WILEY HANDBOOK OF SCIENCE AND TECHNOLOGY FOR HOMELAND SECURITY

Volume 1

WILEY HANDBOOK OF SCIENCE AND TECHNOLOGY FOR HOMELAND SECURITY

Volume 1

Edited by

JOHN G. VOELLER

Black & Veatch

The Wiley Handbook of Science and Technology for Homeland Security is available online at:
http://mrw.interscience.wiley.com/emrw/9780470087923/home/

WILEY

A JOHN WILEY & SONS, INC., PUBLICATION

Published by John Wiley & Sons, Inc., Hoboken, New Jersey
Published simultaneously in Canada

Library of Congress Cataloging-in-Publication Data:

Wiley handbook of science and technology for homeland security / edited by John G. Voeller, Black & Veatch.
p. cm.
Includes bibliographical references and index.
ISBN 978-0-471-76130-3 (cloth : set) – ISBN 978-0-470-13846-5 (cloth : v. 1) – ISBN 978-0-470-13848-9 (cloth : v. 2) – ISBN 978-0-470-13849-6 (cloth : v. 3) – ISBN 978-0-470-13851-9 (cloth : v. 4)
1. Civil defense–Handbooks, manuals, etc. 2. Security systems–Handbooks, manuals, etc. 3. Terrorism–Prevention–Handbooks, manuals, etc. I. Voeller, John G.
UA926.W485 2010
363.34–dc22

2009041798

Printed in Singapore by Markono Print Media Pte Ltd

10 9 8 7 6 5 4 3 2 1

CONTENTS

PREFACE ix

CONTRIBUTORS xiii

INTRODUCTION AND OVERVIEW 1

Policy Development for Homeland Security 3
Threats and Challenges to Homeland Security 21
Terrorist Organizations and Modeling Trends 32
Risk Communication: An Overlooked Tool in Combating Terrorism 45

CROSS-CUTTING THEMES AND TECHNOLOGIES 57

Risk Modeling and Vulnerability Assessment 57

Terrorism Risk: Characteristics and Features 59
Risk Analysis Frameworks for Counterterrorism 75
Risk Analysis and Management for Critical Asset Protection 93
Logic Trees: Fault, Success, Attack, Event, Probability, and Decision Trees 106
Bayesian Networks 117
Using Risk Analysis to Inform Intelligence Analysis 131
Vulnerability Assessment 140
Risk Communication 151
Probabilistic Risk Assessment (PRA) 162
Scenario Analysis, Cognitive Maps, and Concept Maps 186

Time-Domain Probabilistic Risk Assessment Method for Interdependent Infrastructure Failure and Recovery Modeling 197

Risk Transfer and Insurance: Insurability Concepts and Programs for Covering Extreme Events 207

Quantitative Representation of Risk 223

Qualitative Representation of Risk 237

Terrorism Risk 251

Terrorist Threat Analysis 260

Risk Analysis Methods for Cyber Security 279

Defeating Surprise Through Threat Anticipation and Possibility Management 290

Memetics for Threat Reduction in Risk Management 301

High Consequence Threats: Electromagnetic Pulse 309

High Consequence Threats: Nuclear 319

Modeling Population Dynamics for Homeland Security Applications 330

Sensing and Detection 341

Protecting Security Sensors and Systems 343

Threat Signatures of Explosive Materials 359

Radioactive Materials Sensors 371

Knowledge Extraction from Surveillance Sensors 387

RADAR and LiDAR perimeter protection sensors 398

Design Considerations in Development and Application of Chemical and Biological Agent Detectors 411

Sensing Dispersal of Chemical and Biological Agents in Urban Environments 423

Sensing Releases of Highly Toxic and Extremely Toxic Compounds 435

2D-to-3D Face Recognition Systems 468

Eye and Iris Sensors 489

A Tandem Mobility Spectrometer for Chemical Agent and Toxic Industrial Chemical Monitoring 501

Dynamic Load Balancing for Robust Distributed Computing in the Presence of Topological Impairments 512

Passive Radio Frequency Identification (RFID) Chemical Sensors for Homeland Security Applications 523

Protection, Prevention, Response and Recovery 545

Protection and Prevention: An Overview 547

Protection and Prevention: Threats and Challenges from a Homeland Defense Perspective 556

Consequence Mitigation 569

Security Assessment Methodologies for U.S. Ports and Waterways 582
Defending Against Malevolent Insiders Using Access Control 593
Less-Lethal Payloads for Robotic and Automated Response Systems 603
Defending Against Directed Energy Weapons: RF Weapons and Lasers 615
The Sensor Web: Advanced Technology for Situational Awareness 624

Critical Information Infracture Protection *637*

Critical Information Infrastructure Protection, Overview 639
Australia 654
Austria 665
Brazil 675
Canada 686
Estonia 695

PREFACE

The topic of homeland security did not begin with the World Trade Center or the Irish Republican Army (IRA) or the dissidents of past empires, but began when the concept of a nation versus a tribe or kingdom took root and allegiance to people was a choice, not a mandate. The concept of terrorism is part of homeland security but not all of it, as there are other risks to homeland security that come from Mother Nature or our own lack of action, like infrastructure renewal, that have much higher probabilities of creating substantial damage and loss of life than any group of terrorists could ever conceive. Hence, the focus of this Handbook focuses more on the insecurities that can disrupt or damage a nation, its people and economy, and the science and technology (S&T) ideas and tools that can assist in detecting, preventing, mitigating, recovering, and repairing the effects of such insecurities.

The number of S&T topics that are involved in the physical, cyber, and social areas of homeland security include thousands of specialties in hundreds of disciplines, and no single collection could hope to cover even a majority of these. The Handbook was designed to discuss those areas that form a foundation of knowledge and awareness that readers can use to base their understanding on and move to higher levels of sophistication and sensitivity as needed. For example, the many different areas of detection of chemical substances alone could take around 100 volumes to cover, but there is a subset of this knowledge that brings the reader a solid base on which to build a more advanced knowledge view, if desired. Such subsets in each major topic area were the targets of the Handbook.

The Handbook is organized in sections with each addressing a major topic from cyber security to food safety. The articles within each section are designed to range from instructions about fundamentals to some of the latest material that can be shared. Over time, we will add new sections and articles within each to make the Handbook a living entity. John Wiley & Sons has done many such large collections, some being truly massive, and has developed support systems to address such a challenge.

Several key goals were paramount in the creation of this Handbook. First was to gather true experts from all sources to talk about S&T for homeland security, homeland defense, and counterterrorism with very limited control over what was presented. Some of what is done in this vast S&T space has to be classified so as to not "communicate our punches" to our adversaries, which is especially true in a military setting. However, homeland security is largely domestic, and solutions must be available for sale, operation, and maintenance in public infrastructure and networks. Having experts speak in an open channel in the Handbook is important to inform the public, officials, law enforcement, researchers, academics and students so that they can work together and increase our collective national knowledge.

A second goal was to take a portion of the thousands of possible sources of knowledge about the hundreds of S&T topics that are part of homeland security and put them in one location. Moreover, this Handbook increases the opportunity for an expert in one topic to easily find connected, adjacent or codependant topics that would have normally required other searches, references and licenses to access. Homeland security involves so much of cross-discipline action and interdependency examination that this goal was considered especially important.

A third goal was to create a venue where knowledge of different theories, approaches, solutions, and implications could be compared. There are many ways to address homeland security concerns and needs in different disciplines and specialties that nothing less than a multivolume, multiyear project looking for hundreds of authors out of thousands of candidates was required. The Handbook addressed this by the services of some of the best in the world in each major topic area acting as Section Editors. These top experts knew whom to invite, whom could contribute, and most important how much of the overall knowledge in their specialty could be conveyed without drifting into sensitive areas. The Handbook would have been impossible to produce without their incredible efforts in selecting, reviewing, and overseeing their section content.

A fourth goal was to provide a place where even experts in one facet of homeland security could learn about other facets with confidence that the quality of the information would meet their standards. From exceptional discussions about how the European Union views cyber security differently from the United States to massive work on all the different food-safety-detection equipment available, the focus of all contributors was journal quality, peer-reviewed knowledge, with references and links to additional knowledge to allow the reader to go deeper.

A fifth goal was the creation of a substantial enough body of knowledge about the many different facets of homeland security so that policy and decision-makers could get a picture of how much has been done and how much needs to be done to create robust solutions in all the needed areas. Even in places that have dealt with terrorism for over a century, the world still does not have strong, cost-effective solutions to some of the most fundamental problems. For example, we have very limited to no ability to spot a bomb in a car moving toward a building at a sufficient distance to know whether to destroy or divert it before it can damage the target. Even simpler, the ability to spot a personnel-borne improvised explosive device (IED) in a crowd coming into a Las Vegas casino is still beyond our collective capability. The bounding of what we know and don't know that can be applied in a domestic setting needed to be documented at least in part for dozens of major areas in homeland security.

A sixth goal that was not part of the pages of the Handbook was to create a visibility of expertise among all the contributors and reviewers to help them connect with others and enable collaboration. Only a large collection of this type creates such a vast opportunity in known areas of S&T for shared learning and new relationships.

A seventh goal was to present the S&T of homeland security in a way that would allow one of the most important aspects of the economics involved to be considered. This is not the economics of creating or acquiring one solution but rather finding more than one use for a given solution. An inescapable issue in many areas of homeland security S&T is that a fully successful solution applied to only one small market will likely fail because there is insufficient revenue and market to sustain the provider. Building a few hundred detectors for specific pathogens is likely to fail because of lack of volume or will perhaps never see funding as this becomes evident in the original business plan. The solution to this issue is finding multiple uses for each device. For example, a chemical detector looking for contraband or dangerous materials a few days a year may provide continuous service in looking for specific air pollutants related to allergy mitigation in a building. The Handbook provides exposure to the reader in capabilities built for homeland security that might bring benefit in other more frequently needed areas thereby making both applications more viable.

The Handbook authors were asked to contribute material that was instructional or that discussed a specific threat and solution or provided a case study on different ways a problem could be addressed and what was found to be effective. We wanted new material where possible, but given the nature of a handbook we wanted to also bring great work that might already be published in places not easily encountered and with proper permission could be repurposed into the Handbook for broader visibility.

One of the conditions set by the Senior Editor before taking on the project was that the Handbook needed to be published both in print and on the Web. The dynamic online collection will not only allow new articles and topics to be added but also updated when threats, solutions, or methods change. The Senior Editor greatly appreciates John Wiley & Sons for accepting this challenge.

The Section Editors of the Handbook have done a superb job of assembling their authors and topics and ensuring a good balance of foundations and details in their articles. The authors in the Handbook have produced valuable content and worked hard with the Wiley editing staff to enhance quality and clarity. And finally, the Wiley staff has taken on the management of hundreds of contributors with patience and energy beyond measure.

This Handbook was conceived as a living document designed to mutate and grow as the topics presented changed or the capabilities of S&T advanced to meet existing and new threats. We hope readers will consider how they might be able to contribute to the Handbook body of knowledge and consider writing about their special expertise for submission sometime in the future.

Editor-in Chief
John G. Voeller

CONTRIBUTORS

Dulcy M. Abraham, Purdue University, West Lafayette, Indiana, *Consequence Mitigation*

Andrew Anderson, Sionex Corporation, Bedford, Massachusetts, *A Tandem Mobility Spectrometer for Chemical Agent and Toxic Industrial Chemical Monitoring*

George E. Apostolakis, Engineering Systems Division and Department of Nuclear Science and Engineering, Massachusetts Institute of Technology Cambridge, Massachusetts, *Probabilistic Risk Assessment (PRA)*

Bilal M. Ayyub, Center for Technology and Systems Management, Department of Civil and Environmental Engineering, University of Maryland, College Park, Maryland, *Defeating Surprise Through Threat Anticipation and Possibility Management; Memetics for Threat Reduction in Risk Management; Modeling Population Dynamics for Homeland Security Applications; Quantitative Representation of Risk; Risk Analysis Frameworks for Counterterrorism; and Terrorism Risk: Characteristics and Features*

George H. Baker, James Madison University, Harrisonburg, Virginia, *Time-Domain Probabilistic Risk Assessment Method for Interdependent Infrastructure Failure and Recovery Modeling*

Betty E. Biringer, Security Risk Assessment Department, Sandia National Laboratories, Albuquerque, New Mexico, *Defending Against Malevolent Insiders Using Access Control*

Jerry P. Brashear, ASME Innovative Technologies Institute, LLC, Washington, D.C., *Risk Analysis and Management for Critical Asset Protection*

Elgin Brunner, Center for Security Studies (CSS), ETH Zurich, Switzerland, *Australia; Austria; Brazil; Canada; and Estonia*

Donald L. Buckshaw, Innovative Decisions, Inc., Vienna, Virginia, *Logic Trees: Fault, Success, Attack, Event, Probability, and Decision Trees*

Dennis Buede, Innovative Decisions, Inc., Vienna, Virginia, *Bayesian Networks*

David E. Burchfield, Hamilton Sundstrand Corporation, Pomona, California, *A Tandem Mobility Spectrometer for Chemical Agent and Toxic Industrial Chemical Monitoring*

Todd P. Carpenter, Adventium Labs, Minneapolis, Minnesota, *Protecting Security Sensors and Systems*

Myriam Dunn Cavelty, Center for Security Studies (CSS), ETH Zurich, Switzerland, *Critical Information Infrastructure Protection, Overview*

Rama Chellappa, Center for Automation Research and Department of Electrical and Computer Engineering, University of Maryland, College Park, Maryland, *Knowledge Extraction from Surveillance Sensors*

Glenn Coghlan, CPP, Founder and CEO, Coghlan Associates, Inc., Springfield, Virginia, *Vulnerability Assessment*

Kenneth G. Crowther, Center for Risk Management of Engineering Systems, Department of Systems and Information Engineering, University of Virginia, Charlottesville, Virginia, *Scenario Analysis, Cognitive Maps, and Concept Maps*

Michel Cukier, University of Maryland, College Park, Maryland, *Risk Analysis Methods for Cyber Security*

John Cummings, Sandia National Laboratories, Albuquerque, New Mexico, *Protection and Prevention: An Overview*

Robert M. DeBell, Gaithersburg, Maryland, *Qualitative Representation of Risk*

Po-Ching DeLaurentis, Purdue University, West Lafayette, Indiana, *Consequence Mitigation*

Kevin A. Delin, SensorWare Systems, Inc., Pasadena, California, *The Sensor Web: Advanced Technology for Situational Awareness*

Sagar Dhakal, Nortel Networks Inc., Richardson, Texas, *Dynamic Load Balancing for Robust Distributed Computing in the Presence of Topological Impairments*

David Dietz, Department of Electrical and Computer Engineering, University of New Mexico, Albuquerque, New Mexico, *Dynamic Load Balancing for Robust Distributed Computing in the Presence of Topological Impairments*

Robin L. Dillon-Merrill, McDonough School of Business, Georgetown University, Washington, D.C., *Logic Trees: Fault, Success, Attack, Event, Probability, and Decision Trees*

Joseph DiRenzo III, United States Coast Guard, *Security Assessment Methodologies for U.S. Ports and Waterways*

Christopher W. Doane, United States Coast Guard, *Security Assessment Methodologies for U.S. Ports and Waterways*

Larry Drymon, Space and Naval Warfare Systems Center, San Diego, California, *Less Lethal Payloads for Robotic and Automated Response Systems*

Angela M. Ervin, Department of Homeland Security Science and Technology Directorate, Washington, D.C., *Sensing Dispersal of Chemical and Biological Agents in Urban Environments*

Hobart Ray Everett, Space and Naval Warfare Systems Center, San Diego, California, *Less-Lethal Payloads for Robotic and Automated Response Systems*

Robert Finkelstein, Robotic Technology Inc., University of Maryland University College, Adelphi, Maryland, *Memetics for Threat Reduction in Risk Management*

Michael J. Frankel, EMP Commission, Washington, D.C., *Defending Against Directed Energy Weapons: RF Weapons and Lasers; High Consequence Threats: Nuclear; and High Consequence Threats: Electromagnetic Pulse*

Kelly Grant, Space and Naval Warfare Systems Center, San Diego, California, *Less Lethal Payloads for Robotic and Automated Response Systems*

D. Anthony Gray, Syracuse Research Corporation, North Syracuse, New York, *Sensing Releases of Highly Toxic and Extremely Toxic Compounds*

Frank W. Griffin, Sandia National Laboratories, NWS DoD Program Design & Implementation, Albuquerque, New Mexico, *RADAR and LiDAR Perimeter Protection Sensors*

Yacov Y. Haimes, Center for Risk Management of Engineering Systems, Department of Systems and Information Engineering, University of Virginia, Charlottesville, Virginia, *Scenario Analysis, Cognitive Maps, and Concept Maps*

Rida Hamza, Honeywell International, Golden Valley, Minnesota, *Eye and Iris Sensors*

Majeed M. Hayat, Department of Electrical and Computer Engineering and Center for High Technology Materials, University of New Mexico, Albuquerque, New Mexico, *Dynamic Load Balancing for Robust Distributed Computing in the Presence of Topological Impairments*

Christopher D. Hekimian, DXDT Engineering and Research, LLC, Hagerstown, Maryland, *Terrorist Organizations and Modeling Trends*

William Hoffman, Animetrics, Conway, New Hampshire, *2D-To-3D Face Recognition Systems*

Anne E. Hultgren, Department of Homeland Security Science and Technology Directorate, Washington, D.C., *Sensing Dispersal of Chemical and Biological Agents in Urban Environments*

Jeffrey Hunker, Carnegie Mellon University, Pittsburgh, Pennsylvania, *Policy Development for Homeland Security*

Rick Hurley, Sandia National Laboratories, RF and Optics Microsystem Applications, Albuquerque, New Mexico, *RADAR and LiDAR Perimeter Protection Sensors*

Elizabeth M. Jackson, George Mason University School of Law, Arlington, Virginia, *Vulnerability Assessment*

J. William Jones, ASME Innovative Technologies Institute, LLC, Washington, D.C., *Risk Analysis and Management for Critical Asset Protection*

Mark P. Kaminskiy, Center for Technology and Systems Management, Department of Civil and Environmental Engineering, University of Maryland, College Park, Maryland, *Modeling Population Dynamics for Homeland Security Applications; and Quantitative Representation of Risk*

Dan Keller, Sandia National Laboratories, Strategic Business Enterprise Services, Albuquerque, New Mexico, *RADAR and LiDAR Perimeter Protection Sensors*

Greg Kogut, Space and Naval Warfare Systems Center, San Diego, California, *Less Lethal Payloads for Robotic and Automated Response Systems*

Howard C. Kunreuther, Center for Risk Management and Decision Processes, the Wharton School, University of Pennsylvania, Philadelphia, Pennsylvania, *Risk Transfer and Insurance: Insurability Concepts and Programs for Covering Extreme Events*

Mark Lawley, Purdue University, West Lafayette, Indiana, *Consequence Mitigation*

Robert M. Liebe, Innovative Decisions Inc., Vienna, Virginia, *Risk Analysis Frameworks for Counterterrorism*

Suzanne M. Mahoney, Innovative Decisions, Inc., Vienna, Virginia, *Bayesian Networks*

R. M. Mayo, U.S. Department of Energy, National Nuclear Security Administration, Washington, D.C., *Radioactive Materials Sensors*

William L. McGill, College of Information Sciences and Technology, Pennsylvania State University, University Park, Pennsylvania, *Defeating Surprise Through Threat Anticipation and Possibility Management*

Jeffrey D. McManus, The Office of the Secretary of Defense, Washington, D.C., *Protection and Prevention: Threats and Challenges from a Homeland Defense Perspective*

Erwann O. Michel-Kerjan, Center for Risk Management and Decision Processes, the Wharton School, University of Pennsylvania, Philadelphia, Pennsylvania, *Risk Transfer and Insurance: Insurability Concepts and Programs for Covering Extreme Events*

Michael I. Miller, Animetrics, Conway, New Hampshire, *2D-To-3D Face Recognition Systems*

Charles T. C. Mo, Northrop-Grumman IT, Los Angeles, California, *Time-Domain Probabilistic Risk Assessment Method for Interdependent Infrastructure Failure and Recovery Modeling*

William G. Morris, General Electric Global Research Center, Niskayuna, New York, *Passive Radio Frequency Identification (RFID) Chemical Sensors for Homeland Security Applications*

Dale W. Murray, DoD Security System Analysis Department, Sandia National Laboratories, Albuquerque, New Mexico, *Defending Against Malevolent Insiders Using Access Control*

Donald E. Neale, Department of Homeland Security, Washington, D.C., *Vulnerability Assessment*

H. William Niu, Hamilton Sundstrand Corporation, Pomona, California, *A Tandem Mobility Spectrometer for Chemical Agent and Toxic Industrial Chemical Monitoring*

Susmit Panjwani, University of Maryland, College Park, Maryland, *Risk analysis methods for Cyber Security*

Gregory S. Parnell, Department of Systems Engineering, United States Military Academy, West Point, New York, *Logic Trees: Fault, Success, Attack, Event, Probability, and Decision Trees; and Risk Analysis Frameworks for Counterterrorism*

D. Brian Peterman, United States Coast Guard, *Security Assessment Methodologies for U.S. Ports and Waterways*

Jorge E. Pezoa, Department of Electrical and Computer Engineering, University of New Mexico, Albuquerque, New Mexico, *Dynamic Load Balancing for Robust Distributed Computing in the Presence of Topological Impairments*

Radislav A. Potyrailo, General Electric Global Research Center, Niskayuna, New York, *Passive Radio Frequency Identification (RFID) Chemical Sensors for Homeland Security Applications*

Irmak Renda-Tanali, University of Maryland University College, Adelphi, Maryland, *Terrorist Organizations and Modeling Trends*

Edward P. Rhyne, Department of Homeland Security Science and Technology Directorate, Washington, D.C., *Sensing Dispersal of Chemical and Biological Agents in Urban Environments*

David Ropeik, Risk Communication, Ropeik & Associates, Concord, Massachusetts, *Risk Communication—An Overlooked Tool in Combating Terrorism*

Aswin C. Sankaranarayanan, Center for Automation Research and Department of Electrical and Computer Engineering, University of Maryland, College Park, Maryland, *Knowledge Extraction from Surveillance Sensors*

Paul Schuepp, Animetrics, Conway, New Hampshire, *2D-To-3D Face Recognition Systems*

James Scouras, Defense Threat Reduction Agency, Fort Belvoir, Virginia, *Risk Analysis Frameworks for Counterterrorism; and Qualitative Representation of Risk*

Donna C. Shandle, Nuclear, Chemical, and Biological Contamination Avoidance, Aberdeen Proving Ground, Maryland, *Design Considerations in Development and Application of Chemical and Biological Agent Detectors*

Calvin Shipbaugh, Arlington, Virginia, *High Consequence Threats: Nuclear*

Brandon Sights, Space and Naval Warfare Systems Center, San Diego, California, *Less Lethal Payloads for Robotic and Automated Response Systems*

Edward Small, Sacramento Metropolitan Fire District, Sacramento, California and FEMA Urban Search and Rescue Team, CA Task Force 7, Sacramento, California, *The Sensor Web: Advanced Technology for Situational Awareness*

Gary R. Smith, Logical Decisions, Fairfax, Virginia, *Qualitative Representation of Risk*

D. L. Stephens, Pacific Northwest National Laboratory, Richland, Washington, *Radioactive Materials Sensors*

Cheryl Surman, General Electric Global Research Center, Niskayuna, New York, *Passive Radio Frequency Identification (RFID) Chemical Sensors for Homeland Security Applications*

Manuel Suter, Center for Security Studies (CSS), ETH Zurich, Switzerland, *Australia; Austria; Brazil; Canada; and Estonia*

Andrew W. Szumlas, Hamilton Sundstrand Corporation, Pomona, California, *A Tandem Mobility Spectrometer for Chemical Agent and Toxic Industrial Chemical Monitoring*

Joseph A. Tatman, Innovative Decisions, Inc., Vienna, Virginia, *Bayesian Networks*

Lisa Theisen, Contraband Detection, Sandia National Laboratories, Albuquerque, New Mexico, *Threat Signatures of Explosive Materials*

Edward T. Toton, Toton, Incorporated, Reston, Virginia, *Defending Against Directed Energy Weapons: RF Weapons and Lasers*

Monique Mitchell Turner, Center for Risk Communication Research, Department of Communication, University of Maryland, College Park, Maryland, *Risk Communication*

Shawn S. Turner, United States Marine Corps, Arlington, Virginia, *Risk Communication*

Marc Vaillant, Animetrics, Conway, New Hampshire, *2D-To-3D Face Recognition Systems*

Ashok Veeraraghavan, Center for Automation Research and Department of Electrical and Computer Engineering, University of Maryland, College Park, Maryland, *Knowledge Extraction from Surveillance Sensors*

Keith B. Ward, Department of Homeland Security Science and Technology Directorate, Washington, D.C., *Sensing Dispersal of Chemical and Biological Agents in Urban Environments*

David M. Weinberg, Practical Risk LLC, Rio Rancho, New Mexico, *Threats and Challenges to Homeland Security*

Rand Whillock, Honeywell International, Golden Valley, Minnesota, *Eye and Iris Sensors*

Jack F. Williams, Georgia State University, Atlanta, Georgia, *Terrorist Threat Analysis*

Henry H. Willis, RAND Corporation, Pittsburgh, Pennsylvania, *Using Risk Analysis to Inform Intelligence Analysis*

Gordon Woo, Risk Management Solutions, London, United Kingdom, *Terrorism Risk*

INTRODUCTION AND OVERVIEW

POLICY DEVELOPMENT FOR HOMELAND SECURITY

JEFFREY HUNKER
Carnegie Mellon University, Pittsburgh, Pennsylvania

1 INTRODUCTION

In science and technology, five factors make effective and consistent Policy Development for Homeland Security difficult [1].

- The definition and goals of Homeland Security continue to evolve.
- Multiple decision makers and high levels of organizational complexity diffuse decision-making authority and responsibility and make policy prioritization difficult.
- Policy prioritization is further challenged because of the breadth and ambiguity of Homeland Security threats. This, together with highly differentiated interests and levels of support for different projects from the research community challenge policy makers ability to distinguish and invest in the important, not just the interesting.
- Metrics for judging project contribution frequently are difficult to create.
- Distinct roles for key Homeland Security functions—intelligence, prevention, response and reconstruction, and "defend and respond"—overlap with and can be difficult to distinguish from the Nation's overall National Security agenda.

For the practicing policy maker, these characteristics—shifting goals, complex and competing interests, and difficulty in measuring results—are not uncommon. It is the mark of good policy development to overcome these challenges and to produce results that benefit the nation.

2 OVERVIEW OF POLICY DEVELOPMENT

Policy development, in any field, is an art, not a science.

2.1 Defining Policy; Defining Homeland Security

A *policy* is an attempt to define and structure a rational basis for action or inaction [2]. Policy is a long-term commitment; tactics are short-term actions. Tactics and implementation are overlapping concepts in the execution of policy.

Policy also needs to be distinguished from (but overlaps with) *administration* and *politics*. "Administration" is the "management of any office, employment, or organization direction" [3]. Administration is decision making in bounded rationality—making decisions that are not derived from an examination of all the alternatives [2, p. 278]. *Politics*, from the Greek for citizen, is about "exercising or seeking power in governmental or public affairs" [3]. Policy, at least ideally, takes into consideration all alternatives, distinguishing it from administration. A focus, or lack thereof, on power distinguishes policy from politics.

However, policy development is critically constrained by both administration and politics. Political feasibility requires elected officials (or their proxies) to support the policy. Organizational feasibility requires the requisite organizations to support the policy and implement it in a way that makes its success possible [4] (President Kennedy is noted for saying "I like the idea, but I'm not certain that the Government will".).

Homeland Security, the object of policy development for this article, has a shifting definition. *The National Strategy for Homeland Security* (2002) defines it as "a concerted national effort to prevent terrorist attacks within the United States, reduce America's vulnerability to terrorism, and minimize the damage and recover from attacks that do occur" [5]. In practice, however, homeland security now includes protection against and response to natural or accidental manmade disasters, such as hurricanes and toxic spills.

Reflecting this reality, this article principally will address policy development related to terrorism, but will also refer to issues in the prevention and response to natural and accidental disasters.

Homeland Security is thought of in multiple ways even within the narrower confines of protection against terrorism. For example, in protecting key economic and national security assets such as the electric grid, our telecommunication network, and basic utilities, different constituencies will refer to agendas in "critical infrastructure protection (CIP)", "critical information infrastructure protection", or "protection of physical assets". These agendas overlap, but each has its own scientific and political constituency.

The shifting definition of "Homeland Security" as a policy goal prompts three observations. First, prevention and response to natural and accidental disasters is a relatively mature policy agenda in comparison to the terrorism agenda (though provision of insurance for hurricane disasters and perspectives on climate change challenge policy makers and politicians alike). Had not the Federal Emergency Management Agency (FEMA) and the Coast Guard—two principal Federal agencies with responsibilities for natural and accidental disasters—been included in the Department of Homeland Security, it may indeed have been the case that the "mission creep" apparent in the definition of Homeland Security would not have taken place.

However, whether or not natural and accidental disasters are "Homeland Security" issues, policy makers at Department of Homeland Security (DHS) must address these agendas. Their challenge is to integrate and seek synergies in pursuing disparate policy goals. The search for synergies is an important, but oftentimes overlooked, element in policy development. Finally, FEMA's performance, in particular, in responding to Hurricane Katrina highlights the gulf between policy and implementation that policy makers ignore at their peril. The author has reviewed the policies regarding hurricane response in the Gulf of Mexico; on paper they appear more than adequate. Implementation was the problem.

2.2 The Policy Development Process

A common characterization of policy development, useful but inaccurate, lists a series of steps [2, p. 77]:

- *Defining the problem.* What is the context for a policy?
- *Defining the solution.* Who specifies it, the and why?
  - Identifying alternative responses/solutions
  - Evaluating options
  - Selecting the policy option
- *Implementation.* Who implements it, and why? Who follows it, and why?
- *Evaluation.* How is conformity with a policy tracked and evaluated?

This taxonomy is useful in that it describes the steps that any emergent policy follows. However, this taxonomy ignores the real world of policy making, involving interacting cycles of multiple constituencies within government (at many different levels) and outside of government [2, p. 80].

An example of Homeland Security policy development helps to illustrate this observation: In 1999, during the preparation of the first National Plan for Information Systems Protection (the National Plan) [5, 6] a series of informal discussions between two White House offices (the National Security Council and the Office of Science and Technology Policy) and other Executive Branch agencies (the National Science Foundation (NSF) and the Critical Infrastructure Assurance Office (CIAO)) led to the insight that most federally funded cyber security R&D was directed toward mission-specific goals of the funding agencies (e.g. the Defense Advanced Research Projects Agency (DARPA) and the National Security Agency (NSA). Consequently, there were serious gaps in addressing research questions that, although important, did not garner a specific agency constituency. Following several workshops with outside researchers and prolonged internal discussions, a proposal was developed to create a "virtual National Laboratory"—a consortium of US-based research institutions—charged with identifying and addressing the gaps in the Nation's cyber research agenda. This work led to the inclusion in the National Plan of the goal to "establish a new public–private mechanism to coordinate Federal R&D in information systems security with private sector needs and efforts" [6, p. xxi].

Discussions with Congressional members and staff during 1999 evinced considerable interest, but no positive results. Meanwhile, a number of research institutions began vigorously to express interest both to Congress and the Executive Branch in becoming the host institution. That year, Congressional action, independent of Administration's thinking as to possible host institutions, created the Institute for Security Technology Studies (ISTS) at Dartmouth College. With the creation of the DHS, funding and oversight of the ISTS was located in the Science and Technology Directorate. Oversight of ISTS initiatives always has been vigorous, but no quantifiable metrics for performance exist.

There are several lessons from this example. In developing the policy options, there was never a formal development and ranking of alternatives. Consultation with constituencies within and outside the Federal government (Congress, Federal agencies funding cyber R&D, first responders, and outside research institutions) was continuous throughout the policy development process. Events (such as the placement at Dartmouth) were not necessarily planned by the policy makers (though not unwelcome). Quantifiable metrics were never developed; in particular there was never any consideration of cost/benefit analysis.

A final point—of all of the stages of policy development, policy evaluation is perhaps the most difficult. Practicing policy makers often describe policies as "*effective*" or "*ineffective*", yet the policy literature speaks most often of "*efficiency*". A particular allocation of resources is *efficient* if and only if there is no better allocation of those same resources [4, p. 32]. A policy is *effective* if it is adequate to accomplish a purpose, producing the intended or expected result [3]. From a practitioner's perspective, measures of allocative efficiency are rarely meaningful—effectiveness is the most commonly employed heuristic.

To summarize, policy development does not translate easily into the abstract. The context for a policy, who specifies it, who implements it, who follows it, and how conformity of policy is tracked and evaluated, are situation specific. Some generalizations are possible, but not many.

3 CASE EXAMPLES OF POLICY DEVELOPMENT

Three short case examples illustrate the range of issues in developing Homeland Security policy.

3.1 Cyber Security: A Challenge of Defining the Threat and Establishing Incentives

"Cyber Security" means security of our electronic information and communication systems—notably the Internet but also proprietary computer networks (whether used by business or government) including wireless networks [7].

The focus here is on intentional attacks, and mostly on attacks that could affect the "critical functions" that keep a society running well—in commerce, government, and national security/homeland defense.

Following Presidential Decision Directive 63 in May 1998 (CIP) the protection of cyber and information systems against attack has been a national priority. The Department of Defense (DOD), with a focus on protecting its own extensive systems, and DHS, in the Information Analysis and Infrastructure Protection Directorate, have primary Federal responsibility. National Plans and associated Research and Development plans coordinate Federal policy. Private sector participation is key to the policy's effectiveness. In particular, sector specific organizations (e.g. for banking and financial institutions) have been created to both promote private sector cyber security and, very importantly, share information within themselves and with the Federal government about cyber threats and attacks [8].

Our understanding of threats, however, is limited. Proactive anticipation of new threats is difficult because the complexity of software makes *a priori* identification of security vulnerabilities difficult and because new forms of attack (e.g. spear phishing, or distributed denial of service attacks) continually evolve. Publicly available statistics on cyber security are poor. Surveys and compilations of cyber attacks and violations rely on voluntary reporting, and interviews with Chief Information Officers and other officials responsible for security indicate a widespread reluctance to report most intrusions, including even attempted intrusions [9]. With this caveat, the following are examples.

- More than 2,000,000 personal computers are infected and attackers store and serve pornography from them, attack other computers or send out spam using them, or install spy ware to obtain credit card numbers and other personal information.
- Large numbers of sensitive government and contractor computers have been infected with hidden software that records everything done on those computers, and reports back to those that installed that software [8].

General types of threats may include:

- *Cyber-crime* (phishing, extortion, fraud, etc.). This crime is already rampant and is growing in scale and sophistication.
- *Cyber-terror* (attacking a crucial website or a computer-controlled infrastructure (e.g. the electric power grid) or, for example, attacking New York Stock Exchange (NYSE) systems). Many "mischief" attacks of this kind have already been tried and succeeded. They too could easily grow in scale and sophistication—with the potential for use by terrorists.
- *Cyber-warfare* (cyber-terror or cyber-espionage used by one state against another). It appears that this has already been tried at least twice, in the Chinese attempts at reprisals against United States government information networks after the May 1999 accidental bombing of the Chinese embassy in Belgrade, and again by Russian distributed denial of service attacks against Estonian computer networks in May 2007 (both countries deny any involvement).

But key unanswered questions persist. What are the chances that a skilled group of cyber-criminals might hire themselves out as mercenaries for cyber-terror or cyber-warfare? What might they be most likely to attack, and how? Our ability to answer these questions is limited, yet an understanding of where and how threats might materialize is central to building effective policies for protection and response.

Consequently, our security responses, though often quite sophisticated, tend to be piecemeal, ad hoc, and not infrequently focused on the short term.

The possible consequences are not well characterized either. These may include:

- immediate damage or disruption ("planes fall out of the sky", the power grid goes down);
- loss of confidence (e.g. no confidence in NYSE systems, so people begin to take their securities listings and their trading somewhere else);
- general deterioration of an industry or an activity due to constant low-level incidents.

A second major cyber security policy challenge is to create incentives for action. Software developers, for example, are largely immune from tort liability actions challenging the security and reliability of their products. Several states have codified this exemption. The "tragedy of the commons" is also at work in networked systems. The software that acts as "traffic cop" for the Internet—the Border Gateway Protocol (BGP)—is sensitive to accidental (or deliberate) misconfigurations. A decade ago an accidental BGP misconfiguration redirected the entire Internet to a single site in Florida. Although technical solutions to make a repeat of this incident less likely exist, in essence, no single Internet routing point has an incentive to install these solutions. Hence, a decade later, the network still relies upon the good faith and good programming skills of an increasingly large (and increasingly global) community of service providers.

Cyber security presents an example of how although national focus has led to an extensive and detailed policy framework, it has failed to address key foundations. Scientific and understanding the extent and nature of cyber threats, and in technical work in technology solutions (e.g. encryption, firewalls, and intrusion detection) abounds; however, progress in creating risk management systems, and managerial/network imperatives for action are far less advanced.

3.2 Fire: Consistent and Effective Public–Private Partnership

Fire has long been recognized as a serious danger to urban society, commerce, and natural systems. There have been myriad individual homes and businesses destroyed by fire, and occasional large-scale catastrophes—the great London and Tokyo fires of the 17th century, the Chicago fire, and major forest fires such as in Yellowstone Park a decade ago. Though yet to occur, major urban conflagrations, from nuclear or other causes, remain a real, though distant, threat.

Four major outcomes have emerged from our concern with fire.

- Governments, private businesses, and citizens have long worked to *understand* how fires start and spread, how they can be contained and extinguished, and how they can be prevented. Continuous and sustained research has successively addressed new issues, as, for example, when new materials enter into building construction or furnishings, or when new sources of combustion, such as electrical wiring, are introduced. Research takes place at the Federal (e.g. National Institute of Standards and Technology), state, and private sector levels.
- In parallel, common pools of risk knowledge have been created, updated, and perhaps most importantly, widely shared among insurers, risk managers, and researchers. This statistical data provides the necessary foundation for managing the risk of fire.
- The result is a well-developed system in which we have fire codes, fire insurance, agreed-upon standards for products and for fire protection systems, and well-defined procedures and resources in place for calling firefighting companies to the scene of a fire—all backed up by a good knowledge of what the losses could be, in terms of both dollars and human life, and therefore a good way of assessing risk, justifying costs, and compensating for damage.
- For the (fortunately) special case of major conflagrations (forest fires, major urban conflagrations) a well-exercised system of coordinated Federal resources (Department of the Interior, Department of Agriculture, Defense Department (National Guard), DHS (FEMA), and state and local assets) is in place.

The policy response to fire exemplifies an almost three century-long process integrating widespread recognition of the threat together with private and public investments in understanding the threat, working to reduce it, creating systems to respond to fires (large and small) when they occur, and developing sophisticated regulatory and risk management mechanisms to reduce and spread risk. What is most notable is that this policy structure was not created "top–down", but developed from enlightened self-interest and the recognition of a Federal role in two dimensions—research and emergency response and reconstitution. The policy structure is not perfect; for example a comprehensive national fire code has yet to be adopted in place of a myriad of local codes. Nonetheless, it stands as a model of successful policy development.

3.3 Y2K: Top–Down Policy Response to a Specific Threat

From the preparation and execution of Y2K some key lessons can be drawn.

- A clear decision for action was made by the White House, with clear goals and timelines.
- A strong leader, with close ties to the President, and extensive business and government credibility, was chosen.
- Education—of the business community and government agencies—was a major and long term focus.
- Incentives, but not regulation, were used to enhance both action and cooperation among the private sector. For example, the Securities and Exchange Commission (SEC) did not require filing organizations to take action, merely to report publicly in their filings what if any action an organization was taking. National legislation, to promote information sharing and reduce liability for Y2K related actions, was enacted.
- Public–private partnership was emphasized.
- A sophisticated operations' center, coordinating business and government resources and information, was built (the Information Coordination Center); strong leadership (a retired Marine Corps General) led the effort.
- Constant and effective communications kept the press and public informed.
- Extensive and effective outreach to key non-US constituencies, including the UN, helped to ensure that preparation for the Y2K event was, if not global, certainly not exclusively a US priority.
- The core operational team managing the issue was a tight, small, high quality team based at the White House.

The response to the "Y2K bug" illustrates an effective policy development and implementation process. Clear goals (motivated by a pressing threat, though skeptics abounded), strong leadership, effective implementation driven by a subtle combination of "carrot-and-stick", and measurable outcomes (things either worked, or they did not) characterize this initiative.

Some key observations emerge from these case examples. Policies, however detailed, that fail to address fundamental issues reduce their likelihood of being effective (this is sometimes referred to as the "elephant in the drawing room" syndrome—there's an elephant, but no one acknowledges its presence). Policy can be emergent, constructing itself through the uncoordinated actions of various constituencies. Clear goals, strong leadership, and measurable outcomes are critical to successful policy.

4 SELECT RESEARCH AGENDAS AND IMPLICATIONS FOR POLICY DEVELOPMENT

A representative but certainly not exhaustive list of major Homeland Security research topics illuminates some key drivers for policy development.

One taxonomy [10] for research divides scientific challenges into those which have been around for a while and those which have emerged more recently, either in response to new policy concerns (e.g. terrorism, global climate change, and so on) or evolutions in the technology frontier (e.g. greater computational and networking capabilities).

The former includes:

- identification and treatment of known pathogens;
- better technologies for emergency responders;
- blast-resistant and fire-resistant structures;
- air filtering against known pathogens and chemicals;
- decontamination techniques; and
- technologies to enhance security against cyber attacks.

Areas that have emerged more recently include the following.

- creating an intelligent, adaptive electric power grid;
- revising land use and disaster preparedness/response policies in the face of global climate change;
- capturing, analyzing, and assessing useful information for emergency officials and responders with new sensor and surveillance technologies;
- creating a common risk model that allows comparison between and across infrastructures;
- developing methodologies to accurately identify and predict both actors perpetrating and motivations for cyber attacks;
- identifying and predicting paths and methods of currently undetectable food and water alteration;
- developing networks—both physical (e.g. transportation) and electronic (e.g. the Internet) in which security is being imposed as a basic design consideration, not as an add on;
- designing self diagnosing and self repairing systems and facilities; and
- providing a common Homeland Security operating picture available to all decision makers at all levels.

Many other agendas exist. For example the *Draft National Plan for Research and Development in Support of Critical Infrastructure Protection* [11] identifies nine key themes.

- detection and sensor systems;
- protection and prevention;
- entry and access portals;
- insider threats;

- analysis and decision support systems;
- response, recovery, and reconstitution;
- new and emerging threats and vulnerabilities;
- advanced infrastructure architectures and system design;
- human and social issues.

With a mission of "filling gaps" in the Homeland Security R&D agenda, the Institute for Information Infrastructure Protection has identified potentially key R&D grand challenges [12]:

- secure digital healthcare infrastructure;
- value-added infrastructure protection;
- cost-effective CIP through Information Infrastructure Resilience;
- trusted realms;
- national critical infrastructure web for disaster and attack management, analysis, and recovery;
- spatial clustering of information infrastructure—a basis for vulnerability assessment;
- beyond the domain name system (DNS); and
- establishing a national identity.

Implications for Policy Development: These lists of noteworthy projects challenge policy development for Homeland Security in at least four ways.

- There exist numerous and highly differentiated scientific and technical agendas. New challenges with long-standing infrastructures—such as port security—or new issues—like the identification of potentially explosive liquid combinations—continue to emerge. For policy makers, no clear, widely accepted methodology to prioritize initiatives across domains exists.
- Input metrics (e.g. dollars spent) for each initiative are easy to develop; meaningful output metrics (e.g. how much safer are coastal communities from the threat of catastrophic hurricanes, how much safer are US citizens from terrorist threats) largely do not exist.
- The scientific and technical communities demonstrate widely different levels of interest and effort in engaging these topics. For example, of 80 key researchers in Homeland Security at a 2005 conference [13], less than 6 were focused on human and social issues such as insider threats. Detection and sensor systems were the focus of the bulk of the work.
- Some issues of perhaps paramount importance barely appear in the research portfolio. There appears to be a systematic underinvestment in key areas like human social interactions. Interoperability between networked systems was the subject of a recent special session of the IEEE, and is, arguably, a critical element in any system of effective Homeland Security, yet little basic work appears to be taking place [14].

Thus, opportunities in scientific and technical research and deployment for Homeland Security are numerous and varied; this abundance challenges policy makers in establishing clear goals and monitoring and assessing their impact.

5 ORGANIZATIONAL COORDINATION FOR POLICY

The multiplicity of research agendas as well as organizations with a stake in research and development make vital a strong and dynamic integrative framework for communication and cooperation across domains and constituencies, both for policy makers and researchers. Some agendas address issues of immediate concern and impact, while others focus on expanding the frontiers of knowledge. Shortly we will consider in detail an example of such an effective integrative framework, but first we will outline some overall challenge to policy coordination.

5.1 Complexity of the System

Homeland Security should not be thought of as the DHS, but as a system that incorporates a breadth of constituencies—Federal agencies, states and localities, private organizations, individual citizens, and other countries and international organizations.

At least 22 disparate organizations make up the DHS [1, pp. 59–60], [15]. In addition, the FBI, DOD, and the intelligence community are parts of this system. Policy development for science and R&D in this complex system faces several tensions:

- identifying and establishing policies for R&D requirements;
- matching these with the threats;
- resolving organizational conflicts over resources and priorities; and
- measuring progress and success.

Complexity can be viewed on at least two planes. Within the Federal government most agencies have at least some part of the Homeland Security agenda. As an example, the *National Strategy to Secure Cyberspace* engages at least 15 major Federal departments and agencies apart from DHS. Each element brings to bear differing perspectives (law enforcement, National R&D capabilities, new technology policies, and responsibility for economic sectors or citizen concerns) [15, pp. 348–350, 416–419]. Within this framework, the ultimate level of coordinating authority matters. While in the Clinton Administration coordination ultimately rested with a National Coordinator of White House rank, coordination for cyber security policy now resides at a lesser level within DHS.

A more complete, and hence complex, picture of the same agenda (again, only a small part of the Homeland Security agenda) shows how many agents at the first level, firms and their individual actors, at the second, a panoply of legal instruments and national plans (including but not only those of the US), and finally a larger and emerging multinational agenda play a role, each with its own area of focus.

A short (and partial) listing of the published policy plans gives a rough idea of the variety of Homeland Security policies.

- DHS, *Interim National Infrastructure Protection Plan* (2005).
- DHS, *National Response Plan* (2004).
- National Research Council, *Making the Nation Safer: The Role of Science and for Countering Terrorism* (2002).
- Office of Management and Budget (OMB), *2003 Report to Congress on Combating Terrorism* (2003).

- RAND National Defense Research Institute, *The Physical Protection Planning Process*, Proceedings of workshops (2002) sponsored by OSD.
- White House, Homeland Security Presidential Directive 7 *(HSPD-7): Critical Infrastructure Identification, Prioritization, and Protection*, 2003.
- White House, *National Strategy for Homeland Security* (2002).
- White House, *National Strategy for the Physical Protection of Critical Infrastructures and Key Assets* (2003).
- White House, *National Strategy to Secure Cyberspace* (2003).
- White House, NSC-63; *Critical Infrastructure Protection* (1998).

5.2 Coordination of Policy

Overall coordination of these policies takes place in three levels [15].

At its highest level, a Homeland Security Coordination Council, modeled in part on the National Security Council (Cabinet level attendance) provides integration.

For the plethora of plans, several key instruments are used.

- *National Response Plan (NRP)*: The purpose of the NRP is to establish the single comprehensive approach required to enhance US ability to respond to domestic incidents. It provides a framework of incident management protocols to address these threats. Established on the basis of HSPD-5—Management of Domestic Incidents (2003), the NRP applies to high impact events requiring a coordinated and, as appropriate, combined response. As a Response Plan, it does not directly establish science policy, though as a policy document it has a major impact [16].
- An integral component of the NRP is the National Incident Management System. Its purpose is to provide a consistent nationwide approach to prepare for, respond to, and recover from domestic incidents of consequence.
- HSPD-7 assigns responsibility to Sector Specific Agencies' (SSAs) designated for protection activities in specific areas—for example the Department of Energy is responsible for protection of energy assets, including the production, refining, storage, and distribution of oil, gas, and electric power. SSAs report to the DHS on these actions.

As the examples of Y2K and fire protection policy illustrate, numerous and engaged constituencies need not be a barrier to effective policy. However, the evolving definition of what comprises Homeland Security, the long histories of many of the organizations involved, and the sometimes inchoate understanding of what the goals of Homeland Security policy are certainly challenge effective policy making.

6 FEDERAL CYBER SECURITY R&D POLICY: AN EXAMPLE OF EFFECTIVE POLICY DEVELOPMENT

Since 1998, the framework for cyber security R&D has evolved, and now shows great promise of providing an effective framework for decision making. It serves as a good example both of how structures for policy coordination and development evolve over time, and also of how coordination can be achieved by the thoughtful use of metrics and the acquisition of supporting data.

Three themes stand out in this evolution

- focusing the policy making process to incorporate needed cross-cutting and integrative perspectives;
- developing and institutionalizing detailed knowledge of both the "baseline" of R&D projects, and current and projected resource allocations for these projects; and
- Continuous progress to seamlessly integrate cyber security R&D into the CIP agenda, and the even broader homeland security agenda, while also tackling difficult challenges such as technology transfer of R&D results.

As such, federal cyber security R&D policy is a good example for readers of this article.

It is worth noting that federal cyber security programs are relatively small, both in terms of the number of people involved and the dollar amounts. Total federal support for cyber security R&D is of the order of $500 mm, with much of it within the DOD. The number of policy makers engaged is also small. Cyber security R&D is a complex topic, however, and requires a probably unprecedented understanding of and cooperation with the private sector in order to be effective.

6.1 Focusing the Policy Making Process

After PDD 63, the Critical Infrastructure Protection R&D Interagency Working Group (CIP R&D IWG) was formed to coordinate federal R&D policy. The IWG included the principal agencies that performed cyber security R&D work (Defense, National Science Foundation, National Institute of Standards and Technology, and Energy) as well as representatives from agencies charged with working with specific private sectors (energy, information and communications, banking and finance, transportation, vital services, and international). The IWG had a complex reporting structure—a theme that runs through the entire evolution of the policy making process here—and reported to three groups: (1) the Committee on National Security, part of the National Science and Technology Council (NSTC)that in turn was chaired by the White House Office of Science and Technology Policy (OSTP); (2) the Committee on Technology (also a NSTC committee); and (3) the Critical Infrastructure Coordination Group, responsible for coordination all CIP policy, which was chaired by the National Security Council.

The CIP R&D IWG organized its work by sector, and, while important work was done, the sector focuses inadequately addressed at least five challenges [10, p. 4]:

- many different sectors contain infrastructure that is vulnerable to exactly the same threats;
- the majority of the sector specific policies did not address the inherent and broadly applicable interdependencies between infrastructure sectors;
- physical threats and solutions were considered separately from cyber threats and solutions;

- the process was challenged to address simultaneously two different paths toward improved security—special efforts to reduce vulnerabilities and improvements coming from the normal efforts to design new infrastructures for higher performance and quality of service;
- The process was also challenged in evaluating new threats and opportunities coming from new technological advances that might not be readily incorporated into the normal design process.

Along with these challenges, starting in 2002 a number of other changes in the overall policy environment led to a restructuring of the organization and focus of federal cyber security R&D policy. The *Cyber Security Research and Development Act* (Nov 2002) gave responsibility for coordinating cyber security R&D to OSTP, with special charges to NSF and NIST to perform research. The *National Strategy to Secure Cyberspace* was issued in February 2003. The report recommended that OSTP coordinates development of an annual federal cyber security research agenda. Homeland Security Presidential Directive 7 (December 2003) required an annual CIP R&D plan to be developed by OSTP and DHS. A series of outside reports on cyber security R&D—from the National Science Foundation (2002), RAND (2002), the President's Information Technology Advisory Committee (February 2005), and the interagency InfoSec Research Council Hard Problem List (November 2005)—all provided perspective on research priorities, or appropriate strategies, for federal cyber security research.

Following one intermediate reorganization of the policy making process, in mid 2005 the Cyber Security and Information Assurance Working Group (CSIA) was formed to shape federal cyber security R&D policy, reporting to both the NSTC Subcommittee on Networking and Information Technology R&D (NITRD) and the Subcommittee on Infrastructure. Reflecting the continuing theme of complex reporting relationships, these subcommittees in turn report variously to the NSTC Committees on Technology and Homeland and National Security.

Three important and positive changes resulted from this evolution.

- NITRD jointly overseeing CSIA made explicit the recognition that cyber security has a broad impact on the nation's interests beyond just CIP.
- In place of sector-specific policies, initiatives are organized around integrative themes addressing both physical and cyber threats and solutions. In the April 2006 cyber security R&D plan [17] there are eight initiatives:
  - functional cyber security and information assurance;
  - securing the infrastructure;
  - domain-specific security;
  - cyber security and information assurance characterization and assessment;
  - foundations for cyber security and information assurance;
  - enabling technologies for cyber security/information assurance R&D
  - advanced and next-generation systems and architecture;
  - social dimensions of cyber security/information assurance.

- Policy themes and projects are compared and correlated with outside perspectives, starting with the NSF and RAND reports, and also the R&D chapters of the "sector specific" plans developed for the National Infrastructure Protection Plan, and international perspectives from the EU and elsewhere. There is also continued consultation with academia, government labs, and industry. There is a strong match between the themes and projects prioritized by all groups, and recent consultations have surfaced only a few projects that were not already in the plans [18].

6.2 Transparency into the Granularity of Projects and Budgets

A second very important evolution in cyber security R&D policy development has been to create the administrative systems so that decision makers can look at the universe of individual R&D projects and the resources applied to each project.

Previously, there was no comprehensive database of cyber security R&D projects across relevant Federal agencies. A major step forward over the past two years has been to create a very specific database by project—a "program" level perspective is too coarse to provide the needed insight into various efforts—cross-referenced by threat, by sector, by technology, by stage of the project (e.g. basic research), and by agency.

Together with this baseline of projects is a breakout of budget support for cyber security research, starting with the President's FY07 budget submission. Previously, budget amounts for cyber security research were difficult to identify because they were often grouped with noncyber security research in other program areas. While some agencies did not participate in the FY07 NITRD budget breakout for cyber security R&D in the FY07 budget supplement (notably DHS and some elements of the Department of Energy), the Office of Management and Budget's annual budget guidance now requires agencies to submit separate budget amounts for cyber security R&D as part of their annual budget submissions.

These reforms provide two important benefits.

- Decision makers are now able to map R&D priorities against the set of specific projects and their funding, and identify gaps in the national agenda;
- Individual agencies can now identify areas where their individual interests and projects complement or duplicate work going on elsewhere in the Federal government.

6.3 Integrating Cyber Security R&D into Broader Agendas

There is a complex and not universally agreed-upon overlap and integration between the concepts of "cyber security", "CIP", and "homeland security", and this article is, simply put, not the place for an adequate discussion of these issues. Suffice to say that there is a multiplicity of plans addressing some of these different perspectives, as well as a widespread feeling that ultimately cyber security R&D policy needs to be integrated into a comprehensive homeland security R&D policy that also includes consideration and linkages to issues like weapons of mass destruction, and other threats to homeland security. There is also a need to adopt a national perspective—not just a government perspective—that incorporates private sector initiatives and priorities.

Both of these thrust for broader integration are underway. Work is currently being done to integrate cyber and weapons of mass destruction R&D policy, with an explicit goal, as one policy maker said, of "erasing some of these plans" [18]. With "sector coordinating councils" that serve as the forum for dialog between government and the private sector, there is also a forum that appears to be reasonably effective in talking with industry. Hence, the current policy framework shows great promise of being able to not only provide an integrated platform for making effective choices about cyber security R&D policy, but also a way of integrating cyber security with other facets of the broad homeland security R&D agenda across both the government and private sector.

6.4 Challenges

The progress made in creating a framework for effective cyber security R&D policy is by no means complete. One major challenge, for example, is to improve technology transfer from federally funded R&D projects into the hands of users. This is a long-standing challenge, and agencies have adopted various strategies and programs to address it. NSF, for example, largely relies on the project specific researchers to disseminate the results of their work, while the service laboratories in the defense department have technology transfer offices charged with that mission. What is important to note is that this issue is very much a focus of attention by policy makers in OSTP and elsewhere charged with cyber security R&D, and that, while the challenge of tech transfer may never be "solved", considerable improvement can, and most likely will, be made.

To summarize, there is value in looking at instances in which policy system has evolved to provide an ongoing and sustaining framework for better decision making. The evolving structure for Federal cyber security R&D policy provides one such example.

7 LESSONS FOR BETTER POLICY DEVELOPMENT

With a broad set of science and technology research initiatives, the role of Homeland Security policy is to drive, in the national interest, to match policy needs with opportunities. Some key themes for improving policy development for Homeland Security include the following:

7.1 Threats Should Prioritize Policy

Effective Homeland Security policy development is challenged by our incomplete articulation of what we are preparing either to defend against or respond to. The inability to clearly identify threats has at least three significant consequences.

- *Blurring the distinction between policy and tactics.* Policy defines the (longer term) investment interests, tactics relate to more immediate actions, and without a lack of clarity in threats, policy and tactical responses are blurred, and implementation suffers.
- *Impeding organizational coordination.* With multiple and indiscriminate threats, different organizations will focus, without clear metrics, on their perceptions, not on the national needs.
- *Impeding the prioritization of policy goals.* Above all, the lack of a clear structure linking threats to goals tries our ability to prioritize resources to goals of greatest importance.

7.2 Tension, Managed Properly, Makes Good Policy

As an element of good policy development, a tension needs to be managed—but not avoided—between duplication of initiatives on one hand, and on the other hand ensuring a portfolio of projects, perhaps in some cases competitive, but integrated into an operable policy framework.

7.3 Better Metrics Are Needed

Sometimes metrics need not measure direct impacts, but can be proxies for outputs that are inherently difficult to capture. A non-Homeland Security example: ALCOA embarked on a corporate wide and intensive program to improve its safety performance. The genius of this high priority initiative was that a focus on safety was in fact a proxy for a wide range of process improvements within the company and its network of suppliers and customers. A safer workplace was not only a laudable goal in itself, it drove major productivity improvements.

7.4 Implementation Matters

Although policy defines and structures a basis for action, the impact of policy ultimately depends on the actions taken by the plethora of actors—Federal, state, and local agencies; the private sector; and individuals—who are, figuratively or literally, "on the ground". Creating the incentives and structures for assessing effort and impact remains perhaps the single greatest weakness in policy development and implementation—and also the greatest opportunity for improvement.

7.5 Clarifying the Line Between National Security and Homeland Security

Among the major challenges are the existing distinctions between Homeland Security and "National Defense" generally. DOD policies and willingness to engage in homeland defense continue to evolve; a clear set of policies here are needed [1, pp. 213–230]. Secondly, the integration of federal programs and investments with state and local capabilities (both as first responders and as an integral part of ensuring defensive and protective capabilities) is an area for improvement. While integrated communications capabilities, for example, are important, a stronger integration into R&D is needed.

However, an expansion of a single integrative organization—an original conceptualization of DHS—would address this second concern, but does not appear to have much promise given current political realities.

7.6 Leveraging Lessons from the Private Sector

The use of market mechanisms may provide novel insight for more effective policy development, particularly in science and research. Managing key financial and operational risks is central to any organization (e.g. even the United States Government has "Continuity of Government" requirements). Greater use of market mechanisms may prove an important part of better linking policy goals with effective implementation.

7.7 Delegating Responsibility and Dividing the Labor: Who Deals With What?

Ultimately, one who studies Homeland Security policy development is faced with a troubling observation: it remains unclear as to who knows what to do, who manages or drives the policy agenda, and who is in charge of implementation. Ultimately, who terminates projects, and nurtures others? Who reviews the portfolios of investments? Who are the "they" who really will make the decisions?

8 CONCLUSION

As this article indicates, policy development for homeland defense not only supports a vigorous science and technology portfolio but also has room for improvement. Both from a science and technology perspective and as an operational set of activities, significant reforms need to be made. Lessons from our existing post 9/11 experience, from other successful (and less successful) federal agencies, and from non-federal sources can all provide useful insights.

In conclusion, four observations were made

- Policy development for homeland security is highly complex for reasons both of substance and organization.
- Policy making and implementation is fundamentally challenged by the need for effective communication and cooperation—with appropriate metrics to support these policies.
- R&D policy faces a tension between duplication and managing portfolios of competitive initiatives integrated through an operable policy framework;
- Competing interests in conjunction with great organizational and topical complexity can mask or provoke a gap in leadership. Who actually is in charge—both with "big" decisions and smaller projects?

REFERENCES

1. Ranum, M. J. (2004). *The Myth of Homeland Security*, Wiley Publishing Company, Indianapolis, Indiana, pp. 1–50 for a good overview (total pages 1–230).
2. Parsons, W. (1995). *Public Policy: An Introduction to the Theory and Practice of Policy Analysis*, Edward Elgar, Brookfield, Vermont, p. 14 (total pages i–xviii, 1–675).
3. Stein, J. (1966). *The Random House Dictionary of the English Language; The Unabridged Edition*, Random House, New York.
4. Munger, M. C. (2000). *Analyzing Policy: Choices, Conflicts, and Practices*, WW Norton and Co., New York, pp. 14–15 (total pages I–xvii, 1–430).
5. The White House. Office of Homeland Security (2002). *National Strategy for Homeland Security*, The White House, Washington, D.C., July 2002, p. 2. (total pages 1–71).
6. The White House (1999). *Defending America's Cyberspace: National Plan for Information Systems Protection (draft)*, The White House, Washington, D.C., May 1999 (total pages i–xxvi,1–128).

7. Fischer, Eric A. (2005). Creating a National Framework for Cyber Security: An Analysis of Issues and Options. CRS RL 32777, Congressional Research Service, The Library of Congress, February 22, 2005. p. 6, 1–56.
8. *The White House, The National Strategy to Secure Cyberspace.* Washington, DC: The White House, February 2003.
9. Paller, A. (2006). *Research Director, The SANS Institute, Bethesda, Maryland*. Presentation at Carnegie Mellon University, May 2006.
10. *Commentary from Guidance for Writers on Wiley Handbook of Science and Technology for Homeland Security* (2006). John Wiley and Sons, Hoboken NJ.
11. Executive Office of the President, Office of Science and Technology Policy, Department of Homeland Security, Science and Technology Directorate (2004). *The National Plan for Research and Development in Support of Critical Infrastructure Protection*, Washington, DC, pp. 23–67 provides detail in each policy area (total pages 1–81).
12. The Institute for Information Infrastructure Protection www.theI3P.org.
13. *Critical Infrastructure Protection Workshop for Academic and Federal Laboratory R&D Providers* (2005). Science and Technology Directorate, Department of Homeland Security, Washington, DC, June 29, 2005.
14. *IEEE Special Session on Integration and Interoperability of National Security Information Systems* (2006). Cambridge, MA, June 8–9, 2006.
15. Kean, T. H., Hamilton, L. H., Ben-Veniste, R., Kerrey, B., Fielding, F. F., Lehman J. F., Gorelick, J. S., Roemer, T. J., Gorton, S., Thompson, J. R. (2004). *The 9/11 Commission Report: Final Report of the National Commission on Terrorist Attacks Upon the United States*, W.W. Norton and Company, Inc., New York, pp. 423–428 (total pages).
16. U.S. Department of Homeland Security (2005). *Interim National Infrastructure Protection Plan*, Washington, D.C., February 2005, pp. 38–39 (total pages 1–35).
17. U.S. Department of Defense (2005). *Strategy for Homeland Defense and Civil Support*, Washington, D.C., June 2005 pp. 36–38 (total pages 1–40).
18. National Science and Technology Council (2006). Interagency Working Group on Cyber Security and Information Assurance. *Federal Plan for Cyber Security and Information Assurance Research and Development*, National Science and Technology Council, Washington, April 2006.
19. Voeller, J. (2006). OSTP, December 2006.

FURTHER READING

US Department of Justice. *Computer Crime and Intellectual Property Section* www.cybercrime.gov.

US Government Accountability Office (2001). Testimony before the Subcommittee on National Security, Veterans Affairs, and International Relations, House Committee on Government Reform. *Homeland Security: Key Elements of Risk Management* Statement of Raymond J. Decker, Director Defense Capabilities and Management, October 12, 2001. www.house.govInternational CIIIP Directory, based on the G-8 CIIP Experts Initiative. E-mail ciip-directory @niscc.gov.uk for more details.

Other US Government documents: *National Strategy for the Physical Protection of Critical Infrastructure and Key Assets; National Strategy for Homeland Security; National Strategy to Secure Cyberspace.*

David, M. (2002). *Concepts for Enhancing Critical Infrastructure Protection Relating Y2K to CIP Research and Development*, Santa Monica.

National Infrastructure Security Co-ordination Center (NISCC) www.nisc.gov.uk.

THREATS AND CHALLENGES TO HOMELAND SECURITY

David M. Weinberg
Practical Risk LLC, Rio Rancho, New Mexico

1 THREAT SPECTRUM

This survey article is not meant to be exhaustive in detail or citations. Rather, it highlights some conventional threats and challenges and also attempts to tease the reader to consider some less conventional threats. This is done to stimulate the interest of the research community, and to play their role in one of the most complicated issues facing the United States and its people.

Within the context of governmental homeland security, the word *threat* has different meanings to different people and organizations. This article attempts to look at threat in conventional and some unconventional ways. Similarly, the term *challenges* carries much semantic heft, and it too will be considered in terms of conventional ways and otherwise.

Threat is commonly taken to mean that set of activities and purposes aimed at doing harm. Although this definition may be thought to specifically refer to the threat of terrorism, it actually applies to natural hazards and catastrophic accidents as well. A discussion of threat can be broad indeed.

Conventionally, terrorism threat is generally dissected into two components: namely, intent (to perform an act) and capability (resources, including intellectual, to accomplish the act). Recent work by Williams [1, 2] adds a third dimension (or metric), at least to radical jihadist terrorism, namely, authority. Within the Department of Homeland Security (DHS), some workers also break capability into subcomponents such as the intellectual capability to conceive and design what is needed for an attack and the capability to infiltrate the nation, organize all necessary manpower and material logistics, and remain undetected until the attack is executed.

Clearly, the topic of threat includes getting into our adversary's head. This topic is being addressed by the National Consortium for the Study of Terrorism and Responses to Terrorism (START) [3]. Therefore, for the purposes of this article, it is preferable to start this discussion with something a bit simpler than *threat* and examine things that could cause harm in a somewhat more generic sense.

2 TYPES OF THREATS AND CHALLENGES

Terrorism attacks can generally be broken into those that are physical attacks (i.e. 9/11), virtual attacks (i.e. computer hacking and viruses), and a category best described as "other". Physical attacks represent a broad spectrum of possible attack modes (often referred to as *threats* or *threat vectors*) that include the likes of much of what is seen in the media on an all-too-frequent basis. These attacks include improvised explosive devices (IED), a mode faced repeatedly by our troops in Iraq, backpack bombs such as used in the London and Madrid bombings, and suicide vests seen worldwide. An IED's big brother is a vehicle or vessel borne improvised explosive device (VBIED), differing from the IED in its delivery mechanism, size, and potential for destruction. These two attack modes or threat types make up the greatest statistical population of terrorist attacks across the world [4, 5]. Less often experienced within the homeland are other physical attacks that include assassinations and kidnapping, although we have seen these modes perpetrated by terrorists carried out on US citizens abroad.

These conventional physical attacks represent a type-of-attacks spectrum, namely from the somewhat impersonal attack on a group to the very personal attack on an individual. In both cases, there is some individual or group that has conspired to directly harm the homeland and/or its citizens by using a specific designed-for-purpose weapon.

As a class, such threats are fairly predictable in their effect, and to some degree, in their standard practices and procedures. While various types of attacks are "pigeonholed" below for convenience of discussion, it is acknowledged that such summarization may contribute to artificially discretizing what is a continuous, multidimensioned spectrum. For brevity and simplicity, neither multiple attacks, simultaneous or those along a predetermined timeline, are addressed. The reader is referred to other portions of this volume to investigate some of the complications raised by these attack scenarios.

2.1 Conventional Physical Attacks

Attacks can be direct or indirect. Protection and prevention against terrorist acts is a problem not unlike the "inverse problem" in conventional deterministic modeling. Given a result, some (perhaps very large) set of paths exist to go from the initial condition to the observed result (each path representing one determined path). The security problem faced, of course, is that all paths cannot be interdicted, so judgments must be made regarding the various paths and actions taken to disrupt a most likely path.

Evaluation of multiple paths is not unlike the approach taken by law enforcement and counterterrorism by "thinking like the criminal/terrorist", and defining what set of things must be brought together for the act to be realized. It becomes a problem in inductive logic whereby the system of reasoning extends deductive logic to less-than-certain inferences [6]. In this example, a sequence of events leading to the result are believed to support the conclusion, but do not ensure that this conclusion is right. Unfortunately, inductive approaches can miss the unanticipated event [7], sometimes with horrific consequences such as 9/11.

The predictability of such types of direct physical attacks, however, is hampered not only by the number of possible attack paths needed to be considered for interdiction but also by the ingenuity of the adversary. Adversarial ingenuity is demonstrated frequently by their design, and use of less well-known weapons (i.e. peroxide-based explosives, the root cause of our inability to take containers of liquids on airplanes, home-built armor-piercing explosively formed projectiles (EFPs) used in Iraq, and ability to quickly adapt to countermeasures) presents an enormous challenge to the nation.

Subsequent to 9/11, a federal directive was promulgated throughout the rail and chemical sectors to cease shipments of chlorine gas fearing that a rail car might be attacked in a populous area killing or injuring many. A few days later, the directive was lifted because high-density population areas needed chlorine to purify drinking water supplies. Within about 90 days of the attack on the Pentagon, the Blue Plains Wastewater Treatment Plant in Southwest Washington, DC, converted its process so that large tanks of chlorine and sulfur dioxide (an equally hazardous gas) would be essentially eliminated from the plant site and switched to an alternative technology. These examples illuminate preventive actions against what many call indirect attacks because terrorists could use existing infrastructure against the nation. During its first 4 years, the DHS spent significant resources identifying terrorist-created chemical releases as an indirect attack mode with the result that a new organization was created to define and ensure security standards across the chemical industry.

The existence of standards across a sector, however, does not necessarily correlate to security. For instance, chemical contamination of a foodstuff could cause as much damage and panic as the release of a noxious plume from some manufacturing plant. Equally insidious, counterfeit materials (parts or substances) used in sensitive applications can also constitute threats to people, or in some cases, economic well being. In an open society, tracking materials—and people—from origin to endpoint creates a sociological problem, which the nation continues to struggle with.

2.2 Nonconventional Physical Attacks

The attacks described above are classed as being conventional in nature because the means of executing them are reasonably straightforward. Similarly, the tactics used and results obtained from these attacks are conventional. There are, however, less conventional types of attacks of importance to the nation. At the forefront, of course, is that group of attacks termed *weapons of mass destruction/effect* (WMD/E). Those attacks are covered elsewhere in this volume and are not discussed here.

Another unconventional, but not unknown, attack is that class considered denial of use attacks. These scenarios encompass a myriad of agents dispersed into, on, or around infrastructure important to continuity of operations. The anthrax attacks in 2001 using the US Postal Service's Trenton Processing and Distribution Center as a delivery system is one example of such denial of use attack. Unfortunately, in the case of the 2001 attacks, 5 of the 22 citizens exposed to the spores succumbed. Subsequently, then-Senator Tom Daschle's office suite in the Hart Building on Capitol Hill was found to have anthrax contamination causing building evacuation and shutdown of the government mail service until decontamination efforts could clean the premises for occupancy. The Trenton postal facility was not reopened until March 15, 2005, some three and a half years after the contamination was discovered. Had this attack been to a "critical" commercial facility (i.e. one that is essential to the nation and without substitute), it is questionable whether the corporate enterprise or the country could have survived such a lapse in service. Another scenario that could result in denial of use is that of a radiological dispersion device (RDD). In this scenario, radionuclides from any number of sources could be dispersed using explosive or aerosol means and could result in denial of use for years, even decades depending on the material used.

Biological and RDD attacks are not necessarily aimed at creating many casualties. Rather the economic hardship and/or the fear created within the population that works in or near the facility thereby preventing the facility from performing its necessary function may be the true goal. Although such attacks of a neighborhood retail facility may cause no great harm to the nation or inconvenience to the population, there are many facilities that if shut down for extended periods of time can seriously impact the national economy (Wall Street) or national security (single-source for critical military component).

Two other nonconventional attack types being faced by the nation include virtual (cyber) attacks and attacks being staged by hostile nation-states. The former of these is dealt with extensively elsewhere in this volume, and the latter lies outside the scope of the volume. Neither is discussed here.

Other nonconventional attacks that seem farfetched, but nonetheless could wreak havoc throughout America also exist. They are called *attacks* here for the purpose of continuity, but they actually represent broad challenges as well. The first of these types constitutes a form of economic attack by currency, trade, or resource manipulation. These attacks could emerge from nation-states, but could also come from other, even transnational groups bent on controlling some particular part of the commercial or financial market. One example that happened, but was notably nonnefarious, was the over $300b investment in high-profile commercial American real estate by the Japanese in the 1980s. In the early 1990s, market forces reduced the value of these investments by as much as 50% [8]. While this example is one of arguably benign global investing, the question posed becomes "What if intentions are nefarious?"

One such example is clearly illustrated by the 1960 formation of the Organizations of Exporting Petroleum Countries (OPEC) and subsequent withholding of oil exports to the United States in the early 1970s and 1980s. Although academicians continue to argue over the root causes of the embargos, the net result was an energy crisis in the United States that, at least in part, was driven by a political stance taken to punish the alleged wrongdoer. Other technical and geopolitical events eventually nullified the problem, but as a nation, the problem has still not gone away; we are more dependent on foreign oil imports (by over a factor of two) than we were when the embargoes were first exercised 30 years ago. How can the United States protect itself from such economic attacks? "Energy Independence", while making a catchy bumper sticker, is as demonstrably lacking in substance as "Financial Independence". The effects of globalization are rooted deep in American society, and our interdependencies on both external supplies of energy and money create a formidable challenge in a world of highly heterogeneous cultures.

Another nonconventional attack that lays well beyond media headlines constitutes an equally formidable challenge. Simply put, it is the attack, perhaps self-inflicted, that the nation faces with respect to its intellectual infrastructure. Most readers can recall at least one article within the last year chiding "education in this country" for poor scores in science and math, relative to the rest of the world. It is similarly recognized that American colleges and universities are "educating the world". The implications of failing elementary and secondary education for its citizens and excellence at the college level attracting students from across the globe are not straightforward. However, two examples might be useful in stimulating research into how the nation can address this challenge.

Corporate recruiters are always looking for the "best and brightest" regardless of the particular type of expertise they represent. For jobs within the United States, significant resources must be spent if the desired employee is not a US citizen. For jobs within the government that require a security clearance, US citizenship is even more important. Looking at technical fields, the percentage of US graduate students who are US citizens has been decreasing for decades (except for a brief reversal following 9/11 [9]). A recent article [10] states that:

> "International students, especially at the graduate level, are considered an important brainpower infusion to the United States. In certain fields like engineering and physical sciences, foreign students account for more than 40 percent of total students at the graduate level, according to CGS (*Council of Graduate Studies*).
>
> 'There is not a strong domestic pipeline in those disciplines,' said Catharine Stimpson, dean of New York University's Graduate School of Arts and Sciences. 'The U.S. has a strong dependence on international talents.'"

The implications of US dependence on offshore intellectual infrastructure are discussed at length by Canton [11]. As the scientific and technical challenges to homeland security evolve, finding qualified personnel will represent sociological and educational challenges as difficult as anything in engineering or the sciences. Like the national physical infrastructure, our intellectual infrastructure is sufficiently intertwined with that of other nations that makes unilateral solutions (intellectual independence) impossible. From a threat, perspective, denial of access to information or knowledge can be an effective attack not dissimilar to denial of use.

3 ORIGIN OF THREATS

Within the scope of an overview article, exhaustive enumeration of all of the various sources of threats that play a role in homeland security would be redundant to other articles in this volume, and could go on for volumes in themselves. For greatest simplicity, four general types of threat considered here are international terrorism, domestic terrorism and hate groups, natural hazards, and catastrophic accidents. Three are anthropogenic, hence to some degree they can be defended or prevented, but the results of all four must be considered in the context of response and recovery.

3.1 International Terrorists

According to the Memorial Institute for the Prevention of Terrorism, there are over 1200 international terrorist groups [12], all of whom have agendas at odds with normal political intercourse. Although national attention has highlighted Al-Qaeda since 9/11, other groups are also "on the radar". Specific motivational differences between the groups are not of importance to this article. Rather, it is important to understand what kinds of attacks against what kinds of infrastructures may be posed by the transnational terrorists. As mentioned earlier, intent and capability are two venerable types of information needed to judge how realistic a threat from a particular group may be. Also mentioned earlier is the newer concept of authority, at least for radical Muslim jihadists. For more insight into this aspect, the reader is referred to the work of Williams cited below. It may be that his concept could be extended to other groups as well. Simply put, the execution of any particular terrorist event depends on someone effectively saying "Go". Williams shows the role played by fatwas, legal and religious justifications, and speeches given by radical Muslims intent on causing harm. However illogical, that role—choice of target type, what is and is not acceptable behavior during the execution of the attack, and the weapons used (each providing important insights to potential defenders)—can also be seen in historical criminal behaviors (i.e. anecdotal prohibition of violence on family members by the Mafia). Getting this kind of insight is an immense challenge for the nation if only because these reasonings and rationalizations are dynamic even within the groups themselves. Complexity is not a reason to avoid trying to understand these drivers, but developing an institutional understanding of another culture can take decades.

3.2 Domestic Terrorists and Hate Groups

The April 19, 1995, bombing of the Alfred P. Murrah Federal Building in Oklahoma City by disaffected military veterans brought national attention to a threat nexus that had largely been ignored by the public since publicity of the Symbionese Liberation Army, the Black Panthers, and others in the 1970s. Timothy McVeigh and Terry Nichols' attack graphically demonstrated how ill-prepared the country was for dealing with violent acts perpetrated by its own internal terrorists.

Organized domestic terrorist groups such as the Aryan Nation, the Klu Klux Klan, and the New Order reside in the twilight between a "conventional" terrorist group and a "conventional" hate group. The line separating the two may be dim. However, radicalization by Muslim jihadists and others in homeland prisons is a growing and morphing threat, which is not necessarily racially based. Without dwelling on fine distinctions between domestic terror and hate groups, the result of their actions can still terrorize segments of our society or citizens within a particular region. All this compounds the problem of operating cells of transnational groups (e.g. Al-Qaeda, Al-Fuqra, and Aum Shinrikyo) that may form alliances of convenience with domestic groups, including criminal enterprises, possibly with or without their explicit knowledge. Groups such as the Animal Liberation Front and Earth Liberation Front often raise parochial headlines, but are not broadly thought of as national threats.

3.3 Naturally Occurring Challenges

In the simplest terms, natural hazards can be classed into those that are to some extent predictable allowing the population to take some preparatory measures, and those that "come out of the blue". The former would include floods, hurricanes, tornados, some biological events, and wildfires (initiated by lightning strikes). The latter would consist of earthquakes, some biological events, and some volcanic eruptions. Man has been living with and fearing the vicissitudes of Mother Nature for millennia. But, only recently has technology developed to the extent that some of these threats can be prevented (in rare cases) or engineered around to reduce consequences. Medicinal prophylaxis is arguably the most illustrious example of man's ability to prevent a threat from causing harm to health. Certain structures such as levees and dams can mitigate catastrophic impacts but do not prevent threats to them: often making them critical facilities. Similarly, preparations for hurricanes and tornadoes may mitigate impacts, as does buildings designed for earthquakes; but such natural hazards are unique (no two will be exactly alike in consequence or response) and will occur as long as natural processes continue. As demonstrated too well, the national response to Hurricane Katrina was reminiscent of the response to the tragedy of 9/11.

Interruptions to the global integration of economies [13] caused by natural disasters and the continuing interweaving of physical and commercial infrastructure (i.e. chemical feedstocks from Mexico and oil and gas energy from Canada), not only represent serious challenges to homeland security professionals, but also pose a great scientific challenge. Clearly, knowledge-based actions have been shown to have saved lives through weather modeling. Scientific efforts and innumerable data collection efforts have saved lives by evacuating some remote Oregon areas prior to the eruption of Mt St Helens. However, such apparently academic pursuits are rarely seen (or funded) as homeland security efforts; yet the products of these research fields provide much information in the effort to prevent serious consequences of these threats.

3.4 Catastrophic Accidents

In ways similar to natural hazards, catastrophic accidents create impacts that might be indistinguishable from terrorist attacks. Such accidents could include the rupture of rail tank cars filled with toxic chemicals, the core meltdown at a nuclear power plant, a space shuttle crashing, equipment wear/burn-out with catastrophic failure, and so on. Unfortunately, all of these examples did (or nearly did) happen in recent history, but fortunately none occurred in large US population centers. For all intents and purposes, the possibility of the "event that never happened" spawned the field of probabilistic risk assessment (PRA) back in the 1970s when the government and private industry had to develop ways to plan for the risk of such events. The pursuit of PRA and fault-tree analyses by statisticians and engineers over the past three decades has helped reduce the likelihood of such catastrophic events by creating engineering and public safety standards that have prevented Bhopal- or Chernobyl-type events here. These disciplines continue to offer insights into the nation's homeland security.

4 PREVENTION AND PROTECTION

In J. Cummings' article in this volume, he refers to Merriam-Webster's online dictionary for some important definitions [14]. Prevention is defined in several, interlinked ways. Simply put, the DHS seeks to ensure that attacks on the homeland and its people do not occur. Often this is thought to be primarily a function of the intelligence and counterterrorism agencies; those aspects are covered elsewhere in the Handbook. Protection is essentially defined as shielding from an event or attack. Taking these definitions and the threat spectrum discussed above as the context for the technical challenges the nation faces, four activities evolve that provide focus for security professionals, namely; detect the threat, deter the attack, defend against its outcomes, and/or devalue the target. Much is written elsewhere in the Handbook regarding the first three of these, but the last one, devaluing the target (for the attacker) brings into play resiliency and redundancy.

Redundancy is an important and useful way to devalue any given target. However, redundancy is largely an asset-by-asset approach that provides protection from a single-point-of-failure situation. While this approach has been taken by some parts of the private sector, it is not physically or economically feasible to create redundancy for many of the nation's most important infrastructure assets. A large hydroelectric dam is where it is in part because of unique geography. Refineries are extremely expensive and, considering issues as divergent as pipeline connectivity and environmental regulation, cannot easily be duplicated.

Resiliency is a concept that applies to individual assets and to systems or networks of assets. Simply put, resiliency is a design property that allows the asset, network, or system to "fail gracefully", or in such a way as to allow consequences of the failure to be minimized. Consider the automobile tire that you can drive on even after it is ruptured. Self-healing materials and networks are under intense study now, and will continue to play a growing role in homeland security. Greater sophistication in modeling and simulation is also giving rise to designing ways such that systems may actually heal themselves or fail gracefully. However, resiliency must become even broader. We recognize that the interdependencies of the nation's infrastructure are far-reaching and mostly poorly understood. Work in this arena is addressed in the Handbook section titled System and Sector Interdependencies, and the reader is referred to that section for more details.

5 CHALLENGES TO DHS

Some challenges to the DHS and the nation are scattered within the context of the threat spectrum. Many of these challenges are obvious and straightforward, such as sensors for detecting harmful substances or organisms, materials that can provide more and better protection by strengthening facilities while keeping costs reasonable, and software tools to frustrate cyber attacks before they can damage our physical and/or economic infrastructure. Technical challenges related to catastrophic accidents mimic those for natural hazard and terrorism attacks when it comes to physical infrastructure protection. Conventional attacks, by terrorists, nature, or accidents, all require advances in a variety of scientific and engineering endeavors. Less conventional, however, are the security considerations and approaches that will be needed to protect new technologies as they are deployed throughout our infrastructure. There are also two other challenges that the DHS faces as an institution that represents and works for the nation.

5.1 Defining the Unacceptable

In some ways, this problem is reminiscent of the problem faced by the Environmental Protection Agency since its inception "How clean is clean?" Within an attack context, it becomes "How bad (number killed or hurt, dollars lost, people traumatized, etc.) is bad?", and "What constitutes acceptable losses?" As painful as these questions are to contemplate, they must be considered.

Since its inception, the DHS has provided billions of dollars to state, local, tribal, and territorial governments in the form of grants to make the nation safer from terrorism attacks. Both 9/11 and Hurricane Katrina brought public attention to the simple fact that very large-scale events are a national issue requiring a national response. But at what price and for how long? There is no politically correct answer to the question of how many casualties are acceptable, but unfocused funding and unnecessary preventative processes and material are equally unacceptable.

The DHS Secretary, Michael Chertoff stated that "risk management must guide our decision making as we examine how we can best organize to prevent, respond and recover from an attack". To allocate resources, money, material, or personnel, the DHS must prioritize. However, prioritization, like triage, requires that choices be made regardless of how uncomfortable they may be. For many reasons, classical statistics cannot help in the prediction of terrorist attacks although they have proven useful, at least to the insurance industry, to help planning for natural events. There remains, however, the paradox of quantitative (defensible but often technically intricate) versus qualitative (what seems right, albeit possibly quite subjective) solutions within the political environment where there will be winners and losers for federal resources. Making those choices is a significant challenge for the DHS.

5.2 Communicating to the Public

In today's era of 24/7 global news, Edward R. Murrow once said "The newest computer can merely compound, at speed, the oldest problem in the relations between human beings, and in the end the communicator will be confronted with the old problem, of what to say and how to say it." This concept is particularly pertinent to homeland security in general. In simple terms, most people ask two questions: "How likely is something bad to happen?" and "If that bad thing happens, how bad will it be?"

Insight into how the government and private industry has attempted to communicate answers to these questions in the past is, sometimes humorously, documented by Lewis [15]. Most people have great difficulty in fathoming just how likely any number of bad things really are. Schneier said [16], "I think terrorist attacks are much harder than most of us think. It is harder to find willing recruits than we think. It is harder to coordinate plans. It is harder to execute those plans. It is easy to make mistakes. Terrorism has always been rare, and for all we have heard about 9/11 changing the world, it is still rare." Even a casual review of terrorism incidents as compared to violent crimes proves him out. Communicating the risk of both man-made and natural catastrophic events remains a major challenge to the DHS and the nation as a whole.

6 RESEARCH NEEDS

The complexities of our nation's infrastructure belie simple listings of technological needs. The same complexities require bringing together very complicated components, systems, and results. Such complications and the challenges they bring forms most of this Handbook. For this author's part, however, there are three major categories of research needs that will help move us closer to a more secure nation.

The first of these includes more sophisticated modeling and simulation (M&S) of extremely rare events, terrorist systems, and networks, and outcomes from conventional and unconventional attack modes. Thanks to massive increases in computational capabilities, M&S can now be done for problems that only a decade ago were intractable. However, M&S is not reality, nor will it ever replace all of the possibilities that reality represents. That said, M&S does provide important tools into understanding phenomena (physical, virtual, and even psychological) that otherwise could simply not be gathered.

For instance, today's blast models are based on materials with energy equivalent to trinitrotoluene (TNT). The damage done to structures is modeled with a characteristic pressure wave caused by a certain amount of that explosive located at a specified distance from the modeled structure. However, despite the number of plots accomplished and foiled that utilized "bathtub" or peroxide-based explosives, little is known about their explosive characteristics against a variety of target types. It is infeasible to run experiments on all possible combinations of conventional and other explosives and targets. Therefore, more work is needed to better define envelopes of behaviors enabling better-informed protective decisions to be made. Similar statements can be made regarding impact of natural hazards on man-made structures. Some level of experimentation has been done, but many of the historical impacts do not translate directly to today's infrastructure and their interdependencies. In this author's opinion, the M&S of the nation's infrastructure interdependencies is the single greatest and perhaps most difficult M&S infrastructure security challenge facing the nation. It is a problem of such complexity and across so many orders of magnitude that it will take decades to master.

Knowledge management is a second category of research needs. Information overload has become a major challenge in today's technological world. While new sensors and other data are collected (see the Sensing and Detection section of this volume), how we translate the data first into information and then into knowledge are pushing security professionals (and their IT systems) to their limits. Managing all of that in a retrievable way has become a significant and expensive challenge. Within the last decade, IT architectures began evolving from strictly hierarchical to more relational ones. More work needs to be done in this and associated areas in the pursuit of data, information, and knowledge. It is only when easily accessible broad knowledge across many disciplines is fused with judgment that decision makers can plot the best path for their enterprise or the nation.

More research in the social and psychological sciences constitutes the third area of great need for the DHS and the nation. There are two over-arching drivers for these areas to be addressed. First, great good can be accomplished by extensive and excellent scientific advances in all sorts of technologies. While supporting science at large, how these advances can be used to support the making of federal policy, in fact, provides the true return on investment for the government. Second, inasmuch as the government's role is to establish and execute the political will of the nation through policy, gathering data on what the nation wants, needs, and how willing they are to accept it is a supremely difficult task. In some measure, the challenge of communication feeds this research need as well, because policy is fed by communication, which in turn needs to be communicated back to the nation. Within infrastructure protection, a clear understanding of the risks run, and therefore the protection and prevention activities required to address that risk, must be communicated to the consumer, for in the end, it is the consumer that will have to live with the decisions driven by those risks, or less desirably, the perception of those risks. Alfred Hitchcock, who knew something about creating terror in people's minds stated: "There is no terror in a bang, only in the anticipation of it." By being psychologically and socially prepared for the bang, regardless of it being man-made or natural, the impact of the event can be reduced.

7 CONCLUSIONS

The extent of this Handbook's Table of Contents illustrates that homeland security is as complex as life itself. Invigorated by the terror attacks of 9/11, homeland security has expanded to include any and all catastrophic events. Total protection from and prevention of catastrophes is not achievable. However, their impacts to the nation can be partially mitigated by technology, partially by barriers (including regulation and legislation), and to a significant degree by knowing and understanding the risk, which includes threat, the knowledge and understanding of which must be objective, and not be used for fear mongering. In hindsight, the 9/11 attacks are understandable, perhaps even predictable. The perpetrator's ability to execute an attack must be seen as the target of protection and prevention technology. It is within our nation's ability to impact the execution of an event, be it from terrorists or man-made mistakes, and by so doing prevention and protection will make their contribution to homeland security.

REFERENCES

1. Williams, J. F. (2007). Authority and the role of perceived religious authorities under Islamic Law in terrorist operations. *Proceedings Federalist Society—Georgia State University, Atlanta*.
2. Williams, J. F. (2007). Al-Qaida strategic threats to the international energy infrastructure: authority as an integral component of threat assessment. *Proceedings Carlton University—Ottawa Center for Infrastructure Protection, Ottawa Canada*.
3. National Consortium for the Study of Terrorism and Responses to Terrorism (START). http://www.start.umd.edu/, 2008.
4. Memorial Institute for the Prevention of Terrorism. http://www.mipt.org/IncidentTacticModule.jsp, 2007.
5. National Consortium for the Study of Terrorism and Responses to Terrorism. http://209.232.239.37/gtd1/charts/weapon_type_pie.gif andhttp://209.232.239.37/gtd2/charts/weapon_type.gif, 2008.
6. *Stanford Encyclopedia of Philosophy*, http://plato.stanford.edu/entries/logic-inductive/, 2008.
7. Taleb, M. N. (2007). *The Black Swan—The Impact of the Highly Improbable*. Random House, New York, p. 366.
8. Pristin, T. (2005). Commercial real estate; echoes of the 80's: Japanese return to U.S. market. *The New York Times* http://www.nytimes.com/2005/01/26/business/26prop.html.
9. Kujawa, A. (2005). *Foreign Student Enrollment at U.S. Graduate Schools up in 2005*, http://www.america.gov/st/washfile-english/2005/November/20051107160749aawajuk0.8633234.html.
10. Du, W. (2007). *Foreign Student Enrollment Rebounds in U.S.; MSNBC*, http://www.msnbc.msn.com/id/20393318/.
11. Canton, J. (2006). *The Extreme Future —The Top Trends That Will Reshape the World in the Next 20 Years*. Plume, New York, p. 371.
12. Memorial Institute for the Prevention of Terrorism. http://209.232.239.37/gtd2/browse.aspx?what=perpetrator, 2008.
13. Friedman, T. L. (2005). *The World is Flat*. Farrar, Straus, and Giroux, New York, p. 660.
14. Merriam-Webster online dictionary. http://www.merriam-webster.com/, 2008.
15. Lewis, H. W. (1990). *Technological Risk*. WW Norton & Company, New York, p. 353.
16. Schneier, B. (2006). *The Scariest Terror Threat of All*, http://wired.com/politics/security/commentary/securitymatters/2006/06/71152.

FURTHER READING

Chalk, P., Hoffman, B., Reville, R., and Kasupski, A.-B. (2005). *Trends in Terrorism*. RAND Corporation, Santa Monica, CA, p. 75.

Garcia, M. L. (2006). *Vulnerability Assessment of Physical Protection Systems*. Elsivier, Amsterdam, p. 382.

Haimes, Y. Y. (2004). *Risk Modeling, Assessment, and Management*. John Wiley & Sons, New York, p. 837.

Jenkins, B. J., Crenshaw, M., Schmid, A. P., Weinberg, L., Ganor, B., Gorriti, G., Gunartna, R., and Ellis, J. O., Eds. (2007). *Terrorism: What's Coming—The Mutating Threat*. Memo rial Institute for the Prevention of Terrorism, Oklahoma. website: http://www.terrorisminfo.mipt.org/pdf/Terrorism-Whats-Coming-The-Mutating-Threat.pdf9.

Kline, M. (1967). *Mathematics for the Nonmathematician*. Dover Publications, New York, p. 641.

Mueller, J. (2006). *Overblown*. Free Press, New York, p. 259.

Post, J. M. (2005). *The Al-Qaeda Training Manual; USAF Counterproliferation Center, Maxwell Air Force Base*, U.S. Government Printing Office 2005-536-843, p. 175.

Presidential Decision Directive 63: Protecting America's Critical Infrastructures. The White House, May 28, 1998, http://www.fas.org/irp/offdocs/pdd-63.htm.

Ridgeway, J. (2004). *It's All for Sale*. Duke University Press, Durham & London, p. 250.

Roberts, P. (2005). *The End of Oil*. Houghton Mifflin Company, New York, p. 399.

Sauter, M. A., and Carafano, J. J. (2005). *Homeland Security*. McGraw-Hill, New York, p. 483.

Schneier, B. (2006). *Beyond Fear*. Springer, New York, p. 295.

Securing Our Homeland. Department of Homeland Security Strategic Plan, Washington, DC, (http://www.dhs.gov/xlibrary/assets/DHS_StratPlan_FINAL_spread.pdf).

TERRORIST ORGANIZATIONS AND MODELING TRENDS

Irmak Renda-Tanali

University of Maryland University College, Adelphi, Maryland

Christopher D. Hekimian

DXDT Engineering and Research, LLC, Hagerstown, Maryland

1 INTRODUCTION

The US Joint Tactics, Techniques and Procedures (JTTP) for Antiterrorism, Joint Publication 3-07.2 as cited in [1] states: The terrorist organization's structure, membership, resources, and security determine its capabilities and reach". Any method of analysis and understanding that can be directed against the broad threat posed by terrorist organizations (TOs) can contribute to mitigation strategies. Moreover, since TO activities are often covert, and government secrets regarding intelligence pertaining to TOs are closely guarded, knowledge, understanding, and analytical tools may be the only assets that analysts have to direct toward terrorism threat mitigation. Understanding the structures and modes of operation of terrorist groups is a key enabler in the assessment and mitigation of the terrorism threat. Organizational structures of terrorist groups that may appear complex during initial assessments may be more understandable when laid out in systematically modeled formats. This article focuses on existing and ongoing efforts related to terrorist data analysis and modeling aspects that deal with terror risk mitigation.

2 SCIENTIFIC OVERVIEW

The research in support of understanding the construct and operation of TOs can be categorized into (i) studies that focus on definition/conceptual issues; (ii) case studies of particular regions, countries, movements, and events; (iii) counterterrorism and crisis management; (iv) terrorism data analysis and modeling, and other related topics. This article deals with terrorism data analysis and modeling. Discussion of an overview of the seminal thinkers and works on terrorism studies were provided by Hopple in reference 2.

Although there is no universally agreed upon definition of *terrorism*, various definitions exist and have been adopted by organizations worldwide. Therefore it is helpful to disclose the definition up front with the disclaimer that other definitions may or may not be equally valid for the discussion at hand. Key researches on the current bases for classification and categorization of TOs have been summarized in unclassified military documents that are referenced in this article. Other sources on the topic include US Congressional reports and other government and academic reports. The RAND organization provides a large amount of recent research on the operation and function of TOs and has been cited multiple times in this article. A large amount of current research pertaining to the organizational structures of TOs and how those structures tend to affect operations and vulnerabilities are available in military and academic reports and journal articles by Fatur (2005), Shapiro (2005), and Hoffman (2004). There are a wide range of organization modeling methods and scholarly research, including case studies, dissertations, and theses, and articles have been cited in each section of this article. The work of Barry Silverman of University of Pennsylvania, in modeling terrorist behavior, and of Kathleen Carley of Carnegie-Mellon, in network organization modeling, is at the forefront of the advancement of these methods and their application. The reader is encouraged to obtain these documents to find more detailed information on those topics that are beyond the scope of this article.

3 TERRORIST ORGANIZATIONS

3.1 Terrorism Definitions

The definition of what constitutes "terror", "terrorism", and hence a "terrorist" or "terrorist organization", is a matter of significant debate. Some embrace the position that one man's terrorist is another man's freedom fighter. In fact, there is a plurality of reasonable definitions suitable to provide context and focus to discussions on homeland security. For example, a study conducted by the Federal Research Service of the United States Library of Congress [3] presents the following definition for terrorism:

> [T]he calculated use of unexpected, shocking and unlawful violence against noncombatants ... and other symbolic targets perpetrated by a clandestine member(s) of a sub-national group ... for the psychological purpose of publicizing a political or religious cause and/or intimidating or coercing a government(s) or civilian population into accepting demands on behalf of the cause." (Reference 3, p. 12)

Ganor [4] further restricts the definition given above by stipulating that the targets must be civilian and attacked to attain political aims.

Given a definition of terrorism, a terrorist group can be defined as an organizational structure that employs terrorism as a means to further its goals. Terrorist groups can be defined as organizations based on the following criteria set forth by Crenshaw (Reference 5, p. 466):

- The group has a defined structure and processes by which collective decisions are made.
- Members of the organization occupy roles that are functionally differentiated.
- There are recognized leaders in positions of formal authority.
- The organization has collective goals which it pursues as a unit, with collective responsibility for its actions.

A report by the National War College entitled *Combating Terrorism in a Globalized World* [6], states:

> "Collectively, terrorist organizations pose the single greatest threat to American and international peace and prosperity" (Reference 6, p xix). Through links with other TOs, organized crime, drug traffickers, and state and corporate sponsors, TOs constitute a kind of *de facto* nation, complete with the ability to conduct war [6].

The potential targets of terrorist attacks can be summarized as

- the direct victims of the attack;
- members of society who are threatened by the prospect of being victims of similar attacks;
- the wider audience of the act who are intended to receive the message that the TO is a force to be reckoned with;
- government entities whose hand the terrorists are trying to force.

4 TERRORIST ORGANIZATION CONCEPTS

In a broad sense, TOs can be visualized in terms of a set of concentric rings. In the center of the rings is the leadership of the organization. The area just outside the leadership area represents the operations cells, where the responsibility for tactical planning and execution of operations resides. The area outside the operations ring represents the network of those sympathetic to the organization's cause. The sympathizers provide financial support to the organization either directly or indirectly [7].

The following sections describe key concepts associated with TOs, including TO members, TO funding sources, organizational learning for TOs, and TO functions and capabilities.

4.1 TO Members

Members of TOs may typically fall into one of the four general classifications [8]:

1. Leaders, providing direction and policy.
2. Cadres, planning and conducting operations and maintaining logistics, intelligence operations, and communications.
3. Active supporters, engaging in political and fund-raising activities.
4. Passive supporters, sympathizers based on shared end goals or through fear. "[P]assive supporters can be useful for political activities, fund-raising or through unwitting or coerced assistance in intelligence gathering or other nonviolent activities" (Reference 8, p. 3–2).

Members of TOs may progress upward through the power structure by earning the trust of leadership over time or through other factors such as familial or tribal relationships. Trust is likely to be earned through participation in risky operations. After a member has proven to be dedicated to the cause and capable, they are more likely to be rewarded with a leadership role. Typically, leaders are less likely to be involved directly with terrorist tactical operations [9].

4.2 TO Funding

TOs typically rely on any combination of six basic sources of funding [9]:

1. direct contributions from private individuals;
2. donations from charitable institutions;
3. government sponsors;
4. legitimate businesses;
5. contributions from members;
6. profits from criminal enterprises (robbery, kidnapping, hijacking, extortion, trafficking, gambling, black market, etc.).

A TO may be state supported. Sometimes the support exists due to intimidation or extortion. Some governments may support the terrorist's cause ideologically, but disagree with some of the methods employed by the TO. Most financial support for TOs originates from nongovernment sources [10].

4.3 Organizational Learning in TOs

A study of organizational learning within terrorist groups sets forth that in order for terrorist groups to endure, they must adapt to conditions around them (e.g. threats, technology, and societal factors) and within them (e.g. compromise of key organizational elements) [11]. The greater the ability of a TO to learn, the more effective it can be in choosing targets, identifying vulnerabilities for the maximum desired impact of attacks, and avoiding and confounding counterterrorism efforts [11]. Learning within the TO, and the ability to convey knowledge and information in a timely manner, affects the ability of the organization to adapt and survive [11]. The type of organizational structure of a TO and its communication resources will impact the ability of a TO to learn, share knowledge, and adapt. According to Hopmeier [12], this is evolution, which TOs do much better than governments or counterterror organizations, because their response time is much smaller and their "bureaucratic inertia" is less due to the smaller size.

4.4 TO Size

TOs can be of various degrees of maturity and capability. However, the nature of terrorism is such that a large organization is not required to complete a large scale attack that is successful from the terrorist's perspective (e.g. the bombing of the Alfred E. Murrah federal building in Oklahoma City) [8]. TOs are often interconnected such that mutual aid is provided among them. Examples of such aid might be the supply of weapons, ammunition, or training; referral or vetting of personnel; sharing of safe havens; and of course, the exchange of intelligence. In effect, even a small TO may be able to make use of information and resources that they otherwise would not have access to without the support of a greater terrorist community [6].

Emergent terrorist groups can act as proxy or under guidance from larger organizations with more experience and resources. Smaller groups can be absorbed by larger organizations. Several small, hierarchical organizations might coalesce into a larger networked one. Conversely, a smaller organization might splinter off from a larger one. The splintering may occur due to strategic reasons or over disagreements over transitions of power. Each method of formation carries with it implications with respect to the organizational structure, experience level, and capabilities of the resulting organizations [8].

4.5 TO Functions

A 2005 RAND organization report says:

> "In order to act effectively, a TO must be able to organize people and resources, gather information about its environment and adversaries, shape a strategic direction for actions of its members, and choose tactics, techniques and procedures for achieving strategic ends" (Reference 11, p. 95).

Generally, TOs must address certain key functions, including [11]

- training
- logistics
- communications
- fund-raising
- collaboration/interface with other TOs or sponsors
- intelligence
- operational security
- tactical operations
- recruiting
- indoctrination.

Large organizations are also likely to have medical services that are organic to their structure. Well-funded organizations may participate in social services within their regions of influence. Distributing food, providing jobs, and organizing educational and youth activities are all ways of developing and strengthening ties within the communities upon which they rely for cover, support, and new recruits [11].

4.6 TO Categories and Classifications

The military guide to terrorism in the twenty-first century [8] categorizes TOs as follows:

- structure—including hierarchical and networked (such as chain, hub, and flat networks);
- government affiliation—including nonstate supported, state supported, and state directed (operating as an agent of a government);
- motivation—separatist, ethnocentric, nationalistic, revolutionary;
- ideology—including political (for example, right wing, left wing, and anarchist); religious; social (for example, animal rights, abortion, environment, and civil rights);
- international scope—for example, domestic; international (i.e. regional and routinely operational in multiple countries within a specific region); transnational (i.e. transcontinental or global or routinely operational in multiple countries and in multiple regions).

A US Congressional Research Report from 2004 [13] identifies even more characteristics associated with [foreign] TOs. These additional characteristics are included in the following list:

- goals and objectives
- favored tactics
- primary areas of operation
- links with other groups
- business associations
- composition of the organization membership
- nonterror activities.

To understand the motivations and actions of TOs more thoroughly, some researchers have found it useful to categorize them as either *political* or *fanatic* [7]. *Political* TOs tend to use terrorism as a means to achieve political goals. On the other hand, *fanatic* groups tend to be more interested in violence as an end in itself. These groups may have lost sight of their political goals or may be locked in a cycle of revenge, or may have more criminal interests [7].

Most TOs are politically or religiously motivated such that they can benefit from the association with some legitimate or otherwise popular cause [6]. *US Department of State list of Designated Foreign Terrorist Organizations* includes religious as well as various national separatist organizations and ideologically inspired organizations. TOs focusing on racial separatism, opposition to abortion, animal rights, and environmental issues are not uncommon in many of the westernized nations [14].

4.7 Organizational Structures of TOs

The two general categories of structure for TOs are networked and hierarchical. Terrorist groups may be structured as a combination of the two types.

Hierarchical organizations are characterized by well-defined vertical command and control structure. The lower level functional elements of hierarchical organizations are usually specialized (e.g. logistics, operations, and intelligence) as opposed to being stand-alone elements whose capabilities span those same specialties. The latter type is more characteristic of networked organizations [8].

Hierarchical organizational structures are characterized by leadership, that is, centralized in terms of authority. Although the centralized leadership structure provides more organizational control over doctrine, motivation, and operations, these structures are usually more dependent on communication channels, structured logistics, and disciplined membership. These dependencies represent additional vulnerability to successful penetration or counterterror operations [7].

A terrorist network that is of distributed (decentralized) structure tends to be more capable of operation when key leadership is eliminated [15]. However, since terrorist activities are often covert and because modern information and communication systems are susceptible to being intercepted and analyzed, significant challenges to communications and the transfer of funds exist throughout these kinds of TOs. Owing to inexperience, fear of compromise or of leaving an evidence trail, record keeping is likely to be done sparingly or not at all in some cases, adding to the uncertainty and unaccountability of actions within the networked organization [9].

TOs that are bound by broader beliefs, such as religious, environmental, or moral, do not require the type of coordination that politically motivated organizations do. Consequently, networked structures of more or less self-sufficient operational cells distributed geographically are suitable to conduct their operations over a wide area and in cooperation with other like-minded organizations. The leadership of such organizations or of a particular "movement" can set broad goals, and networked TOs can independently choose targets and act against them in a manner that they see fit. The whole organization will expect to benefit in terms of influence and publicity and the attainment of its collective goals [8]. If a network becomes excessively distributed, it tends to lose much of its organizational aspects and instead becomes more of an idea or concept [16].

A correlation has been identified between the general structure of a TO and its ideology or motivating principles [8]. For example, Leninist or Maoist groups tend toward hierarchical structure (implying centralized leadership). Hierarchical groups are better suited for coordination and synchronization with political efforts. Larger organizations tend to adopt a networked, cellular structure at some point to reduce the risk of security breaches and counterintelligence failures [8].

4.8 TO Enabling Factors

According to the National War College report, the "most prominent contributing factors that enable terrorism to flourish" are (Reference 6, p. 54)

- poverty and economic and social inequities;
- poor governance with economic stagnation;
- illiteracy and lack of education;

- resentment to the encroachment of western values;
- unpopular foreign policies among potential target countries.

5 MODELS

A current trend in terrorist threat mitigation is to employ technology in the form of analytical tools as models, simulations, and data mining software to derive understanding about TOs where hard intelligence resources are limited or nonexistent.

A general knowledge of the prevalent models of terrorist organizational structures can be expected to lead to a better understanding of the threat, functionality, capabilities, and vulnerabilities of the organization [8]. The following sections discuss, in general terms, the most current analytical methods employed against the modeling and analysis of TOs.

5.1 Network Models

To conduct network analysis on a terrorist group, one typically represents the members of the group as nodes and the links between the nodes are representative of associations such as chain of command or resource dependencies [17]. The relative number of links emanating from a node tends to suggest a leadership position within a network, or otherwise, a key resource node [17]. When there are many short paths passing through a member, a gatekeeper role is likely. A gatekeeper acts as a facilitator between subgroups of a network [17]. Nodes (members) that are not linked are likely to exist in separate subgroups [7].

Organizational network modeling programs are available that can automatically identify the links per node of a network and present the results graphically in a top-down (hierarchical) fashion or in a rose form where the most influential nodes are located in the center of the diagram. The same programs can be used to identify subgroups within the network [18].

The *NetBreaker* modeling and analysis tool developed by Argonne National Laboratory [19] takes as input a list of known organization members (and their functions, if known), along with any unknown members and any known or hypothesized interactions involving the group. The interim analysis result is a set of all the possible terrorist networks that could include the input set. The interim analysis is based on validated network formation rules. Subsequent questions and rules are applied to reduce the size of the interim solution set, thereby honing in on the most likely actual structure of the organization. This kind of analysis is useful for identifying key functionaries in the network and for identifying vulnerabilities so that counterterror efforts can be more keenly focused.

The information required as input to network modeling tools is more likely to be found in a centralized terror network. Compromised elements of a centralized terrorist network will tend to lead, ultimately, to other elements. However, centralized structures can be expected to operate through well-established leadership chains and have well-organized communication and logistics channels. Distributed networks tend to be more difficult to identify or eliminate since leadership communication and logistics channels can be expected to be shorter. For distributed networks where elements act with more autonomy and with greater independence, it tends to be more difficult to identify dependencies between network elements.

Network modeling methods can be useful for determining what subgroups exist within a network. Moreover, the following information may be uncovered [20]:

- Whether subgroups are subordinate to one another.
- Whether the subgroups exist within a common logistics chain.
- Whether the subgroups have members in common.
- Whether the subgroups rely on one another for operational or financial support.
- Whether the overall network is centralized or distributed in form.
- What roles do members or subgroups play?

Clues to the structure of TOs can be uncovered that may lead to insights as to where limited counterterrorism resources can be directed for the most effect. For example, if a network is found to be more of a centralized structure, penetrating or destroying the nucleus of the network would tend to offer the greatest impact against the network as a whole. Similarly, when chain-like dependencies and linkages to subgroups are identified, whole operational cells (subgroups) could be effectively cut off and temporarily isolated with a "surgical" application of counterterror operations [17].

5.2 Network Influence Models

Influence models are derived from network models. They are based on an assumption that for the most part, members with more links attached to them have influence over those members with fewer links. The degree of influence is taken as a degree of importance of an individual to the organization as a whole [18]. Influence diagrams are intended to capture the interrelationship of factors pertinent to a given decision at a snapshot in time. Therefore, unlike causal and Bayesian models that are discussed in the following section, they have the weakness of being insensitive causal factors and decision-making processes [21].

5.3 Causal and Markov Modeling

Causal modeling of TOs is a method of identifying precursor conditions and/or actions that lead to some other condition or action on behalf of the TO. Some of the questions that causal modeling would address might be as follows [2]

- What conditions lead a TO to evolve from a nationally focused one to a transnational organization?
- How do national characteristics manifest in TOs?
- How do large events, including natural disasters, likely to affect TOs?
- What is the relationship between political activity and terrorism activities?

Causal models can be built based on a Markov chain construct where actions, conditions, and decision points are modeled in a flow chart fashion. Transition from one node in the Markov chain to another will occur based on a probability determined by the current state of model (i.e. what conditions are currently prevailing within the TO), and not based on precursor conditions that led the TO to the current state. Known information can be compared with a validated causal model to identify the patterns associated with specific terrorist activities and threats [22].

5.4 Bayesian Models

Bayesian models built on the Markov technique are used to answer high-level questions regarding a TO based on more conditions that can affect the transition of state. The types of questions that are answered might include the following: Will the organization merge with another? Will it attack a specific target? Will it escalate an attack? The Bayesian aspect of the modeling method addresses the decision-making processes and reactions within the organization that are conditioned upon previous actions and the current state of affairs. The Markovian aspect of the model defines the basic processes associated with operating a TO or planning or carrying out a terrorist attack. Bayesian (probabilistic) decisions are derived at different states along a chain of Markov-modeled events based on the plurality of conditions. The combined result of the Bayesian and Markov modeling is a complex model that can be used as a test bed for antiterrorism policy [23] and as a foundation for agent-based models such as those described in Section 5.6.

5.5 Dynamic Organizational Theory

Although the structure of TOs may hold clues to the strengths and/or vulnerabilities of it, understanding the dynamic aspects of the organization is also of great interest. The dynamic aspects might reveal under what conditions certain key functions such as training, recruiting, and funding become critically challenged or significantly enabled. Any probabilistic rules governing the likely responses of the organizational behavior to counterterror, bureaucratic, or societal stimuli are of interest to those planning antiterror strategies or conducting risk mitigation [24].

DeGhetto sets forth that organization theory (i.e. the study of organizational dynamics) and, specifically, organizational decline theory can be used effectively against TOs [25]. The agent-based modeling (ABM) methods described in the following section provide a means for testing counterterror strategies such as those outlined in DeGhetto's thesis [25].

Terrorist group decline factors, as identified by Kent Layne Oots, are the lack of entrepreneurial leadership, recruitment, ability to form coalitions with other groups, political and financial outside support, internal and external competition, and internal cohesiveness [26]. Preemption, deterrence, backlash, and burnout are the main factors for terrorist group decline, as identified by Gurr and Ross [27]. Another factor might be the failure of legitimate or illegitimate commercial ventures that the organization might be involved in.

5.6 Agent-Based Models and Complex Adaptive Systems

A system modeled as a set of independently simulated, interacting, and adaptive "agents" is referred to as a *complex adaptive system (CAS)*. Modeling a TO as a CAS is often effective in bringing out the dynamic aspects of the organization. The agents that comprise a CAS are themselves models of dynamic entities such as people or other groups or organizations.

The rules that govern agent behaviors are typically based on a large set of empirical and/or random variables [24]. Basic agent rules might govern movement, trading behavior, combat, interaction with the environment, cultural behaviors, and interaction between sexes and noncombatants [28].

In a sense, with ABM, a model of a relevant portion of the world, with as many relevant factors and conditions represented as possible, is developed. Within that world, a TO is modeled as a CAS comprising many free-acting agents (perhaps sharing the same goal or motivations) that are programmed to behave and respond like real people. The combined result of the agents responding independently to conditions, other agents, and stimuli is an emergent and unpredictable higher level organizational behavior [24].

ABM in the context of a dynamic network model allows internal reactions and regrouping of a TO to be anticipated when one or more members are compromised or eliminated. The capability also can be used to help identify terrorists or to identify hidden dependencies on critical personnel or resources [29].

ABM provides a kind of "flight simulator" functionality that can serve as a test bed for tying various tactical and policy approaches in response to the terror threat and under a wide range of conditions [30]. Simulations based on ABM are also useful to determine the limits of an organization's capabilities.

5.7 Human Behavior Models

In ABM, modeled agents can have individual human characteristics, including personality traits such as temperament, dedication to the group, and ambition. These traits provide input to behavioral models. The actions and roles of the agents are subject to rules of social interaction and broader guiding principles [19].

A human behavior model developed by Barry Silverman et al., University of Pennsylvania, includes, for example, over 100 interdependent submodels of anthropological, physiological, medical, societal, cultural, religious, and political factors. The models have been incorporated into sophisticated, game-like simulations with life-like avatars, each with specific personalities and motivations. The models can be used to train in counterterror operations and to help identify terrorists based on interactions with others and patterns of behavior [29].

5.8 Population Dynamics Models

High-level modeling of TOs in terms of the size of the organization is taking place at the University of Maryland, Center for Technology and Systems Management (CTSM). Terrorist population dynamics (TPD) models rely on data pertaining to the growth and contraction of terrorist network population over a given time interval to estimate factors such as current terrorist population size, typical rates of growth and contraction of the TO, and correlations of TO size with activities and societal forces acting outside of the TO [31].

6 RESEARCH DIRECTIONS

The effectiveness of the modeling and analytical methods described in this article is limited by the quality and accuracy of information that the models are provided with and are based on. Increasingly, models and historical data are turned to fill the gaps of knowledge about TOs that are the result of otherwise poor intelligence. Models can be expanded but the ability to validate the models based on known facts about TOs will continue to be a challenge.

Case studies directed toward validation of the methods will always be valuable. A common set of metrics is needed to base evaluations of models and their specific applications. These metrics will allow a host of model and analytical techniques to be evaluated against each other in the context of a wide range of questions, TOs and conditions.

Areas for continued research include the hybridization of some of the methods described in the article. Review of the literature indicates that TO dependencies on resources such as arms, real properties, various kinds of communications, and transportation can be more rigorously modeled, perhaps revealing new insights or points of vulnerability. The flow of specific commodities within a TO can provide clues to the timing, nature, and scale of pending attacks.

A recurring theme in the literature is that TOs inevitably persist under challenged conditions that are often exclusive to covert, illegal, and largely unpopular organizations. The notion that TOs do not face at least the same problems with other large organizations, including bureaucracy, conflict, fraud, poor morale, attrition, and financial hardship, is not founded based on the research. Consequently, the opportunity exists to aggravate and exploit some of these factors to mitigate the threat posed by TOs [25] [9].

REFERENCES

1. US Joint Chiefs of Staff. *Joint Tactics, Techniques and Procedures (JTTP) for Antiterrorism*, U.S. Government Joint Chiefs of Staff 3-07.2. (Revised first draft). 2004 Apr 9. (FOUO-Referenced in [7], pp. 3–1.
2. Hopple, G. W. (1982). Transnational terrorism: prospectus for a causal modeling approach. *Terrorism Int. J.* **6**(1), 73–100.
3. Library of Congress, Federal Research Center. (1999). *The Sociology and Psychology of Terrorism: Who Becomes a Terrorist and why*. Report. Washington (DC), 1999 Sept. 186. There are other standard definitions. One compendium is provided by the Terrorism Research Center Inc, URL: http://www.terrorism.com.
4. Ganor, B. (2002). Defining terrorism: is one man's terrorist another man's freedom fighter? *Police Pract. Res.* **3**(4), 287–304.
5. Crenshaw, M. (1985). An organizational approach to the analysis of political terrorism. *Orbis* **29**(3), 465–489.
6. National War College Student Task Force on Combating Terrorism. (2002). *Combating Terrorism in a Globalized World*. Report. National War College, Washington, DC, 2002 Nov. 88 pages.
7. Franck, R. E., and Melese, F. (2004). Exploring the structure of terrorists' WMD decisions: a game theory approach. *Def. Secur. Anal.* **20**(4), 355–372.
8. U.S. Army Training and Doctrine Command. (2005). *A Military Guide to Terrorism in the Twenty-First Century; TRADOC DCSINT Handbook*, Number 1 Chapter 3: Terrorist group organization, Leavenworth, KS, 3-1–3-12. Available from http://www.fas.org/irp/threat/terrorism/index.html; Internet; accessed Jan. 28, 2007.
9. Shapiro, J. (2005). The greedy terrorist: a rational-choice perspective on terrorist organizations' inefficiencies and vulnerabilities. *Strateg. Insights* **4**(1), 13.
10. Mickolus, E. (2005). How do we know if we are winning the war against terrorists? Issues in measurement. *Stud. Conflict Terrorism* **25**(3), 151–160.
11. Jackson, B. A., Baker, J. C., Cragin, K., Parachini, J., Trujillo, H. R., and Chalk, P. (2005). *Aptitude for Destruction: Volume 2: Case Studies of Organizational Learning in Five Terrorist*

Groups. RAND Corporation, Santa Monica, CA, p. 216, available from: http://www.rand.org/pubs/monographs/2005/RAND_MG332.pdf, accessed 2007 Feb. 24.

12. Hopmeier, M.. Unconventional. (2007). *Terrorism Expert*, Interview by phone. 2007 Mar. 18.
13. Cronin, A. R., Aden, H., Frost, A., and Jones, B.. Congressional Research Service [CRS]. (2004). Foreign terrorist organizations. Report for Congress. Library of Congress; 2004 Feb. 6. 111. Available from: http://www.fas.org/irp/crs/RL32223.pdf, accessed ~2007 Feb. 24.
14. National Defense University (US) [NDU]. (2002). *Chemical, Biological, Radiological, and Nuclear Terrorism: the Threat According to the Current Unclassified Literature*. Center for the Study of Weapons of Mass Destruction. ISN Publishing House, p. 46, available from: http://www.isn.ethz.ch/pubs/ph/details.cfm?v21=94077&lng=en&id=26595, accessed 2007 Feb 24.
15. Fatur, R. B.. (2005). *Influencing transnational terrorist organizations: using influence nets to prioritize factors*, [masters thesis]. Air Force Institute of Technology Wright-Patterson AFB OH School of Engineering and Management, 2005 June. 94 p. A523634.
16. Hoffman, B. (2004). The changing face of Al Qaeda and the global war on terrorism. *Stud. Conflict Terrorism* **27**(6), 549–560.
17. Xu, J., and Chen, H. (2005). Criminal network analysis and visualization. *Commun. ACM* **48**(6), 101–107.
18. Brams, S., Mutlu, H., and Ramirez, S. L. (2006). Influence in terrorist networks: from undirected to directed graphs. *Stud. Conflict Terrorism* **29**(7), 679–694.
19. North, M. J., Macal, C. M., and Vos, J. R.. (2004). Terrorist organizational modeling. *Argonne National Laboratory: NAACSOS Conference*, Pittsburgh, PA, 2004 June 27; n.d., p. 4 http://www.casos.cs.cmu.edu/events/conferences/2004/2004_proceedings/North_Michael.doc., accessed Feb 24, 2007.
20. McAndrew, D. (1999). The structural analysis of criminal networks. In *The Social Psychology of Crime: Groups, Teams, and Networks, Offender Profiling Series, III*, D. Canter, and L. Alison, Eds. Darthmouth, Aldershot.
21. Clemen, R. T., and Reilly, T. (2001). *Making Hard Decisions with Decision Tools*. Duxbury Resource Center, Belmont, CA, p. 752.
22. Coffman, T. R., and Marcus, S. E.. (2004). Dynamic classification of groups through social network analysis and HMMs. *IEEE: Aerospace Conference 2004*, BigSky, MO, 2004 Mar. 6, IEEE, 2004, p. 8.
23. Tu, H., Allanach, J., Singh, S., Pattipati, K. R., and Willett, P.. (2005). *Information Integration via Hierarchical and Hybrid Bayesian Networks [Internet]*. Storrs, CT: [cited 2007 Feb. 24]. p. 14, available from: http://servery.engr.uconn.edu/cyberlab/Satnam/docs/HHBN.pdf.
24. Elliott, E., and Kiel, L. D. (2004). A complex systems approach for developing public policy toward terrorism: an agent-based approach. *Chaos Solitons Fractals* **20**, 63–68.
25. DeGhetto, T. H. (1994). *Precipitating the decline of terrorist groups: a systems analysis*, [master's thesis]. Naval Postgraduate School, Monterey, CA, Mar. 24. 89 p.
26. Oots, K. L. (1989). Organizational perspectives on the formation and disintegration of terrorist groups. *Terrorism* **12**(3), 139–152.
27. Ross, J. I., and Gurr, T. R. (1989). Why terrorist subsides: a comparative study of Canada and the United States. *Comp. Polit.* **21**(4), 405–426.
28. Epstein, J. M. (1989). Agent-based computational models and generative social science. *Complexity* **4**(5), 41–60.
29. Goldstein, H.. (2006). *Modeling Terrorists*. IEE Eng Spectrum [serial on the Internet]. 2006 Sept. [cited 2007 Jan. 30]; Available from: http://spectrum.ieee.org/print/4424.
30. Holland, J. H. (1995). *Hidden Order: How Adaptation Builds Complexity*. Helix Books, Reading, MA.

31. Kaminskiy, M., and Ayyub, B. (2006). Terrorist population dynamics model. *Risk Anal.* **26**(3), 747–752.

FURTHER READING

Ackoff Center for Advancement of Systems Approaches. (2007). Available from: http://www.acasa.upenn.edu/. See for more information on agent-based social behavior models at University of Pennsylvania.

Center for Computational Analysis of Social and Organizational Systems (CASOS). (2007) http://www.casos.cs.cmu.edu/terrorism/projects.php. See for more information on social network modeling efforts at Carnegie Mellon University.

Farey, J. D. (2003). Breaking Al Qaeda cells: a mathematical analysis of counterterrorism operations (a guide for risk assessment and decision making). *Stud. Conflict Terrorism* **26**, 399–411.

Gunaratna, R. (2005). The prospects of global terrorism. *Society* **42**(6), 31–35.

Gunaratna, R. (2005). Responding to terrorism as a kinetic and ideological threat. *Brown J. World Aff.* **11**(2), 243.

Johnston, R. (2005). *Analytic culture in the U.S. intelligence community*. The Center for the Study of Intelligence. CIA, Pittsburgh, PA, p. 184, available from: http://www.fas.org/irp/cia/product/analytic.pdf, accessed~n.d.

Klerks, P. (2001). The network paradigm applied to criminal organizations: theoretical nitpicking or a relevant doctrine for investigators? Recent developments in the Netherlands. *Connections* **24**(3), 53–65.

Krebs, V. E. (2001). Mapping networks of terrorist cells, *Connections* **24**(3), 43–52.

Newman, M., Barabasi, A. L., and Watts, D. J. (2006). *The Structure and Dynamics of Networks*. Princeton University Press, Princeton, NJ.

RISK COMMUNICATION—AN OVERLOOKED TOOL IN COMBATING TERRORISM

David Ropeik

Risk Communication, Ropeik & Associates, Concord, Massachusetts

1 THE NEED

The terrorist attacks on September 11, 2001, killed approximately 3000 people, directly. But the death toll was higher. 1018 more Americans died in motor vehicle crashes

October through December 2001 than in those 3 months the year before, according to researchers at the University of Michigan's Transportation Research Institute. As those researchers observe " ... the increased fear of flying following September 11 may have resulted in a modal shift from flying to driving for some of the fearful" [1]. 1018 people died, more than one-third the number of people killed in the attacks of September 11, in large part because they perceived flying to be more dangerous and driving less so, despite overwhelming statistical evidence to the contrary.

As much as 17% of Americans outside New York City reported symptoms of post-traumatic stress two months after the September 11, 2001, attacks [2]. Even 3 years later, a significant number of Americans were still suffering serious health problems as a result of that stress. In a random sample of 2000 Americans, people who reported acute stress responses to the 9/11 attacks, even if they only watched the events on television, had a 53% increased incidence in doctor-diagnosed cardiovascular ailments like high blood pressure, heart problems, or stroke for up to 3 years following the attacks. The impact was worse among those who continued to worry that terrorism might affect them in the future. These people were three to four times more likely to report a doctor-diagnosed cardiovascular problem [3].

The Oxford English Dictionary defines terrorism as "the action or quality of causing dread". But that definition is inadequate. The dread caused by terrorism is just an intermediate outcome. More important are the health effects that result from such fear. Terrorism injures and kills both directly—from the attacks themselves—and indirectly, from what has been called the social amplification of risk, from the behaviors and stress that our worries produce [4]. Risk communication is an underutilized tool for combating those effects and minimizing the harm that terrorism can cause.

2 RISK COMMUNICATION DEFINED

The term *risk communication* arose largely as a result of environmental controversies in the 1970s, when public concern was high about some relatively low threats to human and environmental health. Scientists, regulators, and the regulated community described this public concern as irrational, and in their frustration they looked for ways to make people behave more rationally (as defined by those experts), especially about issues such as air and water pollution, nuclear power, and industrial chemicals. The goal of early risk communication was rarely to enlighten people so that they might improve their health. It was frequently to reduce conflict and controversy, an effort to talk people out of opposing some product or technology of which they were afraid. One researcher defined risk communication as "a code word for brainwashing by experts or industry" [5].

But risk communication has evolved. This article will use the following definition:

> "Risk communication is a combination of actions, words, and other messages responsive to the concerns and values of the information recipients, intended to help people make more informed decisions about threats to their health and safety."

That definition attempts to embody the ways that risk communication has matured over the past two decades. The consensus among experts in the field now rejects the one-way "We'll teach them what they need to know" approach. A National Research Council effort to move the field forward produced this definition in 1989. "*Risk communication is an interactive process of exchange of information and opinion among individuals, groups, and institutions. It involves multiple messages about the nature of risk and other messages, not strictly about risk, that express concerns, opinions, or reactions to risk messages or to legal and institutional arrangements for risk management*" [6]. In other words, risk communication should be considered a dynamic two-way street. Both sides get to talk, and both sides have to listen, and respond to input from the other.

More fundamentally, and intrinsic to the idea of the two-way street, is the growing acceptance among risk communication experts that risk means something different to the lay public than to scientists and regulators. "Risk" is perceived as more than a science-based rational calculation by the general public. Other attributes, like trust, dread, control, and uncertainty, also factor into the judgments people make about what they are afraid of.

As risk communication has evolved, more and more experts in the field agree that both the science-based view of experts and the affective view of risk among the general public are valid, and both must be respected and incorporated if communications about risk is to be effective.

This evolution is summed up in Risk Communication and Public Health, edited by Peter Bennett and Kenneth Calman:

> "... there has been a progressive change in the literature on risk:
>
> - *from* an emphasis on 'public *misperceptions*', with a tendency to treat all deviations from expert estimates as products of ignorance or stupidity
> - *via* empirical investigation of what actually concerns people and why
> - *to* approaches which stress that public reactions to risk often have a rationality of their own, and that 'expert' and 'lay' perspectives should inform each other as part of a two-way process" [7].

The evidence that illuminates what actually concerns people and why, requires discussion at some length. A solid body of careful research from a number of fields has established that the lay public's perception of risk is based on a dual process of fact-based analysis *and* intuitive, affective factors. The Greek Stoic philosopher Epictetus said "People are disturbed, not by things, but by their view of them." Understanding the roots of what shapes those views allows the true dialogue of modern risk communication to take place.

3 THE BIOLOGY OF FEAR

Neuroscientists have found that what we consciously describe as fear begins in a subcortical organ called the amygdala. Critically for risk communication, in very simplified terms, information is processed in the amygdala, the part of the brain where fear begins, *before* it is processed in the cortex, the part of the brain where we think. We fear first and think second [8]. That alone suggests that risk communication that merely attempts to communicate the facts, without factoring in the emotional issues involved, will not be as successful.

There is also neuroscientific evidence suggesting that as we process information, we fear *more*, and think *less*. Neural circuits have been identified that lead from the amygdala to parts of the cortex, circuits which, in essence, trigger a "fight or flight" response (accelerated heart rate, hormonal responses, etc.). The pathways coming back into the amygdala from the thinking "rational" cortex have also been identified. And there are more circuits out of the amygdala, the organ that stimulates a fear response, than there are circuits coming back in from the "thinking" brain, which could moderate that response.

So when we encounter information that might pose a threat, we generally fear first and think second, and fear more and think less. This basic description of the way the human brain is physically wired has fundamental implications for risk communication and dramatically reinforces the importance of findings from social science, which explain why risk means one thing to experts and another to the lay public.

4 RISK PERCEPTION PSYCHOLOGY

Some of what we are commonly afraid of seems instinctive: snakes, heights, the dark, and so on. But how do we subconsciously "decide" what to be afraid of, and how afraid to be, when the threat does not trigger an instinctive reaction; when we hear about a new disease, product, or technology, or when we try to gauge the risk of something against its benefits, or when we witness an act of terrorism? How does the human mind translate raw data into our perceptions of what is risky and what is not?

The answers can be found in two literatures, both critically relevant to risk communication. The first is the study of how people generally make judgments of any kind, including judgments about risk, under conditions of uncertainty. The second is the specific study of the psychology of risk perception, which has identified more than a dozen affective attributes that tend to make some threats feel more worrisome than others, even when our apprehension is not consistent with the scientific data.

4.1 General Heuristics and Biases

The discovery of systematic heuristics and biases—mental shortcuts—that we use to make choices under uncertainty, when we do not have all the facts, or all the time we need to get all the facts, or all the intellectual ability to fully understand the facts we have, was led by, among others, Daniel Kahneman, who was awarded the 2002 Nobel Gold Medal in Economics for his work. Kahneman and others identified a number of mental processes that simplify decision making when time or complete information is not available. This field has direct relevance for risk communication, as noted in a seminal paper on risk perception: "When laypeople are asked to evaluate risks, they seldom have statistical evidence on hand. In most cases, they must make inferences based on what they remember hearing or observing about the risk in question." "These judgmental rules, known as heuristics, are employed to reduce difficult mental tasks to simpler ones" [9].

Here are a few of the heuristics and biases relevant to risk perception, and therefore to risk communication.

- *Availability.* " ... people assess the ... the probability of an event by the ease with which instances or occurrences can be brought to mind" [10]. The risk of terrorism in the United States is statistically quite low. But apprehension has been elevated since September 11, 2001, in part because such an event is more "available" to our consciousness. The availability heuristic explains why, when a risk is in the news (flu vaccine issues, an outbreak of food poisoning, child abduction, etc.), it evokes more fear than when the same risk is around, at the same level, but just not making headlines.

- *Framing.* The way a choice is presented can distort the judgment that results. Imagine you are the mayor of a city of 1 million people and a fatal disease is spreading through your community. It is occurring mostly, but not exclusively in one neighborhood of 5000 residents. With a fixed amount of money, you can either (i) save 20% of the 5000 residents in that neighborhood, or (ii) save 0.2% of the entire city of 1 million. What do you do?

  A sizable number of people in risk communication classes I teach choose option (i), which produces a greater percentage effectiveness, but condemns 1000 people to death. Reframed, the choice would be: you can spend a fixed amount of money and save 1000 people or 2000. Presented that way, the choice is obvious. But the framing of the question in terms of percentages skews the judgment. Understanding the importance of framing is a key to better risk communication.
- *Anchoring and adjustment.* People estimate probabilities based on an initial value and adjusting from there. In one experiment, two groups of high school students estimated the sum of two numerical expressions that they were shown for just 5 s, not long enough for a complete computation. The first group was shown $9 \times 8 \times 7 \times 6 \times 5 \times 4 \times 3 \times 2 \times 1$. Their median estimate was 2250. The median estimate for the second group, shown the same sequence, but in ascending order—$1 \times 2 \times 3 \times 4 \times 5 \times 6 \times 7 \times 8 \times 9$—was 512 [11]. Knowledge of the anchoring effect is another tool for better risk communication.
- *Representativeness.* This is "the tendency to regard a sample as a representation of the whole, based on what we already know" [12]. Consider two people:
  - A white woman who is shy and withdrawn, with little interest in people, a strong need for order and structure, and a passion for detail.
  - A young man of middle-eastern complexion who is passionate, but sullen, quick to anger, bright, and unconcerned with material possessions.

Which one is the librarian, and which one is the terrorist? Without complete data by which to make a fully informed choice, the representativeness heuristic gives you a simple mental process by which to take the partial information and fit it into the preexisting category it represents. This suggests that risk communication must consider the patterns of knowledge and information people already have, on which they will base their response to what the communicator says.

4.2 Risk Perception Characteristics

Work in a related field, the specific study of the perception of risk, has identified a number of attributes that make certain risks feel more worrisome than others.

These risk perception factors are essentially the personality traits of potential threats that help us subconsciously "decide" what to be afraid of and how afraid to be. They offer powerful insight into why "risk" means different things to the lay public than it does to experts. A few of these factors have particular relevance to terrorism.

- *Trust.* When we trust *the people informing us* about a risk, our fears go down. When we trust *the process* deciding whether we will be exposed to a hazard, we will be less afraid. When we trust *the agencies that are supposed to protect us*, we will be less afraid. If we do not trust the people informing us, the process determining our exposure to a risk, or the people protecting us, we will be more afraid.

  Trust comes from openness, honesty, competence, accountability, and respecting the lay public's intuitive reasoning about risk.
- *Risk versus Benefit.* The more we perceive a benefit from any given choice, the less fearful we are of the risk that comes with that choice. This factor helps explain why, of more than 400,000 "first responders" asked to take the smallpox vaccine in 2002, fewer than 50,000 did. They were being asked to take a risk of about one in a million—the known fatal risk of the vaccine—in exchange for ZERO benefit, since there was no actual smallpox threat. Imagine, however, there was just one confirmed case of smallpox in a US hospital. The fatality risk of the vaccine would still be one in a million, but the benefit of the shot would suddenly look much greater
- *Control.* If you feel as though you can control the outcome of a hazard, you are less likely to be afraid. This can be either physical control as when you are driving and controlling the vehicle, or a *sense* of control of a process, as when you feel you are able to participate in policy making about a risk through stakeholder involvement, participating in public hearings, voting, and so on.

  This is why, whenever possible, risk communication should include information not just about the risk ("Terrorists have attacked the food supply"), but also offer information about what people can do to reduce their risk ("Boil milk before you drink it"). Specifically as regards food-related terrorism, information about how people can participate in a food recall is of particular value, by giving people a sense of control.
- *Imposed versus voluntary.* We are much less afraid of a risk when it is voluntary than when it is imposed on us, as is the case in terrorism, agricultural, or otherwise.
- *Natural versus human-made.* If the risk is natural, we are less afraid. If it is human-made, we are more afraid. A radiologically contaminated conventional explosive—a "dirty bomb"—will evoke much more fear than radiation from the sun, which will cause far more illness and death. A natural foodborne pathogen such as *E. coli* O157:H will likely produce less concern than a "militarized" pathogen such as anthrax, regardless of their scientific risk profiles.
- *Dread.* We are more afraid of risks that might kill us in particularly painful, gruesome ways than risks that kill us in more benign fashion. Ask people which risk sounds worse, dying in a fiery plane crash or dying of heart disease, and they are likely to be more afraid of the plane crash, despite the probabilities.

  This factor helps explain why the United States has a "War on Cancer", but not "War on Heart Disease". Cancer is perceived as a more dreadful way to die, so it evokes more fear, and therefore more pressure on government to protect us, thought heart disease kills far more people annually.
- *Catastrophic versus chronic.* We tend to be more afraid of things that can kill a lot of us in one place at one time, such as a plane crash, than heart disease or stroke or chronic respiratory diseases or influenza, which cause hundreds of thousands more

deaths, but spread out over time and location. This factor makes foodborne illness outbreaks much more frightening than the chronic presence of foodborne illness, which sickens one American in four per year.

- *Uncertainty.* The less we understand about a risk, the more afraid we are likely to be, as is the case with terrorism, particularly a terrorist attack on the food supply, where there will likely be many unknowns. When uncertainty exists because all the facts are not in, the fear that results must be acknowledged and respected.
- *Is the risk personal.* Understandably, a risk that we think can happen to us evokes more concern than a risk that only threatens others. As a demonstration of this, consider how the attacks of September 11 made terrorism a risk not just to Americans living somewhere else, but to Americans at home. Suddenly we realized "this could happen to ME!" We began referring to the United States as "The Homeland". We could probably take the "H" and the "O" out of the word. What we are really saying is that now terrorism could happen in the "MEland".

  This factor explains why numbers alone are ineffective as risk communications. One in a million is too high if you think you can be the one.
- *Personification.* A risk made real by a person/victim, such as news reports showing someone who has been attacked by a shark or a child who has been kidnapped, becomes more frightening than one that is statistically real, but only hypothetical.

There are a few important general qualifications about the heuristics and biases mentioned earlier, and the risk perception factors listed immediately above. Often, several of these factors are relevant for any given risk. A terrorist attack on the food supply will certainly evoke issues of trust, dread, and control, among other factors. The availability heuristic will certainly affect how afraid we are.

Also, while the research suggests that these tendencies are universal, any given individual will perceive a risk uniquely depending on his or her life circumstances, that is, age, gender, health, genetics, lifestyle choices, demographics, education, and so on. This means that although it is good risk communication practice to consider the emotional concerns of the audience, not everyone in a large audience shares the same concerns. As the National Research Council report suggests, "For issues that affect large numbers of people, it will nearly always be a mistake to assume that the people involved are homogeneous . . . . It is often useful to craft separate messages that are appropriate for each segment" [13].

5 RECOMMENDATIONS

In general, by understanding and respecting the psychological reasons for people's concerns (or lack of concerns in the case of terrorism preparedness), risk communication strategies can be devised that take these factors into account and shape messages that are more resonant with people's perceptions. That in turn, increases the likelihood that the messages will be more trusted, better-received, which increases the impact they will have.

However, as the National Research Council report noted, " . . . there is no single overriding problem and thus no simple way of making risk communication easy" [14]. So although this article provides suggestions on fundamentals, it cannot offer a detailed how-to guide to risk communication.

But there are several widely accepted general recommendations:

Include risk communication in all risk management policy making and action. Far more is communicated by what you do than what you say. "Risk communication ... must be understood in the context of decision making involving hazards and risks, that is, risk management" (NRC) [15]. Consider the example cited a few pages ago of the failed Bush administration smallpox vaccination policy. Had the risk perception factor of "risk versus benefit" been considered when the policy was being discussed, officials might not have chosen a policy unlikely to meet its objectives since it asked people to take a risk (albeit low) for ZERO benefit. In other words, the policy itself, not the press releases about it, carried implicit, but very clear risk communication information that had a lot to do with how people responded.

Information that affects how people think and feel about a given risk issue is conveyed in nearly all of the management actions an agency or a company or a health official takes on that issue. All risk management should include consideration of the risk perception and risk communication implications of any policy or action under review. Quite specifically, this means that *organizations should include risk communication in the responsibilities of senior managers, not just of the public relations or communications staff*. As the NRC report suggests, risk managers cannot afford to treat risk communication as an afterthought that comes at the end of the process after risk assessment has been done and policy set.

Recognize that *the gaps between public perception and the scientific facts about a risk can lead to behaviors that can threaten public health. These gaps are part of the overall risk that must be managed*. Whether people are more afraid of a risk than they need to be or when they are not afraid enough, this perception gap is a risk in and of itself and must be included in dealing with any specific risk issue and in all risk management and public health efforts.

Consider the example or the fear of flying post 9/11. One of the messages of the federal government was, paraphrasing, "Live your normal lives or the terrorists win. Go shopping." Had they considered the importance of the feeling of control to people's perceptions, perhaps the message might have suggested "Live your normal lives or the terrorists win. For example, flying seems scary right now. But if you choose not to fly and drive instead, because having a sense of control makes driving safer, remember that driving is much riskier, and if you die behind the wheel, the terrorists have won." Such a message might have saved the lives of some of those who made the choice to drive instead of fly.

Trust is fundamentally important for effective risk communication, and it is on the line with everything you do. " ... messages are often judged first and foremost not by content but by the source: 'Who is telling me this, and can I trust them?' If the answer to the second question is 'no', *any* message from that source will often be disregarded, no matter how well-intentioned and well delivered" (Bennett and Calman) [16].

Trust is determined in part by who does the communicating. When the anthrax attacks took place in the fall of 2001, the principal government spokespeople were the Attorney General, the Director of the FBI, and the Secretary of Health and Human Services, and not the head of the CDC or the US. Surgeon General—doctors likely to be more trusted than politicians. Had risk communication been included in the considerations of senior managers as the anthrax issue was beginning to develop, and incorporated into the deliberations of how to manage the overall anthrax risk, the more trusted officials would have done the majority of the public speaking, which might have done more to help the public keep their concern about the risk of bioterrorism in perspective. This lesson should be applied to any risk communication in connecting with agroterrorism.

But trust is more than just who does the talking. Trust also depends on competence. If people believe that a public health or safety agency is competent, they will trust that agency to protect them, and be less afraid, than if they doubt the agency's ability. When the first mad cow case in the United States was found in 2003, the US Department of Agriculture and Food and Drug Administration were able to point to a long list of regulatory actions they had taken for years to keep the risk low. So the *actions* taken by those agencies, years before the news conferences and press releases about that first case, had risk perception implications by establishing trust and thus affecting the public's judgment about the risk and their behavior. This helps explain why beef sales in the United States after that first case was discovered were effectively unchanged.

Trust is also heavily dependent on honesty. Of course, honesty means many things. In some instances, it can mean apologizing or taking responsibility for mistakes. When leaks developed in underground tunnels that are part of a major transportation project in Boston, press attention and public criticism focused on the contractor responsible for the tunnels until the chairman of the company said at a tense public hearing "We apologize for our mistakes" [17] (Note that the apology was made 'sincere' by the fact that it came from the head of the company, and the fact that the company offered to pay for repairs.). Criticism of the company dropped substantially thereafter.

Another example of honesty is avoiding the desire to over-reassure. Again, the way the USDA handled mad cow disease illustrates one example. In the years prior to that first sick cow being found, top officials never promised there was ZERO risk of mad cow disease, either in animals or in humans, just that the risk was very low. Had they followed the initial inclination of some senior USDA officials and promised that the risk was ZERO, that single first case would probably have provoked more public concern because people might have feared that the government's overassurance was not honest and could not be trusted.

And, obviously, honesty means not covering things up or telling untruths or half-truths. Being caught keeping secrets is almost always worse than revealing the information, even if damaging, first. Remember the framing heuristic mentioned above. How people think about an issue is based in part on the first way it is presented. Even if information is damaging, revealing it first gives the communicator the opportunity to "paint the first picture" of how people will think about the matter.

Adopting risk communication into intrinsic risk management requires fundamental cultural change. Sharing control, admitting mistakes, acknowledging the validity of the public's intuitive risk perception, not keeping secrets, being open and honest ... these are all countercultural to political, legal, and scientific organizations and people, the kinds of organizations and people who will be in charge of dealing with terrorist threats to the food supply. These are countercultural suggestions in a litigious society. They are countercultural to the myth of the purely rational decision-maker. As risk communication researcher and practitioner Peter Sandman has observed, "What is difficult in risk communication isn't figuring out what to do; it's overcoming the organizational and psychological barriers to doing it" [18].

Nonetheless, countless examples demonstrate how adoption of the principles of risk communication are in the best interests of most organizations, public safety officials, politicians, as well as the interest of public health. In the case of terrorism, they help officials with more effective risk management to protect public health. They increase support for an agency's overall agenda or a company's brand and products, political support for a candidate or legislation, and they reduce controversy and legal actions. While these benefits may not be readily quantifiable, and only realized over the long term, they are real, well-supported by numerous examples, and argue strongly for the cultural change necessary for the adoption of best practice risk communication principles.

Finally, *if at all possible within constraints of time and budget, any specific risk communication should be systematically designed and executed, including iterative evaluation and refinement*. "We wouldn't release a new drug without adequate testing. Considering the potential health (and economic) consequences of misunderstanding risks, we should be equally loath to release a new risk communication without knowing its impact" [19].

Risk communication messages and strategies specific to each plausible terrorist scenario should be developed in advance, and tested and revised to maximize effectiveness. Being prepared for purposeful contamination of the food supply, with various agents, at various points of entry in the farm-to-fork system, is vital to protecting public health in such events.

6 CONCLUSION

The human imperative of survival compels us to make rapid decisions about the threats we face. But this decision-making process is almost always constrained by a lack of complete information, a lack of time to collect more information, and a lack of cognitive abilities to understand some of the information we have. In response, humans have evolved a dual system of reason and affect to rapidly judge how to keep ourselves safe. In many cases these judgments work to protect us. But sometimes they can lead to behaviors that feel right, but actually raise our risk, whether we are more afraid of a relatively low risk or not afraid enough of a relatively big one. Great harm to public health can occur in such cases. To mitigate this threat, it is critical that an understanding of risk perception and its application to effective risk communication become an intrinsic part of how organizations deal with the threat of terrorism.

REFERENCES

1. Sivak, M., and Flanagan, M. (2004). Consequences for road traffic fatalities of the reduction in flying following September 11, 2001. *Trans. Res. Part F* **7**(4-5), 301–305.
2. Silver, R. C., Holman, E. A., McIntosh, D., Poulin, M., and Gil-Rivas, V. (2002). Nationwide longitudinal study of psychological responses to September 11. *JAMA* **288**, 11235–11244.
3. Holman, E. A., Silver, R. C., Poulin, M., Andersen, J., Gil-Rivas, V., and McIntosh, D. (2008). Terrorism, acute stress, and cardiovascular health, a 3-year study following the September 11th attacks. *Arch. Gen. Psychiatry* **65**(1), 73–80.
4. Pidgeon, N., Kasperson, R., and Slovic, P., Eds. (2003). *The Social Amplification of Risk*, Cambridge University Press, Cambridge, UK.
5. Jasanoff, S. (1989). Differences in national approaches to risk assessment and management. *Presented at the Symposium on Managing the Problem of Industrial Hazards: the International Policy Issues*, National Academy of Sciences, Feb. 27.
6. *Improving Risk Communication*, (1989). National Research Council, National Academy Press, p. 21.

7. Bennett, P., and Calman, K., Eds. (1999). *Risk Communication and Public Health*, Oxford University Press, New York, p. 3.
8. This very simplified synthesis of LeDoux's work comes from Ledoux, J. (1998). *The Emotional Brain: the Mysterious Underpinnings of Emotional Life*, Simon and Schuster, New York.
9. Slovic, P., Fischhoff, B., and Lichtenstein, S. (2001). A revised version of their original article appears. In *Judgment Under Uncertainty: Heuristics and biases*, D. Kahneman, P. Slovic, and A. Tversky, Eds. Cambridge University Press, Cambridge, UK, pp. 463–489.
10. Kahneman, D., Slovic, P., and Tversky, A., Eds. (1982). *Judgment Under Uncertainty: Heuristics and biases*, Cambridge University Press, Cambridge, UK, pp. 11–12.
11. Kahneman, D., Slovic, P., and Tversky, A., Eds. *Judgment Under Uncertainty: Heuristics and biases*, Cambridge University Press, Cambridge, UK, pp. 14–15.
12. Kahneman, D., Slovic, P., and Tversky, A., Eds. (1982). *Judgment Under Uncertainty: Heuristics and biases*, Cambridge University Press, Cambridge, UK, p. 24.
13. *Improving Risk Communication*, (1989). National Research Council, National Academy Press, p. 132.
14. *Improving Risk Communication*, (1989). National Research Council, National Academy Press, p. 3.
15. *Improving Risk Communication*, (1989). National Research Council, National Academy Press, p. 22.
16. Bennett, P., and Calman, K. (1991). *Risk Communication and Public Health*, Oxford University Press, Oxford, UK, p. 4.
17. *Big Dig Firm Apologizes, Considers Fund for Repairs*, (2004). Boston Globe, Dec. 3, p. 1.
18. Sandman, P. *The Nature of Outrage (part1)*, www.psandman.com.
19. Morgan Granger, M., Fischhoff, B., Bostrom, A., and Altman, C. (2002). *Risk Communication A Mental Models Approach*, Cambridge University Press, Cambridge, UK, p. 180.

CROSS-CUTTING THEMES AND TECHNOLOGIES

RISK MODELING AND VULNERABILITY ASSESSMENT

TERRORISM RISK: CHARACTERISTICS AND FEATURES

BILAL M. AYYUB
Center for Technology and Systems Management, Department of Civil and Environmental Engineering, University of Maryland, College Park, Maryland

1 INTRODUCTION

Risk is associated with all projects, business ventures, and activities taken by individuals and organizations regardless of their sizes, natures, and time and place of execution and utilization. Acts of violence including terrorism can be considered as an additional hazard source. These risks could result in significant losses, such as economic and financial losses, environmental damages, budget overruns, delivery delays, and even injuries and loss of life. In broad context, risks are taken even though they could lead to adverse consequences because of potential benefits, rewards, survival, and future return on investment. Risk taking is a characteristic of intelligence for living species since it involves decision making that is viewed as an expression of higher levels of intelligence. The chapter defines and discusses terrorism risk and its characteristics and features.

2 TERMINOLOGY

Definitions that are needed for risk analysis are presented herein [1].

Several definitions are available for the term *terrorism*, though without a globally accepted one. The following are selected definitions:

- US Code of Federal Regulations: "... the unlawful use of force and violence against persons or property to intimidate or coerce a government, the civilian population, or any segment thereof, in furtherance of political or social objectives" (28 C.F.R. Section 0.85).
- Current US national security strategy: "premeditated, politically motivated violence against innocents".

- United States Department of Defense: the "calculated use of unlawful violence to inculcate fear; intended to coerce or intimidate governments or societies in pursuit of goals that are generally political, religious, or ideological".
- British Terrorism Act 2000 defines terrorism so as to include not only attacks on military personnel but also acts not usually considered violent, such as shutting down a website whose views one dislikes.
- 1984 US Army training manual says "terrorism is the calculated use of violence, or the threat of violence, to produce goals that are political or ideological in nature".
- 1986 Vice-President's Task Force: "Terrorism is the unlawful use or threat of violence against persons or property to further political or social objectives. It is usually intended to intimidate or coerce a government, individuals, or groups or to modify their behavior or politics."
- Insurance documents define terrorism as "any act including, but not limited to, the use of force or violence and/or threat thereof of any person or group(s) of persons whether acting alone or on behalf of, or in connection with, any organization(s) or government(s) committed for political, religions, ideological or similar purposes, including the intention to influence any government and/or to put the public or any section of the public in fear".

A *hazard* is an act or phenomenon posing potential harm to some person(s) or thing(s), that is, a source of harm, and its potential consequences. For example, uncontrolled fire is a hazard, water can be a hazard, and strong wind is a hazard. In order for the hazard to cause harm, it needs to interact with person(s) or thing(s) in a harmful manner. Hazards need to be identified and considered in projects' life cycle analyses since they could pose threats and could lead to project failures.

Threat is any indication, circumstance, or event with the potential to cause the loss of or damage to an asset. Threat can also be defined as the intention and capability of an adversary to undertake actions that would be detrimental to assets.

Reliability can be defined for a system or a component as its ability to fulfill its design functions under designated operating and/or environmental conditions for a specified time period. This ability is commonly measured using probabilities. Reliability is, therefore, the occurrence probability of the complementary event to failure.

For a failure event, *consequences* can be defined as the degree of damage or loss from some failure. Each failure of a system has some consequence(s). A failure could cause economic damage, environmental damage, injury or loss of human life, or other possible events. Consequences need to be quantified in terms of failure—consequence severities using relative or absolute measures for various consequence types to facilitate risk analysis.

Risk originates from the Latin term *risicum* meaning the challenge presented by a barrier reef to a sailor. The Oxford dictionary defines risk as the chance of hazard, bad consequence, loss, and so on. Also, risk is the chance of a negative outcome. Formally, risk can be defined as the potential of losses for a system resulting from an uncertain exposure to a hazard or as a result of an uncertain event. Risk should be identified based on risk events or event scenarios. Risk can be viewed as a multidimensional quantity that includes event-occurrence probability, event-occurrence consequences, consequence significance, and the population at risk; however, it is commonly measured as a pair of the probability of occurrence of an event, and the outcomes or consequences associated with the event's occurrence. Another common representation of risk is in the form of an exceedence probability function of consequences.

Probability is a measure of the likelihood, chance, odds, or degree of belief that a particular outcome will occur. A conditional probability is the probability of occurrence of an event based on the assumption that another event (or multiple events) has occurred.

An asset is any person, environment, facility, physical system, material, cyber system, information, business reputation, or activity that has a positive value to an owner or to society as a whole.

The occurrence probability (p) of an outcome (o) can be decomposed into an occurrence probability of an event or threat (t) and the outcome-occurrence probability given the occurrence of the event ($o|t$). The occurrence probability of an outcome can be expressed as follows using conditional probability concepts:

$$p(o) = p(t)p(o|t) \tag{1}$$

In this context, threat is defined as a hazard or the capability and intention of an adversary to undertake actions that are detrimental to a system or an organization's interest. In this case, threat is a function of only the adversary or competitor, and usually cannot be controlled by the owner of the system. The adversary's intention to exploit his capability may, however, be encouraged by vulnerability of the system or discouraged by an owner's countermeasures. The probability $p(o|t)$ can be decomposed further into two components: success probability of the adversary and a conditional probability of consequences as a function of this success. This probability $p(o|t)$ can then be computed as the success probability of the adversary times the conditional probability of consequences given this success. The success probability of the adversary is referred to as the *vulnerability of the system* for the case of this threat occurrence. *Vulnerability* is a result of any weakness in the system or countermeasure that can be exploited by an adversary or competitor to cause damage to the system and result in consequences.

The *performance* of a system or component can be defined as its ability to meet functional requirements. The performance of an item can be described by various elements, such as speed, power, reliability, capability, efficiency, and maintainability. The design and operation of system affects this performance.

A *system* is a deterministic entity comprising an interacting collection of discrete elements and commonly defined using deterministic models. The word *deterministic* implies that the system is identifiable and not uncertain in its architecture. The definition of the system is based on analyzing its functional and/or performance requirements. A description of a system may be a combination of functional and physical elements. Usually functional descriptions are used to identify high information levels on a system. A system can be divided into subsystems that interact. Additional details in the definition of the system lead to a description of the physical elements, components, and various aspects of the system. Methods to address uncertainty in systems architecture are available and can be employed as provided by [3].

Risk-based technologies (RBT) are methods or tools and processes used to assess and manage the risks of a component or system. RBT methods can be classified into risk management that includes risk assessment/risk analysis and risk control using failure prevention and consequence mitigation, and risk communication. Risk assessment consists of hazard identification, event-probability assessment, and consequence assessment. Risk control requires the definition of acceptable risk and comparative evaluation of options and/or alternatives through monitoring and decision analysis. Risk control also includes failure prevention and consequence mitigation. Risk communication involves perceptions of risk, which depends on the audience targeted. Hence, it is classified into the media, the public, and the engineering community.

Safety can be defined as the judgment of risk tolerance (or acceptability in the case of decision making) for the system. Safety is a relative term since the decision of risk acceptance may vary depending on the individual making the judgment. Different people are willing to accept different risks as demonstrated by different factors such as location, method or system type, occupation, and lifestyle. The selection of these different activities demonstrates an individual's safety preference despite a wide range of risk values. It should be noted that *risk perceptions* of safety may not reflect the actual level of risk in some activity.

Risk assessment is a technical and scientific process by which the risks of a given situation for a system are modeled and quantified. Risk assessment can require and/or provide both qualitative and quantitative data to decision makers for use in risk management. Risk analysis is the technical and scientific process to breakdown risk into its underlying components. Risk assessment and analysis provide the processes for identifying hazards, event-probability assessment, and consequence assessment. The risk assessment process answers three basic questions: (i) What can go wrong? (ii) What is the likelihood that it will go wrong? (iii) What are the consequences if it does go wrong? Answering these questions requires the utilization of various risk methods as discussed in this section. A summary of selected methods is provided in Table 1. A typical overall risk analysis and management methodology can be expressed in the form of a workflow or block diagram consisting of the following primary steps:

1. definition of a system based on a stated set of analysis objectives;
2. hazard or threat analysis, definition of failure scenarios, and hazardous sources and their terms;
3. data collection in a life cycle framework;
4. qualitative risk assessment;
5. quantitative risk assessment; and
6. management of system integrity through countermeasures, failure prevention, and consequence mitigation using risk-based decision making.

Methods to support these steps are described in various articles of this section on "Risk Modeling and Vulnerability Assessment".

Risk can be assessed and presented using matrices for preliminary screening by subjectively estimating probabilities and consequences in a qualitative manner. A risk matrix is a two-dimensional presentation of likelihood and consequences using qualitative metrics for both the dimensions as given in Tables 2–4 and Figure 1 with risk subjectively assessed as high (H), medium (M), and low (L). The articles on "Quantitative representation of risk" and "Qualitative representation of risk" describe other methods for representing risk.

A countermeasure is an action taken or a physical capability provided whose principal purpose is to reduce or eliminate one or more vulnerabilities or to reduce the frequency of attacks. *Consequence mitigation* is the preplanned and coordinated actions or system

TABLE 1 Risk Assessment Methods

Method	Scope
Safety/review audit	Identifies equipment conditions or operating procedures that could lead to a casualty or result in property damage or environmental impacts
Checklist	Ensures that organizations are complying with standard practices
What-If	Identifies hazards, hazardous situations, or specific accident events that could result in undesirable consequences
Hazard and operability study (HAZOP)	Identifies system deviations and their causes that can lead to undesirable consequences and determine recommended actions to reduce the frequency and/or consequences of the deviations
Preliminary hazard analysis (PrHA)	Identifies and prioritizes hazards leading to undesirable consequences early in the life of a system. It determines recommended actions to reduce the frequency and/or consequences of the prioritized hazards. This is an inductive modeling approach
Probabilistic risk analysis (PRA)	Quantifies risk, and was developed by the nuclear engineering community for risk assessment. This comprehensive process may use a combination of risk assessment methods
Failure modes and effects analysis (FMEA)	Identifies the components (equipment) failure modes and the impacts on the surrounding components and the system. This is an inductive modeling approach
Fault tree analysis (FTA)	Identifies combinations of equipment failures and human errors that can result in an accident. This is an deductive modeling approach
Event tree analysis (ETA)	Identifies various sequences of events, both failures and successes that can lead to an accident. This is an inductive modeling approach
The Delphi Technique	Assists to reach consensus of experts on a subject such as project risk while maintaining anonymity by soliciting ideas about the important project risks that are collected and circulated to the experts for further comment. Consensus on the main project risks may be reached in a few rounds of this process [3].
Interviewing	Identifies risk events by interviews of experienced project managers or subject-matter experts. The interviewees identify risk events based on experience and project information
Experience-based identification	Identifies risk events based on experience including implicit assumptions
Brain storming	Identifies risk events using facilitated sessions with stakeholders, project team members, and infrastructure support staff

features that are designed to reduce or minimize the damage caused by attacks (consequences of an attack), support and complement emergency forces (first responders), facilitate field-investigation and crisis management response, and facilitate recovery and reconstitution. Consequence mitigation may also include steps taken to reduce short- and long-term impacts, such as providing alternative sources of supply for critical goods and services. Mitigation actions and strategies are intended to reduce the consequences (impacts) of an attack, whereas countermeasures are intended to reduce the probability that an attack will succeed in causing a failure or significant damage.

TABLE 2 Likelihood Categories for a Risk Matrix

Category	Description	Annual Probability Range
A	Likely	≥0.1 (1 in 10)
B	Unlikely	≥0.01 (1 in 100) but <0.1
C	Very unlikely	≥0.001 (1 in 1,000) but <0.01
D	Doubtful	≥0.0001 (1 in 10,000) but <0.001
E	Highly unlikely	≥0.00001 (1 in 100,000) but <0.0001
F	Extremely unlikely	<0.00001 (1 in 100,000)

TABLE 3 Consequence Categories for a Risk Matrix

Category	Description	Examples
I	Catastrophic	Large number of fatalities and/or major long-term environmental impact
II	Major	Fatalities and/or major short-term environmental impact
III	Serious	Serious injuries and/or significant environmental impact
IV	Significant	Minor injuries and/or short-term environmental impact
V	Minor	Only first aid injuries and/or minimal environmental impact
VI	None	No significant consequence

TABLE 4 Example Consequence Categories for a Risk Matrix in 2003 Monetary Amounts (US $)

Category	Description	Cost
I	Catastrophic loss	≥$10,000,000,000
II	Major loss	≥$1,000,000,000 but <$10,000,000,000
III	Serious loss	≥$100,000,000 but <$1,000,000,000
IV	Significant loss	≥$10,000,000 but <$100,000,000
V	Minor loss	≥$1,000,000 but <$10,000,000
VI	Insignificant loss	<$1,000,000

Probability category							
	A	L	M	M	H	H	H
	B	L	L	M	M	H	H
	C	L	L	L	M	M	H
	D	L	L	L	L	M	M
	E	L	L	L	L	L	M
	F	L	L	L	L	L	L
		VI	V	IV	III	II	I
	Consequence category						

FIGURE 1 Example risk matrix.

Risk management entails decision analysis for a cost-effective reduction of risk within available resources. The benefit of a risk mitigation action can be assessed as follows:

$$\text{Benefit} = \text{unmitigated risk} - \text{mitigated risk} \tag{2}$$

The benefit minus the cost of mitigation can be used to justify the allocation of resources. The benefit-to-cost ratio can be computed, and may also be helpful in decision making. The benefit-to-cost ratio can be computed as

$$\text{Benefit-to-cost ratio } (B/C) = \frac{\text{Benefit}}{\text{Cost}} = \frac{\text{Unmitigated risk} - \text{Mitigated risk}}{\text{Cost}} \tag{3}$$

The cost in Eq. (3) is the cost of the mitigation action or countermeasure. Ratios greater than one are desirable. In general, the larger the ratio, the better the mitigation action. Internal rate of return can be used instead of benefit-to-cost ratios [1]. Assuming B and C random variables with normal probability distributions, a benefit–cost index ($\beta_{B/C}$) can be defined as follows:

$$\beta_{B/C} = \frac{\mu_B - \mu_C}{\sqrt{\sigma_B^2 + \sigma_c^2}} \tag{4}$$

where μ and σ are the mean and standard deviation. In the case of lognormally distributed B and C, benefit–cost index can be computed as

$$\beta_{B/C} = \frac{\ln\left(\frac{\mu_B}{\mu_C}\sqrt{\frac{\delta_C^2+1}{\delta_B^2+1}}\right)}{\sqrt{\ln[(\delta_B^2 + 1)(\delta_C^2 + 1)]}} \tag{5}$$

where δ is the coefficient of variation. The probability of not realizing the benefits (P) can be computed as

$$P = 1 - \Phi(\beta_{B/C}) \tag{6}$$

where Φ is the cumulative distribution function of the standard normal. In the case of mixed distributions or cases involving basic random variables of B and C, the advanced second moment method or simulation method can be used [1]. In cases where benefit is computed as revenue minus cost, benefit might be correlated with cost requiring the use of other methods [1].

The following are four primary ways available to deal with risk within the context of a risk management strategy:

- risk reduction or elimination;
- risk transfer, for example, to a contractor or an insurance company;
- risk avoidance; and
- risk absorbance or pooling.

Risk communication can be defined as an interactive process of exchange of information and opinion among stakeholders such as individuals, groups, and institutions. It often involves multiple messages about the nature of risk or expressing concerns, opinions, or reactions to risk managers or to legal and institutional arrangements for risk management. Risk communication greatly affects risk acceptance and defines the acceptance criteria for safety.

3 TERRORISM RISK ANALYSIS

Terrorism can be assessed at various levels, starting with a facility and its assets, a region, a sector, and a national territory. For example, considering risk analysis for an asset requires examining several threat-asset scenarios [4, 5]. Each individual risk for a scenario can be evaluated as the product or combination of a specific defined consequence, expressed as a numerical range (e.g. dollars) and the probability range that the specific consequence will occur. This results in a mean value for the risk, and can be reevaluated to produce upper and lower bounds using uncertainty analysis techniques [3]. For asset screening, qualitative measures of probability and consequence can be combined on a risk matrix. For most postulated adversary actions, there is a spectrum of possible outcomes (consequences), each with an associated probability range. Terrorism risk analysis can be performed using a scenario-based approach. The probability range associated with a specific consequence range is the product of the frequency of attacks by adversaries and a series of individual conditional probability ranges associated with the chain of events that occur after the attack. These probability ranges are determined based on the various capabilities that the asset has for dealing with or avoiding the adversary action. These capabilities may be consequence mitigation systems or threat response or avoidance actions (countermeasures), which can be pictured as nodes along an event tree. The consequence ranges associated with a successful attack by an adversary on an asset, combined with the probability range associated with each consequence, define the risk range. The risks associated with each of these probability range/consequence range pairs can be added (given that they are measured in the same units, e.g. dollars) to obtain the risk for a postulated adversary action. Similarly, the overall risk for a spectrum of postulated adversary actions is the sum of the risks for each individual action. Figure 2 schematically shows the relationship of risk, in terms of the above features, and the spectrum of potential countermeasures and consequence mitigation strategies. Figure 3 presents a primary structure of risk analysis for critical assets. The figure shows a threat block on the left that targets asset or sector vulnerabilities protected by existing countermeasures. If a threat materializes and succeeds, it might lead to failures that might pass through existing consequence mitigation measures leading to consequences.

4 UNIQUE FEATURES OF TERRORISM RISK ANALYSIS

Terrorism risk analysis includes some features of typical probabilistic risk analysis (PRA) methods, in which probability and consequence ranges are determined numerically. However, it relies heavily on many of the features of traditional qualitative approaches

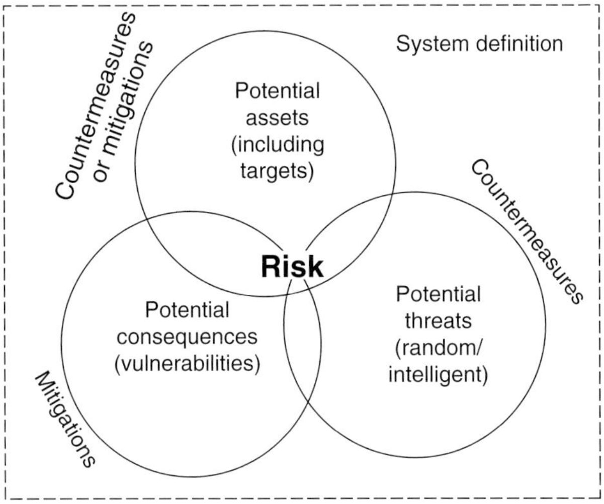

FIGURE 2 Schematic of the approach to risk analysis.

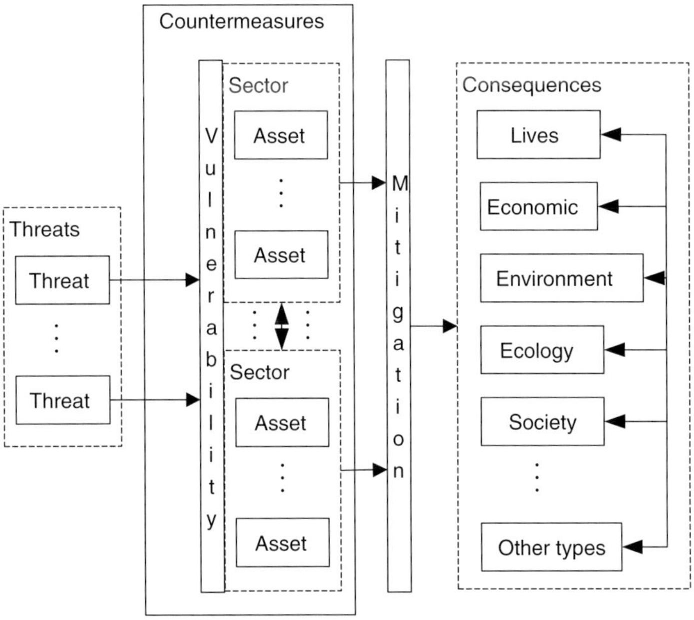

FIGURE 3 A primary structure for risk analysis.

to balance the time and resources required to do the analysis with the need for numerical risk measures that can be used to inform resource allocation decisions. Some of the unique features of risk analysis related to assets, threats, and countermeasures and consequence mitigation are summarized in Table 5.

TABLE 5 Unique Features of Risk Analysis for Asset Protection from Adversary Threats

Features	Unique Characteristics
Risk analysis framework	Should be performed accounting for the perspectives of adversaries as well as the perspectives of defenders. And as a multilevel analysis ranging from an asset, to multiassets, to a sector, and to multisectors to sufficiently account for interdependencies that may affect the risks pertinent to the decision being made
Asset (target) features	Include attractive assets, critical assets, soft assets, assets with vulnerabilities that are sufficiently known to adversaries
Assets (targets) selected by adversaries	Include high consequence assets (or scenarios) with high probability of success
Threat features	Include the dynamic nature of threats, threat types and probabilities; their nonrandomness but deliberateness using design basis threats; possibly being of unknown or unknowable types
Threat–asset dependencies	Include dynamically responding to asset protection using countermeasures and consequence mitigation
Ingenuity of adversaries	Include converting assets to threats by capitalizing on the efficiency of infrastructures, for example: Transportation efficiency by converting airplane assets into explosive weapons Mail efficiency by using mail items for bioagent delivery Other efficient infrastructure systems include power and information systems
Capabilities of adversaries	Include the ability to select targets and accurately deliver the weapon to them and the ability to adapt to countermeasures to redirect the weapon to another target
Asset vulnerabilities	Include identifying targets outside the system boundaries to exploit system vulnerabilities through system dependencies
Consequences	Are broadly defined to include public health, economic loss, loss of vital commodities, interruption of government operation, and national psyche
Asset and sector interdependencies	Include interdependencies in functionality and subsequently in consequences
Decision analysis	Includes trade-offs based on national security, safety, and economics
Information flow	Is a two-way flow of defenders acquiring knowledge about the adversaries; adversaries acquiring knowledge about the assets, countermeasures, and consequence mitigation plans
Countermeasures	Include countermeasures at the asset level and metacountermeasures at the multiasset, sector, and multisector levels. Countermeasures reduce the probability of selection of an asset as well as the probability of success of an attack
Consequence mitigation plans	Include mitigation at the local level and metamitigations at the state, regional, and national level. Mitigation actions reduce consequences

5 ANALYSIS OBJECTIVES, ASSETS, AND SYSTEM BOUNDARIES

One of the first steps in a risk analysis is to define the objectives. The objectives may include

- understanding the nature of the risks to define and optimize the allocation of resources for countermeasures and consequence mitigation strategies;
- maximizing threat risk reduction benefits and utility to stakeholders; and/or
- understanding the nature of the risks for better communication with stakeholders.

Risk analysis requires a systems framework in order to achieve the stated objectives. Such a view of risk analysis requires structuring and formulating a problem or approaching a design with the following in mind:

1. The structure should be within a systems framework.
2. The approach should be systematic and should capture all critical aspects of the problem or decision situation.
3. Uncertainties should be assessed and considered.
4. An optimization scheme should be constructed for the utilization of available resources including maximizing benefits and utility to stakeholders. The definition of a system is therefore driven by a well-stated description of the objectives of the risk analysis.

After the objectives have been defined, the next step is to identify potential critical assets for further screening. This is done using judgment and experience. Taxonomy of potentially critical assets is provided in Table 6 for illustration purposes. The taxonomy is provided under selected headings and could be expanded hierarchically according to asset categories and sectors.

Identifying critical assets requires defining the features that define criticality. Categories of critical assets are relatively broad and inclusive. The criticality of an asset should be based on features such as the impact of total destruction of or significant damage to an asset on

- public service and the operation of government;
- the local, regional, and national economy;
- surrounding population;
- national security; and
- environment.

Note that critical assets are identified primarily on the basis of the consequences of a successful attack by an adversary rather than the probability that the attack will be successful. However, other asset features that should be considered include

- asset softness, that is, accessibility and inability to limit it;
- softness of targets within an asset; and
- other specific features of these targets.

TABLE 6 Asset Taxonomy

Agriculture and Food	Information and Telecommunications	Banking and Finance
Supply	Public switched telecommunications network (PSTN)	Physical facilities (buildings)
Processing	Internet	Operations centers
Production	Switch/router areas	Regulatory institutions
Packaging	Access tandems	Physical repositories
Storage	Fiber/copper cable	Telecommunications networks
Distribution	Cellular, microwave, and satellite systems	Emergency redundancy service areas
Transportation	Operations, administration, maintenance and provisioning systems	**Chemical/Hazardous Materials Industry**
Water	Network operations centers	Manufacturing plants
Dams, wells, reservoirs, and aqueducts	Underwater cables	Transport systems
Transmission pipelines	Cable landing points	Distribution systems
Pumping stations	Collocation sites, peering points, and telecom hotels	Storage/stockpile/supply areas
Sewer systems	Radio cell towers	Emergency response and communications systems
Treatment facilities	**Energy**	**Postal and Shipping**
Storage facilities	**Electricity (Nonnuclear)**	Processing facilities
Public Health	Hydroelectric dams	Distribution networks
National strategic stockpile	Fossil-fuel electric power generation plants	Air, truck, rail, and boat transport systems
National institutes of health	Distribution systems	Security
State and local health departments	Key substations	**National Monuments and Icons**
Hospitals	Communications	National parks
Health clinics	**Oil and Natural Gas**	Monuments
Mental health facilities	Offshore platforms	Historic buildings
Nursing homes	Refineries	**Nuclear Power Plants**
Blood supply facilities	Storage facilities	Commercial owned/operated
Laboratories	Gas processing plants	Government owned/operated
Mortuaries	Product terminals	Physical facilities
Pharmaceutical stockpiles	Pipelines	Spent fuel storage facilities
Emergency Services	**Transportation**	Safety/security systems
Fire houses	Aviation	**Dams**
Rescue	Railways	Large
Emergency medical services	Highways	Small
Law enforcement	Trucking	Government owned
Mobile response	Busing	Private/corporate owned
Communications systems	Bridges	**Government Facilities**

TABLE 6 (*Continued*)

Agriculture and Food	Information and Telecommunications	Banking and Finance
Defense Industry Base	Tunnels	**National Security Special Events**
Supply systems	Borders	**Commercial Assets**
	Seaports	Prominent commercial centers
	Pipelines	Office buildings
	Maritime	Sports centers/arenas
	Mass transit	Theme parks
		Processing/service centers

The definition of the system and the establishment of system boundaries should be based on an analysis of its functional and/or performance requirements. A description of a system may be a combination of functional and physical elements. For example, the system to be considered for a particular highway bridge should include both its structural design characteristics (e.g. a steel span across a river that is a quarter mile wide) and its functional capability (e.g. carry 1000 vehicles per hour). A system may be divided into subsystems that interact with each other. For example, the elements of a petroleum refinery are highly interdependent, such that each unit might constitute a subsystem within a system under analysis. Additional detail leads to a description of the physical elements, components, and various aspects of the system. The analysis objectives drive the system definition and boundaries. Buffer zones should be included in the definition of the systems, so that their effect on threat success and consequence mitigation can be appropriately assessed.

6 UNIQUE FEATURES OF THREAT

The analysis of threats is generally the most uncertain part of risk analysis for homeland security. When considering events such as equipment failures, human errors, or natural disasters an extensive body of historical experience exists that can be used to establish the frequency or probability range of various initiating events. Although, unfortunately, there is also an extensive history of terrorist and other adversarial acts, the nature of these events is constantly changing, so historical experience provides less guidance in trying to predict the future. The following characteristics make terrorist and other adversarial threats unique relative to other risk contributors:

- Terrorist and other adversarial threats are focused rather than random events. This is often characterized as "intelligent versus random threats". Accidents and natural disasters occur in a random pattern that is often statistically predictable. On the other hand, prediction of the frequency range of a specific adversary action against a specific target should be based less on historical data and more on an analysis of factors such as

  - prevailing political situation;
  - objectives and motivations of adversaries with access to or near the target;
  - attractiveness of the target to the adversaries;
  - number and type of adversaries with sufficient access to the target to carry out the threat;
  - weapons and other capabilities available to the adversaries;
  - local security surveillance level;
  - quality of intelligence information.
- Unlike accidents and natural disasters, adversaries are able to adapt to changing circumstances. This is often characterized as "dynamic versus static threats". For example,
  - hardening of a target by improving countermeasures that adversaries are aware of can drive them to select another target;
  - changing perceptions of the impact of damage to various targets and the effectiveness of various types of attacks can lead to changes in terrorist and other adversarial strategies;
  - terrorists/adversaries typically try to accumulate more material and capability than the minimum necessary to achieve their desired consequences.
- Even "unsuccessful" attacks can have a significant consequence. For example, an attempt to shoot down a commercial airliner in the United States could have a significant impact on the airlines and the US economy even if the attack is not successful.

Efficient systems for creating threats, delivering threats, and propagating consequences should be identified, such as transportation systems, mail systems, computer networks, and power systems. Adversaries could capitalize on the efficiencies of infrastructure systems to create/advance/transmit/propagate threats. Threats could be classified as threats by the system, through the system, and to the system.

Threats can be classified by type as shown in Table 7. Analyzing threats over a spectrum from low to high enables analysts to focus their attention and resources. Figure 4 shows such a threat spectrum based on threat magnitude measured in terms of its potential impact. Design basis threats can be used to address threats for which an individual asset owner is expected to provide countermeasures and/or consequence mitigation; whereas regional or national countermeasures are necessary for large-scale threats, for example, weapons of mass destruction. For the region in between these two extreme categories, risk-informed decision making should be used to decide on appropriate countermeasures and consequence mitigation strategies.

7 SECTION SCOPE AND OUTLINE

This article starts with a presentation of overall frameworks for analyzing and assessing risks for homeland security purposes, and for critical infrastructure and key resource protection. The article also includes various risk analysis methods, such as logic trees: event, fault, success, attack, probability, and decision trees; scenario analysis, cognitive maps,

TABLE 7 Threat Types

Threat Type	Delivery Mode	Weapon/Agent	Quantity/Quality
Chemical	Outdoor dispersal	Ricin	Potent
		Mustard gas	Potent
	Crop duster	VX	Potent
		Chlorine gas	Potent
	Propelled missile	Any of the above	Potent
	Postal mail	Ricin	Potent
Biological	Outdoor dispersal	Anthrax	Potent
		SARS	Potent
	Postal mail	Anthrax	Potent
	Food buffets	Hepatitis	Potent
		Salmonella	Potent
	Missile	Any of the above	Potent
Radiological	Standard deployment	Dirty bomb	Strong
		Radiological release	Strong
Nuclear	Standard deployment	Improvised nuclear device	2 kt
		Strategic nuclear weapon	100 kt
Explosive	Standard deployment	Backpack bomb	10 lb Trinitrotoluene (TNT)
		Propelled missile	50 t
	Truck	Fertilizer bomb	200 lb
			500 lb
			1000 lb
			4000 lb
	Boat	C4	200 lb
	Airplane	Jet fuel	5000 ga
Sabotage	Physical	Cut power cable	Not applicable
		Cut bolts	Not applicable
		Improper operation or maintenance	Not applicable
	Cyber	Providing unauthorized access	Obvious
Cyber	Physical	Cut SCADA cable	Not applicable
		Magnetic weapons	Power units
	Cyber	Worm virus	Obvious

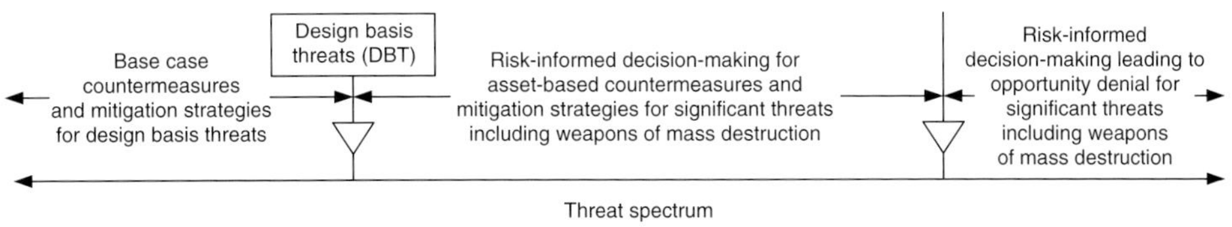

FIGURE 4 Threat spectrum.

and concept maps; Bayesian networks; probabilistic risk assessment; game theory; and representation of risk. National interdependence models of infrastructure are discussed. High consequence threats, such as electromagnetic pulse, nuclear, biological, chemical, and high-grade explosives are discussed. Also, special chapters on bioterrorism, cyber security, risk perception, and soft failure modes are provided. Regulations and standards relating to risk analysis are discussed and their features are summarized. Also, methods for policy development are discussed.

The article covers various analytical steps needed for performing threat analysis, vulnerability assessment, consequence analysis with infrastructure interdependencies and social and psychological issues, systems analysis, multiobjective decision analysis including risk-based prioritization, and risk communication. Uncertainty modeling and analysis are also introduced and discussed. Countermeasures, including robustness, resilience and security, and consequence mitigation methods are discussed with a special presentation relating to deterrence and defeating surprise.

Experiences from terrorism risk analysis from the insurance industry including the economics of terrorism and risk transfer are discussed. Regional risk analysis and protection are also covered.

Terrorism risk analysis requires data and information. Data scarcity and unavailability require the use of expert opinion elicitation. Data sources for threat analysis are needed [2]. Treads in threats and their organizations and emergence, and terrorism databases are also discussed and covered. The legal aspects of information security needed for risk analysis are discussed.

REFERENCES

1. Ayyub, B. M. (2003). *Risk Analysis in Engineering and Economics*, Chapman and Hall/CRC Press, FL.
2. Ayyub, B. M., and Klir, G. J. (2006). *Uncertainty Modeling and Analysis in Engineering and the Sciences*. Chapman and Hall/CRC Press, Boca Raton, FL.
3. Ayyub, B. M. (2001). *Elicitation of Expert Opinions for Uncertainty and Risks*. CRC Press, Boca Raton, FL.
4. Ayyub, B. M., McGill, W. L., and Kaminskiy, M., (2007). Critical asset and portfolio risk analysis for homeland security: an all-hazards framework, *Risk Anal. Int. J. Soc. Risk Anal.*, **27**(3), 789–801.
5. McGill, W. L., Ayyub, B. M., and Kaminskiy, M., (2007). A quantitative asset-level risk assessment and management framework for critical asset protection, *Risk Anal. Int. J. Soc. Risk Anal.*, **27**(5), 1265–1281.

FURTHER READING

Kumamoto, H., and Henley, E. J. (1996). *Probabilistic Risk Assessment and Management for Engineers and Scientists*, 2nd ed. IEEE Press, New York.

Modarres, M. (1993). *What Every Engineer Should Know About Reliability and Analysis*. Marcel Dekker, Inc., New York.

Modarres, M., Kaminskiy, M., and Krivstov, V. (1999). *Reliability Engineering and Risk Analysis: A Practical Guide*, Marcel Decker Inc., New York.

RISK ANALYSIS FRAMEWORKS FOR COUNTERTERRORISM*

James Scouras

Defense Threat Reduction Agency, Fort Belvoir, Virginia

Gregory S. Parnell

Department of Systems Engineering, United States Military Academy, West Point, New York

Bilal M. Ayyub

Center for Technology and Systems Management, Department of Civil and Environmental Engineering, University of Maryland, College Park, Maryland

Robert M. Liebe

Innovative Decisions Inc., Vienna, Virginia

1 INTRODUCTION

Although there are numerous definitions of risk concepts in use in the risk analysis community, the great majority of these definitions recognize the same common elements. We focus on these essential elements to develop straightforward definitions of risk, risk analysis, risk assessment, risk management, threat, vulnerability, and consequence. To further clarify these terms, we also discuss the relationships among them. We also expand some of the classic questions addressed by risk assessment and risk management to more explicitly address terrorism.

A framework is a conceptual or procedural structure used to address complex issues. A risk framework can facilitate both internal and external risk communication and enable risk analysis involving diverse threats and multiple participants. We analyze a selection of risk frameworks to develop a set of tasks that would be included in a comprehensive framework and against which a particular framework could be evaluated: (i) identify goals and objectives, (ii) define system, (iii) assess threats, (iv) assess vulnerabilities, (v5) assess consequences, (vi) assess baseline risk, (vii) identify risk management options, (viii) analyze benefits and costs, (ix) make decisions, (x) communicate risks, (xi) implement risk management actions, and (xii) monitor risk management actions.

The United States is engaged in a global war on terrorism with domestic and international battlefronts. US military and civilian leaders have defined this conflict as the "Long War" to convey the expectation that the terrorist threat will be with us for years—possibly decades—and the war will require significant resources and resolve to win [1].

There is a consensus among analysts and policy makers that risk management provides the proper framework for defending against terrorism. We need to understand the threats, reduce our vulnerabilities, and prevent terrorists from achieving the

*The views expressed in this article represent those of the authors; they do not necessarily represent the views of any governmental or commercial entity.

consequences they seek with attacks on the United States and worldwide. The US Government Accountability Office (GAO) advocates adopting a risk management approach universally across the federal government [2]:

> After threat, vulnerability, and criticality assessments have been completed and evaluated in this risk-based decision process, key actions can be taken to better prepare ourselves against potential terrorist attacks. Threat assessments alone are insufficient to support the key judgments and decisions that must be made. However, in conjunction with vulnerability and criticality assessments, leaders and managers will make better decisions based on this risk management approach. If the federal government were to apply this approach universally and if similar approaches were adopted by other segments of society, we could more effectively and efficiently prepare in-depth defenses against acts of terrorism against our country.

This article focuses on risk frameworks, a first step toward coordinated risk management approaches. We begin with discussions of essential risk terminology and the relationships among basic concepts, as well as several important practical considerations in risk analysis. We then introduce the concept of a risk framework and summarize our research on 17 diverse frameworks. We illustrate two types of risk frameworks: procedural and methodological. On the basis of our research on the risk frameworks, we identify 12 risk framework elements. We then describe each of these elements and how they can be used to evaluate risk frameworks.

2 DEFINITIONS OF RISK ANALYSIS TERMS

It is important to begin with a clear understanding of the terminology of risk analysis. Here, we limit ourselves to definitions of risk, the major components of risk analysis, and the elements of risk assessment. There are a multitude of definitions in use for each of these terms. Our analysis of these definitions leads us to the observation that much effort can be expended in debating the nuances of alternative definitions to little avail. The definitions we propose are intended to be simple in that they convey the essential elements of the term, with as little extraneous baggage as possible.

Risk is the potential for loss or harm due to the likelihood of an unwanted scenario and its potential adverse consequences. There are two essential features of risk that are embedded in this definition. First, the consequences that contribute to risk are negative. (There is no short-term financial risk associated with winning the lottery.) Secondly, both the scenario (sequence of future states) and its associated consequences are uncertain. (There is no risk that governments will abandon taxation.) We use the term *likelihood* (rather than *probability*) to acknowledge that risk can be represented qualitatively or quantitatively.

Risk assessment is a systematic analytic process for describing the nature and magnitude of risk associated with a scenario, including consideration of relevant uncertainties. When we consider natural hazards (e.g. hurricanes) and engineered systems (e.g. nuclear reactor accidents), the risk assessment objective is to provide, to the extent practical, a scientific and analytically sound basis for answering the following questions: What can go wrong? How can it happen? What are the potential consequences if it does happen? How likely is it to happen? However, these questions do not emphasize the unique human aspect of the terrorist threat: terrorists are intelligent and motivated adversaries who can adapt to their experiences, environment, and anticipations of the future. Thus, when we consider terrorism, we propose a set of more explicit questions:

1. What adversaries threaten US interests?
2. What are their motivations, capabilities, and intentions?
3. What vulnerabilities could be exploited in an attack?
4. What are the potential consequences of an attack?
5. How likely is an attack to occur?

Risk management is the process of constructing, evaluating, selecting, implementing, and monitoring actions to alter levels of risk. For natural and engineered system hazards, it addresses the questions:

1. What can we do?
2. What should we do?
3. What are the results of our actions? Again, the terrorist is an intelligent adversary, which requires we make explicit the need to answer an additional question:
4. What can be done to account for the response of an adaptive, intelligent adversary?

The goal of risk management is scientifically sound, cost effective, integrated actions, including providing information (i.e. risk communication) that transfer, mitigate, or accept risks while taking into account social, cultural, ethical, political, economic, and legal considerations. Because of these additional considerations, which can significantly constrain the risk management actions available and their evaluations, we refer to homeland security decisions as *risk informed*, rather than *risk based*.

It is worth noting that some risk professionals use the term *risk management* to encompass *risk assessment* and *risk control*, where *risk control* is what we have defined as risk management in this article [3]. This is a useful taxonomy of concepts, but (unfortunately) not as widely used in the community of risk analysts and practitioners.

As with many terms in English, *risk analysis* has more than one meaning. One definition has risk analysis encompassing the full spectrum of activities, processes, and phenomena related to risk, including all of risk assessment and risk management, as depicted in Figure 1a. This was the meaning intended when the term was used in this article. Although risk communication is often called out as an additional separate element of risk analysis, we do not do so here because it is integral to both risk assessment and risk management.

A related but distinct definition holds risk analysis to be the process of separating the whole of risk into its component parts (e.g. threat, vulnerability, and consequence). As such, it supports both risk assessment and risk management as depicted in Figure 1b.

A *threat* is a scenario that could result in loss or harm. Terrorist threats are intentional. More generally, the term *hazard* encompasses terrorist and other intentional threats, as well as unintentional events such as accidents (e.g. nuclear power plant failures) and natural phenomena (e.g. hurricanes). Risk management of counterterrorism must include consideration of all hazards (AH) because (i) nonterrorist threats might not pose the greatest risks or the greatest opportunities for risk mitigation with limited resources and (ii) many risk management actions (e.g. stockpiling vaccines) will reduce both terrorist and nonterrorist risks and should be evaluated on that broader basis.

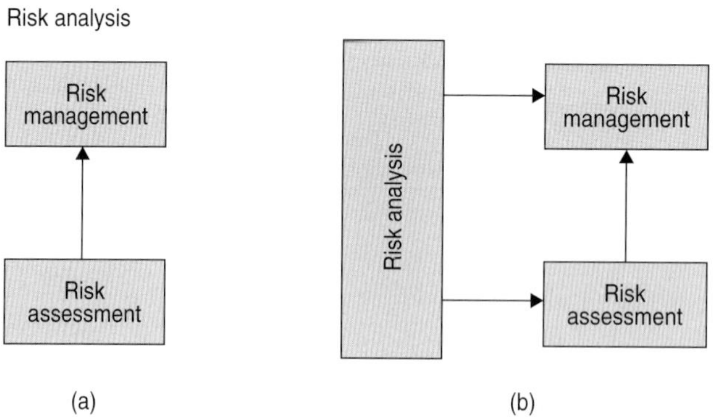

FIGURE 1 Alternative taxonomies of risk analysis, risk assessment, and risk management: (a) risk analysis *encompasses* risk assessment and risk management and (b) risk analysis *supports* risk assessment and risk management.

A *vulnerability* is an attribute of a system that could be exploited by an adversary to cause loss or harm or, similarly, a weakness of a system that could result in loss or harm in the event of a terrorist attack, an accident, or a natural phenomenon.

A *consequence* is an adverse outcome. Consequences include the tangible and intangible, the quantifiable and unquantifiable: mortality, morbidity, economic loss, psychological and societal damage, and myriad other forms of loss and harm. Consequences can cascade through interdependent economic infrastructures and can persist, or even increase, far into the future.

3 RELATIONSHIPS AMONG THREAT, VULNERABILITY, AND CONSEQUENCE

Two depictions of the relationships among threat, vulnerability, consequence, and risk are shown in Figures 2 and 3.

Figure 2 shows risk in a Venn diagram at the intersection of threat, vulnerability, and consequence. This widely used representation is meant to convey that if any one of these elements is absent, there is no risk. For example, even with a recognized vulnerability and definite severe consequences of an attack that exploits that vulnerability, without a plausible threat there is no risk. As an illustration, there is (essentially) no risk of a UK nuclear attack against the United States, even though we are defenseless against such an attack and the consequences would be horrific.

Note that while our definition of risk has two components, likelihood and consequences, Figure 2 shows three—threat, vulnerability, and consequence. These are reconciled in probabilistic risk analyses with the following definitions:

$$\text{Risk} = \text{threat} \times \text{vulnerability} \times \text{consequence} \tag{1}$$

where threat is the probability of an attempted attack, vulnerability is the probability of successful attack, given an attempted attack, and consequence is the consequences of a successful attack.

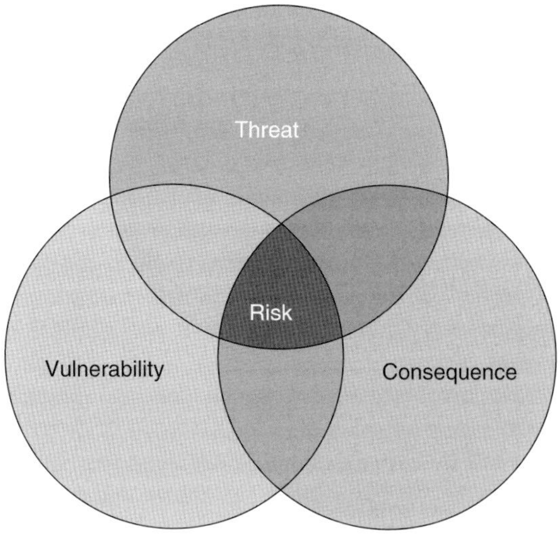

FIGURE 2 Venn diagram representation of threat, vulnerability, consequence, and risk.

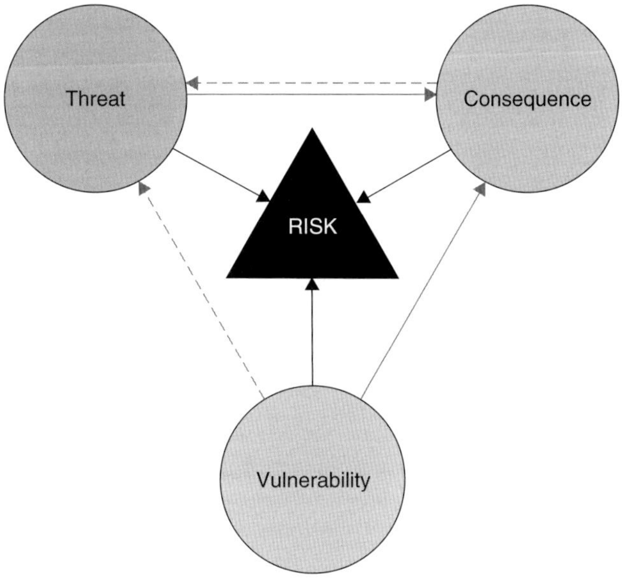

FIGURE 3 Information flow among threat, vulnerability, consequence, and risk.

successful attack. With these assignments, we can identify likelihood as the probability of a successful attack, the product of threat and vulnerability.

The Venn diagram representation of risk can be misleading. By its symmetry, it suggests that threat, vulnerability, and consequence are equal and independent elements of risk. Figure 3 breaks this symmetry by depicting the flow of information among threat, vulnerability, consequence, and risk, thereby making explicit their interdependencies.

The black arrows from threat, vulnerability, and consequence to risk indicate that all three elements contribute to risk. The solid red arrows indicate that both threat and vulnerability contribute to consequence. The dashed red arrows indicate that intelligent adversaries' (IA) perceptions of vulnerability and consequence influence their thinking, and thus the threat. (Readers are requested to refer the online version for color indication.) These arrows would not be present in an analysis of unintentional hazards.

4 INTRODUCTION TO RISK FRAMEWORKS

With this set of definitions and discussion, we now turn to the utility of, indeed *need for*, a risk framework to help assess and manage risk. We use the term *framework* to mean a conceptual or procedural structure used to address complex issues.

There is no doubt that the question of how to address the risks of terrorism qualifies as a complex issue. Terrorism is characterized by multiple adversaries with opaque motivations, diverse threats, learning, and adaption—all of which are evolving in uncertain ways over time. Counterterrorism must address multiple stakeholders with differing perceptions and priorities; profound uncertainties in threats; vulnerabilities of complex interdependent systems; myriad direct and indirect consequences; and the challenges of risk communication to diverse stakeholders.

In general terms, conceptual frameworks are simpler than procedural frameworks. Although both conceptual and procedural frameworks identify the major risk analysis tasks, conceptual frameworks may or may not explicitly specify an order to these tasks or the flow of information among them, and do not specify techniques or provide specifics with respect to data and measures. Procedural frameworks are more detailed and prescriptive than conceptual frameworks. Procedural frameworks do indicate the order of the tasks, the flow of information among them, can specify the techniques used to perform each task, and may or may not prescribe the data and measures used. Of course, many risk frameworks employ a mixture of features of both types of frameworks.

A risk framework makes a number of contributions to risk analysis. In particular, both conceptual and procedural frameworks serve as vehicles for risk communication. Internal risk communication among analysts, stakeholders, and decision makers requires a common terminology and understanding of the risk analysis process often undertaken by multiple participants from diverse disciplines across many public and private organizations. Even for highly conceptual frameworks, the process itself of developing a risk framework can be very helpful in hammering out differences in terminology and in refining the objectives, scope, and methodology of a risk analysis.

Moreover, for risk management decisions regarding policy, procedures, and the allocation of resources to gain acceptance and support, the results of risk analyses need to be communicated to stakeholders, the media, and the public. A graphical representation of a conceptual or procedural risk framework is an essential means toward this end.

The second major contribution of a risk framework applies more to procedural frameworks. The means of terrorist attack are diverse (e.g. biological, chemical, explosive, nuclear, and radiological), as are targets and consequences of attack. Since the goal of risk management is to effectively and efficiently allocate limited resources, risk analysis needs to compare risks across the full spectrum of possibilities. This requires the development of a risk framework that specifies the data to be collected, the techniques to be employed, and the measures to be used.

Finally, we observe that risk frameworks are either widely perceived to be useful or there is an unexplained compulsion to develop them. The ubiquity of risk frameworks is addressed in the following section.

5 SURVEY OF RISK FRAMEWORKS

In this section, we analyze a selection of common risk frameworks. These frameworks should be considered as an illustrative, rather than a representative sampling of all risk frameworks. One of our key screening criteria was clear evidence of use of the risk framework by public or private organizations. We offer the general observation that the variety of risk frameworks suggesting the development of a common risk framework for use throughout the homeland security enterprise is neither wise nor feasible.

Table 1 provides the following information about each risk framework: a reference number that includes the category of the framework, the framework name, the application area(s), the framework sponsor and/or developer, the user(s), and a reference (the full citation is provided in the references section). We have categorized the risk frameworks by the type of hazards they are designed to address: unintentional hazards (H), IA, and AH. The Coast Guard's Risk-Based Decision Making (RBDM) Guidelines is different from the other frameworks. RBDM has a list of 12 risk analysis techniques and 4 risk management techniques and provides information about when each of the techniques is appropriate.

Table 2 provides our analysis framework. We identify each task included in each of the frameworks and record the terminology used to describe it. By this process we have identified 12 distinct tasks that span the elements used in the 17 risk frameworks:

1. *Identify goals and objectives.* Identifying the goals and objectives of the risk framework is a useful task in problem framing. This can be achieved with a general objective (reduce risk of computer network attack) or multiple qualitative and quantitative objectives (see MORDA).
2. *Define system.* Another important framing or scoping task is defining the system. Because of the complex interconnections of systems, the system boundary may not be clear. For example, any component connected to the internet may be vulnerable to a cyber attack. Because of the system complexity, this task sometimes is required for critical asset identification. The assumption is that not all of the system is required to be considered—only the most critical assets.
3. *Assess threats.* This is a very important bounding task. Threats can be identified by specific sources (e.g. terrorist group T) or by classes of threats (domestic terrorist, foreign terrorist group, rogue nation–state, etc.). A common technique is the use of threat scenarios.
4. *Assess vulnerabilities.* Assessing the vulnerabilities is a very common task. This task usually requires significant understanding of the system (or at least the critical assets) and the full spectrum of the threats.
5. *Assess consequences.* Assessing consequences is a very common task. Identifying the potential types of consequences and their relationships is problematic. For example, will terrorist attacks of airplanes cause economic consequences of reduced discretional flying. If so, how much and for how long?

TABLE 1 Surveyed Risk Frameworks

Reference Number	Name	Application(s)	Sponsor/Developer	User(s)	Reference
H-1 RBDM	Risk-Based Decision Making (RBDM) Guidelines	Marine safety prevention, preparedness, and response	Coast Guard	Public and private organizations (12 risk analysis tools and 4 risk management techniques)	Coast Guard, 2008 [4]
H-2 IRGC	IRGC Risk Governance Framework	Global risks (environmental, social, etc.)	International Risk Governance Council	International agencies	Renn, 2005 [5]
H-3 HVA	Highway Vulnerability Assessment	Highway vulnerabilities	Federal Highway Administration/SAIC	Highway administrators	SAIC, 2002 [6]
H-4 RFRM	Risk Filtering and Ranking Method	Highway infrastructures	State of Virginia/UVA	Highway administrators	Haimes et al., 2004 [7]
H-5 MRA	Revised Framework for Microbial Risk Assessment	Microbial risk assessment	United Nations/ILSI	Environmental protection offices	ILSI, 2000 [8]
H-6 DOT	DOT Hazardous Materials Transport	Transportation of hazardous materials	DOT/ICF Consulting	Public and private hazardous materials transporters	ICF, 2000 [9]
H-7 NASA	NASA	Safety and reliability of engineered systems	NASA	NASA risk analysts and risk managers	NASA, 2002 [10]
IA-1 RAMCAP	RAMCAP	Infrastructure critical asset protection	DHS/ASME	Public and private organizations	ASME, 2004 [11]
IA-2 FEMA	FEMA 452	Terrorist threats against buildings	FEMA	Building developers and owners	FEMA, 2006 [12]

IA-3 DS	Digital Sandbox	Ranks relative risks for assets at a facility	Digital Sandbox	Public and private facility owners	Ware et al., 2002 Digital Sandbox, 2005 [13]
IA-4 ODP	ODP	Terrorist threat of WMD to facilities	Office of Domestic Preparedness	Federal, state, and local community organizations	ODP, 2003 [14]
IA-5 GAO	GAO Risk Management	Ports and other critical infrastructure	GAO	GAO to assess government frameworks	GAO, 2005 [15]
IA-6 MORDA	MORDA	Critical information systems	DoD/Innovative Decisions Inc.	DoD information assurance organizations	Buckshaw et al., 2005 [16]
IA-7 COMSEC	Canada COMSEC	Communications security	Canada	Information assurance organizations	Canada, 1999 [17]
AH-1 Orange Book	Orange Book	Risks to achieving organizational objectives	Her Majesty's Treasury	Public organizations in the United Kingdom	Her Magesty's Treasury, 2004 [18]
AH-2 CAPRA	CAPRA	Variety of hazards (safety, reliability, and security)	State of Maryland/University of Maryland	Public and private organizations	Ayyub, 2006 [19]
AH-3 Army	Army FM 100-14	Operational missions and daily tasks	US Army	Army commanders and staff officers	Army, 1998 [20]

TABLE 2 Comparison of Risk Frameworks

Reference Number	Identify Goals and Objectives	Define System	Assess Threats	Assess Vulnerabilities	Assess Consequences	Assess Baseline Risk	Identify RM Options	Analyze Benefits and Costs	Make Decisions	Communicate Risks	Implement RM Actions	Monitor RM Actions
H-1 RBDM	Some techniques	Some techniques	Boating safety hazards	Most techniques	Most techniques	Most techniques	Most techniques	Some techniques	Risk informed decisions	Not addressed	Four techniques identified	
H-2 IRGC		Problem framing	Hazard identification	Exposure and vulnerability	Concern assessment (risk perception)	Risk characterization	Risk management	Option evaluation	Decision making	Center of framework	Implementation	
H-3 HVA		Not addressed	Critical assets	Not addressed	Assess vulnerabilities	Assess consequences	Not addressed	Identify CM	Estimate CM cost	Review operational security planning	Not addressed	
H-4 RFRM		Scenario identification and filtering	Bicriteria filtering Quantitative ranking	Multicriteria evaluation	Bicriteria filtering Quantitative ranking	Bicriteria filtering Quantitative ranking	Risk management		Filtering	Interview survey	Operational feedback	
H-5 MRA	Water and foodborne pathogens	Not addressed	Pathogen characterization	Occurrence and exposure analysis	Health effects and dose response analysis	Exposure and host–pathogen profile	Not addressed					
H-6 DOT	Party involved	Hazmat transport activities	Conduct risk analyses with probabilities and consequences					Risk control points	Establish priorities, analyze cost/benefits, and decide	Party involved	Implement	Verify

H-7 NASA	Identify				Analyze		Plan			Communicate and document	Track	Control
IA-1 RAMCAP	Critical assets: prepare for study	Define scope	Analyze threats	Analyze vulnerabilities	Analyze consequences	Analyze risks	Identify action strategies	Analyze benefits and costs	Make informed decisions	Communicate results for all steps	Implement results	Monitor risks
IA-2 FEMA	Terrorist attacks against buildings	Buildings	Threat identification and rating	Vulnerability rating	Asset value rating	Threat* asset value * vulnerability[a]	Mitigation options	Prioritize	Select mitigation options	Not addressed		
IA-3 DS	Facility security	Asset assessment	Threat discovery and assessment	Vulnerability assessment	Asset assessment	Risk assessment	Mitigation option analysis		Operations and security planning	Not addressed	Remediation program management	Trend analysis and auditing
IA-4 ODP	Not addressed	Criticality assessment	Threat assessment	Vulnerability assessment	Impact assessment	Risk assessment	Not addressed					
IA-5 GAO	Strategic goals, objectives, and constraints		Risk assessment				Alternatives evaluation		Management selection	Not addressed	Implementation and monitoring	
IA-6 MORDA	Value models	Information system	Threat categories	Attack trees	Value models	Baseline risk	CM design options	Cost–benefit analysis	Optimization and decision making	Not addressed	Implement CM design options	Not addressed

(continued overleaf)

TABLE 2 (*Continued*)

Reference Number	Identify Goals and Objectives	Define System	Assess Threats	Assess Vulner-abilities	Assess Con-sequences	Assess Baseline Risk	Identify RM Options	Analyze Benefits and Costs	Make Decisions	Communicate Risks	Implement RM Actions	Monitor RM Actions
IA-7 COMSEC	Planning	Identify assets	Threat analysis	Identify vul-nerabilities	Assess con-sequences	Risk analysis	Identify safeguards	Not addressed	Avoid, transfer, accept, reduce	Not addressed	Implement and certify	Operations and main-tenance
AH-1 Orange Book	Not addressed	Risk environment/ context, extended enterprise	Identify risks	Assessing risks			Addressing risks			Communica-tions and learning	Not addressed	
AH-2 CAPRA	Identify	Define	Assess	Assess	Assess	Assess	Identify	Analyze	Benefit–cost analysis	Communicate	Evaluate	
AH-3 Army	Army mission objectives	Missions	Identify Hazards	Assess prob-abilities	Assess severity	Assess risk level and mission risk	Develop controls	Assess controls	Make decision	Not addressed	Implement	Supervise and evaluate

[a]CM, countermeasures; RM, risk management.

6. *Assess baseline risk.* Assessing the baseline risk is not as common. However, without a baseline risk assessment, it is difficult to compare the costs and benefits of risk management options.
7. *Identify risk management options.* This task is used in all risk frameworks that address risk management. The quality and quantity of the risk management options are important considerations.
8. *Analyze benefits and costs.* Benefit–cost analysis is a common task. Organizations have limited budgets. Resources spent on risk management are not spent on providing products and services to generate profit (private organization) or provide value to citizens (public organization). Benefits and costs can be assessed qualitatively or quantitatively.
9. *Make decisions.* Decision making is a common risk management task. For all public organizations and many private organizations, decision makers are accountable to other individuals and organizations to obtain the most benefit for the resources.
10. *Communicate risks.* Risk communication requires the identification of stakeholders and development plans to provide timely information in a form that stakeholders will understand. Stakeholder participation earlier in the process tends to build understanding and confidence in the results.
11. *Implement risk management actions.* Implementing risk management options is not always included. Implementation requires the support of key stakeholders who have important roles in implementation.
12. *Monitor risk management actions.* Monitoring of risk management options to evaluate the extent to which they achieve risk mitigation objectives is seldom included in risk frameworks, but seems like an important task to ensure that benefits and costs were correctly assessed. In addition, this task is required to help identify the need for future actions.

These tasks constitute a comprehensive risk analysis framework. Although we do not advocate a one-size-fits-all common framework for the entire homeland security enterprise, we do believe that there is potential utility in developing a risk analysis framework with all these elements that could then be tailored to specific applications. Such tailoring could involve combining or eliminating tasks, alternative terminology, providing greater detail on certain tasks, and so on. At the very least, this list provides a basis to evaluate frameworks. If some tasks are not addressed, a justification should be provided.

We make the following additional observations:

Risk problem identification. In Table 2 we have two risk problem identification tasks: identify goals and objectives (of the study) and define the system. In contrast to decision analysis and systems engineering, stakeholder analysis is *not* a first task in the risk frameworks. Only one framework explicitly addresses the risk environment (UK Orange Book). We found that risk analysis goals and objectives are explicitly included in most, but not all, of the frameworks. In some frameworks, the scope is obvious by the name of the framework (COMSEC) or the framework application area (Federal Emergency Management Agency (FEMA) terrorist attack on buildings). In other frameworks, the scope and objectives must be determined (GAO, Orange Book, and Army). The most comprehensive frameworks seem to be the frameworks with the broadest scopes (Risk Governance Framework, RAMCAP, and Critical Asset and Portfolio Risk Analysis (CAPRA)).

Risk assessment. Most of the frameworks assess threats, vulnerabilities, and consequences. However, the definitions of these terms and the order of assessment is not the same. Several of the frameworks focus on critical assets; some have a methodology to identify the critical assets (Risk Filtering and Ranking Method, RFRM). Only one framework explicitly addresses dependencies (RAMCAP). A couple of frameworks emphasize the importance of learning about threats and vulnerabilities (Orange Book and IRGC).

Risk management. There is no common terminology or approach for risk management. Different terms are used including options, countermeasures, actions, plans, and strategies. The risk management tasks and techniques can also be quite different. Cost–benefit analysis is included in some frameworks (RAMCAP, DOT, and MORDA. Several of the frameworks address implementation and monitoring of risk management actions (IRGC, RFRM, DOT, NASA, DS, GAO, COMSEC, CAPRA, and Army). Only one framework (IRGC) explicitly addresses risk perceptions. Seven of the 17 frameworks included risk communication.

6 EXAMPLE RISK FRAMEWORKS

In this section we present an example of a conceptual and a procedural framework.

6.1 A Conceptual Framework: The Government Accountability Office Risk Management Framework

The GAO uses a procedural risk framework (Figure 4) with five tasks [15, 21]:

1. strategic goals, objectives, and constraints
2. risk assessment
3. alternatives evaluation
4. management selection
5. implementation and monitoring.

Although there is an ordering of these tasks, the cyclical representation in the GAO framework suggests an iterative process. This framework does not specify the specific analysis techniques or the outputs of the analysis. The GAO uses this framework to evaluate homeland security risk management programs.

We are particularly fond of this conceptual framework because it is relatively straightforward to draw a correspondence between its 5 steps and 11 of the 12 tasks that span all 17 frameworks surveyed. The one task not explicitly included in the GAO Risk Management Framework is *Communicate Risks*. We believe this framework would be improved were this task shown supporting the other five tasks from the center of the oval.

6.2 A Procedural Framework: Critical Asset and Portfolio Risk Analysis (CAPRA)

CAPRA provides a quantitative approach for all-hazards risk analysis [22, 23]. As we see in Figure 5, CAPRA specifies the analysis tasks, the analysis techniques, and the output of each task.

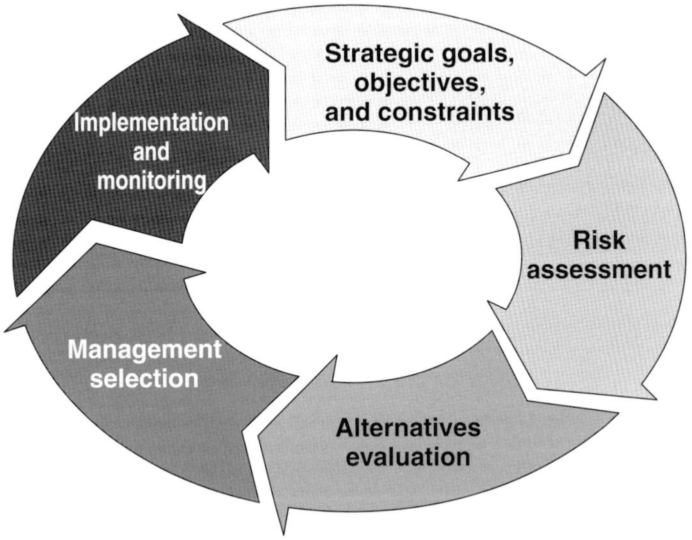

FIGURE 4 GAO risk management framework.

In general, CAPRA is a five-phase process that identifies hazard scenarios that are relevant to the region or asset of interest, assesses the losses for each of these scenarios given they were to occur, allows for consequence-based screening, assigns a probability of success given a hazard as one, assesses the annual rate occurrence for each scenario, and provides results suitable for benefit–cost analysis. CAPRA produces *risk assessments* that can form the basis for identifying alternative risk mitigation strategies and evaluating them for their cost-effectiveness, affordability, and ability to meet risk reduction objectives.

The following provides a description of each phase:

- *Scenario identification.* characterizes the missions applicable to a region and identifies hazard and threat scenarios that could cause significant regional losses should they occur. For natural hazards, this phase considers the estimated annual rate of occurrence, and screens out infrequent scenarios. For security threats, this phase identifies relevant scenarios based on the inherent susceptibilities of a region's mission and lifeline services to a wide spectrum of threat types. The product of this phase is a complete set of hazard and threat scenarios that are relevant to the region under study.
- *Consequences and criticality assessment.* assesses the loss potential for each scenario identified for the region by considering the maximum possible loss, physical vulnerability of key missions and services, and effectiveness of consequence-mitigation measures to respond to and recover from a scenario. The results of this phase provide estimates of potential loss for each hazard and threat scenario, which are used to screen scenarios and determine those that warrant further analysis.
- *Security vulnerability assessment.* assesses the effectiveness of measures to deny, detect, delay, respond to, and defeat an adversary determined to cause harm to a

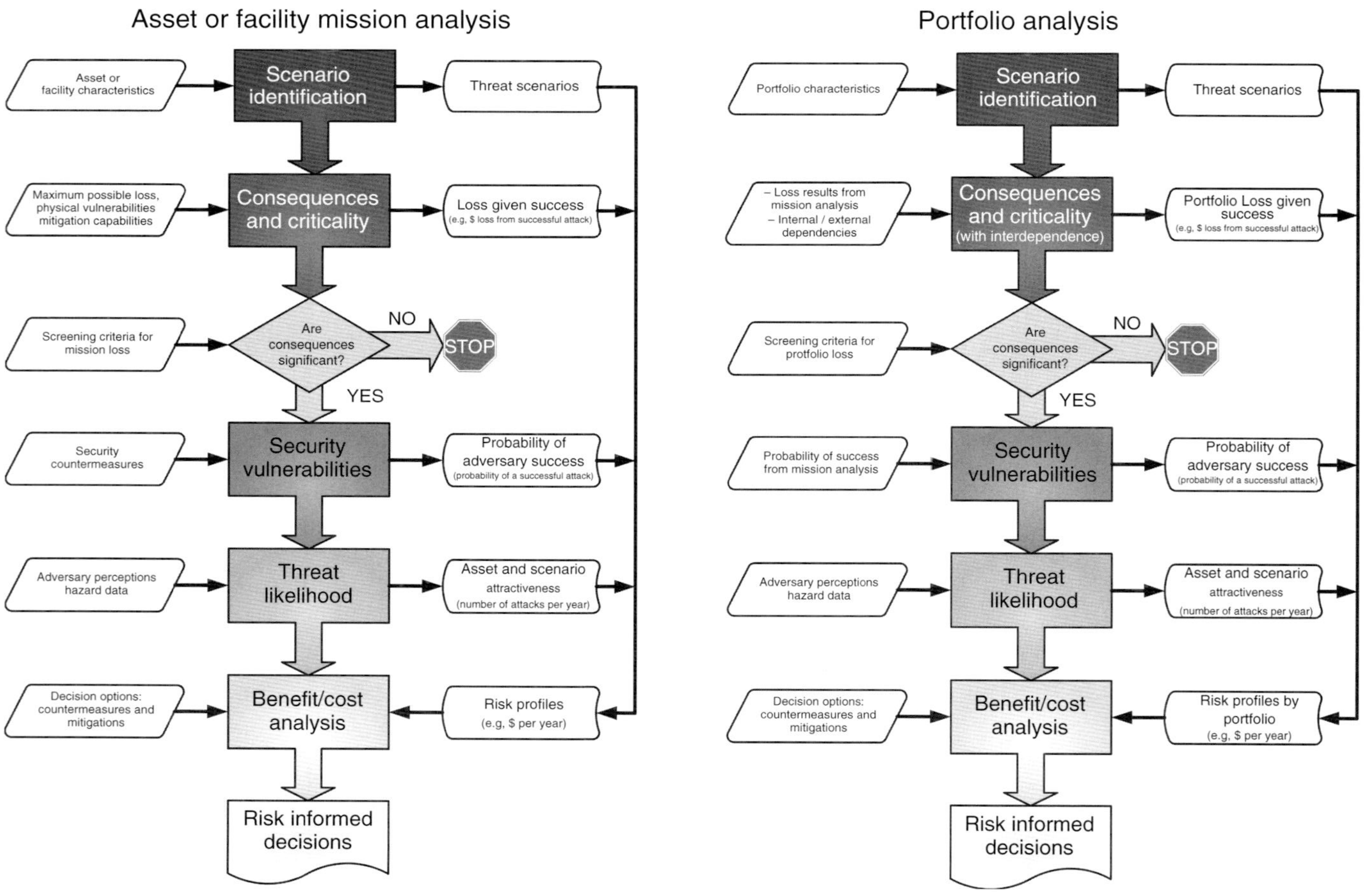

FIGURE 5 Critical asset and portfolio risk analysis (CAPRA) [4].

region. The results from this phase provide estimates of the probability of adversary success for each threat scenario, which combined with loss yields an estimate of conditional risk.

- *Threat likelihood assessment.* assesses scenario attractiveness from the adversary's point of view. The results from this phase provide estimates of the annual rate of occurrence for each threat scenario.
- *Benefit/cost analysis.* assesses the cost-effectiveness of proposed countermeasures and consequence-mitigation strategies produced from the developing of strategy tables. The results from this phase provide benefit-to-cost ratios for each proposed risk mitigation strategy, which are used to inform resource allocation decisions.

7 A CONCLUDING THOUGHT

It is no exaggeration to characterize the terrorist threat, from a national perspective, as *existential*. The consequences of terrorist attacks, and potentially ineffective, inefficient, and counterproductive responses to those attacks could threaten our socicty in fundamental ways. In countering the terrorist threat, we must be at least as intelligent, resourceful, and adaptive as our adversaries have proven to be. Among other things, this requires the best risk analyses we can develop. Greater rigor in the use of terminology and risk frameworks is a useful first step.

REFERENCES

1. Graham, B. and White, J. (2006). Abizaid credited with popularizing the term 'Long War'. *The Washington Post*, A08.
2. U.S. General Accounting Office (GAO) (2001). GAO-02-150T. *HOMELAND SECURITY Key Elements of a Risk Management Approach*, Washington, DC.
3. Ayyub, B. M. (2003). *Risk Analysis in Engineering and Economics*, Chapman and Hall/CRC Press, p. 39.
4. U.S. Coast Guard (2008). *Risk-based Decision-making Guidelines*. Coast Guard web site: http://www.uscg.mil/hg/g-m/risk/e-guidelines/RBDMGuide.htm, accessed January 21, 2008.
5. Renn, O. (2005). *Risk Governance: Towards an Integrative Approach*, International Risk Governance Council, Geneva, Switzerland.
6. Science Applications International Corporation (SAIC) (2002). *A Guide to Highway Vulnerability Assessment*, Vienna, VA.
7. Haimes, Y., Lambert, J., Horowitz, B., Kaplan, S., Pikus, I., Leung, F., and Mosenthal, A. (2004). *Risk assessment and management of critical highway infrastructure, Report to Virginia Transportation Research Council*, University of Virginia, Charlottesville, VA.
8. International Life Science Institute (ILSI) (2000). *Revised Framework for Microbial Risk Assessment*. An ILSI Risk Science Institute report. Washington, DC.
9. ICF Consulting (2000). *Risk Management Framework For Hazardous Materials Transportation*, report to US Department of Transportation, Fairfax, VA.

10. National Aeronautics and Space Agency (NASA) (2002). *NPR 8000.4 NASA Risk Management Directive*, Office of Safety and Mission Assurance, Washington, DC.
11. American Society of Mechanical Engineers (ASME) (2004). *Risk Analysis And Management For Critical Asset Protection: General Guidance*.
12. Federal Emergency Management Agency (FEMA) (2006). *FEMA 452. Risk Assessment: How-to Guide to Mitigate Potential Terrorist Attacks Against Buildings*, Washington, DC.
13. Ware, B. S., Beverina, A. F., Gong, L., Colder, B. (2002). *A Risk-Based Decision Support System For Antiterrorism*. Digital Sandbox web site: http://www.dsbox.com/Documents/MSS A Risk-Based Decision Support System for Antiterrorism.pdf, accessed January 21, 2008.
14. Office of Domestic Preparedness (ODP) (2003). *Special Needs Jurisdiction Toolkit for Official Use Only*, Washington, DC.
15. U.S. Government Accountability Office (GAO) (2005). GAO-06-91. *RISK MANAGEMENT Further Refinements Needed to Assess Risks and Prioritize Protective Measures at Ports and Other Critical Infrastructure*, Washington, DC.
16. Buckshaw, D. L., Parnell, G. S., Unkenholz, W. L., Parks, D. L., Wallner, J. M., and Saydjari, O. S. (2005). Mission oriented risk and design analysis of critical information systems. *Mil. Operat. Res*. **10**(2), 19–38.
17. Government of Canada (1999). *Threat and Risk Assessment Working Guide*, Communications Security Establishment, Ottawa, Ontario, Canada.
18. Her Majesty's Treasury (2004). *The Orange Book: Management of Risk Principles and Concepts*, London, UK.
19. Ayyub, B. M. (2006). *Guide on the protection of critical infrastructure and key resources for homeland security*, Center for Technology and Systems Management, University of Maryland.
20. Department of the Army (1998). *Risk Management. Field Manual 100-14*, Washington, DC.
21. U.S. Government Accountability Office (GAO) (2007). GAO-07-386T. *HOMELAND SECURITY Applying Risk Management Principles to Guide Federal Investments*, Statement of William O. Jenkins, Jr., Director Homeland Security and Justice Issues.
22. Ayyub, B. M., McGill, W. L., Kaminskiy, M. (2007). Critical asset and portfolio risk analysis for homeland security: an all-hazards framework. *Risk Anal. Int.J., Soc. Risk Anal*., **27** (3), 789–801. DOI: 10.1111/j.1539-6924.2007.00911.x.
23. McGill, W. L., Ayyub, B. M., and Kaminskiy, M. (2007). A quantitative asset-level risk assessment and management framework for critical asset protection. *Risk Anal. Int. J., Soc. Risk Anal*. **27**(5), 1265–1281. DOI: 10.1111/j.1539-6924.2007.00955.x.

FURTHER READING

Ayyub, B. M. (2003). *Risk Analysis in Engineering and Economics*, Chapman and Hall/CRC Press.

Haimes, Y. Y. (2004). *Risk Modeling, Assessment, and Management*, 2nd ed., John Wiley & Sons, Inc., Hoboken, NJ.

Parnell, G. S., Dillon-Merrill, R. L., and Bresnick, T. A. (2005). Integrating risk management with homeland security and antiterrorism resource allocation decision-making. In *The McGraw-Hill Handbook of Homeland Security*, D. Kamien, Ed., New York, pp. 431–461.

Scouras, J., Cummings, M. C., McGarvey, D. C., Newport, R. A., Vinch, P. M., Weitekamp, M. R., Colletti, B. W., Parnell, G. S., Dillon-Merrill, R. L., Liebe, RM, Smith, GR, Ayyub, B. M., and Kaminskiy, M. P. (2005). *Homeland Security Risk Assessment*. Volume I, An Illustrative Framework. RP04-024-01a. Homeland Security Institute, Arlington, VA.

RISK ANALYSIS AND MANAGEMENT FOR CRITICAL ASSET PROTECTION

Jerry P. Brashear and J. William Jones
ASME Innovative Technologies Institute, LLC, Washington, D.C.

1 INTRODUCTION

The current economic crisis and events of 9/11, Hurricane Katrina, terrorist attacks and natural disasters at home and abroad have heightened the nation's awareness of the risks to critical infrastructures. This awareness has stimulated the requirement that risks and risk-reduction options be assessed in ways permitting the direct comparisons needed for rational allocation of resources. Numerous risk methodologies are in use by individual firms and industries, but their results are generally not comparable with other firms or industry sectors or, in some cases, not even with other facilities within the sector. Most are qualitative or ordinal only, producing relative results that can be compared only locally, if at all. Moreover, several of the available methods require the assistance of specialized consultants and/or considerable amounts of time, money and personnel resources, which discourages their use and makes them costly to use on a regular basis. The RAMCAP Plus process—through the cost-effective application of common and consistent terminology, processes and metrics—provides an objective, repeatable basis for assessing risk, resilience, and the benefits and costs of improvements in a transparent, consistent, quantitative, and directly comparable manner.

2 ORIGIN AND DEVELOPMENT

Following the attacks of September 11, 2001, the American Society of Mechanical Engineers (ASME) convened more than one hundred industry leaders, at the request of the White House, to define and prioritize the requirements for protecting our nation's critical infrastructure. Their primary recommendation was to create a risk analysis and management process to support decisions allocating resources to initiatives that can reduce risk. This process would necessitate quantitative objectivity; common terminology; common metrics; and consistent processes for analysis and reporting, often tailored to the technologies, practices and cultures of the respective industries. This commonality would permit direct comparisons within and across industrial sectors, scales of analysis from asset to region to nation, and time for measuring trends and effectiveness as well as maintaining accountability. Such direct comparisons are seen as *essential* to supporting rational decision-making in allocating limited private and public resources to reducing risk and enhancing resilience of critical infrastructures.

Published by permission of ASME Innovative Technologies Institute, LLC.

In response to this recommendation, ASME assembled a team of distinguished risk assessment experts from industries and universities to develop a suitable methodology. The team defined a seven-step methodology that enables asset owners to perform assessments of their risks and risk-reduction options, relative to specific attacks. A series of reviews with infrastructure executives and engineers added the design criterion: to be useful, acceptable and useable by personnel at facilities of concern; the methodology must be *appropriate for self-assessment by on-site staff, in a relatively short period of time*. The original version was simplified and streamlined to meet this criterion. Throughout the experience with RAMCAP^{TM}, balancing practicality with scientific rigor has been a challenge—and still is.

The simplified version of RAMCAP [1] served as the basis for consistent sector-specific guidance documents for (i) nuclear power generation; (ii) spent nuclear waste transportation and storage; (iii) chemical manufacturing; (iv) petroleum refining; (v) liquefied natural gas offloading terminals; (vi) dams and navigational locks; and (vii) water and wastewater systems. In addition, ASME-ITI has prepared a version for higher education campuses, that is currently being tested. Experience in field testing these tailored processes, the devastation caused by recent natural disasters, and a growing appreciation of the range of threats to critical infrastructures caused the simplified process to evolve into the present RAMCAP Plus.

3 RISK AND RESILIENCE DEFINED

Consistent with the widely held definition that risk is the expected value of the consequences of an adverse event, that is, the combination of the event's likelihood and consequences, the National Infrastructure Protection Plan [2], and RAMCAP Plus [3] split the likelihood term into event likelihood and conditional vulnerability, given the event:

$$\text{Risk} = (\text{Threat}) \times (\text{Vulnerability}) \times (\text{Consequence}) \text{ or } R = T \times V \times C \quad (1)$$

where

Risk = The probability of loss or harm due to an unwanted event and its adverse consequences. When the probability and consequences are expressed as numerical point estimates, the expected risk is computed as the product of those values.

Threat (T) = The likelihood that an adverse event will occur within a specified period, usually one year. The event could be anything with the potential to cause the loss of, or damage to, an asset or population.

Vulnerability (V) = The probability that, given an adverse event, the estimated consequences will ensue.

Consequence (C) = The outcomes of an event occurrence, including immediate, short and long-term, direct and indirect losses and effects. Loss may include human fatalities and injuries, economic damages and environmental impacts, which can generally be estimated in quantitative terms, and less tangible, nonquantifiable effects, including political ramifications, decreased morale, reductions in operational effectiveness or military readiness, etc. RAMCAP Plus estimates economic losses to the infrastructure owner and to the community served, and can readily be extended to states, multistate regions or the nation.

A second, closely related concept, resilience, is not an element in the risk equation, but is central to the purposes of risk management for critical infrastructures. *Resilience* is defined as the ability of an asset, system or facility to withstand an adverse event while continuing to function at acceptable levels or, if functioning is diminished, the speed by which an asset can return to the acceptable level of function (or a substitute function or service provided) after the event. Resilience as a concept is still being formalized, but candidate metrics include reductions in the duration and severity of service denial and/or economic losses to the community due to service denial. For the purposes of this article, resilience is defined in different ways for the asset owner and the community.

For the asset owner, the level of resilience for a particular asset–threat combination is

$$\text{Resilience}_{\text{Owner}} = \text{Lost net revenue} \times \text{Vulnerability} \times \text{Threat} \tag{2}$$

where

Lost revenue = The product of the *duration* of service denial (in days) and the *severity* of service denial (in physical units per day) and pre-event price of the service less variable costs avoided (in dollars per unit), all of which are essential parts of estimating the owner's financial loss, that is

$$\begin{aligned}\text{Lost net revenue} &= \text{Duration of denial} \times \text{Severity of denial}\\ &\quad \times (\text{Unit price} - \text{Variable costs})\end{aligned} \tag{3}$$

For the community, the level of resilience for a particular asset–threat combination is

$$\text{Resilience}_{\text{Community}} = \text{Lost community economic activity} \times \text{Vulnerability} \times \text{Threat} \tag{4}$$

where

Lost economic activity in the community = The amount of decreases in both the losses of income, direct and indirect, throughout the economy of the metropolitan region due to denial of service. It is usually estimated as a function of the asset's lost revenue and the duration of the service denial, while using a static application of basic regional economic data and an input–output model, modified to reflect the resilience of the respective business sectors. Impacts on the number of jobs and employment levels are also often estimated in the same model [4–9].

The constituent elements of risk and resilience are treated as independent, single-point, "best" estimates; they are not means of underlying distributions of the estimates. More complete treatment of uncertainty and dependencies is being considered for the future.

4 THE RAMCAP PLUS PROCESS

The RAMCAP Plus process comprises seven steps (Figure 1). Taken as a whole, these steps provide a rigorous, objective, replicable, and transparent foundation for data-collection, interpretation, analysis, and decision-making.

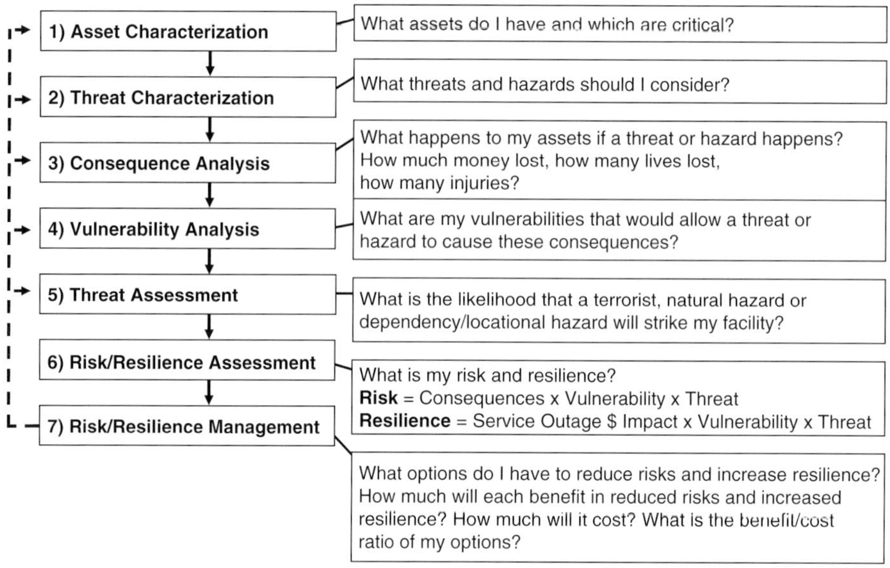

FIGURE 1 The RAMCAP Plus process.

The figure also shows the iterative nature of the RAMCAP process. The feedback arrows imply that the assessment of risk-reduction and resilience enhancement benefits is a reiteration and modification of some or all of the same logical steps as the initial, baseline risk estimate. Enhancing security and resilience requires that the options being considered reduce consequences (including duration of service denial), vulnerability, and/or the likelihood of occurrence. The process estimates the changes attributable to a countermeasure or mitigation option.

Benefits are defined as the change in risk and/or resilience (the result of changing the elements in Eqs 1–3); and *costs* include the investment and operating costs of the option. With these estimates, the net benefit (benefit less costs) and benefit–cost ratio can be used to rank the options by the magnitude and efficiency of security or resilience improvement per dollar of cost. Reductions of other consequences (e.g. fatalities) can be either converted to dollar values using the value of a statistical life, or can be maintained as a separate indicator.

The feedback arrows also imply that the process is iterated for three additional concepts: (i) for each relevant threat for a given asset; (ii) for each asset critical to the mission of the organization; and (iii) over time as part of continuous improvement and evaluating periodic progress (e.g. annually) or as needed based on changing threat circumstances.

Step 1. Asset characterization analyzes the organization's mission and operational requirements to determine which assets, if damaged or destroyed, would diminish the facility's ability to meet its mission. Critical assets are identified and a preliminary estimate is made of the gross potential consequences from various threats or hazards, in ordinal terms (e.g. "very small" to "very large" in five to seven intervals). The assets evaluated include those that are directly engaged in performing the most important missions or functions, the assets that support these and the infrastructures on which they depend. These assets may include physical plants, cyber systems, knowledge bases, human resources, customers or critical off-site suppliers.

Since the number of assets owned by an organization can be substantial, the assessment team conducts an initial ranking to identify the high priority assets, screening out the rest. The term "asset" means components of an organization's system.

The assets that directly perform the organization's mission are usually fairly obvious, but the assets and systems on which they depend may be less so. For example, a water plant has systems through which water flows for treatment and distribution and many of these are critical, but these systems require electricity, chemicals, automated monitoring, water testing, skilled labor, and so on, which can also be critical because the assets directly performing the mission cannot operate without them.

The supporting assets, in turn, may be dependent on yet other assets, which are then seen as critical, for example, the electricity substation from which the plant draws its power. Whenever an alternative source of critical support is independently available, the supporting asset may not be critical, for example, an emergency generator with sufficient fuel to last through an event would make the substation noncritical. Noncritical assets are not considered further.

Step 2. Threat characterization is the identification and description of reference threat scenarios in enough detail to estimate vulnerability and consequences. As summarized in Table 1, there are a wide variety of threat scenarios. Each is specified in more detail in actual application [3].

One key to comparability of results is the use of a common set of reference threats. These threat scenarios are not "design basis threats," which imply that the organization must take steps to withstand the threat to continue operations. Rather, these are "benchmark" or "reference" threats that span the survivable range of possible threats across all critical infrastructure sectors. Five distinct types of reference threats have been defined as follows:

1. *Terrorism.* Attacks by enemies, as suggested by the U.S. Department of Homeland Security (DHS) based on analyses by DHS and others as an understanding of the means, methods, motivations and capacities of terrorists.
2. *Natural hazards.* Currently including hurricanes, floods, tornadoes and earthquakes, based on the physical location of the facility and federal data.
3. *Product or waste stream contamination.* Suggested by the water sector and also applicable to food and pharmaceuticals, to address concerns of intentional or accidental contamination.
4. *Supply chain hazards.* Immediate dependencies, mostly supply chain issues such as suppliers, labor, and customers included as an initial step toward dealing with dependencies on other organizations for critical elements of the organization's mission.
5. *Proximity hazards.* Potential to become collateral damage from events at nearby sites.

The organization decides which of the defined scenarios represent possible physical threats for the facility; some, such as a major marine attack in a desert, may be impossible. For those threats which are possible, the organization should summarily assess the consequences of a successful attack by each threat against each asset earlier defined as critical. A convenient way to do this is to array a matrix of the critical assets identified in the first step versus the possible threats and estimating ordinally according to a five- or seven-point ordinal scale (e.g. very low, low, moderate, high, and very high).

TABLE 1 Summary of *RAMCAP Plus* Reference Threat Scenarios

Attack Type	Tactic/Attack Description			
Marine	M1 Small boat	M2 Fast boat	M3 Barge	M4 Deep draft shipping
Aircraft	A1 Helicopter	A2 Small plane (Cessna)	A3 Medium, regional jet	A4 Large plane long-flight jet
Land-based vehicle	V1 Car	V2 Van	V3 Mid-size truck	V4 Large truck (18 wheeler)
Assault team	AT1 1 Assailant	AT2 2–4 Assailants	AT3 5–8 Assailants	AT4 9–16 Assailants
Sabotage	S(PI) Physical-Insider	S(PU) Physical-Outsider	S(CI) Cyber-Insider	S(CU) Cyber-Outsider
Theft or diversion	T(PI) Physical-Insider	T(PU) Physical-Outsider	T(CI) Cyber-Insider	T(CU) Cyber-Outsider
Product contamination	C(C) Chemical	C(R) Radionuclide	C(B) Biotoxin	C(P) Pathogenic
	C(W)—Weaponization of waste disposal system			
Natural hazards	N(H) Hurricanes	N(E) Earthquakes	N(T) Tornadoes	N(F) Floods
Dependency and proximity hazards	D(U) Loss of utilities	D(S) Loss of suppliers	D(S) Loss of employees	DI Loss of customers
	D(T) Loss of transportation		D(P) Proximity to other targets	

This establishes the sequence by which asset–threat pairs will be analyzed: examine the highest ranked and proceed to lower ranked until the consequences are acceptable.

Step 3. Consequence analysis is the identification and estimation of the *worst reasonable consequences* generated by each specific asset–threat combination. This step examines facility design, layout and operation in order to estimate fatalities, serious injuries and economic impacts.

RAMCAP Plus defines "economic impacts" for risk management at two levels: (i) the financial consequences to the organization and (ii) the economic consequences to the regional metropolitan community the organization serves. Economic consequences for communities larger than the metropolitan area, for example, the state, multistate region, or nation, may also be estimated using the same methods, as needed by decision-makers. For many critical infrastructures and facilities, interdependencies make the metropolitan region most relevant to decision-makers.

Financial consequences to the organization include all necessary costs to repair or replace damaged buildings and equipment, abandonment and decommissioning costs, site and environmental cleanup, net revenue losses (including fines and penalties for failing to meet contractual production levels but excluding avoided variable costs) while service is reduced, direct liabilities for casualties on and off the property and environmental damages. These costs are reduced by applicable insurance or restoration grants and must be corrected to account for tax effects for tax-paying organizations.

The primary concern for the public or community is the length of time, quantity and sometimes quality of critical service denied, and the direct and indirect economic consequences of service denial [5, 6]. When the service denial is short and/or customers are able to cope by such actions as conservation, substitution, redundancies, making up lost production later, the region is said to be "resilient" [7]. The public's objective is to enhance the resilience of critical infrastructures on which they depend.

RAMCAP Plus estimates the direct and indirect losses to the regional community by a modified input–output algorithm. While recognizing the classical critiques of input–output modeling of a major disruption of critical infrastructures, it remained necessary to quantify at least roughly the community impact, to guide public choices. To minimize the methodological problems without adding inordinate complexity, RAMCAP Plus adopted a model originally developed to fill a gap in the computational ability of HAZUS-MH [8], the Federal Emergency Management Agency's loss estimation software referred to as a *HAZUS patch* [9]. The algorithm can be applied to any estimate of infrastructure service disruption to compute direct and indirect losses of regional output, income and jobs.

Other consequences are identified and described qualitatively, and include impact on iconic structures, governmental ability to operate, military readiness, and citizen confidence in the organization, product, or the government.

Step 4. Vulnerability Analysis estimates the conditional likelihood that the estimated consequences will occur, *given* the occurrence of the specific threat or hazard. Vulnerability analysis involves an examination of existing security capabilities and structural components, as well as countermeasures and their effectiveness.

A variety of rigorous tools can be used to estimate vulnerability, such as those described in Table 2.

Direct elicitation often seems to be easier and less time-consuming, but the time to reason through each threat–asset pair can lead to long discussions and it is difficult to maintain logical consistency across a number of such judgments. Some RAMCAP sector-specific guidance documents provide prespecified structure of vulnerability logic, event or decision trees for users to populate with estimates of the required elements to enhance comparability and reliability.

Step 5. Threat assessment estimates the probability that a particular threat–terrorist, natural, contamination, dependency, or proximity–will occur in a given time frame (usually one year). The approach differs depending on the type of hazard, as characterized in Table 3.

Terrorism likelihood (and its contribution to contamination, proximity, and even dependency hazards) is the most difficult to estimate and is still being refined. In its most advanced formulation, it recognizes that terrorists are cognizant, near-optimizing adversaries in a contest perhaps best modeled by game theory. Because of RAMCAP's specification to keep the process simple and brief, however, simpler techniques of

TABLE 2 Frequently Used Vulnerability Tools

Method	Description
Direct expert elicitation	Members of the evaluation team discuss the likelihood of success and their reasoning for their estimates; in its more formal form, a statistical "Delphi" processor Analytical Hierarchy Process can be used to establish a consensus.
Vulnerability logic diagrams (VLDs)	Plot of the flow of events from the time an adversary approaches the facility to the terminal event in which the attack is foiled or succeeds, considering obstacles and countermeasures that must be surmounted, with each terminal event associated with a specific likelihood estimate. This is frequently complemented with an estimate of the reaction time of a counterforce once the attack has been detected
Event trees (also called "failure trees")	Tree with branches for the sequence of events between the initiation of the attack and the terminal events. The evaluation team estimates the probability of each outcome. Multiplying the probabilities along each branch, from the initiating event to each terminal event, calculates the probability of each unique branch, while all branches together sum to 1.0. The sum of the probabilities of all branches on which the attack succeeds is the vulnerability estimate
Decision trees	Very similar to event trees except that the decisions by the adversary are modeled at each node in the unfolding tree to capture the adaptive behavior of the adversary; a sophisticated variant is to model the tree as a two-player game
Hybrids of these	Often used by the more sophisticated assessment teams

TABLE 3 Estimation of Hazard Likelihood

Hazard Type	Likelihood/Probability Estimation
Terrorist attack	Based on the terrorists' objectives and capabilities, (generally provided by intelligence and law enforcement agencies), and the attractiveness of the facility relative to alternative targets, the asset's expected value (vulnerability x consequences), and the cost/effectiveness of the attack
Natural hazards	Based on the historical federal frequency data for various levels of severity at the specific location of the asset. Can be adjusted if there is reason to believe that the future frequency or severity will differ from the past
Dependency hazard	Based on local historical records for the frequency, severity and duration of service denials as a baseline estimate of "business as usual," incrementally increased if they may be higher due to terrorist activity or natural events on required supply chain elements. Confidential conversations with local utilities and major suppliers can inform these estimates
Product contamination	Treated the same as terrorism and dependency likelihood, except additional consideration is given to accidental contamination of inputs and the vulnerability of critical processes to accidents
Proximity hazard	Based on asset's location relative to other assets that may incur adverse events leading to collateral damage, using the same logic in estimating terrorist and natural hazard threats

approximation based on observable or previously estimated factors are used. RAND Corporation has contributed relative likelihood of attack based by metropolitan region and asset type [10]. The previously estimated conditional risk (consequences, times, vulnerability) aptly characterizes the expected value to the terrorist of the asset–threat pair, while the asset's size and prominence relative to other assets of the same type in the region can indicate attractiveness. The adversary might also consider the likelihood of preattack detection and the "cost" in resources. This approach is explained in reference [11].

Two additional analyses can assist in appraising the realism of this approach to terrorism likelihood:

1. *Comparison of terrorism risk with natural hazard risk.* uses a natural hazard risk that is accepted by the organization to deduce a terrorism threat likelihood equating the two risks. The analyst and decision-maker then judge whether the deduced likelihood is reasonable or not. If the likelihood in the deduced risk is equal to or less than the judged reasonable level, then the terrorism risk is as tolerable as the natural hazard risk and the likelihood is moot. If, on the other hand, the likelihood in the deduced risk is greater than the accepted level, the judgment of the reasonable level sets a minimum and the asset–threat pair's risk justifies taking the next steps.
2. *Investment break-even.* assumes the decision-maker's choices are simple "go/no-go" on individual options. This method can only be applied as part of Step 7 because it requires the calculation of a baseline risk, conceptual design and cost estimation of an investment option to materially reduce the risk, and an assessment of the risk with the option in place. Given the reestimated consequences and vulnerability and the option cost, the calculated "break-even" likelihood is the one that yields a net benefit of exactly zero and a benefit–cost ratio of 1.0. The decision-maker can then judge whether the "break-even" likelihood is plausible or not. If the decision-maker believes the actual likelihood exceeds the break-even, the option has value and results in a "go" decision, and vice versa.

Step 6. Risk and resilience assessment creates the foundation for prioritizing and selecting among risk-reduction and resilience enhancement options. The risk assessment step is a systematic and comprehensive evaluation of the previously developed estimates. The risk for each threat for each asset is calculated from the risk relationship expressed in Eq. (1).

Resilience, the ability to function despite and during a traumatic event or to restore functionality in very short time, is defined in different ways for the asset owner (Eq. 2) and community (Eq. 3), respectively, for each asset–threat pair.

Step 7. Risk and resilience management is the step that actually reduces risk and increases resilience. Having determined the risk and resilience of each important asset–threat pair, this step defines new security countermeasures and consequence-mitigation resilience options, and evaluates them to achieve a portfolio that yields an acceptable level of risk and resilience at an acceptable cost. The 10 actions described in Table 4 constitute this crucial step.

In essence, the value or benefit of the options is estimated by revisiting Steps 3, 4 and/or 5 and reestimating the (reduced) threat likelihood, vulnerability or consequences to calculate a new risk and resilience with the option in place. The reduction in risk and the increase in resilience are the benefit or value of the option, which can be compared

TABLE 4 Risk and Resilience Management Actions

Activity Title	Activity Description
1. Acceptance level	Establish whether the risk/resilience level is acceptable
2. Design	Design potential countermeasures and consequence-mitigation options that would reduce risk and/or enhance resilience
3. Costs	Estimate the investment and operating costs of each option
4. Reestimation	Reestimate consequences, threat likelihood, and/or vulnerability, whichever is affected by the option
5. Benefits	Recalculate risk and resilience, given the option, and subtract it from the risk without the option (the "do nothing" baseline option) to define the *benefit* of the option
6. Combinations	Combine the options that affect multiple asset–threat pairs, for example, if a higher fence changes the vulnerability for an attack by one assailant, it may do the same for two to four. Add the benefits of the asset-pairs to compute the total benefit of the option
7. Key metrics	Calculate the net benefits (less costs) (value) and the benefit–cost ratio (efficiency) of the option
8. Rank and select	Select the options that have the highest net benefits and/or benefit–cost ratios and the lives saved, injuries avoided, considering both risk and resilience until resources are fully committed (less any reserved amounts)
9. Manage	Manage the implementation and operation of the selected options, evaluate their effectiveness and make mid-course corrections for maximum effectiveness
10. Recycle	Repeat the risk analysis cycle periodically or as needed given intelligence or changing circumstances, for example, new technologies and new facilities

to the cost of implementing it and to the benefits of other options. Taking no action is always a baseline option against which all others are compared.

Net benefits measure the magnitude of the value added by the option, while the benefit–cost ratio measures of the amount of risk reduction per unit of cost, an efficiency test. For fatalities and serious injuries, examine the gross reductions and the expected number required to make the needed trade-offs. The full set of options should be as a portfolio to establish if equity and balance are maintained. Financial, human, and other resources are then allocated to implement and operate the selected options.

Choices among the options are virtually never made with a single metric, but rather a set of difficult trade-off decisions must be made. Some organizations apply explicit preferences to establish an initial portfolio of options and then adjust the selections as needed to balance the portfolio or program of risk-reduction and resilience enhancement measures. It is common to estimate a "value of statistical life" to roll human casualties into the dollar-denominated benefits. When this is done, RAMCAP Plus calls for displaying the casualty estimates separately as well, for decision-makers to consider.

Once these decisions are made, risk management extends to implementation of the chosen options, monitoring their effectiveness and taking corrective actions as needed. The risk management process is the essential part of continuous security and resilience improvement, repeated periodically (e.g. annual budget process) or as necessitated by changes in the threats, vulnerabilities, consequences, technologies or the evolving development of the organization's systems.

In addition to investing in these options, risk can also be managed by acquiring insurance, entering into cooperative agreements, or simply accepting the calculated risk when it compares favorably with other risks such as financial or investment alternatives. Ideally, the organization would consider all these risk-reduction and resilience enhancement options collectively as a mixed portfolio of risk and resilience management.

5 BENEFITS OF USING THE RAMCAP PLUS PROCESS

Use of the RAMCAP Plus process generates a number of benefits or advantages to the organization using it, the sector or industry that adopts it, the communities served and the public policy toward infrastructure security and resilience. These are summarized in Table 5.

Several of the entries in Table 5 mention benefits that occur if the process becomes a voluntary consensus standard. As this article is being written, three voluntary consensus American National Standards are being developed based on RAMCAP Plus: one, an overarching standard applicable to any asset-based industry and the others, specifically tailored for water–wastewater utilities and higher education campuses. Three additional RAMCAP Plus standards are under consideration. These standards and others that will follow provide for continuous improvement of the process, *while* maintaining consistency and comparability. They cost the federal government little or nothing other than perhaps sector-specific guidance (SSG) development because they are maintained by volunteers in officially designated standards development organizations, of which ASME and ASME-ITI are two. These benefits result in dynamic, effective risk and resilience management that is driven by the private and public infrastructure organizations in true partnership with all other stakeholders' interests, including public and nonprofit concerns.

In summary, use of the RAMCAP Plus process yields significant benefits to the asset owners and industries who use it, to the communities they serve, and to the local, regional and/or national economies to which they contribute.

6 LOOKING TO THE FUTURE

RAMCAP Plus is a "living" tool, now in its third version [3], as new challenges and improved methods are incorporated. Even as this is published, a number of enhancements are being developed or evaluated, including the following:

- sector-specific guidance for additional sectors;
- software for more systematic, efficient application, and facilitation of more sophisticated techniques;
- explicit treatment of uncertainties in, and dependencies among, the key terms of risk and resilience instead of point estimates;
- adding wear, aging, technological obsolescence, and rise in sea level in coastal areas to the hazard set;
- augmenting the analysis of dependencies and interdependencies on the regional scale to track multistage "cascading failures" explicitly;

TABLE 5 Using RAMCAP Plus and Its Standards Benefits All Levels of Decision-making

Beneficiaries	Benefits
Infrastructure organizations	• Cost-effective enhancement of security and resilience • Rational allocation of resources across assets, facilities, sites, and lines of business • More efficient management of capital and human resources • Consistently quantified risk and resilience levels, potential net benefit and benefit–cost ratios of investment options • Repeated application over time measures progress and trends while enabling accountability for execution • Enhanced reliability in performance of the mission • Ability to define risk and resilience levels quantitatively at the *community* level enables partnering with other firms and public agencies for large-scale solutions • If adopted as industry voluntary consensus standard, it becomes the vehicle for incentives, such as preferred supplier status, lower insurance costs, higher credit ratings, and lower liability exposure
Whole industry or sector	• Ability to identify the assets with the greatest need and value of improvement • Cross-facility comparisons reveal industry-wide vulnerabilities for collective action (e.g. R&D, new technology and standards) • Direct comparison of the sector's risk and resilience level to other sectors for higher level resource allocation and policy-making • If sector-specific guidance becomes a consensus standard, additional benefits can be incurred, e.g., ◦ preferential treatment by insurers, financial rating services and customers ◦ potential affirmative defense in liability cases ◦ ability to substitute self-regulation by standards for bureaucratic regulation, and direct participation in federal regulatory, procurement or other federal actions
Metropolitan regional community	• Ability to estimate value of security and resilience investments to the region, a salient criterion in both private and public decisions • Consistent terminology provides common language for meaningful dialogue between private organizations and government agencies • Identification and valuation of "public goods" and shared-benefit programs; encourages public–private partnerships • Cooperative decision-making based on comparability of risk, resilience, and benefit estimates for rational regional trade-offs • If consensus standards become available, communities can designate the standards as the local codes of expected practice

TABLE 5 (*Continued*)

Beneficiaries	Benefits
	• Repeated application over time measures progress and trends while enabling accountability for regional execution
State, multistate regions and/or federal agencies	• All the metro regional community benefits, above • Consistency, transparency and direct comparability needed to evaluate major public infrastructure and program investments • Methods used to estimate economic losses to metropolitan regions can be extended to whatever scales are relevant to the decisions to be made—states, multistate regions or the national economy—in the same, directly comparable terms • Rational allocation of resources to maximize the security and resilience enhancement within a finite budget • If consensus standards are developed, the industry can self-regulate with public compliance audits; maintenance of the standards costs government nothing, *for as long as there is demand for the standard*

- enhanced estimation of economic impacts on metropolitan, state, multistate, and national scales;
- modification to include valuing both existing and new infrastructures required by a growing population and economy in the same framework;
- adding new metrics of merit such as the upside gains from improved infrastructure, socioeconomic distribution of benefits, quantitative environmental impacts (including effects on global climate), and employment (during construction and subsequently as the improvements in infrastructure contribute to the economy);
- portfolio analysis to exploit correlations and dependencies in selecting collections of investment options.

As these and other enhancements are made, they will be incorporated into the tools and standards that constitute the RAMCAP program.

REFERENCES

1. ASME Innovative Technologies Institute, LLC. (2006). *RAMCAP: The Framework, Version 2.0*. ASME-ITI, Washington, DC.
2. U.S. Department of Homeland Security. (2006). *National Infrastructure Protection Plan*. DHS, Washington, DC.
3. ASME Innovative Technologies Institute, (LLC). (2009). *All-Hazards Risk and Resilience: Prioritizing Critical Infrastructures Using the RAMCAP Plus^{SM} Approach*. ASME Press, New York.
4. Rose, A. (2006). Economic resilience to disasters: toward a consistent and comprehensive formulation. In *Disaster Resilience: An Integrated Approach*, D. Paton, and D. Johnston, Eds. Charles C. Thomas, Springfield, IL, pp. 226–248.

5. Rose, A. (2004). Economic principles, issues, and research priorities in natural hazard loss estimation. In *Modeling the Spatial Economic Impacts of Natural Hazards*, Y. Okuyama, and S. Chang, Eds. Springer, Heidelberg, pp. 13–36.
6. Rose, A., and Liao, S. (2005). Modeling regional economic resilience to disasters: a computable general equilibrium analysis of water service disruptions. *J. Reg. Sci.* **45**(1), 75–112.
7. Rose, A., Oladosu, G., and Liao, S. (2007). Business interruption impacts of a terrorist attack on the water system of Los Angeles: customer resilience to a total blackout. In *Economic Costs and Consequences of Terrorist Attacks*, H. Richardson, P. Gordon, and J. Moore, Eds. Edward Elgar, Cheltenham, pp. 291–316.
8. Federal Emergency Management Agency (FEMA). (2006). *HAZUS-MH: Multi-hazard Loss Estimation Methodology*. National Institute of Building Sciences, Washington, DC.
9. Multi-Hazard Mitigation Council (MMC). (2005). *Mitigation Saves: The Benefits of FEMA Hazard Mitigation Grants*. National Institute of Building Sciences, Washington, DC.
10. Willis, H. H., LaTourrett, T., Kelly, T. K., Hickey, S., and Neil, S. (2007). *Terrorism risk modeling for intelligence analysis and infrastructure protection*, l, RAND Center for Terrorism Risk Policy.
11. Brashear, J. B. (2009). *Approximating Terrorism Threat Likelihood*. ASME Innovative Technologies Institute, LLC, Washington, D. C. (in press).

LOGIC TREES: FAULT, SUCCESS, ATTACK, EVENT, PROBABILITY, AND DECISION TREES

ROBIN L. DILLON-MERRILL

McDonough School of Business, Georgetown University, Washington, D.C.

GREGORY S. PARNELL

Department of Systems Engineering, United States Military Academy, West Point, New York

DONALD L. BUCKSHAW

Innovative Decisions, Inc., Vienna, Virginia

1 INTRODUCTION

This article provides an introduction to logic trees. Six types of logic trees are described, compared, and illustrated in this article: fault trees, success trees, attack trees, event trees, probability trees, and decision trees. Probabilistic risk analysis (PRA) models may include fault trees, success trees, attack trees, and event trees [1]. Decision analysis models generally include probability trees and/or decision trees [2]. We illustrate the different models using bioterrorism examples. Table 1 provides a summary of the different logic tree models. The table includes the uses, mathematical foundation, data required, advantages, and limitations.

2 FAULT, SUCCESS, AND ATTACK TREES

A fault tree is a graphical probabilistic risk assessment (PRA) technique whereby an undesirable event (called the *top event*) is postulated and the possible ways for this top event to occur are systematically deduced for combinations of initiating and intermediate events [3]. The events are generally binary (or Boolean), that is, events may or may not occur. System components are either in parallel or in series, so combinations of events that lead to failure are identified with logic gates. Figure 1 shows two fault tree diagrams in which the failure of the system depends on two Boolean events. Figure 1a portrays a system that fails when events A *or* B occur, and Figure 1b portrays a system that fails when events C *and* D occur. The Fail block is known as the *top event*, and events A, B, C, and D are basic events. In more complex trees, events in between the top event and the basic events are called *intermediate events*.

Bell Telephone Laboratories developed fault tree analysis in 1961 to support the US Air Force in development of the Minuteman missile system [4]. Others realized the benefits of fault tree analysis and began using the technique for analyzing failures of complex systems. In 1981, the Nuclear Regulatory Commission published the Fault Tree Handbook [5] which remains a valuable resource today. Many others have contributed to development of theory and tools to enable fault tree analysis; Ericson [6] provides a timeline of significant individuals and their contributions up to 1999. Fault trees are useful for assessing risks in almost any uncertain situation, and applications have included software design, space system design, nuclear safety, project management, and information assurance.

Derivatives of fault trees include success trees and attack trees. Fault trees have historically been used to analyze the failure of a system (e.g. the auxiliary water feed system in a nuclear power plant) where the basic events are failures of system components (e.g. tanks, pumps, etc.) and/or acts of nature. A success tree is the complement of a fault tree and models the combination of events that lead to success. Events in attack trees [7] are defined by hostile actions from an adversary against a system. Attack trees (also called *vulnerability trees* [8]) can be used to determine the probability of success, given an attack where the top event is a successful system attack. Additionally, an attack tree used to determine the probability of an attack is often referred to as a *threat tree* [9]. All of these techniques use the basic fault treelike structure and probabilistic relationships between components or events to determine the likelihood of the undesired, top event. Boolean logic and probability remain the foundation of all these models.

Sometimes attack trees can be a mixture of fault and event trees (event trees are described in the next section). The fault tree portion enumerates the cut sets—the smallest number of components that, if they all fail, will lead to system failure—in an efficient manner. Once you have the cut sets, the attacks are treated in a manner similar to the event tree, where sequence matters and the probabilities of events are conditional on the

TABLE 1 Comparison of Logic Trees

Logic Tree	Use	Mathematical Foundation	Data Required	Advantages	Limitations
Fault tree	Calculate the probability of failure Determine the cut sets	Boolean logic Probability theory Reliability theory	System knowledge failure modes, and probabilities	Focus on components and failure modes	Specialized software
Success tree	Calculate the probability of success Determine the cut sets	Boolean logic Probability theory Reliability theory	System knowledge Success modes and probabilities	Focus on success modes	Specialized software
Attack tree	Calculate the probability of successful attack Determine the cut sets	Boolean logic Probability theory	System knowledge Adversary knowledge Attack steps and probabilities	Focus on adversary actions	Specialized software
Event tree	Calculate the probability of scenarios and consequences	Probability theory	Events Sequencing Outcome spaces	Multiple outcomes Conceptually simple to develop and solve	Binary outcomes Do not appropriately model terrorist decisions.
Probability tree	Calculate the probability of any uncertain event in a joint probability distribution	Probability theory Expected value Dominance Bayes' theorem	Events sequencing Outcome spaces Probabilities Consequences	Multiple outcomes Conceptually simple to develop and solve	Large trees are difficult to understand, display, and solve Do not appropriately model terrorist decisions.
Decision tree	Identify the best decision strategy under uncertainty	Expected value Dominance Bayes' theorem Bellman optimality principle Utility theory axioms	Events Sequencing Outcome spaces Probabilities Alternatives Consequences	Conceptually simple to develop and solve Can model terrorist decisions	Large trees are difficult to understand, display, and solve

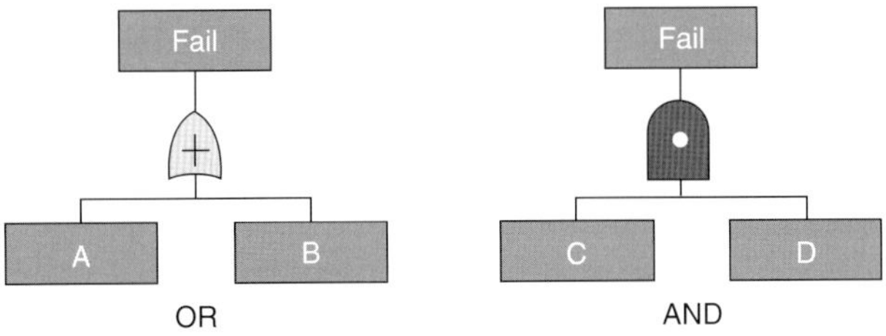

FIGURE 1 Example fault trees.

previous steps [10]. But, there are many dead-end branches of the event tree that cannot happen given a failure at an early step.

Fault trees are constructed using deductive logic starting at the top event (the failure of all or part of the system) and diagramming the relationships and interactions of events that can constitute a system failure until the system is decomposed into a set of basic events [11]. Each fault tree specifically addresses one top-level system failure; defining this failure is the first step of fault tree construction. The tree is then expanded using "gate" and "event" symbols to logically show the possible events that would lead to the top-level failure. Event symbols represent specific occurrences within the system, such as a component failure; gate symbols describe how lower-level events can be integrated up to higher-level events.

While other logic gates are available, the two basic gate symbols used for fault tree construction are the AND gate and the OR gate. The AND gate demonstrates that the higher-level event (the gate's output) will occur only if all immediate lower-level events (the gate's inputs) occur. Logically, OR gates demonstrate that only one of the gate's inputs must occur for the gate's output to occur. Note that although a gate can have many inputs, gates have only one output.

Fault tree diagrams should be detailed enough to satisfy the scope of the analysis, yet they should be presented such that the relationships between the components are easily understood. Qualitative analysis of fault trees requires determining the minimal cut set(s). This involves a Boolean manipulation of events. The most common algorithm to determine the minimal cut sets for complex fault trees is the successive substitution method (see [5] for algorithm details). In Figure 1a, there are two minimal cut sets with one event each: {A} and {B}. In Figure 1b, there is a single minimal cut set that comprises both events: {C, D}. From this minimal cut set determination, one may qualitatively assess the most important components of a system (i.e. a qualitative ranking of components contributing to system failure) and possible common failure causes.

To conduct any quantitative analysis of the system's reliability, it is necessary to have data on the reliability (i.e. probability of failure) of the system's components, which can then be used to calculate the reliability of the overall system. A quantitative evaluation of a fault tree evaluates the likelihood that the system will fail due to any of the cut sets and all their respective events occurring. Based on a quantitative analysis, one may rank the contributors to system failure and examine the system's sensitivity to component failure data.

For example, in the system in Figure 1a, if A and B are independent events, the system will fail with probability:

$$\text{P(System Fail)} = \text{P(A)} + \text{P(B)} - \text{P(A)P(B)}$$

Simplistically, if P(A) is greater than P(B), one may conclude that prevention of event A is more important than prevention of event B in reducing the likelihood of system failure. If C and D are independent events, the system in Figure 1b will fail with probability:

$$P(\text{System Fail}) = P(C)P(D)$$

In the second example, an effective strategy could emphasize making one event extremely unlikely regardless of the likelihood of the other event.

Fault trees (and success and attack trees) have several advantages over purely qualitative techniques for risk assessment. These models are designed to handle complex logical relationships between system components. Where human logic and computational abilities fail to comprehend component interactions when multiple subsystems and components are introduced, fault trees provide concise information about the state of the system and its components and allow rapid calculation of the effects of component changes or changes in system design. As a technique with wide applicability to scenarios requiring decomposition and analysis of system components and sources of failure, fault trees can be valuable additions to any risk assessment task. Some fault tree software packages allow embedding of fault trees within decision analysis models.

In conclusion, fault trees, success trees, and attack trees impose logic and calculate probabilistic interactions between component states to support the analysis of failure, success, and attack scenarios. For complex systems, specialized software is required to support these modeling techniques.

2.1 Fault, Success, and Attack Tree Example

Consider an anthrax attack where the delivery mechanism is the US Postal Service. For the attack to be successful, the anthrax package must not be detected.

Suppose the postal service detects anthrax packages either by physical observation or by electronic sensors (see Figure 2). If anthrax packages are to bypass physical protections they must not show any outward threat (i.e. leaking powdery substance) or no one can detect the visible threat. For packages to bypass electronic detection, the package may

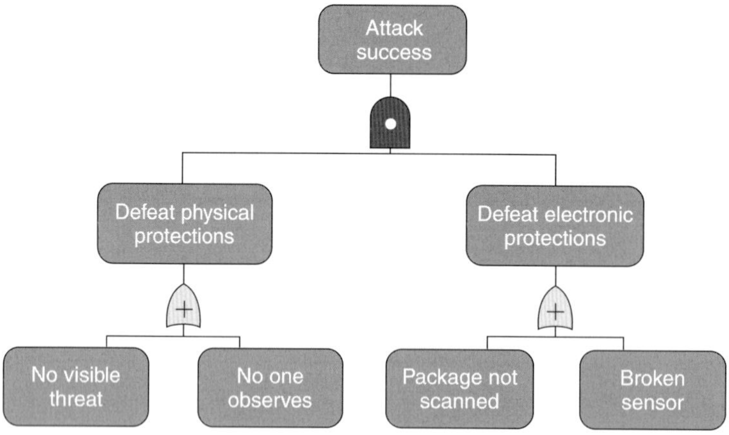

FIGURE 2 Bioterrorism attack example.

either not be scanned by electronic sensors or the sensors may not be functioning properly. Given a set of event probabilities, we can easily calculate the probability of attack success.

3 EVENT TREES AND PROBABILITY TREES

An event tree is a logic tree model for illustrating the sequence of outcomes which may arise after the occurrence of an initiating event. An event tree diagrams the sequences of random events where the chain or path through the tree represents a particular failure scenario and each node represents a binary outcome of success or failure for each event in the scenario. A probability tree is a similar but more general tool than an event tree because in a probability tree, event nodes can have more than binary branches.

Event trees are generally constructed using deductive logic, starting with an initiating event, and then considering the occurrence or nonoccurrence of other possible events, to determine the outcome of each possible failure scenario [11]. The probabilities associated with each additional event (or node) are conditional on all previous outcomes in the tree.

Event trees are useful for structuring failure scenarios because building an event tree addresses all three risk assessment questions (What can happen? What is the likelihood? What are the consequences?). Separate trees can be built for each possible initiating event, where end-path outputs document the chain of events that starts with an event such as an attack and progresses through intermediate events (e.g. loss of a critical infrastructure asset) to determine outcomes and consequences. Event tree analysis was developed in conjunction with fault tree analysis by the nuclear industry [12, 13], and was used to study the operability of nuclear power plants (to include identifying accident sequences in the Three Mile Island-2 accident).

Event trees are commonly integrated with fault trees in a comprehensive risk analysis. In these cases, the left side of the event tree connects to a top event identified by a fault tree, and the right side of the tree with a damage state model (e.g. a plume dispersion model for a dirty bomb). Each node in the tree models the branching probabilities that can be obtained from a system analysis. In contrast to the fault tree, the event tree can easily capture the issue of timing (or evolution) of events. The initiating events used to model event trees are those commonly identified as top events from the fault tree analysis.

Event trees are mainly used in consequence analysis for preincident and postincident application. Because risk assessment frequently requires the systematic identification of all failure (or accident or attack) scenarios, event tree structures provide convenient tools for analyzing families of such scenarios. To quantitatively evaluate an event tree, an event's probability must be assessed conditionally on all prior event outcomes. The probability of any path is the joint probability distribution for the events on that path.

3.1 Event and Probability Tree Example

Three nodes are denoted in the example for a bioterrorism event in Figure 3. For such an event to succeed, the terrorist needs to be able to acquire the agent, deliver the agent, and successfully contaminate the surrounding population. In this simple event tree, the probability of a successful bioterrorism attack in a fixed period of time would be calculated as 0.045 (0.05 * 0.30 * 0.50). Consequence models could be used in conjunction with an event tree to model the potential health impacts if contamination is successful.

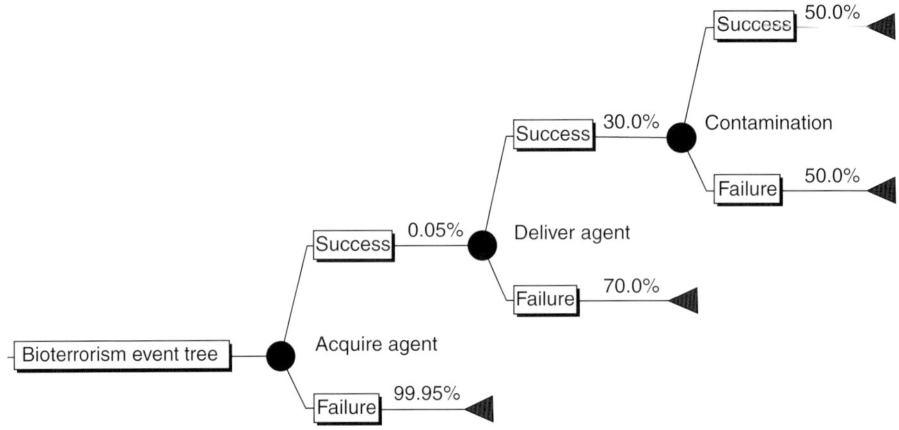

FIGURE 3 Event tree for bioterrorism event.

4 DECISION TREES

Decision trees extend probability trees by including decision nodes among the event nodes [2], where a decision node's branches represent the choices at that node. The decisions and events are logically sequenced in time. To solve a decision tree, one must specify the relevant decision alternatives, the relevant branch probabilities, and the outcome values (sometimes called *consequences*) associated with each path. A decision tree is solved by starting at the right end and working backwards to the base. At each uncertainty node, the expected value of its branches is found, and at each decision node, the branch that maximizes/minimizes the expected value/expected cost is chosen. By taking this approach, the best decisions and their expected values are found. Decision trees work the same way for expected utility [2].

4.1 Decision Tree Example

Decision trees expand the modeling capabilities of simple event trees by including decisions such as which agent and/or which target an adversary could choose (See Figure 4). The remaining event nodes (acquire agent, deliver agent, and contamination) would be assessed conditional on the agent and target. In an alternative formulation, we could model the adversary actions as uncertain and add decision nodes interdiction, tests, warning, and treatment decisions. Decision trees would generally include consequences at the terminal nodes of the trees. This allows the risk analyst to calculate the probability distributions for each decision strategy and identify the nondominated decision strategies [2].

5 SEEING THE FOREST FOR THE TREES

In this section we discuss the advantages and limitations of each type of logic tree used in decision and risk analysis. Table 1 summarizes the key comparisons of each logic tree based on uses, mathematical foundations, data requirements, advantages, and limitations. We focus our discussion here on the advantages and limitations.

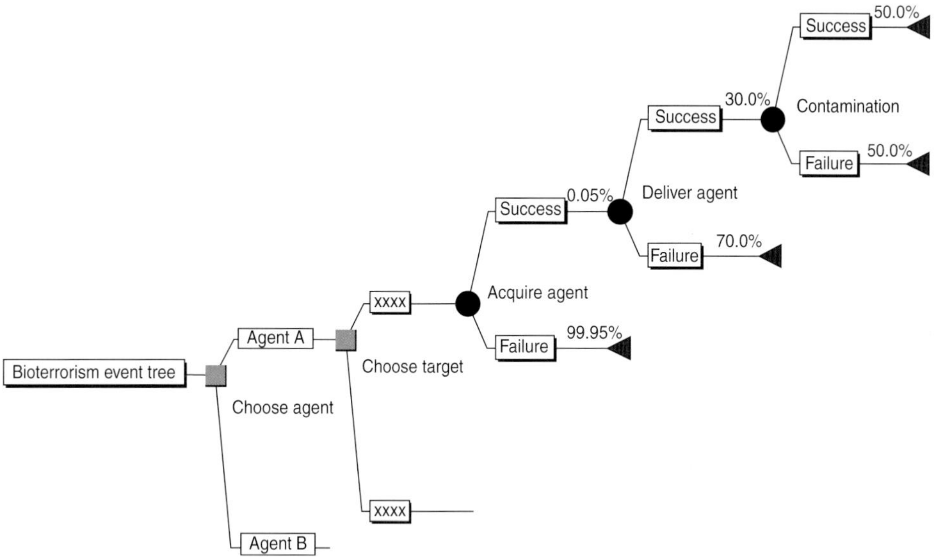

FIGURE 4 Decision tree for terrorist bioterrorism decisions and events.

5.1 Advantages

For event and decision trees: (i) analysis focuses upon paths or scenarios that lead to failure; (ii) analysts can concentrate upon one specific scenario at a time; (iii) results provide a graph that depicts system weaknesses; and (iv) trees serve other risk analysis techniques, such as PRA. These advantages are also true for fault trees. However, unlike fault trees, probability trees can capture complex dependent temporal events that need not be binary. In addition, probability trees can include more than two outcomes and independence assumptions are not required.

5.2 Limitations

For event, decision, and fault trees: (i) possibility of overlooking a significant source of failure when building the fault tree; and (ii) difficulties of eliciting probability estimates, particularly where human reliability is concerned [14]. Eliciting adversary attack type and attack target probabilities is problematic. A recent National Research Council study (NRC 2008) recommends not using event trees for terrorist decisions. Another disadvantage of decision trees is that a few decision nodes can quickly make the tree large and complex. For these reasons, an influence diagram [2] is often a better representation of a failure scenario. The requirement for a specialized analysis tool is the primary disadvantage to using fault trees for risk assessment. Use of specialized fault tree software tools typically requires an experienced risk analyst to develop an appropriate model and represent the probabilistic relationships consistently. The necessity to include all relevant factors requires input from multiple stakeholders and many areas of expertise may be time-consuming. Haimes [14] identifies other limitations such as the implausibility of an assumption of independence between component failures.

6 RESEARCH CHALLENGES

Logic trees for decision and risk analysis offer several research challenges for homeland security applications: probability elicitation of expert data, appropriate models for intelligent adversary decisions, software to solve large problems and perform value of information for risk analysis, tools to allow collaboration between risk analysts, techniques to integrate logic tree analysis results for resource allocation, and ability to compare risk analysis between risk areas.

6.1 Probability Elicitation of Expert Data

Probability elicitation of expert data is a common challenge in decision and risk analysis. Tversky and Kahneman show that subject-matter experts are subject to several biases which can lead to poor probability estimations [15]. Fischhoff suggests elicitation support techniques that might be promising in helping experts avoid many of the biases [16]. But few have shown any value in minimizing the biases. Of more concern is a study by Weiss and Shanteau [17] that showed that intelligence analyst probability estimates were better than flipping a coin but their judgments could be highly suspect and should be used with caution. Buede et al. [18] reviewed the literature and recommended best practices for probability elicitation and aggregation of multiple expert assessments; however, more work is required to test the recommendations through controlled experimentation.

6.2 Models for Intelligent Adversary Decision Making

Intelligent adversaries determine their actions based on their capabilities, the defensive system capabilities, the risk they are willing to take, and the consequences they hope to achieve. Modeling terrorists involves the types of adversaries (e.g. nation state, international terror organization, or individuals), the capabilities of adversaries, the type of attack, the probability of attack, the type of target, the attack location(s), and the timing of attack. Given the challenges of eliciting expert data for the states of nature, the challenges of Modeling intelligent actors are much more problematic. Promising approaches to modeling intelligent adversaries include terrorist decision analysis models, attacker-defender-attacker optimization models, game theory [19], and red teams (NRC 2008).

6.3 Software to Solve Large Logic Trees

Homeland security problems involve complex systems with many interactions and many ways that intelligent adversaries may attempt to defeat the system defenses and achieve the consequences they desire. To defeat these potential attacks, homeland security analysts will need to understand the adversaries' capabilities, the potential targets, target vulnerabilities, consequences of these attacks, and potential mitigation strategies. The PRA software should integrate consequence models with decision trees and should be able to perform value of information calculations [2] on uncertain information.

6.4 Software to Solve Large PRA Combined with Decision Analysis

Homeland security decision makers need to make resource allocation decisions. They will need to evaluate several countermeasures to assess their potential to reduce the likelihood of attack, reduce the vulnerabilities, and/or mitigate the consequences. We may need to develop new software to solve the large decision trees and determine the best portfolio of design changes and countermeasures for the resources available. An information assurance example of this capability is the Microsoft Excel and Access implementation of the mission-oriented risk and design analysis (MORDA) process used by the Department of Defense [10] that can be improved through professional software development.

6.5 Collaborative Risk Analysis

Many of our critical infrastructure are interdependent. Information technology connects and enables the efficient operation of many of our infrastructure. For example, Supervisory Control and Data Acquisition (SCADA) systems are used to manage many of our infrastructure; and information enabled transportation systems are critical for the efficient operation of many infrastructure. Independent risk analysis wastes resources (e.g. multiple organizations analyzing the same systems); may miss the risk associated with interdependent systems due to lack of system knowledge; may lead to suboptimal decisions (e.g. one organization selecting an inefficient countermeasure when collaboration could result in a more efficient use of resources); and may lead to risk transfers (e.g. a simple reprioritization of the adversaries target selection). We will need to develop a standard risk analysis lexicon and allow collaborative risk analyses. Collaborative risk analyses would share information and allow for the most efficient risk analysis (both risk assessments and risk management actions). Collaborative risk analyses will require the sharing of information between private and public organizations and between government agencies, including federal, state, and local governments.

6.6 Ability to Compare Risk Analyses between All-hazards

The Department of Homeland Security (DHS) has the responsibility of securing the homeland against all hazards. In order for the DHS to make an efficient allocation of risk assessment and risk management resources, the risk analysis results must be communicated in a manner that allows the DHS and oversight organizations to assess the relative risk for all hazards and the cost-effectiveness of mitigation actions. A critical first step would be to require all risk assessment results to be presented in comparable formats (e.g. the risk of bioterrorism and the risk of chemical spills would be comparable). The second critical step would be to require that all risk management actions be compared with the reduction in risk per dollar spent.

REFERENCES

1. Ayyub, B. M. (2003). *Risk Analysis in Engineering and Economics*, Chapman and Hall/CRC, London.
2. Clemen, R. T. (1996). *Making Hard Decisions*, 2nd Ed., Duxbury Press. Belmont, CA.
3. Modarres, M. (1993). *What Every Engineer Should Know About Reliability and Risk Analysis*, Marcel Dekker, Inc., New York.
4. Watson, H. A. (1961). *Launch Control Safety Study*, Section VII, Volume 1, Bell Labs, Murray Hill, NJ.
5. U.S. Nuclear Regulatory Commission (1981). *Fault Tree Handbook. NUREG-81/0492.*

6. Ericson II, C. A. (1999). *Course No. 9Sv276 Fault tree analysis student workbook, D&SG Employee Training and Development*, The Boeing Company. Seattle, Washington.
7. Schneier, B. (1999). Attack trees: modeling security threats. *Dr. Dobbs J. Softw. Tools* **24**(12), 21–29.
8. Unkenholz, W. (1989). *Vulnerability Tree Analysis Method*, Unpublished Manuscript.
9. Amoroso, E. (1994). *Fundamentals of Computer Security Technology*, Prentice-Hall, Englewood Cliffs, NJ.
10. Buckshaw, D. L., Parnell, G. S., Unkenholz, W. L., Parks, D. L., Wallner, J. M., and Saydjari, O. S. (2005). Mission oriented risk and design analysis of critical information systems. *Mil. Operat. Res.* **2**(10), 19–38.
11. Paté-Cornell, M. E. (1984). Fault trees vs. event trees in reliability analysis. *Risk Anal.* **4**(3), 177–186.
12. U.S. Nuclear Regulatory Commission (1975). *Reactor safety study, WASH-1400 (NUREG-75/014)*, Washington, DC.
13. U.S. Nuclear Regulatory Commission (1983). *PRA procedures guide (NUREG/CR-2300)*, Washington, DC.
14. Haimes, Y. Y. (1999). *Risk Modeling, Assessment, and Management*, John Wiley and Sons, Inc., Hoboken, NJ.
15. Tversky, A., and Kahneman, D. (1974). Judgment under uncertainty: heuristics and biases. *Science*. **185** (1124–1131).
16. Fischhoff, B. (2002). Heuristics and biases. In *Application in Heuristics and Biases: The Psychology of Intuitive Judgment*, Gilovich, Griffin and Kahneman, Eds. Cambridge University Press. New York.
17. Weiss, D. J. and Shanteau, J. (2003). The vice of consensus and the virtue of consistency. In *Psychological explorations of competent decision making*, J. Shanteau, P. Johnson, and C. Smith, Eds. Cambridge University Press, New York.
18. Buede, D. M., Mahoney, S. M., Ulvila, J. W., and Smith T. C. (2006). *Review of Probability Elicitation and Examination of Approaches for Large Bayesian Networks. U.S. Department of Defense*, Unpublished Manuscript.
19. Hausken, K. (2002). Probabilistic risk analysis and game theory. *Risk Anal.* **22**(1). Available at http://www1.his.no/vit/oks/hausken.
20. Report on Methodological Improvements to the Department of Homeland Security's Biological Agent Risk Analysis: A Call for Change, Committee on Methodological Improvements to the Department of Homeland Security's Biological Agent Risk Analysis, National Research Council of the National Academies, The National Academy Press, Washington, DC, (2008).

FURTHER READING

Bedford, T. Cooke R. (2001). *Probabilistic Risk Analysis: Foundations and Methods*, Cambridge University Press, Cambridge, UK.

Kumamoto, H. and Henley, E. J. (1992). *Probabilistic Risk Assessment: Reliability Engineering, Design, and Analysis*, IEEE Press. New York.

Parnell, G. S., Dillon-Merrill, R. L., Bresnick, T. A. (2005). *Integrating Risk Management with Homeland Security and Antiterrorism Resource Allocation Decision-making*, D. Kamien, Ed. The McGraw-Hill Handbook of Homeland Security, New York, pp. 431–461.

von Winterfeldt, D., and Rosoff, H. (2005). *Using Project Risk Analysis to Counter Terrorism*, Symposium on Terrorism Risk Analysis, Los Angeles, CA.

BAYESIAN NETWORKS

DENNIS BUEDE, SUZANNE M. MAHONEY, AND JOSEPH A. TATMAN
Innovative Decisions, Inc., Vienna, Virginia

1 INTRODUCTION

Bayesian networks (BNs) may be used to address Department of Homeland Security (DHS) needs ranging from identifying and tracking suspect individuals to analyzing vulnerabilities and providing warning of attack. BNs factor complex problems into manageable networks of random variables. The graphical interface of BNs facilitates explanation while the local nature of the probabilistic relationships facilitates assessment. Computational algorithms provide rapid evaluation of available evidence to support decision-making in complex and evolving situations.

Section 2 provides a brief introduction to BNs while Section 3 provides some insight into the worldwide research community. The next section describes the application of BNs to meet DHS' critical needs. In the final section, research goals relevant to meet DHS's critical needs are summarized.

2 SCIENTIFIC OVERVIEW

This section presents the basics of BNs. This is followed by a section on inference (or computation). The issue of obtaining the probabilistic inputs for a model is described under knowledge acquisition.

2.1 Modeling with Bayesian Networks

A BNs is a factorization of a joint probability distribution. An acyclic directed graph specifies the network's structure. Nodes stand for random variables. Directed arcs indicate probabilistic dependence. Lack of an arc indicates probabilistic independence. A BNs stores parameters with each node in the form of a conditional probability table (CPT). A CPT contains distributions of a dependent/child variable given possible combinations of its conditioning/parent variables. Given its parents, each random variable is independent of its nondescendants [1].

Practitioners often use BNs to model uncertain situations. Frequently, these models follow the pattern shown in Figure 1. *Domain uncertainty* applies to hypotheses and their observables. Four types of random variables represent the domain of interest.

- *Context* that is broadly defined includes the random variables whose values are usually known when the problem arises, but not when the model is built. Context potentially influences all other variables in a network. In a medical application, context may include a patient's gender and age. In an information assurance application, context would include operating system and software.

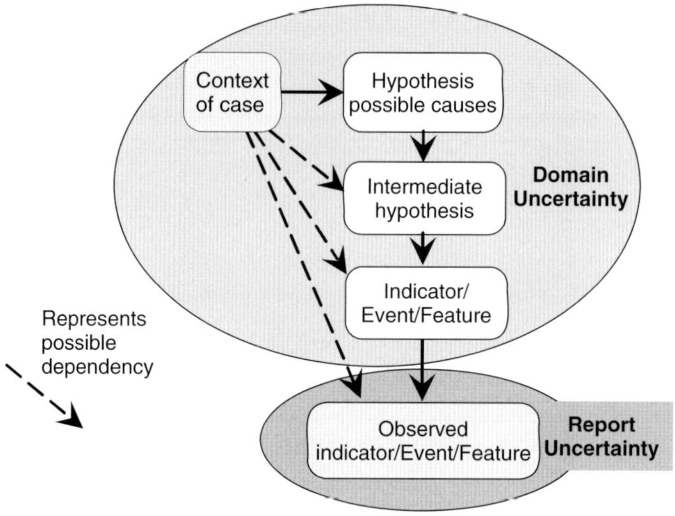

FIGURE 1 Pattern for a Bayesian network.

- *Hypothesis/Possible Causes* variables are the variables of interest in a network whose purpose is diagnosis. Examples include diseases in a medical diagnostic network, possible information attack plans in an information assurance network, faults in an equipment repair application, and types of bombs in a bomb detection application.
- *Intermediate hypotheses* provide support to one or more of the hypotheses. Examples include plan steps of an information attack and detectable components of a dirty bomb.
- *Indicator/Events/Features* are those elements of a situation that are observable. Examples include symptoms of diseases, equipment readings (e.g. level of γ rays), and airline passenger behavior. These are often the variables of interest in networks whose purpose is prediction.

Observations are subject to *report uncertainty*. In addition to depending upon the context of the domain and the indicator/event/feature being observed, observations also depend upon characteristics of the observer. For example, a sensor is limited by its inherent sensitivity (whether it can detect an event when the event is present) and specificity (how often it detects an event when the event is not present). Human reports are influenced by the ability of the observer to make the observation, whether the observer actually believes his/her observations and whether the observer is lying.

2.2 Applying Bayesian Networks

Inference in a BNs flows throughout the network. Consequently, a network may be used for multiple purposes. For example, in a medical application one uses test results and symptoms to diagnose the disease. In this case one makes inferences about the hypotheses, given the observations of the disease indicators. This is the typical diagnostic application. At the same time, one may make inferences about unobserved disease indicators or how a probable disease is likely to progress over time. In this case, one is using the same network for prediction or what-if evaluation.

The BN's structure and parameters represent a knowledge base about a problem of interest. For example, the structure of a disease network shows that colds and allergies share some of the same symptoms. The parameters of the network indicate how much more likely a fever is with a cold than with an allergy.

In applying BNs to real world applications the evidence or observations may come from a variety of sources. Evidence may be stored in a database or generated by a simulation or entered in real-time by a human observer or received from a sensor. Because BNs provide a general framework for representing uncertainty, they are excellent models for multisensor fusion.

2.3 Example

Figure 2 shows a small BNs that collects information about passengers boarding a public conveyance. The network's purpose is to decide whether a passenger is "of interest" and should therefore be subject to further screening.

The node in the upper right hand corner supplies the context "passenger screening". The nodes above the line across Figure 2 reflect domain uncertainty. The "passenger" node is the hypothesis. There are no intermediate hypotheses in this network. The three other nodes above the line are observable features. Representing report uncertainty, the node "reported behavior" is an observation of the "suspicious behavior" node. The "observer quality" node conditions how well an observer's report matches the observed behavior.

Each node displays the random variable's possible values or states with their associated probability (displayed as percents in the nodes of Figure 2). As evidence about nodes in the network is accumulated the probabilities are updated by the inference algorithm.

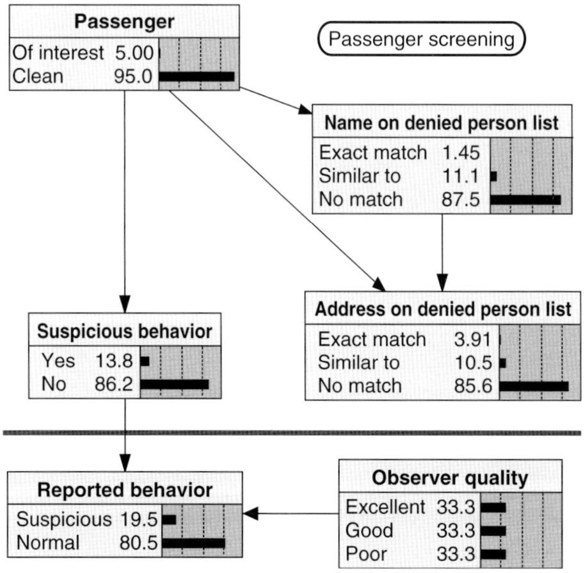

FIGURE 2 Passenger screening network.

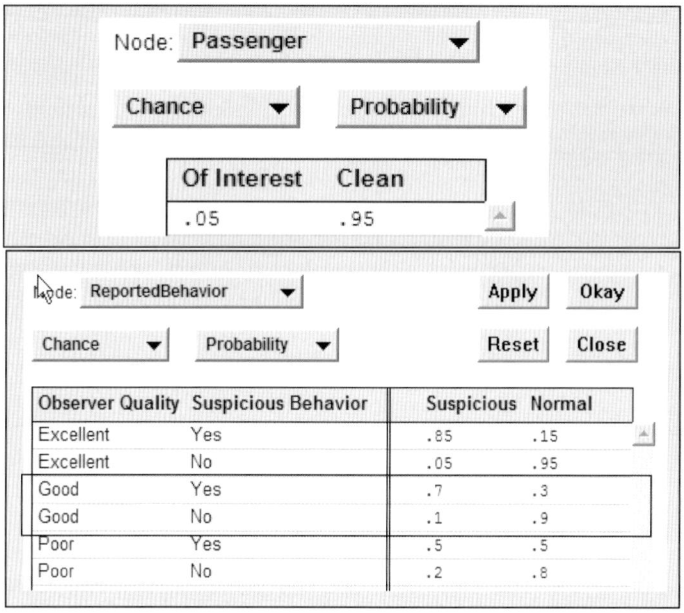

FIGURE 3 Conditional probability tables for the passenger screening network.

Figure 3 shows two of the passenger screening network's CPTs. The top CPT, belonging to the "passenger" node, displays a single distribution indicating that about 5% of passengers are of interest. The bottom CPT is for the node "reported behavior"; it depends on two other nodes. The states of these other two nodes are displayed on the left side of the CPT. The CPT represents the probability of an expert of a specified quality making a particular observation given the true state of the observed behavior. The portion of the CPT outlined by the box displays the distributions associated with a good observer. For passengers exhibiting suspicious behavior, the good observer is correct 70% of the time while, for passengers behaving normally, the observer is correct 90% of the time.

Figure 4 shows two instances of the passenger screening network. Each shows some nodes set to specific values. Compare the displayed probability distributions of these networks with the ones in Figure 2. The instance on the left shows how the probabilities in the unobserved nodes change when we note that a passenger's name and address are similar to ones on the denied persons list. This evidence not only changes the probabilities of the hypothesis node "passenger" but also the probabilities of the nodes "suspicious behavior" and "reported behavior". This reflects our belief that persons who may be on the denied person list are more likely to exhibit suspicious behavior. At the same time, this knowledge about the passenger does not change our belief about the quality of the observer. Many people have names similar to those in the database, so we also observe the passenger's behavior before making a decision to subject the passenger to further screening. As shown in the network on the right side of Figure 4, the observation of suspicious behavior by a good quality observer raises the probability of the passenger being of interest to 69%. In this case, the quality of the observer makes a difference in the probabilities of the unobserved nodes. A poor observer would produce a probability of only 54% that the passenger is of interest.

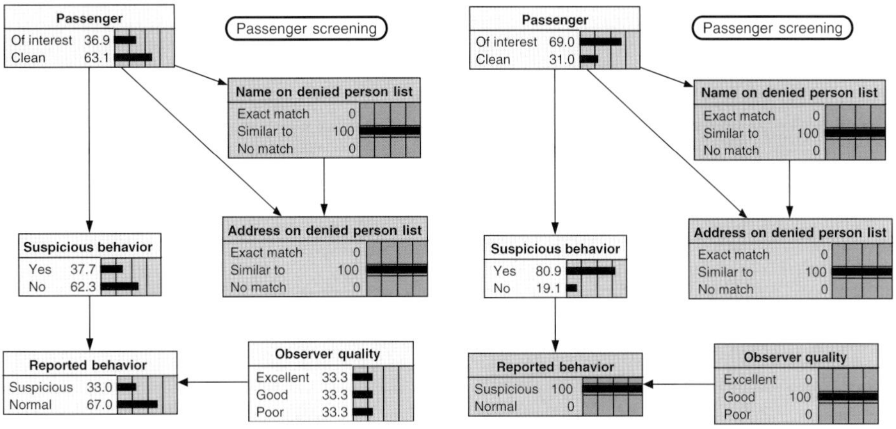

FIGURE 4 Passenger screening network with observations.

2.4 Bayesian Network Structures

For many applications of BNs, networks similar to those in Figure 1 are reused for different entities (passengers). With each reuse the evidence of previous uses is cleared and new evidence is entered. This is an example of a template model. When in use, the network's structure and parameters remain fixed and unchanging. Only the evidence regarding the current circumstance differs from one use to the next. Other usage examples are dynamic Bayesian networks (DBNs) and situation-specific networks. DBNs [2] are specifically designed to model cases in which the situation changes over time in a predictable fashion. Situation-specific BNs [3] are constructed by pulling network fragments from a knowledge base in response to evidence and context about the situation being modeled. Unlike template networks, DBNs and situation-specific BNs extend the structure of a BNs to include additional variables. Like template networks, the parameters associated with the variables remain unchanged.

2.5 Inference

A BNs factors the joint probability distribution so that any query about joint/marginal probability distributions in a BNs can be computed by the product of all CPTs found in the BNs followed by the marginalization of the variables not in the query. For example the joint probability distribution of the network shown in Figure 2 is computed from the following product:

$$
\begin{aligned}
&P(\text{Passenger, Name on Denied Person List, Addresson Denied Person List, Suspicious}\\
&\text{Behavior Observer Quality, Reported Behavior})\\
&= P(\text{Passanger}) \times P(\text{Name on Denied Person List} \mid \text{Passenger})\\
&\quad \times P(\text{Name on Denied Person List} \mid \text{Name on Denied Person, Passenger})\\
&\quad \times P(\text{Suspicious Behavior} \mid \text{Passenger})\\
&\quad \times P(\text{Observer Quality})\\
&\quad \times P(\text{Reported Behavior} \mid \text{Suspicious Behavior, Observer Quality}).
\end{aligned}
$$

Pearl's algorithm [1] applied a message passing approach to a BNs whose structure was a polytree. In a polytree there is exactly one path from one node to another. To solve BNs, whose graphs are not polytrees, researchers have taken two different approaches to convert such networks to polytrees. Graph theoretic approaches use the topology of the graph to cluster nodes into a junction tree [4], while algebraic approaches consider the nature of the query and evidence [5].

The worst-case complexity of any exact inference algorithm is known to be NP-hard [6]. Hence, researchers have also turned to approximate inference algorithms. Existing approximate BNs inference algorithms fall into three major categories.

- *Monte Carlo sampling* algorithms estimate a posterior probability distribution (the target distribution) by sampling from another computationally simpler probability distribution (the sampling distribution). Sampling algorithms include: logic sampling, Gibbs sampling, likelihood weighting, and adaptive importance sampling [7]. For DBNs, particle filtering [8] algorithms represent the past expression, a summary of all past observations, by a set of samples called *particles*.
- *Dependency reduction* algorithms weaken or ignore some of the dependencies among nodes. The iterative belief propagation algorithm, also known as *loopy belief propagation* [9], applies Pearl's belief propagation algorithm for singly connected BNs to loopy networks. For DBNs, Boyen and Koller [10] factorize the past expression into a set of smaller tables by ignoring the dependencies among some interface variables.
- *Variational* methods fit an approximating distribution to the true posterior [11].

Many approximate algorithms exhibit an anytime property, which offers the user the control of the trade-off between the cost and accuracy. The complexity of the approximate inference algorithms to obtain the desired accuracy is also known to be NP-hard [12].

2.6 Knowledge Acquisition

Knowledge acquisition includes the following activities: identify the structure of the BNs; assess its parameters; and evaluate the resulting network. Sources of knowledge include both subject matter experts and available data. One advantage of a BNs is that both sources can be combined by treating expert knowledge as an equivalent number of data cases.

2.7 Eliciting Knowledge from Experts

When obtaining knowledge from experts, the facilitator is presented with a number of challenges. Experts arrive with different amounts and types of expertise. Furthermore, the elicitation method may be unfamiliar to the experts, making communication between the elicitor and the experts difficult. In addition, humans have biases that may skew the elicitation process. Among these are overconfidence, extrapolating population statistics from small samples, and adjusting probabilities too little given an initial value. Reliability studies have found that experts dealing with scientific areas are the most reliable while those working with human activities are the least reliable.

Methods for eliciting probabilities and distributions have been researched for more than 40 years. Initially, the emphasis was on eliciting probabilities for single events.

With the introduction of BNs, populating CPTs became a concern. Generally, a facilitator elicits probabilities from experts. However, the requirement to fill numerous CPTs in large BNs and the capability to develop graphical user interfaces have led to the use of computer-supported graphical tools.

Eliciting the structure of BNs has received little research attention. Causal structures have been recommended as a way of minimizing network complexity. Others recommend an iterative object-oriented approach. Various researchers have noted patterns that appear regularly in BNs.

2.8 Learning from Data

The problem of learning a BNs from observations can be decomposed into the tasks of searching over possible structures, estimating the CPTs given the structure, and scoring the structure given the data [13].

1. Search efficiently over the large number of possible structures to find a subset of good structures.

   There are an exponentially large number of possible structures for a given set of variables. The K2 algorithm [14] applied a simple heuristic hill-climbing search. It compared the posterior probability of the structures that differed by a single arc added to the network. More recent approaches use genetic algorithms and Markov Chain Monte Carlo (MCMC) to search over possible structures.
2. Estimate the local probability tables for a given structure.

   Cooper and Herskovits [14] developed the first algorithm for joint estimation of structure and parameters for the discrete variable case. The discovery of more general estimation methods for mixtures of graphical models is an active area of research.
3. Estimate the relative posterior probabilities of any subset of structures.

   Most algorithms assume a uniform probability across all structures. The structure score is an estimate of how well it explains the data. The posterior probability of the structure is calculated from

   $$P(D) = \sum_{S_c} P(D/S_c).P(S_c)$$

   where S_c is the structure being considered and D is the data. The term $P(D|S_c)$ is calculated directly from a proposed network.

In cases of incomplete or missing data, one commonly applied approach is the expectation maximization (EM) algorithm [15]. EM applies when data are "missing at random". The EM algorithm is guaranteed to converge to a local optimum. An alternative option is to simulate missing observations along with structures in an MCMC algorithm.

3 RESEARCH AND FUNDING DATA

Research into BNs began in the middle of the 1980s. There have been at least 20 books written on this topic since 1988. There are active research programs on the topic at the most prestigious universities (e.g. MIT, Stanford University, Harvard University, University of Cambridge, and Oxford University) as well as many other universities around the world. There are several major conferences that address BNs: Uncertainty in Artificial Intelligence (http://www.auai.org/), Florida Artificial Intelligence Research Society (http://www.flairs.com/), and Artificial Intelligence and Statistics (http://www.stat.umn.edu/~aistat/index.html). Finally the major IT companies around the world are investing heavily in the use of BNs for their product development: Microsoft, Yahoo, IBM, Boeing, Google, and Intel.

Current research topics reported across several conferences and journals include the following.

- Advanced algorithms for difficult networks, for example compiling graphical models.
- Advanced modeling techniques, for example dealing with non-Gaussian continuous variables.
- Advanced software implementations, for example sensitivity analysis and user friendliness for naïve users.
- Advanced learning techniques.
- Improved elicitation of expert's probabilities.
- Development of ontologies for probabilistic reasoning.
- Causal reasoning on the basis of probabilistic modeling.

More applied research deals with applications of BNs to

- user modeling;
- cognitive agents and practical robots for network management, unmanned robots, computer games, educational games, and sensor management;
- sensor fusion;
- forensic science;
- reliability models;
- spam filters;
- biologic systems;
- cyber attacks and user behavior;
- decision support to military commanders;
- gene expression;
- combining BNs and social networks to address predicting organizational activities and responses to changes.

Funding sources for research into the theory and application of BNs includes corporations, government agencies and foundations. Current corporate funding comes from

- Microsoft
- Yahoo
- IBM
- Boeing

- Google
- Intel
- Siemens
- Toyota
- SAIC.

Funding from government agencies in the United States comes from:

- ONR
- NSF
- Air Force
- JPL/NASA
- DARPA
- NASA
- Federal Aviation Administration
- National Library of Medicine
- National Institute of Health.

Funding from governments around the world comes from:

- Academy of Finland
- Finnish Funding Agency for Technology and Innovation
- Netherlands Organization for Scientific Research
- German Research Foundation
- Norwegian Academy of Science and Letters
- Israel Science Foundation
- US–Israel Binational Science Foundation
- National Sciences and Engineering Research Council of Canada
- Canada Foundation for Innovation
- Ministry of Education, Culture, Sports, Science, and Technology of Japan
- Swedish Research Council
- Swedish Foundation for Strategic Research
- Information Societies Technologies (IST) Programme of the European Commission
- Dutch Ministry of Economic Affairs.

Foundations supporting BNs research are:

- Leverhulme Trust, UK
- Rothschild Foundation
- Sumitomo Foundation
- Kayamori Foundation of Information Science Advancement
- Japan Society for the Promotion of Science
- James S. McDonnell Foundation Causal Learning Collaborative

- Sloan Foundation
- Gatsby Charitable Foundation
- Minerva Foundation
- Ridgefield Foundation

4 CRITICAL NEEDS ANALYSIS

Two major classes of BNs are relevant for DHS. Diagnostic networks are useful for a broad array of problems involving a set of hypotheses and a collection of evidence that distinguishes among the hypotheses. They provide a probability distribution over the set of hypotheses given any combination of evidence. An example is the passenger screening network shown in Figure 2. The "passenger" node provides the hypotheses. Its probability distribution is updated as findings are entered into the evidence nodes such as "name on denied person list". Predictive networks have one or more output nodes with probabilistic relationships to a set of input nodes as in Figure 5. The CPT of the output node accounts for dependencies among the states of the input nodes. In Figure 5, the output node, "structural failure", is predicted by the combination of possible values of the other nodes in the network.

The next paragraphs discuss four critical needs of DHS. These needs are relevant to two major functions of DHS, counterterrorism and border control.

4.1 Indications and Warnings

The DHS strategic plan states that "The prevention of terrorist attacks is the first priority in securing our homeland." Key to preventing an attack is obtaining a warning that such

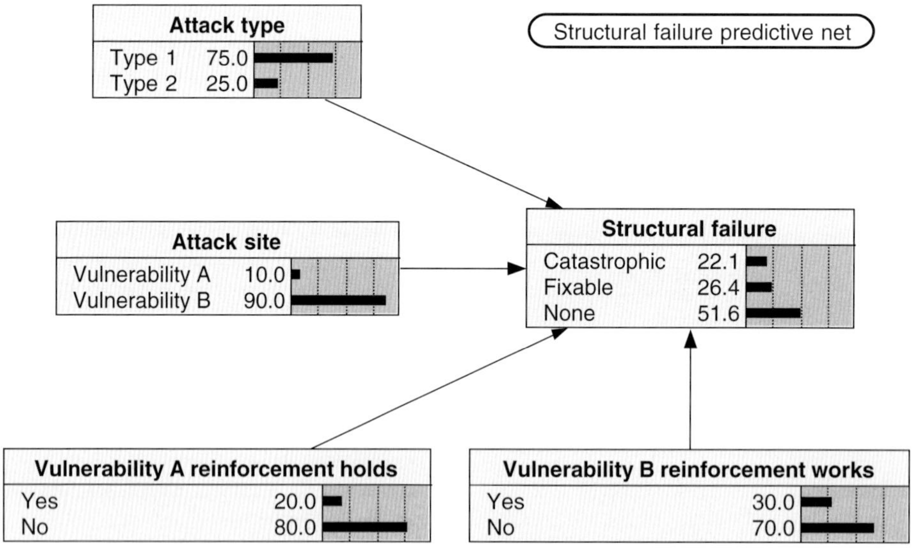

FIGURE 5 Predictive bayesian network.

an attack is planned. To do so, one must combine many disparate indicators in a way that lends support to whether or not such an attack is planned.

The diagnostic class of BNs offers an effective, systematic framework for integrating diverse sets of data. The hypotheses are whether or not an attack is being planned and the stage of development of the plan. The indicators include events such as the purchase of specific types of materials, communication between persons who may have terrorist leanings and known terrorists, experimentation by suspected terrorists with specific items that could be used for weapons, and anomalous behavior of suspected terrorists.

In actual applications, BNs have been applied to understanding adversary intent in international crises, detecting deception by an adversary, and in decision support tools that predict enemy attacks for battlefield commanders. These examples demonstrate that indications and warning is a high impact, high probability of success application of BNs.

4.2 Assessment of Information Quality

The DHS strategic plan recognizes the need for integrating information received from diverse sources ranging from humans to electronic sensors and coming from all levels of government. BNs effectively address two problems here. The first problem is the need to combine information from diverse types of sources (as in the "indications and warning" discussion). The second problem is the reliability of human reporting.

The diagnostic class of BNs is relevant for both problems. Such a BNs has a hypothesis node that represents the event being reported. One evidence node (of possibly many) is the human report. In this case the link between hypothesis and evidence (or report) is decomposed into nodes representing the competence and credibility of the source. It further breaks down credibility into sensitivity, objectivity, and veracity, see Figure 6. The BNs facilitates calculation for this complex problem. BNs provide a methodology for integrating information not only from multiple human sources but also from traditional

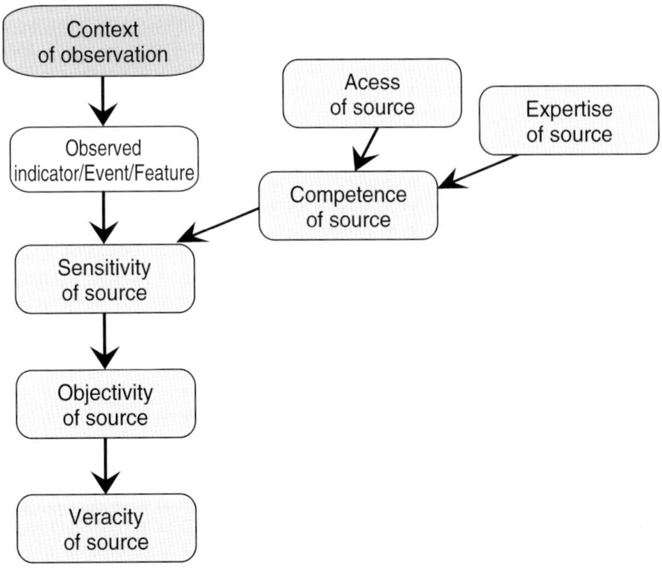

FIGURE 6 Source credibility model.

sensors. Their application to the assessment of information quality is rigorous and proven in real world applications [16].

4.3 Analyze Risks and Vulnerabilities

A top priority in implementing the DHS strategic plan is a framework for continuously assessing what can go wrong, the likelihood of occurrence, and the potential consequences. Also, a strategic objective asserts that risk-based analytic tools will be used to help prioritize projects to protect critical infrastructure.

Such risk assessment and analysis is a natural application for the predictive class of BNs. The variable of interest in a risk application could be the level of damage to a critical building as in Figure 5. This level of damage depends on many random events or variables, such as the nature of the attack, the location of the attack, or the degree of training of security personnel.

BNs have been successfully applied to the risk analysis of major engineering development projects. The impact of many uncertain variables on the timing and cost of successful project completion had to be considered. In another application, assessments performed by a large number of experts on the safety level of a critical software-intensive system were integrated [17].

The level of complexity and the factors modeled in these applications indicate that they can be extended to meet DHS's vulnerability and risk assessment needs.

4.4 Identify and Track Suspect Persons

Critical to counterterrorism is the ability to identify and track individuals. Identifying the suspect in a scene is a natural application for a diagnostic BNs. The network is used to assimilate many bits of evidence about the suspect to make an identification. On the other hand, a predictive BNs can be applied to track a person. The variable of interest may be the person's location during a specific time period. Such a location would be a probabilistic function of the person's location during a previous time period and other uncertain factors such as the receipt of a cell phone call.

One application automatically tracks a varying number of people exposed to an advertisement, and determines whether they looked at the ad. A hybrid dynamic BNs simultaneously infers the number of people in the scene and their body and head locations [18]. Another application is a tracking algorithm that can identify and track distinct individuals in a scene and recover when it loses track [19].

These applications, while limited, provide a starting point in attacking this difficult problem.

4.5 Summary

A broad collection of DHS critical needs may be addressed using BNs. BNs are a natural, effective solution for problems involving uncertainty and the integration of evidence from diverse sources. They provide a powerful probability calculator for problems requiring the calculation of risk. Finally, BNs balance mathematical rigor with a capability to analyze messy real world problems involving uncertainty, while providing an intuitive modeling front-end for most decision makers.

5 RESEARCH DIRECTIONS

Homeland security is exemplified by intelligent adversaries and evolving science and technology. These characteristics mean that what is true or likely today may not be true or likely tomorrow or next week and certainly not true or likely next year. Such a domain is difficult for any human reasoning, whether qualitative or quantitative. Those approaches that use modular building blocks that can be swapped in and out relatively easily are the most likely to provide rapid and useful support. BNs certainly fall into this category. But achieving the benefits of this type of an approach is not easy or readily available.

A sample of research directions for dealing with domains that have dynamic expertise are:

- extend the theory behind model fragments;
- develop adaptive learning algorithms that make use of limited data;
- develop graphical user interfaces to facilitate the capture of expertise from humans.

In addition to homeland security having a dynamic base, there is a tremendous amount of complexity. This complexity includes technology interactions such as incompatibilities among communication devices, as well as finding the right sequence of activities during a catastrophe that will save the most lives. Most of this complexity is present because of the need to harness hundreds or thousands of humans together to work as a team when they have never performed as a team before in dealing with a situation never before envisaged. Automated reasoning tools (including BNs) can be used both to increase the productivity of individuals as well as to manage the team effort more effectively.

A sample of research directions for dealing with large, complex domains are:

- develop more adaptable connections between BNs and other modeling tools, for example simulation packages and spread sheets;
- embed BNs in radios, personal digital assistants (PDAs), and so on.

REFERENCES

1. Pearl, J. (1988). *Probabilistic Reasoning in Intelligent Systems: Networks of Plausible Inference*, Morgan Kaufmann, San Mateo, CA.
2. Murphy, K. (2004). Dynamic Bayesian networks. Proposed chapter in *Probabilistic Graphical Models*, M. Jordan, Ed. http://www.cs.ubc.ca/~murphyk/Papers/dbnchapter.pdf.
3. Wellman, M. P., Breese, J. S., and Goldman, R. P. (1992). From knowledge bases to decision models. *Knowl. Eng. Rev.* **7**(1), 35–53.
4. Lauritzen, S., and Spiegelhalter, D. (1988). Local computations with probabilities on graphical structures and their application to expert systems. *J. R. Stat. Soc.* **B 50**, 157–224.
5. Dechter, R. (1996). Bucket Elimination: A unifying framework for probabilistic inference. *Proceedings of UAI-96*. Portland Oregon, August. http://www.ics.uci.edu/~dechter/publications/.
6. Cooper, G. F. (1990). The computational complexity of probabilistic inference using Bayesian belief networks. *Int. J. Artif. Intell.* **42**, 393–405.
7. Shachter, R. D., and Peot, M. A. (1989). Simulation approaches to general probabilistic inference on belief networks. *Proceedings of the Fifth Workshop on Uncertainty in Artificial Intelligence*. Santa Cruz, CA, pp. 311–318.

8. Doucet, A., deFreitas, N., Murphy, K., and Russell, S. (2000). Rao-Blackwellised particle filtering for dynamic BNs, *UAI-00*. www.foi.se/fusion/fusion24.ps.
9. Murphy, K., Weiss, Y., and Jordan, M. (1999). Loopy-belief propagation for approximate inference: an empirical study. *Proceedings of UAI-99*. http://citeseer.ist.psu.edu/murphy99loopy.html.
10. Boyen, X., and Koller, D. (1998). Tractable inference for complex stochastic processes. *Proceedings of the Conference on Uncertainty in AI*. http://ai.stanford.edu/~koller/papers/nips98.html.
11. Jaakkola, T. S., and Jordan, M. I. (1996). Recursive algorithms for approximating probabilities in graphical models. In *Advances in Neural Information Processing Systems*, 9. http://people.csail.mit.edu/people/tommi/inf.html.
12. Dagum, P., and Luby, M. (1993). Approximating probabilistic inference in Bayesian belief networks is NP-hard. *National Conference on Artificial Intelligence*. Washington, DC.
13. Heckerman, D. (1999). A tutorial on learning with Bayesian networks. In *Learning in Graphical Models*, M. Jordan, Ed. MIT Press, Cambridge, MA.
14. Cooper, G. F., and Herskovits, E. (1992). A Bayesian method for the induction of probabilistic networks from data. *Mach. Learn.* **9**, 309–347.
15. Lauritzen, S. L. (1995). The EM algorithm for graphical association models with missing data. *Comput. Stat. Data Anal.* **19**, 191–201.
16. (a) Schum, D. (1989). Knowledge, probability, and credibility. *J. Behav. Decis. Mak.* **2**, 39–62; (b) Schum, D. (1992). Hearsay from a layperson. *Cardozo Law Rev.* **14**(1), 1–77.
17. Bouissoum, M., and Nguyen, T. (2002). Decision making based on expert assessments: use of belief networks to take into account uncertainty, bias, and weak signals. *Proceedings of the Lambda Mu 13/ESREL Conference*. perso-math.univ-mlv.fr/users/bouissou.marc/ExpertsAndBN_ESREL02.pdf.
18. Smith, K., Ba, S., and Gatica-Perez, D. (2006). Tracking the MultiPerson wandering visual focus of attention. *International Conference on Multimodal Interfaces*. www.idiap.ch/~smith/SMITH_ICMI06.pdf.
19. Ramanan, D., and Forsyth, D. A. (2003). Finding and tracking people from the bottom up. *Proceedings Computer Vision and Pattern Recognition (CVPR)*. www.cs.berkeley.edu/~ramanan/papers/trackingpeople.pdf.

FURTHER READING

Howson, C., and Urbach, P. (2005). *Scientific Reasoning: The Bayesian Approach*, Open Court Publishing Company, Chicago, IL.

Jensen, F. V. (2002). *Bayesian Networks and Decision Graphs*, Springer, NY.

Jordan, M., Ed. (1999). *Learning in Graphical Models*, MIT Press, Cambridge, MA.

Korb, K. B., and Nicholson, A. E. (2003). *Bayesian Artificial Intelligence*, Chapman & Hall/CRC Press, Boca Raton, FL.

Neopolitan, R. E. (2004). *Leaning Bayesian Networks*, Pearson Prentice Hall, Upper Saddle River, NJ.

Pearl, J. (1988). *Probabilistic Reasoning in Intelligent Systems: Networks of Plausible Inference*, Morgan Kaufmann, San Mateo, CA.

Pearl, J. (2000). *Causality: Models, Reasoning, and Inference*, Cambridge University Press, New York.

Zdziarski, J. A. (2005). *Ending Spam: Bayesian Content Filtering and the Art of Statistical Language Classification*, No Starch Press, San Francisco, CA.

USING RISK ANALYSIS TO INFORM INTELLIGENCE ANALYSIS

HENRY H. WILLIS
RAND Corporation, Pittsburgh, Pennsylvania

1 INTRODUCTION

The goal of intelligence is to produce guidance on the basis of available information within a time frame that allows for purposeful action. In efforts to combat terrorism, actionable guidance could come in many forms.

Sometimes guidance is needed to shape strategy. For example,

- the federal government must decide whether to maintain stockpiles to enhance emergency preparedness, or
- state and local governments must choose for which scenarios to develop response plans and train.

Sometimes guidance is needed to inform operational decisions. For example,

- if the federal government decides to use stockpiles, it must decide what to put in them and how to preposition them; or
- airports must decide how to deploy technologies and modify operations to enhance security.

Sometimes the required guidance is on a tactical level. For example,

- law enforcement must know when to deploy additional surveillance around a building or for an event;
- law enforcement is interested in who may be planning an attack; or
- critical infrastructure owners and operators need to know when greater security is required.

All of these examples require different information, but have one thing in common. They all require that the information be appropriate for the intended use.

The concept of the intelligence cycle provides a structure to the process of producing this guidance. The intelligence cycle (Figure 1) begins with the direction of intelligence collection (step 1). This results in collection of new information (step 2) that must be processed (step 3), analyzed (step 4), and disseminated and used (step 5). Use of the intelligence products creates new information through either active (i.e. new directed intelligence collection) or passive (i.e. observance of resulting events) means. The intelligence cycle is completed by feeding this new information back into this process [1].

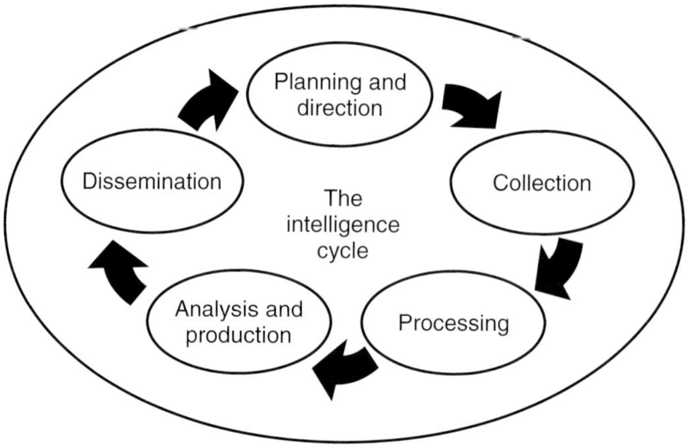

FIGURE 1 The intelligence cycle (adapted from Krizan 1999) [1].

Time is critical to the success of this process because adversaries also carry out their own intelligence processes to identify promising opportunities for attack and target vulnerabilities. Intelligence is more valuable if the intelligence cycle operates faster than the opponent's. More rapid intelligence enables faster recognition of new threats and adaptation to shifts in opponents' strategies. Thus, methods to improve the accuracy and speed of the process provide a strategic advantage in efforts to combat terrorism.

This article discusses how risk analysis can help intelligence analysts assess threats of terrorism. The discussion leads to two conclusions. First, risk analysis can be used to sharpen intelligence products. Second, risk analysis can be used to prioritize resources for intelligence collection. However, it is important that practitioners applying risk analysis recognize its limitations to ensure that results are appropriate for the purpose and that its use does not blind the analyst to potential surprises.

The remainder of the article is organized as follows. The next section describes intelligence analysis as an input–output process and maps risk analysis to this process. Following this description is an introduction of challenges to the successful application of risk analysis to intelligence analysis. The article closes with a summary of how risk analysis can best serve the intelligence analysis community.

2 INTELLIGENCE ANALYSIS AS AN INPUT–OUTPUT PROCESS

The analysis function of the intelligence cycle in Figure 1 can be considered an input–output process in which raw intelligence is the input and intelligence products are the outputs. Within this framing, Willis et al. [2] described how risk analysis can be connected to the intelligence cycle (Figure 2).

The process outlined in Figure 2 represents an interaction between the intelligence community and the intelligence customer. In the case of terrorism risk, it is the managers who are responsible for implementing homeland security policies and programs. In the same way that the intelligence cycle must be an iterative process, the intelligence community and the homeland security community must interact closely and at many stages throughout the process of collection, processing, and analysis.

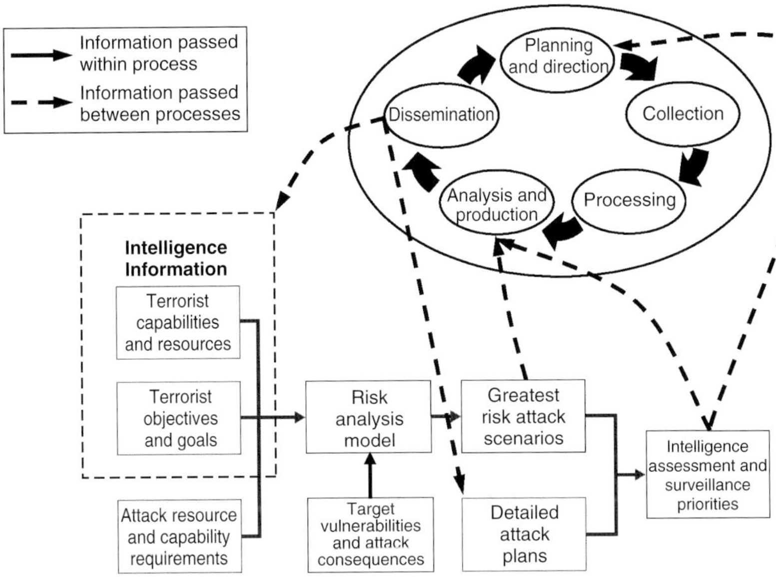

FIGURE 2 Connections between risk analysis and the intelligence cycle (adapted from Willis et al., 2006).

Collection activities produce intelligence that can be used to assess the magnitude and nature of terrorism threats. Here, relevant information is that which describes terrorist capabilities to carry out attacks of different complexity, fiscal and personnel resources to support such attacks, goals in pursuing terrorist activities, and objectives associated with any particular attack plan. This information alone has been used to assess the range of terrorist threats that exist and how they are adapting to evolving security postures. Analyzing all of these factors is important as threat does not exist unless a group or an individual has both the intent and capability to conduct an attack [3, 4].

However, terrorism risk is not determined by threat assessment alone. Risk from terrorism only exists when there is a credible threat of attack that could cause harm to a target that is vulnerable [5]. Risk analysis provides a framework for considering threat, vulnerability, and consequences of potential terrorist attacks and developing strategies to manage these risks most effectively with limited resources.

As depicted in Figure 2, risk analysis combines intelligence about the objectives and capabilities of terrorist groups with assessments of the required capabilities and resources to complete an attack successfully, assessments of the vulnerabilities of targets to different attack modes, and assessments of consequences of different types of attacks on different targets. The product of this analysis is identification of terrorism scenarios that present the greatest risk.

This result itself can be used by the intelligence cycle. Thus, at its most basic form, risk analysis can be integrated into the intelligence cycle as part of the analysis and production step. However, assessments of the relative risks of different attack scenarios may only be adequate as intelligence products to support for strategic and operational analysis.

Often more specificity than relative risks of different attacks is required about where an attack will occur, when the attack will occur, or who will try to attempt the attack. In particular, law enforcement agencies attempting to prevent future terrorist attacks need to have guidance of where to conduct further surveillance and who to target with such efforts. This information requires an understanding of the detailed steps and time lines associated with the planning and orchestrating of an attack. Some aspects of this are specific to the target and attack mode being considered. Other aspects are specific to how the group that is planning the attack operates.

Detailed information about attack planning can be developed through surveillance of terrorist groups, investigations into foiled or successful plans for attacks, and red-teaming studies facilitated by tools of risk analysis. Rosoff and von Winterfeldt [6] have demonstrated how probabilistic risk analysis can be used in this way within the context of scenarios for detonating a radiological device at the ports of Los Angeles and Long Beach. In this study, probabilistic risk analysis was used to decompose the attack scenario into its component steps and explore how defensive countermeasures directed at each step could reduce the risks of attack.

By combining the results of risk analysis with detailed assessments of the planning stages of terrorist attacks, the intelligence process can provide directed intelligence products and refine future planning and direction of intelligence collection.

3 ANALYST'S CHALLENGES TO APPLYING RISK ANALYSIS PRODUCTIVELY

The productive application of risk analysis to support intelligence analysis must address four methodological challenges: (i) developing methods that can be supported with obtainable information; (ii) matching resolution of results with the problem; (iii) applying the best practices of risk analysis; and (iv) avoiding the potential for blinding analysts to the possibility of surprise.

3.1 Basing Analysis on Obtainable Information

While discussions of terrorism risk have received more attention recently, the methods of risk analysis are supported by decades of development and application, which include the study of risks of terrorism to critical infrastructure [7, 8]. This creates a strong methodological foundation on which to build and a pool of expertise on which to draw.

However, it also creates the potential that well-intentioned risk analysts could develop tools for which required input data are not obtainable in an effort to bring their capabilities to bear on the issue *de jure*. This problem of pushing tools in a manner for which they cannot be used can be avoided by asking and answering four questions at early stages of a risk analysis (Figure 3).

First, simply consider what data are needed. Asking this question initiates the process of considering what information is required and what information is available.

Second, determine whether the risk assessment requires additional data collection. In some cases, if the data are not already available, financial or time considerations will preclude going further. In other words, if the data are not obtainable, the analytic approach is dead on arrival. In other cases, learning that new data are needed initiates a process of refining future planning and direction of intelligence collection or developing innovative approaches to obtaining proxies for required data elements.

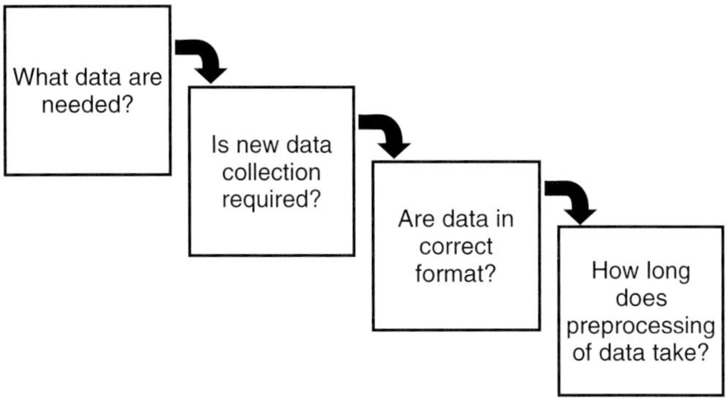

FIGURE 3 Considerations regarding availability of information to support risk analysis.

Third, analysts must consider whether data are in the correct format and resolution for the analysis. Typically, the data that have been already collected will not have been assembled with risk analysis in mind. The data could be in an incorrect format, could be incomplete because of missing data elements, or could also be internally incoherent because of conflicting reports or assessments of particular data elements. In any of these cases, the data may require cleaning and processing before they can be used in risk analysis. This is important because of the fourth question: How long does any required preprocessing take to complete?

It may be the case that new data are required and that new or existing data must be preprocessed before the data can be used in a risk analysis. However, the time required to do this affects how and whether the information can be used to support intelligence analysis. Tactical decision making is generally very time critical and allows little time for analysis. Strategic and operational decision making is generally less urgent. Ideally, all required information is immediately available and its collection and processing does not lengthen the time of the intelligence cycle. To the extent more time is required, the information may become less useful for supporting tactical operations where the value of intelligence information is measured in its ability to inform decisions that are made in a matter of hours if not a few days.

These four questions are all basic and seemingly second nature to analysts familiar with informing real decisions. However, if not carefully considered when developing or proposing applications of risk analysis, the results can be a process of little or no value.

3.2 Matching Resolution of Analysis to the Problem

There is no single risk assessment tool that fits the demands of all problems. Each problem has unique aspects that determine requirements for the spatial and temporal resolution of results [9].

Risk assessments intended to support design or performance assessment for security need to be tuned to a specific threat or target, but not necessarily to a specific time. For example, consider an assessment used to buttress physical security at a nuclear power plant. Designs need to be focused on specific types of attacks on specific parts of the plant, but it is much less important whether the attacks would occur this year or next.

Risk assessments for strategic planning need to be specific about what types of attacks could occur, but do not require specificity of when or where the attacks will occur because strategic assessments need only reflect the range of threats of terrorism. For example, think of an analysis to support the division of resources among control of nuclear proliferation and border security. Here the decision making does not require distinctions between specific places or which attack will happen first. It is important instead that the planning consider the correct range of attacks.

Finally, risk assessments to support tactical decisions require both spatial and temporal specificity. They will be used to help commanders decide what actions to take and when to take them. For example, consider an analysis that would help local law enforcement determine where and how many officers to deploy in security around a national political convention and when and where to supplement security in response to specific threats. Assessments of risks based on capabilities terrorists had last year are irrelevant if the way the group operates has changed dramatically.

Figure 4 presents the results of a comparative risk assessment that one user may see as having little value but which another could see as insightful. This figure presents an estimate of the distribution of the relative risk of terrorism across Manhattan in terms of casualty costs associated with workers' compensation claims following a terrorist attack. In this figure, darker shaded areas reflect regions of higher risk. A first order conclusion drawn from this figure is that terrorism risk is greatest in midtown Manhattan and the financial district of lower Manhattan.

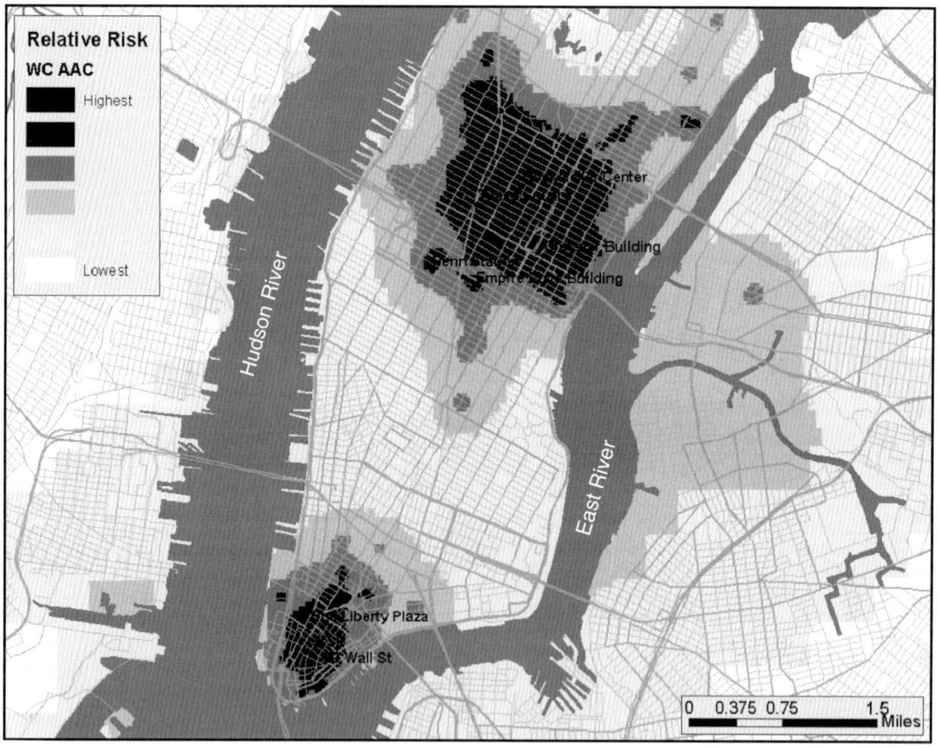

FIGURE 4 Graphical depiction of assessment of relative terrorism risk in Manhattan from the risk management solutions terrorism risk model expressed as workers' compensation (WC) average annual consequences (AAC). Source: Willis et al. (2006).

To the New York Police Department, such information is likely to have little value. The local community enters the challenge of protecting New York with a strong understanding of where the greatest vulnerabilities exist, and which events or locations represent particular value targets. For them, risk analysis must have a much sharper resolution to be useful. In the effort to prevent future terrorist events, local law enforcement requires help in determining which intersections to patrol, what questions to ask detainees to crack terrorist networks, and when to step up security because threats seem more imminent.

For state or federal officials, the analysis may have more value. These groups may not have the same entrenched knowledge of local vulnerabilities and targets. They may also be solving different problems. For example, federal officials are responsible for allocating resources to combat terrorism across the United States in proportion to terrorism risk [5]. Also, when faced with new threat information, federal officials may need to ascertain quickly for which communities the threat information is relevant [2]. In each of these cases, there is value in having the capability to access quickly or conduct studies of relative risks of terrorism across multiple cities using a common approach with consistent assumptions and data. In such cases, results like those presented in Figure 4 could be useful if accompanied by similar analyses for other communities across the nation.

3.3 Applying the Best Practices of Risk Analysis

The best practices of risk analysis recognize that risk is a social construct and that risk analysis requires an analytic and deliberative process [10]. For terrorism risk, these characteristics can be refined to provide further definition of a good assessment.

3.3.1 Analytic. An analytic terrorism risk assessment must address all three factors that determine terrorism risk: (i) threat, (ii) vulnerability, and (iii) consequence [2]. When feasible, this should be done quantitatively, using qualitative methods to fill in data gaps were necessary and appropriate. To the extent qualitative methods are used, risk analysis will be more useful for sorting high risks from low ones than for optimizing or fine tuning risk management strategies.

Risk assessments must be repeatable so all parties can replicate, analyze, and understand them. Risk assessment will require standard definitions within an analysis or methodology to ensure that results are consistent among analysts. It may be impossible to develop consistent definitions across all risk analyses. However, that is not necessary so long as consistency is applied within an analysis and when results are compared from one analysis to another. It is most important that the analysis is transparent such that definitions are clearly documented and appropriate for the problem on hand.

Because of uncertainties associated with terrorism risk, standard expected value decision-making tools that focus on the average or best estimate of risks may not be appropriate [8]. In particular, the most significant uncertainty surrounds assessment of terrorism threat. There is little consensus about when terrorists will attack next, how severe such an attack will be, and how quickly terrorist threats are evolving as terrorist groups attempt to obtain weapons of mass destruction and governments adopt tighter security. In light of this tremendous uncertainty, approaches that consider a very broad range of plausible threats may be necessary as well as adoption of decision support tools that help identify strategies that perform well across a wide spectrum of these plausible scenarios (see [11] as an example of such an approach).

3.3.2 ***Deliberative.*** A deliberative process is necessary because the notion of a cold, actuarial risk assessment is unrealistic. Although one might think risk analysis could be performed only on the basis of data about threat, vulnerability, and consequences, it is not possible to assess risks without considering individual values and judgments about risks and risk exposures. As a result, risk analyses must include deliberative processes that make it possible to take these judgments into account. A transparent analytic process, as outlined above, is necessary to support the deliberative process. This is the only way to address credibly trade-offs between risks to people from risks to property and risks from a conventional bomb, nuclear attack, biological attack, or even hurricane or other natural disaster. Applications of risk analysis to terrorism have to date focused more on the analytic components of risk than on the deliberative dimensions that require a difficult discussion of priorities and a judgment of which risks shall be tolerated.

3.4 Avoiding Blinding Analysts to Surprise

The strength of applying risk analysis to intelligence analysis is that it provides a set of tools for translating available information about terrorist motivations, terrorist capabilities, infrastructure vulnerabilities, and attack consequences into a set of common metrics that can be used to develop strategies to protect communities from attack effectively. However, this is also the root of one weakness of the approach.

The results of a risk analysis are bounded by the information and assumptions that go into it. Only those attacks that are envisioned will be assessed and only those targets that are considered relevant will be considered. As a result, the potential exists that risk analysis could lead the intelligence community to place too much attention on events that are presumed to be likely, thus only reinforcing prior beliefs about terrorist threats and risks and not revealing new insights or trends.

To counter this potential bias, it is necessary to state explicitly the principal assumptions built into terrorism risk assessments and homeland security plans, fully explore uncertainties around terrorism risk [5], adopt methods that allow analysts to consider the extent to which information they are assessing could explain alternatives that they are not considering [12], and consider institutional structures designed to prevent myopic policies with respect to the possibility of surprise [13].

4 SUMMARY

This article describes how methods of risk analysis can be integrated into the intelligence cycle used to produce terrorism warnings and threat assessments. This connection reveals two ways by which risk analysis can be of potential value to the intelligence community.

Risk analysis can be a tool that can help intelligence practitioners sharpen their conclusions by providing analytic support for identification of scenarios of greatest concern. Risk analysis can also be used to direct future collection efforts on information that appears to be most relevant to refining existing estimates of terrorism risks.

However, risk analyses must be conducted to meet challenges of information availability, matching resolution of results to the problem, reflecting risk as the social construct that it is, and not ignoring the possibility of surprise.

ACKNOWLEDGMENTS

This research was supported by the United States Department of Homeland Security through the Center for Risk and Economic Analysis of Terrorism Events (CREATE) under Contract N00014-05-0630 (Office of Naval Research). Any opinions, findings, conclusions, or recommendations in this document are those of the authors and do not necessarily reflect views of the United States Department of Homeland Security. I would also like to thank my colleagues Brian Jackson, Terrence Kelly, Don Kleinmuntz, Tom LaTourrette, Jack Riley, Errol Southers, Greg Treverton, Detlof von Winterfeldt, and Mike Wermuth at RAND and USC for their informal comments on my work in this research area.

REFERENCES

1. Krizan, L. (1999). *Intelligence Essentials for Everyone*, Occasional Paper Number Six. Joint Military Intelligence College, Washington, DC. Available online at http://www.dia.mil/college/pubs/pdf/8342.pdf (accessed February 5, 2007).
2. Willis, H. H., LaTourrette, T., Kelly, T. K., Hickey, S. C., and Neill, S. (2007). *Terrorism Risk Modeling for Intelligence Analysis and Infrastructure Protection*. TR-386-DHS. RAND Corporation, Santa Monica, CA.
3. Cragin, K., and Daly, S. (2004). *The Dynamic Terrorist Threat: An Assessment of Group Motivations and Capabilities in a Changing World*, MR-1782-AF. RAND Corporation, Santa Monica, CA.
4. Chalk, P., Hoffman, B., Reville, R., and Kasupski, A.-B. (2005). *Trends in Terrorism: Threats to the United States and the Future of the Terrorism Risk Insurance Act*, MG-393. RAND Corporation, Santa Monica, CA.
5. Willis, H. H., Morral, A. R., Kelly, T. K., and Medby, J. J. (2005). *Estimating Terrorism Risk*, MG-388-RC. RAND Corporation, Santa Monica, CA.
6. Rosoff, H., and von Winterfeldt, D. (2007). A risk and economic analysis of dirty bomb attacks on the ports of Los Angeles and Long Beach. *Risk Anal.* **27**(3), 533–546.
7. Garrick, J. B. (2002). Perspectives on the use of risk assessment to address terrorism. *Risk Anal.* **22**(3), 421–423.
8. Haimes, Y. Y. (2004). *Risk Modeling, Assessment, and Management*, 2nd ed. John Wiley & Sons, Hoboken, NJ.
9. Willis, H. H. (2005). *Analyzing Terrorism Risk*. Testimony presented before the House Homeland Security Committee, Subcommittee on Intelligence, Information Sharing, and Terrorism Risk Assessment, on November 17, 2005.
10. International Risk Governance Council (2005). *White Paper on Risk Governance: Towards an Integrative Approach*. International Risk Governance Council, Geneva. Available online at http://www.irgc.org/IMG/pdf/IRGC_WP_No_1_Risk_Governance_reprinted_version_pdf (accessed February 5, 2007).
11. Lempert, R., Popper, S. W., and Bankes, S. C. (2003). *Shaping the Next One Hundred Years: New Methods for Quantitative, Long-Term Policy Analysis*, MG-1626-RPC. RAND Corporation, Santa Monica, CA.
12. McGill, W., and Ayyub, B. (2006). Quantitative methods for terrorism warnings analysis. Presentation at the *Annual Meeting of the Society for Risk Analysis*, December 6, 2006, Baltimore, MD.
13. Posner, R. A. (2005). *Preventing Surprise Attacks: Intelligence Reform in the Wake of 9/11*. Rowman & Littlefield Publishers, Lanham, MD.

VULNERABILITY ASSESSMENT

DONALD E. NEALE
Department of Homeland Security, Washington, D.C.
ELIZABETH M. JACKSON
George Mason University School of Law, Arlington, Virginia
GLENN COGHLAN, CPP
Founder and CEO, Coghlan Associates, Inc., Springfield, Virginia

1 DEFINING TERMINOLOGY

The term *vulnerability assessment* is commonly misunderstood. Surveys and inspections are regularly labeled as vulnerability assessments, possibly due to the fact that surveys, inspections, and vulnerability assessments all have a security focus and use a checklist to maintain their scope. Although these terminologies are often used interchangeably, the following descriptions delineate their differences.

Survey. In general terms, a survey identifies current security posture and helps ascertain what countermeasures/protective measures exist or need to exist. The survey process is used to gather information about a site and its countermeasures/protective measures. The information is then analyzed in order to make recommendations for the security design of a site, whether an individual building, an entire complex with multiple buildings or facilities, or an event location.

Inspection. An inspection is conducted to verify compliance with pre-established security standards and procedures as described in policy and regulation in an effort to help maintain security posture. Inspection teams use checklists to compare a site's security posture to that outlined by an authoritative or regulatory body, the results of which are detailed in after-action reports provided to the responsible party.

Vulnerability assessment. A vulnerability assessment is an aspect or component of risk analysis. Essentially, it is the systematic examination of a security posture to identify vulnerabilities or gaps in security or supporting infrastructures that may allow an adversary to exploit a site's weaknesses. A vulnerability assessment also helps to determine the appropriate countermeasures/protective measures necessary to reduce risk.

A vulnerability assessment represents a holistic look at security. Its goal is to identify any overlooked details in a site's security posture; details may be disregarded due to underdeveloped procedures or due to lack of understanding of the capabilities of existing technologies or of an adversary to exploit vulnerabilities within security posture to compromise the site and surrounding population. Ensuring that a qualified team conducts a vulnerability assessment can present a broader perspective and greater knowledge of key security issues to a site's owner/operator. Assessment teams complete assessment reports that document a site's vulnerabilities and recommend specific countermeasures/protective measures to address those vulnerabilities.

A good analogy of a vulnerability assessment is the instance of a home buyer who examines the house he intends to purchase, but overlooks a popped nail on the roof partly holding down a shingle. During high winds and rain, the shingle may be knocked loose, leading to water damage inside the house. If the water damage is excessive, the house could be lost. Similarly, if a vulnerability is overlooked in an assessment, that vulnerability could be exploited and lead to catastrophic damage to the site and surrounding population, as well as significant cascading effects. A common maxim encapsulates this situation, "for want of a nail ... the kingdom was lost ... "—attributed to a Benjamin Franklin poem in 1757s *Poor Richard's Almanac*.

One example of exploited vulnerabilities is illustrated by the terrorist attacks of September 11, 2001. Terrorists are trained to find what has been overlooked, whether by surveillance or open source research; such training was apparent in these attacks. Terrorists identified a vulnerability at airport security checkpoints and exploited that vulnerability by defeating X-ray technology and its operators, thereby enabling them to successfully hijack airliners in an attack resulting in the deaths of thousands of Americans. Additional countermeasures/protective measures and training programs relative to airport screening were subsequently implemented, and a heightened tolerance for increased security screening was instilled in the American people.

2 VULNERABILITY ASSESSMENT METHODOLOGIES

Many hybrids of vulnerability assessments have been developed over the years. The majority of vulnerability assessment methodologies concentrate on one of three primary focal points: population protection, site and systems security effectiveness of high-value/high-risk assets, and mission/service survivability. While most assessment methodologies integrate these concentrations, outlined below, each will typically have a defined scope.

Population protection. Population protection is sought through the combination of security initiatives protecting the population both inside and around a site from attack. Using an integrated approach, population protection coordinates countermeasures/protective measures to mitigate the effects of an incident, whether it is the product of a natural or man-made disaster or terrorism. This process is illustrated in Figure 1.

Site and systems security effectiveness (high-value/high-risk assets). Site and systems security effectiveness implies a comprehensive vulnerability assessment of a layered security system within security postures. Utilizing "design basis threat", it dictates the evaluation of security system effectiveness to deter, detect, defend against, and respond to adversarial capabilities. Following evaluation, specific system vulnerabilities may be identified. Modeling and estimation of success rates are used to gauge the appropriate countermeasures/protective measures for implementation. This process is illustrated in Figure 2.

Mission/service survivability. Mission/service survivability centers on the protection, robustness, and assurance of mission or service continuation, as well as the protection of workforce essential to mission/service execution. It considers a full-threat,

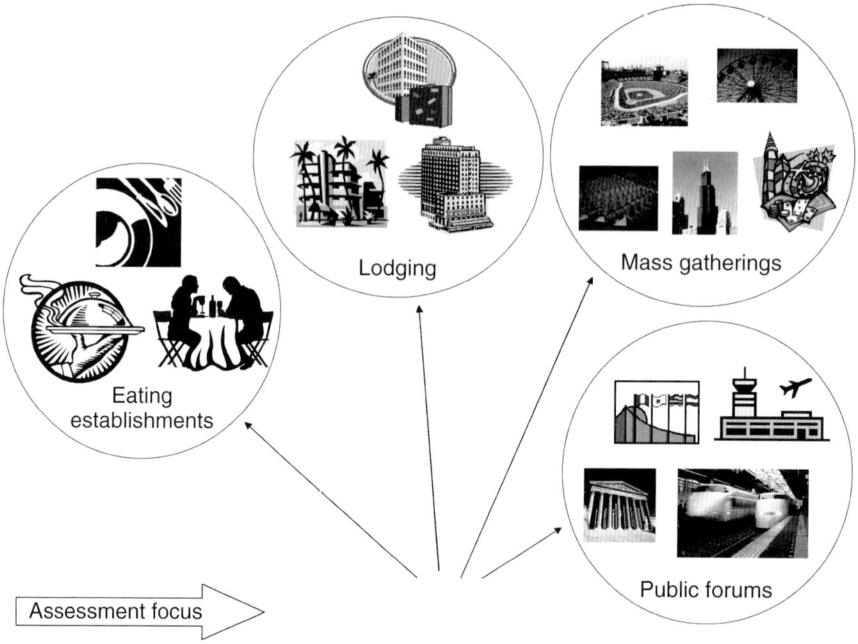

FIGURE 1 Population protection.

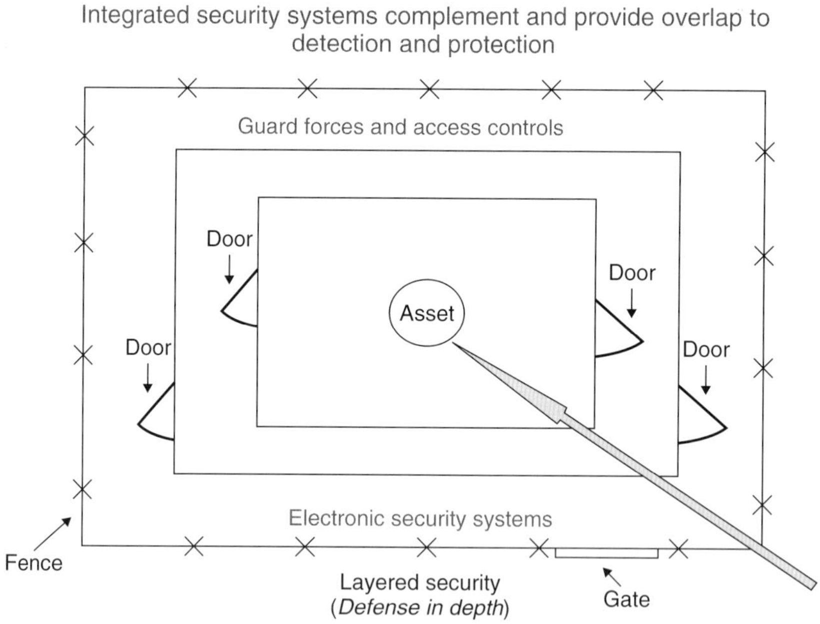

FIGURE 2 Site and systems security effectiveness.

all-hazards spectrum, looking at site criticality (internal) in conjunction with assessing the criticality of outside supporting infrastructures (external). In doing so, it:

- normally starts from the center of the site and moves to the fence line and beyond, to at least the first node of critical failure of any supporting infrastructure;
- produces suggestions to reduce strategic infrastructure vulnerabilities;
- identifies any cascading effects stemming from a systems failure;
- identifies interoperability, interdependencies, redundancies, and diversity of infrastructure, to include the resiliency of specific critical infrastructure;
- incorporates site experts/systems engineers and security specialists; and
- considers the operating parameters, realistic threats, and acceptable loss and funding of the site being assessed.

This process is illustrated in Figure 3.

There is no single methodology that adequately encapsulates all three concentrations in totality. However, no one methodology is better than another; the results of a vulnerability assessment are predominantly affected by the composition of the assessment team and the expertise of other individuals within the vulnerability assessment process. Given this, many methodologies attempt to employ the same skill sets. It must also be noted that vulnerability assessment methodologies regardless of their focus or scope, more often than not, use a similar common sense, systematic approach.

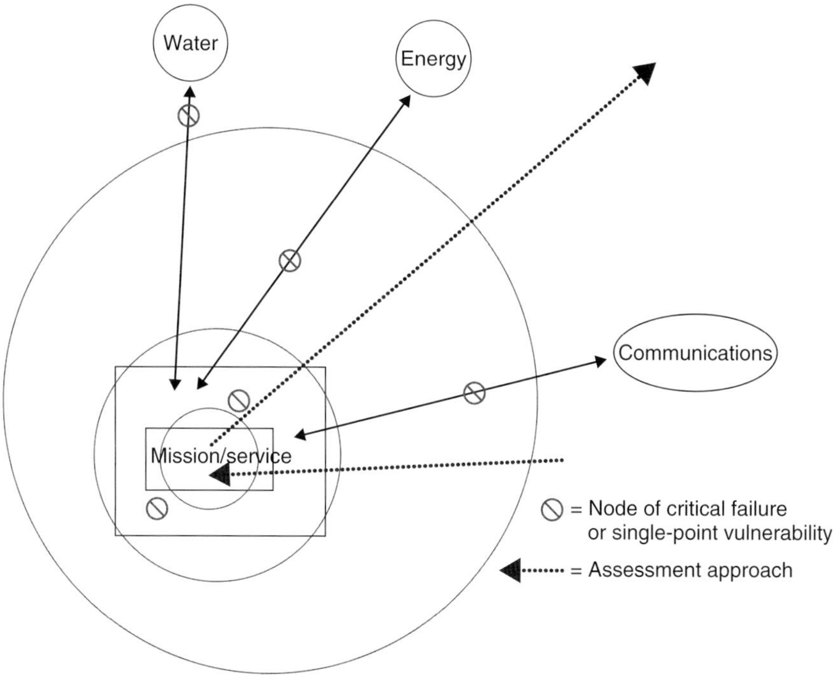

FIGURE 3 Mission/service survivability.

3 VULNERABILITY ASSESSMENT TEAM COMPOSITION

The focus of the assessment and a site's uniqueness will dictate specific vulnerability assessment team composition. The following list of selected protection elements and disciplines, each of which can encompass a vast array of specific concentrations or areas of expertise, may be considered in the makeup of an assessment team:

- Technical integration (Assessment Team Lead)
- Physical security
- Information operations
  - Cyber security
  - Information security
  - Operations security
- Personnel security
- Surveillance operations
- Site expert/systems engineer
- Communications
- Supervisory control and data acquisition (SCADA) systems
- Utilities
  - Energy
  - Water
- Structural engineering (building hardening)
- Emergency management and response
- Fire protection
- Restoration/continuity of operations
- Weapons of mass destruction
- Electromagnetic and radio frequency protection

The assessment team will follow a systematic vulnerability assessment process similar to that outlined below. The Assessment Team Lead will oversee and integrate the various aspects of the assessment, while each assessment team member will conduct a comprehensive assessment of his/her respective protection element or discipline. *Typically, these subject-matter experts will break down the assessed area into systems, subsystems, and components (or categories, topics, and subtopics) to obtain a clearer, more detailed picture of the vulnerabilities and gaps in overall security posture.*

4 VULNERABILITY ASSESSMENT PROCESS

The assessment process stresses a common sense approach to protecting and securing systems, components, and personnel necessary for the continued operation of a site's mission (or services) and is executed by the assessment team in three phases: pre-assessment, on-site assessment, and post-assessment. Established regulations, policies, and standards should neither be used as guiding principles nor reviewed for compliance as, in some instances, such guidance has been found to cause mission vulnerabilities. This being said, assessment team members must have a basic understanding of the regulatory requirements that a site's owner/operator uses when establishing site security guidelines.

The following explanation of the execution of an all hazards vulnerability assessment focusing on mission/service survivability can help the reader understand and recognize the basic framework of vulnerability assessments, as well as elements that can be tailored for a specific need. *The assessment described does not detail specific protection elements and disciplines considered during the assessment; rather, it depicts an overall systematic approach used by all methodologies*.

4.1 Phase I: Pre-assessment

Phase I generally consists of two primary sets of activities: customer (responsible party, owner/operator or otherwise) orientation and coordination and assessment planning. This phase normally begins weeks prior to Phase II (the actual on-site assessment) because of the complexities and unique requirements of each assessment. During this phase, assessment team members conduct the essential groundwork to underwrite assessment success.

Working with the assessment team according to standard operating procedures, the Assessment Team Lead will select a team task organized for the site's mission, and liaise with the appropriate parties to pass clearances, make lodging and travel arrangements, and settle other administrative matters. During Phase I, the Assessment Team Lead will coordinate with secondary assessment sites (external) as appropriate, relative to the support of the primary site (internal) and its mission. Unless restrictions are put in place by the customer, assessment team members should initiate contact with their customer counterparts as soon as the Assessment Team Lead obtains points of contact (POCs) for specific protection elements and disciplines.

During any assessment phase, assessment team members should notify the Assessment Team Lead of any significant problems or issues that arise, and provide recommended actions to remedy such problems or issues. The Assessment Team Lead shall provide guidance to the assessment team on submission deadlines and procedures.

4.1.1 Customer Orientation and Coordination Activities. Key customer orientation and coordination activities will be performed by the Assessment Team Lead or by assessment team members in close coordination with the Lead. These activities should include, but are not limited to, the following tasks:

1. Identify the site(s) to be assessed. For each site, identify:
   - its assigned mission or services it provides;
   - an initial list of other key sites or infrastructures (e.g. host site or other tenants) that provide vital support to the site to be assessed; and
   - the customer's goals and objectives for the assessment.
2. Obtain the site briefing and organizational chart.
3. Coordinate the scope of the assessment:
   - identify any constraints or limitations relative to the site (e.g. site boundaries, security classification, or special access restrictions);
   - ensure the customer understands that recommended countermeasures/protective measures will not solve all problems, meaning that there is no 100% guarantee for site protection; and

- ensure the customer understands that periodic assessment reviews and a risk management program should supplement any vulnerability assessments.

4. Identify the customer's perception of the primary threats to the site's ability to execute its assigned mission.
5. Coordinate the assessment's timeline with the site POC.
6. Conduct a site pre-visit.
   - Provide the customer, key personnel, and representatives of key supporting infrastructures with an assessment presentation that, at a minimum:
     - addresses customer goals, objectives, intended scope, envisioned timeline with major milestones or events, and any known constraints or limitations;
     - identifies or confirms specifics for milestones, such as the date, time, location, and attendees for the in-brief and out-brief;
     - provides an overview of the planned assessment methodology;
     - describes the assessment team's functional composition and POCs for specific protection elements and disciplines, and provides team contact information;
     - details requirements for the assessed site to coordinate with other host, tenant, or local supporting sites or infrastructures;
     - describes assessment team Phase I and II information and documentation requirements for each protection element and discipline;
     - emphasizes that information and documentation is required for the assessment team to effectively conduct its assessment, and that the team's timely access to required documentation is critical to assessment thoroughness and efficiency; and
     - describes assessment team on-site administrative and logistics support requirements, such as office space, parking, security containers, telephone access, information on personnel POCs and POC contact information (e.g. telephone numbers and e-mail addresses), and any other support issues.

4.1.2 Assessment Planning Activities. Primary assessment planning activities that take place during Phase I include, but are not limited to, the following tasks:

1. Conduct an initial review and analysis of the site's mission and supporting documents, as well as other information obtained during Phase I, to identify vulnerabilities and to focus Phase II efforts. Site documents and information for review include:
   - briefing and organizational chart;
   - critical mission and mission support functions;
   - relative priorities or importance over time;
   - interrelationships and interdependencies;
   - logical, physical, and functional network and system diagrams, blueprints, or user manuals;
   - concept of operations (CONOPs);
   - previous vulnerability, threat, or risk assessment results and/or reports;

- previous exercise after-action reports;
- training program curricula, standard operating procedures, and relevant policies or directives to include applicable laws and regulations;
- continuity of operations, contingency, disaster recovery, and backup plans or procedures; and
- details on sites or infrastructures that support critical mission and mission support functions (e.g. SCADA, telephone/fiber-optic line connectivity, and utilities).

2. Particularly for large, complex sites, develop an assessment timeline with major milestones or events for assessment Phase II activities to maximize the efficiency of on-site activities and minimize the impact on the site and its personnel.
3. Conduct open source research for information to characterize the site being assessed, critical mission and mission support functions, and supporting infrastructures.
4. Obtain preliminary, relevant information on:
   - the key site personnel perceptions of the primary threats to the site's critical mission and mission support functions;
   - the type of threat: insider (access, knowledge, and privileges), hacker (novice or professional), organized crime or terrorist, foreign or industrial intelligence, weapon of mass destruction, and etc.; and
   - anticipated threat tactics: social engineering or collusion, stealth, brute force, physical attack or damage, open source research, third-party usage, and etc.
5. Identify the site's POC for each assessment protection element or discipline.
6. Contact site POCs and set up tentative interview schedules (dates, times, and locations). Doing this prior to Phase II is a courtesy to the site personnel in that they can schedule the interviews around other commitments, minimize impact on their jobs, and ensure that key personnel are appropriately involved.
7. Conduct intra-assessment team coordination to integrate efforts, thus minimizing impact on on-site personnel.

4.2 Phase II: On-Site Assessment

This phase consists of all on-site activities, such as site personnel interviews, site tours, data collection and analyses, and observations to develop and verify ground truth about the site's mission survivability. Specific assessment Phase II activities include, but are not limited to, the following tasks:

- Document mission and mission support survivability activities and security posture through observation of systems, assets, or personnel in action; interviews; analyses of documentation; and assessment tool and/or modeling usage, as authorized.
- Identify, obtain, review, and analyze additional, relevant customer information and documentation.
- Work with intelligence agencies and local law enforcement to discuss potential and known threats directed toward the site.
- Observe demonstrations and/or exercises related to security and survivability.

4.2.1 ***Assessment In-brief.*** The assessment team will normally meet its key personnel counterparts, those with knowledge in the specific protection elements and disciplines considered, at the assessment in-brief. Similar to the pre-visit conducted by the Assessment Team Lead, the assessment in-brief should, at a minimum, provide the customer and key personnel with:

- identification or confirmation of specifics for major milestones or events, such as the date, time, location, and attendees for the assessment out-brief;
- an overview of the assessment methodology;
- a description of the assessment team's functional area composition and protection element and discipline POCs;
- requirements for the site to coordinate with other host, tenant, or local support sites and infrastructures;
- an overview of information and documentation requirements; and
- requirements for site authorization documentation, such as that needed for photography, special badges, or access.

4.2.2 ***Site Tours.*** After the in-brief, assessment team members typically receive a group tour of the site. This initial tour should be regarded as a site overview. It is appropriate for team members to ask general questions; however, detailed questions regarding specific protection elements and disciplines are best noted and asked later. Assessment team members should schedule detailed follow-up tours of site and supporting infrastructures of direct interest as necessary.

4.2.3 ***Interviews.*** Ideally, site POCs are obtained during the site pre-visit, and initial interviews are scheduled before Phase II begins. If not, scheduling interviews with key personnel should be a priority for the assessment team once it arrives on site.

The objective of the interviews is to learn the reality of how operations occur, capturing information on both official and unofficial procedures and processes used. Where applicable, data gained via this means should be compared with official policies, directives, or standard operating procedures.

Key considerations to conducting successful interviews are as follows:

- Ask questions to learn, not to interrogate.
- Be prepared, focused, and organized (with core questions written or thought out in advance).
- Maintain control of the interview.
- Use two assessment team members, when possible: one to focus on questioning and the other to focus on note taking.

Questions to consider asking during interviews include the following:

- What are the cascading effects resulting from loss or damage to the site's critical mission and mission support functions?
- What redundancies exist if the site's critical mission and mission support functions are lost or damaged?
- What are your worst fears?
- What are your perceived vulnerabilities and penetration points?

- What are your recommendations for the customer to improve the security posture of the site?
- What do you consider critical, relative to your specific discipline(s), and why?

4.2.4 Data Collection and Analysis. The assessment team should seek to obtain, review, and analyze all available documentation and information on the site's critical mission and mission support functions, processes, and procedures. Ideally, this data collection and analysis will begin during Phase I.

The assessment team should characterize critical mission and mission support functions and supporting infrastructures to identify vulnerabilities. This includes evaluating and assessing security posture and vulnerabilities, with emphasis on identifying single-point vulnerabilities, for the impact on the site's mission, specifically if systems are lost or degraded.

The assessment team should assess the impact of the loss or degradation of each critical mission and mission support function, its critical components, and its critical dependencies. Specific issues to consider include the following:

- What actions, components, factors, or assumptions are required for critical mission accomplishment or success?
- What are the high-value nodes, components, or systems?
- What actions, components, or factors could result in loss of life or personnel injuries?
- What actions, components, or factors could result in damage to property or equipment?
- What actions, components, or factors could present a potential liability to the site?
- What actions, components, or factors could result in mission failure, disruption, delay, or denial?
- What amount of time is required to recover from an incident?
- What redundancies or backup capabilities exist to ensure mission survivability?
- What contingency and continuity of operations plans exist? How often are they reviewed, exercised, or tested?

4.2.5 Assessment Out-brief. At the conclusion of the assessment team site visit, the team should present an out-brief to the customer and key personnel. The out-brief should be focused on and address all single-point and major vulnerabilities, to include any interdependencies between them. The assessment out-brief should not attempt to address all minor vulnerabilities or issues that will be contained in the assessment final written report. It should reflect validated, coordinated, and deconflicted information regarding specific protection elements and disciplines. Other areas for consideration include the following:

- What are the cascading effects resulting from loss or damage to the site's critical mission and mission support functions?
- What redundancies exist if the site's critical mission and mission support functions are lost or damaged?
- What are the current protection strategies and plans?
- What are the gaps in current protection strategies and plans?
- Do any single-point or major vulnerabilities exist? If so, what are they? Do interdependencies between them exist? If so, what are they?

- What can be done to mitigate existing vulnerabilities through improved policies, resources (personnel, equipment, training, and funding), procedures, or processes in either the near or long term?

4.3 Phase III: Post-assessment Activities

This phase includes final report writing, delivering the report to the customer, and conducting follow-up actions as needed to complete the assessment process. This phase is highly intensive, time consuming, and requires extensive collaboration among assessment team members.

5 KEY CONSIDERATIONS FOR VULNERABILITY ASSESSMENTS

The aforementioned descriptions of vulnerability assessment methodology focal points, systematic approach, and assessment execution touch upon the fundamental components of vulnerability assessments. Nonetheless, additional considerations must be taken into account when conducting a vulnerability assessment. These considerations may drive funding for countermeasures/protective measures, as well as overall security posture. They include, but are not limited to, the following:

- Political and social tolerance for loss.
  - For example, the common, prudent person on the street would favor spending the money necessary to protect nuclear weapons from theft. Due to the zero tolerance for loss of these weapons systems, more money will be spent on security and the implementation of stringent countermeasures/protective measures.
- Costs versus benefits of added countermeasures/protective measures.
  - Recommended countermeasures/protective measures must be commensurate with risk. For instance, is the value of a site less than the cost of a countermeasure/protective measure with limited benefits?
- Agreement of safety issues and security.
  - Recommended countermeasures/protective measures must work in conjunction with life safety issues. For instance, do the recommended countermeasures/protective measures comply with fire safety regulations?

6 CONCLUSION

Vulnerability assessments represent systematic approaches to identifying weaknesses in security postures, procedures, or infrastructure configuration. With no one methodology better than another, vulnerability assessments are distinguished by their focus—population protection, security system effectiveness, or mission survivability. Importantly, the results of a vulnerability assessment are affected, either positively or negatively, by the expertise of those conducting the assessment. Institutions are attempting to develop a single vulnerability assessment methodology that ensures consistent results. However, this feat is virtually impossible due to the subjectivity involved in assessment execution and report development. Rather than trying to develop such a methodology, efforts should be focused on standardizing expertise criterion and required background experience for assessment team members in every protection element and discipline. Experienced vulnerability assessment professionals can coalesce assessment concentrations and apply tailored assessment approaches based on specific needs to enhance the assessment and the assessment team's findings. Vulnerability assessment methodologies are consistently based on the same framework, using the same approach; the human element is where strengths or weaknesses lie.

END NOTE

The views expressed herein are those of the authors and may not represent the views of their respective organizations.

FURTHER READING

Broder, J. F. (2000). *Risk Analysis and the Security Survey*, 2nd ed., Butterworth-Heinemann, Boston.

Center for Chemical Process Safety. (2003). *Guidelines for Analyzing and Managing the Security Vulnerabilities of Fixed Chemical Sites*, American Institute of Chemical Engineers, New York.

Cumming, N. (1992). *Security: A Guide to Security System Design and Equipment Selection and Installation*, 2nd ed., Butterworth-Heinemann, Boston.

Garcia, M. L. (2001). *The Design and Evaluation of Physical Protection Systems*. Butterworth-Heinemann, Boston.

Ramo, S., St. Clair, R. K. (1998). *The Systems Approach: Fresh Solutions to Complex Civil Problems Through Combining Science and Practical Common Sense*, KNI, Incorporated, Anaheim.

RISK COMMUNICATION

Monique Mitchell Turner

Center for Risk Communication Research, Department of Communication, University of Maryland, College Park, Maryland

Shawn S. Turner

United States Marine Corps, Arlington, Virginia

1 INTRODUCTION

In 1987, University of Oregon Psychology Professor, Paul Slovic, wrote "in recent decades, the profound development of chemical and nuclear technologies has been accompanied by the potential to cause catastrophic and long-lasting damage to the earth and the life forms that inhabit it. The mechanisms underlying these complex technologies are unfamiliar and incomprehensible to most citizens" (p. 280). Slovic's words are still poignant. The recent rash of natural and man-made disasters and the ever present threat of terrorist attack abroad and at home demonstrate the ever increasing need for organizations charged with communicating risk to be more proactive in their planning, more strategic in their approach, and more educated about risk communication. Social scientists and communication professionals have long studied how shifts in the ease of availability of information coupled with the credibility of the source and the emotional state of the receiver impact individual behaviors and perceptions of events. Unfortunately, findings from this research are often slow to reach frontline practitioners. For organizations involved in communicating risks to publics, these shifts are particularly significant because they are part and parcel to a dynamic communication environment in which our understanding of what constitutes truly effective communication is rapidly evolving. When risks present themselves, communication professionals face formidable challenges in collecting and assessing vital information, designing effective communication strategies, and assessing public's response. In this article, we will present some of the challenges that risk communicators face, best practices for risk communicators, and promising directions for the future of risk communication scholarship.

2 CHALLENGES TO EFFECTIVELY COMMUNICATING RISK

Risk communication is a challenging enterprise. Many of those challenges are in no small part due to the lack of clear understanding regarding the nature of risk, particularly as it relates to scientific innovations and public safety. Elucidating the difficulties associated with risk communication starts with understanding the word "risk". To us, risk is the probability of negative consequences of a hazard occurring and the magnitude of those consequences. So, risk deals with the severity of any given threat (i.e. hazard and risk). But, it is also tied to susceptibility to that threat [1]. Moreover, it is important that we note we are talking about *perceived* risk. Certainly, there are risk assessors who can calculate our risk based on a series of factors—but, humans rarely perceive an objective amount of risk. In fact, there are often huge discrepancies between "real" risk and "perceived risk" [2]. Hence, the notion of uncertainty, pros and cons, as well as doubt and ambiguity are inherent in the term risk. Therefore, when communicators talk about risk, they have to recognize the difficulty of the decisions people must make. We must be mindful of the fact that an outcome cannot be guaranteed. There is no such thing as zero risk and there are numerous psychological and emotional factors affecting the way key audiences perceive risk. We highlight some of the specific challenges in communicating risk.

2.1 Uncertainty

The notion that science and technology is constantly evolving, coupled with the dynamic nature of public security threats, creates and compounds risk related uncertainty. Lay audiences are often unclear about the risks associated with new innovations and overwhelmed by the amount and complexity of public safety information. Misinterpretations and seemingly contradictory assessments of scientific findings can lead to both over- and underreactions, and create feelings of anger and resentment toward the communicating organization. Moreover, the act of communicating uncertain information, that is, communicating information on topics about which the full implications are unknown, represents unique challenges for practitioners. A real-world example of this phenomenon can be seen in public's perception of the risk associated with the avian bird flu. According to the Centers for Disease Control, the risk from avian influenza is generally low to most people, because the viruses do not usually infect humans. But, a study released at the World Economic forum entitled [3] claimed that avian bird flu was a more significant threat than terrorism and that, given the right circumstances, could kill between 40 and 50 million people worldwide. Such contradictory information highlights the complexity of the risk, and will also lead to a great deal of frustration and doubt on the part of the public. The goal is to reduce uncertainty by communicating in a confident, credible, and assertive manner while ensuring that key audiences respond appropriately to new and evolving information.

2.2 Experts and Consumers Define Risk Differently

Slovic [2, 4] reviewed the biases that laypeople employ in order to make risk assessments. He examined the complex and subtle meanings that people have when they say that something is "risky". To do so, he examined past psychometric data of particular risks with the goal of developing a taxonomic scheme from which we can understand the distinctions among risks. A factor analysis indicated that most risks break down into two continuous factors: dread risk and unknown risk. High dread risk is defined by a perceived lack of control, dread, catastrophic potential, fatal consequences, high risk to future generations, not easily reduced, involuntary, increasing in riskiness, and the inequitable distribution of risks and benefits. Nuclear weapons, nerve gas incidents, nuclear weapons fallout, and radioactive waste among others scored high on dread risk. Unknown risk, at its high end, is defined by hazards that are unobservable, unknown to those exposed, unknown to scientists, new, and delayed in their manifestation of harm. Chemical technologies score high on unknown risk as did cadmium usage and radioactive waste.

These data have been used to predict the kinds of risk that will lead to public outcry. Risks that are perceived to be high on dread risk are perceived to be bigger threats. Risks that are high in both dread and unknown factors will be the most likely to garner attention by the public. One way that experts have attempted to deal with public's rising concerns with particular risks is to educate and inform people about the true nature of the risk. For example, although experts ranked nuclear power twentieth out of 30 potential risks, college students and the league of women voters (examples of lay groups) rated it first; revealing the discrepancy in knowledge about the risk among experts and publics. Yet, simply communicating facts and figures about risks will not decrease public's concerns.

2.3 Use of Heuristics in Decision Making

People often use heuristics, simple decision-making tactics, when making decisions that revolve around risk. Often, use of these heuristics can lead to uninformed decisions. Risk decisions are often affected by the availability heuristic. Risks that are more memorable or readily accessible are believed to be riskier. Unfortunately, if the media catch hold of an especially interesting or controversial risk, the story is likely to gain attention of the nation. Media attention can lead to increased memorability on the part of receivers—exacerbating the availability heuristic. Problematically, the availability heuristic leads people to overestimate the frequency of rare events [5, 6].

How a risk is framed will also affect the decisions that people make [6, 7]. Describing the costs of accepting a risk (i.e. a negative frame) will lead to different decisions than will describing the benefits of rejecting a risk (i.e. a positive frame). Research indicates that when an event is communicated in terms of its risks, a negative frame will lead to more attitude change in the direction of the message, but when that same event is communicated in terms of prevention, a positive frame will be more effective.

Extant data also illustrates the omission bias, which reveals that people believe that no action is less harmful than action [8, 9]. For example, parents who are concerned about the risks of vaccination are more likely to withhold vaccinating their children because they believe that their action will be more harmful than no action.

3 BEST PRACTICES IN RISK COMMUNICATION

Given the potential negative repercussions of unskilled risk communication, it is important that practitioners be well trained and capable of developing strategies that are informed by relevant communication research. Often, communicators believe that more risk communication equates to better risk communication. However, it is quality and not quantity that lays the foundation for effective risk communication. A significant body of research has led to the development of several risk communication best practices. These practices are reviewed here and some advice on implementing these practices in your organization communication plans are also provided.

3.1 Risk and Crisis Communication is an Ongoing Process

Given the uncertainties that are inextricably tied to risk, it is imperative that communicators do their job in communicating the evolving nature of science to community members. Communicators should be clear that they are using the most up-to-date science in their assessments, and as they get more information they will update the public. This way, when updates do arise people will not mistake this as "the so-called experts have no idea what they are talking about". When possible, communicate with the public about why uncertainties exist, that you are collaborating with teams of scientists who are still working, and/or the difficulties in establishing a certain outcome. It is important to be clear about when updates will be communicated and how they will be communicated. People are more accepting of the evolving nature of risk when communicators tell them to expect it.

3.2 Listen to Public's Concerns and Understand Your Audience

The discrepancy between true, objective risk and perceived risk is clear. But, until communicators take the time to understand the community members experiencing the risk, we will not know if their perceived risk is exaggerated, too low, or accurate.

When perceived risk is exaggerated the goal is to calm the audience down. Communicating empathetically is an important skill in this regard and is discussed later. The way that people act when their risk is inflated depends upon the emotion they are experiencing [10]. Turner's [11] Anger Activism Model suggests that when people are angry about an issue and they feel they can do something to ameliorate (what they think is) the problem, they are likely to begin engaging in activist behaviors. This can be troublesome for some organizations if consumers begin protesting, picketing, or engaging in a sit-in. In such cases, the communicators must discuss the risk in ways that address audiences' concerns.

When our perceived risk is too low, people are unlikely to engage in precaution behaviors (e.g. emergency preparedness). In such situations, it is imperative to communicate about the severity of the problem and audience's susceptibility to encounter the risk. Fear appeal theory [1, 12] reveals that when people perceive high severity and high susceptibility, they will perceive an imminent threat in their environment. If the audience feel efficacious in their ability to handle the issue (efficacy is discussed in a subsequent section), then they will engage in preventative behaviors.

Hence, in high stress situations, people must perceive high trust and credibility and they must feel efficacious. But, we also understand that the anxiety created by stress can lead to deficiencies in information processing [13]. To help audiences understand the issue, risk communicators have to create targeted messages. Using clear, nontechnical language is advisable when you are communicating about the nature, severity, or magnitude of risk [15]. The US Department of Health and Human Services provides the following tips:

- Use consistent names, denominators, and other terms throughout the risk (or crisis) situation. For example, do not switch from person per million to person per billion. This may throw people off, and they might not notice that you changed your unit of analysis.
- Use graphics and pictures to help the audience understand the risk.
- Avoid jargon and acronyms.
- Answer not only the question "how much?" but also "will it hurt me?" to ensure that people get relevant information.
- Use familiar frames (or metaphors) to explain how much, how small, or how many. Try to create a mental picture for the audience.

3.3 Demonstrate Honesty, Candor, and Openness

Covello's [14] research indicates that during low stress situations, public's perception of trust in communicator is based on their perceived level of expertise. So, in low stress situations people will look to someone's competencies, job title, education, and so on. However, in high stress situations, people base their trust on perceptions of listening, caring, and empathy.

Communicating trust and credibility is the first and foremost important skill of the risk communicator. Decades of research in source credibility help lay the groundwork for what makes a communicator appear credible [16]. Credibility is made up of multiple components, in particular, trust, dynamism, and expertise. Dynamism is not covered in this article, but researchers have studied what people think of when they use the word "*expert*". This research indicates that experts are those who have or communicate the following:

Training. Experts have an advanced knowledge and/or degrees in the area being spoken about.

Skill. Experts have specialized skills.

Informed. Experts stay up to date on advanced research and are well informed on the current information about his/her topic.

Authoritative. Experts speak with authority. Experts act assured in their knowledge.

Ability. Experts have the ability to take action.

Intelligence. Experts have generalized intelligence.

Studies show that in low-trust, high-concern situations, credibility is assessed using these four measures: empathy and/or caring (50%, assessed in the first 30 s), competence and expertise (15–20%), honesty and openness (15–20%), and commitment and dedication (15–20%). Nonverbal cues have also been documented to impact trust and credibility [14].

Often, risk communication teams will have their scientists as the public spokesperson during times of risk and crisis because the public believe that a scientist knows what is going on. In fact, successful risk communication does require expertise in conveying understandable and usable information to all interested parties. Risk managers and technical experts may not have the time or skill to do all of the complex risk communication jobs such as responding to the media, the public, industry, and so on. People with real expertise should be involved as early as possible. This expertise was likely developed by training and experience.

Trustworthiness is the second dimension of credibility. It is truly not enough to be an expert—who cares if you are an expert if people do not believe in your proclivity to tell the truth? Also, when an issue is personal, frightening, or serious, we look for trustworthy speakers. The risk communication literature identifies the following three factors that determine whether or not the public will perceive a communicator as trustworthy:

- empathy and caring,
- honesty and openness,
- dedication and commitment.

When communicators make certain to relate their message to audience's perspectives, emphasizing relevant information to any practical steps that they can take, the message will be more effective. Use clear and plain language. Make sure that you clearly state the existence of uncertainty, and avoid trivializing the concern. Covello et al. [18, 19] also offer the following guidelines:

- be balanced;
- focus on a specific issue;
- pay attention to what the audience already knows;
- be respectful in tone and recognize that people have legitimate feelings and thoughts;
- be honest about the limits of scientific knowledge;
- consider and address the broader social dynamics in which risks are embedded;
- be subjected to careful evaluation;

The single most important determinant of gain or loss of trust is whether the communicator is subsequently proven to be correct or incorrect, and that the communicator is demonstrated to be unbiased. Trust is contextual; that is, whether you are seen as trustworthy may depend on the audience. Take into consideration whether the audience is made up of scientists, members of the military, or government employees, for example.

Speakers should always act calmly. Do not hesitate when speaking or appear nervous. Do not give the impression that you are holding back information, or that you do not like what you are saying. If you are a nervous speaker, you should practice speaking. Be willing to admit that you do not know everything. If someone asks you a question that you do not know the answer to, say so. Or, you may say "*I do not know, but I will find out the answer to that*". When you are honest about not knowing something, people will assume that you are honest about other things as well. And, it appears that you are not simply trying to look good.

- Trust is linked with perceptions of accuracy and expertise—these qualities work in concert.
- Admit to uncertainty. Facilitate public understanding that risk science is a process.
- Be forthcoming with information and involve the public from the outset.
- Avoid secret meetings.

3.4 Communicate with Compassion, Concern, and Empathy

Select a messenger who can really connect with the audience and help them understand that the risk affects their lives personally. This is known as *creating outcome involvement* [19]. Outcome involvement refers to how much the audiences feel that the issue at hand will directly affect their lives. Provided that message, receivers are cognitively able to process information; the more involved they feel, the more they will pay attention to the arguments presented. An audience who feels involved is a thinking audience. The communicator should ask the community what their concerns are. When speaking to the community, use opening remarks to solicit their concern and issues. Address those issues with sincerity. Here are some other practical pieces of advice:

- Stay late after your talk. Show the audience that you are there to answer their questions. This communicates your commitment to their concerns. This principle holds for showing up early too.
- If you make a promise, keep it.
- Provide contact information—give your audience your phone number and/or e-mail address.
- Listen to what various groups have to tell you.

3.5 Provide Messages that Foster Efficacy

One of the most common mistakes that risk communicators make is limiting their communication to the susceptibility and severity of the risk. For example, we often communicate that the audience is at high risk, but take no other steps. When people perceive high risk, they are likely to become fearful and anxious. Although this is a natural and effective human response to threat, when people are afraid the "fight or flight" mode is activated.

Bandura [20–22] argued that perceptions of self-efficacy influence thought patterns, actions, and emotional arousal. Self-efficacy refers to the perception that one has the personal capability to do the things necessary to avert the threat. Bandura also noted, "perceived self-efficacy helps to account for such diverse phenomena as changes in coping behavior produced by different modes of influence" (p. 122). But, self-efficacy is only one part of this picture; audiences must also perceive that the recommendations you are giving to them will work to avert the threat. This latter issue is known as *creating response efficacy*. When individuals have both response and self-efficacy, they will be able to cognitively process information about the risk and engage in "danger control". Danger control refers to behaviors that work at preventing the threat versus simply avoiding the threat.

3.6 Conduct Precrisis Planning

Deciding when to communicate is also an important part of strategic risk communication. Scherer [23] describes two types of strategies:

1. Reactive
2. Proactive.

Reactive strategies essentially describe communication that is in reaction to an event or occurrence. This type of strategy does not call attention to a particular risk, but waits until there is already considerable public and media attention about a risk issue.

Advantage
- allows the public to vent about the issue.

Disadvantages
- science may be less relevant when issues become highly emotionally charged;
- places communicator in defensive position;
- people may not believe information that is delayed;
- people may unknowingly be exposed to risk.

Proactive strategies represent a more ongoing risk communication effort. Rather than waiting until an event happens or is subsequently discovered by the public, this type of strategy calls attention to a risk issue, both potential and existing; suggests the agenda for discussion; and provides mechanisms for information exchange.

Advantages
- may alert people to something of which they are not aware;
- allows for a much more meaningful discussion of risk;
- generates more balanced discussion about risk.

Disadvantage
- may alert people to something of which they are not aware.

4 CRITICAL NEEDS ANALYSIS AND RESEARCH DIRECTIONS

There is much still to be discovered about risk communication.

4.1 Communicating Uncertainty

It was made clear in this article that the biggest difficulty with communicating risk is the uncertainty involved. Data does indicate what does not work well: using ambiguous terms such as "probably", "remote", or "almost certainly" [24]. But, it is less clear *how to* communicate uncertain data. In particular, the importance of considering your audience and messenger cannot be overstated. Fessenden-Raden et al. [25, p. 100] wrote,

> No matter how accurate it is, risk information may be misperceived or rejected if those who give information are unaware of the complex, interactive nature of risk communication and the various factors affecting the reception of the risk message.

Researchers must begin to focus on better ways to communicate uncertainty without frustrating the audience. We suggested earlier using graphic or visual evidence, but even here knowledge is limited. What kinds of visuals are most effective? Is effectiveness moderating by personality factors such as learning styles?

4.2 Emotion and Risk

Emotion plays an important role in risk perception. Emotions prompt responses, which facilitate individuals' ability to deal quickly with problems in the environment [10, 26]. And, negative emotions (e.g. anger, fear, and guilt) cause distinct cognitive patterns. Fischhoff et al. [27] found that relative to fear, anger activated optimistic perceptions of terrorism. Lerner et al. [28] found that compared to neutral feelings, anger activated more punitive attributions and more heuristic processing. Turner et al. [13] found that when people perceive that a threat is high (high severity and high susceptibility), their anxiety related feelings increased. This increase in anxiety motivated people to engage in information seeking about the risk, but decreased their ability to correctly process the information. Despite this understanding, much is still unknown. For example, most of the attention has been paid on anger and fear. Yet, guilt is a relevant emotion surrounding risk. Pregnant women, for example, may experience a great deal of anticipated guilt if they are led to believe that a future risk decision could harm their unborn child. Parents generally may feel guilty if they believe their decisions affected their children.

Also, our risk communication can induce emotional responses. There is a great deal of research on fear appeals [1, 12], but little is known about other kinds of emotion appeals such as guilt, anger, disgust, or even hope appeals. Turner [11] has focused attention on anger messages and has found that when the anger appeal is proattitudinal and communicates efficacy, persuasiveness linearly increases. But, we know relatively little about how emotional appeals can be used to motivate emergency preparedness and risk prevention. In situations such as preparing for a terrorist attack, which is emotion laden to begin with, what emotions might be used to motivate people?

5 CONCLUSION

Communicating risk is filled with opportunities. When risk is correctly communicated, we have the ability to help people make informed and effective decisions. Effective communication helps ensure that people understand the positive and negative aspects of risk, and that they can process that information without negative emotions overloading their cognitive capacity. We continue to live in an increasingly dangerous world where national security has become an important focus of our attention. Scientists continue to develop new technologies to thwart such threats, but these technologies lessen some risks and create new risks. Consumers must understand, weigh pros and cons, and make effective decisions about how they deal with such things. To do so, they must be communicated with appropriately and sensitively.

REFERENCES

1. Witte, K. (1992). Putting the fear back in to fear appeals: the extended parallel process model. *Commun. Monogr*. **59**, 329–349.
2. Slovic, P. (1987). Perception of risk. *Science* **236**, 280–285.
3. Global Risks (2006). (n.d.) *A world economic forum report*. Retrieved October 18, 2008, from http://www.weforum.org/pdf/CSI/Global_Risk_Report.pdf.
4. Slovic, P. (1992). Perceptions of risk: reflections on the psychometric paradigm. In *Risk Communication*, J. C. Davies, V. Covello, and F. Allen, Eds. The Conservation Foundation, Washington, DC.
5. Carroll, J. S. (1978). The effect of imagining an event on expectations for the event: an interpretation in terms of the availability heuristic. *J. Exp. Soc. Psychol*. **14**, 88–96.
6. Tversky, A., and Kahneman, D. (1974). Judgment under uncertainty: heuristics and biases. *Science* **185**, 1124–1131.
7. Tversky, A., and Kahneman, D. (1981). The framing of decisions and the psychology of choice. *Science* **211**, 453–458.
8. Ritov, I., and Baron, J. (1990). Reluctance to vaccinate: omission bias and ambiguity. *J. Behav. Decis. Mak.* **3**, 263–277.
9. Spranca, M. D., Minsk, E., and Baron, J. (1991). Omission and commission in judgment and choice. *J. Exp. Soc. Psychol.* **27**, 76–105.
10. Lerner, J. S., and Tiedens, L. Z. (2006). Portrait of the angry decision maker: how appraisal tendencies shape anger's influence on cognition. *J. Behav. Decis. Mak.* **19**, 115–137.
11. Turner, M. M. (2007). Using emotion to prevent risky behavior: the anger activism model. *Public Relat. Rev.* **33**, 114–119.
12. Witte, K. (1994). Fear control and danger control: a test of the extended parallel process model. *Commun. Monogr.* **61**, 113–134.
13. Turner, M. M., Rimal, R. N., Morrison, D., and Kim, H. (2006). The role of anxiety in seeking and retaining risk information: testing the risk perception attitude framework in two studies. *Hum. Commun. Res.* **32**, 130–156.
14. Covello, V. T. (1992). Risk communication: an emerging area of health communication research. In *Communication Yearbook 15*, S. Deetz, Ed. Sage Publications, Newbury Park and London, pp. 359–373.
15. U.S. Department of Health and Human Services (2002). *Communicating in a Crisis: Risk Communication Guidelines for Public Officials*.
16. McCroskey, J. C. (1966). Scales for the measurement of ethos. *Speech Monogr.* **33**, 65–72.
17. Covello V. T., Sandman P., and Slovic, P. (1988). *Risk Communication, Risk Statistics and Risk Comparisons: A Manual for Plant Managers*. Chemical Manufacturers Association, Washington, DC.
18. Covello, V. T. and Allen, F. (1988). *Seven Cardinal Rules of Risk Communication*. US Environmental Protection Agency, Office of Policy Analysis, Washington, DC.

19. Johnson, B. T., and Eagly, A. H. (1989). Effects of involvement on persuasion: a meta-analysis. *Psychol. Bull.* **106**, 290–314.
20. Bandura, A. (1969). *Principles of Behavior Modification*. Holt, Rinehart & Winston, New York.
21. Bandura, A. (1971). *Social Learning Theory*. General Learning Press, New York.
22. Bandura, A. (1986). *Social Foundations of Thought and Action*. Prentice-Hall, Engelwood Cliffs, NJ.
23. Scherer, D.C. (1991). Strategies for communicating risks to the public. *Food Technol.* **45**, 110–116.
24. Cooke, R. M. (1991). *Experts in Uncertainty: Opinion and Subjective Probability in Science*. Oxford University Press, New York.
25. Fessenden-Raden, J., Fitchen, J. M., and Heath, J. S. (1987). Providing risk information in communities: factors influencing what is heard and accepted. *Sci. Technol. Human Values* **12**(3 & 4), 94–101.
26. Frijda, N. H. (1986). *The Emotions*. Cambridge University Press, Cambridge.
27. Fischhoff, B., Gonzales, R. M., Lerner, J. S., and Small, D. A. (2005). Evolving judgments of terror risks: foresight, hindsight and emotion. *J. Exp. Psychol.* **11**, 124–139.
28. Lerner, J. S., Goldberg, J. H., and Tetlock, P. E. (1998). Sober second thought: the effects of accountability, anger and authoritarianism attributions of responsibility. *Pers. Soc. Psychol. Bull.* **24**, 563–574.

FURTHER READING

Fearn-Banks, K. (2002). *Crisis Communications: A Casebook Approach*, 2nd ed. Lawrence Erlbaum, Mahwah, NJ.

Fisher, A., Pavlova, M., and Covello, V. (Eds.) (1991). *Evaluation and Effective Risk Communications: Workshop Proceedings*. Interagency Task Force on Environmental Cancer and Heart and Lung Disease. EPA/600/9-90/054.

Grunig, J., and Hunt, T. (1984) *Managing Public Relations*. Holt, Rinehart and Winston.

Kasperson, R. E., Golding, D., and Tuler, S. (1992). Social distrust as a factor in siting hazardous facilities and communicating risks. *J. Soc. Issues* **48**(4), 161–187.

Meyer, P. (1988). Defining and measuring credibility of newspapers: developing an index. *Journal. Q.* **65**, 567–574, 588.

National Research Council (1989). *Improving Risk Communication*. National Academy Press, Washington, DC.

Plough, A., and Krimsky, S. (1987). The emergence of risk communication studies: social and political context. *Sci. Technol. Human Values* **12**(3 & 4), 4–10.

Powell, D., and Leiss, W. (1997) *Mad Cows And Mother's Milk: The Perils Of Poor Risk Communication*. McGill University Press, Montreal and Kingston.

Presidential/Congressional Commission on Risk Assessment and Risk Management (1997). *Framework for Environmental Health Risk Management*. Final Report, Volume 1. Available at http://www.riskworld.com/Nreports/1997/risk-rpt/html/epajana.htm.

Renn, O., and Levine, D. (1991). Credibility and trust in risk communication. In *Communicating Risks to the Public: International Perspectives*, R. E. Kasperson, and P. J. M. Stallen, Eds. Vol. 4. Kluwer Academic Publishers, Dordrecht, pp. 175–218.

Sandman, P. M. (1993). Definitions of risk: managing the outrage, not just the hazard. Paper presented at *Regulating Risk: The Science and Politics of Risk*. Washington, DC.

PROBABILISTIC RISK ASSESSMENT (PRA)

George E. Apostolakis

Engineering Systems Division and Department of Nuclear Science and Engineering, Massachusetts Institute of Technology Cambridge, Massachusetts

1 INTRODUCTION

Probabilistic risk assessment (PRA) is a scenario-based analytical methodology that has been developed to manage the risks of complex technological systems such as nuclear power plants (NPPs) [1], chemical agent disposal facilities [2], and space systems (e.g. the International Space Station (ISS) [3]). It is also called *probabilistic safety assessment (PSA)* and *quantitative risk assessment (QRA)*. In general terms, PRA answers the following three questions [4]: What can go wrong? What are the consequences? How likely is it? The principal PRA results are the probabilities of various consequences of accidents, the identification of the most likely scenarios (event sequences) that may lead to these consequences, as well as the most important (from a risk perspective) structures, subsystems, and components. This information is very valuable to operators, designers, and regulators because the responsible parties can focus resources on what is really important to the safe operation of the system.

For a given system, PRA answers the above three questions by proceeding as follows:

1. A set of undesirable *end states* is defined.
2. For each end state, a set of disturbances to normal operation is defined, which, if uncontained or unmitigated, can lead to this end state. These are called *initiating events (IEs)*.
3. Sequences of events that start with an IE and end at an end state are identified. Thus, *accident scenarios* are generated.
4. The probabilities of these scenarios are evaluated using all available evidence, primarily past experience, and expert judgment.
5. The scenarios are ranked according to their contributions to the frequencies of the end states.
6. Systems, structures, and components are also ranked according to their contributions to the frequencies of the end states.

2 SCENARIO IDENTIFICATION

The identification of scenarios that may defeat the built-in defenses of the system and lead to undesirable system states requires intimate knowledge about the system and its operation. These scenarios are the potential failure modes of the system. The development of the scenario list is usually facilitated by the use of computer codes such as SAPHIRE, RISKMAN, and RiskSpectrum [5].

2.1 End States

The end states are any undesirable states that are of interest to the risk manager (decision maker). For example, the two end states that are used routinely in NPP risk assessments are *reactor core damage* and *large, early release of radioactivity to the atmosphere* ("early" means that the release occurs before any evacuation of the surrounding population can be effected). For the ISS, the two end states that have been analyzed are *loss of crew and vehicle* and *evacuation of the ISS*.

2.2 Initiating Events

A PRA requires a good understanding of the system and its normal operation. There is usually a set of functions that must be performed for normal operation. For example, in an NPP, the reactor core produces large amounts of heat that must be removed by the cooling systems. Disturbances to the heat production or removal may start a sequence of failures that may ultimately lead to the end states. These disturbances are the IEs. Examples are the loss of off-site electric power and various sizes of loss-of-coolant accidents. For the ISS, the station functions (e.g. propulsion) and the functions needed for the crew's health (e.g. removal of CO_2 from the station's atmosphere) are the basis for defining the IEs.

2.3 Accident Scenarios

The question now is how an IE could lead to an end state, that is, what additional events (failures) must occur for the end state to materialize. The logic diagram that is used for this purpose is the event tree (*see* Logic Trees: Event, Fault, Success, Attack, Probability, and Decision Trees), which is a decision tree without decision nodes.

The technological systems to which PRAs are applied are well defended. This means that sufficient redundancy and diversity are built into their design to prevent IEs from happening and to mitigate their consequences. Let us suppose that two protective systems, PS1 and PS2, have been designed to mitigate the consequences of a particular IE. Either system can mitigate the consequences. Figure 1 shows how the accident sequences associated with this IE are determined. The top line is the sequence $IE \cap PS1 \cap PS2$, that is, the IE has occurred and both protective systems work. The result is OK, that is, no damage. The sequence leading to *damage level 1* is $IE \cap \overline{PS1} \cap PS2$, that is, the IE has occurred, PS1 has failed, and PS2 has worked. Similarly, *damage level 2* results from the sequence $IE \cap \overline{PS1} \cap \overline{PS2}$, in which both the protective systems have failed.

An event tree may be developed at a high level in which general protective functions are listed, for example, *electric power available*, or at lower levels in which individual systems or subsystems are listed, for example, *off-site power available* and *diesel generator A available*. In the latter case, the number of sequences increases significantly. It is customary to start with high-level event trees and to proceed to develop more detailed trees as the accident sequences are becoming better understood. It is evident that the term *accident sequence* is not well defined since it may be a high-level sequence involving functions or a very detailed sequence involving components. It is common practice to call those sequences involving system failures as accident sequences.

The development of event trees requires detailed knowledge about system function, operation, and intersystem dependencies. Various tools have been developed to facilitate the collection and understanding of the relevant information. For example, event sequence diagrams can be developed to show the sequence of actions required by procedures as various events occur, and dependency matrices to display the dependencies among systems [6, 7].

The next step is the investigation of the failure modes of the systems/functions that appear in the event tree, that is, the protective systems in Figure 1. The question that the analysts ask is "how can PS1 fail?" This question is to be contrasted with the question that is asked when the event tree is developed: "how can this IE lead to an end state?"

The failure modes of individual systems can be identified using fault-tree analysis (*see* Logic Trees: Event, Fault, Success, Attack, Probability, and Decision Trees). The analysis includes hardware failures and human errors during routine operations such as test and maintenance [8]. Human actions during the evolution of an accident sequence are usually placed in the event tree. The unique failure modes of a system resulting from fault-tree analysis are usually called *minimal cut sets*. A minimal cut set is a set of failures whose occurrence guarantees the failure of the system. The word "minimal" means that, if a failure is removed from this set, the system will not fail.

There are still models being developed for human actions that occur post-IE, that is, during the accident [9–12]. The focus is the identification of the contexts within which the human actions occur and the evaluation of the probabilities of errors given a specific context. The context is defined both by the accident sequence that is occurring and various cognitive and organizational crew performance shaping factors. The evaluation of the probabilities is done by utilizing expert judgments supported by extensive discussions of the contexts, the results of simulator exercises, and other supporting information.

In developing the accident sequences and system failure modes, it is important to capture the possible dependencies among systems and components. System functional dependencies are usually included in the event trees. For example, if "availability of ac power" is one of the event tree headings, then all subsequent systems or components that depend on electric power will be assumed failed if electric power has failed. Other major dependencies result when fires, earthquakes, tornadoes, and other similar events occur. These "external" events can cause an IE and, at the same time, affect the performance

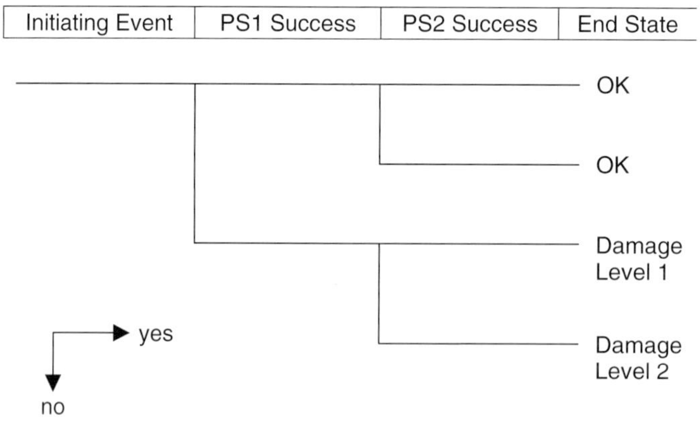

FIGURE 1 Example of an event tree.

of safety systems. These events and the resulting dependencies are usually investigated separately [13, 14].

The various events that appear in event and fault trees are binary (true–false). We associate an indicator variable X_j with event j so that the event is true (failure) when $X_j = 1$ and false when $X_j = 0$. Then the result of the whole analysis is represented by a logic function that expresses the logical relationship between the end state k and the failures that may lead to it, that is,

$$X_{\mathrm{ES}k} = \varphi(X_1, \cdots, X_n) \equiv \varphi(\underline{X}) \tag{1}$$

where $X_{\mathrm{ES}k}$ is the indicator variable for end state k and $(X_1, \ \ldots \ , X_n)$ is the set of (primary or basic) failures that have been input to the logic diagrams. Eq. (1) contains all the information that the logic diagrams have produced. It is called the *structure function of the system*.

Let AS_i be an accident sequence (or a minimal cut set in the case of fault trees). Eq. (1) is equivalent to

$$X_{\mathrm{ES}k} = \sum_{i=1}^{N} \mathrm{AS}_i - \sum_{i=1}^{N-1} \sum_{j=i+1}^{N} \mathrm{AS}_i \mathrm{AS}_j + \cdots + (-1)^{N+1} \prod_{i=1}^{N} \mathrm{AS}_i \tag{2}$$

where N is the total number of accident sequences and

$$\mathrm{AS}_i = \prod_{X_m \in \mathrm{AS}_i} X_m \tag{3}$$

that is, the indicator variable of accident sequence AS_i is the intersection of all the indicator variables of the failure events that belong to this accident sequence.

The binary modeling of systems and components (success–failure) that is employed in event trees and fault trees is not appropriate when phenomenological (physical, chemical, and biological) processes must be considered in the evaluation of the end states. For example, the analysis of fires requires models for the fire plume and the radiative and convective heat transfer processes that may damage systems and components [14]. These analyses result in accident scenarios that consist of the sequences of the phenomenological processes which, in combination with binary component failures, may lead to the end states.

Risk assessments for nuclear waste geologic repositories [15, 16] are completely dominated by phenomenological processes and the concept of binary accident scenarios is of little use. The end state is usually the amount of radioactivity released to the accessible environment at a particular time (e.g. 10,000 years in the future). The so-called base-case scenario models the evolution of the repository during very long times using a suite of computer programs that model the interdependent physical and chemical processes that are expected to occur. Similarly, when an IE is considered, such as human intrusion via exploratory drilling for natural resources, the evaluation of the end state is again done using the computer programs. The formulation of Eqs. (1–3) does not apply to these "performance assessments" (as risk assessments for repositories are called). Accident scenarios as defined in Eq. (3) cannot be identified (a proposal for an alternative scenario definition appropriate for performance assessments is presented in [17]). We will see later that a similar situation arises when infrastructures are investigated in the sense that computer programs appropriate to the specific infrastructure must be used to determine its response to external disturbances.

3 PROBABILITIES

The probability distribution of an end state requires the probability distributions of the primary failure events that are input to the event and fault trees. In mathematical terms, the probability distributions of the indicator variables X_i are propagated through Eq. (2), the logic of the system, to produce the probability distribution of the indicator variable for the end state $X_{\mathrm{ES}k}$. This leads us to the following equations:

$$R_k \equiv \Pr(X_{\mathrm{ES}k}) = \sum_{i=1}^{N} \Pr(\mathrm{AS}_i) - \sum_{i=1}^{N-1} \sum_{j=i+1}^{N} \Pr\left(\mathrm{AS}_i \mathrm{AS}_j\right) + \cdots + (-1)^{N+1} \prod_{i=1}^{N} \Pr(\mathrm{AS}_i) \tag{4}$$

where

$$\Pr(\mathrm{AS}_i) = \Pr\left(\prod_{X_m \in \mathrm{AS}_i} X_m\right) \tag{5}$$

The probabilities of the accident sequences $\Pr(AS_i)$ are produced from the probabilities of the primary failures that are input to the event and fault trees. In Eq. (5), the probability of the accident sequence is not, in general, the product of the primary failure probabilities; the dependencies among the primary failures must be taken into account [6, 7].

The primary failure probabilities are generally low (<0.1), and so the resulting probabilities of the end states are very low ($<10^{-2}$). In other words, we are in the realm of rare events. This is a consequence of the fact that, as stated earlier, these technological systems are well defended, that is, sufficient redundancy and diversity are built into their design to reduce their failure probabilities to acceptable levels.

As the failure probabilities become small, the uncertainties become large because the statistical evidence from system operation becomes weak (this is, of course, also true for "new" designs for which there may be no operational experience). This means that the analysts must rely on expert judgment in evaluating probabilities. Such judgments are subject to biases [18], thereby increasing the uncertainties. This state of affairs creates the need to examine interpretation of the concept of probability.

3.1 Interpretation of the Concept of Probability

The theory of probability, as expounded in textbooks, is an axiomatic mathematical theory that needs no interpretation. In engineering practice, however, it is useful to have an interpretation in mind. In risk assessment, in particular, the rarity of the events of interest makes it a necessity [19, 20].

The usual interpretation of probability is as the limit of relative frequencies. This interpretation is very restrictive for PRA. It allows the use of statistical evidence only in probability evaluations, whereas, as stated above, the use of expert judgment is inevitable when rare events are analyzed. The uncertainties in the probability estimates are expressed in terms of confidence intervals, which are impossible to propagate in any meaningful way through the structure function of the system (Eq. 1) to produce a statement on the uncertainties of the probabilities of the end states. There are no PRAs that have been conducted with this interpretation in mind.

Probability is interpreted as a measure of degree of belief [21]. We accept that it is meaningful to say that one judges an event to be more (equally, less) likely than another event. Probability is simply a numerical expression of this judgment. When we say that event A has probability 0.6, we mean that A is more likely than any other event whose probability is less than 0.6 and it is less likely than all events whose probability is greater than 0.6. This primitive notion of likelihood, along with several other axioms, allows the development of a rigorous mathematical theory of subjective probability. Both the frequentist and subjectivist interpretations are consistent with the mathematical theory of probability.

It is useful to discuss how this concept is applied to PRA. The analysis begins by constructing a "model of the world", that is, a mathematical model of the system that may include the development of its structure function using event and fault trees as well as models for physical processes such as hear transfer. The "world" is defined as *the object about which the person is concerned* [22]. Occasionally, we refer to the model of the world as simply the model or the mathematical model. Constructing and solving such models is what most physical scientists and engineers do.

There are two types of models of the world, deterministic and probabilistic. The former include event and fault trees for the system logic and mechanistic models for physical processes such as convective heat transfer. Although it would be highly desirable to have the complete PRA done using deterministic models, we soon realize that many important phenomena cannot be modeled deterministically. For example, the failure time of a component while running (assuming a successful start) exhibits variability that we cannot eliminate; it is impossible for us to predict when the failure will occur. We, then, construct a model of the world that reflects this uncertainty. This model is usually the exponential distribution whose probability density function (pdf) is

$$f(t/\lambda, M) = \lambda e^{-\lambda t} \qquad (6)$$

This expression shows explicitly that this probability is conditional on our knowing the numerical value of the parameter λ (the failure rate) and accepting the assumption M that the exponential model is appropriate, that is, its fundamental assumption that the failure rate is constant is valid. This observation is valid for deterministic models as well. For example, event trees are developed making assumptions regarding the success criteria of the protective systems. As another example, the development of deterministic models for fires requires assumptions regarding the fire plume. The numerical values of parameters of such models like the burning rate of the fuel must be known for the models to be used.

The uncertainty described by the model of the world is sometimes referred to as *randomness* or *stochastic uncertainty*. Recently, the term *aleatory models* was adopted because the preceding terms are used in many contexts in probabilistic analyses.

As stated above, each model of the world is conditional on the validity of its assumptions, M_i, and on the numerical values of its parameters. In general, the model has a number of parameters, which can be represented in vector form as θ. Since there is usually uncertainty associated with M_i and θ, we introduce the *epistemic* probability model that reflects in quantitative terms our state of knowledge (degree of belief) regarding the validity of M_i (and hence the results of the model of the world) and the numerical values of θ. It is important to bear in mind that the model of the world deals with observable quantities, such as failure times, and the epistemic model with parameters and assumptions that are not observable.

In its landmark Regulatory Guide 1.174 that established the framework for risk-informed regulatory decision making [23], the US Nuclear Regulatory Commission staff considers three categories of uncertainties: parameter, model, and completeness. Although we could consider the last two as part of model uncertainties, it may be useful in some cases to consider completeness separately. The Regulatory Guide states: "Completeness is not in itself an uncertainty, but a reflection of scope limitations. The result is, however, an uncertainty about where the true risk lies. The problem with completeness uncertainty is that, because it reflects an unanalyzed contribution, it is difficult (if not impossible) to estimate its magnitude.... There are issues, however, for which methods of analysis have not been developed, and they have to be accepted as potential limitations of the technology. Thus, for example, the impact on actual plant risk from unanalyzed issues such as the influences of organizational performance cannot now be explicitly assessed." Completeness is expected to be a major issue in terrorism studies.

Parameter uncertainties are usually represented by pdf's $\pi(\theta/M_j)$ that are produced using statistical evidence and expert judgments, as it is seen later. Model uncertainties are handled in a variety of ways, for example, using qualitative arguments, sensitivity studies, and expert judgments. Completeness uncertainties are handled in the context of decision making. For example, qualitative arguments may be made that completeness is not a significant issue for the decision to be made. Alternatively, additional safety-related requirements may be imposed that are not based on the PRA results (in other words, the decision-making process is risk *informed* rather than risk *based*).

3.2 The Epistemic Distributions and Expert Judgment

There are two kinds of evidence that are utilized to develop the epistemic pdf for the parameters: event-specific statistical evidence, E_S, and generic information, E_G, from other sources. To make the discussion concrete, we will deal with one parameter, the failure rate λ of the exponential model (Eq. 6). Let $\pi(\lambda/E_G, M)$ be the pdf developed from generic information. The epistemic pdf for the failure rate is produced by combining E_G with E_S using Bayes' theorem:

$$\pi'(\lambda/E_{\mathrm{S}}, E_{\mathrm{G}}, M) = \frac{\pi(\lambda/E_{\mathrm{G}}, M)L(E_{\mathrm{S}/\lambda}, M)}{\int_0^\infty \pi(\lambda/E_{\mathrm{G}}, M)L(E_{\mathrm{S}}/\lambda, M)d\lambda} \tag{7}$$

The term $L(E_S/\lambda, M)$ is the likelihood of the evidence E_S assuming that λ is known and is calculated using the model of the world. In the case of component failures, the evidence E_S may consist of the failure times of n components, $t_1, \ldots, t_n$. Then, the likelihood function is

$$L(E_S/\lambda, M) = \lambda^n \exp\left[-\lambda\left(\sum_1^n t_j\right)\right] \tag{8}$$

Expert judgment is prevalent in this formulation. Determining the evidence E_S itself is usually not straightforward. Deciding what constitutes a "failure" and whether the operating conditions of the failed component were "identical" to those of the component of interest requires the exercise of judgment on the part of the analysts. In many cases, however, the impact of this judgment is overwhelmed by the judgment required to develop the generic pdf $\pi(\lambda/E_G, M)$.

Generic information may take many forms. For example, some generic sources may report "point" values of the failure rate while others may report a pdf. The degree of applicability of the information provided may also be an issue. For example, information on dependent failures in NPPs may be utilized to develop pdfs for such failures in space systems in which both the operating environment and the personnel culture are different. Widely acceptable methods for handling such situations are not available.

In cases of large model uncertainties and when the issue is of such importance that appropriate resources are available, the elicitation and processing of expert judgments have been formalized [13, 24, 25]. The purpose of the elicitation process is to reduce the biases as much as possible and to train the experts on how to express their judgments in terms of probabilities. The processing of the elicited judgments may be done in a number of ways all having advantages and disadvantages.

4 PRA RESULTS AND RISK MANAGEMENT

The epistemic distributions of the probabilities of the primary inputs to the PRA model are propagated through the model usually via Monte Carlo simulation to produce an epistemic distribution for the probability of each accident sequence, Eq. (5), and ultimately for each end state, Eq. (4). Monte Carlo simulation is also used when physical and chemical processes are modeled as in performance assessments. Because of the long running times of the computer programs involved, Latin hypercube sampling is employed to reduce the number of Monte Carlo trials [26].

Figure 2 shows the result for the end state *latent cancer fatalities* for an NPP [27]. The risk curves show the frequency of a given number of fatalities or greater. As an example, the frequency of 100 or more cancer fatalities has epistemic uncertainty that ranges from about 4×10^{-7} per reactor year (fifth percentile) to about 3×10^{-5} per reactor year (95th percentile). This uncertainty is the result of all the epistemic uncertainties associated with the primary inputs to the PRA (IE frequencies, component failure rates, human error rates, phenomenological uncertainties associated with severe accidents, and others) propagating through the PRA model. The mean and median frequency can also be found in the figure. An examination of the accident sequences that contribute the most to cancer fatalities shows that the dominant sequences are initiated by station blackout (all ac power lost) and sequences that bypass the containment. The complete accident sequences initiated by these events are detailed in the PRA report.

For the end state *core damage*, the results for the same plant are fifth percentile, 1.2×10^{-5}; median, 3.7×10^{-5}; mean, 5.7×10^{-5}; and 95th percentile, 1.8×10^{-4}

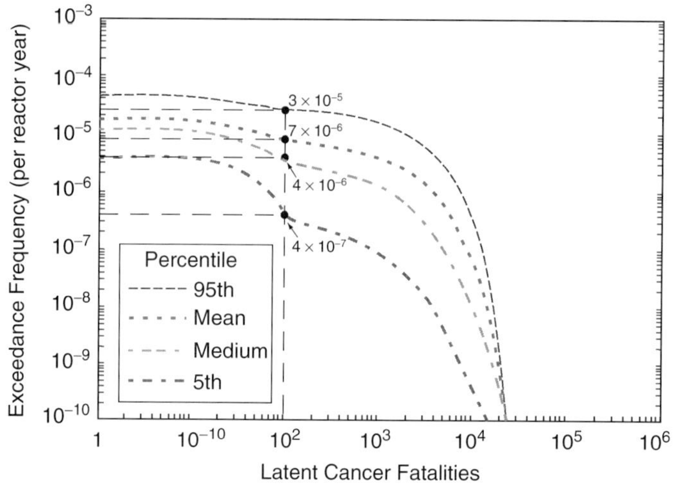

FIGURE 2 Example of risk curves.

per reactor year. The dominant accident sequences for this end state are loss-of-coolant accidents and station blackout.

These results serve two purposes: they allow a determination as to whether the frequencies of the end states are "acceptable" by the society (or its representatives, that is, the relevant regulatory agencies) and, if it is found that the risk is "unacceptable" or if risk reduction is cost effective, measures can be taken to reduce the frequency of the dominant sequences. The risks that society "accepts" or "tolerates" (see, for example, [28]) are usually the basis for deciding what risks are "acceptable". The Nuclear Regulatory Commission has adopted quantitative heath objectives that define an acceptable level of risk from NPP operations that is 0.1% of all other risks to which an average individual of the public is exposed [29]. The United Kingdom Health and Safety Executive considers three regions of risk [30]. Risks to an individual of the general public greater than 10^{-4} per year (probability of death) are generally unacceptable and risks smaller than 10^{-6} per year are considered to be broadly acceptable. The region between these two limits is the tolerability region and represents risks "from activities that people are prepared to tolerate in order to secure benefits". The risks in this region are kept as low as reasonably practicable.

When the uncertainties in the frequencies of the end states are so large that they inhibit meaningful risk management, it may be possible to define subsidiary objectives for which the uncertainties are smaller. This is the case with NPPs. Because the uncertainties associated with the evaluation of individual risks are very large, the Nuclear Regulatory Commission staff is using a goal of 10^{-4} for the frequency (per year) of reactor core damage and 10^{-5} for the frequency (per year) of a large release of radioactivity from the containment building before evacuation of the surrounding population can be effected. These values are compared with the epistemic means of the frequencies of the corresponding end states.

Additional information useful to risk management is provided by assessing the relative importance of the thousands of events that are included in the PRA. This is done using importance measures [31]. One such measure is the risk reduction worth (RRW) of an event. It is defined as follows:

$$\mathrm{RRW}_i^k = \frac{R_k}{R_k^{-i}} \tag{9}$$

where RRW_i^k is the risk reduction worth of event i with respect to end state k and R_k^{-i} is the risk metric R_k (Eq. 4) with the ith event assumed to always be a success (zero probability of failure).

This importance measure is particularly useful for identifying improvements in the reliability of elements, which can most reduce the risk. Using this importance measure, [31] finds that, with respect to the end state *core damage*, reducing the frequency of very small loss-of-coolant accidents could lead to a core damage frequency reduction of up to 38% in a particular case. Similarly, the improvement of the operator ability to control sprays during a small loss-of-coolant accident could reduce the core damage frequency by up to 37%.

Other importance measures used in practice are the Fussell–Vesely (FV) and the risk achievement worth (RAW) measures. The FV measure is defined as follows:

$$\mathrm{FV}_i^k = \frac{R_k - R_k^{-i}}{R_k} = 1 - \frac{R^{-i}}{R_k} = 1 - \frac{1}{\mathrm{RRW}_i^k} \tag{10}$$

As Eq. (10) shows, this measure can be derived from RRW_i^k; therefore, it does not provide any additional information.

RAW is defined as follows:

$$\mathrm{RAW}_i^k = \frac{R_k^{+i}}{R_k} \tag{11}$$

where R_k^{+i} is the risk metric R_k (Eq. 4) with the ith event assumed to be always a failure (probability of failure equal to unity). RAW is a measure of the "worth" of a basic event in "achieving" the present level of risk and indicates the importance of maintaining the current level of reliability for this basic event.

Importance measures have found a variety of applications in risk management. They have been used to establish the maintenance strategy for systems and components at facilities so that specific reliability targets can be met. This allows maintenance to be performance based rather than prescriptive. Importance measures have also been used to identify systems and components on which special quality assurance requirements are imposed that go beyond the normal industrial standards. An application of this concept to infrastructures is seen later in this article.

The PRA results (probabilities of end states, dominant accident sequences, and important events) are extremely useful to risk management. It is important to point out, however, that decisions made by responsible authorities are never risk based. There are too many judgments and uncertainties (especially those due to models and completeness) for decisions to be risk based [23, 32]. The risk results are submitted to a deliberative process in which the known and unknown uncertainties are discussed and the assumptions behind the analyses are scrutinized. The decisions are usually based on a combination of risk insights, qualitative engineering analyses, and traditional safety measures that are precautionary in nature [33]. Such an "integrated decision-making process" [23] relies on the experience and judgment of the risk managers and usually leads to decisions that are more conservative than the risk numbers would suggest. This approach is similar to the analytic-deliberative process for decision making proposed by the National Research Council [34]. The prioritization of accident sequences and events that PRA produces is the starting point for the deliberation among the risk managers (or, more generally, the stakeholders [35]). Risk management for terrorism is also expected to be risk informed rather than risk based.

5 TERRORISM

5.1 Challenges

After September 11, 2001, protection against terrorism has become a US and international focus and is likely to remain one in the foreseeable future. Large amounts of resources are expended to prevent terrorist acts and to mitigate their consequences if they occur. Given the many forms, terrorist acts can assume and the openness of democratic societies, it is evident that attempting to protect against all such acts is impossible. As an example, [36] offers numerous recommendations for vulnerability reduction in several diverse infrastructures. Implementing all of them would impose a considerable financial burden on the United States and would ignore the relative importance of these vulnerabilities.

It is evident that a prioritization of the national needs for protection against terrorism is needed. Since PRA has been proven to be an effective analytical tool for prioritizing accident sequences in complex technological systems, it is natural to explore the possibility of using it in combating terrorism. We must acknowledge, however, that the application of PRA to terrorism poses challenges that must be overcome.

Before we proceed to discuss challenges specific to terrorism, we state some major PRA limitations that exist even for applications to technological systems [32, 37, 38]. As mentioned earlier (Section 1.3), models are still being developed for human actions that occur post-IE, that is, during the accident [9–12]. At this time, there is no universally accepted model for such human actions, especially for errors of commission (the crew does something that worsens the situation). A related topic is that of the influence of organizational and managerial factors. Although they are recognized as having an important impact on personnel performance [39], they are not included in PRAs explicitly. It is stated in [32] that it is doubtful that the inclusion of organizational factors in PRA will be achieved any time soon. Finally, probabilistic models for digital software systems that are embedded in the system are in their infancy [40]. These challenges will, of course, be present when PRA methodology is applied to terrorism. Additional challenges due to some unique features of assessments that involve malevolent acts are discussed below.

The end states in technological system PRAs are very few (two or three) and are focused on health and safety impacts. Because they are very few, the results can be used easily by risk managers in combination with other requirements in their "integrated decision-making process" (deliberation) that ultimately leads to action. For specific assets and for specific purposes, the number of end states can be kept small even in the case of terrorism. As an example, the US Nuclear Regulatory Commission focuses on public health and safety and states [41]: "For the facilities analyzed, the results confirm that the likelihood of both damaging the reactor core and releasing radioactivity that could affect public health and safety is low. Even in the unlikely event of a radiological release, due to a terrorist use of a large aircraft against an NPP, the studies indicate that there would be time to implement the required on-site mitigating actions."

For terrorism, the set of end states must be expanded considerably. As we stated earlier, the end states are determined by the risk manager (decision maker) because they are judged to be important to the decision. Decisions involving terrorism are usually based on more than just health and safety impacts. As stated in [42], "the end states must reflect initial, cascading, and collateral consequences (or levels of damage)". This approach leads to a large number of candidate end states. These include end states related to health and safety such as fatalities and injuries; economic losses such as the value of the assets attacked and replacement/repair costs; environmental impacts on flora and fauna; national security such as impacts on government functions; and many others. An additional complication is that some of the impacts may be immediate and others delayed. It is evident that not all of these end states are equally important or relevant to all situations. The end states to be evaluated depend on the assets under attack, the nature of the threat, and the interests of the risk managers (the decision makers).

The construction of scenarios leading to the end states is also a challenge. An important feature of technological systems to which PRAs have been applied is their spatial compactness. By focusing on NPPs and limiting the number of end states, the US Nuclear Regulatory Commission has been able to evaluate scenarios initiated by aircraft attacks using traditional PRA methods, as stated above. A major concern in terrorist studies is the vulnerability of infrastructures. These are interconnected across systems and spread out geographically [43]. Further, societal infrastructures have overlapping ownership and responsibility in private organizations and local, state, and national government. Therefore, technical complexity may be compounded by sociopolitical complexity. The views of stakeholders must be taken into account.

A traditional PRA requires the probabilities of IEs. The evaluation of the probability of a terrorist attack is a major challenge. Even if one resorts to expert opinion elicitation methods as discussed above in the context of model uncertainties, the uncertainties will be very large. Questions such as who the experts are and whether they have access to all available information contribute to these uncertainties. A major complicating factor is the fact that terrorist groups, unlike accidents, are intelligent and adaptable adversaries.

The challenges we have discussed have created a fertile area of research. Several investigators have proposed ideas and approaches to the management of terrorism risks that build on the principles of risk assessment and management. Thus the authors of [42] provide a detailed review of PRA methods and discuss how they would apply to terrorism. Traditional event trees are used to analyze a sample regional grid and the relevant probabilities are assumed to be produced by expert judgment. One of the authors' conclusions is: "It is with respect to catastrophic attacks that the principles and practices of quantitative risk assessment have their greatest value. The structuring of catastrophic attack scenarios could be one of the most important short-term benefits of quantitative risk assessment."

The observation that terrorist groups are intelligent adversaries has prompted researchers to explore the applicability of game theory. The authors of [44] develop an "overarching model" whose objective is to organize the mass of available information and to bring together various types of threat scenarios, various groups of attackers, and the amount of damage they could inflict. Decision analysis is employed to prioritize the threats and game theory helps in analyzing the dynamics of the events and the moves and countermoves of the United States and the perpetrators. A conclusion is: "In designing and implementing strategies of response to potential terrorist attacks, it is essential to think beyond the reoccurrence of the last event.... The model presented here involves, in particular, probabilistic dependencies, and it uses a forward-looking approach to generate a set of possibilities and scenarios." The authors argue that " ... in its practical application, the analysis must include the dynamics of the events and the moves and countermoves of both sides (the United States and the perpetrators)".

Game theory, optimization methods, and reliability analysis are applied in [45] to simple series and parallel systems of components to develop insights into the nature of optimal defensive investments for managing terrorist attacks. The authors reach the following interesting conclusion: "Our results suggest that some high-value targets with a low probability of being successfully attacked may not merit investment, while other (less valuable but more vulnerable) targets may merit defending".

Finally, game theory has been applied to the security analysis of computer networks [46]. The authors determine the Nash equilibria in a two-player (the attacker and the administrator) stochastic game and argue "a Nash equilibrium gives the administrator an idea of the attacker's strategy and a plan for what to do in each state in the event of an attack".

5.2 A Proposed Framework for Infrastructures

A framework that combines decision analysis [47] and PRA methods has been proposed and implemented in several infrastructure studies [48–50]. Decision analysis allows us to focus on the interests of the risk managers and stakeholders and to include the end states of concern to them. The objective is to rank the infrastructure elements according to their value to the decision maker. A further objective is to identify specific issues that arise when risk assessment methods are applied to infrastructures. The general framework is shown in Figure 3 and explained below.

The methodology begins by identifying assets and components of the infrastructure whose failure may lead to undesirable consequences. These failures may be random, in which case a traditional PRA is performed for the infrastructure, or due to a malevolent act, in which case terrorism or vandalism is analyzed. The element failures are the IEs in this case. The physical consequences of these failures are determined by analyzing the response of the infrastructure and human intervention. A central concept is that each of the infrastructure elements is characterized by a "capacity". This concept can represent a carrying capacity, production, or consumption of goods. Any infrastructure has some characteristic that can be considered as its capacity. The units of this capacity might be the amount of water produced by treatment plants and transported by pipelines in a water-supply system [49] or the electric power generated by generators and transmitted by transmission lines in a power-supply system [50].

There is a physical limit on the amount of goods that can be generated/processed/transported/consumed by each element. The assumed failure of an element leads to a rearrangement of the amounts transported and consumed possibly leading to cascading failures. This analysis requires appropriate computer programs such as the Sandia AC load flow simulation model [51] for a bulk power infrastructure that was used in [50]. The consideration of the capacity of the infrastructure elements makes this analysis different from traditional PRAs, in which the performance of each element is modeled as a success or failure, thus leading to Boolean expressions such as those of Eqs. (1–3). This analysis resembles that conducted in performance

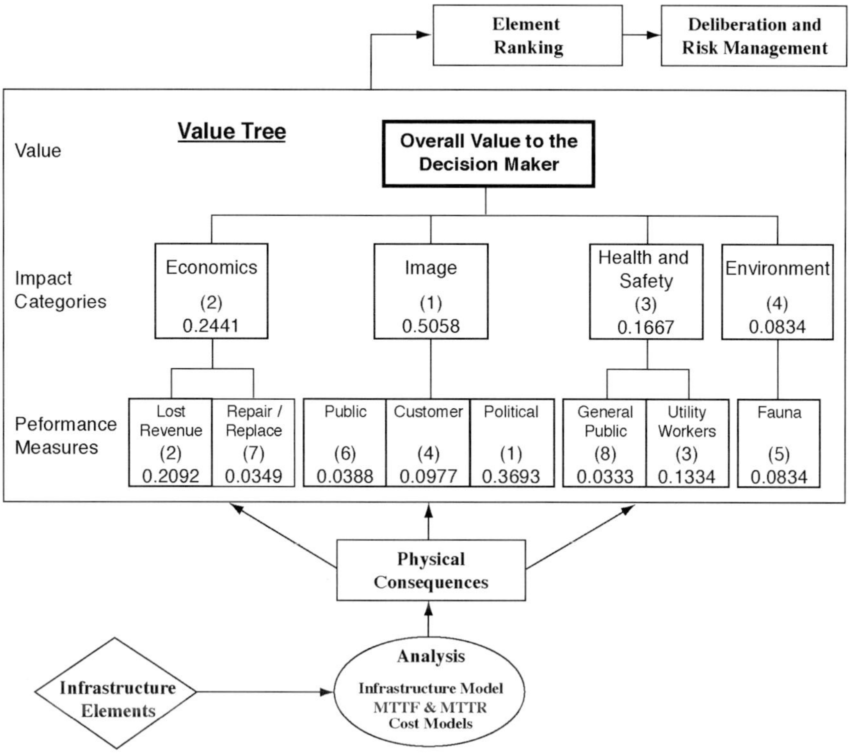

FIGURE 3 A general framework for the evaluation of infrastructure elements.

assessments in which simulation models are used for the relevant physical and chemical processes.

The physical consequences of the failure of an infrastructure element include the numbers and types of customers affected (industrial users and individual households) and the duration of the denial of service, for example, power outages and no water supplies. The evaluation of these consequences requires inputs, such as the mean time to failure (MTTF) and mean time to repair (MTTR) of individual elements, cost estimates, and other inputs as appropriate for the problem at hand.

The physical consequences are input to a value tree that incorporates the decision maker's views on possible impacts. The value tree is then used to determine the impact the consequences have on the decision maker. The amount of impact a component represents to the decision maker is its value. Each component value is combined with its susceptibility to failure or attack. The combination of value and susceptibility is used to rank the components according to their risk significance.

In order to be able to assess the value of the physical consequences, the value system of the decision maker needs to be modeled explicitly. The value tree, also called *objectives hierarchy* [47], is based on multiattribute utility theory (MAUT) and provides a hierarchical view of the impact each physical consequence may have on the stakeholders. The value tree generally consists of three levels in which the top level is the overall impact, or value, of a failure scenario (Fig. 3). The second level breaks this overall impact into broad categories that we call impact categories. In the third level, these categories are expressed as specific attributes, called *performance measures (PMs)*, that specifically describe the various ways physical consequences result in impacts to the decision maker.

In order to be able to rank the element failures, we need a single index that combines all the impacts (PMs) into a single number. This index includes the relative weights the decision maker assigns to the PMs and a number representing the degree of impact of each physical consequence on each PM (the "disutility" of that level).

The relative weights of the PMs can be determined using one of several methods that exist in the literature [47]. We have found the analytic hierarchy process (AHP) [52] particularly useful. The AHP is a hierarchical method of pairwise comparisons in which preferences among objectives are converted into numerical weights. Since the decision maker has already structured the value tree in a hierarchical arrangement, the application of AHP is fairly straightforward. In performing AHP, the decision maker first makes pairwise comparisons of the impact categories with respect to the overall goal. After making these comparisons and establishing the weights of the categories, the decision maker compares the next level of objectives (the PMs) to the objective above it. It is important to remember that the comparisons of objectives are not made in an absolute sense; they are made, instead, with respect to the decision context at hand. This means that the decision maker does not compare, for example, *health and safety* to *image* in general; rather, he/she compares them with respect to the *overall value to the agency* in the context of the problem that is being investigated. Therefore, the decision maker must be mindful of the ranges of the corresponding PMs. An advantage of the AHP is that it offers an estimation of the consistency of the judgments of the decision maker. We point out that the AHP is used here as a means for producing relative weights from stakeholder input and not as a decision-making method. The latter use has been criticized in the literature, for example, in [53, 54].

Figure 3 shows the relative weights for a particular decision maker. The relative rankings of the various elements of the value tree are shown in parentheses. Thus the decision maker ranks the impact category *image* as the most important and *environment* as the least important. This is a good example of our earlier comment that the evaluation is made in the context of the specific decision that the decision maker must make. In the present case, the decision maker was aware of the fact that failures of infrastructure elements would have very minor physical consequences. However, the political consequences of even the smallest failure would be major.

In several applications, we have found that the impact categories shown in Figure 3 capture the stakeholder concerns. Of course, the relative weights vary according to the application.

The PMs are described by constructed scales, that is, discrete levels of potential impacts on each PM. As an example, Table 1 shows the constructed scale and associated disutilities for the PM *image with the general public*. The disutilities are developed in close collaboration with the decision maker using pairwise comparisons of the levels of the constructed scale. Figure 3 shows the relative weights and rankings of the PMs in our example. Again, the most important PM for the decision maker in this problem is *political*.

The average performance index (PI) of a failure scenario is determined using the following equation:

TABLE 1 Constructed Scale for the Performance Measure "Image with the General Public"

Level	Constructed Scale	Disutility
4	International interest from the media	1.0
3	Repeated articles in the national media, mention in the international media	0.4092
2	Repeated articles in the local media, appearance in the national media	0.1363
1	Single appearance in the local media	0.0374
0	No negative image with the general public	0.0

$$\overline{\mathrm{PI}}_j = \sum_{i=1}^{k_{\mathrm{PM}}} w_i \overline{d}_{ij} \tag{12}$$

where $\overline{PI}_j$ is the expected performance index of the scenario initiated by the failure of element j, w_i is the weight of the performance measure i, $\overline{d_{ij}}$ is the expected disutility of performance measure i and failure scenario j, and k_{PM} is the total number of PMs.

An intermediate step not shown in Eq. (12) is the evaluation of the physical consequences of scenario j, which are converted to the levels d_{ij} of the constructed scales of the corresponding PM.

The additive independence assumption made in establishing Eq. (12) means that the preferences among the PMs can be assessed independently of one another. The analysts should work with the decision maker to confirm that the PMs are reasonably independent. The independence assumption is very strict and is rarely satisfied fully in practice. Nevertheless, it is widely used in applications. As stated in [47]: "Even if used only as an approximation, the additive utility function takes us a long way toward understanding our preferences and resolving difficult situations." Reference 47 also states: "In extremely complicated situations with many attributes, the additive model may be a useful rough-cut approximation". An additional safeguard is that the results of the formal analysis are subjected to deliberation, in which the decision maker evaluates the results of the analysis for reasonableness.

The physical consequences correspond to the end states of a traditional PRA. Unlike the framework described above, the values of the decision maker are not stated explicitly but, rather, are taken into account in the deliberation that is part of the integrated decision-making process. This is possible because the end states are part of a single impact category (that is, they are all related to health and safety) and their number is kept small. A proposal to apply the framework presented here to NASA systems is presented in [55].

5.3 Risk Ranking

The average performance index $\overline{PI}$ is the metric that should be used to rank the infrastructure elements. The inequality $\overline{PI}_j > \overline{PI}_m$ means that the decision maker assesses that element j is of higher disutility, that is, its failure leads to less desirable consequences, than element m. From a risk management perspective, element j should attract more attention than element m.

As shown in Eq. (12), the theory demands that the average PI should be calculated. This means that all the uncertainties should be accounted for in the analysis. These include the uncertainties in the failure rates of the elements. When random failures are considered, these uncertainties can be evaluated using past experience and expert judgment, as is done in traditional PRAs. In the case of terrorism or vandalism, however, the probabilities of successful attacks are difficult to evaluate in a meaningful way. Reference 42 argues that expert opinions can be used in this evaluation. This approach would require access to intelligence information that is generally unavailable and also raises the question of whether experts on this issue actually exist.

An alternative to the evaluation of the probabilities of a successful attack is to calculate a conditional PI assuming that the threat exists. Similarly, the disutilities of the various consequences are assessed conservatively so that probabilities are not needed. The conditional PI is then

$$\mathrm{PI}_j = \sum_{i=1}^{k_{\mathrm{PM}}} w_i d_{ij} \tag{13}$$

where PI_j is the performance index of the scenario initiated by the failure of element j and d_{ij} is the conservative value of the disutility of performance measure i and failure scenario j.

Even though the probability of attack has been excluded from Eq. (13), the structuring of the scenarios that lead to physical consequences depends on the nature and magnitude of the assumed threat. In the applications of the methodology presented in [48–50], the threat is assumed to be "minor". This means that major attacks that fail several infrastructure elements at the same time are not considered. Minor threats (vandalism) emanate from individuals or small groups of individuals and not from organized terrorist organizations.

Although Eq. (13) excludes the probability of a successful attack, we can add more information to the ranking process by including the susceptibility to attack of the infrastructure elements. As Table 2 shows, we consider six levels of susceptibility to malevolent acts ranging from completely secure (the lowest level) to completely open (the highest level). These susceptibility levels are combined subjectively with the numerical results for the PI to place a scenario into one of five color categories as shown in Table 3. Table 4 shows how the categories were developed by combining susceptibility and PI in one particular case. The decision maker should develop a similar table for the problem in hand. This subjective evaluation, which may lead to results inconsistent with those

TABLE 2 Susceptibility Levels of Infrastructure Elements

Level	Description
5—Extreme	Completely open, no controls, no barriers
4—High	Unlocked, non-complex barriers (door or access panel)
3—Moderate	Complex barrier, security patrols, video surveillance
2—Low	Secure area, locked, complex closure
1—Very low	Guarded, secure area, locked, alarmed, complex closure
0—Zero	Completely secure, inaccessible

TABLE 3 Vulnerability Categories

Vulnerability Category	Description
Red	This category represents a severe vulnerability in the infrastructure. It is reserved for the most critical locations that are highly susceptible to attack. Red vulnerabilities are those requiring the most immediate attention.
Orange	This category represents the second priority for counter terrorism efforts. These locations are generally moderate to extreme valuable and moderately to extreme susceptible.
Yellow	This category represents the third priority for counter terrorism efforts. These locations are normally less vulnerable because they are either less susceptible or less valuable than the terrorist desire.
Blue	This category represents the fourth priority for counter terrorism efforts.
Green	This is the final category for action. It gathers all locations not included in the more severe cases, typically those that are low (and below) on the susceptibility scale and low (and below) on the value scale. It is recognized that constrained fiscal resources is likely to limit efforts in this category, but it should not be ignored.

TABLE 4 Determination of the Vulnerability Category from the Performance Index and Susceptibility Level

	Susceptibility levels					
Value Levels	Zero	Very Low	Low	Moderate	High	Extreme
0.0000–0.0049	G	G	G	G	G	G
0.0050–0.0299	G	B	B	B	B	B
0.0300–0.0499	G	B	B	Y	Y	Y
0.0500–0.0999	G	B	Y	Y	O	O
0.1000–0.2499	G	Y	Y	O	O	R
>0.2500	B	Y	O	R	R	R

of a rigorous quantitative analysis [56], is the price the decision maker pays for not quantifying the probabilities of successful attacks.

The results of the analysis presented above should be subjected to deliberation among the stakeholders to assess their reasonableness and to evaluate the assumptions on which they are based [34]. The deliberation should be supported by extensive sensitivity analyses. The analytical results should not be the sole basis for decision making. As stated earlier, the decision-making process should be risk informed, not risk based.

In [49], it was found that the water treatment plants were the highest ranked vulnerabilities of the water-supply network both with respect to random failures and vandalism. This result was deemed to be reasonable because these plants are the sources of water for the entire infrastructure. In [50], the transmission lines are found to be the highest vulnerabilities due to their openness and remote locations (i.e. their extreme susceptibility to attacks).

5.4 Multiple Stakeholders

The evaluation of the PI, which is the central metric for risk ranking, depends on the decision maker in very important ways. The structure of the value tree and the corresponding numerical assignments (relative weights and disutilities) reflect the decision maker's values and preferences. In decision analysis, there are methods for eliciting preferences that aid an individual decision maker in making the necessary judgments [47].

In decisions involving infrastructures, the decision maker may be a private organization that owns the infrastructure or a government agency charged with protecting the national interest. In these cases of "societal" decision making the situation is much more complex.

A practical way of developing a consensus value tree is to consult with a few managers of the decision-making agency and use the analytical results as the basis for sensitivity analyses and deliberation. As an example, such an approach was employed in the evaluation of alternative sites for the disposal of nuclear waste in which the decision-making agency was the Office of Civilian Radioactive Waste Management of the US Department of Energy [57]. Another example involves the management of minor incidents in NPPs [58]. In [49], a small group of engineers developed the consensus value tree that was presumed to represent the water utility's values.

If, for whatever reason, consensus is not achieved, a number of evaluations using different value trees can serve as the basis for deliberation. In [50], five representatives of a regional electric utility were consulted. The group included two senior members of the management division. The stakeholders agreed on the structure of the value tree but differed in the weights they assigned to the PMs. Five rankings of the infrastructure elements were produced using the five different inputs. It turned out that the differences in the element rankings were minimal.

The value of the evaluations that we have discussed is better appreciated when we remember that the results are used to inform the decision-making process. The final decision should not be based on these results. These results and the appropriate sensitivity analyses should be input to a deliberative process that will lead to the final decision. A structured way for using analytical results from risk assessment to inform the deliberation is presented in [35].

5.5 Spatial Dependence

As discussed earlier, a useful PRA result, in addition to the ranking of accident sequences, is the ranking of individual events using importance measures. The latter may also be used in infrastructures to determine critical locations, that is, locations which, if attacked successfully, would affect multiple infrastructures. An example using the campus of the Massachusetts Institute of Technology (MIT) is presented in [59].

The concept of RAW, Eq. (11), has been extended to networks in [60]. For multiple Monte Carlo simulations, RAW is defined as

$$\mathrm{RAW}_{ykj} = \frac{U^{+}_{ykj}}{U_{kj}} \tag{14}$$

where U_{kj} is the percent of simulations in which there is no path connecting the user j to an infrastructure k source and U^{+}_{ykj} is the percent of simulations in which element y (node or arc) of infrastructure k *is failed* and there is no path connecting the user j to an infrastructure k source.

The importance measure RAW_{ykj} describes the ratio of risk to user j for infrastructure k when element y is always unavailable to the base-case risk to the user. Reference 59 combines this importance measure with the PI to define the valued worth of element y of infrastructure k as follows:

$$\mathrm{VW}_{yk} \equiv \sum_{j}\left[\mathrm{RAW}_{ykj} \cdot \sum_{i=1}^{k_{\mathrm{PM}}} w_i d_{ijk}\right] \tag{15}$$

The valued worths represent the RAW-scaled potential disutility they evoke to all users of their respective infrastructure. The higher the valued worth, the more important the element y of infrastructure k is to the decision makers. It is important to note that these values represent the potential of an element's unavailability to fail the system and cause a certain amount of disutility; the failure of a high valued worth item may or may not lead to the loss of infrastructure service.

Geographic information systems (GIS) are programs that display geospatial information stored in a database in graphical form. They allow us to perform a network analysis and determine the valued worths for the elements of each infrastructure independently, and then, through GIS algorithms, we can find spatially close nodes/arcs, for example, parallel pipes within a certain distance from each other. Doing this "intersection analysis" after the network analysis allows us to easily change the "intersection distance", that is, how close different elements must be to be considered spatially coincident.

First, we develop a generic grid to be laid across the map of all the infrastructures. The side of each grid space (hexagon) is the size of the threat's radius of influence. We use a hexagonal closely packed grid across the entire region of analysis. Then, we use an internal GIS function to take the maximum valued worth of all elements of the same infrastructure that are located within each hexagon. We sum the maximum valued worth elements from each infrastructure for each hexagon and define the geographic valued worth (GVW) as follows:

$$\mathrm{GVW}_{xz} = \sum_{k} \max\left(\mathrm{VW}_{y_{xz}k}\right) = \sum_{k} \max\left[\sum_{j}\left(\mathrm{RAW}_{y_{xz}kj} \cdot \sum_{i=1}^{k_{\mathrm{PM}}} w_i d_{ijk}\right)\right] \tag{16}$$

where GVW_{xz} is the geographic valued worth of the grid space at coordinates (x, z) and $\max\left(VW_{y_{xz}k}\right)$ is the maximum valued worth element y out of all the elements of infrastructure k that pass through grid element (x, z)

An example of the use of the GVW is shown in Figure 4 in which the lightest gray represents the lowest numerical GVW values and solid black the highest. These GVWs are conditional on a threat that destroys everything within a 7-m radius. The GVWs were calculated on a grid of hexagons with the height and width of two times the radius of influence. We observe that there is a high-GVW "loop" that services the users in the figure. There are lines for electric power, steam, water, and natural gas that all run under the same streets very close to each other to service the dorms that are present in this part of the MIT campus.

Conditional risk maps similar to that in Figure 4 are given to the decision maker who must, then, decide whether the susceptibility of the infrastructure elements in the dark areas of the map is such that the area may be declared a critical location. It is important

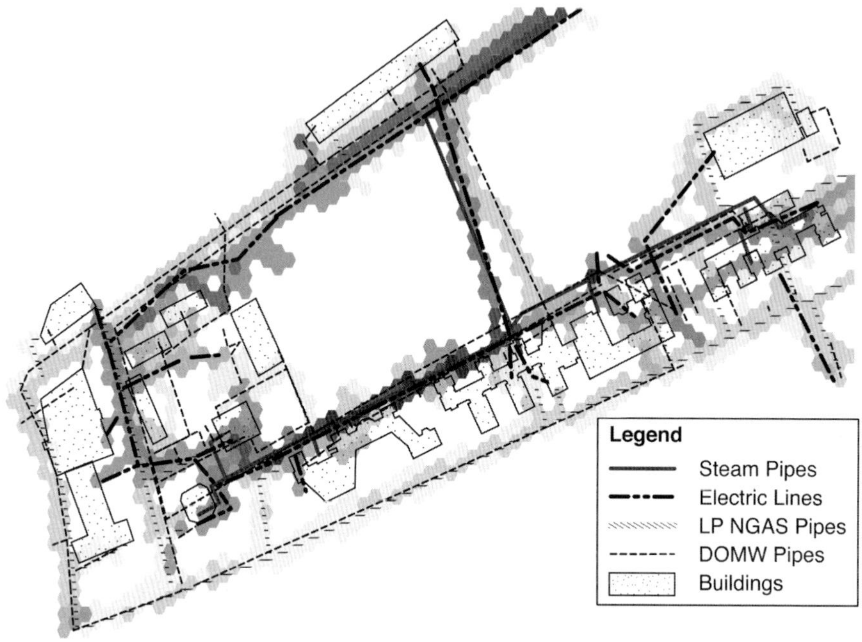

FIGURE 4 GVW conditional risk map based on GVW for part of the MIT campus (reprinted from [59] with permission from Elsevier).

to note that locations of moderate importance to the networks, individually, may be of high importance if all infrastructures are considered together.

5.6 Future Research Directions

The rankings of the infrastructure elements that have been produced using the proposed framework are conditional on the assumption of a "minor" threat, for example, vandalism. For other kinds of threats, multiple elements may be damaged and the methodology needs to be extended to cover these cases.

The element rankings are relative. The absence of the probability of attack inhibits decision making. Just because an infrastructure element is ranked as the most important, it does not necessarily follow that something should be done about it. The absolute value of the infrastructure risk may already be very low therefore not justifying hardening this particular element. In other words, even assuming that society is willing to determine levels of "acceptable" or "tolerable" terrorism risks (a very strong assumption), we do not currently have the analytical tools that allow the quantification of the risk. Yet, resources must be allocated to counterterrorism measures; therefore, the need for developing such tools is high.

REFERENCES

1. US Nuclear Regulatory Commission. (1990). *Severe Accident Risks: An Assessment for Five US Nuclear Power Plants*, Report NUREG-1150, Washington, DC.

2. National Research Council. (1997). Risk assessment and management at deseret chemical depot and the tooele chemical agent disposal facility. *Committee on Review and Evaluation of the Army Chemical Stockpile Disposal Program*. National Academy Press, Washington, DC.
3. Futron Corporation. (2001). *Probabilistic Risk Assessment of the International Space Station. Phase III–Stage 12A.1 Configuration*, Washington, DC.
4. Kaplan, S., and Garrick, B. J. (1981). On the quantitative definition of risk. *Risk Anal.* **1**, 11–27.
5. *SAPHIRE* is available at: http://saphire.inel.gov. RISKMAN is available at: http://www.absconsulting.com/riskmansoftware/index.html. RiskSpectrum is available at: http://www.relcon.se/.
6. Stamatelatos, M. G., Apostolakis, G., Dezfuli, H., Everline, C., Guarro, S., Moieni, P., Mosleh, M., Paulos, T., and Youngblood, R. (2002). *Probabilistic Risk Assessment Procedures Guide for NASA Managers and Practitioners*, Version 1.1, Office of Safety and Mission Assurance. NASA Headquarters, Washington, DC.
7. Nuclear Regulatory Commission. (1982). *PRA Procedures Guide*, Report NUREG CR-2300, Washington, DC.
8. Swain, A. D., and Guttmann, H. E. (1983). *Handbook of Human Reliability Analysis with Emphasis on Nuclear Power Plant Applications*, Report NUREG/CR-1278. Nuclear Regulatory Commission, Washington, DC.
9. Forester, J., Bley, D., Cooper, S., Lois, E., Siu, N., Kolaczkowski, A., and Wreathall, J. (2004). Expert elicitation approach for performing ATHEANA quantification. *Reliab. Eng. Syst. Saf.* **83**, 207–220.
10. Nuclear Regulatory Commission. (2000). *Technical Basis and Implementation Guidelines for A Technique for Human Event Analysis (ATHEANA)*, Report NUREG-1624, Washington, DC.
11. Moieni, P., and Spurgin, A. J. (2004). Advances in human reliability analysis methodology. *Reliab. Eng. Syst. Saf.* **44**, 27–66.
12. Julius, J., and Grobbelaar, J. (2006). Integrating Human Reliability Analysis Approaches in the EPRI HRA Calculator. *8th Probabilistic Safety Assessment and Management Conference (PSAM 8)*. sponsored by the International Association for Probabilistic Safety Assessment and Management, New Orleans, LA, (http://www.iapsam.org/).
13. Budnitz, R. J., Apostolakis, G., Boore, D. M., Cluff, L. S., Coppersmith, K. J., Cornell, C. A., and Morris, P. A. (1997). *Recommendations for Probabilistic Seismic Hazard Analysis: Guidance on Uncertainty and Use of Experts*, Report NUREG/CR-6372, Vols. 1 and 2, Nuclear Regulatory Commission, Washington, DC.
14. Nuclear Regulatory Commission and Electric Power Research Institute. (2005). *Fire PRA Methodology for Nuclear Power Facilities*, Report NUREG/CR-6850 (EPRI 1011989), vols. 1 and 2, Washington, DC/Palo Alto, CA.
15. Helton, J. C., Anderson, D. R., Basabilvazo, G., Jow, H.-N., and Marietta, M. G. (2000). Conceptual structure of the 1996 performance assessment for the waste isolation pilot plant. *Reliab. Eng. Syst. Saf.* **69**, 151–165.
16. Thompson, B. G. J., and Sagar, B. (1993). The development and application of integrated procedures for post-closure assessment, based upon Monte Carlo simulation: The Probabilistic Systems Assessment (PSA) approach. *Reliab. Eng. Syst. Saf.* **42**, 125–160.
17. Ghosh, S. T., and Apostolakis, G. E. (2006). Extracting risk insights from performance assessments for HLW repositories. *Nucl. Technol.* **153**, 70–88.
18. Tversky, A., and Kahneman, D. (1974). Judgments under uncertainty: heuristics and biases. *Science* **185**, 1124–1131.
19. Apostolakis, G. (1990). he concept of probability in safety assessments of technological systems. *Science* **250**, 1359–1364.

20. Winkler, R. L. (1996). Uncertainty in probabilistic risk assessment. *Reliab. Eng. Syst. Saf.* **54**, 127–132.
21. De Finetti, B. (1974). *Theory of Probability*, Vols. 1 and 2. John Wiley & Sons, New York.
22. Savage, L. J. (1972). *The Foundations of Statistics*. Dover Publications, New York.
23. Nuclear Regulatory Commission. (2002). *An Approach for Using Probabilistic Risk Assessment in Risk-Informed Decisions on Plant-Specific Changes to the Current Licensing Basis*, Regulatory Guide 1.174, Revision 1, Washington, DC.
24. Keeney, R. L., and von Winterfeldt, D. (1991). Eliciting probabilities from experts in complex technical problems. *IEEE Trans. Eng. Manag.* **38**, 191–201.
25. Budnitz, R. J., Apostolakis, G., Boore, D. M., Cluff, L. S., Coppersmith, K. J., Cornell, C. A., and Morris, P. A. (1998). Use of technical expert panels: applications to probabilistic seismic hazard analysis. *Risk Anal.* **18**, 463–469.
26. Helton, J. C., Johnson, J. D., Sallaberry, C. J., and Storlie, C. B. (2006). Survey of sampling-based methods for uncertainty and sensitivity analysis. *Reliab. Eng. Syst. Saf.* **91**, 1175–1209.
27. Nuclear Regulatory Commission. (1990). *Severe Accident Risks: An Assessment for Five US Nuclear Power Plants*, Report NUREG-1150, Washington, DC, available at http://www.nrc.gov/.
28. Wilson, R., and Crouch, E. A. C. (2001). *Risk-Benefit Analysis*. Harvard University Press, Cambridge, MA.
29. Nuclear Regulatory Commission. (1986). *Safety Goals for the Operations of Nuclear Power Plants*. Federal Register, vol. 51, p. 30028, August 4.
30. United Kingdom Health and Safety Executive. (2001). *Reducing Risks, Protecting People. HSE's Decision-Making Process* . Her Majesty's Stationery Office, Norwich, available at http://www.hse.gov.uk/risk/theory/r2p2.pdf
31. Cheok, M. C., Parry, G. W., and Sherry, R. R. (1998). Use of importance measures in risk-informed regulatory applications. *Reliab. Eng. Syst. Saf.* **60**, 213–226.
32. Apostolakis, G. E. (2004). How useful is quantitative risk assessment? *Risk Anal.* **24**, 515–520.
33. Sorensen, J. N., Apostolakis, G. E., Kress, T. S., and Powers, D. A. (1999). On the role of defense in depth in risk-informed regulation. *Proceedings of PSA '99, International Topical Meeting on Probabilistic Safety Assessment*, Washington, DC, August 22–26. American Nuclear Society, La Grange Park, IL, pp. 408–413.
34. National Research Council. (1996). *Understanding Risk: Informing Decisions in a Democratic Society*. National Academy Press, Washington, DC.
35. Apostolakis, G. E., and Pickett, S. E. (1998). Deliberation: integrating analytical results into environmental decisions involving multiple stakeholders. *Risk Anal.* **18**, 621–634.
36. National Research Council. (2002). *Making the Nation Safer. The Role of Science and Technology in Countering Terrorism*. National Academy Press, Washington, DC.
37. Bley, D., Kaplan, S., and Johnson, D. (1992). The strengths and limitations of PSA: where we stand. *Reliab. Eng. Syst. Saf.* **38**, 3–26.
38. Bier, V. M. (1999). Challenges to the acceptance of probabilistic risk assessment. *Risk Anal.* **19**, 703–710.
39. Reason, J. (1997). *Managing the Risks of Organizational Accidents*. Ashgate, Aldershot, Hampshire, UK.
40. Aldemir, T., Miller, D. W., Stovsky, M. P., Kirschenbaum, J., Bucci, P., Fentiman, A. W., and Mangan, L. T. (2006). *Current State of Reliability Modeling Methodologies for Digital Systems and Their Acceptance Criteria for Nuclear Power Plant Assessments*, Report NUREG/CR-6901, Nuclear Regulatory Commission, Washington, DC.

41. Nuclear Regulatory Commission. *Protecting Our Nation*. Available at www.nrc.gov.
42. Garrick, B. J., Hall, J. E., Kilger, M., McDonald, J., McGroddy, J., O'Toole, T., Probst, P., Rindskopf Parker, E., Rosenthal, R., Trivelpiece, A., Van Arsdale, L., and Zebroski, E. (2004). Confronting the risks of terrorism: making the right decisions. *Reliab. Eng. Syst. Saf.* **86**, 129–176.
43. Haimes, Y. Y. (2002). Roadmap for modeling risks of terrorism to the homeland. *J. Infrastruct. Syst.* **8**, 35–41.
44. Paté-Cornell, M. E., and Guikema, S. (2002). Probabilistic modeling of terrorist threats: a systems analysis approach to setting priorities among countermeasures. *Mil. Oper. Res.* **7**, 5–20.
45. Bier, V. M., Nagaraj, A., and Abhichandani, V. (2005). Protection of simple series and parallel systems with components of different values. *Reliab. Eng. Syst. Saf.* **87**, 315–323.
46. Lye, K., and Wing, K. M. (2005). Game strategies in network security. *Int. J. Inf. Secur.* **4**, 71–86.
47. Clemen, R. T. (1996). *Making Hard Decisions: An Introduction to Decision Analysis*, 2nd ed. Duxbury Press, Belmont, CA.
48. Apostolakis, G. E., and Lemon, D. M. (2005). A screening methodology for the identification and ranking of infrastructure vulnerabilities due to terrorism. *Risk Anal.* **25**, 361–376.
49. Michaud, D., and Apostolakis, G. E. (2006). Methodology for ranking the elements of water-supply networks. *J. Infrastruct. Syst.* **12**, 230–242.
50. Koonce, A. M., Apostolakis, G. E., and Cook, B. K. (2008). Bulk power grid risk analysis: ranking infrastructure elements according to their risk significance. *Int. J. Electr. Power Energy Syst*. **30**: 169–183.
51. Richardson, B. (2005). *Sandia Load Flow Simulation Model*. Sandia National Laboratories, Albuquerque, NM.
52. Saaty, T. L. (2000). *Fundamentals of Decision Making and Priority Theory*, Vol. VI. RWS Publications, Pittsburgh, PA.
53. Holder, R. D. (1990). Some comments on the analytic hierarchy process. *J. Oper. Res. Soc.* **41**, 1073–1076.
54. Forman, E. H. (1996). Facts and fictions about the analytic hierarchy process. In *The Analytic Hierarchy Process*, T. L. Saaty, Ed. RWS Publications, Pittsburgh, PA.
55. Stamatelatos, M., Dezfuli, H., and Apostolakis, G. (2006). A proposed risk-informed decision-making framework for NASA. Presented at the *8th International Conference on Probabilistic Safety Assessment and Management (PSAM 8)*. New Orleans, LA, 15–18 May. (www.iapsam.org).
56. Cox, L. A., Jr., Babayev, D., and Huber, W. (2005). Some limitations of qualitative risk rating systems. *Risk Anal.* **25**, 651–662.
57. Merkhofer, M. W., and Keeney, R. L. (1987). A multiattribute utility analysis of alternative sites for the disposal of nuclear waste. *Risk Anal.* **7**, 173–194.
58. Pagani, L., Smith, C., and Apostolakis, G. (2004). Making decisions for incident management in nuclear power plants using probabilistic safety assessment. *Risk Decis. Policy* **9**, 271–295.
59. Patterson, S. A., and Apostolakis, G. E. (2007). Identification of critical locations across multiple infrastructures for terrorist actions. *Reliab. Eng. Syst. Saf.* **92**: 1183–1203.
60. Zio, E., Podofillini, L., and Zille, V. (2006). A combination of Monte Carlo simulation and cellular automata for computing the availability of complex network systems. *Reliab. Eng. Syst. Saf.* **91**, 181–190.

SCENARIO ANALYSIS, COGNITIVE MAPS, AND CONCEPT MAPS

Kenneth G. Crowther and Yacov Haimes
Center for Risk Management of Engineering Systems, Department of Systems and Information Engineering, University of Virginia, Charlottesville, Virginia

1 INTRODUCTION

In her book *Pearl Harbor: Warning and Decision* [1962] recently cited by the 9/11 Commission Report [1], Roberta Wohlstetter made the following comment [2]:

> It is much easier after the event to sort the relevant from the irrelevant signals. After the event, of course, a signal is always crystal clear; we can now see what disaster it was signaling since the disaster has occurred. But before the event it is obscure and pregnant with conflicting meetings.

Those same words, which describe the Japanese attack on Pearl Harbor in 1941, could be used 60 years later to describe many modern disasters. Nassim Taleb [3], in his highly provocative essay *The Black Swan: The Impact of the Highly Improbable*, confirms Roberta Wohlstetter's insight with one additional caveat—the hindsight understanding produced by assessing signals *after* an event is usually not general enough to yield insight, missing causal understanding, and often lacks meaningful impact on decisions. "Clear" hindsight without inventive foresight is insufficient to cope with risks of improbable and elusive events to the homeland. Moreover, it is the improbable, elusive events that result in the greatest impact (e.g. unimagined terrorist attacks with passenger planes in New York; implausible combinations of hurricane, infrastructure, and social demographics in New Orleans; unlikely underwater earthquakes and resulting tsunami in southeastern Asia; etc.).

By their definition, disasters constitute extreme and catastrophic events; thus, their probabilities and associated consequences defy any common expected value representation of risk. Many analysts and decision theorists are beginning to recognize a simple yet fundamental philosophical truth—in the face of such unforeseen or elusive calamities as bridges falling, dams bursting, airplanes crashing, tsunamis washing, and hurricanes landing with great force, we must acknowledge the importance of studying "extreme" events [4, 5].

2 OVERVIEW—SCENARIO ANALYSIS

Lowrance [6] described *risk* as "a measure of the *probability* and *severity* of *adverse effects*". Kaplan and Garrick [7] were the first to formalize a theory of quantitative risk assessment with the triplet:

What can go wrong?
What is the likelihood?
What are the consequences?

Risk assessments that answer the above triplet questions hinge on the ability to identify risk scenarios and organize them according to some understanding of their approximated consequences and perceived frequency. To begin such a risk assessment process, the analyst must clearly define the as-planned or success scenario [8]. The risk scenarios or *what-can-go-wrong* events are then defined as any deviation from the as-planned scenario. Kaplan et al. [9] present a Theory of Scenario Structuring (TSS) based on a systemic set of methods to decompose a system with respect to time, operation, process components, or other means to create a complete set of risk scenarios as a foundation of analysis. TSS is a generalized method into which most common risk assessment methods can be described, including [10]: (i) failure modes and effects analysis (FMEA), which divides a system into functional parts and derives risk scenarios from the failures of functional components (e.g. MIL-STD-1629A; Benbow et al. [11]), (ii) hazard and operations analysis (HAZOP), which divides a plant into physical sections and assesses the impacts of risk scenarios that result from various extreme operational conditions (e.g. Kletz [12]), (iii) event trees (ET), which seeks to assess cascades of errors or risk scenarios through time that derive from an initiating event (e.g. Smith [13]), (iv) fault trees (FT), which seeks the causes of predefined, undesirable end states or risk scenarios (e.g. Hoyland and Rausand [14] and US Nuclear Regulatory Commission [15]), (5) anticipatory failure determination (AFD), which assesses what an operator or agent can make go wrong to cause a risk scenario to occur [9], and (vi) hierarchical holographic modeling (HHM), which decomposes a system into hierarchal overlapping categories of operation and function on the basis of fundamental system structure and develops risk scenarios relevant to each overlapping category [5, 16, 17]. All of these methods are a form of scenario analysis—prescribing a systematic way of analyzing a system under a systemic set of conditions. The following subsections will focus on AFD and HHM as example resources in scenario analysis. Many of the other methodologies are supported by these methods and are described in other articles of this book.

2.1 Scenario Analysis—Anticipatory Failure Determination (AFD)

The TRIZ method, whose acronym in Russian stands for "Theory of Inventive Problem Solving", is a scenario structuring method developed to assist in stimulating the creation of new inventions and finding new solutions for real world problems [18]. An off-shoot tool of TRIZ is AFD method, which is used for failure analysis. Unlike traditional failure analysis methods, such as FMEA, which look for a cause of failure, AFD views failure as an intended consequence [19]. Analysts then try to devise ways of ensuring that failures always happen, thereby generating a list of causes. AFD asks the question "If I wanted to create a particular failure, how can I do it in the most effective way?" [10]. AFD provides a structure to the use of "red teaming" to identify weaknesses, which seems well documented [20–22].

2.2 Scenario Analysis—Hierarchical Holographic Modeling (HHM)

Real large-scale systems are complex in nature. Many large-scale systems exhibit characteristics, daunting to modelers, such as a large number of subsystems, a hierarchical structure, multiple decision makers, and elements of risk and uncertainty. It may be impractical to use a single-model analysis to describe such large-scale systems, particularly in the context of identifying potential threat scenarios and other sources of risk. HHM [5, 16] is a holistic philosophy and methodology which captures and represents the essence of the inherent diverse characteristics and attributes of a system—its multiple aspects, perspectives, facets, views, dimensions, and hierarchies—and is suitable for describing complex, large-scale systems. Specifically, HHM has the capability of presenting risk scenarios from multiple overlapping perspectives, consistent with the overlapping view of the system [10].

The term *holographic* refers to the multiple perspectives with which a system can be described (as opposed to a single view or a flat planar image). The term *hierarchical* in HHM refers to describing the myriad levels of a system hierarchy—structurally or organizationally. Because it examines the system from multiple perspectives and decomposes those perspectives hierarchically into subsystems, HHM is a valuable tool for identifying the myriad sources of risks to a system. Sources of risk can be categorized by a number of perspectives, or HHM head-topics, such as hardware, software, organizational, temporal, and geographical. Sources of risk can manifest themselves at macroscopic and microscopic levels of a system or at various management levels of an organization, each described by HHM subtopics and further.

Recent applications of HHM in identifying sources of risk include defining types of hardening to reduce vulnerabilities in water systems partially shown in Figure 1 [17], enumerating the sources of risk in large-scale software acquisition [23], identifying risk scenarios as a tool in the TSS framework [10], identifying risks to information assurance [24], capturing risk scenarios for military operations other than war [25], cataloging risk issues encountered in space missions [5], and identification of possible transportation system disruptions [26].

2.3 Scenario Analysis—Collaborative Adaptive Multiplayer HHM (CAM-HHM)

If determining what can go wrong is the thesis (e.g. through HHM) and determining how can we make the system fail is the antithesis (e.g. through AFD), then their synthesis combines the efficacy of both for holistic analysis of risk scenarios. A recent extension of HHM, the collaborative adaptive multiplayer hierarchical holographic modeling (CAM-HHM), better establishes the collection of data from multiple perspectives and sources. Haimes and Horowitz [27] introduced the concept as the adaptive two-player HHM game, which allowed two teams (e.g. Red and Blue) with very different perspectives to generate risk scenarios for the purpose of intelligence collection and analysis. Developing two independent structures of risk scenarios provides (i) a more holistic view of the system created when the two models are aggregated, (ii) valuable benchmark information on the depth and breadth of the assessment, and (iii) greater self-understanding and knowledge of the opponent [27]. HHMs are created by each team and combined in an iterative process until convergence on a final, systemic HHM is generated.

The technique was designed to encourage collaboration among several experts from disparate backgrounds for identifying sources of risk to the system. The fundamental difference between the red team and blue team is their knowledge base, perspective, and experience. The Red Team approaches the problem from a terrorist perspective: What kind of threats can I create from publicly available or stolen intelligence data that will defeat the system? The Blue Team approaches the problem from a defensive perspective:

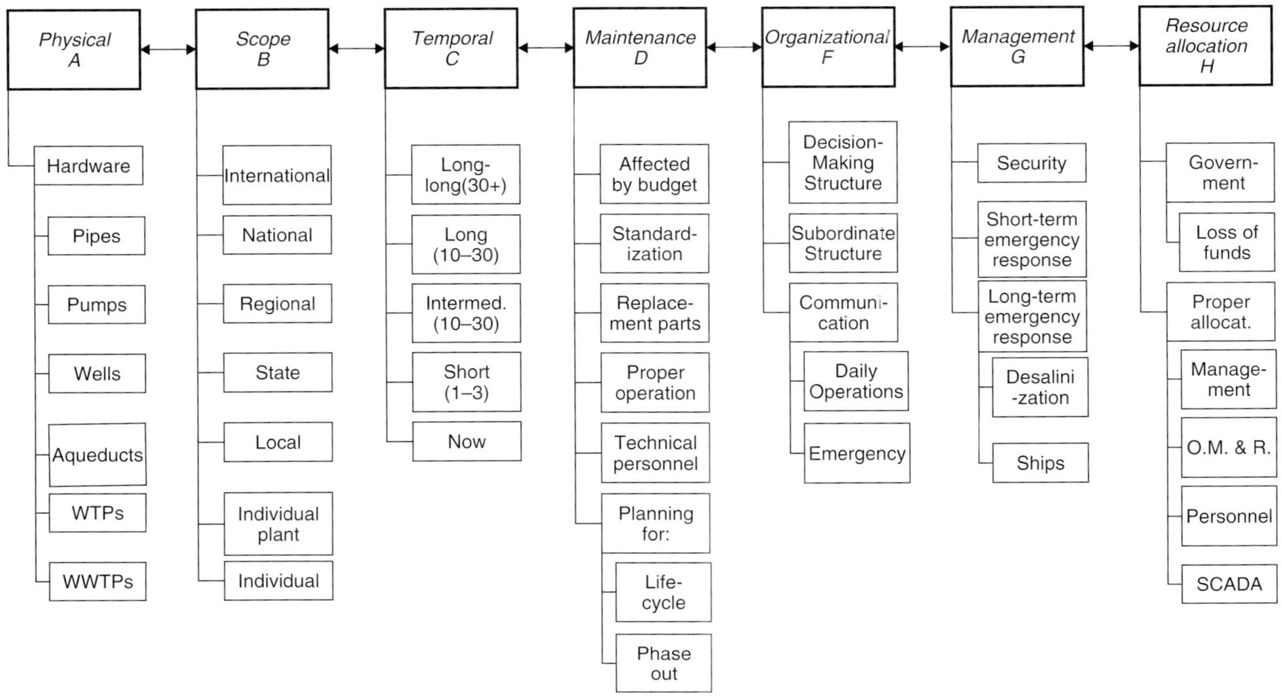

FIGURE 1 Partial HHM structuring multiple perspectives on the hardening of the water-supply system [4, p. 121].

Using our internal or classified information, what weaknesses or vulnerabilities are present? [27] Collaboration across teams provides a more holistic collection of data from many perspectives and multiple hierarchical levels.

The two-player concept was extended to account for multiple Red Team and Blue team perspectives with the CAM-HHM. One of the first deployments of the CAM-HHM tool was in capturing a board perspective on risk in oil and gas infrastructures due to cyber threats for the Institute for Information Infrastructure Protection (I3P) [28]. In that example, sources of risk to supervisory control and data acquisition (SCADA) systems were gathered from four perspectives: attackers and hackers (representing Red Team interests), SCADA owners and operators (representing Blue Team interests), SCADA vendors, and stakeholders. In four parallel 40-min sessions, they created a taxonomy of 148 classes of industrial cyber risks [26].

A more recent deployment of the CAM-HHM tool dealt with identifying sources of risk to the 2006 Virginia gubernatorial inauguration [29]. Three teams were involved in the CAM-HHM game: officials from the transportation sector, officials with health care experience and others from the private sector, and members with experience in creating security plans (representing Red Team interests). The CAM-HHM game collaboratively produced 272 sources of risk of interest to security officials in planning for the inauguration. The results of the CAM-HHM fit into a larger scenario analysis framework, where risk scenarios are identified, qualitatively and quantitatively assessed, and risk management plans are developed to reduce or eliminate the sources of risk.

2.4 Scenario Analysis—Risk Filtering, Ranking, and Management (RFRM)

The objective of risk scenario structuring is to generate a comprehensive set of perspectives with which to view a system and its numerous sources of risks, potentially producing an unmanageable number of risk scenarios. Often, it is impractical to perform scenario analysis on hundreds of sources of risk. There is a need to discriminate among them in a systematic fashion according to the likelihood and severity of occurrence. To serve this purpose, the risk filtering, ranking, and management (RFRM) method was developed [5, 30].

The RFRM methodology systematically evaluates a comprehensive set of risk scenarios and then differentiates them according to the likelihood and severity of their consequences. The methodology effectively considers a variety of risk scenarios and integrates empirical and conceptual, descriptive and normative, as well as quantitative and qualitative methods and approaches to complete the scenario analysis process. The RFRM process is aimed at providing priorities in analyzing assets in the decision makers' domain of interest. This means exploring the more urgent sources of risks or scenarios first; it does not imply ignoring the sources of risks that have been filtered out earlier.

The RFRM process is systematically carried out in eight phases, as described below.

Phase I: scenario identification. An HHM is developed to describe the system in a comprehensive way, including all relevant risk scenarios.

Phase II: scenario filtering. The risk scenarios enumerated in the HHM produced in Phase I are filtered according to the interests and scope of the analysis.

Phase III: bicriteria filtering and ranking. The set of risk scenarios is further filtered with qualitative relative likelihoods and consequences.

Phase IV: multicriteria evaluation. The abilities of the remaining risk scenarios to defeat the defenses of the system are evaluated with a set of 11 criteria, including duration of effects, cascading effects, and complexity and emergent behaviors.

Phase V: quantitative ranking. Filtering and ranking of scenarios continues with a quantitative and qualitative matrix of likelihoods and consequences.

Phase VI: risk management. Risk management options are identified to deal with the sources of risks. Included are issues of costs and benefits in risk reduction.

Phase VII: safeguarding against missing critical items. The performance of the risk management options selected in Phase VI is evaluated against the scenarios previously filtered out during Phases II through V.

Phase VIII: operational feedback. The scenario filtering and decision processes of previous phases are refined using the experience gained in applying the risk management options.

The RFRM framework has been applied in several risk and scenario filtering contexts, including focusing limited resources on the most risky and uncertain sources of risk to military operations other than war [25], filtering over 900 sources of risk to US Army telecommunications systems information assurance [24], prioritizing the protection of transportation assets against terrorist attacks [31], and identifying and prioritizing risks to US Army critical assets relative to organizational goals [32]. In the 2006 VA Gubernatorial inauguration exercise, the original 272 HHM topics identified were prioritized into 77 scenarios based on their likelihoods and severity assessments. The prioritized set of risk scenarios then become the focus of resource allocation and other risk management actions and more detailed quantitative modeling.

2.5 Scenario Analysis—Risk Management

Haimes [4, 5] presents a second set of triplet questions that outline the fundamental tasks of risk management (see Phase IV of RFRM):

What can be done and what options are available?

What are the trade-offs in terms of costs, benefits, and risks?

What are the impacts of current decisions on future options?

Answers to these questions critically enhance the capacity to make appropriate decisions about acceptable levels of risk. Haimes [5] provides an extensive discussion of quantitative risk assessment and management. No single model can capture all the dimensions necessary to adequately evaluate the efficacy of risk assessments and management activities because of the impossible task of identifying all relevant state variables and their substates for adequately representing large and multiscale systems [5, 16, 33].

Scenario analysis legitimizes the exploration and experimentation of out-of-the-box and seemingly "crazy" ideas and ultimately discovers insightful implications that otherwise would have been completely missed and dismissed [26]. To illustrate, consider a hypothetical coastal region that clearly lies in a flood zone. Now consider several dimensions of the system that might be juxtaposed to result in adverse regional consequences, including rare hazards such as hurricanes, uncertain forecasts of the hurricanes, and demographic tension (such as racial). A scenario analysis focusing on a rare hurricane that could potentially destroy a coastal region would cost millions, and might justify the appointment of an individual who constantly communicates with the national weather services. However, the uncertainty might result in a last minute capability to evacuate the region. Such uncertainty in the forecast would justify the utilization of an otherwise unused assets, such as school buses to aid in the evacuation from the coastal region. However, a focus in the scenario analysis that focus on demographic tensions might result in the justification of contracting charter buses through a prehurricane memorandum of understanding for last minute evacuation at a known cost. The output of the scenario structuring methods is a taxonomy of identified risk scenarios that are constructed from the multiple perspectives of a system for modeling (e.g. hazard, infrastructure, information, and demographics). Moreover, the output of the scenario analysis process is a justification or rationalization of mitigation against the improbable event, and investment in preparedness or learning activities to protect against critical forced changes or emergent risks—investment that might not otherwise have been approved. Through logically organized and systemically executed analyses, scenario analysis provides a reasoned experimental modeling framework with which to explore and thus understand the intricate relationships that characterize the nature of multiscale infrastructure and organizational systems. This philosophy rejects dogmatic problem solving that relies on a single modeling approach structured on one school of thinking. Rather, its modeling schema builds on the multiple perspectives gained through generating multiple scenarios. This leads to the construction of appropriate models to deduce tipping points as well as meaningful information for logical conclusions and future actions. Currently, models assess what is optimal, given what we know, or what we think we know. Scenario analysis extends this mind-set framework to answer other questions, such as (i) What do we need to know? (ii) What value might appear from risk reduction results having more precise and updated knowledge about complex systems, and (iii) Where is that knowledge needed for acceptable risk management and decision making?

Masterpieces are not usually created on the first painting, but are usually created by selecting themes and exploring those themes to develop knowledge and understanding the final product of which can then be carefully designed based upon what is learned through experience. Scenario analysis is a modeling paradigm that is congruent with and responsive to the uncertain and ever-evolving world of emergent systems. In this sense, it serves as an adaptive process, a learn-as-you-go modeling laboratory, where different scenarios (e.g. generated through the CAM-HHM, introduced in a previous section) of needs and developments for emergent systems can be explored and tested. In other words, to represent and understand the uncertain and imaginary evolution of a future emergent system, we need to deploy an appropriate modeling technology that is equally agile and adaptive.

FEMA's HAZUS-MH for hurricanes [34] is an example of the type of tool that has emerged to support scenario analysis. The tool construction has resulted from the integration of databases, models, and simulation tools that have been developed across many disciplines over the last several decades., As an integrated tool it can be used to study the impact of various hurricane scenarios on regions and their system states. At the basic modeling level there are databases of buildings, businesses, essential facilities, and other basic structural and regional knowledge that characterize the various system dimensions of the region under study. These databases are editable to enable explorations of agile properties of structures that may change the impact of hurricanes. Scientific models from decades of structural research estimate probabilistic structural damage from wind gusts to various structural vulnerabilities. Finally, there is a hazard model to estimate peak wind gust given historical or user-defined catastrophes. The integration of databases, causal damage models, and flexible hazard simulations results in a tool that enables regions to fully explore ranges of improbable situations.

3 SCENARIO ANALYSIS, COGNITIVE MAPS, AND CONCEPT MAPS

Scenario analysis is a methodological process for developing such analyses that will have the flexibility to capture emergent behavior of both regional vulnerabilities and the threats. Moreover, these solutions to a scenario process result in a method to trace changes in problem definitions, critical variables, critical metrics, available data, and so on, in a way that enables us to measure learning, changing, and improvement in risk management activities over time. However, effective scenario analyses rely on the capability to ascertain the decision processes of a region, namely what assumptions are needed to make decisions, how decision makers connect information and map it to decisions, and how do groups of decision makers, which govern a region, assimilate new information to adapt their decision-making processes. Cognitive maps support our ability to understand how decision makers map understanding of the world. Concept maps have general use. In this article, we describe how concept maps can be used to represent and capture the cognitive maps in order to customize the filtering of the scenario analysis.

4 OVERVIEW—COGNITIVE MAPS

Cognitive maps is supplement and complement scenario analyses. Kitchin [35] synthesizes the use of the term *cognitive maps* through several decades of literature. Tolman [36] is credited as the earliest to use the term. Tolman [36] created the term to articulate his opposition to a commonly held viewpoint that rats learn through physiological changes, and proposed instead that rats form mental or psychological notions of their environment (i.e. cognitive maps) that evolve and adapt as they gain experience. Kitchin [35] describes how multiple uses of the terminology have developed over time to have explicit, analogical, metaphorical, and hypothetical meanings. For example, Lieblich and Arbib [37], describe parts of the brain that actually store topologies of the environment (i.e. explicit definition), whereas Downs [38] describes how our actions are consistent with a model of environmental processes *as if* we stored a mental map (i.e. metaphorical definition). Despite all these meanings, every person forms models of their environment through acquisition, storage, analysis, and synthesis of perceived environmental information. Developing cognitive maps are useful tools for analysis of factors that influence decision processes.

Axelrod [39] was the earliest to develop cognitive maps as a means of explaining phenomena that are caused by cognition and decisions. Walsh [40] provides an extensive review and organization of approximately 400 books and peer reviewed journal articles to describe how cognitive maps have been developed from multiple fields of psychology and management. Weick [41] describes cognitive mappings as a well-established technique for capturing the thinking of decision makers about situations, problems, and adverse events.

There are several approaches that have been used to elicit and depict cognitive maps. They are typically customized to the background of the individual researcher and the topic. See Walsh [40] for several examples. This article will focus on the utilization of concept maps to guide the elicitation and representation of cognitive maps.

5 OVERVIEW—CONCEPT MAPS

Concept maps are graphical tools to organize and represent knowledge. They are constructed from a focus question (i.e. a decision) and are constructed by identifying concepts and concept relationships that are believed to result in a capability to answer the focus question. Triplets of concepts and their relationships are propositions or semantic units of understanding. Data can be acquired to provide evidence concerning a concept or semantic unit on the map. Such evidence over time will result in a learning process that connects actions and occurrences to answers and questions. If the concept map represents the group cognitive map of a process, then the concept map is able to channel the occurrences to update the set of decisions that results from the scenario analyses. Novak and Musonda [41] describe that concept maps were developed in 1972 from a study of the growth of children's knowledge of science.

6 EXAMPLE APPLICATION OF CONCEPT MAPS TO CYBER SECURITY

To illustrate how one might capture cognitive maps through concept maps consider the following cyber security example. A team of executives, information technology managers, and security managers meet together and perform a scenario analysis that results in a demonstration that high leverage, low cost solutions (e.g. antivirus, intrusion detection, and patch management) are highly desired. Moreover, the scenario analysis identifies potential justification of encryption and key management technologies depending on the actual likelihood that intellectual property will be stolen. In order to provide a solution that implements a learning strategy in addition to high leverage, low cost security, the decision team constructs a concept map that captures the major assumptions and data needs that might be required to make a final justification decision for encryption and key management.

Figure 2 shows the focus question on the left, namely whether a certain portion of the IT budget should be invested in encryption and key management or in other portions of the IT budget that improve productivity. Across the team the decision depends on four general categories of knowledge that include the character of the risk of cyber theft, the effectiveness of encryption to mitigate the threat, the nominal market share of the corporation, and the effectiveness of other IT investment options in increasing productivity. All of these general categories are then further characterized with more specific concepts that define the categories. For example, the risk of cyber theft is defined by the probability that the particular theft might occur and the consequences if the theft is successful. The consequence of theft is the reduction in nominal market share of the company. Other relationships among the concepts can be defined as shown in Figure 2.

This concept map will not hold for all teams or all companies that face similar decisions, but rather represents the method of cognition of the individual team that creates it. The result is a set of concepts that represent the data requirement for making a decision concerning an investment related to a tipping point identified by the scenario analysis. If the HHM methodology (described in a previous section) is adapted for the scenario structure component of the scenario analysis, then the subcategories of the system will become a pool of concepts that can be utilized in the construction of concept maps.

This process applies broadly to homeland security and counterterrorism, where systems are large and complex and where the threats against the system are emergent and largely not well defined.

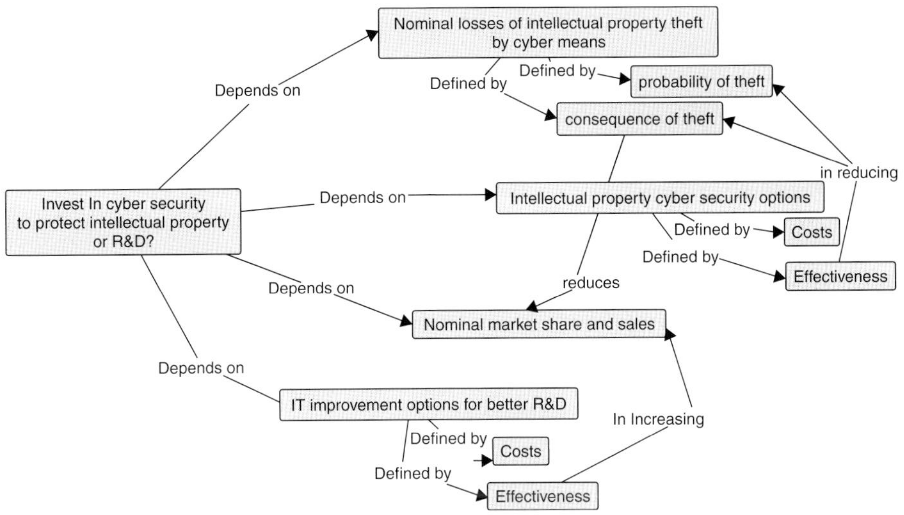

FIGURE 2 Concept map for a cyber security investment question.

REFERENCES

1. 9/11 Commission (2004). *The 9/11 Commission Report: Final Report of the national Commission on Terrorist Attacks upon the United States*, W.W. Norton and Company, New York.
2. Wohlstetter, R. (1962). *Pearl Harbor: Warning and Decision*, Stanford University Press, Stanford, CA.
3. Taleb, N. N. (2007). *The Black Swan: The Impact of the Highly Improbable*, Random House, New York.
4. Haimes, Y. Y. (1991). Total risk management. *J. Risk Anal.* **11**(2), 169–171.
5. Haimes, Y. Y. (2004). *Risk Modeling, Assessment, and Management*, 2nd ed., Wiley, Hoboken, NJ.
6. Lowrance, W. W. (1976). *Of Acceptable Risk: Science and the Determination of Safety*, William Kaufmann, Inc., Los Altos, CA.
7. Kaplan, S., and Garrick, B. J. (1981). On the quantitative definition of risk. *Risk Anal.* **1**(1), 11–27.
8. Kaplan, S. (1991). The general theory of quantitative risk assessment. In *Risk Based Decision Making in Water Resources V*, Y. Y. Haimes, D. A. Moser, and E. Z. Stakhiv, Eds. American Society of Civil Engineers, New York, pp. 11–39.
9. Kaplan, S., Haimes, Y. Y., and Garrick, B. J. (2001). Fitting hierarchical holographic modeling into the theory of scenario structuring and a resulting refinement to the quantitative definition of risk. *Risk Anal.* **21**(5), 807–819.
10. Benbow, D. W., Berger, R. W., Elshennaway, A. K., and Walker, H. F. (2002). *The Certified Quality Engineer Handbook*, ASQ Quality Press, Milwaukee, WI.
11. Kletz, T. A. (1999). *Hazop and Hazan*, 4th ed., CRC Press, Boca Raton, FL.
12. Smith, D. J. (2005). *Reliability, Maintainability, and Risk: Practical Methods for Engineers including Reliability Centered Maintenance and Safety-Related Systems*, 7th ed., Elsevier, New York.
13. Hoyland, A., and Rausand, M. (1994). *System Reliability Theory, Models, and Statistical Methods*, Wiley, New York.

14. US Nuclear Regulatory Commission (1981). *Fault Tree Handbook*, NUREG-81/0492, U.S. Nuclear Regulatory Commission, Washington, DC.
15. Haimes, Y. Y. (1981). Hierarchical holographic modeling. *IEEE Trans. Syst. Man Cybern.* **11**(9), 606–617.
16. Haimes, Y. Y., Matalas, N. C., Lambert, J. H., Jackson, B. A., and Fellows, J. F. R. (1998). Reducing the vulnerability of water supply systems to attack. *J. Infrastruct. Syst.* **4**(4), 164–177.
17. Kaplan, S. (1996). *An Introduction to TRIZ: The Russian Theory of Inventive Problem Solving*, Ideation International, Inc., Southfield, MI.
18. Clarke, D. W. Sr. (2000). Inventive troubleshooting. *Mach. Des.* **15**, 78–80.
19. Metz, S., and Johnson, D. V. II (2001). *Asymmetry and U.S. Military Strategy: Definition, Background and Strategic Concepts*, http://www.au.af.mil/au/awc/awcgate/ssi/asymetry.pdf. Accessed April 25, 2007.
20. Schneider, W. J., Gold, T., and Hermann, B. (2003). *The Role and Status of DoD Red Teaming Activities*, Defense Science Board, Washington, DC.
21. Bier, V. M., Nagaraj, A., and Abhichandani, V. (2005). Protection of simple series and parallel systems with components of different values. *Reliab. Eng. Syst. Saf.* **87**(3), 315–323.
22. Lambert, J. H., Haimes, Y. Y., Li, D., Schooff, R. M., and Tulsiani, V. (2001). Identification, ranking, and management of risks in a major system acquisition. *Reliab. Eng. Syst. Saf.* **72**(3), 315–325.
23. Haimes, Y. Y., Longstaff, T. A., and Lamm, G. A. (2002b). Balancing promise and risk with information assurance in Joint Vision 2020. *Mil. Oper. Res.* **7**(3), 31–46.
24. Dombroski, M., Haimes, Y. Y., Lambert, J. H., Schlussel, K., and Sulcoski, M. (2002). Risk-based methodology for support of operations other than war. *Mil. Oper. Res.* **7**(1), 19–38.
25. Haimes, Y. Y. (2007). Phantom system models for emergent multiscale systems. *J. Infrastruct. Syst.* **13**(2), 81–87.
26. Haimes, Y. Y., and Horowitz, B. M. (2004). Adaptive two-player hierarchical holographic modeling game for counterterrorism intelligence analysis. *J. Homeland Secur. Emerg. Manage.*, **1**(3), 302.
27. Haimes, Y. Y., Santos, J. R., Crowther, K. G., and Henry, M. H. Lian, C., and Yan, Z. (2007). Analysis of Interdependencies and Risk in Oil and Gas Infrastructure Systems, *Institute for Information Infrastructure Protection Research Report No. 11, June 2007*. Published online http://www.thei3p.org/docs/publications/researchreport11.pdf.
28. Agrawal, A. B. (2006). Integrated Risk Assessment and Management for the 2006 Virginia Gubernatorial Inauguration. *MS Thesis, University of Virginia, Charlottesville, VA*.
29. Haimes, Y. Y., Kaplan, S., and Lambert, J. H. (2002a). Risk filtering, ranking, and management framework using hierarchical holographic modeling. *Risk Anal.* **22**(2), 381–395.
30. Leung, M., Lambert, J. H., and Mosenthal, A. (2004). A risk-based approach to setting priorities in protecting bridges against terrorist attacks. *Risk Anal.* **24**(4), 963–984.
31. Anderson, C. W., Barker, K., and Haimes, Y. Y. (2008). Assessing and prioritizing critical assets for the United States Army with a modified RFRM methodology. *J. Homeland Secur. Emerg. Manage.* **5**(1), 5.
32. Haimes, Y. Y. (1977). *Hierarchical Analyses of Water Resources Systems: Modeling and Optimization of Large-Scale Systems*, McGraw-Hill, New York.
33. Federal Emergency Management Agency. (2003). *Multi-Hazard Loss Estimation Methodology Hurricane Model: HAZUS-MH Technical Manual*, Department of Homeland Security, Emergency Preparedness and Response Directorate, Federal Emergency Management Agency, Mitigation Division, Washington, DC.

34. Kitchin, R. M. (1994). Cognitive maps: what are they and why study them? *J. Environ. Psychol.* **12**(1), 1–19.
35. Tolman, E. C. (1948). Cognitive maps in rats and men. *Psychol. Rev.* **55**(4), 189–208.
36. Lieblich, I., and Arbib, M. A. (1982). Multiple representations of space underlying behaviour and associated commentaries. *Behav. Brain Sci.* **5**(4), 627–660.
37. Downs, R. M. (1976). Cognitive mapping and information processing: a commentary. In G. T. Moore, and R. G. Golledge, Eds. *Environmental Knowing*, Dowden, Hutchingson, and Ross, Stroudsburg, PA, pp. 67–70.
38. Axelrod, R. (1976). *Structure of Decision*, University of Princeton Press, Princeton, NJ.
39. Walsh, J. P. (1995). Managerial and organizational cognition: Notes from a trip down memory lane. *Organ. Sci.* **6**(3), 280–321.
40. Weick, K. E. (1995). *Sensemaking in Organizations*, Sage, Thousand Oaks, CA.
41. Novak, J. D., and Musonda, D. (1991). A twelve-year longitudinal study of science concept learning. *Am. Educ. Res. J.* **28**(1), 117–153.

TIME-DOMAIN PROBABILISTIC RISK ASSESSMENT METHOD FOR INTERDEPENDENT INFRASTRUCTURE FAILURE AND RECOVERY MODELING

George H. Baker
James Madison University, Harrisonburg, Virginia
Charles T. C. Mo
Northrop-Grumman IT, Los Angeles, California

1 INTRODUCTION

The report of the President's Commission on Critical Infrastructures [1, 2] concluded that the nation's physical and economic security depend on critical energy, communications, and computer infrastructures. The Department of Homeland Security has issued national strategy documents for the protection of physical and cyber infrastructures that call for vulnerability assessments of critical infrastructure systems [3, 4]. Even in a world free of malicious activities, such assessments are exceedingly useful to help predict and hopefully prevent outages and costs resulting from natural disasters and normal accidents. Critical infrastructure networks and facilities are subject to many different failure modes. It is important to anticipate these modes, the likelihood of their occurrence, and the relative seriousness of their consequences.

Failures may be due to many causes, intentional and nonintentional, including aging, accidents, and sabotage from insiders or external malefactors. Failures can propagate such that seemingly minor problems may lead to complete functional failure. Of particular concern is the presence of "single-point failure" locations in many known facilities and systems. A question of concern is which failure points or point combinations would lead to the most serious and "most to-be-avoided" consequences. Some serious failure modes may be counterintuitive. Assessments provide an important basis for determining the most serious failure modes, implementing cost-effective countermeasures, and planning for reconstitution [5].

It is worth noting that modeling infrastructure system interdependencies emerged at the top of National Science Foundation's research objectives at their 2006 Workshop on Resilient and Sustainable Infrastructures [6].

2 THE TIME-DOMAIN PROBABILISTIC RISK ASSESSMENT METHOD OVERVIEW

Most critical infrastructure modeling capabilities address the initial failure of critical systems. A major recent push has been developing the capability to model cascading failure of interdependent, interconnected infrastructures [1]. An important capability of interest that has not received wide attention, but is important for understanding and improving infrastructure resiliency is the ability to model the postfailure, recovery phase of infrastructure debilitation.

Infrastructure systems are functionally complex and their system-wide failure probabilities, modes, and consequences are often not obvious. A useful technique for assessing the infrastructure failure modes and their respective likelihoods has been probabilistic risk assessment (PRA) [7]. The technique admittedly does not enable study of detailed system operation that physics-based or agent-based system simulation tools offer. On the other hand, the method requires fewer input parameters and has a faster-turnaround making it a good choice for scoping or screening studies.

To analyze and quantify survivability, conventional PRA methods provide a snapshot of potential failure modes at a single point in time for certain initiating conditions. We have developed a method that extends PRA into the time domain. The method computes the evolution of overall system functionality in time by evaluating initial failure probabilities, effects onset times, and system repair/reconstitution times for single or combinations of critical systems. Using this technique, we can determine which types of failures have the highest probability of putting a critical system off-line for the longest period of time or for a time window of interest. The output thus measures the "resilience" of a system, or system of systems. Results are useful for decision-makers to determine where scarce resources may be invested for replacement parts, service personnel, and so on, to lessen risks and improve system resiliency. The technique provides information useful in weighing the advantages of buying protection to reduce initial failure probabilities (often a costly proposition) or accepting high initial failure probabilities and relying on emergency response contingencies.

The time-domain PRA technique enables a comparative evaluation of potential functional debilitation mode consequences, using realistically available system information

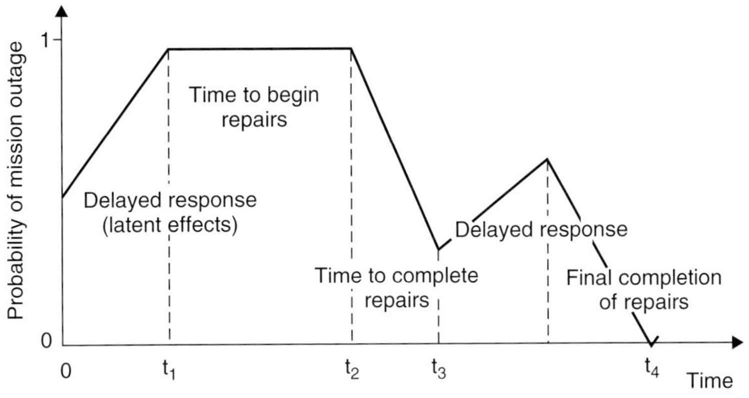

FIGURE 1 Notional output of the time-domain probabilistic risk assessment (PRA) technique. Standard PRA techniques would provide the probability of effect value at $t = 0$.

augmented by reasonable engineering assumptions. The technique is designed to quantify the probability of mission outage (P_e) and outage longevity. The formulation provides output to the user in a simple plot of P_e versus time. A notional output plot is shown in Figure 1.

The tool starts with the building blocks of traditional PRA. Mission outage modeling is accomplished by building a fault diagram. This construct includes a tree structure plus any closed loops and failure linkages among tree elements. The structure provides the hierarchy of systems and subsystems necessary for an infrastructure facility to perform its mission.

As an aid for developing fault trees for infrastructure facilities, it is helpful to divide facility systems into three categories according to their function as diagrammed in Figure 2: (i) Operations function, (ii) Protection function, and (iii) Support function. "Operation systems" perform mission-specific functions such as communication, manufacturing, storage, or materials/energy/transport inventory. "Protection systems" function to provide physical security, information security, fire protection, emergency services, and so on. "Support systems" provide the environment control, electric power, or communications necessary for the operation of mission and protection systems.

At first blush, the operational/mission systems appear to be the most critical to system performance. However, experience shows that the debilitation of support systems often has the most widespread effects on facility functionality [7]. If a saboteur is familiar with subsystem fault trees and interdependencies, he may be able to bring down an entire facility by attacking one subsystem of a support system. Mission systems tend to have more protection/security, making support systems the most accessible targets.

To address system resilience, the technique accounts for system and subsystem reconstitution times and constraints. Subsystems vary tremendously in their repair and replacement times. Some subsystem failures may also result in cascading effects exhibiting latent time delays. For example, the failure of a facility's environmental control supervisory control and data acquisition (SCADA) may cause room temperature to rise to a point in time where electronics begin to overheat. For this reason the formulation is designed to account for three subsystem time factors: (i) effects' onset delays, (ii) damage repair/replacement times, and (iii) repair commencement time bases on resource availability and system priority.

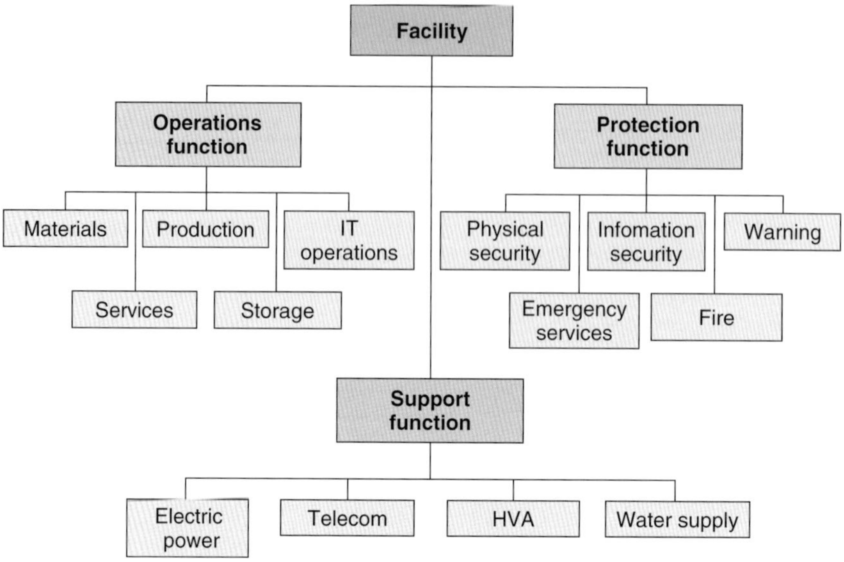

FIGURE 2 Infrastructure facility generic functional diagram.

3 TIME-DOMAIN PRA MATHEMATICAL FORMULATION

Input information includes subsystem fault and function diagrams, scenario stresses on critical subsystems, subsystem component strengths (damage or disruption thresholds), subsystem effects onset probabilities and times, effects propagation paths and time lengths, and condition-constrained repair times. These inputs drive the computation of the probability of mission outage versus time. The method compares stresses to subsystem strengths, to determine the probability of effect at individual components. In cases where subsystem component's P_e's are known, they can be input directly. These probabilities can be evaluated at any system point. They can also be calculated for a specified time of performance and for altered function diagrams. A diagram of the basic process flow is shown in Figure 3.

3.1 Probability of Effects Calculation at the Subsystem Level

Consider a system consisting of a number of subsystems with each subsystem consisting of many components. Let the components be the basic units for which we compare an environment-induced stress, **S**, with the component's physical damage threshold, **T**. The stress occurs at the weapon onset time ($t = 0$). **T** can be either a scalar quantity or a K-dimensional vector, where K is the number of components. If a component j sees $\mathbf{S} \geq \mathbf{T}$, defined as:

$$\{\min(\mathbf{S}_{\mathrm{jk}} - \mathbf{T}_{\mathrm{jk}}) \geq 0, k = 1 \ldots K\}$$

then it is counted as having suffered physical damage and will, at a latent time delay $\mathbf{TL}_{\mathrm{j}}$ after the stress onset, not perform its function. Component j is then defined as being in a state of "functional outage." Notice that the overall component mission state at any given

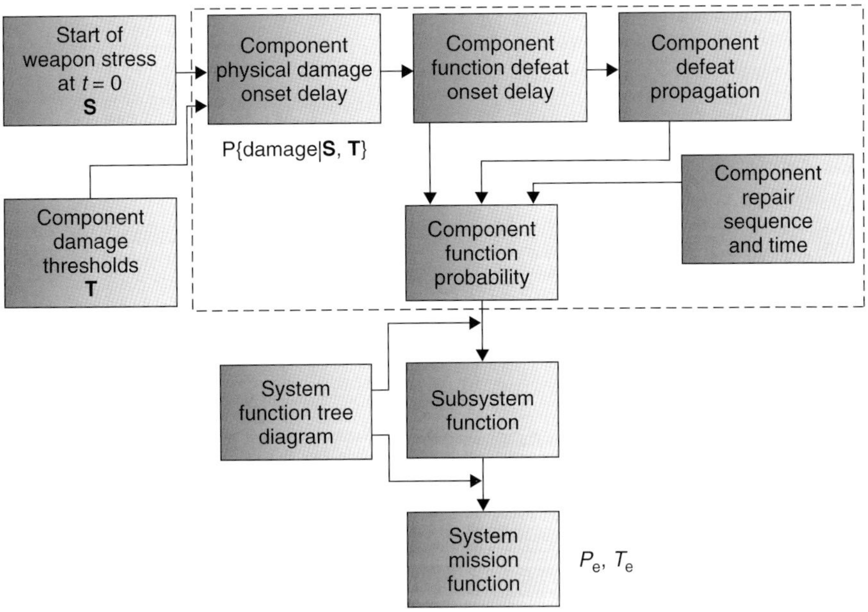

FIGURE 3 Computational process flow.

time is modeled as a binary *functional* XOR *not-functional*. Gradual mission function degradation is very difficult to model due to the inherent nonuniqueness of associated system states and problems associated with aggregating such failures. Such degradation is beyond the capabilities of current fault-tree-based models.

The method uses parameter, $\mathbf{P}_{oj}$, to represent the probability that component j is physically damaged as a consequence of the stress $\mathbf{S}_j$ exceeding the component's threshold $\mathbf{T}_j$ at $t = 0$. In mathematical notation:

$$\mathbf{P}_{oj} = Pr\{\mathbf{S}_j \geq \mathbf{T}_j\}$$

S and **T** are modeled as random variables, each with its own probability distribution determined on engineering and physical bases. These distributions, in practice, represent all estimated uncertainties and known variations. For different components, $\mathbf{S}_j$ may be statistically dependent due to physical proximity or other physical factors.

The model defines a functional outage status index variable, $\mathbf{I}_{dj}$. $\mathbf{I}_{dj} = 1$ if, and only if, component j is in a state of functional outage and $\mathbf{I}_{dj} = 0$ if, and only if, the component is functional. This binary function status is a simplification. The variable could be graduated into "shades of gray" in the interval [0,1] quite straightforwardly, if required. The model includes a latent time delay, $\mathbf{TL}_j$ to account for the time duration necessary for component j's physical damage to result in loss of its functionality. The model allows for uncertainty in latent time delay, by treating $\mathbf{TL}_j$ as a random variable with a probability distribution $f_j(t)$. This probability distribution may be scenario dependent.

At the subsystem level, any one of many sets of prespecified combinations of nonfunctional components serves to prevent subsystem function. And, any one set of prespecified combinations of subsystem outages constitutes overall debilitation of the system's mission.

3.2 Top-Level Functional Outage Computation

In order to quantify the infrastructure mission outage as a time varying probability, we define a state vector for each component j, $\mathbf{I}_j = (\mathbf{I}_{dj}, \mathbf{I}_{cj})$, where $\mathbf{I}_{dj}$ is the outage status index and I_{cj} tracks the outage time condition for component j. $\mathbf{I}_j = (0{,}0)$ if there is no physical damage and $\mathbf{I}_j = (0, TL)$ if it is damaged at effects onset time, but will manifest a latent functional outage at a later time. At the functional outage onset time, $\mathbf{t}_s$, $\mathbf{I}_j = (1, \mathbf{t}_s)$. Note that $\mathbf{t}_s$ is itself a function of time because a component can be damaged and repaired and damaged again due to functional outage propagation paths of other system components. If damaged, component j may be repaired in time $\mathbf{TR}_j$, where $\mathbf{TR}_j$ is modeled as a random variable with a scenario-dependent probability distribution.

The repair start time may not commence immediately after a subsystem/component outage. The model accommodates situations where a repair sequence is needed, that is, certain components must be repaired before others can be fixed. The repair sequence is an explicit input and the model checks the sequence for preceding components' operational status and allows component j to be repaired only if preceding components are functional.

It is also possible to model situations when, even if a subsystem is not damaged directly at $t = 0$, it can suffer a functional outage due to the accumulated outage of other "upstream" subsystems. An example would be electronics overheating due to the failure of heating, ventilation and air conditioning (HVAC) subsystems. Such outage propagation paths and times, TP are accommodated by an additional layer of random variables with their uncertainty distributions as explicit inputs.

The probabilistic nature of system failure is implied by the input component stresses and thresholds, which are themselves described probabilistically. Likewise, the latent time delay, the propagated outage time, and the start and duration of repair times are described using probability distributions. Such distributions provide bounds for both the uncertainties in the exact physical description of a specific single system of interest, and system-to-system variations if more than one system is involved. The time factors result in a time varying probability of effects.

Theoretically, the fault-diagram representation maps each system state point $P(X)$, of the 2^N possible points in the component function space (where N is the total number of components in the system), into a system function measure, 0 XOR 1. As time goes on, the state point, $P(X, t)$ traces a trajectory in the component function state space with its path influenced and controlled by all the physical constraints including damage, delay, propagation, and repair (see Section 2). The time changes occur in discrete time steps. Thus, the system functional state is a piecewise step function of time, with value 1 or 0. Because these physical constraints, including their initial values, are probabilistic, the method indicates the probability that the system is in a state of mission outage up to time t based on the specified stress scenario.

The equivalent mathematical problem of coupled finite difference equations with stochastic "forces" is much more challenging to solve analytically in closed forms [8]. This method does it numerically, invoking a basic Monte Carlo simulation.

3.3 Practical Factors: Approximations and Uncertainties

With perfect knowledge of a system and its environment, one can predict certain system outages or functionalities, deterministically, as a function of time. Then there would be no need for a probabilistic formulation. Except in obvious, trivial cases, it is practically infeasible to develop a rigorous, deterministic, analytical closed-form solution to quantify functional outage probabilities and their mathematical-statistical inference from data. This is particularly true if a classical frequency probability approach is used. At the other extreme, if too wide a range of weapon scenarios or facility variations are included in a probabilistic formulation, the quantification loses practical meaning. Also, purely subjective formulations lack consistency and credibility.

We have attempted to achieve a balanced approach, retaining the frequency interpretation as much as possible and using probability theory as a tool to treat and propagate the uncertainties. If necessary, the model can accept expert estimates, both explicitly and numerically.

Infrastructure modeling using any approach is nontrivial. With the time-domain PRA approach the following factors require particular care.

- The mission function must be specified and the necessary and sufficient component subset disablement to achieve functional outage must be determined.
- Determining the level of detail, and basic components to include in functional diagrams requires care.
- **S** and **T** are nonscalar in nature. Stresses at different components, and their uncertainties, are often not independent.
- There is often a latent time delay between component physical damage and the onset of functional outage.
- The propagation path and associated time delay from one component to another must be included.
- Component and subsystem repair times and inherent repair sequences must be specified. Also, the effect of human resource constraints and scenarios must be factored into the repair/reconstitution ability.
- The uncertainties associated with all input values and assumptions must be specified

The difficulty of specifying the joint probability distribution of mutually dependent and time-dependent stresses, thresholds, internal damage propagation, and repairs makes an analytically explicit solution infeasible. Nevertheless, the model is set up to provide a logical and defendable range of results based on available system information and reasonable engineering approximations.

3.4 Method Validity and Limitations

Incomplete knowledge of the facility/system functional components, their interrelationships and outage mechanisms can limit the accuracy of the calculation in two ways:

1. If an in-parallel functional component, which contributes to system performance redundancy, is omitted from the model, then the true effect on system operation is less than or equal to that assessed. In this case, the assessed effect measure is defense-conservative.
2. If an in-series functional component exists, which contributes to system vulnerability, but is omitted from modeling due to lack of information, then the true effect measure may be more serious than assessed. That is, the assessed results are offense-conservative.

Thus, the completeness of the set of identified model components results in uncertainties in the output probability in both directions of under- and overestimating system effects. This property can be helpful in performing sensitivity studies. The method can be used to investigate sensitivity to model completeness by selectively omitting from and adding to a system's component ensemble.

A critical part of the system assessment is to determine which components or combination of components must be nonfunctional to negate the system's mission capability. In some PRA literature this is referred to as a *cut set*, in that cutting off any one of its members, the set is invalidated as a sufficient cause of mission outage. Another concept useful for ensuring assessment realism involves identifying the minimum sufficient set of components that ensures functional performance if all its members perform their respective functions. This is equivalent to an "all component in series" functional diagram and is sometimes referred to as a *path set* in PRA literature, because a functional path through all of the set members constitutes a system function.

A theoretically trivial, but important and sometimes tricky consideration is the avoidance of double counting. This is a special case of correlated-cause component dependencies. In cases where one component's operational status appears in different sets, the model must ensure that the component is modeled as one and the same.

Model validation can be performed using logical implication checks. This involves running self-consistency and special limiting-case checks. Because of the general inability to do multiple system copy pass–fail experiments, validation must be indirect and relative. This is a general problem for all PRA models of complicated systems.

4 CRITICAL NEEDS ANALYSIS

It is very important to understand the risk of cascading failures and interacting infrastructures. Understanding the interactions and failure correlations among critical systems allows us to understand the consequences of failure of any one of the systems. Analysis of past high consequence infrastructure failures suggests that the risk of failure in one system can be significantly affected by fairly weak coupling with other systems.

Furthermore, it is not sufficient to merely predict the risk of system failure. Understanding system resilience requires the ability to predict system recovery timelines. Because it is cost-prohibitive to completely prevent system failures, understanding recovery processes and timelines emerges as arguably the most important aspect of assuring system resiliency. The time-domain PRA approach incorporates the ability to model system risk of failure at the initial time of stress and the subsequent time-line for system recovery.

There is a need for models at a high level of system abstraction for screening studies, because they make it possible to look at many different scenarios and long-time system interaction dynamics including system failure and recovery processes. The time-domain PRA approach offers a reduced, relatively simple approach to system modeling to complement full-physics system simulation models. Full-physics system simulations tend to be impractical for problems in which multiple scenarios need to be examined. Large, multiple-trial parameter studies can be computationally intractable in full-physics and agent-based models. Simplified system models that do not incorporate all system details offer a useful, complementary approach.

The time-domain PRA method offers the ability to evaluate the probability and duration of infrastructure system mission outages with a relatively limited set of system information. Required input information includes system fault trees, effect "stress" and subsystem/component "strength," and subsystem repair times and repair sequences. The approach also allows specification of latent time delay effects associated with the onset of damage. The method allows specification of variances for all input parameters. Input data may be empirical or estimated based on reasonable engineering approximations. Scenarios may include intentional physical or cyber attacks, Radio Frequency (RF) weapons, natural disasters, or normal accidents. The model enables comparison of the effectiveness of alternative mitigation strategies and techniques, and enables location of most serious failure points from the standpoint of highest probability of effect and longest duration functional outage. Output is provided in the form of probability of effects versus time output facilitates estimation of recovery times.

5 RESEARCH DIRECTIONS

The time-domain PRA approach has received attention only relatively recently. A simple, basic prototype of the method has been formulated using MATHCAD software. The method has been successfully demonstrated by modeling Electromagnetic Pulse (EMP) effects on a simple communication facility and blast effects on a national command center (classified study). These problems demonstrated the usefulness of the approach both computationally and as a "gedanken framework" for understanding prime system failure points and mechanisms.

Future research is needed to build on and generalize this introductory work, including:

- development of a fast-running production version of the method for application to larger infrastructure systems and networks;
- demonstration of the method on a problem of national interest, for example, interdependencies between the power grid and telecommunications grid [9]; and
- development of outage cost analysis algorithms based on system debilitation timelines.

On a broader scale, this approach should be beneficial for the development of:

- more complex "global" functional failure propagation and causation of interdependent infrastructures;
- techniques for system performance self-diagnosis and repair prioritization;
- dynamic sensitivity studies to develop more robust designs of complex systems and networks; and
- quantification of continuous instead of binary system performance degradation.

REFERENCES

1. Rinaldi, S. (2004). Modeling and simulating critical infrastructure interdependencies. *Proceedings of the 37th Annual Hawaii Conference on System Sciences*, Hawaii.

2. Robert T. Marsh (1997). *Critical foundations: Protecting America's Infrastructures*. Report of the President's Commission on Critical Infrastructure Protection (PCCIP), Chairman, Washington, DC, 1997.
3. U.S. Department of Homeland Security. (2003). *The National Strategy for the Physical Protection of Critical Infrastructures and Key Assets*, U.S. Department of Homeland Security, February 2003.
4. U.S. Dept of Homeland Security. (2003). *The National Strategy to Secure Cyberspace*, U.S. Dept of Homeland Security, February 2003.
5. Baker, G. (2005). *A Vulnerability Assessment Methodology for Critical Infrastructure Facilities*, Department of Homeland Security R&D Partnerships Symposium, Boston, MA.
6. National Science Foundation. (2006). *Resilient and Sustainable Infrastructures Workshop*, Arlington, VA, 4-5 December 2006.
7. Kumamoto, H. and Henley, E.. (1996). *Probabilistic Risk Assessment and Management for Engineers and Scientists*, 2nd ed. IEEE Press, New York.
8. Baker, G. (2005). *A Vulnerability Assessment Methodology for Critical Infrastructure Facilities*, Department of Homeland Security R&D Partnerships Symposium, Boston, MA.
9. Kohlberg, I., Clark, J., Morrison, P.. (2007). Dynamics of recovery of coupled infrastructures following a natural disaster or malicious insult. *Proceedings of the James Madison University/Federal Facilities Council Symposium on Cascading Infrastructure Failures*. National Research Council, Washington, D.C., May 2007.

FURTHER READING

Alliance, L. (2006). *Power Systems, Transportation and Communications Lifeline Interdependencies*, FEMA-National Institute of Building Sciences, March 2006, Washington, DC.

Amin, M. (2000). National infrastructures as Complex Interactive Networks. In *Automation, Control, and Complexity: An Integrated Approach*, Samad, T. and Weyrauch, J., Eds. John Wiley and Sons, New York, pp. 263–286.

Bruneau, M., Chang, S., Eguchi, R., Lee, G., O'Rourke, T., Reinhorn, A., Shinozuka, M., Tierney, K., Wallace, W., and von Winterfeldt, D. (2003). A framework to quantitatively assess and enhance the seismic resilience of communities. *Earthquake Spectra* **19**(4), 733–752.

Gnedenko, B., Belyayev, Y., and Solovyev, A. (1969). *Mathematical Methods of Reliability Theory*. Academic Press, New York.

Henley, E. and Kumamoto, H. (1992). *Probabilistic Risk Assessment*, IEEE Press, New York.

Koubatis, A., Schonberger, J. (2005). Risk management of complex critical systems. *Int. J. Crit. Infrastructures* **1**(2/3), 195–215.

National Research Council. (2002). *Making the Nation Safer: The Role of Science and Technology in Countering Terrorism* . The National Academies Press, Washington, DC.

Nozick, L. and Turnquist, M. (2004). Assessing the performance of interdependent infrastructures and optimizing investments. *Proceedings of the 37th Hawaii International Conference on System Sciences*, Hawaii.

Peerenboom, J. (2001). *Infrastructure Interdependencies: Overview of Concepts and Terminology*. Argonne National Laboratory Pamphlet.

Rinaldi, S. (2004). Modeling and simulating critical infrastructures and their interdependencies. *Proceedings of the 37th Hawaii International Conference on System Sciences,* Hawaii.

U.S. Department of Homeland Security. (2003). *The National Strategy for the Physical Protection of Critical Infrastructures and Key Assets*, February 2003.

Wolthusen, S. D. (2004). Modeling critical infrastructure requirements. *Proceedings of the 2004 IEEE Workshop on Information Assurance and Security,* June 2004, Charlotte, NC.

RISK TRANSFER AND INSURANCE: INSURABILITY CONCEPTS AND PROGRAMS FOR COVERING EXTREME EVENTS

HOWARD C. KUNREUTHER AND ERWANN O. MICHEL-KERJAN
Center for Risk Management and Decision Processes, the Wharton School, University of Pennsylvania, Philadelphia, Pennsylvania

1 INTRODUCTION

The United States has extensive experience with natural catastrophes. But the hurricanes that occurred in the Gulf Coast during the 2004 and 2005 seasons have changed the landscape forever. Coupled with the terrorism attacks of September 11, 2001, there is recognition by both the public and private sectors that one needs to rethink our strategy for dealing with these low probability but extreme consequence events.

The 2002 White House National Strategy defines homeland security as "the concerted effort to prevent attacks, reduce America's vulnerability to terrorism, and minimize the damage *and* recover from attacks that do occur". To succeed, homeland security must be a national and comprehensive effort. Moreover, that definition must apply to technological and natural disasters as well.

While protecting residential and commercial construction and critical infrastructure services (transportation, telecommunications, electricity and water distribution, etc.) in risky areas may limit the occurrence and/or the impacts of major catastrophes, we know that major disasters will still occur. In these situations one must provide adequate emergency measures and rapidly restore critical services. The question as to who will provide financial protection to victims (residents and commercial enterprises) will take center stage. The insurance infrastructure will then play a critical role [1].

This article discusses some fundamentals of the operation of insurance as well as some of the insurance programs that have been established in the United States to cover economic losses due to large-scale catastrophes.

2 HOW DOES INSURANCE WORK AND DOES NOT WORK

2.1 Determining Premiums and Coverage

2.1.1 ***Basic Concepts.*** The insurance business, like any other business, has its own vocabulary. A *policyholder* is a person who has purchased insurance. A *premium* is the amount that a policyholder pays in return for the promise of a payment from the insurer should he suffer a loss covered by his policy. The term *benefit* denotes the payment by the insurer to the policyholder given that he has suffered a reduction in wealth due to a loss. A *claim* means that the policyholder is seeking to recover financial payments from the insurer for damage covered by the policy. A claim will not result in a payment by the insurer if the amount of the insured's financial loss is below the stated *deductible* (i.e. the amount or proportion of an insured loss that the policyholder agrees to pay before any recovery from the insurer) or if the loss is subject to policy exclusions (e.g. war or insurrection). However, insurers will still incur expenses for investigating the claim.

Insurer *capital* represents the net worth of the company (assets minus liabilities). Capital enables the insurer to pay any losses above those that were expected. It serves as a safety net to support the risk that an insurer takes on by writing insurance and helps ensure that the insurer will be able to honor its contracts. As such, it supports the personal safety nets of homeowners, business owners, workers, dependents of heads of households, and others who rely on insurance to provide financial compensation to rebuild their lives and businesses after covered losses occur. Insurer capital is traditionally referred to as *policyholders' surplus*. Despite the connotation of the term *surplus*, there is nothing superfluous about it—it is, in fact, an essential component supporting the insurance promise. The cost of that capital is an insurer expense that must be considered in pricing insurance, along with expected losses, sales, and administrative expenses for policies written.[1]

The capital needed by an insurer varies directly with the risk that the insurer takes on. If an insurer wishes to take on more risk, it must have capital to support that risk. *Insurance regulators* and *rating agencies* in their efforts to assure policyholders that insurers will be able to pay their losses, devote significant efforts toward evaluating the adequacy of insurer capital relative to the amount and types of risk they are taking on. Holding an adequate level of capital is critical to the continued viability of an insurer.

Insurance markets function best when the losses associated with a particular risk are independent of each other and the insurer has accurate information on the likelihood of the relevant events occurring and the resulting damage. By selling a large number of policies for a given risk, the insurer is likely to have an accurate estimate of claim payments it expects to make during a given period of time. To illustrate this point with a simple example, consider an insurer who offers a fire insurance policy to a set of identical homes each valued at \$100,000. Based on past data, the insurer estimates that the likelihood that the home will be destroyed by fire next year is 1/1000 and that this is the only loss that can occur. In this case the expected annual loss for each home would be \$100 (i.e. 1/1000 × \$100,000).

If the insurer issued only a single policy to cover the full loss from a fire, then there would be a variance of approximately \$100 associated with its expected annual loss.[2] As the number of policies issued, n, increases, the variance of the expected annual loss, or the mean loss per policy, decreases in proportion to n. Thus, if $n = 10$, the variance of the mean loss will be approximately \$10. When $n = 100$ the variance decreases to \$1, and with $n = 1000$ the variance is \$0.10. It is thus not necessary to issue a very large number of policies to reduce significantly the variability of expected annual losses

[1]Consider, for example, insurance for property damage caused by hurricanes. An insurer's expected losses are relatively low, because in a typical year, the policyholder will not suffer a hurricane loss. However, it is possible that losses will be quite high—far in excess of those expected at the time policies are priced. In the event of a serious hurricane, a substantial portion of the loss must be paid from insurer capital. For terrorism coverage, maximum losses are extremely high relative to expected losses, so the capital issue is critical.

[2]The variance for a single loss L with probability p is $Lp\,(1-p)$. If $L = \$100{,}000$ and $p = 1/1000$, then $Lp(1-p) = \$100{,}000\,(1/1000)(999/1000)$, or \$99.90.

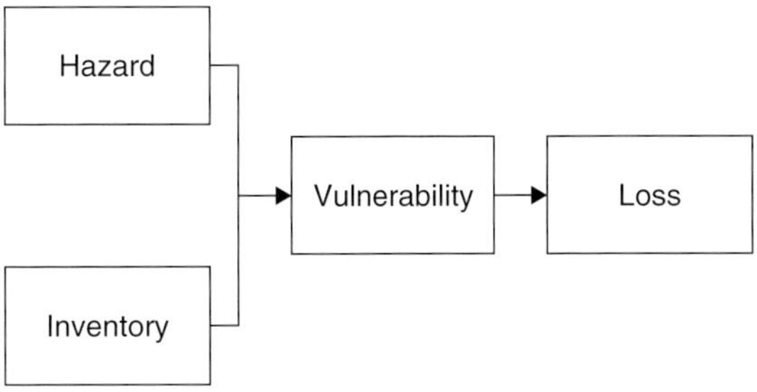

FIGURE 1 Structure of catastrophe models.

per policy if the risks are independent. This model of insurance works well for risks such as fire, automobile, and loss of life where the above assumptions of independence and ability to estimate probabilities and losses are satisfied. As will be shown below, terrorism risk does not satisfy the above conditions, so it is more problematic to insure.

***2.1.2 Catastrophe Models*[3].** Before insurance providers are willing to offer coverage against an uncertain event they feel they must be able to identify and quantify, or at least partially estimate the chances of the event occurring and the extent of losses likely to be incurred. Such estimates can be based on past data (e.g., loss history of the insurer's portfolio of policyholders, loss history in a specific region) coupled with data on what experts know about a particular risk through the use of catastrophe models.

The four basic components of a catastrophe model are hazard, inventory, vulnerability, and loss, as depicted in Figure 1, and illustrated for a natural hazard such as a hurricane. First, the model determines the risk of the *hazard* phenomenon, which in the case of a hurricane is characterized by its projected path and wind speed. Next, the model characterizes the *inventory* (or portfolio) of properties at risk as accurately as possible. This is done by first assigning geographic coordinates such as latitude and longitude to a property based on its street address, zip code, or another location descriptor, and then determining how many structures in the insurer's portfolio are at risk from hurricanes of different wind speeds and projected paths. For each property's location in spatial terms, other factors that characterize the inventory at risk are the construction type, the number of stories in the structure, and its age.

The hazard and inventory modules enable one to calculate the *vulnerability* or susceptibility to damage of the structures at risk. In essence, this step in the catastrophe model process quantifies the physical impact of the natural hazard phenomenon on the property at risk. How this vulnerability is quantified differs from model to model. On the basis of this measure of vulnerability, the *loss* to the property inventory is evaluated. In a catastrophe model, loss is characterized as direct or indirect in nature. Direct losses include the cost to repair and/or replace a structure. Indirect losses include business interruption impacts and relocation costs of residents forced to evacuate their homes.

Catastrophe models were introduced in the mid-1980s but did not gain widespread attention until after Hurricane Andrew hit southern Florida in August, 1992, causing insured losses of over $21.5 billion (in 2004 prices). Until 9/11 this was the largest single loss in the history of insurance. Nine insurers became insolvent as a result of their

[3]This section is based on [2].

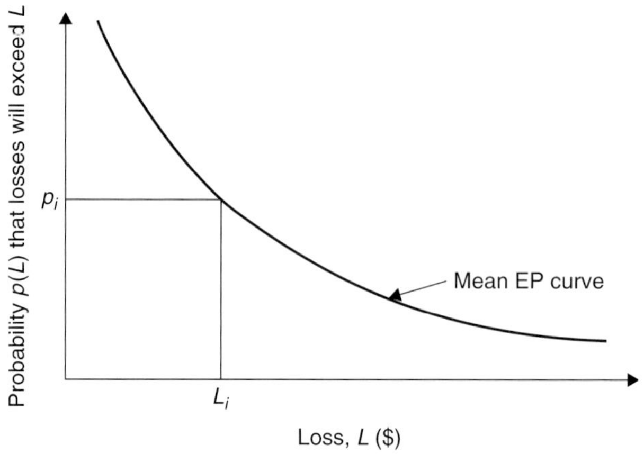

FIGURE 2 Sample mean exceedance probability curve.

losses from Hurricane Andrew. Insurers and reinsurers thought that, in order to increase the chances of remaining in business, they needed to estimate and manage their natural hazard risk more precisely. Many companies turned to the modelers of catastrophe risks for decision support.

***2.1.3 Exceedance Probability Curves*[4].** On the basis of the outputs of a catastrophe model, the insurer can construct an exceedance probability (EP) curve that specifies the probabilities that a certain level of losses will be exceeded. The losses can be measured in terms of dollars of damage, fatalities, illness, or some other unit of analysis.

To illustrate with a specific example, suppose one were interested in constructing an EP curve for an insurer with a given portfolio of insurance policies covering wind damage from hurricanes in a southeastern US coastal community. Using probabilistic risk assessment, one would combine the set of events that could produce a given dollar loss and then determine the resulting probabilities of exceeding losses of different magnitudes. On the basis of these estimates, one can construct a mean EP curve such as the one depicted in Figure 2. The x-axis measures the loss to insurer in dollars and the y-axis depicts the probability that losses will exceed a particular level. Suppose the insurer focuses on a specific loss L_i. One can see from Figure 2 that the likelihood that insured losses exceed L_i is given by p_i.

An insurer utilizes its EP curve for determining how many structures it will want to include in its portfolio given that there is some chance that there will be hurricanes causing damage to some subset of its policies during a given year. More specifically, if the insurer wanted to reduce the probability of a loss from hurricanes that exceeds L_i to be less than p_i it will have to determine what strategy to follow. The insurer could reduce the number of policies in force for these hazards, decide not to offer this type of coverage at all (if permitted by law to do so) or increase the capital available for dealing with future hurricanes that could produce large losses.

Federal and state agencies may want to use EP curves for estimating the likelihood that losses to specific communities or regions of the country from natural disasters in the coming year will exceed certain levels in order to determine the chances that it will

[4]This section is based on material in [3].

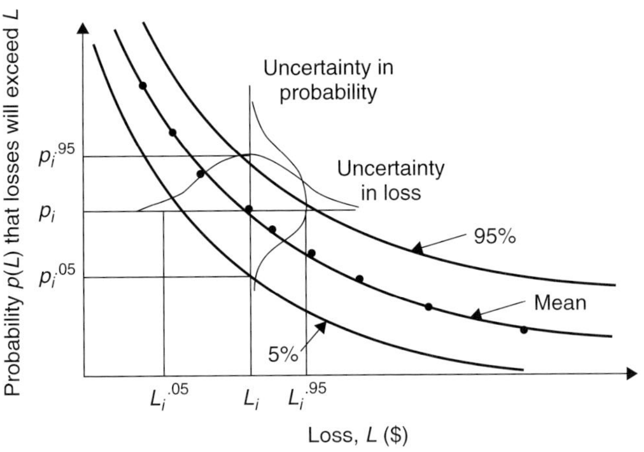

FIGURE 3 Confidence intervals for a mean exceedance probability (EP) curve.

have to provide disaster assistance to these stricken areas. At the start of the hurricane season in 2004, Florida could have used an EP curve to estimate the likelihood of damage exceeding $23 billion. Although this probability would have been extremely low, we now know that a confluence of events (i.e., Charley, Frances, Ivan, and Jeanne) produced an outcome that exceeded this dollar value.

The uncertainty associated with the probability of an event occurring and the magnitude of dollar losses of an EP curve is reflected in the 5 and 95% confidence interval curves in Figure 3. The curve depicting the uncertainty in the loss shows the range of values, $L_i^{0.05}$ and $L_i^{0.95}$ that losses can take for a given mean value, L_i, so that there is a 95% chance that the loss will be exceeded with probability p_i. In a similar vein one can determine the range of probabilities, $p_i^{0.05}$ and $p_i^{0.95}$ so that there is 95% certainty that losses will exceed L_i. For low probability-high consequence risks, the spread between the 5 and 95% confidence intervals depicted in Figure 3 shows the degree of indeterminacy of these events.

The EP curve serves as an important element for evaluating risk management tools. It puts pressure on experts to make explicit the assumptions on which they are basing their estimates of the likelihood of certain events occurring and the resulting consequences.

2.2 Determining Whether to Provide Coverage

On the basis of their knowledge of likelihood and outcome, an insurer has to make a decision as to whether to cover the risk (unless they are required to do so by law). In his study on insurers' decision rules as to when they would market coverage for a specific risk, Stone [4] develops a model whereby firms maximize expected profits subject to satisfying a constraint related to the survival of the firm.[5] An insurer satisfies its survival constraint by choosing a portfolio of risks with an overall expected probability of total claims payments greater than some predetermined amount (L^*) that is less than some threshold probability, p_I. This threshold probability reflects the trade-off between the

[5]Stone also introduces a constraint regarding the stability of the insurer's operation. Insurers have traditionally not focused on this constraint in dealing with catastrophic risks but reinsurers have, as discussed in the next article.

expected benefits of another policy and the costs to the firm of a catastrophic loss that reduces the insurer's surplus by L^* or more. This threshold probability does not necessarily correspond to what would be efficient for society. The value of L^* is determined by an insurer's concern with insolvency and/or a sufficiently large loss in surplus that will lead a rating agency to downgrade its credit rating.

A simple example illustrates how an insurer would utilize its survival constraint to determine whether a particular portfolio of risks is insurable with respect to hurricanes. Assume that all homes in a hurricane-prone area are identical and equally resistant to damage such that the insurance premium, P, is the same for each structure. Furthermore assume that an insurer has S dollars in current surplus and wants to determine the number of policies it can write and still satisfy its survival constraint. Then, the maximum number of policies, n, satisfying the survival constraint is given by Eq. (1):

$$\text{Probability [claims payments } (L^*) > (n \cdot P + S)] < p_1 \qquad (1)$$

The insurer will use the survival constraint to determine the maximum number of policies it is willing to offer, with possibly an adjustment in the amount of coverage and premiums, and/or a transfer of some of the risk to others in the private sector (e.g. reinsurers or capital markets). It may also rely on state or federal programs to cover its catastrophic losses.

Following the series of natural disasters that occurred at the end of the 1980s and in the 1990s, insurers focused on the survival constraint to determine the amount of catastrophe coverage they were willing to provide because they were concerned that their aggregate exposure to a particular risk did not exceed a certain level. Rating agencies, such as A.M. Best, focused on insurers' exposure to catastrophic losses as one element in determining credit ratings, so insurers paid attention to this risk.

2.3 Setting Premiums

If the insurer decides to offer coverage, it needs to determine a premium rate that yields a profit and satisfies its survival constraint given by Eq. (1). State regulations often limit insurers in their rate-setting process. Competition can play a role as well as to what premium can be charged in a given marketplace. Even in the absence of these influences, an insurer must consider problems associated with the *ambiguity of the risk*, asymmetry of information (*adverse selection* and *moral hazard*), and degree of *correlation* of the risk in determining what premium to charge. We briefly examine each of these factors in turn.

2.3.1 Uncertainty of the Risk. The infrequency of major catastrophes in a single location implies that the loss distribution is not well specified. The ambiguities associated with the probability of an extreme event and with the outcomes of such an event raise a number of challenges for insurers with respect to pricing their policies. As shown by a series of empirical studies, actuaries and underwriters are averse to ambiguity and want to charge much higher premiums when the likelihood and/or consequences of a risk are highly uncertain than if these components of risk are well specified [5].

Figure 4 illustrates the total number of loss events from 1950 to 2000 in the United States for three prevalent hazards: earthquakes, floods, and hurricanes. Events were selected that had at least \$1 billion of economic damage and/or over 50 deaths [6].

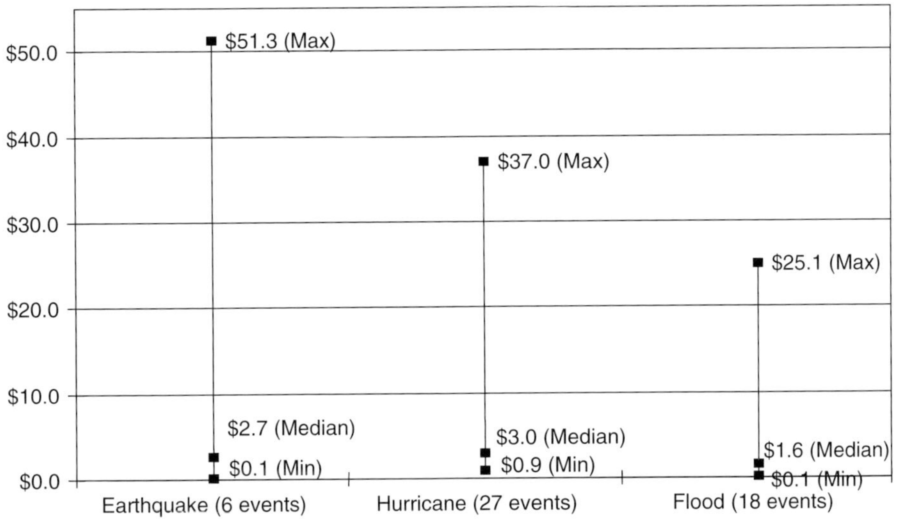

FIGURE 4 Historical economic losses in $ billions versus type of significant US natural disaster for the 50-year period from 1950 to 2000.

Looking across all the disasters of a particular type (earthquake, hurricane, or flood), for this 50-year period, the median loss is low while the maximum loss is very high. Given this wide variation in loss distribution, it is not surprising that insurers are concerned about the uncertainty of the loss in estimating premiums, or even providing any coverage in certain hazard prone areas.

The 2004 and 2005 seasons have already dramatically changed the upper limits in Figure 4. Hurricane Katrina is estimated to have caused between $150 billion and $170 billion in economic losses, more than four times higher than the most costly hurricane between 1950 and 2000. On the other hand, no hurricane hit the US landfall this year, despite predictions earlier in the year indicated higher than normal intensive season.

2.3.2 Adverse Selection. If the insurer cannot differentiate the risks facing two groups of potential insurance buyers and each buyer knows his/her own risk, then the insurer is likely to suffer losses if it sets the same premium for both groups by using the entire population as a basis for this estimate. If only the highest risk group is likely to purchase coverage for that hazard and the premium is below its expected loss, the insurer will have a portfolio of "bad" risks. This situation, referred to as *adverse selection*, can be rectified by the insurer charging a high enough premium to cover the losses from the bad risks. In so doing, the good risks might purchase only partial protection or no insurance at all because they consider the price of coverage to be too expensive relative to their risk[6].

This was the argument made by private insurers regarding the noninsurability of flood risk that led to the creation of the National Flood Insurance Program (NFIP). Indeed, insurers thought that family who had lived in a specific flood-prone area for many years had a much better knowledge of the risk than any insurer would have unless it invested in costly risk assessment tools.

[6]For a survey of adverse selection issues, see [7].

In the context of hurricane, however, it is not clear whether there is any adverse selection. Indeed, there is no evidence that those at risk have an informational advantage over the insurer. In fact, the opposite might be true: if insurance companies spend a lot of resources estimating the risk (what they actually do) they might gain an informational advantage over their policyholders who cannot afford or want to do so. Over the past 5 or 6 years, there has been a growing literature studying the impact of insurers being *more* knowledgeable about the risks than the insured themselves. Research in this field reveals that insurers might want to exploit this "reverse information asymmetry", which results in low risk agents being optimally covered while high risks are not [8].

2.3.3 ***Moral Hazard.*** This refers to an increase in the expected loss (probability or amount of loss conditional on an event occurring) caused by insurance-induced changes in the behavior of the policyholder. An example of moral hazard is more careless behavior *vis-à-vis* natural hazards or other types of risk as a result of purchasing coverage. Providing insurance protection may lead the policyholder to change behavior in ways that increase the expected loss from what it would have been without coverage. If the insurer cannot predict this behavior and relies on past loss data from uninsured individuals to estimate rates, the resulting premium is likely to be too low to cover losses.

Even after the insurer is aware that people with insurance have higher losses, its inability to observe loss-enhancing behavior may create problems of moral hazard. The introduction of specific deductibles, coinsurance or upper limits on coverage can be useful tools to reduce moral hazard by encouraging insureds to engage in less risky behavior, as they know they will have to incur part of the losses from an adverse event.

2.3.4 ***Correlated Risks.*** For extreme events, the potential for high correlation between the risks will have an impact on the tail of the distribution. In other words, at a predefined probability p_i, the region below the EP curve is likely to expand for higher correlated risks covered by insurers. This requires additional capital for the insurer to protect itself against large losses. Insurers normally face spatially correlated losses from large-scale natural disasters. State Farm and Allstate Insurance paid \$3.6 billion and \$2.3 billion in claims, respectively, in the wake of Hurricane Andrew in 1992 due to their high concentration of homeowners' policies in the Miami/Dade County area of Florida. Given this unexpectedly high loss, both companies began to reassess their strategies of providing coverage against wind damage in hurricane-prone areas [9].

Hurricanes Katrina and Rita that devastated the US Gulf Coast in August and September 2005 impacted dramatically on several lines, including life, property damage, and business interruption. Edward Liddy, chairman of Allstate, which provided insurance coverage to 350,000 homeowners in Louisiana, Mississippi, and Alabama, declared that "extensive flooding has complicated disaster planning ... and the higher water has essentially altered efforts to assess damage. We now have 1100 adjusters on the ground. We have another 500 who are ready to go as soon as we can get into some of the most-devastated areas. It will be many weeks, probably months, before there is anything approaching reliable estimates" [10].

2.3.5 ***Role of Capital Costs.*** The importance of capital and its need to secure an adequate rate of return is often not sufficiently understood. In particular, the prices charged for catastrophe insurance must be sufficiently high to cover the expected claims costs and other expenses, but must also cover the costs of allocating risk capital to underwrite this risk. Moreover, because the large amounts of risk capital are needed to underwrite catastrophe risk relative to the expected liability, the capital cost built into the premium is high, often dominating the expected loss cost. Thus, an insurer usually needs to charge high prices relative to its loss expenses, simply to earn a fair rate of return on equity and thereby maintain its credit rating.

To illustrate, we construct a hypothetical example that is somewhat conservative by ignoring taxes. Consider a portfolio that has 1000 in expected liabilities, $E(L)$. Since actual losses will not equal expected losses, the insurer needs to hold considerable capital. Here we will assume that \$1 of capital is needed for each \$1 of expected liability to maintain the insurer's credit rating. Thus the insurer needs capital, $E(L) = 1000$. In addition to paying claims, the insurer has an additional 200 in upfront-expected expenses that include commissions to agents and brokers, and underwriting expenses. Moreover, given the risk characteristics of the portfolio, investors require a rate of return (ROE) of 15% on their investment to compensate for risk. The insurer invests its funds in lower risk vehicles that yield an expected return, r, of 10%. What premium P would the insurer have to charge to secure a return of 15% for its investors?

The formula for the premium can be expressed as a function of the cash flows, the return on investment and the required return on equity (k is the ratio of equity to expected losses)

$$P = \frac{X(1+r) + E(L)}{(1+r) - k(ROE - r)}$$

$$P = \frac{200(1+0.05) + 1000}{(1+0.05) - 1(0.15 - 0.05)} = 1210$$

Premiums need to be 1210 to generate this required 15% return on equity.

This calculation is very sensitive to the ratio of capital to expected liability, k, needed to preserve credit. In the above example, the ratio was one dollar of capital for one dollar of expected liability. This ratio is in the ballpark for many property liability insurers for their combined books of business. However, for catastrophe risk, with its very high tail risk (which severely affects credit risk), the capital to liability ratio needs to be considerably higher. Indeed, the capital to liability ratio depends on volatility (particularly, the downside or tail risk) of the catastrophe liability and its correlation with the insurer's remaining portfolio. For higher layers of cat risk, the expected loss is often quite low and the volatility very high. At these layers, the required capital to liability ratio can be considerably greater than unity as shown in this example. An increase in the capital to liability ratio will increase the premium required to generate a fair return on equity.

A second issue with catastrophe risk is that it can be expensive to underwrite since it requires extensive modeling. Many companies will buy commercial models and/or use their own in-house modeling capability. If we rework the above premium calculation now with the transaction costs set at 100% of expected losses and 4:1 capital to expected liability ratio the required premium is now 3154:

$$P = \frac{1000(1+0.05) + 1000}{(1+0.05) - 4(0.15 - 0.05)} = 3154$$

For very high layers even more capital may be needed, thus further increasing the premium. There are other considerations that can dramatically leverage upwards the capital cost, notably the impact of double taxation [11] have simulated the tax burden over many parameterizations and show that tax costs alone can reasonably be as much as the claim cost and lead to further increases in premiums. When we account for all these factors (i.e. high capital inputs, transaction costs, and taxes), catastrophe insurance premiums often are several multiples of expected claims costs.

2.4 Role of Rating Agencies

During the past few years, rating agencies have played increasing attention to the impact that catastrophic risks will have on their view of the financial stability of insurers and reinsurers. The rating given to a company will affect their ability to attract business and hence their pricing and coverage decisions.

To illustrate how ratings are determined consider A.M. Best. It undertakes a quantitative analysis of an insurer's balance sheet strength, operating performance, and business profile. Evaluation of catastrophe exposure plays a significant role in the determination of ratings, as these are events that could threaten the solvency of a company. Projected losses of disasters occurring at specified return periods (a 100-year windstorm/hurricane or a 250-year earthquake) and the associated reinsurance programs to cover them are two important components of the rating questionnaires that insurers are required to complete.

For several years now, A.M. Best has been requesting such information for natural disasters. Their approach has been an important step forward in the incorporation of catastrophe risk into an insurer's capital adequacy requirements. Up until recently the rating agency has been including probable maximum loss (PML) for only *one* of these severe events (100-year windstorm/250-year earthquake, depending on the nature of the risk the insurer was mainly exposed to) in its calculation of a company's risk-adjusted capitalization. In 2006 A.M. Best introduced a second event as an additional stress test. The PML used for the second event is the same as the first event in the case of hurricane (a 1-in-100 year event; the occurrence of one hurricane is considered to be independent of the other one). If the main exposure facing the insurer is an earthquake, the second event is reduced from a 1-in-250 year event to a 1-in-100 year event [12]. These new requirements have increased the amount of risk capital that insurers have been forced to allocate to underwrite this risk and have made them more reluctant to provide this coverage unless they are able to raise premiums sufficiently to reflect these additional costs.

In March 2006, Standard and Poor's, another rating agency, indicated that it would revise criteria for measuring cat risk which has traditionally been based on premium charges. But the new criteria will measure catastrophe risk based on exposure of the insurer. This will include an exposure-based capital charge for insurers similar to what it does for reinsurers based on net expected annual aggregate property losses for all perils at 1-in-250 year return period. There will be a 6–12 month phase period to allow companies to adjust risk profiles [13].

2.5 Role of Market State Regulation

In the United States, insurance is regulated at the state level with the principal authority residing with insurance commissioners. Primary insurers are subject to solvency regulation and rate and policy form regulation, whereas domestic reinsurers are subject only to solvency regulation (the price and terms of reinsurance transactions are not subject to regulation). Solvency regulation addresses the question as to whether the insurer or reinsurer is sufficiently capitalized to fulfill its obligations if a significant event occurs and inflicts major losses on its policyholders.

Insurance commissioners regard solvency as a principal objective even if it means requiring higher premiums or other insurer adjustments (e.g. reducing their catastrophe exposures). On the other hand, insurance regulators face political pressure to keep insurance premiums "affordable" and coverage readily available. In balancing solvency and consumer protection goals, insurance regulators are required by state laws to ensure that rates are adequate but not excessive and not unfairly discriminatory. Regulators' assessment of insurers' rates and other practices involves some degree of subjectivity that can result in rate restrictions that reduce the supply of insurance or cause other market problems and distortions. "Parameter uncertainty" and different opinions on the level of risk of loss can lead to disagreements between insurers and regulators over what constitutes adequate rates and appropriate underwriting practices.[7]

State legislatures, governors, and the courts also play a significant role in the regulation of insurers and insurance markets. Consequently, insurance regulators are subject to a number of constraints on their authority and discretion and the other branches of state government may impose their preferences on how state laws, regulations, and policies govern insurers and insurance markets. Ultimately, all elected officials and their appointees are subject to the will of the voters—if government officials act contrary to the preferences of voters, they are subject to being replaced by people who will obey the voters, even if their actions are economically unsound.

3 US FEDERAL AND STATE CATASTROPHE PROGRAMS

We now turn to the important role that the federal and state governments in the United States play in supplementing or replacing private insurance with respect to natural disasters, nuclear accidents, and other catastrophic losses. This section provides a brief overview of several of these programs to illustrate the types of public–private partnerships that have been implemented in the past.

3.1 Flood and Hurricane Insurance

3.1.1 Flood. Insurers have experimented over the years with providing protection against water damage from floods, hurricanes, and other storms. After the severe Mississippi Floods of 1927, they concluded that the risk was too great, and refused to provide private insurance again. As a result, Congress created the NFIP in 1968, whereby homeowners and businesses could purchase coverage for water damage. Private insurers market flood policies, and the premiums are deposited in a federally operated Flood Insurance Fund, which is then responsible for paying claims. The stipulation for this financial protection is that the local community makes a commitment to regulate the location and design of future floodplain construction to increase safety from flood hazards. The federal government established a series of building and development standards for floodplain construction to serve as minimum requirements for participation

[7]See [14–16] for more detailed discussions of insurance regulatory policies in general and specific to natural disaster risk.

in the program. The creation of the Community Rating System in 1990 has linked mitigation measures with the price of insurance in a systematic way [17].

The number of claims paid by the NFIP differs from year to year, but between 1980 and 2002 was never higher than 62,400, which was the count in 1995. The severity of flood losses from the 2004 hurricane season led to 75,000 claims, a new record in the history of the program. The breach in the New Orleans levees from Hurricane Katrina coupled with the flood losses from Hurricanes Katrina, Rita, and Wilma triggered some 239,000 claims with the NFIP in 2005, 80% of which were from Hurricane Katrina. It is estimated that the NFIP will pay in excess of $23 billion in flood claims for the 2005 hurricane season, the equivalent of 10 years of premiums. This raises major questions regarding the future of the program. Two bills are currently being discussed in Congress as to how modify its operation so it fits better with this new loss dimension.

3.1.2 Hurricane Insurance. The need for hurricane insurance is most pronounced in the state of Florida. Following Hurricane Andrew in 1992, nine property-casualty insurance companies became insolvent, forcing other insurers to cover these losses under Florida's State Guaranty Fund. Property insurance became more difficult to obtain as many insurers reduced their concentrations of insured property in coastal areas. During a special session of the Florida State Legislature in 1993 the Florida Hurricane Catastrophe Fund (FHCF) was created to relieve pressure on insurers to reduce their exposures to hurricane losses. The FHCF, a tax-exempt trust fund administered by the State of Florida, is financed by premiums paid by insurers that write policies on personal and commercial residential properties. The fund reimburses a portion of insurers' losses following major hurricanes (above the insurer's retention level) and enables insurers to remain solvent [10]. The four hurricanes that hit Florida in the fall of caused an estimated $29 billion in insured losses, with only about $2.6 billion paid out by the fund. Each hurricane was considered a distinct event, so that retention levels were applied to each storm before insurers could turn to the FHCF.

As this article goes to press, the future of hurricane and flood insurance in the United States is being analyzed as part of a research initiative between the Wharton School, Georgia State University, and the Insurance Information Institute, in partnership with over 15 insurers and reinsurers, trade associations, and federal agencies.

3.2 Earthquake Insurance

The history of earthquake activity in California convinced legislators that this risk was too great to be left in the hands of private insurers alone. In 1985, a California law required insurers writing homeowners coverage on one to four unit residential buildings to also offer earthquake coverage. Since rates were regulated by the state, insurers felt they were forced to offer coverage against older structures in poor condition, with rates not necessarily reflecting the risk. Following the 1994 Northridge earthquake, huge losses on insured property created a surge in demand for coverage. Insurers were concerned that if they satisfied the entire demand, as they were required to do by the 1985 law, they would face an unacceptable level of risk and become insolvent following the next major earthquake. Hence, many firms decided to stop offering coverage or restricted the sale of homeowners' policies in California.

In order to keep earthquake insurance alive in California, the State legislature authorized the formation of the California Earthquake Authority (CEA) in 1996. The CEA is a state-run insurance company that provides earthquake coverage to homeowners. The innovative feature of this financing plan is the ability to pay for a large earthquake while committing relatively few dollars up front. There is an initial assessment of insurers of $1 billion to start the program and then contingent assessments to the insurance industry and reinsurers following a severe earthquake. Policyholders absorb the first portion of an earthquake through a 15% deductible on their policies [18]. However, 8 years after the creation of the CEA, the take-up rate for homeowners is about 15%, down from 30% when the California State Legislature created the CEA [19]. It is questionable how effective this program will be in covering losses should a major earthquake occur in California.

3.3 Nuclear Accident Insurance[8]

The Price-Anderson Act, originally enacted by Congress in 1957, limits the liability of the nuclear industry in the event of a nuclear accident in the United States. At the same time, it provides a ready source of funds to compensate potential accident victims, which would not ordinarily be available in the absence of this legislation. The Act covers large power reactors, small research and test reactors, fuel reprocessing plants, and enrichment facilities for incidents that occur through plant operation as well as transportation and storage of nuclear fuel and radioactive wastes.

Price-Anderson sets up two tiers of insurance. Each utility is required to maintain the maximum amount of coverage available from the private insurance industry—currently $300 million per site. In the United States, this coverage is written by the American Nuclear Insurers, a joint underwriting association or "pool" of insurance companies. If claims following an accident exceed that primary layer of insurance, all nuclear operators are obligated to pay up to $100.59 million for each reactor they operate payable at the rate of $10 million per reactor, per year. As of February 2005, the US public had more than $10 billion of insurance protection in the event of a nuclear reactor incident. More than $200 million has been paid in claims and costs of litigation since the Price-Anderson Act went into effect, all of it by the insurance pools. Of this amount, approximately $71 million has been paid in claims and costs of litigation related to the 1979 accident at Three Mile Island.

In February 2003, Congress extended the law for power reactors licensed by the Nuclear Regulatory Commission (NRC) to the end of 2003.[9] Coverage for facilities operated by the Department of Energy has been extended until the end of 2006 in a separate legislative action. Congress is now considering further extension of the law as part of comprehensive energy legislation.

3.4 Federal Aviation Administration Third Party Liability Insurance Program

Since the terrorist attacks of September 11, 2001, the US commercial aviation industry can purchase insurance for third party liability arising out of aviation terrorism. The

[8]For more details on nuclear accident insurance see Nuclear Energy Institute "Price-Anderson Act Provides Effective Nuclear Insurance at No Cost to the Public", February 2005.

[9]Although the existing law has technically expired, its provisions are "grandfathered" and continue to apply to all existing NRC licensees, that is to say, to power reactor operators with operating licenses issued prior to the expiration date. Personal Correspondence with John Quattrocchi July 21, 2005.

current mechanism operates as a pure government program, with premiums paid by airlines into the Aviation Insurance Revolving Fund managed by the Federal Aviation Administration (FAA).

As the program carries a liability limit of only $100 million, losses paid by government sources in the event of an attack will almost surely exceed those available through the current insurance regime. In that case, either the government would need to appropriate additional disaster assistance funds as it did in the aftermath of September 11th, or victims would be forced to rely on traditional sources of assistance [20].

3.5 International Terrorism Risk Insurance Program (TRIA)

Although the United States has been successful since 9/11 in preventing terrorist attacks on its own soil, the impact on the economy of another mega-attack or series of coordinated attacks will cause serious concerns to the government, the private sector and citizenry [21, 22]. With security reinforced around federal buildings, the commercial sector constitutes a softer target for terrorist groups to inflict mass casualties and stress on the nation. These threats require that the country as a whole develops strategies to prepare for and recover from a (mega-)terrorist attack. Insurance is an important policy tool for consideration in this regard.

Quite surprisingly, even after the terrorist attack on the World Trade Center in 1993 and the Oklahoma City bombing in 1995, insurers in the United States did not view either international or domestic terrorism as a risk that should be explicitly considered when pricing their commercial insurance policy, principally because losses from terrorism had historically been small and, to a large degree, uncorrelated. Thus, prior to September 11, 2001, terrorism coverage in the United States was an unnamed peril included in most standard all-risk commercial and homeowners' policies covering damage to property and contents.

The terrorist attacks of September 11, 2001, killed over 3000 people from over 90 countries and inflicted insured losses currently estimated at $32.5 billion that was shared by nearly 150 insurers and reinsurers worldwide. Reinsurers (most of them European) were financially responsible for the bulk of these losses. These reinsurance payments came in the wake of outlays triggered by a series of catastrophic natural disasters over the past decade and portfolio losses due to stock market declines. Having their capital base severely hit, most reinsurers decided to reduce their terrorism coverage drastically or even to stop covering this risk.

In response to such concerns, the Terrorism Risk Insurance Act (TRIA) of 2002 was passed by Congress and signed into law by President Bush on November 26, 2002.[10] It constitutes a temporary measure to increase the availability of risk coverage for terrorist acts [23]. TRIA is based on risk sharing between the insurance industry, all policyholders (whether or not they have purchased terrorism insurance) and the federal government (taxpayers) up to $100 billion of insured losses on US soil. President Bush signed into law a 2-year extension of TRIA on December 22, 2005, the Terrorism Risk Insurance Extension Act (TRIEA) that expanded the private sector role and reduced the federal share of compensation for terrorism insured losses. Since TRIA was passed prices have stabilized in most industries and take-up rate continuously increased. Today, over 60%

[10]The complete version of the Act can be downloaded at: http://www.treas.gov/offices/domestic-finance/financial-institution/terrorism-insurance/claims_process/program.shtml.

of large commercial firms have some type of terrorism insurance coverage [24]. As we write this article in January 2007, it is unclear what type of long-term terrorism insurance program, if any, will emerge at the end of 2007 for dealing with the economic and social consequences of terrorist attacks.

4 CONCLUDING REMARKS

The past 5 years have demonstrated that the United States can suffer today catastrophic losses from disasters that far exceed those from events that occurred prior to 2000. Insurance has played a critical role in the recovery process, but these recent catastrophes have raised questions as to under which conditions the private sector can continue to provide coverage (and will want to) and the role that the public sector should play in dealing with these events. One limiting factor of the programs we discussed in this article is that the premiums they charge are often not based on risk, reducing the economic incentive to invest in cost-effective mitigation measures. Implementing reforms that will have premiums based on risk is certainly one way to start.

Also, so far, the country has responded by developing specific programs for each type of catastrophes. Whether this is the best way to go remains an open question. Other countries have developed some coverage that protects homeowners and businesses against all types of disasters: wind, flood, terrorist attacks, and so on [25]. This idea has been proposed for the United States many years ago [26] and has been resurrected following Hurricane Katrina[11]. Whether a risk-based all-hazard disaster insurance is now more appropriate given the events of the past 5 years is an issue for the new US Congress to study in more detail.

REFERENCES

1. Auerswald, P., Branscomb, L., LaPorte, T., and Michel-Kerjan, E., Eds. (2006). *Seeds of Disasters, Roots of Response. How Private Action Can Reduce Public Vulnerability*. Cambridge University Press, New York, p. 504.
2. Grossi, P. and Kunreuther, H., Eds. (2005). *Catastrophe Modeling: A New Approach to Managing Risk*, Chapter 2. Springer, New York.
3. Kunreuther, H. Meyer, R., and van den Bulte, C. (2004). *Risk Analysis for Extreme Events: Economic Incentives for Reducing Future Losses*. NIST Monograph GCR 04-871.
4. Stone, J. (1973). A theory of capacity and the insurance of catastrophic risks: part I and part II. *J. Risk Insur.* **40**, 231–243, (Part I); 339–355, (Part II).
5. Kunreuther, H., Meszaros, J., Hogarth, R., and Spranca, M. (1995). Ambiguity and underwriter decision processes. *J. Econ. Behav. Organ.* **26**, 337–352.
6. American Re. (2002). *Topics: Annual Review of North American Natural Catastrophes 2001*.
7. Dionne, G., Doherty, N., and Fombaron, N. (2000). Adverse selection in insurance markets. In *Handbook of Insurance*, G. Dionne, Ed. Chapter 7. Kluwer, Boston.
8. Henriet, D. and Michel-Kerjan, E. (2006). *Optimal Risk-Sharing under Dual Reversed Asymmetry of both Information and Market Power: A Unifying Approach*, Working Paper,

[11]For a more detailed discussion of this proposal see [27].

Center for Risk Management and Decision Processes, The Wharton School, Philadelphia, PA.

9. Lecomte, E. and Gahagan, K. (1998). Hurricane insurance protection in Florida. In *Paying the Price: The Status and Role of Insurance Against Natural Disasters in the United States*, H. Kunreuther and R. Roth Sr., Eds. Joseph Henry Press, Washington, DC, pp. 97–124.
10. Francis,T. (2005). CEO says allstate adjusts storm plan. Interview of Edward Liddy. *Wall St. J.* C1–C3.
11. Harrington,S. E. and Niehaus,G. (2001). Government insurance, tax policy, and the affordability and availability of catastrophe insurance. *J. Insur. Regul.* **19**(4), 591–612.
12. Best, A. M. (2006). *Methodology: Catastrophe Analysis in AM Best Ratings*, April.
13. Insurance Journal. (2006). *S&P to Implement New Way to Assess Insurer Cat Risk*, March 31, 2006.
14. Klein, R. (1995). Insurance regulation in transition. *J. Risk Insur.* **62**, 263–404.
15. Grace, M., Klein, R., and Liu, Z. (2005). Increased hurricane risk and insurance market responses. *J. Insur. Regul.* **24**, 2–32.
16. Klein, R. W. (2006). *Catastrophe Risk and the Regulation of Property Insurance Markets*, working paper.Georgia State University, November for a more detailed discussion of insurance regulatory policies in general and specific to natural disaster risk.
17. Pasterick, E. (1998). The national flood insurance program. In *Paying the Price: The Status and Role of Insurance Against Natural Disasters in the United States*, H. Kunreuther and R. Roth Sr., Eds. Chapter 6. Joseph Henry Press, Washington, DC.
18. Roth, R., Jr. (1998). Earthquake insurance protection in California. In *Paying the Price: The Status and Role of Insurance Against Natural Disasters in the United States*, H. Kunreuther and R. Roth Sr., Eds. Chapter 4. Joseph Henry Press, Washington, DC.
19. Risk Management Solutions. (2004). *The Northridge, California Earthquake. A 10-Year Retrospective*.
20. Strauss, A. (2005). *Terrorism Third Party Liability Insurance for Commercial Aviation, Federal Intervention in the Wake of September 11*. The Wharton School, Center for Risk Management and Decision Processes, Philadelphia, PA.
21. Kunreuther, H. and Michel-Kerjan, E. (2004). Challenges for terrorism risk insurance in the united states. *J. Econ. Perspect.* **18**(4), 201–214.
22. Kunreuther, H. and Michel-Kerjan, E. (2005). *Insurability of (mega)-Terrorism, Report for the OECD Task Force on Terrorism Insurance, in OECD (2005)*, July 5, Terrorism Insurance in OECD Countries, Organization for Economic Cooperation and Development, Paris.
23. U.S. Congress. (2002). *Terrorism Risk Insurance Act of 2002*. HR 3210.Washington, DC, November 26.
24. Michel-Kerjan, E. and Pedell, B. (2006). How does the corporate world deal with mega-terrorism? Puzzling evidence from terrorism insurance markets. *J. Appl. Corp. Finance* **18**(4), 61–75.
25. Michel-Kerjan,E. and deMarcellis-Warin,N. (2006). Public-private programs for covering extreme events: the impacts of information distribution and risk sharing. *Asia Pac. J. Risk Insur.* **1**(2), 21–49.
26. Kunreuther, H. (1968). The case for comprehensive disaster insurance. *J. Law Econ.*
27. (a) Kunreuther, H. (2006). Has the time come for comprehensive natural disaster insurance? In *On Risk and Disaster*, R. Daniels, D. Kettl, and H. Kunreuther, Eds. University of Pennsylvania Press, Philadelphia, PA; (b) Kunreuther, H. and Michel-Kerjan E. (in press). Improving homeland security in the wake of large-scale disasters: would risk-based all-hazard disaster insurance help in the post-Katrina world? *Economic and Risk Assessment of Hurricane Katrina*, Routledge.

QUANTITATIVE REPRESENTATION OF RISK

Bilal M. Ayyub and Mark P. Kaminskiy
Center for Technology and Systems Management, Department of Civil and Environmental Engineering, University of Maryland, College Park, Maryland

1 INTRODUCTION

Risk representation is defined as the description of risk in an appropriate manner for a decision-making situation. The representation of risk that would retain an appropriate level of information is essential to decision makers, and should be tailored for specific applications and types of decision-making situations under consideration.

Risk is defined as the potential for loss or harm to systems due to the likelihood of an unwanted event and its adverse consequences. The term *potential* is used in the definition to imply uncertainty as a central notion to risk. Uncertainties are inherent in both the likelihood of the unwanted event and in the consequences including their nature and severity. Loss or harm includes all negative consequences, including both tangible effects such as human casualties and/or financial losses and less tangible impacts such as political instability, decreased morale, and reduced operational effectiveness. The term *event* represents the occurrence that triggers scenarios and the consequences. While consequences of an event can be both advantageous and/or adverse, risk considers only the adverse consequences—meaning that risk is a function of perspective. Risk to us is opportunity to our adversaries. The term *likelihood* refers to both the occurrence of the event and its potential adverse consequences. Probability is a quantitative measure of likelihood [1].

Each of homeland security risks can be classified into three components of threat, vulnerability, and potential consequences that should be analyzed in a systems framework. Analytically and computationally, risk assessment requires the following estimates:

- threat analysis to define and assess the likelihood of an attack by an adversary according to some threat scenario;
- vulnerability analysis to assess the likelihood that the adversary attack is successful (i.e. overcoming the security and physical protection systems) given an attack; and
- consequence analysis to assess the consequences given a successful adversary attack.

Risk assessment can involve a range of qualitative and quantitative methods [1]. For quantitative analysis, the first two likelihoods may be assessed as probabilities and multiplied to calculate the probability of an adversary's successful attack. In risk applications (e.g. natural hazards) and homeland security applications, combining these three elements into a single measure of risk would deprive a decision maker from the full information that might be critical for rational decision making.

The representation of risk that would retain an appropriate level of information is essential to decision makers, and should be tailored for specific applications and types of decision-making situations under consideration.

The above scenario can be viewed as a cause and, if it occurs, may result in consequences with severities. The representation of risk is essential for risk communication and decision making. The objective of this article is to provide a summary of risk representation methods, including random variables relating to losses, occurrence rates, and moments and parameters. The methods include qualitative and quantitative representations of risk. Loss exceedance probability (EP) distributions, loss exceedance rates, probability density functions (PDFs), and descriptive point estimates are included for this purpose. Computational examples are used to illustrate these methods. The selection of an appropriate method should be based on the decision-making situation under consideration.

2 RISK MEASURES AND REPRESENTATION

Risk results from an event or sequence of events, called a *scenario*, with occurrence likelihood. A scenario can be viewed as a cause and, if it occurs, may result in consequences with severities, called *losses*. A risk measure accounts for both the probability of occurrence of a scenario and its consequences. Both the probability and consequences could be uncertain. This section provides fundamental cases for representing risks to prepare readers for subsequent sections.

2.1 Fundamentals of Risk Representation

The representation or display of risk may include risk matrices (or tables), risk plots (or graphs), and probability distributions of adverse consequences in the form of cumulative probability distributions or EP distributions [1]. The choice of representation techniques depends on the type of analysis (qualitative or quantitative) and stakeholder/decision maker preferences. The risk display becomes the baseline for comparison of the effectiveness of risk management alternatives. It is important to recognize that the probability of the event is not plotted as a function of its potential adverse consequences. Rather, the two elements of risk are plotted separately on their own axes. Uncertainties in both the elements of risk are represented by line segments, which form a cross that depicts the risk of the event.

2.2 Probability Trees for Defining Scenarios

Probability trees can be used to develop scenarios and associated branch probabilities (p_i) and consequences (i.e. losses) L_i. In this section, simple cases are used to illustrate the development of scenarios. Figure 1 shows a generic probability tree related to homeland security applications. A sequence of events constitutes a scenario in this tree. By following a series of branches under each of the headings shown in the figure, a scenario can be identified (or developed). The scenario probability as indicated in the figure can be evaluated as the product of the conditional probabilities of the branches appearing in the scenario. The conditional probability for a branch is the probability of occurrence of the branch under the assumed conditions that all the branches leading to this particular branch in the scenario have occurred. This product can be viewed as the best (or point)

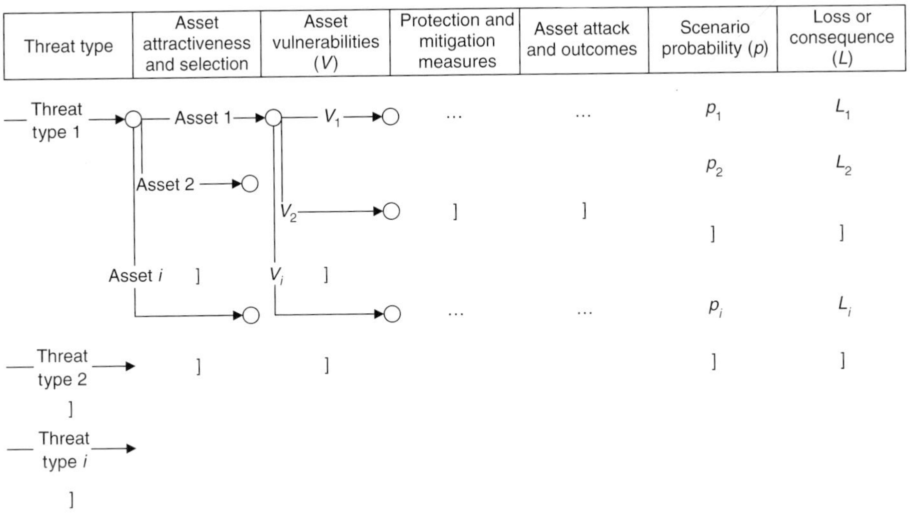

FIGURE 1 Construction of a generic probability tree.

estimate of the scenario probability. The loss or consequence associated with a scenario is also conditional on the occurrence of the scenario, and can be provided either as a best (i.e. point) estimate or a probability distribution. A percentile interval can be used instead, and converted into a probability distribution once a distribution type is assumed.

2.3 One-Scenario Representation

For a scenario i, the risk pair (p_i, L_i) can be represented in any of the forms provided in Figure 2 reflecting the type of data available: (i) point estimates, (ii) interval estimates, and/or (iii) probability distributions. A percentile interval can be used, and converted to a probability distribution once a distribution type is assumed.

2.4 Multiscenario Representation

Probability trees similar to Figure 1 lead to multiscenario cases, which is a simple generalization of the case of one-scenario shown in Figure 2. All these scenarios are based on point estimate data types. For other data types, that is, intervals or probability distributions, they can be converted to point estimates. Uncertainty modeling can be used at the end of any analytical process, such as probability exceedance distributions presented in the following section, to propagate these uncertainties based on the data types to assess their effects on the outcomes of the analytical process.

3 EXCEEDANCE PROBABILITY DISTRIBUTIONS

3.1 Definitions

Risk can be represented using *EP distributions* (*or curves*) [2]. The EP curve gives the probabilities of specified levels of loss exceedance. The notion "losses" can be expressed in terms of dollars of damage, number of fatalities, casualties, and so on.

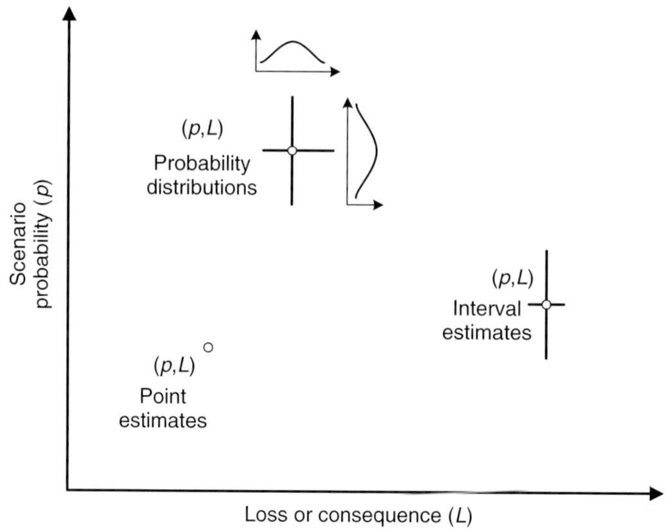

FIGURE 2 Risk plots and data types.

The construction of an EP curve begins with the data that might be empirically obtained or produced using simulation methods. An example by Kunreuther et al. [2] illustrates the empirical construction of an EP curve based on a set of loss-producing events. In this example, an EP curve is constructed for a portfolio of residential earthquake policies in Long Beach, California, and based on dollar losses to homes in Long Beach from earthquake events. The objective is to combine these loss-producing events, and then to determine respective return periods and annual probabilities of exceeding losses of different magnitudes. On the basis of these estimates, the EP, or mean EP, is developed as shown in Figure 3. According to this figure with a particular loss L_i, the curve provides the probability that the loss as a random variable exceeds L_i with the respective y-axis value p_i. Thus the x-axis measures the loss value in given units, and the respective y-axis value is the probability that the loss exceeds a given value.

Using an EP curve, the effects of countermeasures and mitigation strategies can be examined based on the shifts of the EP curve downward. In other words, the EP curves can be used to estimate benefit–cost effect of these strategies.

The EP curves can also express uncertainty associated with the probability of occurrence of an undesirable event and the magnitude of the respective loss as a result of uncertainties in specified values as inputs. Such uncertainties can be expressed using the *percentile* EP curves. For example, the 5th and 95th percentile EP curves depict uncertainties associated with losses as well as the uncertainties associated with respective probabilities. In our case, the EP curve depicting uncertainties in losses would show the interval $(L_i^{0.05}, L_i^{0.95})$, which can include the loss related to a given mean value L_i associated with probability p_i. Similarly, the EP curve depicting uncertainties in probabilities shows the percentiles $(p_i^{0.05}, p_i^{0.95})$ associated with loss mean value L_i.

It should be noted that due to data availability, constructing EP curves is much easier for the problems dealing with natural disasters (such as earthquakes and floods) compared to the risk assessment problems relating to homeland security where data are limited or nonexistent.

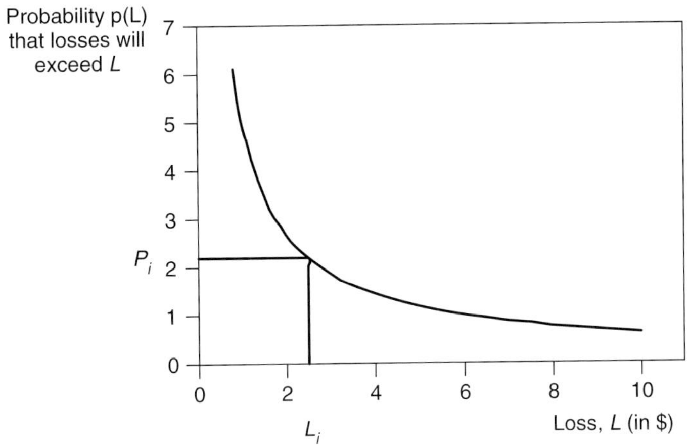

FIGURE 3 Sample mean exceedance probability curve [2].

3.2 Constructing Exceedance Probability Distributions

The available data are assumed to be collected as a set of n disaster events, E_i, $i = 1, 2, \ldots, n$ with respective annual probabilities of occurrence p_i. Also, estimates of the respective losses (L_i) associated with these events are made after the occurrence of these events. Coming back to the example discussed in the previous section, the events are earthquakes with magnitudes that can be physically measured in the form of ground accelerations. If one deals with floods, the respective events can be physically defined as well, for example, water-level elevation. An example of such earthquake data including 15 events is given in Table 1.

Some of the entries in Table 1 are for events having equal (or almost equal) values of losses, but with respective annual probabilities that are different, such as for $Event_{10}$, and $Event_{11}$. These two events correspond to earthquakes of different magnitudes occurring, perhaps, at different locations with different populations at risk and producing the same loss estimates.

In order to apply the notions of annual probability and random variable, the disaster events must be, to an extent, *repeatable*, which is, to an extent, true in the case of natural disasters such as earthquakes, floods, hurricanes, and so on. On the other hand, it should be noted that identification of events is not necessarily straightforward in the case of terrorist actions. Nevertheless, the respective events can be identified. For example, they might be defined as explosions committed in public communication systems, like metro, railway stations, and so on.

The loss associated with a given disaster event can be treated as a continuous random variable, whereas the number of events occurring in some specified period of time, such as a year, can be treated as a discrete random variable. Table 1 provides loss data estimates for these events. These loss values can be considered as best estimates for the respective events, and can be treated as central-tendency point estimates of the continuous random variable.

For a set of natural disaster events, E_i, $i = 1, \ldots, n$, each event has an annual probability of occurrence, p_i, and an associated loss estimate, L_i. The number of events per year is not limited to one; numerous events can occur in the given year. Fifteen such

TABLE 1 Constructing Exceedance Probability Curves

Event (E_i)	Annual Probability of Occurrence (p_i)	Loss (L_i)	Exceedance Probability ($EP(L_i)$)	$E(L) = (p_i L_i)$	Rate of Occurrence ($\lambda(L_i)$)
$Event_1$	0.002	25,000,000	0.0020	50,000	0.0020
$Event_2$	0.005	15,000,000	0.0070	75,000	0.0070
$Event_3$	0.010	10,000,000	0.0169	100,000	0.0170
$Event_4$	0.020	5,000,000	0.0366	100,000	0.0373
$Event_5$	0.030	3,000,000	0.0655	90,000	0.0677
$Event_6$	0.040	2,000,000	0.1029	80,000	0.1086
$Event_7$	0.050	1,000,000	0.1477	50,000	0.1598
$Event_8$	0.050	800,000	0.1903	40,000	0.2111
$Event_9$	0.050	700,000	0.2308	35,000	0.2624
$Event_{10}$	0.070	500,000	0.2847	35,000	0.3351
$Event_{11}$	0.090	500,000	0.3490	45,000	0.4292
$Event_{12}$	0.100	300,000	0.4141	30,000	0.5346
$Event_{13}$	0.100	200,000	0.4727	20,000	0.6400
$Event_{14}$	0.100	100,000	0.5255	10,000	0.7455
$Event_{15}$	0.283	0	0.6597	0	1.0779

events are listed in Table 1, ranked in descending order of the amount of loss. Event 15 was defined to be encompassing of all other zero-loss events so that the set of all events is collectively exhaustive. In order to keep this example simple, these events were chosen so that the set is exhausted, that is, the sum of the probabilities for all the events equals one.

The events in Table 1 are assumed to be independent Bernoulli random variables with the following probability mass functions:

$$P(E_i \text{ occurs}) = p_i \tag{1a}$$

$$P(E_i \text{ does not occur}) = 1 - p_i \tag{1b}$$

The expected loss (E) for a given event E_i is

$$E(L) = p_i L_i$$

Indexing the events in reverse order of their losses (i.e. $L_i \geq L_{i+1}$), and assuming that only one disaster can occur during a given year, the mean (expected) EP for a given loss $EP(L_i)$ can be found as

$$\begin{aligned} EP(L_i) &= P(L > L_i) = 1 - P(L \leq L_i) \\ &= 1 - \prod_{j=1}^{i} (1 - p_j) \end{aligned} \tag{2}$$

Equation (2) shows that the resulting annual EP that the loss exceeds a given value L_i is one minus the probability that losses below or equal the value L_i have not occurred.

The *EP* curve based on the data from Table 1 is shown in Figure 3. In general, the summation of the probabilities of Events 1–14 can exceed one, since they are independent Bernoulli events.

3.3 Curve Fitting to Exceedance Probabilities

Procedures to fit a curve to EPs, that is, mainly applied to insurance problems, are not considered by Kunreuther et al. [2]. In contrast to the insurance applications, for the problems related to the risk representation for homeland security, fitting of EP curves is essential due to data unavailability or limited availability that might represent the results of expert opinion elicitation processes.

In this section, a special approach to fitting the EP curves is suggested, assuming that limited data is available in the form discussed in the previous section, that is, as a set of couples of p_i and L_i.

The suggested choice of EP curve functional form is based on meeting general trends expected for probabilistic reasoning. The function used for EP curve fitting must be simple and concave upward (Figures 3 and 4). Such function should be positive and limited from the above by unity. These requirements are satisfied for the survivor function of Pareto distribution, which (for the random variable L) can be written as

$$P(L) = \left(\frac{d}{L}\right)^{c} \tag{3a}$$

where c and d are the positively defined shape and scale parameters, respectively. It should be noted that the Pareto distribution has support on $L > d$, which is very important from the standpoint of the parameters estimation. Another form of the Pareto distribution having support that $L > 0$ is given by the following survivor function [3, 4]:

$$P(L) = \left(1 + \frac{L}{d}\right)^{-c} \tag{3b}$$

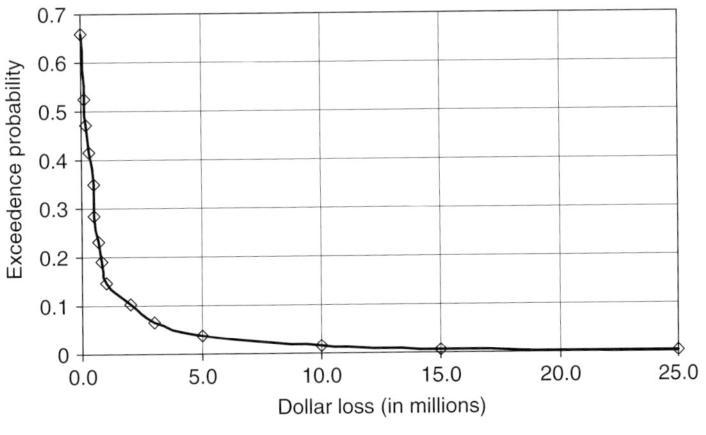

FIGURE 4 Mean exceedance probability curve based on data from Table 1.

The Pareto distribution is used to model variables relating to a random time to failure. For example, consider a population of components with an exponential time to failure (T) distribution. The parameter of the exponential distribution (λ) is supposed to be random and distributed according to the gamma distribution with parameters α and k. A new random variable closely related to the time to failure T is introduced as $\tau = (T/\alpha + 1)$. It can be shown that the introduced variable τ is distributed according to the Pareto distribution with the scale parameter equal to 1, and the shape parameter equal to k.

It is also interesting to note that Eq. (3a), at least formally, coincides with the model used for fitting of the so-called S–N or Wohler curves, where S is stress amplitude and N is time to failure in cycles, such that:

$$N = kS^{-b} \tag{4}$$

where b and k are material parameters estimated from the data. Because of the probabilistic nature of fatigue life, one has to deal with not only one S–N curve, but with a family of S–N curves so that each curve is related to a probability of failure as the parameter of the model. These curves are called S–N–P curves or curves of constant probability of failure.

Figure 5 displays the results of Eq. (3a) loglinear least squares (LS) fitting based on the data given in Table 1. The results of Eq. (3a) nonlinear LS fitting using Gauss–Newton method for the same data are shown in Figure 6. Finally, the results of fitting the model provided in Eq. (3b) nonlinear LS fitting using Gauss–Newton method for the same data are shown in Figure 7. The estimates of the models parameters and R^2 statistics

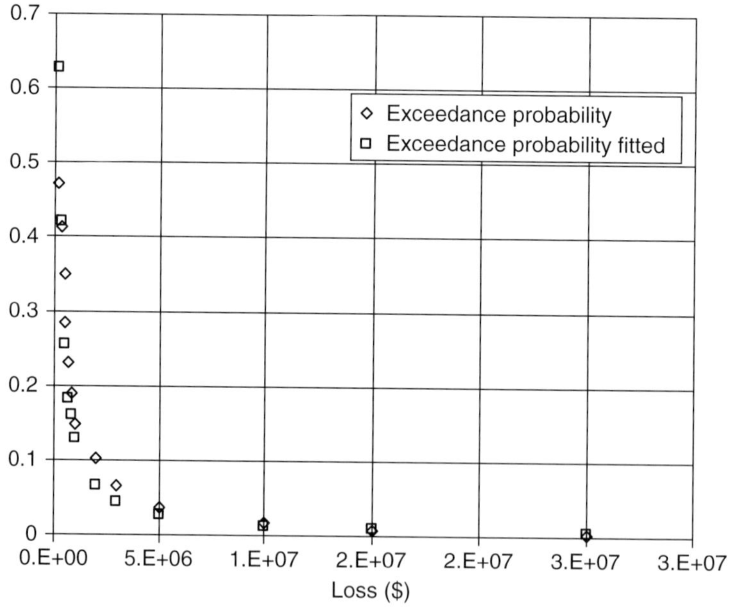

FIGURE 5 Results of Eq. (3a) loglinear LS fitting for data given in Table 1.

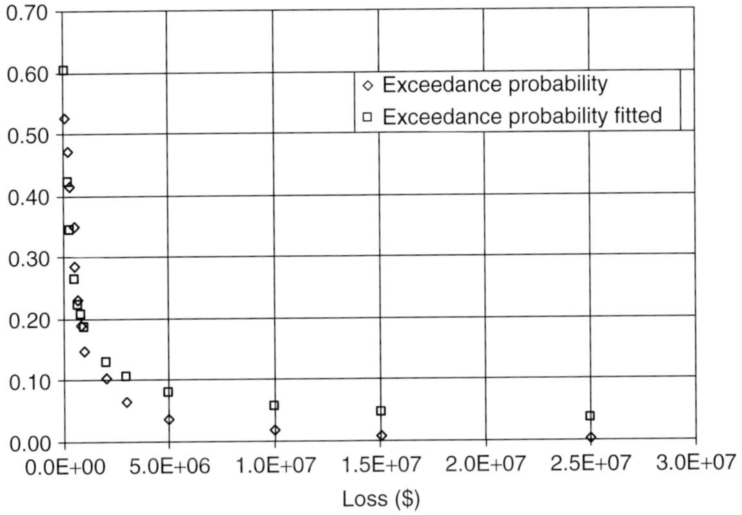

FIGURE 6 Results of Eq. (3a) nonlinear LS fitting using Gauss–Newton method for data given in Table 1.

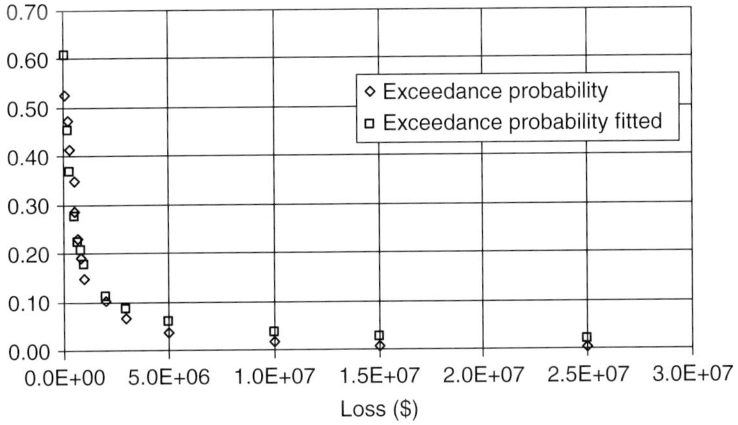

FIGURE 7 Results of Eq. (3b) nonlinear LS fitting using Gauss–Newton method for data given in Table 1.

(representing the fraction of variation explained by the fitted model) for each fitted model are given in Table 2. By analyzing the table, one might draw a conclusion that Eq. (3b) provides the best fit.

4 OCCURRENCE RATES OF EVENTS

The objective of this section is to relate the EP curves to the rates of occurrence of loss events that are of particular interest [1, 5].

TABLE 2 Parameters Estimates of Models

Model	Fitting Procedure	Estimates of Parameters		R^2
		c	d	
Eq. (3a)	Loglinear LS	0.984	124,514	0.930
Eq. (3a)	Nonlinear LS	0.512	37,256	0.934
Eq. (3b)	Nonlinear LS	0.724	101,412	0.959

4.1 Nonrandom Rates

At this point, it should be noted that any point of the mean EP curve $EP(L)$ can be related to an event resulting in loss $>L$. These events could be related to specific real events.

In order to assess how the losses are evolving in time, one needs to recall that, strictly speaking, the considered *EPs* are, in essence, *annual EPs*. The annual probability is considered as the probability (equal to EP) that a given event occurs in the time interval (0, 1], where 1 is 1 year. The *annual probabilities of occurrence* (p) can be related to the *rate of occurrence of disaster events* resulting in a loss exceeding L in a stochastic process. In the situation considered, the only reasonable model for the stochastic process is the homogeneous Poisson process.

It should be noted that the homogeneous Poisson process is a widely used model for the insurance and security applications. It is reasonable to assume that the cumulative distribution function (CDF) is exponential that is the time between events, each resulting in loss $>L$, is distributed as a function of time according to the exponential distribution. For any value of loss L_i the parameter λ of this distribution can be simply evaluated as

$$\lambda(L_i) = -(1/t)\, ln(1 - EP(L_i)) \tag{5}$$

where $t = 1$ year. The parameter $\lambda(L)$ can be considered as the rate of occurrence. Its numerical values for the above discussed example are exhibited in the right column of Table 1. One can notice that the values of $\lambda(L_i)$ are very close to p_i for the small values of p_i. The higher the value of p_i, the wider the difference between p_i and $\lambda(L_i)$. It can be easily explained using the two-term Tailor expansion: $EP = 1 - \exp(-\lambda t) = 1 - \exp(-p)$. For $p \to 0$, one gets $EP \approx 1 - 1 + p = p$.

The occurrence rate of the events for which loss lies in a two-sided interval can be found in the following way. The situation that the rate λ related to the event associated with a loss interval (L_l, L_u) is considered. By introducing the width of the respective probability interval $\Delta = EP(L_u) - EP(L_l)$, Eq. (5) takes the form

$$\begin{aligned}\lambda(\Delta) &= -(1/t)\ \ln(1 - (EP(L_u) - EP(L_l)) \\ &= -(1/t)\ \ln(1 - \Delta)\end{aligned} \tag{6}$$

where $t = 1$ year for the annual probabilities considered throughout this section.

In order to better reveal the probabilistic meaning of Eq. (6), let us make a rather realistic assumption that Δ is small enough (e.g. $\Delta = 0.1$) such that the Taylor expansion for $\ln(1 - \Delta)$ can be applied. The respective approximation yields

$$\lambda(\Delta) \approx \Delta/t \tag{7a}$$

$$\approx EP(L_l)/t - EP(L_u)/t \tag{7b}$$

$$\approx \lambda(L_l) - \lambda(L_u) \tag{7c}$$

Finally, for the complementary event that the loss does not exceed a given value L_i, that is, $NEP(L_i) = P(L < L_i)$, one can get the following equation for the respective occurrence rate:

$$\lambda(L_i) = -(1/t)\ln(1 - (1 - EP(L_i))) \tag{8}$$

4.2 Accumulated Loss as a Function of Occurrence Rate

The rate (λ) was initially considered as a nonrandom quantity. Randomness in the rate can be added to the model as provided by Ayyub [1].

Now let us assume that the number of event occurring during a nonrandom time interval is governed by homogeneous Poisson process. The loss associated with each event is modeled using a continuous random variable S with the CDF $F_S(s)$. The CDF of the accumulated damage (loss) during a nonrandom time interval $[0, t]$ is given by

$$F(s; t, \lambda) = \sum_{n=0}^{\infty} \mathrm{e}^{-\lambda t} \frac{(\lambda t)^n}{n!} F_S^{(n)}(s) \tag{9}$$

where $F_S{}^{(n)}(s)$ is the n-fold convolution of $F_S(s)$. In other words, $F_S{}^{(n)}(s)$ is the probability that the total loss accumulated over n events (during time t) does not exceed s. For $n = 0$, $F_S{}^{(0)}(s) = 1$ and for $n = 1$, $F_S{}^{(1)}(s) = F_S$. For $n = 2$, the twofold convolution $F_S{}^{(2)}(s)$ can be evaluated using conditional probabilities as

$$F_S^{(2)}(s) = \int_0^{\infty} F_S(s - x)\, \mathrm{d}F_S(x) \tag{10}$$

In the case of a normal probability distribution, the twofold convolution $F_S{}^{(2)}(s)$ can be evaluated as follows:

$$F_S^{(2)}(s) = P(S + S < s) = F_S(s; 2\mu, \sqrt{2}\sigma) \tag{11}$$

where P is the probability and F_S $(s; 2\mu, \sqrt{2}\,\sigma)$ is the CDF of $S + S$ that can be evaluated using normal CDF of S with a mean value of 2μ and a standard deviation of $\sqrt{2}\,\sigma$ for independently and identically distributed losses. For other distribution types, the distribution of the sum $S + S$ needs to be used. In general for the case of $S + S$, the following special relations can be used:

- $S + S$ is normally distributed if S is normally distributed;
- $S + S$ has a gamma distribution if S has an exponential distribution; and
- $S + S$ has a gamma distribution if S has a gamma distribution.

The threefold convolution $F_S^{(3)}(s)$ is obtained as the convolution of the distributions of $F_S^{(2)}(s)$ and $F_S(s)$ for independently and identically distributed losses represented by a normal probability distribution as follows:

$$F_S^{(3)}(s) = P(S + S + S < s) = F_S(s; 3\mu, \sqrt{3}\sigma) \tag{12}$$

Higher-order convolution terms can be constructed in a similar manner for the independently and identically distributed losses represented by a normal probability distribution as follows:

$$F_S^{(n)}(s) = P(\overbrace{S + S + \cdots + S}^{\text{n times}} < s) = F_S(s; n\mu, \sqrt{n}\sigma) \tag{13}$$

Therefore, Eq. (9) can be written for independently and identically distributed losses represented by a normal probability distribution as follows:

$$F(s; t, \lambda) = \sum_{n=0}^{\infty} e^{-\lambda t} \frac{(\lambda t)^n}{n!} F_S(s; n\mu, \sqrt{n}\sigma) \tag{14}$$

If λ is random with the PDF $f_\lambda(\lambda)$, Eq. (9) can be modified to

$$F(s; t) = \int_0^\infty \left(\sum_{n=0}^{\infty} e^{-\lambda t} \frac{(\lambda t)^n}{n!} F_S^{(n)}(s) \right) f_\lambda(\lambda)\, d\lambda \tag{15}$$

where $F_S^{(n)}(s)$ is the n-fold convolution of $F_S(s)$.

5 RISK DESCRIPTORS

5.1 Probability Distributions

As it was discussed in Section 2, an EP curve can be treated as the survivor function of the respective distribution of losses. Thus, one can write the following equation for the CDF of losses, $F(L)$,

$$F(L) = 1 - S(L) \tag{16}$$

where $S(L)$ is the fitted survivor function of losses.

It should be noted that this distribution of losses is related to 1 year, that is, it is *annual distribution* of losses. On the basis of the CDF of losses, $F(L)$, the respective PDF can be easily found.

The Pareto distribution is further used to illustrate the suggested methodology. In this case, the CDF of losses $F(L)$ takes on the following form:

$$F(L) = 1 - \left(\frac{d}{L}\right)^c \tag{17}$$

where $L > d$. The corresponding PDF of losses, $f(S)$, is given by

$$f(L) = \frac{cd^c}{L^{c+1}} \tag{18}$$

It should be noted that the above distribution parameters d and c are obtained as a result of the respective EP curve fitting (Section 2.3). These point estimates of the loss distribution parameters are used in the following section for evaluation of the main risk descriptors.

5.2 Risk Descriptors and Functions

By evaluating the PDF of losses, one can find the main loss descriptors. For example, the *mean value of loss* (*expected loss*), $E(L)$, can be evaluated as

$$E(L) = \int_0^\infty l f(l)\, \mathrm{d}l \tag{19}$$

For the Pareto distribution, mean value of loss is given by the following equation:

$$E(L) = \frac{cd}{c-1} \tag{20}$$

The *variance of loss* $Var(L)$ is evaluated as

$$Var(L) = \int_0^\infty [l - E(L)]^2 f(l)\, \mathrm{d}l \tag{21}$$

For the considered Pareto distribution, this variance is given by

$$Var(L) = \frac{cd^2}{(c-1)^2(c-2)} \tag{22}$$

Other useful descriptors such as median, percentiles, and interpercentile ranges can be obtained in a similar way.

The nonrandom hazard rates λ associated with one-sided and two-sided loss intervals are considered in Section 4.1 and the respective formulas are given by Eqs. (5–7).

Having the annual distribution of losses (i.e. Eq. 16) available and using Eq. (9), one can estimate the accumulated random damage during time t, introduced in Section 3.2, as a function of time t (in years) and the respective annual occurrence rate λ.

In this case, the annual occurrence rate λ can be estimated using the Apostolakis–Mosleh estimate suggested for the rate of the precursor events of the core damage in nuclear power plants [6]

$$\lambda = \frac{N(T)}{T} \tag{23}$$

where in the case considered, $T = 1$ year, and $N(T)$ is given by

$$N(T) = \sum_{i=1}^{n} p_i$$

where n is the number of events observed during time T (a year) and p_i are their respective probabilities.

It should be noted that the probabilities of events included in the above sum are related to the same events, which were used for constructing the respective EP curve. For example, based on Table 1, the annual occurrence rate λ is estimated as 0.717 events per year.

If the annual occurrence rate λ is treated as *random*, and its PDF $f(\lambda)$ is available, for example, as a result of expert opinion elicitation, the accumulated random damage can be evaluated using Eq. (15).

ACKNOWLEDGMENTS

The authors are please to thank the anonymous reviewer and Dr Bruce Colletti for their thoughtful and constructive comments and suggestions.

REFERENCES

1. Ayyub, B. M. (2003). *Risk Analysis in Engineering and Economics*. Chapman & Hall/CRC Press, London.
2. Kunreuther, H., Meyer, R., and Van den Bulte, C. (2004). *Risk Analysis for Extreme Events: Economic Incentives for Reducing Future Losses*, NIST Report GCR 04-871 National Institute of Standards and Technology (NIST).
3. Hoyland, A., and Rausand, M. (1994). *System Reliability Theory*. John Wiley & Sons, New York.
4. Frees, E., and Valdez, E. (2001). Understanding relationships using copulas. *N. Am. Actuar. J.* **2**(1), 1–25.
5. Ayyub, B. M., and McCuen, R., (2003). *Probability, Statistics and Reliability for Engineers and Scientists*, 2nd ed. Chapman & Hall/CRC Press, Boca Raton, FL.
6. Apostolakis, G., and Mosleh, A., (1979). Expert opinion and statistical evidence: an application to reactor core melt frequency. *Nucl. Sci. Eng.* **70**, 135.

FURTHER READING

Modarres, M., Kaminskiy, M., and Krivtsov, V. (1999). *Reliability Engineering and Risk Analysis: A Practical Guide*. Marcel Dekker, New York, Basel.

Modarres, M., Martz, H., and Kaminskiy, M. (1996). The accident sequence precursor analysis: review of the methods and new insights. *Nucl. Sci. Eng.*, **123**, 238–258.

Pate-Cornell, E. (2004). On signals, response, and risk mitigation. In *Accident Precursor Analysis and Management*, J. R. Phimister, V. M. Bier, H. C. Kunreuther Eds. The National Academic Press, Washington, DC.

QUALITATIVE REPRESENTATION OF RISK*

Gary R. Smith
Logical Decisions, Fairfax, Virginia

James Scouras
Defense Threat Reduction Agency, Ft. Belvoir, Virginia

Robert M. DeBell
Gaithersburg, Maryland

1 INTRODUCTION

A risk representation is an attempt to describe or clarify a particular risk. A qualitative risk representation describes or clarifies without the explicit use of numbers. A qualitative risk representation can be used to communicate the results of a qualitative risk assessment or summarize a quantitative risk assessment. Qualitative risk representations can be used to describe individual risks or to compare the relative severity of different risks.

Risk can be defined as the potential for injury or loss. This definition has two parts: the potential (or likelihood) and the loss (usually called a *consequence in risk analysis*). Quantitative representations describe likelihood with probabilities and loss using numeric scales such as dollars or fatalities. Alternatively, qualitative representations can be used for the likelihood, the loss, or for the risk as a whole.

All risk representations are subject to misinterpretation. This can be due to poor definition of the risk being represented or lack of caveats regarding the underlying analysis. Qualitative risk representations are further subject to misinterpretation due to the imprecision inherent in language and due to imperfect alignment between qualitative representations and underlying quantitative assessments. It is also important to avoid the temptation to do inappropriate mathematical manipulations with qualitative risk representations. For example, one should not arbitrarily establish a rule that two "low" risks add to become a "medium" risk.

Qualitative risk representation is not a formal field of scientific study. Disciplines that touch on qualitative risk representation include risk analysis, measurement theory, psychology, sociology, and linguistics. More specialized researchers investigate risk communication, but seldom distinguish between qualitative and quantitative representations.

*The views expressed in this article represent those of the authors; they do not necessarily represent the views of any governmental or commercial entity.

2 VERBAL RISK REPRESENTATIONS

Verbal representations describe risk or risk components with words. We use verbal here in the sense of "communicated with words" rather than "spoken out loud". Verbal risk representations are the most common type of qualitative risk representation, but not the only one. Other types of qualitative risk representations, such as pictorial, graphical, and cartographic, are not discussed here.

The simplest verbal representations are warnings such as "smoking may be hazardous to your health". Warnings can be useful, but generally provide little information about the degree of risk or how the risk compares or relates to other risks.

More sophisticated verbal risk representations use two general approaches—scales and risk comparisons. Scales are a standardized way of describing individual risks by assigning to each risk one of a predefined set of descriptions. Risk comparisons directly compare risks with one another or with standardized reference risks.

3 RISK REPRESENTATIONS USING SCALES

Stevens [1] classifies scales of measurement (both verbal and numeric) into four different types—nominal, ordinal, interval, and ratio.

The simplest scale of measurement is a nominal scale or categorization. Categorizations describe the nature of the risk without attempting to describe its magnitude. For example, "chemical, biological, radiological, nuclear, and explosive" is a common risk categorization. The major limitation of categorizations is that they do not provide any ordering of risks. It is not possible to use a categorization to say that one risk is greater than another.

The next simplest scale of measurement is an ordering. More formally, orderings are called *ordinal scales*. The archetypical ordinal scale is "high, medium, and low". By definition, ordinal scales provide a partial ordering of risks. What they do not provide is an indication of the relative degree of risk. There is no way to tell how much worse a high risk is than a medium risk or whether the change from low to medium is greater or lesser than the change from medium to high. In addition, the ordering is only partial since there is no way to determine the order of two risks assigned the same scale point.

Adjusting an ordinal scale such that each scale degree represents the same amount of change results in an interval scale. Interval verbal scales have almost all of the properties of integer numbers, except that they do not have an explicit zero point. Practically, it is very difficult to develop a purely qualitative interval scale. It will usually be necessary to tie a verbal scale to an underlying numeric scale to ensure that the changes between scale points are equal.

If one of the descriptions in an interval risk scale represents zero risk, then the scale is a ratio scale. The integers are an example of a ratio scale. While the Fahrenheit and Centigrade temperature scales are interval scales since their zero points do not correspond to a complete absence of temperature, the Kelvin temperature scale is a ratio scale since its zero point *can* be interpreted that way. Ratio scales allow statements like "risk A is twice as severe as risk B".

Risk presenters are often not as careful as they should be about the type of scale they are using and the operations and inferences that are valid for each scale type. Valid

TABLE 1 Valid Operations and Results for Scale Types

Scale	Valid Operations	Results
Nominal	Count	Frequency distribution
Ordinal	Greater than Less than Equal to	Rank Median
Interval	Add Subtract Multiply by a constant	Mean Standard deviation
Ratio	Divide elements	Ratio

operations and results for each scale type are shown in Table 1. (The operations for simpler types are all allowed for more complex types.)

Nominal and ordinal scales should generally be verbal, since the normal rules of arithmetic cannot be applied and the use of numbers would be confusing. Interval and ratio scales should generally be numeric since arithmetic operations are allowed. Verbal interval and ratio scales would typically be tied to an underlying numeric scale and used principally for risk communication.

When using verbal scales, imprecision in the scale point descriptions can make it difficult to determine which scale point to assign. For example, *moderate* consequences to one person might be classified as *severe* by another, leading to disagreements on what should go where. Thus, verbal scales require more elaborate—sometimes quite elaborate—definitions to allow consistent assignment and interpretation.

3.1 Verbal Likelihood Scales

Verbal descriptions are often used to represent the first component of risk—likelihood. If the consequence is clearly defined (e.g. death), then the verbal representation of likelihood is equivalent to a verbal representation of risk.

One widely used likelihood scale was developed by Kent [2] for use in intelligence analysis. The ordinal scale, shown in Table 2, associates verbal descriptions with probability ranges. An interesting feature of this scale is that it is asymmetric, with, for example, 63–87% represented by "probable" and 20–40% represented by "probably not." Curiously, the Kent scale has some gaps in probability, notably between the categories "almost certainly not" and "probably not."

Though the Kent scale is associated with numbers, it is not an interval scale, since the scale points do not represent equal changes. Another commonly used verbal likelihood scale, the Juster scale [3], spreads its scale points much more evenly. Originally developed for use in marketing studies, the Juster scale combines an 11-point interval scale (with minor exceptions in the first and last categories) with a set of verbal descriptions and associated probabilities, as shown in Table 3. If the top and bottom scale points were redefined as "certain (100 in 100 chance)" and "no chance (0 in 100 chance)", respectively, the Juster scale would be a ratio scale.

TABLE 2 The Kent Scale

Range of Probability	Verbal Description
100%	Certainty
93% give or take about 6%	Almost certain
75% give or take about 12%	Probable
50% give or take about 10%	Chances about even
50% give or take about 10%	Chances about even
30% give or take about 10%	Probably not
7% give or take about 5%	Almost certainly not
0%	Impossibility

TABLE 3 The Juster Scale

Score	Verbal Description
10	Certain, practically certain (99 in 100 chance)
9	Almost sure (9 in 10 chance)
8	Very probable (8 in 10 chance)
7	Probable (7 in 10 chance)
6	Good possibility (6 in 10 chance)
5	Fairly good possibility (5 in 10 chance)
4	Fair possibility (4 in 10 chance)
3	Some possibility (3 in 10 chance)
2	Slight possibility (2 in 10 chance)
1	Very slight possibility (1 in 10 chance)
0	No chance, almost no chance (1 in 100 chance)

Other verbal representations that are not explicitly linked to probability numbers have been used in various settings. A difficulty with these representations is the wide range of interpretations of various terms used to describe likelihood. Numerous studies have documented this. (See Hillson [4] and Wallsten and Budescu [5] for reviews of this literature.) Hillson presents survey results from over 500 respondents that link verbal terms with the probabilities perceived to be associated with them. The results summarized in Figure 1 demonstrate a wide range of variation in interpretation of verbal probability terms. Note also some apparent inconsistencies in the responses. "Better than even", is interpreted by some respondents, to be associated with probabilities lower than 50%. Also, "definite" and "impossible" are associated with probabilities less than 100% and greater than 0, respectively. Note that these interpretation difficulties may persist even for scales such as the Juster scale that explicitly tie words to numbers. It is not clear if users of the scale will interpret it using the numbers or using their preexisting interpretation of the words in the scale.

Verbal representations of the likelihood of a terrorist attack have also been developed. The most well known of these is the Homeland Security Advisory System administered by the US Department of Homeland Security (DHS) [6]. This system, introduced in March 2002 [7], is an ordinal scale of five colors (threat levels) and associated text descriptions, as shown in Figure 2. Although the scale refers to various levels of "risk", this is not

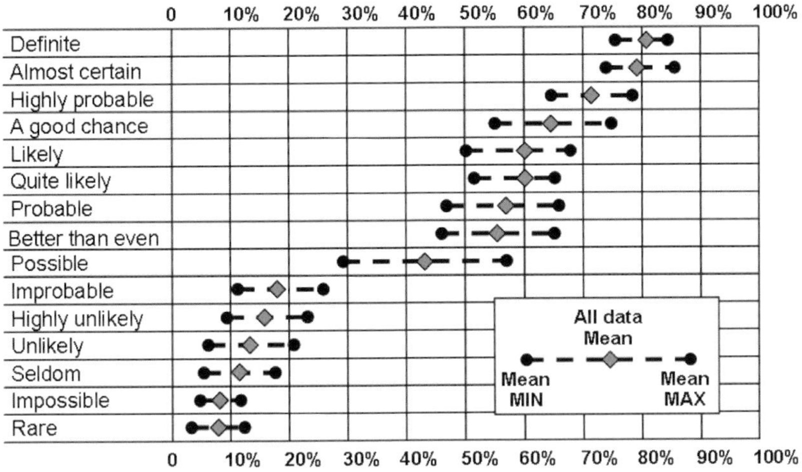

FIGURE 1 Perceptions of probability represented by common words. Source [4].

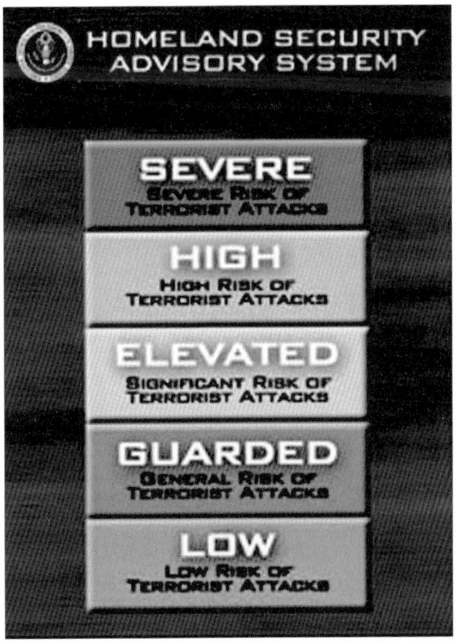

FIGURE 2 DHS Homeland Security Advisory System.

an accurate usage in the sense of the definition provided above. Rather, as used in the scale, risk refers to the likelihood of the associated consequence being “terrorist attacks”. There are no published guidelines that establish how the threat level is determined, but DHS has developed guidelines for government agencies [8] and civilians [9] on how to respond to different announced threat levels. The United Kingdom has developed a similar five-point ordinal scale, though it is not color coded [10].

3.2 Verbal Consequences Scales

Verbal representations are also frequently used for describing consequences. These types of scales are perhaps most often used in association with natural disasters such as earthquakes and hurricanes.

The modified Mercalli scale [11] shown in Table 4 qualitatively describes the consequences of earthquakes. This 12-point ordinal scale is composed of a category number, a brief category description, and a longer discussion of consequences. One advantage of this type of scale is that historical earthquakes that occurred before the development of the quantitative Richter scale can be estimated on the modified Mercalli scale using contemporary descriptions. Similar intensity scales have also been developed for hurricanes (the Saffir–Simpson scale developed in 1969 by Herbert Saffir and Bob Simpson) and tornados (the Fujita scale developed in 1971 by Dr Ted Fujita [12]).

Qualitative consequences scales have also been developed in the computer industry. Microsoft [13], for example, developed the ordinal scale shown in Table 5 to characterize the potential consequences of reported vulnerabilities in Microsoft products. A consequences scale developed by Symantec is discussed in the next section.

A common problem with consequence scales for terrorist attacks is that the full spectrum of consequences includes effects on life and health, the economy, social institutions, and individual psychology. Some of these elements are quantifiable; others are not. Verbal scales with rigorous definitions are difficult in both cases. Combining these effects into a single scale is an even more daunting challenge. Thus, often only one element of consequences—typically the most readily quantifiable element such as lives lost or economic damage—is considered in a terrorism risk assessment.

3.3 Verbal Scales that Combine Likelihood and Consequences

Verbal representations of risk combine likelihood and consequences into a single description. This is commonly done either with a single scale or in a risk matrix with likelihood on one dimension and consequences on the other. The two methods are discussed further in the following sections.

3.4 One-Dimensional Combined Risk Scales

One-dimensional verbal risk representations combine likelihood and consequences information into a single scale. This is often done by developing text descriptions that include both likelihood and consequences information. An excellent example of this is the Torino impact hazard scale originally developed by Dr Richard Binzel [14, 15]. This scale indicates the risk posed by an asteroid that could potentially collide with earth. The representation consists of 11 verbal descriptions organized into five color-coded categories, as shown in Figure 3. The scale descriptions indicate the likelihood of an event, the potential consequences, and the actions that are recommended to be taken.

Similar scales have been developed for information technology (IT) risks. For example, Symantec Corporation developed a security response threat severity assessment scale [16] that evaluates the risk from threats such as viruses and worms. The scale addresses likelihood by examining the extent to which the threat is already present "in the wild" and the rate at which it spreads. The scale addresses consequences by describing the damage that the threat could cause. These three risk aspects are combined into a five-point risk scale that includes brief descriptions (very low, low, moderate, severe, and very severe),

TABLE 4 The Modified Mercalli Scale for Earthquakes

Category	Category Description	Level of Damage
I	Instrumental	Not felt except by a very few under especially favorable conditions
II	Feeble	Felt only by a few persons at rest, especially on upper floors of buildings. Delicately suspended objects may swing
III	Slight	Felt quite noticeably by persons indoors, especially on the upper floors of buildings. Many do not recognize it as an earthquake. Standing motor cars may rock slightly. Vibration similar to the passing of a truck. Duration estimated
IV	Moderate	Felt indoors by many, outdoors by few during the day. At night, some awakened. Dishes, windows, doors disturbed; walls make cracking sound. Sensation like heavy truck striking building. Standing motor cars rocked noticeably. Dishes and windows rattle
V	Rather strong	Damage negligible. Small, unstable objects displaced or upset; some dishes and glassware broken
VI	Strong	Damage slight. Windows, dishes, glassware broken. Furniture moved or overturned. Weak plaster and masonry cracked
VII	Very strong	Damage slight moderate in well-built structures; considerable in poorly built structures. Furniture and weak chimneys broken. Masonry damaged. Loose bricks, tiles, plaster, and stones will fall.
VIII	Destructive	Structure damage considerable, particularly to poorly built structures. Chimneys, monuments, towers, elevated tanks may fail. Frame houses moved. Trees damaged. Cracks in wet ground and steep slopes.
IX	Ruinous	Structural damage severe; some will collapse. General damage to foundations. Serious damage to reservoirs. Underground pipes broken. Conspicuous cracks in ground; liquefaction.
X	Disastrous	Most masonry and frame structures/foundations destroyed. Some well-built wooden structures and bridges destroyed. Serious damage to dams, dikes, embankments. Sand and mud shifting on beaches and flat land.
XI	Very disastrous	Few or no masonry structures remain standing. Bridges destroyed. Broad fissures in ground. Underground pipelines completely out of service. Rails bent. Widespread earth slumps and landslides.
XII	Catastrophic	Damage nearly total. Large rock masses displaced. Lines of sight and level distorted.

more detailed text descriptions, and references to the threat's rating on its three individual risk components.

3.5 Representing Consequence and Likelihood Scales Using a Risk Matrix

The risk matrix is a technique commonly employed to display risk. Its essential feature is that it does not combine the two elements of risk—likelihood and consequences. Rather,

TABLE 5 The Microsoft Severity Rating System

Rating	Definition
Critical	A vulnerability whose exploitation could allow the propagation of an internet worm without user action
Important	A vulnerability whose exploitation could result in compromise of the confidentiality, integrity, or availability of users' data, or of the integrity or availability of processing resources
Moderate	Exploitability is mitigated to a significant degree by factors such as default configuration, auditing, or difficulty of exploitation
Low	A vulnerability whose exploitation is extremely difficult, or whose impact is minimal

Normal (Green Zone)	1	A routine discovery in which a pass near the earth is predicted that poses no unusual level of danger. Current calculations show the chance of collision is extremely unlikely with no cause for public attention or public concern. New telescopic observations very likely will lead to re-assignment to Level 0.
Meriting Attention by Astronomers (Yellow Zone)	2	A discovery, which may become routine with expanded searches, of an object making a somewhat close but not highly unusual pass near the Earth. While meriting attention by astronomers, there is no cause for public attention or public concern as an actual collision is very unlikely. New telescopic observations very likely will lead to re-assignment to Level 0.
	3	A close encounter, meriting attention by astronomers. Current calculations give a 1% or greater change of collision capable of localized destruction. Most likely, new telescopic observations will lead to re-assignment to Level 0. Attention by public and by public officials is merited if the encounter is less than a decade away.
	4	A close encounter, meriting attention by astronomers. Current calculations give a 1% or greater chance of collision capable of regional devastation. Most likely, new telescopic observations will lead to re-assignment to Level 0. Attention by public and by public officials is merited if the encounter is less than a decade away.
Threatening (Orange Zone)	5	A close encounter posing a serious, but still uncertain threat of regional devastation. Critical attention by astronomers is needed to determine conclusively whether or not a collision will occur. If the encounter is less than a decade away, governmental contingency planning may be warranted.
	6	A close encounter by a large object posing a serious but still uncertain threat of a global catastrophe. Critical attention by astronomers is needed to determine conclusively whether or not a collision will occur. If the encounter is less than three decades away, governmental contingency planning may be warranted.
	7	A very close encounter by a large object, which if occurring this century, poses an unprecedented but still uncertain threat of a global catastrophe. For such a threat in this century, international contingency planning is warranted, especially to determine urgently and conclusively whether or not a collision will occur.
Certain Collisions (Red Zone)	8	A collision is certain, capable of causing localized destruction for an impact over land or possibly a tsunami if close offshore. Such events occur on average between once per 50 years and once per several 1000 years.
	9	A collision is certain, capable of causing unprecedented regional devastation for a land impact or the threat of a major tsunami for an ocean impact. Such events occur on average between once per 10,000 years and once per 100,000 years.
	10	A collision is certain, capable of causing a global climatic catastrophe that may threaten the future of civilization as we know it, whether impacting land or ocean. Such events occur on average once per 100,000 years, or less often.

FIGURE 3 The Torino Impact Hazard Scale.

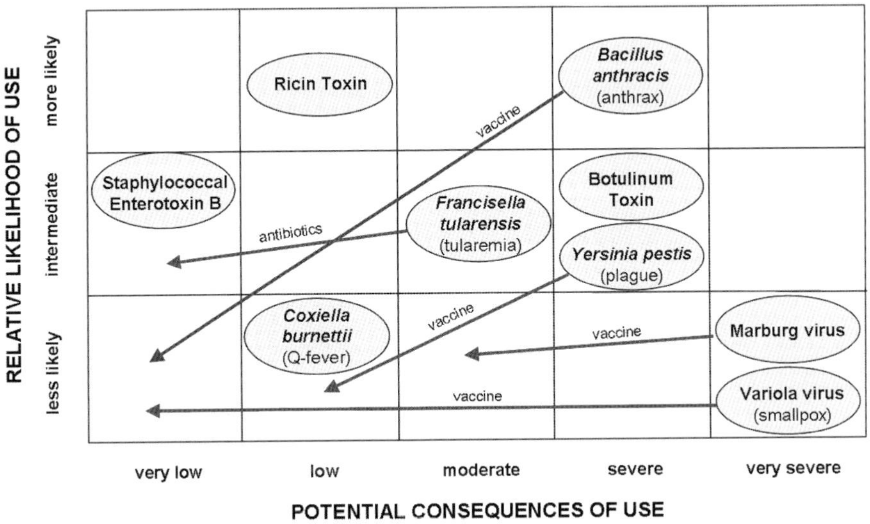

FIGURE 4 Risk assessment for selected biological agents.

these elements define the two dimensions of the risk matrix. An example of risk matrix is presented in Figure 4 to provide focus for the discussion in this section. It displays the risks of a terrorist attack with a biological weapons agent. The practical issues in developing a risk matrix, and advantages and limitations of the resultant display are discussed.

In contrast to a quantitative risk graph that would utilize continuous numerical scales for both dimensions, a qualitative risk matrix divides both dimensions into bins defined by verbal scales. As discussed above, verbal scales often require detailed descriptions to ensure consistency.

For example, both likelihood and consequences of terrorist use of biological agents have multiple factors that need to be combined into a single scale to define the dimensions of the risk matrix.

Some of the factors that affect consequences include agent stability and persistence in the environment, infectivity, communicability, availability of clinical therapies, and duration of effects and lethality. In combining these factors into a scale, there is a trade-off between fewer and more scale points. Fewer scale points make the assessment job easier, but are less discriminating. More scale points can describe more variations in the factors, but make the matrix more complex. In practice, risk matrices typically have 3–6 rows and columns.

We have combined the consequence factors into the following five-point scale:

Very low. A biological agent preparation that would likely be incapacitating with little or no associated lethality. Such an agent would not likely be communicable, not persistent for more than a few days in the environment, and cause no permanent or long-lasting effects. Effective clinical treatments would be available to limit or eliminate disease effects.

Low. An incapacitating agent with a low level of associated lethality (less than 10%). Clinical treatments for such an agent would be available, but possibly difficult to distribute. The agent would not likely be communicable, but could be persistent in the environment and cause only rare instances of permanent effects.

Moderate. A lethal agent with an associated moderate level of lethality (at least 10%, but less than 35%). Clinical treatments would likely be available, but limited by quantities or logistics for distribution. This agent could be communicable through direct contact with an infected individual, especially for prolonged periods. The agent could be stable in the environment and persistent either because it may be protected as in the case of spores or because carried by vectors.

Severe. An extremely lethal agent preparation with no or very limited clinical therapies available. The agent would be communicable through direct contact with infected individuals or families, and would likely be stable in the atmosphere, but not necessarily persistent for long periods.

Very severe. An extremely lethal agent preparation with no available effective clinical therapy. The agent would likely be highly communicable, stable, and persistent in the atmosphere, with a low infectivity.

Even this attempt to precisely define a consequences scale is inadequate since we have not covered all the different possible combinations of factors. In addition, most agents do not fit every aspect of these definitions for any single category, so the analyst will have to decide which scale point provides the closest fit for each agent. Also note that quantitative elements have sometimes crept into the qualitative descriptions.

The other dimension of the risk matrix—likelihood—has a similar set of definitional problems. The factors that affect likelihood of use for a given agent as a biological weapon include availability of materials, financial resources, expertise needed, safety precautions necessary, ease of cultivation, production and storage, practicality of dissemination, and anticipated consequences of use. Note that since the anticipated consequences will influence likelihood of use, the likelihood and consequences scales are not independent.

Continuing our example of a terrorist attack with a biological weapons agent, the following three-point scale for relative likelihood of use based on these factors is developed:

More likely. Agents are easy to acquire and commonly available almost anywhere in the world. These agents would also be quite easy to produce in desired quantities and would not require high-level containment. Individuals could employ simple protocols for production and would use protective clothing and available therapies such as vaccines for protection. These agents would not be communicable. The consequences for the use of such agents would vary depending on the efficacy of the preparation and the method for dissemination.

Intermediate. Agents are easy to acquire as these may have a wide distribution, but not necessarily global. Individuals would not have great difficulties acquiring such agents, although the agents most frequently would occur in endemic areas. The protocols needed to produce such agents would be technically challenging, but with adequate equipment and sufficient expertise, needed quantities of agent could be produced. Because clinical therapies may or may not be available and because such agents could be communicable, containment facilities would be needed. The consequences for the use of such agents would have a wide range of variation depending on the agent, as well as the ability to produce stable, effective preparations.

Less likely. Agents are difficult to find in the environment and difficult to produce in usable quantities. These agents would likely produce serious diseases for which there may not be adequate therapies. Thus, these agents would require high-level containment and high-level expertise to be able to manage production or cultivation. These agents could be highly communicable, and the consequences for the use of such agents might be very severe.

Although a considerable effort is made to ensure that our scales are comprehensive and accurate, we note some shortcomings. Many of the agents are hybrids of these categories. For example, in Figure 4 *Yersinia pestis*, the causative agent of plague, is shown to be an intermediate threat on the matrix. There is no vaccine available to prophylactically protect individuals, but effective antibiotics do exist. If the plague organism is successfully disseminated, and because it is communicable from person-to-person and animal-to-person, it could potentially cause severe consequences. However, actual consequences could be significantly less, depending on whether antibiotics could be effectively distributed to the affected population. Thus, defining the risk likelihood is difficult and problematic.

Note that the likelihood scale of the risk matrix depicts relative, rather than absolute, likelihood. Although an agent assigned to a bin immediately above another means that the agent is assessed to be more likely to be used, it does not indicate *how much* more likely. Nor does the scale imply that the bins are equally separated in likelihood. Thus the likelihood scale in the risk matrix is an ordinal scale. The same is true of the consequences scale.

Once we have overcome all the difficulties in defining our risk matrix and we have placed our selected biological agents on the matrix, how can it be used? The primary value of a risk matrix is a communication tool. It provides an overview of a broad-spectrum of diverse threats and allows quick identification of the most severe relative risks (which are those in the upper right-hand corner of the matrix). This can provide a starting point for discussions on how to mitigate the risks by either reducing the consequences or reducing the likelihood of individual risks. For example, potential risk reduction measures (use of vaccines and antibiotics, which do not apply to toxins) are shown on the matrix as arrows linking the cells to an item that would be located in before and after application of the measure.

Note that we have not assigned the matrix cells to levels on an overall risk scale that combines likelihood and consequences. The reasons for this are discussed further in the next section.

3.6 Caveats about Combining Qualitative Risk Scales

Many qualitative risk analyses develop qualitative scales for likelihood and consequences, and then combine those scales to arrive at an overall qualitative risk scale [17, 18]. A typical approach is shown in Table 6, where three-level qualitative scales for likelihood and consequences are combined into a three-level qualitative scale for overall risk. Of course, information is necessarily lost when two pieces of data are combined into one. This is especially problematic in risk assessment since the underlying concepts of likelihood and consequences are more easily comprehended than the abstract concept of risk.

In the table, the level of the likelihood scale is determined by the row and the level on the consequences scale is determined by the column. The word in each cell is the level on the overall risk scale. As was demonstrated by Cox et al. [19], even in this simple case reasonable assignments of qualitative labels to likelihoods and consequences can

TABLE 6 Example of Combining Two Qualitative Scales into an Overall Qualitative Risk Scale

		Consequences		
		Low	Medium	High
Likelihood	High	Medium	High	High
	Medium	Low	Medium	High
	Low	Low	Low	Medium

result in the same label for risk being assigned to quite different underlying situations. One remedy to this problem is to use formal elicitation techniques to identify the relative rankings of the different combinations of likelihood and consequences.

4 RISK COMPARISONS

Many individuals find it difficult to intuitively understand probability and risk estimates. Because of this, estimates are often presented as comparisons to other, more familiar and therefore presumably better known and understood, risks. Comparisons are an ordinal representation of risk that can be considered qualitative since ordering is the only operation that is used. Risk comparisons can be completely verbal without specifying the underlying risk. For example, the risk of dying from lung cancer is greater for smokers than nonsmokers. In practice, however, the risks are generally presented as part of the comparison if they are known.

Hoerger [20] states that providing a sense of perspective about risks being estimated through analogy is a desirable attribute of a risk assessment report. Fiering and Wilson [21] describe various methods used for estimating risk by comparisons with other similar risks. Common examples are the use of results in laboratory animals to estimate the analogous risk in humans and the use of past history as an analogy for the future. They state that the process of using the analogy introduces another layer of uncertainty in representing risk.

Comparative risk representations can be problematic if the comparisons are not appropriate. Inappropriate comparisons include comparing voluntary risks (such as skydiving) with nonvoluntary risks (such as air pollution) and comparing risks with personal control (such as driving) with uncontrollable risks (such as flying in a commercial aircraft) [22].

Covello et al. criticize commonly used risk comparisons such as comparisons to the risk of death in traffic accidents [23]. They developed five categories, shown in Table 7, that order comparisons from most to least suitable.

The usefulness and acceptability of a risk comparison depend on the rank of the comparison as defined in Table 7, the context of the comparison, and the audience of the comparison. If the comparison is made simply to give a feel for the magnitude of the numbers, then wider latitude should be granted. If the comparison is made for decision-making purposes, for example, to determine acceptability or to prioritize mitigation measures, a greater effort to ensure that the risks are truly comparable should be made.

TABLE 7 Categories for Risk Comparisons Developed by Covello et al [23]

First-Rank Risk Comparisons (most acceptable)
a. Comparisons of the same risk at two different times
b. Comparisons with a standard
c. Comparisons with different estimates of the same risk

Second-Rank Risk Comparisons (less desirable)
a. Comparisons of the risk of doing and not doing something
b. Comparisons of alternative solutions to the same problem
c. Comparisons with the same risk as experienced in other places

Third-Rank Risk Comparisons (even less desirable)
a. Comparisons of average risk with peak risk at a particular time or location
b. Comparisons of the risk from one source of a particular adverse effect with the risk from all sources of that same adverse effect

Fourth-Rank Risk Comparisons (marginally acceptable)
a. Comparisons of risk with cost or of one cost/risk ratio with cost/risk ratio
b. Comparisons of risk with benefit
c. Comparisons of occupational with environmental risks
d. Comparisons with other risks from the same source
e. Comparisons with other specific causes of the same disease, illness, or injury

Fifth-Rank Comparisons (rarely acceptable—use with extreme caution!)
a. Comparisons of unrelated risks

5 CONCLUSIONS

Natural language is the medium used for most qualitative risk representations. This provides both advantages and disadvantages. The advantages include simplicity and widely used terminology. However, care must be exercised in the use of qualitative risk representations. Audiences often have differing understandings of common risk and probability terms and misunderstandings will occur unless, and perhaps even if, the precise meaning of the terms used is explained.

Useful verbal scales have been developed for both likelihood and consequences. These scales must be tailored to the particular risk being discussed. When developing scales, a balance must be struck between fully describing multiple relevant factors and conciseness. Ordinal verbal scales are common, but conclusions drawn from them are limited to those based on counting and rank ordering.

The risk matrix is a useful technique for presenting likelihood and consequence in a single display without aggregating them. Aggregation of consequence and likelihood scales is not recommended unless care is taken to ensure the aggregated scale does not obscure relevant information.

Risk comparisons are commonly used to place unfamiliar risks in the context of more familiar risks. When comparing risks, the comparison must be as close as possible for results to be accepted. In particular, comparisons of risks with very different levels of acceptability (e.g. voluntary vs. involuntary risks) are less likely to be valid.

REFERENCES

1. Stevens, S. S. (1946). On the theory of scales of measurement. *Science* **103**, 677–680.
2. Kent, S. (1964). Words of estimative probability. *Stud. Intell.* **8**, 49–65.
3. Juster, F. T. (1966). *Consumer Buying Intention and Purchase Probability*. National Bureau of Economic Research, Columbia University Press, Los Angeles, CA.
4. Hillson, D. A. (2005). Describing probability: the limitations of natural language. *Proceedings of the EMEA PMI Global Congress*, Edinburgh, UK.
5. Wallsten, T. S., and Budescu, D. V. (1995). A review of human linguistic probability processing: general principles and empirical evidence. *Knowl. Eng. Rev.* **10**, 43–62.
6. U.S. Department of Homeland Security. (2006). Web site, http://www.dhs.gov/dhspublic/display?theme=29.
7. Ridge, T. (2006). Remarks quoted in White House press release. March 12, 2002.
8. U.S. Department of Homeland Security. (2006). Web site, http://www.dhs.gov/dhspublic/interapp/press_release/press_release_0046.xml.
9. U.S. Department of Homeland Security. Citizen Guidance on the Homeland Security Advisory System. Retrieved from http://www.dhs.gov/xlibrary/assets/CitizenGuidanceHSAS2.pdf, October 16, 2008.
10. Reid, J. (2006). *Threat Levels: the System to Assess the Threat from International Terrorism*. The Stationary Office, Norwich, UK.
11. Wood, H. O., and Neumann, F. (1931). Modified Mercalli intensity scale of 1931. *Bull. Seismol. Soc. Am.* **21**, 277–283.
12. Fujita, T. T. (1971). *Proposed Characterization of Tornadoes and Hurricanes by Area and Intensity*, Satellite and Mesometeorology Research Project Report 91, The University of Chicago.
13. Microsoft Corporation. (2006). Web site, http://www.microsoft.com/technet/security/bulletin/rating.mspx.
14. Binzel, R. P. (1995). A near-earth object hazard index. Near-Earth Objects, the United Nations International Conference. *Proceedings of the International Conference held April 24–26*, New York; Remo, J. L. Eds. *Ann. N.Y. Acad. Sci.* 822, 545.
15. Morrison, D., Chapman, C. R., Steel, D., and Binzel R. P. (2004). *Impacts and the Public: Communicating the Nature of the Impact Hazard. Mitigation of Hazardous Comets and Asteroids*, M. J. S. Belton, T. H. Morgan, N. H. Samarasinha and D. K. Yeomans, Eds. Cambridge University Press.
16. Symantec Corporation. (2006). Web Site, http://www.symantec.com/avcenter/threat.severity.html.
17. FDA. (2003). *Guidance for Industry 152 – Evaluating the Safety of Antimicrobial New Animal Drugs with Regard to their Microbiological Effects on Bacteria of Human Health Concern*, October 23, 2003 Available at http://www.fda.gov/cvm/guidance/fguide152.pdf.
18. American Petroleum Institute. (2004). *Security Vulnerability Assessment Methodology for the Petroleum Industry*, 2nd ed., October 2004.
19. Cox, L. A., Babayev, D., and Huber, W. (2005). Some limitations of qualitative risk rating systems. *Risk Anal.* **25**(3), pp. 651–662.
20. Hoerger, F. (1990). Presentation of risk assessments. *Risk Anal.* **10**(3), pp. 359–361.

21. Fiering, M., and Wilson, R. (1983). Attempts to establish a risk by analogy. *Risk Anal.* **3**(3), pp. 207–216.
22. Starr, C. (1969). Social benefit versus technological risk. *Science* **165**(3899), pp. 1232–1238.
23. Covello, V. T., Sandman, P. M., and Slovic, P. (1988). *Risk Communication, Risk Statistics, and Risk Comparisons: A Manual for Plant Managers*. Chemical Manufacturers Association, Washington, DC.

TERRORISM RISK

GORDON WOO
Risk Management Solutions, London, United Kingdom

1 SCIENTIFIC OVERVIEW

Einstein remarked that nature is subtle, but not malicious. There is no universal definition of terrorism, but all such acts are recognized as being malicious. Also, not all terrorist campaigns are deadly and enduring, but these are the words used by the director general of the British security service, Manningham-Buller, [1] to categorize the global Jihadi threat, at a time when the MI5 perceived Britain to be Al Qaeda's prime target [2].

The purpose of this article is to describe methods for modeling this source of terrorism risk, and to identify research directions, especially in analysis on a global scale. In the latter regard, aviation and maritime risks are given prominence, because of their significance in border protection. Skeptics of terrorism risk modeling may perceive terrorism to be simply a Manichaean struggle between good and evil, or imagine that terrorists are stupid and crazy. On the contrary, in reality, capable terrorists are both rational and intelligent. Terrorists have to be intelligent in order to make an impact in asymmetric warfare. Atwan [3] has warned the West not to underestimate the intellectual prowess of the Al Qaeda leadership. Osama bin Laden honored Khalid Sheikh Mohammed, the 9/11 mastermind, with the title "mukhtar", meaning "the brain" [4]. Indeed, it may be argued that the most powerful biological weapon in the terrorist's arsenal is not any deadly virus but the human brain itself.

But is it rational for a Jihadi to undertake a suicide mission? Yes, according to the seventeenth century French philosopher, Blaise Pascal. Given the promise of eternal paradise after a martyr's death, and a nonzero likelihood of this promise being actually realized, it is perfectly rational for a Jihadi to accept Pascal's wager, and bet on this outcome of a martyrdom mission. It is known that some terrorists have followed this line of philosophical thought. In the words of one Palestinian, "If you want to compare it to the life of Paradise, you will find that all of this life is like a small moment. You know, in mathematics, any number compared with infinity is zero" [5].

Building on the understanding that Islamist militants are rational and intelligent, an overview of the principles of terrorism risk modeling that govern the frequency and severity of Jihadi terrorist attacks, the choice of weaponry, and the selection of targets is presented.

1.1 Terrorist Targeting

A cornerstone of terrorist targeting is target substitution [6]: if a designated target is too hard, an alternative softer target may be substituted. If, however, there is no available alternative of a similar status, efforts at striking the original target may be redoubled. Target substitution operates on all spatial scales, and is a phenomenon very familiar to criminologists (Yezer, [7]). At the street level, the Bali cafés bombed on October 1, 2005, had been chosen for their inferior security. At the town level, the British embassy in Istanbul was bombed on November 20, 2003, in preference to the fortress-like US embassy [4]. At the national level, the IRA switched a truck bomb attack from London to Manchester, when the border security around the city of London was tightened. At an international level, Jemaah Islamiyah switched an embassy attack from Manila to Singapore because of the difficulty in the Philippine targets [8].

The malice underlying terrorism suggests applying game theory, as Sandler and coworkers did long before 9/11 [9], and others have done since. Game theory models with just a few variables can be analyzed in elaborate mathematical detail. For example, Kardes [10] has provided solutions to some illustrative stochastic game problems with a small number of states. A basic model of a territory with a handful of terrorist targets or weapon systems, may be mathematically tractable, but the US homeland is dense with terrorist targets. As with models of a few targets, insights into what to protect may be gained from considering a simple model with numerous targets, where all, or almost all, of the targets are equally valuable to the defender (Bier [11]).

To develop a full-scale practical model for the entire US homeland, some simplifying assumptions are clearly necessary. These may be coarse, but they can be substantiated from knowledge of the terrorist modus operandi, and validated against information on planned attacks. Al Qaeda operatives are trained to be meticulous over target surveillance [12], and are very sensitive to changes in security. One may reasonably assume that through diligent surveillance, they effectively immunize the attack loss potential against changes in security (Major [13]), in the mathematical sense that, to first order, the expected loss from an attack is invariant against such changes.

For various terrorist organizations such as the IRA, killing large numbers of civilians has very low utility for the terrorists, since such attacks would severely erode their popular support base. Therefore, a target such as a packed sports stadium, had a much lower utility for the IRA than for the defenders. However, for Islamist militants, enraged by many thousands of civilian Muslim deaths in conflict zones, mass fatalities are perceived as a legitimate attack objective. In his statements [14], Osama bin Laden has indeed encouraged such attacks: "It is the duty of Muslims to prepare as much force as possible to attack the enemies of God".

The chief Al Qaeda strategist, Dr Ayman Al Zawahiri, has explained the influence of the Umma, (the global community of Muslims), in their targeting strategy: "Al Qaeda wins over the Umma when we choose a target that it favors" [15]. Opinion poll surveys show that a significant proportion of Muslims around the world condone terrorist attacks against Western targets as a reprisal for Western indifference to the loss of Muslim lives. The Leeds clique of British–Pakistanis who bombed London on 7/7 had held a celebration party, immediately after 9/11 [16]. Cities with international recognition are collectively favored by the Umma, as is affirmed by the list of postcard cities attacked since 9/11: Bali, Mombasa, Casablanca, Riyadh, Amman, Madrid, and London. In contrast, towns are not favored by the Umma if they are unknown to most Americans, let alone on the Arab street.

The Jihadi preparedness to use maximum force, and the Umma's perception of Crusaders' pain as Islam's gain, suggests that, as a first approximation, target utility valuations based on economic and symbolic value, and casualty potential, are broadly similar for Jihadi attackers and Western defenders. If only Jihadis were motivated to attack a Western target that had low defender utility. A cartoon in Punch magazine once depicted London's Royal Albert Memorial as being an ideal honey-pot IRA target. To the regret of architectural aesthetes, this undefended monument to Victorian garishness was never targeted by the IRA. The Statue of Liberty is a prime example of an iconic terrorist target that is well protected, even though its economic value and casualty potential are comparatively modest. Unlike the undefended Albert Memorial, it is known that the defended Statue of Liberty is a Jihadi target.

If it is posited, as with the Statue of Liberty, that defensive resources commensurate with their utility are applied to protect targets, a parsimonious targeting model is then derived (Woo [17, 18]). This is a phenomenological model: the few parameters are estimated through elicitation of the judgment of international terrorism experts. Cities are grouped into discrete tiers by the experts, for example, New York and Washington D.C. comprise the first tier. A similar style of discrete target tiering is developed for types of targets, for example, hotels, government offices, airports and so on.

The model automatically yields targeting likelihoods akin to a Pareto 80/20 rule, in that it forecasts that the great majority of attacks will be against a minority of potential targets. This focusing of attacks against a small proportion of targets is consistent with the historical experience of major prolonged foreign terrorist campaigns, such as the IRA bombing campaign in England, which concentrated terror in its leading cities: London, Manchester, and Birmingham.

This baseline model can be updated with site-specific information on whether security is relatively better or worse than the norm for a potential target of a particular ranking. The essential characteristics of terrorist targeting are captured by the model:

- Terrorists may substitute one target with another, according to the relative security of the targets.
- Local security enhancement transfers threat elsewhere: excessive site protection may be undesirable on a societal level. For example, some government buildings may be protected against vehicle bombs well beyond terrorist capability, whereas there may be insufficient expenditure on protecting private offices, or public infrastructure.
- There is safety in numbers: increasing the number of targets of a specific type dilutes the individual risk. By contrast, where there are few targets of a specific type, for example, tourist hotels in a city, then the risk is increased, as with the Amman, Jordan, bombings on November 9, 2005.
- Terrorist attacks are geographically focused, with attack likelihood decreasing logarithmically for descending target tiers. This is exemplified by the IRA's choice of

the English city to attack: the first tier being London; the second tier comprising Manchester and Birmingham, and so on.

1.2 Weapon Attack Modes

Rationality pervades the operational modus operandi of terrorists. The handbook of all guerrilla movements is Sun Tzu's "The Art of War", which identifies optimal modes of combat: "Now an army may be likened to water, for just as water avoids heights, and hastens to the lowlands, so an army avoids strength and strikes weakness". This dictum echoes the principle of following the path of least resistance that governs the dynamics of the universe, including terrorist activity [19]. This elegant principle was first enunciated by Pierre de Maupertuis: "The great principle is that, in producing its effects, Nature acts always according to the simplest paths".

In hydrology, the principle of minimum energy expenditure governs the pattern of river drainage networks. Similar to the flow of water, the flow of terrorist activity is towards weapon modes and targets, against which the technical, logistical, and security barriers to mission success are least. Since 9/11, the counterterrorism environment for the development of new weapons and planning complex strategic operations has become oppressive for Al Qaeda. Accordingly, it inclines towards off-the-shelf, ready-to-use weapons, (such as man-portable air-defense systems (MANPADS), mortars, hijacked aircraft, and propane tankers), or improvised conventional explosive devices, which do not involve intricate and potentially failure-prone technological development. Al Qaeda is known to be highly adaptive in learning from past terrorist successes and failures from all terrorist organizations around the world [4]. Neural network simulation models can represent the social learning process [20].

The logistical burden of alternative weapon systems can be evaluated in terms of the terrorist demands on finances, equipment, material, trained personnel, and sleeper cell support. Calibration against actual experience is possible for conventional attack modes, but not for exotic attack modes, such as chemical, biological, radiological, and nuclear (CBRN). Event-tree methods have been devised to elaborate the alternative foreign and domestic pathways by which such unconventional weapons can be manufactured or procured.

1.3 Frequency of Successful Macro-Terror Attacks

Macro-terror attacks are acts of terrorism, such as the one perpetrated on 9/11, that aim to cause substantial loss, and require significant logistical resources and considerable time for planning and preparation. Al Qaeda is renowned for meticulous detail over its centrally organized attack planning that involves diligent reconnaissance, surveillance, and rehearsal. Most notably, Al Qaeda has developed a long-term strategy, and is extremely patient in planning its military campaign over decades. Indeed, patience is half of the Islamic concept of faith, that covers action, (for which gratitude is due to Allah), and abstinence from action, (for which patience is demanded). Dr Ayman Al Zawahiri [15] has explained the Al Qaeda goal: "We must inflict the maximum casualties against the opponent, for this is the language understood by the West, no matter how much time and effort such operations take".

It turns out that the number of operatives involved in planning and preparing attacks has a tipping point with respect to the ease with which the dots might be joined by counterterrorism forces. The opportunity for surveillance experts to spot a community of terrorists, and gather sufficient evidence for courtroom convictions, increases nonlinearly with the number of operatives—above a critical number, the opportunity improves dramatically. This nonlinearity emerges from analytical studies of networks, using modern graph theory methods (Derenyi et al. [21]). Below the tipping point, the pattern of terrorist links may not necessarily betray much of a signature to the counterterrorism services. However, above the tipping point, a far more obvious signature may become apparent in the guise of a large connected network cluster of dots, which reveals the presence of a form of community. The most ambitious terrorist plans, involving numerous operatives, are thus liable to be thwarted. As exemplified by the audacious attempted replay in 2006 of the Bojinka spectacular, too many terrorists spoil the plot (Woo, [22]).

Intelligence surveillance and eavesdropping on terrorist networks thus constrain the pipeline of planned attacks that logistically might otherwise seem almost boundless. Indeed, such is the capability of the Western forces of counterterrorism, that most planned attacks, as many as 80–90%, are interdicted. For example, in the three years before the 7/7 London attack, eight plots were interdicted. Yet, any noninterdicted planned attack is construed as a significant intelligence failure. The public expectation of flawless security is termed the *90-10 paradox*. Even if 90% of plots are foiled, it is by the 10% which succeed that the security services are ultimately remembered.

Thanks to the diligence of the security services, which deter the planning of large numbers of attacks, and interdict most of those that are planned, the frequency of successful terrorist attacks is kept low. Only a small proportion of attacks succeed, and these attacks tend to be those involving fewer active operatives. Through this control process, which comes at the cost of some personal civil liberty, the uncertainty in the frequency of successful terrorist attacks is constrained. As happened after 9/11 and 7/7, after each major terrorist attack, democracies will respond by rebalancing the desire for liberty with the need for security.

2 RESEARCH ON TERRORISM RISK

Terrorism is a global phenomenon. Where activists cannot effect political change through peaceful means, they coerce through political violence. Terrorism research has a wide international agenda: the structure of terrorist organizations, weaponry, targeting, vulnerability and security, counterterrorism and so on. Terrorism risk assessment forms part of the research agenda, and is shaped and directed by the practical needs of the public and private sectors.

The insurance industry operates worldwide. Accordingly, global models of terrorism risk have been developed to assist risk managers of insurance companies and captive insurers. These models vary in resolution from one country to another, depending on the degree of commercial interest. Insurance risk management requires control of aggregate loss potential. Accordingly, models cover all terrorist groups and all plausible attack modes. Although their adoption is discretionary, risk models are used quite widely by insurers, in recognition of which the insurance industry has funded a considerable amount of research on terrorism risk. Notable is the research carried out by the RAND Center for Terrorism Risk Management Policy, which is a joint project of the RAND Institute for Civil Justice, RAND Public Safety and Justice, and Risk Management Solutions (RMS).

Besides the insurance industry, the US government has a direct stake in assessing terrorism risk on a national scale, with purposes to improve counterterrorism resource allocation among others. Any implementation of a risk-based allocation procedure must address concerns over uncertainty about terrorist targeting. A detailed RAND study (Willis et al. [23]), based on the RMS terrorism risk model, has developed an approach for making allocation decisions, robust against uncertainty in model parameterization.

A considerable volume of academic terrorism risk research has been undertaken to support national public policy, notably at the University of Southern California's Center for Risk and Economic Analysis of Terrorism Events (CREATE), a DHS University Center of Excellence. At spatial scales below the national level, terrorism risk assessment is potentially useful for state and local public officials, for managers of transport infrastructure, utilities, critical facilities, as well as major buildings. Absent prescriptive terrorism mitigation standards, a risk-based approach to decision-making on security improvement is relevant. For major transportation systems, cost–benefit analyses for project prioritization and resource allocation have been conducted (e.g. King and Isenberg [24]).

The challenge of terrorism risk assessment varies according to scale. On one hand, the more restricted the geographical scope, the narrower is the spectrum of possible targets. On the other hand, the dynamical coupling of terrorism risk across regional boundaries and between diverse targets is lost. The same applies to research focusing on specific weapon systems: the more restricted the military scope, the harder it is to address the coupling and switching between alternative choices of weaponry. An analogy with weather forecasting is apposite, since meteorological conditions are dynamically coupled across regions. Local weather forecasting is possible, but only if a large-scale model of the atmosphere is used to define regional boundary conditions.

The problem of scale manifests itself particularly acutely in frequency estimation. An absolute measure of frequency cannot be generated internally from within the confines of a research project that has a limited scope. However, setting aside discussion of the absolute probability of an attack against a specific target, risk assessments can usefully focus on the conditional probability of an attack.

2.1 Aviation and Maritime Risk Assessment

Just as the fighter jet is symbolic of militarism, so is the passenger jet of terrorism. Civil aviation is a vulnerable link in the global economy. The continuous adaptation of attack modes to evade security enhancements is a conundrum of aviation risk assessment. A wide range of airport security measures have been analyzed by Martonosi [25] from an operational research perspective. The cost-effectiveness of MANPAD countermeasures has been investigated by von Winterfeldt and O'Sullivan [26], using decision analysis techniques. Their analyses show that measures to deflect SAM missiles might be cost-effective if the probability of an attack exceeds 0.5 in 10 years; the losses are large ($100 billion); and the countermeasures are relatively inexpensive (<$15 billion). It turns out that the RMS estimate of the 10-year MANPAD attack probability for shooting down a plane in the United States falls around the threshold criterion. Also, given that the scale of economic losses and the cost of countermeasures might also straddle the criterion levels, any decision on implementing countermeasures remains awkward, bearing in mind the inevitable prospect of the terrorist threat shifting to another alternative aircraft attack mode. Since 9/11, there has been an intermittent series of foiled plots against civil aviation, several of which have involved the smuggling of explosives.

Like aircraft, ships can be attacked or converted into floating bombs. Apart from these maritime dangers, US ports face a considerable challenge in preventing illicit terrorist cargo or personnel from entering the United States, while allowing the free flow of maritime commerce. Security standards are tightening at foreign ports, but the possibility exists that an inbound vessel might have terrorist connections. It is impractical for the coast guard to search more than a small percentage of inbound vessels, so an intelligence-led risk-based procedure is needed to improve the search selection.

With access to a global database of all commercial vessel movements, (such as maintained by the Lloyds Marine Intelligence Unit), a threat-ranking system can be devised, based on public information on ownership, flag of registration, crew nationality, last ports of call, and so on. Like any profiling system, it can be subverted by an adaptive terrorist intent on surprise, except that intelligence leads may give reason to revise this threat-ranking. A risk-based methodology for incorporating available intelligence leads, and treating surprising data of dubious reliability, has been devised by Woo [27] using the conceptual framework of Bayesian epistemology.

3 CRITICAL NEEDS ANALYSIS

Of the various measures that may be taken to protect the US homeland against terrorism, securing national borders is a priority. The particular contribution made by specific border security measures, such as the US VISIT program, which involves the gathering of personal identity information about people in transit to the United States, or at US border posts, is somewhat indeterminate. A strict regime of personal identity checking casts a virtual security net around the entire United States that takes advantage of the small-world phenomenon of social networks: terrorists may be separated by thousands of miles, but connected instantaneously by a computer database of suspected or known terrorist links.

The potential to unravel a terrorist network may have very substantial deterrent value. A sizeable community of Jihadis involved in planning a large-scale technically complex operation, for example, a CBRN attack, would come under heavy counterterrorism pressure. Given that a major US operation carries a significant risk of network unraveling, there should be a strong impetus for terrorists either to lower the attack scale to reduce their detectability footprint, or to switch the attack to a softer country within the Western alliance. Through its unwavering support for US foreign policy, Britain is seen by Islamists as forming an axis of evil with the United States and Israel. The United Kingdom is an obvious substitute target: the United Kingdom is politically almost exactly aligned with the United States on the war on terrorism, has a far more radicalized Muslim community, and yet is far behind on biometric border security. Already, the terrorist threat to London is comparable with that to New York or Washington. Because Al Qaeda has a global franchise, and terrorist target substitution operates on a transnational scale, there is a critical need for risk assessment which is conducted on a global basis [28].

3.1 Global Scale of Terrorism Risk Modeling

Global terrorism models are needed to comprehend the terrorist threat to the US homeland from Islamist militancy. This is obvious for international aviation and maritime transportation attacks, and it is true for all threat modes, since foreign policy is a prime driver of Muslim discontent and Jihadi recruitment. Further research on global terrorism risk modeling will benefit efforts on focusing counterterrorism action. Terrorist social networks, weapon procurement and transport, and the security of US assets abroad are important concerns that merit analysis from an international perspective.

The Umma has been galvanized collectively by Al Qaeda to seek redress for the plight of Muslim brothers and sisters in conflict zones. The global social networks of Muslims provide a support framework for radicalism, extremism, and terrorism. A crucial role in facilitating terrorism in the Western democracies is played by Muslim diaspora communities: Pakistanis in United Kingdom, Algerians in France, Moroccans in Spain, Somalis in Italy, and so on.

Improved models need to be developed for the evolution of terrorist support. Transcending the boundary between sociology, social psychology, and physics, is the modern discipline of sociophysics. This offers a scientific methodology for incorporating the social dynamics of diaspora communities into terrorism risk assessment. Basic rules of social interaction, akin to the dynamics of physical systems, are capable of enhancing understanding of complex social behavior. The formation of ghettos is a classic paradigm of sociophysics; one which is relevant to terrorism risk. "Jihad" and "Osama bin Laden" are the favorite Google keywords in some British Muslim ghettos. It is no surprise that as many as 100,000 British Muslims considered the 7/7 London bombings to be justified [1].

With the same objective, large-scale agent-based computer simulations of the collective behavior of Islamist militants have been developed (e.g. MacKerrow [29]), but there is much room for advancement in the understanding of Islamist militancy. The discourse on this subject is beset at the highest level by political autism—the inability to think what others are thinking. Mathematical psychologists like Vladimir Lefebvre have found mathematical ways of encoding the recursive sequence of thinking what others think. Terrorism risk assessment will be distorted if its political components misrepresent the underlying root causes of terrorism (Richardson [30]).

4 RESEARCH DIRECTIONS

Terrorism risk analysis is no longer a fledgling discipline. The core principles that govern terrorist actions are subject to observation and empirical validation, and, with the passage of time, the database of planned attacks is becoming more amenable to quantitative analysis. What is already clear is that, in the asymmetric warfare waged by the Western democracies against militant Islam, the forces of counterterrorism are achieving considerable success in controlling terrorism. The record is not perfect, as illustrated by the rail bombings of Madrid in March 2004 and London in July 2005. As the authorities remind their citizens: itis not a question of if, but when the next attack will occur. But at least, the question is thankfully not: how many major attacks will there be?

Research into global counterterrorism is needed to elucidate important aspects of the dynamics of the terrorism control process. The close cooperation between security services in the G8 countries contributes significantly to the successful interdiction of the great majority of planned terrorist attacks against these countries by Islamist militants, and to restricting their international operations. The dynamics of terrorism control depend on the global political environment, which is easily destabilized, and subject to the law of unintended consequences, even if politicians would hope otherwise. In order to be meaningful for homeland security decision-making, cost–benefit analysis should be broadened in scope from having a restricted internal domestic frame of reference to having global coverage of US interests and actions. As terrorist threat is integrated across the globe, terrorism risk research must also be directed globally.

REFERENCES

1. Manningham-Buller, E. (2006). *The International Terrorist Threat to the UK*, Speech at Queen Mary College, London.
2. Rusbridger, A. (2006). The Guardian newspaper. Manchester, October 19.
3. Atwan, A. B. (2006). *The Secret History of Al Qaida*, Saqi books, London, pp. 256.
4. Gunaratna, R. (2005). *Terrorism Risk Briefing*, Washington, DC.
5. Oliver, A. M., and Steinberg, P. (2005). *The Road to Martyr's Square*, Oxford University Press, Oxford, pp. 214.
6. Drake, C. J. M. (1998). *Terrorists' Target Selection*. Macmillan Press, London, pp. 272.
7. Yezer, A. (2006). Terrorist attacks and their consequences. Presentation at the CREATE workshop on benefit methodologies for Homeland Security Analysis, June 8–9, Washington, D.C.
8. Bell, S. (2005). *The Martyr's Oath*. John Wiley & Sons, Ontario, pp. 254.
9. Sandler, T., and Lapan, H. (1988). An analysis of terrorists' choice of targets. *Synthèse* **76**, 245–261.
10. Kardes, E. (2005). *Robust stochastic games and applications to counter-terrorism strategies*. CREATE report, 64.
11. Bier, V. (2005). *Choosing what to protect*. CREATE report, 27.
12. Gunaratna, R. (2002). *Inside Al Qaeda*. Hurst & Co., London, pp. 282.
13. Major, J. A. (2002). Advanced techniques for modeling terrorism risk. *Journal of Risk Finance* **4**, 15–24.
14. Lawrence, B., Ed. (2003). *Message to the World: The Statements of Osama bin Laden*, Verso, London, pp. 224.
15. Zawahiri A. *Knights under the prophet's banner*, London, 2002.
16. Rusbridger, A. (2006). The Guardian newspaper. Manchester, June 24.
17. Woo, G. (2002). Quantitative terrorism risk assessment. *Journal of Risk Finance* **4**, 7–14.
18. Woo, G. (2003). Insuring against Al Qaeda. *NBER Insurance Workshop Presentation*, Cambridge, Mass.
19. Ranstorp, M. (2006). *In the service of Al Qaeda*. In preparation.
20. Carley, K. (2003). *Destabilizing terrorist networks*, CASOS Working Paper.
21. Derenyi, I., Palla, G., and Vicsek, T. (2005). Clique percolation in random networks. *Phys. Rev. Lett.* **94**, 160202.
22. Woo, G. (2006). *Small World Constraints on Terrorism Attack planning*, Vol. 5. RUSI/Jane's Homeland Security & Resilience Monitor, London.
23. Willis, H. H., Morral, A. R., Kelly, T. K., and Medby, J. J. (2005). *Estimating terrorism risk*, RAND Corporation, Report from Center for Terrorism Risk Management Policy, Santa Monica, pp. 66.
24. King, S., and Isenberg, J. (2005). *Assessment of Urban Transportation Infrastructure for Terrorism Risk Management*. ICOSSAR, Millpress, Netherlands, pp. 2773–2780.
25. Martonosi, S. E. (2005). An operations research approach to aviation security. PhD thesis, M.I.T, 163.

26. von Winterfeldt, D., and O'Sullivan, T. M. (2005). *A decision analysis to evaluate the effectiveness of MANPAD counter-measures*. CREATE report No. 05–30, 24.
27. Woo, G. (2005). Institutionalizing imagination to prevent surprise. Invited talk at the *IDSS National Security Conference*, Singapore.
28. Woo, G. (2006). The terrorist threat to the US from abroad. Presentation at the CREATE workshop on benefit methodologies for Homeland Security Analysis. June 8–9, Washington D.C.
29. MacKerrow, E. (2003). Understanding why-dissecting radical Islamist terrorism with agent-based simulation. *Los Alamos science*, **28**: 184–191.
30. Richardson, L. (2006). *What terrorists want*. John Murray, London, pp. 288.

FURTHER READING

Wiktorowicz, Q. (2005). *Radical Islam rising: Muslim extremism in the West*. Rowman & Littlefield, New York, pp. 288.

TERRORIST THREAT ANALYSIS

JACK F. WILLIAMS
Georgia State University, Atlanta, Georgia

1 INTRODUCTION

Threat analysis is the art of wasting information, that is, the art of peeling away layers of irrelevant information, incorrect information, and disinformation, in order to expose a state or states of relevant and credible information related to a question of interest to the client. Analysts rarely have too little information to form an opinion. Rather, the analyst is often inundated with information culled from classified sources, unclassified but private or sensitive sources, or open sources [1]. After all, there are two ways in which to hide a needle: in a haystack or in a stack of needles. Threat assessment more closely resembles the latter task. Threat analysis is a nuanced and complex inquiry. Central to its understanding is that the most relevant threat information is usually in the hands of the adversary [2]. Thus, we are constantly guessing about threat, and our adversary is constantly guessing about how we are guessing about threat. In order to understand threat, conventional wisdom segregated threat into two components: (i) capability and (ii) intent [3]. Thus, an analyst could study an adversary's prior attack events, training activity, and the like to ascertain the capability to undertake and carry out an attack. Additionally, an analyst could study an adversary's expressed or implied intent regarding specific targets and operational modes. Synthesizing the two components, an analyst could describe past and present threat profile and develop future threat profiles given a known adversary. Yet, these traditional threat profiles were lacking in a key ingredient: authority. Particularly with religiously motivated adversaries (but certainly not limited to such), moral and theological authority is a necessary condition to make the transition from attack planning to execution [4, 5]. Furthermore, the concept of authority has a direct effect on target selection and tactics, and lends legitimacy to the group among nonviolent, but sympathetic, populations [4–6].

2 TRADITIONAL THREAT ASSESSMENT

Conventional wisdom holds that threat must be mapped to a given adversary and consist of that adversary's capability to carry out an event and that adversary's intent to conduct such operations [3, 7]. Thus, traditionally, threat (T) is a product of capability (C) and intent (I) of a *specific nation-state*:

$$T = f(C \times I) \tag{1}$$

The role of the analyst was to develop a threat profile centered on these two attributes. Under the Soviet intelligence paradigm, the adversary was generally self-evident: The Soviet Union, Eastern Bloc Countries, or Soviet Union Surrogate Countries (Vietnam, Syria, etc.). Sources of information to flesh out threat included classified, unclassified, and open-source material. Among analysts, it was well accepted and well understood that the golden source was classified material followed far behind by unclassified material [1, 8]. Open source was rarely considered and, in fact, regularly dismissed out-of-hand by analysts [1, 8]. Notwithstanding a heated division among US intelligence analysts, open source has gained currency, aligning US intelligence with allies such as the United Kingdom, Canada, Australia, Europe, Israel, Morocco, and Jordan, to name a few [8–11].

2.1 An Adversary's Capability

An analyst uses several tools to assess an adversary's capability. Initially, analysts subdivided capability into three components: operational, technical, and logistical [12]. An assessment of operational capability focuses on the question whether an adversary had sufficient leadership, command and control, intelligence function, financing, experience, and expertise in carrying out a specific terrorist event. A traditional inquiry would consider an adversary's past events (both operational successes and failures), training centers, centers of gravity, access to various weapons platforms, expertise, indicators and warnings, sources of financing, and alliances with other organizations. An assessment of technological capability would focus on whether an adversary had either internal or external access to the required technology to carry out an event. An analyst would consider an adversary's prior events, research and development programs, and alliances. Careful attention would be paid to indicators of in-house technological development and attempts to secure such technology from outside sources. After considering the relevant information, an analyst may develop operational grids that identify indicators and precursors of operational and technological mastery and advancement, creating a composite profile of an adversary's capability. An assessment of logistical capability would focus on the ability of a given adversary to sustain a threat on a regional or global level, differentiating terrorism threat from criminal attempt. An analyst would consider an adversary's leadership structure, recruitment, training, educational, social outreach, communications, propaganda, political, and financial footprints.

Conceptually, all three subcomponents of threat capability could be mapped to the foundation stones of global terror, those attributes of a terrorist organization necessary for the organization to sustain itself as a continuing regional or global threat. These attributes include territory, infrastructure, and finance [13]. Territory and infrastructure operate together to provide sanctuary for a terror group, an essential attribute in order to pose a sustainable regional or global threat. Moreover, the closer to the theater of operations a terrorist group can establish sanctuary, the more effective the terrorist activity becomes. This is in part why insurgencies are so effective; the sanctuary and theater of operations are one and the same.

In prior writings, I have argued that, in addition to territory, infrastructure, and finance, for a terrorist group to pose a strategic threat, that group must operate within recognized authority [4, 5]. Authority serves two distinct but closely related purposes. First, *internal authority* is absolutely necessary to conduct operations, recruit, and replenish [4, 5]. For example, within al-Qaeda, authority in the form of the issuance of a *fatwa* or a series of *fatawa* elevates what may appear to be murder and suicide, which are acts strictly forbidden by the Quran, Hadith, Sharia and other Islamic religious sources, into a sacramental duty by embracing principles of *jihad* and martyrdom [4, 5, 14]. Without a carefully reasoned *fatwa* by a recognized religious authority providing religious cover, a suicide bomber is stripped of the patina of religious authority behind his act and exposed as an *irhabi* (wanton criminal or terrorist) or *mufsidoon* (an evildoer) engaged in *fitna* (the sowing of societal chaos) whose proper end is eternity in hellfire [15]. Secondly, external authority is necessary in order for the audience to which religious terrorists speak to acquiesce, accept, or better yet embrace the terrorist group's tactics and goals. External authority is better understood as conferred legitimacy. For terrorist organizations like al-Qaeda, the intended audience is the *umma* (greater Muslim community) and not the West. Jihadist groups like al-Qaeda, who base their operational philosophy on a loose, cobbled collection of *Salafia* authorities [16], *Qutbist* writings [17], and contemporary Jihadist jurists and theologians, engage in what they perceive to be *dawa* (the "call") or the teaching and invitation to non-Muslims and non-devout Muslims to the Muslim faith [16]. Without a solid perception of external legitimacy among the *umma*, the *dawa* fails to resonate; once the *dawa* fails, the educational function of a movement like al-Qaeda dries up; once the educational function dries up, the organization cannibalizes itself and will cease to exist [18, 19]. A terror group's burn rate is exceptionally fast.

2.2 An Adversary's Intent

An adversary's intent may be subdivided into (i) general intent to commit certain acts directed at general target categories and (ii) specific intent to commit specific acts directed at specific targets [20]. Each subdivision may be further divided into strategic, operational, and tactical elements. An analyst employs several techniques to ascertain an adversary's intent. Primarily, the analyst peruses vast amounts of information, carefully teasing out an adversary's beliefs, motives, and intents, by exploring historical interest, past attacks (both successes and failures), current adversarial interest in a site or asset, current surveillance, documented or authenticated threat, desired level of consequence, ideology, religious beliefs, historical authenticity of attack, and ease of attack [21]. Absent a declared intent by an adversary, such as Iranian President Ahmadinejad's intent toward Israel, this usually requires a mastery of the tradecraft and mind-set of an adversary. Analysts draw on cultural, religious, philosophical, ideological, legal, historical, geographical, and linguistic sources to portray the adversary's thought-process and inclinations. The analyst then patiently and deliberately moves from an adversary's general intent to harm, to specific intent to engage in potential acts toward specific targets. That migration is not always graceful, reliable, or accurate.

Often, an analyst may discover that an adversary's intent does not match that adversary's capability. Such a preliminary conclusion could mean a number of things. First, it could mean that present intent telegraphs future capability. An analyst then considers intelligence (or requests intelligence) that might suggest increased research and development by an adversary in a given expertise or the quest by an adversary to find someone or some group that would presently possess that expertise and join ventures with the subject adversary. This often can be accomplished, for example, through training the members of one terrorist group by subject-matter and technical experts from another terrorist group, often for remuneration [22]. Second, it could mean that the manifestation of intent that we discovered or captured is part of a denial and deception ("D & D") campaign. Analysts have developed several heuristics and tools in hedging against being misled by deception, including a technique known as the analysis of competing hypotheses (ACH) [23–29]. Historically, most counterdeception techniques did not work well when confronted by a sophisticated deception program, where, among other things, the deceivers did not even know that they were part of a deception. Stech and Elsaesser have done excellent work in developing robust counterdeception techniques, including an ACH model building on cognitive task analysis (CTA) of the deception detector, courses of action (COA), and Bayesian Belief Networks, which have advanced D & D detection and counterdeception strategies [29, 30].

2.3 Synthesizing Capability and Intent

An analyst would think of threat as the product of capability times intent. That way, if either variable is zero, there is zero threat. During the Cold War, for example, the United States worried a lot about Soviet submarines off our coasts loaded with nuclear missiles; however, the United States harbored no concern about British submarines similarly loaded. Although capability was the same, the British intent against us was zero. The goal of synthesizing capability and intent is to develop an adversary's threat profile mapped to a specific target and operational means [8]. Traditionally, this requires the analyst to think of capability and intent as two vectors ranging, for example, from no capability to conduct an event, to complete confidence that our adversary could do so. An analyst then gauges capability and intent somewhere along the relevant vectors. In synthesizing capability and intent, an analyst may employ classification, description, and prediction techniques [8].

3 CULTURAL THREAT SPACE AND THE NON-STATE ACTOR

September 11, 2001, is a hinge date in the history of intelligence analysis. The terrorist attacks on the World Trade Center in New York City, the Pentagon in Arlington, Virginia, and a field outside Shanksville, Pennsylvania, thrust the analyst into a more visible light. Although many of us had assessed threat from non-state actors well prior to 9/11, the dramatic paradigm shift from a state-centric focus to a non-state one was undeniable. With that focus came a new and honest assessment of how analysts assessed threat [9, 11]. In this article, I unpack a new frame of reference in assessing threat by a non-state actor. In particular, I assert that threat cannot be understood without a cultural awareness and assessment of the adversary [31–33]. Although many good analysts employ informal cultural assessments in their threat analyses, we can all gain from a more explicit assessment and more robust understanding of how culture affects threat. Furthermore, I assert that the present threat formula must be expanded to include another variable, authority, in order to capture a more robust and accurate profile of threat.

3.1 Cultural Space

One cannot understand threat without a sophisticated assessment of the culture within which an adversary operates [31, 33]. Failure to comprehend this basic step in a threat assessment leads well-meaning and highly intelligent analysts to run afoul of the Napoleonic caution to avoid "making pictures of the enemy." For a terrorist movement similar to al-Qaeda, this requires an assessment of both, the sociopolitical and religious attributes of the movement, and the universe within which it operates and seeks to influence [34–36]. Figure 1 offers a basic depiction of the sociopolitical universe of operation and influence. Note that the smaller the circle within the nesting, the tighter the bond and the greater sense of belonging. These nesting circles include Tribe or Village; Other Tribes or Villages (such other tribes historically posed a threat to the tribe); Western-Influenced Nation-States which has not completely displaced tribal influence as the United States has learned in Operation Enduring Freedom (OEF) and Operation Iraqi Freedom (OIF); Arab Nations; and Muslim Nations (which forms the periphery marking the transition from the so-called "Arab World" to the "Rest of the World"); and the Rest of World [37–39]. Groups like al-Qaeda engage the West through this lens, as an inherent adversary, an outsider, a continuing threat [40].

Figure 2, adapted from the Combating Terrorism Center's Militant Ideology Atlas [16], depicts, again through the use of nesting circles, a Jihadist organization's religious world view. In this world view, at its periphery are non-Muslims. Groups like al-Qaeda have essentially rejected the distinction, well documented in the Quran, Hadith, and so on, between "Pagans" (idol worshippers or those who reject monotheism) and the "People of the Book" (Jews and Christians who are in spiritual if not historical possession of prior revelations). Jihadist religious authorities have collapsed in an unprincipled manner the religious distinction between these two groups and have rejected the relatively favored treatment of "People of the Book." Moving toward the center, the next circle encompasses all Muslims. This circle would include the full range of Muslim attitudes toward religion as part of politics. Thus, this circle includes secularists, progressives, reformists, and fundamentalists [41]. These members would generally self-identify as Muslims, even with no religious institutional affinity. At their core, those that would consider themselves

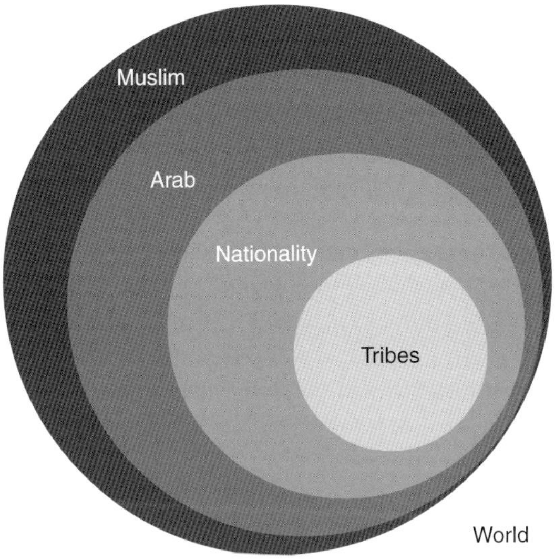

FIGURE 1 Sociopolitical world view.

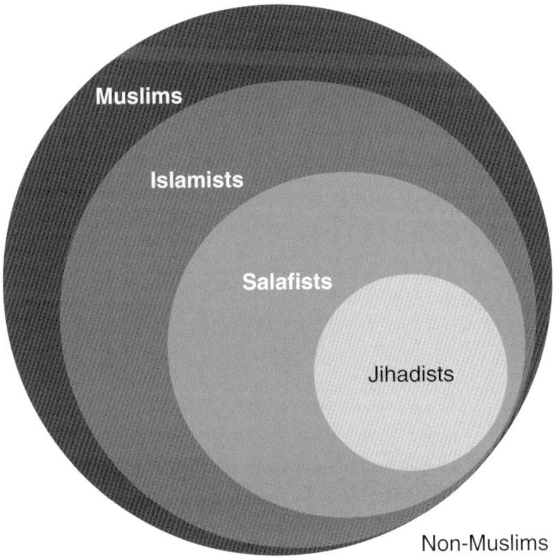

FIGURE 2 Religious-political world view.

observant of religious practices (for Islam is an orthoproxic religious movement), recognize the five pillars of Islam. The five pillars include the *shahadah* (confession of faith that there is no G-d but G-d and that Mohammed is His Prophet (the final seal of the prophets); *salat* (prayer five times a day at designated times); *zakat* (the giving of charity as a percentage of one's wealth and not simply income); *sawm* (fasting in the month of *Ramadan*); and *hajj* (the pilgrimage to Mecca in one's life time if one can afford it) [42]. These Muslims are the followers of G-d and effectuate His will by following the words of the Quran (a collection of chapters increasing in length) of the revelations of G-d to Mohammed over a number of years through the Angel *Jibril* (Gabriel) and the examples of the Prophet Mohammed collected in the Hadith [43, 44]. Within this circle one would find the *Sunni*, *Shia*, and other movements within Islam.

The next circle includes Islamists. This is better understood as a political movement within Islam. This is not to suggest that religious sources, doctrine, traditions, and practices do not influence Islamist, they all do, but that political ambition and goals are elevated over the role of traditional religion. Islamists are people who want Islamic law to be the primary source of law and cultural identity within a Muslim state. We often think of Islamists as fundamentalists, a term actually coined in reference to Christian theology and of little analytical use in understanding this Islamic political phenomenon. They are not. In fact, one may document case after case where Islamists (and Jihadists) regularly stray from the well-established teachings of Islam reflected in Holy Sources as a matter of political expediency. The Muslim Brotherhood is probably the most well-known Islamist organization [45].

The next circle includes Salafists. These people are *Sunni* Muslims who want to establish and govern Islamic states based solely on a revivalist interpretation of the Quran and Hadith as understood by the first generation of Muhammad's followers [16]. The most influential Salafists are Saudi religious authorities [16].

The tightest circle is the target audience to which al-Qaeda's teaching resonates. This final circle includes Jihadists. This is the militant core led by a cluster of Salafist thinkers, including al-Maqdisi of Jordan, al-Tartusi and Abu Qatada of England, Ab'd al Azziz of Egypt, and several Saudi religious authorities, who maintain that the only religiously legitimate way for Islam to engage the West is through a violent manifestation of *jihad* until the West submits to the superiority of Islam and either converts or pays the *jizah* (tax) and, according to some authorities, subjects itself to political capitulation and humiliation [16]. This is the culture within which our present adversary operates. Assessing capability, intent, or authority without an awareness of an adversary's cultural space leads to analytical drift, subconscious projection, and superficial threat assessment.

Scholars have not reached a consensus on the terms to be used to describe the Muslims that have engaged in acts of terrorism in the name of Islam. Some scholars have used the terms "Jihadists," "Militant Jihadists," "Salafists," "Islamists," "Muslim Extremists," "Salafi-Jihadists," and "Militant Muslims," or some variation thereof. Others have referred to movements like al-Qaeda as "Militant Islam," "Extremist Islam," "Radical Islam," "Fundamentalist Islam," "Neo-Fundamentalist Islam," "Neo-Wahhabist," "Jihadist Islam," and "Islamist." What makes this endeavor doubly difficult is that some scholars will employ one label for a particular movement, while another scholar will employ that same label for another movement. Moreover, the US government has not applied a consistent terminology. Its use of certain terms to classify movements within Islam in the 9/11 Commission Report was not illuminating. In an excellent essay entitled *"What's in a Name?" The Use of Terminology When Dealing with Islam and Muslims*, collected in First Impressions: American Muslim Perspectives on the 9/11 Commission Report (October, 2004), Maliha Balala made this point clear. For example, the lack of consultation on choice of terms with American Muslims and experts of Islam led to the use of words, such as "Islamist," to qualify the militant groups that constitute a serious threat to America, without distinguishing between nonviolent "Islamists" (a word that could apply easily to all Muslims and sounds to the public like "Islamic") and militant groups. The use of such terms runs the danger of including as a threat to America those Islamists who seek to foster sociopolitical changes in their societies through peaceful means. In this article, I will use the terms *radical*, *fundamentalist*, or *militant* movements within Islam, where appropriate, to mean distinct but potentially overlapping movements. *Radical* is used to refer to movements within Islam that seek major political or ideological change, particularly major changes in the form of government. These changes in form of government may be through nonviolent or violent means, depending on the methods of any given radical movement. *Fundamentalist* is used to refer either to a political movement within Islam or to an approach to religious sources. As a political movement, fundamentalists may be radical (when they seek the *nonviolent* change of governmental structure) or nonradical (when they attempt to work within a given governmental structure). As far as an approach to religious sources goes, fundamentalists believe in a return to the fundamentals of Islam and a strict interpretation of the sources. *Militant* or *Jihadist* is used to refer to those Muslims and movements within Islam that seek radical change through *violence*, often directed at the West and mainstream Muslims.

3.2 Authority and Legitimacy

As previously discussed, the traditional threat model is deficient. In addition to an assessment of an adversary's capability and intent, an analyst must consider authority or legitimacy. Thus, the threat equation should be expanded to include authority (A) and cultural influences (Z) and may be expressed as follows:

$$T = [(C \times I \times A) + Z] \tag{2}$$

Empirical research suggests that moral and theological authority is a necessary condition for religiously motivated terrorists to make the transition from attack planning and development to execution [4, 5, 14, 15]. As previously discussed, authority serves two distinct but related purposes: internal authority for operational and recruiting purposes and external legitimacy for recruiting, financial, and other support purposes. Furthermore, this concept of authority has a direct effect on target selection and tactics, and lends legitimacy to the group among nonviolent, but sympathetic, populations [4, 5]. The typical and traditional manner by which this authority is communicated is through the issuance of *fatawa*, or religious verdicts by religious authorities sympathetic to Jihadist causes [14, 15, 41]. A *fatwa* is a legal judgment or learned interpretation that a qualified jurist (*mufti*) may give on issues pertaining to Sharia (Islamic law) [46].

Originally, only a jurist possessing a number of qualifications and carefully trained in the Holy Sources and, among other things, the techniques of *ijtihad* (personal reasoning), was allowed to issue a legal opinion or interpretation of an established law. Ijtihad is a potential source of Islamic law after the Quran, the Sira or the Prophet Muhammad's life, the Hadith or his collected words and deeds, the Sunna (custom), and Ijma (consensus). Later, all trained jurists were permitted to issue *fatawa* [46]. More recently, even those who self-profess to be knowledgeable or those with little training (either formal or informal since both methods are historically permissible) have taken to issuing *fatawa* [46]. Furthermore, the great universities of *fiqh* (jurisprudence) and *sharia*, such as Jami'at al-Qarawiyyin in Fez, Morocco, and al-Azhar in Egypt, have lost stature, respect, and influence, and no longer serve the role of containing and isolating poorly reasoned *fatawa*.

A *fatwa* is generally nonbinding, and a Muslim may seek another opinion. Many *fatawa* of famous jurists are collected in books and may be used as precedents. Since 1991, at Georgia State University, we have been engaged in a project to collect and analyze *fatawa* related to Jihadist activity. These *fatawa* may include justifications for attacks against the West, those targets by category or type that are *hallal* (permissible) or forbidden (*haram*), those attack means that are *hallal* or *haram*, the permissibility of collateral damage, the general prohibition and exceptions to the spilling of sacrosanct (Muslim) blood, the destruction of natural resources both outside and within Muslim lands, the justification of prior attacks, etc.

For Jihadist organizations, the *fatawa* are rules of engagement that must be followed in order to perform a legitimate act of *jihad*. We have identified four types of *fatawa* relevant for threat purposes. These four types include the following:

1. *By fiat.* Authority bound within the issuer itself;
2. *By narrative.* Authority is housed in accuracy of the narrative through assessment of transmitters;
3. *By analogy.* Authority rests on strength of classifications, a classical approach to Sharia;
4. *By logic.* Authority rests on soundness of premise and power of logic employed, an approach recently influenced by Western intellectual thought.

Virtually all *fatawa* of interest to an analyst would fall within Types 2, 3 or 4. Absent a centralized *Sunni* Islamic authority structure, multiple persons in many locations issue *fatawa* [46]. These issuances and Muslims' reactions to them reflect the author's relative position within the collection of Islamic scholars and spectrum of Islamic doctrine. Therefore, *fatawa* carry differing degrees of validity and influence, and must be assessed individually [46]. After identifying the type of *fatwa*, we then assess the *fatwa* according to the following protocol:

1. The *fatwa's* author
2. The author's teacher
3. The author's religious, political, and intellectual alliances
4. The author's students
5. The subject matter of the *fatwa*
    - Legitimacy
    - Necessity
    - Proportionality
6. The response to the *fatwa*
    - Internal among religious thinkers
    - External among potential followers
7. *Fatwa* issuer's connection to terror groups and terror acts

Each *fatwa* is then scored as unpersuasive, persuasive, or compelling, depending on both the objective application and subjective evaluation of the protocols.

In addition to the influence of a *fatawa*, this threat methodology incorporates the influence of other religious symbols, statements, songs (*nasheed* or *anasheed*), and propaganda to describe terrorist target sets and operational parameters more completely.

Thus, the analyst should pay closer attention to religious decrees and cultural signs and symbols for clues about potential targets and attack means.

3.3 Synthesis

After one populates the threat model by assigning information to bins of capability, intent, and authority, one can then assess and evaluate the data and the information's credibility, reliability, relevance, inferential force, and adversary purpose. Credibility is a measure of belief and authenticity [8]. That measure may be either quantitative (for example, a source has an acceptable track record) or qualitative (subjective indicators suggest believability) or both. Reliability is a measure of replicability [8]. That measure focuses on consistency and coherence. Relevance is a measure of fit. That measure seeks a determination of whether the information proves or disproves a fact of interest to the analyst and the ultimate client. Inferential force is a measure of weight [8]. Not all information is of equal value in developing and supporting a conclusion [1]. Clark identifies seven pitfalls in measuring inferential force [8]. These include what he calls (i) vividness weighting (information experienced directly is given too much weight); (ii) weighing based on the source (for example, the snobbery of downplaying open-source information versus classified information, the belief that the information from spies is more valuable than refugees or defectors, or the belief that intercepted communications are the gold standard); (iii) favoring the most recent evidence (even though an informational signal tends to degrade over time); (iv) favoring or disfavoring the unknown (absence of evidence problem); (v) trusting hearsay (information from a third-party source) about what someone else said or did though the source may be biased; (vi) trusting expert opinions (too much deference, little attempt to gauge objectivity, etc.); and (vii) premature closure and philosophical predisposition (forming an opinion early in the process and then looking only for evidence that supports that opinion). Adversary purpose is a measure of informational intent. Recall that most threat information is initially in the hands of an adversary [2, 8]. Thus, an analyst must consider whether the information, in fact, may be disinformation playing a part in a D & D program [29, 30]. This D & D assessment is as such difficult to make, and is made even more difficult where the adversary's own actors are unaware that they are playing out a D & D program. Important hedges against D & D programs include consideration of alternative hypotheses; perspective shifts; a focus on collection content and not quantity; infusion of randomness into information collection; and development of an efficient feedback loop among all constituent parties to the intelligence process [8, 29, 30]. Another important hedge is something called "Red Teaming," that is, the use of an independent team, made of subject-matter experts, whose job is to try to debunk an analysis, invalidate assumptions, disprove hypotheses, etc. A good Red Team should be a hedge against complacency and "group think" that often send analyses down a wrong path.

The next step in the threat process is to identify information as either divergent or convergent. Clark suggests that multiple items of information are divergent if they favor different conclusions [8]. Clark further subdivides divergent information into two subclasses of note: (i) divergent information where multiple strands are true but support different conclusion and (ii) divergent information that is contradictory because these strands suggest logically opposing conclusions [8]. Clark then suggests that multiple items of information are convergent if they suggest or favor the same conclusion. He notes that convergent evidence may be redundant. As Clark observes, "redundancy is one way to improve the chances of getting it right" [8]. Redundant information may be further subdivided into (i) corroborative redundancy and (ii) cumulative redundancy. Both forms of redundancy are measures of the increased weight and credibility of the information.

The next step in the process of synthesizing threat information is to combine information across the threat variables. The following diagram illustrates this step and suggests that the highest threat would come from the intersection of capability, intent, and authority in the intersection space labeled "1" (Fig. 3).

3.4 Analysis

After synthesizing the information, the penultimate step in the threat assessment process is to analyze the information in order to make an informed estimate of what may happen in the future. Threat assessments are always predictive. As Clark cogently remarks, "Describing a past event is not intelligence analysis, it is history [8]." Here, an analyst

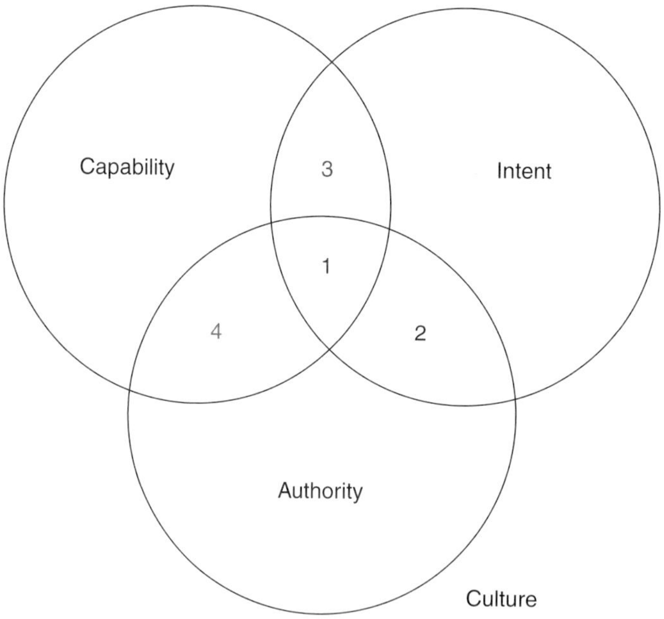

1. Capability AND Intent AND Authority = Represents High Threat
2. Intent AND Authority BUT NOT Capability = Warrant Close Monitoring of Proliferation for Capability Attainment
3. Capability AND Intent BUT NOT Authority = Warrant Close Monitoring of Fatawa for Authority Attainment
4. Capability AND Authority BUT NOT Intent = May be Perpetrating a Threat Deception to Enhance Negotiation or Increase Influence

FIGURE 3 Culture-centric threat model.

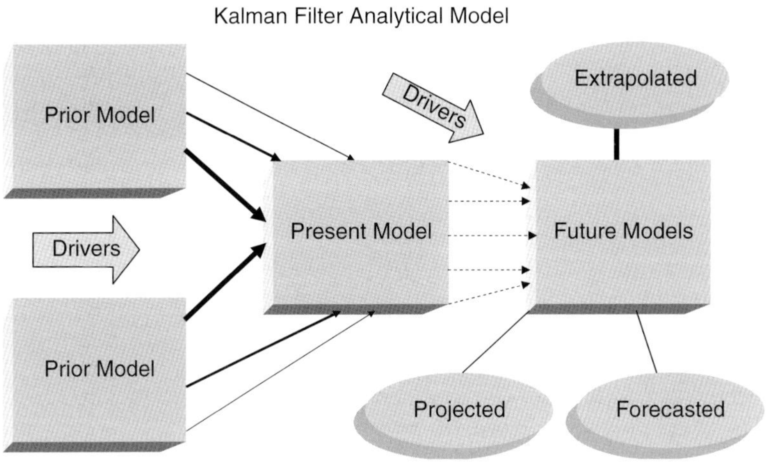

FIGURE 4 Kalman Filter analytical model.

might begin with a relevant Venn diagram that depicts, in a conceptual frame, what we know, what we do not know, what has happened in the past, and what is the current state.

Threat is a string of conditional probabilities. The Venn diagram above sums up the idea of conditional probabilities, but we may be able to go further in the analysis. Intelligence analysts have used several analytical techniques in order to combine vast amounts of information in complex systems. Among these is the Kalman Filter, developed by Rudolf Kalman [47]. Used by engineers and adapted by financial and restructuring advisors in financially distressed business situations, the Kalman Filter estimates the state of a dynamic system from a series of incomplete or "noisy" measurements (Fig. 4).

The method begins with a thorough description of a past and present threat model [8]. That description may be quantitative, qualitative, or both. After past and present threat models are constructed, the next step is to identify and analyze the drivers (or forces) that acted on the past model to drive it to its present state. Drivers may be strong or weak, direct or indirect, reliable or unreliable, certain or uncertain. Through the use of extrapolation, projection, and forecasting, the analyst may then develop a set of scenarios in order to construct future threat models [8]. Extrapolation holds the drivers constant among past, present, and future stages of the threat model; projection accounts for dynamic drivers among the threat models; and forecasting accounts not only for dynamic drivers but also for new drivers that may emerge and former drivers, may become irrelevant. During each step of the process, the analyst must assess information credibility, determine convergent or divergent information, and render an opinion as to the analyst's confidence in the models, drivers, and predictions.

Analysts use scenarios to identify relatively large drivers. According to Clark, analysts use four types of scenarios [8]: (i) demonstration scenarios (imagine an end state and then describe a plausible path to achievement); (ii) driving-force scenarios (identify and manipulate forces at work on the process, typically holding them constant or at some predetermined rate of change, to derive multiple futures); (iii) system-change scenarios (a cross-impact analysis that accounts for changing forces in a probabilistic manner and emerging synergies and degradations to derive multiple futures); and (iv) slice-of-time scenarios (an analysis that jumps to a future state without an explanation of how that state was derived).

The Kalman Filter technique, through the use of driving-force or system-change scenario modeling, leads us back to the beginning of this article, to the original drivers of threat that exist within a cultural space: (i) capability, (ii) intent, and (iii) authority (external/internal). These general drivers are then further subdivided into more specific ones.

Threat Drivers

- *Capability*
  - Operational
  - Technological
  - Logistical
- *Intent*
  - General
- Strategic
- Operational
- Tactical
  - Specific
- Strategic
- Operational
- Tactical
- *Authority*
  - Existence of *fatwa*
  - Acceptance of *fatwa*
  - Contra-*fatawa* (a fatwa that counters, refutes, or distinguishes the fatwa of interest)
  - Other sources of authenticity, authority, resonance
- Treatises
- Symbols
- Signs
- Song
- *Cultural*
  - Social dynamics
  - Demographic trends
  - Economic trends
  - Political (internal/external)
  - Communications
  - Education
  - Religion
  - Song
  - Story
  - Myth

For example, based on this analysis, we have determined that al-Qaeda has never conducted an operation that had not been previously sanctioned by a *fatwa*. Thus, a *fatwa* is a strong, direct, and reliable indicator of threat. In fact, we have also learned that al-Qaeda has called off at least one operation where no *fatwa* permitted the operation and that lack of a *fatwa* played an indirect role in terminating an al-Qaeda operative where he had regularly exceeded his authority.

These observations lead us to the final step in the threat assessment process, that is, monitoring. Analysts must be active participants in the intelligence loop. For example, the use of the culture-centric threat model identified above where we have identified the existence of a *fatwa* as a strong, direct, and reliable driver has led us to the development of various weapons of mass destruction (WMD) scenarios. The planned attack by an al Qaeda cell that was called off involved a chemical attack in a subway system in the United States. Such an unprecedented attack mode (at least for Jihadists) with the potential for heavy collateral damage to non-Muslims who are noncombatants and Muslims who may use the subway led Ayman al-Zawahiri to request the opinion of a Saudi religious authority. Shortly after that request, Nasser bin Hamed al-Fahd, a prominent Saudi *Salafi* scholar, in an Islamic ruling published in May 2003, approved the use of WMDs against America [48]. He based his indictment on the principle of retaliation, and argued that Muslims have the right to kill 10 million Americans in response to the crimes of their government against the Muslim nation [48]. Al-Fahd elaborated the circumstances under which it is religiously permitted to kill noncombatant Americans: during a military operation when it is hard to distinguish between soldiers and civilians and according to military needs or considerations [48]. Ascribing great importance to the military considerations, he asserted that the military leaders who are responsible for the execution of *jihad* have the authority to make the decisions concerning what types of weapons to use against the infidels. If they decided to use WMDs based on military need, it would be an obligation under Islamic law to use them [48]. Although a thorough analysis of this *fatwa* is beyond the scope of this article, it is important to note that the attack means permitted the use of WMDs particularly chemical agents (and except, possibly the use of biologics), in targeting mass transportation in the United States. This is what we label a dangling *fatwa*, one that exists without a corresponding terrorist event as of yet. Based on our analysis, the intelligence analyst may want to monitor closely any technological developments in the processing or dispersing of chemical agents, access to WMD technology, and the emergence of alliances between al-Qaeda and various right-wing groups located in the Continental United States and Europe with chemical manufacturing and dispersion expertise.

4 OBSERVATIONS

Intelligence analysis may be divided into three separate but interdependent stages: (i) collection, (ii) management, and (iii) understanding. Robust threat assessments call on the skills necessary to discharge the three steps in the analytical process at a high level of sophistication. As we migrate from a threat space where we envision threat from a nation-state to threat from a non-state actor, we must refocus our efforts on a more robust threat model that accounts not only for the traditional drivers of capability and intent, but also incorporates authority and cultural drivers in our efforts to build meaningful scenarios that may help us in constructing a predictive model of threat. Yesterday's threat is history; tomorrow is where we confront threat head-on. Throughout this process, we must remain cognizant of one sobering condition that permeates threat assessments and humbles analysts and academics alike. Many of us were taught deep in our youth that "the truth shall set you free." For an analyst, that translates into a quest for the truth so that we may meaningfully inform our intelligence client. The humbling fact is that there is someone just as committed to obfuscate or hide the truth from us. The attainment of truth involves a constant struggle in an ever-changing environment with a cunning adversary who is fighting back. As intelligence analysts, we will never know the truth until it is too late. We may only approximate it. However, truth is not our ultimate goal, it is not an end in itself; it is our means toward victory.

REFERENCES

1. Shulsky, A. N., and Schmitt, G. J. (2002). Silent warfare. *Understanding the World of Intelligence 1–30*, 3rd ed. Brassey's, Inc., Washington, DC.
2. Khalsa, S. (2004). *Forecasting Terrorism 1–59*. Scarecrow Press, Inc., Lanham, MD.
3. Department of Homeland Security (2006). *National Infrastructure Protection Plan 39*, USA.
4. Williams, J. F. (2007) al-Qaeda strategic threats to the international energy infrastructure: authority as an integral component of threat assessment. *Proceedings of Carlton University–Ottawa Center for Infrastructure Protection*.
5. Williams, J. F. (2007) The nature of the terrorist threat. *Proceedings of Conference Board of Canada–Targeting the World's Transportation Systems*.
6. Weinberg, D. M., Coplon, G. H., and Williams, J. F. (2008). Understanding terrorism risk and its possible impacts. *Oil Gas J.*
7. Pattakos, A. N. (Pat) (1998). Threat analysis: defining the adversary. *Competitive Intelligence Review*, Vol. 9(2). John Wiley & Sons.
8. Clark, R. M. (2004). *Intelligence Analysis: A Target-Centric Approach*. CQ Press, Washington, DC, pp. 99, 105, 108–110, 113, 137–138, 166–173, 200–201, 204–205, 214–215, 262–263.
9. *Commission on the Intelligence Capabilities of the United States Regarding Weapons of Mass Destruction; Report to the President of the United States 378–379 (2005)*, USA.
10. Director of National Security (effective July 11, 2006). *Intelligence Community Directive Number 301*. National Open Source Enterprise.
11. National Commission on Terrorist Attacks Upon the United States (9-11 Commission) (2004). *Final Report of the National Commission on Terrorist Attacks Upon the United States*, 413.
12. Cragin, K., and Daly, S. D. (2004). *The Dynamic Terrorist Threat: An Assessment of Group Motivations and Capabilities in a Changing World*. RAND Project Air Force.
13. Lesser, I. O., Hoffman, B., Arquilla, J., Ronfeldt, D., Zanini, M. (1999). *Countering the New Terrorism*. RAND, Santa Monica, CA.
14. Hoffman, B. (1999). Terrorism trends and prospects 19–20 and n.33. In *Countering the New Terrorism*, I. O. Lesser, et al., Ed. RAND, Santa Monica, CA.
15. Bar, S. (2006). *Warrant for Terror: The Fatwas of Radical Islam, and the Duty of Jihad*. Rowman & Littlefield, Stanford, CA.
16. McCants, W., Ed. (2006). *Combating Terrorism Center, U.S. Military Academy; Militant Ideology Atlas*. Combating Terrorism Center, West Point, NY.
17. E.g., Qutb, S. (1964). *Ma'alim fi al-Tariq ("Signposts on the Road" or "Milestones")*.
18. Hoffman, B. (1998). *Inside Terrorism 162*. Columbia University Press, New York.
19. Kepel, G. (2003). *The Trial of Political Islam*. Belknap Press.

20. Santos, E.Jr., and Negri, A. (2004). Constructing adversarial models for threat/enemy intent prediction and inferencing. *5423 Enabling Technologies for Simulation Sciences VIII: Int'l Society for Optical Engineering*.
21. Bringer, B. E., et al. (2007). *Security Risk Assessment and Management 11*. John Wiley & Sons.
22. Hunter, T. (1997). *Bomb School International Training Camps*. *Janes Intelligence Review*.
23. Heuer, R. J. (1999). *Psychology of Intelligence Analysis*. Central Intelligence Agency Center for the Study of Intelligence.
24. Heuer, R. J. (1981). Strategic deception and counterdeception: a cognitive approach. *Int. Stud. Q.*, **25**(2), 294–327.
25. Jones, R. V. (1995). Enduring principles: some lessons in intelligence. *CIA Stud. Intell.*, **38**(5), 37–42.
26. Jones, R. V. (1989). *Reflections on Intelligence*. Mandarin, London.
27. Whaley, B., and Busby, J. (2002). Detecting deception: practice, practitioners, and theory. In *Strategic Denial and Deception: The Twenty-First Century Challenge*, R. Godson, and J. J. Wirtz, Eds.
28. Johnson, P. E., et al. (2001). Detecting deception: adversarial problem solving in a low base-rate world. *Cogn. Sci. Multidisciplinary J.* **25**(3), 355–392.
29. Stech, F. J., and Elsaesser, C. (2005). *Deception of Detection by Analysis of Competing Hypotheses*. The MITRE Corporation.
30. Stech, F. J., and Elsaesser, C. (2004). *Midway Revisited: Detecting Deception by Analysis of Competing Hypotheses*. The MITRE Corporation.
31. McFate, M., and Jackson, A. (2005). An organizational solution for DoD's cultural knowledge needs. *Mil. Rev.*
32. United States Marine Corps (USMC) (1940). *United States Marine Corps Small Wars Manual*. Sunflower University Press, Manhattan, KS.
33. Kipp, L., et al. (2006). The human terrain system: a CORDS for the 21st century. *Mil. Rev.*
34. Lewis, B. (2003). *The Crisis of Islam: Holy War and Unholy Terror*. Random House, New York. (A concise treatment of the role religion may play in militant movements within Islam with an elegant treatment of Islamic history in context).
35. Esposito, J. L. (2002). *Unholy War: Terror in the Name of Islam*. Oxford University Press, New York. (Another important treatment of the role or, to be more precise, the abuse of religion by militant movements within Islam, but from a different, but equally persuasive, perspective than that of Professor Lewis).
36. Kepel, G. (2004). *The War for Muslim Minds*. Harvard University Press, Cambridge, MA. (Among one of the most gifted scholars of political Islam or Islamism, the book presents an insightful treatment of militant Islam from a cultural perspective with an excellent critique of the post 9-11 al-Qaeda leadership and organizational structure).
37. Patai, R. (2002). *The Arab Mind*, Rev. ed. Hatherley Press, New York. (Dismissed by some as outdated or insensitive; to others, a treasure trove of information through its comprehensive treatment of the subject, especially its careful selection of Arab authors addressing several sensitive cultural subjects; you may disagree with Dr. Patai, but any serious treatment of Arabic culture must confront his work head on).
38. Pryce-Jones, D. (2002). *The Closed Circle: An Interpretation of the Arabs*. (Thoughtful treatment of how tribal, religious, and cultural traditions drive the Arabs in their dealings with each other and with the West).
39. Nydell, M. K. (2006). *Understanding Arabs: A Guide for Modern Times*, 4th ed. Intercultural Press, Yarmouth, ME. (From the perspective of a linguist, this book does a good job of covering the cultural waterfront of Arabs in the Middle East; although she takes issue with Dr. Patai

on a number of points, her observations, with several notable exceptions, are quite consistent with Dr. Patai's own observations).

40. Ruthven, M., and Nanji, A. (2004). *Historical Atlas of Islam*. Harvard University Press, Cambridge. (One of the best short introductions to Islamic history with engaging but concise text, a beautiful use of maps, and splendid pictures).
41. Williams, J. F. (2005). *The Battle Within Islam: Militant Movements in Islam and the Threat They Pose to the West and to Mainstream Muslims*. The MITRE Corporation.
42. Ruthven, M. (2000). *Islam: A Short Introduction*. (one of the best short introductions to Islam as religion, culture, and politics).
43. Algar, H. (2000). *The Sunna: Its Obligatory and Exemplary Aspects*. Islamic Publications International, New York.
44. Ali, M. M. (2001). *A Manual of Hadith*.
45. Mitchell, R. P. (1993). *The Society of the Muslim Brothers*. Oxford University Press, New York.
46. Masud, M. K., Messick, B., and Powers, D. S., Eds. (1996). *Islamic Legal Interpretation: Muftis and Their Fatwas*. Harvard University Press, Cambridge, MA.
47. Kalman, R. E. (1960). A new approach to linear filtering and prediction problems. *Trans. ASME J. Basic Eng.* **82**, 35–45.
48. Hamd Al-Fahd, N. B. (2003). *A Treatise on the Legal Status of Using Weapons of Mass Destruction Against Infidels (May 2003 [Rabi' I 1424])*.

FURTHER READING

Al-Qaeda Training Manual.

CIA. (2007). *The World Fact Book*.

Proceedings of the Kuala Lumpur International Forum on Islam. (2002).

The Holy Qur'an.

The Koran. (N. J. Dawood trans. 1956 and updated 1999).

The Noble Qur'an.

Time Almanac (2008).

United Nations Arab Human Development Report (2002).

BOOKS

1. Ajami, F. (1998). *Dream Palace of the Arabs: A Generation's Odyssey*.
2. Algar, H. (2000). *The Sunna: Its Obligatory and Exemplary Aspects*.
3. Algar, H. (2002). *Wahhabism: A Critical Essay*.
4. Algar, H. (1997). *Surat Al-Fatiha: Foundation of the Qur'an*.
5. Ali, M. M. (2001). *A Manual of Hadith*.
6. Anonymous. (2002). *Through our Enemies' Eyes: Osama bin Laden, Radical Islam, and the Future of America*.
7. Anonymous. (2004). *Imperial Hubris: Why the West is Losing the War on Terror*.
8. Armstrong, K. (2000). *The Battle for God: A History of Fundamentalism*.
9. Armstrong, K. (2000). *Islam: A Short History*.
10. Armstrong, K. (2001). Was it inevitable? In *How Did This Happen? Terrorism and the New War*, J. F. Hoge, and G. Rose, Eds.

11. Bakhtiar, L. (1996). *Encyclopedia of Islamic Law: A Compendium of the Major Schools*.
12. Beckett, I. F. W. (2001). *Modern Insurgencies and Counter-Insurgencies*.
13. Benjamin, D., and Simon, S. (2002). *The Age of Sacred Terror*. Random House, New York.
14. Bergen, P. L. (2001). *Holy War, Inc.*
15. Carr, C. (2002). *The Lessons of Terror*.
16. Cohen, M. (1994). *Under the Crescent and the Cross: The Jews in the Middle Ages*
17. Doi, A. R. I. (1984). *Shariah: The Islamic Law*.
18. El Fadl, K. A. (2001). *Rebellion and Violence in Islamic Law*.
19. El Fadl, K. A. (2001). *Speaking in God's Name: Islamic Law, Authority and Women*.
20. El Fadl, K. A., (with Ali, T., Viorst, M., and Esposito, J., et al. (2002). *The Palace of Tolerance in Islam*.
21. Esposito, J. (2000). *The Oxford History of Islam*.
22. Esposito, J. (1999). *The Islamic Threat: Myth or Reality?* 3rd ed.
23. Esposito, J. (2004). *Islam: The Straight Path*.
24. Esposito, J. L. (2002). *Unholy War: Terror in the Name of Islam*.
25. Fuller, G. (2003). *The Future of Political Islam*.
26. Habeck, M. (2006). *Knowing the Enemy: Jihadist Ideology and the War on Terror*.
27. Hitti, P. (1967). *History of the Arabs from Earliest Times to the Present*, 9th ed., (emphasis on Islamic culture, arts, and sciences).
28. Huband, M. (1999). *Warrior of the Prophet: The Struggle for Islam*.
29. Huntington, S. (1996). *The Clash of Civilizations and the Remaking of World Order*.
30. Jones, R. V. (1978). *Most Secret War*.
31. Juergensmeyer, M., Ed. (1992). *Violence and the Sacred in the Modern World*.
32. Juergensmeyer, M. (1993). *The New Cold War?*
33. Juergensmeyer, M. (2000). *Terrorism in the Mind of God*.
34. Kent, S. (1966). *Strategic Intelligence for American World Policy*.
35. Kepel, G. (2004). *The War for Muslim Minds*.
36. Kramer, M. (1987). *The Moral Logic of Hizbullah*.
37. Laqueur, W. (1977). *Terrorism*.
38. Laqueur, W. (1979). *The Terrorism Reader*.
39. Laqueur, W. (1999). *The New Terrorism: Fanaticism and the Arms of Mass Destruction*.
40. Ledeen, M. A. (2003). *The War Against the Terror Masters*.
41. Lewis, B. (2003). *The Crisis of Islam: Holy War and Unholy Terror*.
42. Lewis, B. (2002). *What Went Wrong? Impact and Middle Eastern Response*.
43. Lewis, B. (1960). *The Arabs in History*.
44. Lewis, B. (1968). *The Assassins: A Radical Sect in Islam*.
45. Lewis, B. (1991). *The Political Language of Islam*.
46. Lewis, B. (1995). *The Middle East: A Brief History of the Last 2000 Years*.
47. Lewis, B. (2001). *Music of a Distant Drum: Classical Arabic, Persian, Turkish and Hebrew Poems*.
48. Lewis, B. (2001). *Islam in History: Ideas, People, and Events in the Middle East*.
49. Lewis, B. (2004). *From Babel to Dragomans: Interpreting the Middle East*.
50. Marty, M. E., and Appleby, S. R. (1995). *Fundamentalism Comprehended*.
51. O'Neill, B. E. (1990). *Insurgency & Terrorism: Inside Modern Revolutionary Warfare*.
52. Pipes, D. (1996). *The Hidden Hand: Middle East Fears of Conspiracy*.

53. Pipes, D. (2002). *Militant Islam Reaches America*.
54. Qutb, S. (Hamid Algar trans. 2000). *Social Justice in Islam*.
55. Ramadan, T. (1999). *To Be a European Muslim*.
56. Roolvink, R. (1958). *Historical Atlas of the Muslim People*, (Best collection of maps of Muslim world at important times in history).
57. Roy, O., and Volk, C. (1998). *The Failure of Political Islam*.
58. Sageman, M. (2004). *Understanding Terror Networks*.
59. Schanzer, J. (2005). *Al-Qaeda's Armies: Middle East Affiliate Groups and the Next Generation of Terror*.
60. Schacht, J. (1982). *An Introduction to Islamic Law*.
61. Shanor, C. A., and Hogue, L. L. (2003). *National Security and Military Law*.
62. Sprinzak, E. (1999). *Brother Against Brother*.
63. Thompson. L. (2002). *The Counter Insurgency Manual*.
64. Tibi, B. (1998). *The Challenge of Fundamentalism: Political Islam and the New World Disorder*.
65. Townshend, C. (2002). *Terrorism: A Very Short Introduction*.
66. Treverton, G. (2003). *Reshaping National Intelligence for an Age of Information*.
67. Trimble, P. R. (2002). *International Law: United States Foreign Relations Law*.
68. Von Grunebaum, G. E. (K. Watson trans. 1970). *Classical Islam*.
69. Ye'or, B. (2002). *Islam and Dhimmitude: Where Civilizations Collide*.

ARTICLES AND ESSAYS

1. Vlahos, M. (2002). *Terror's Mask: Insurgency Within Islam*. Johns Hopkins University Applied Physics Laboratory, Laurel, MD.
2. Aboul-Enein, Y. H., and Zuhur, S. (2004). *Islamic Rulings on Warfare*. Strategic Studies Institute, Carlisle, PA.
3. Anspach, M. (1991). Violence against violence: Islam in historical context. *Terrorism and Political Violence 3, 3*.
4. Brei, W. S. (1996). *Getting Intelligence Right: The Power of Logical Procedure*. Joint Military Intelligence College, Washington, DC.
5. Cordesman, A. H. (2004). *The Intelligence Lessons of the Iraq War(s)*. Center for Strategic Studies.
6. Halliday, F. (2000). Terrorisms in historical perspective. *Nation and Religion in the Middle East*. Lynne Rienner Publishers, Boulder, CO.
7. Merari, A. (1990). The readiness to kill and die: suicidal terrorism in the middle east. In *Origins of Terrorism: Psychologies, Ideologies, States of Mind*, W. Reich, Ed.
8. Rapoport, D. (1984). Fear and trembling: terrorism in three religious traditions. *Am. Polit. Sci. Rev.* **78**(3), 658–677.
9. Schwartz, S. (2004). *Rewriting the Koran. The Weekly Standard 19 Sep 27 2004*.

OPEN-SOURCE INTERNET SITES

1. http://www.odl.state.ok.us/usinfo/terrorism/911.htm (Collections of annotated bibliographies by Oklahoma Department of Libraries)

2. http://www.nyazee.com/islaw/Islamic%20Law%20Infobase.html (Islamic Law Infobase)
3. http://www.odci.gov/cia/publications/factbook/ (*CIA World Factbook* 2007)
4. http://www.bartleby.com/65/ (*Columbia Encyclopedia*)
5. http://www.arches.uga.edu/%7Egodlas/maps.html (populations, maps, and countries of Muslim world)
6. http://www.lib.utexas.edu/maps (maps)
7. http://www.umr.edu/~msaumr/topics/ (English translation and interactive topical index to Quran)
8. http://www.acs.ucalgary.ca/~elsegal/I_Transp/IO6_Shia.html (sectarian discussion of Sunni/Shia)
9. http://www.fordham.edu/halsall/islam/islamsbook.html (Islamic history)
10. http://www.acs.ucalgary.ca/~elsegal/I_Transp/Ilist.html (class notes on Islam)
11. http://www.quran.org/ (authenticated English translations [and other languages] of the Quran)
12. http://www.carm.org/islam/faith_imam.htm (Introduction to Islam from nonMuslim source with chart comparison to Christianity)
13. http://al-islam1.org/laws/ (Islamic law)
14. http://www.mb-soft.com/believe/txo/wahhabis.htm (Wahhabism)
15. http://www.wsu.edu/~dee/ISLAM/UMAY.HTM (Islamic history)
16. http://www.eppc.org/ (Contains transcripts of conversations by leading scholars on Political Islam, Iraq, and Terrorism)
17. www.arabia.com
18. www.islamfortoday.com
19. www.memri.org (Middle East Media Research Institute)
20. www.saudiinstitute.org (Saudi reformers)

RISK ANALYSIS METHODS FOR CYBER SECURITY

Michel Cukier and Susmit Panjwani
University of Maryland, College Park, Maryland

1 INTRODUCTION

Executive Order 13010 defines the nation's critical infrastructure as "telecommunications, electrical power systems, gas and oil storage and transportation, banking and finance, transportation, water supply systems, emergency services (including medical, police, fire, and rescue), and continuity of government" [1]. Traditionally, the nation's critical infrastructure assets were considered independently from the information assets. However, the development of an information-based economy and the wide proliferation of the Internet have changed the way these critical infrastructure assets are accessed, maintained, and used. The critical infrastructures are now exposed to a different threat profile raised by the interdependence of information assets and critical infrastructure assets.

This is clearly illustrated in the Clinton Administration's Policy on Critical Infrastructure Protection (CIP). Presidential Decision Directives (PDD) 62 and 63, released on May 22, 1998 by President Clinton address the new and nontraditional "cyber-security" threats against critical infrastructure [2, 3]. PDD 63 is the key directive focusing on CIP from both the physical and cyber security perspective [3, 4].

On October 16, 2001, President Bush announced Executive Order 13231, entitled "Critical Infrastructure Protection in the Information Age" [5]. In this executive order, President Bush explicitly stated that "the information technology revolution has changed the way business is transacted, government operates, and national defense is conducted" [6, 7]. Executive Order 13231 established the President's Critical Infrastructure Protection Board [5].

2 SECURITY RISK ASSESSMENT

The aforementioned directives emphasize the gravity of the new and evolving risk associated with cyber-security. A security risk assessment framework is needed to quantify this risk. Research has been conducted to define guidelines and frameworks. These frameworks should be leveraged to drive the investment decisions regarding critical infrastructure assets.

The Office of Management and Budget (OMB) Circular A-130, "Security of Federal Automated Information Resources", mandated that the federal agencies must "consider risk when deciding" which security countermeasures to implement [8, 9]. According to the circular, a "risk-based approach" should be adopted to determine the adequate level of security [8, 9]. This circular encourages agencies to consider "major risk factors, such as the value of the system or application, threats, vulnerabilities, and the effectiveness of safeguards" [8, 9]. The OMB Director further emphasized this risk-based approach to security when he issued Memorandum 99-20, "Security of Federal Automated Information Resources", explicitly stating the importance of "continually assessing the risk" associated with information assets to maintain adequate levels of security [9, 10].

Recognizing the gravity of the cyber-security threat, the National Institute of Standards and Technology (NIST) also released guidelines on security risk assessment, contained in "An Introduction to Computer Security: The NIST Handbook" [11] and "Generally Accepted Principles and Practices for Securing Information Technology Systems" [12].

The need for implementing cost-effective, risk-based information security programs was further emphasized by The Federal Information Security Management Act of 2002 [13]. The act was meant to strengthen computer security within the federal government and affiliated parties (such as government contractors) by mandating yearly audits.

3 A QUANTITATIVE APPROACH TO CYBER-SECURITY RISK ASSESSMENT

The main aspect of making a risk-based decision is the significance of the data used to make the decision. Often this data is collected in the form of expert opinion or the attack trend observed on the Internet. Because of the continuous evolution of the Internet, the attack trend changes rapidly, and it is important to quantify this evolving attack threat to develop appropriate countermeasures. Moreover, the attack trend observed by an organization may differ from the attack trend observed on the Internet. Indeed, the complexity and technology used by organizations to build their information infrastructure may differ significantly from the networks seen on the Internet. Hence, it is crucial that organizations measure the attack trend at the organization level to make more accurate decisions.

In this article, we present a quantitative approach to security risk assessment that can be used to make risk-based decisions using data representative of the organization's infrastructure. The core of the quantitative risk assessment is the ability to form and evaluate critical hypotheses based on attack data and to perform attack trend analysis. In this article, we describe how quantitative data can be leveraged for (i) hypothesis evaluation and (ii) attack trend analysis. The research test-bed that was used to collect the data and two empirical experiments are also described.

4 DATA COLLECTION AND ANALYSIS

The experimental test-bed used in this research consisted of two target computers used just for the purpose of being attacked. Other computers were used to closely monitor the target computers, however, attackers were unaware that they were observed. This architecture is similar to the one developed by the Honeynet project [14]. Differences between the two architectures and details on the deployed architecture are presented in Ref. 15.

The selected subnet for the target computers is unmonitored; IP addresses are assigned to users dynamically. Both target computers ran Windows 2000 and had the same services (i.e. IIS, FTP, Telnet, and NetTime [16]) and the same vulnerabilities maintained throughout the data collection period. Twenty-five vulnerabilities, shown in Table 1, were selected to cover a broad range of discovery dates (from 2000 to 2004), various services (i.e. RPC, LSA, IIS, FTP, HTTP, and Telnet), and different levels of criticality (i.e. Critical, Important, Moderate, and Low). Note that most UDP traffic was filtered at the gateway level, which is why the analysis focused on ICMP and TCP traffic. The outbound traffic was limited to 10 TCP connections per hour, 15 ICMP connections per hour, and 15 other connections per hour.

The traffic was filtered at multiple stages before being analyzed. The data consisted of (i) malicious traffic from the Internet and (ii) management traffic like spanning tree protocol (STP) traffic generated by the bridge, DNS resolutions, and NTP queries. The data were parsed into a format that could be stored in a database. The data were then parsed based on a protocol to filter out the remaining management traffic. Traffic not directed toward either target computer was also filtered.

TABLE 1 List of Vulnerabilities Remaining on the Target Computers

Year	Bulletin Number	Service	Criticality	Vulnerability Description
2004	MS04-012	RPC/DCOM	Critical	Race condition
2004	MS04-012	RPC/DCOM	Important	Input vulnerability
2004	MS04-012	RPC/DCOM	Low	Buffer overflow
2004	MS04-012	RPC/DCOM	Low	Input vulnerability
2004	MS04-011	LSA	Critical	Buffer overflow
2004	MS04-011	LSA	Moderate	Buffer overflow
2003	MS03-010	RPC endpoint mapper	Important	Vulnerability not clearly specified
2003	MS03-018	IIS	Important	Memory vulnerability
2003	MS03-026	RPC interface	Critical	Buffer overflow
2003	MS03-049	Workstation service	Critical	Buffer overflow
2003	MS03-039	RPC	Critical	Buffer overflow
2002	MS02-062	IIS	Moderate	Memory vulnerability
2002	MS02-018	FTP	Critical	Vulnerability not clearly specified
2002	MS02-018	HTTP	Critical	Buffer overflow
2002	MS02-004	Telnet	Moderate	Buffer overflow
2001	MS01-041	RPC	Not rated	Input vulnerability
2001	MS01-044	IIS	Not rated	Input vulnerability
2001	MS01-026	IIS	Not rated	Vulnerability not clearly specified
2001	MS01-026	FTP	Not rated	Memory vulnerability
2001	MS01-026	FTP	Not rated	Vulnerability not clearly specified
2001	MS01-016	IIS WebDAV	Not rated	Input vulnerability
2001	MS01-014	IIS exchange	Not rated	Input vulnerability
2000	MS00-086	IIS	Not rated	Vulnerability not clearly specified
2000	MS00-078	IIS	Not rated	Input vulnerability
2000	MS00-057	IIS	Not rated	Input vulnerability

5 HYPOTHESIS EVALUATION: ARE PORT SCANS PRECURSORS TO AN ATTACK?

In this section, we show how quantitative data can be utilized to evaluate security related hypotheses. The security community continues to debate if port scans are considered to be precursors to an attack. The case study described in this section (i.e. a short version of experiment described in Ref. 15) shows how the data collection architecture previously described to quantify such an assumption can be leveraged.

5.1 Data Filtering

In order to determine the link between scans and attacks, we filtered the malicious traffic collected into four categories: ICMP scans, port scans, vulnerability scans, and attacks. ICMP scans can be easily recognized by their protocol. Regarding port scans, as reported by Ref. 17, three packets are sufficient to finish a TCP handshake and establish a connection. Information on open ports and services can be gathered by using as few as two packets. We showed empirically using Nmap [18] in Ref. 15 that in 99.8% of the cases, port scans were defined as connections with four or fewer packets. Regarding vulnerability scans, we ran NeWT 2.1 [19] and showed empirically that 99.9% of the vulnerability scans consisted of connections having between 6 and 12 packets. Based on these results:

- port scans were characterized as connections with less than five packets;
- vulnerability scans were characterized as having connections between 5 and 12 packets; and
- attacks were characterized as connections with more than 12 packets.

5.2 Data Analysis

The goal of the experiment was to analyze the link between scans and attacks. Therefore, we only kept unique scans and attacks. For example, if multiple port scans were launched from one specific source IP address toward one of the target computers, we recorded that this source IP address had launched at least one port scan without recording the actual number of port scans. Similarly, if one source IP address launched several attacks (of the same type or of different types) against one of the target computers, we recorded that at least one attack had been launched from that source IP address against the target computer. The link between the 22,710 connections of malicious activity (collected over a period of 48 days) and unique ICMP scans, port scans, vulnerability scans, and attacks is shown in Table 2.

For each of the 760 attacks from different source IP addresses, we checked if a scan or combination of scans was linked to the attack (from the same source IP address toward the same target computer). The number and percentage of direct attacks (i.e. attacks not linked to a scan) and attacks linked to different scans and combinations of scans are provided in Table 3. We observed that more than 50% of the attacks were not linked to a scan. However, over 38% of the attacks were linked to a vulnerability scan. Port scans and combinations of port and vulnerability scans were linked to 3–6% of the attacks.

5.3 Interpretation of Results

These experimental results indicated that the majority of the attacks were not linked to a scan. When scans were linked to attacks, the most frequent ones were (i) a vulnerability scan, (ii) a combination of port and vulnerability scans, and (iii) a port scan.

One explanation for these observations might be the large number of automated attacks captured by the honeypot-based test-bed. These attacks were developed in a manner to fingerprint the vulnerability rather than the machine they were trying to compromise.

TABLE 2 Distribution of Malicious Activity: Nonunique and Unique Records

Malicious Activity	No. of Records	No. of Unique Records
ICMP scans		3007
Port scans	8432	779
Vulnerability scans	2583	1657
Attacks	2035	760
Total	22,710	6203

TABLE 3 Distribution of Scans Linked to an Attack

Type of Scan	No. Attacks Linked to a Scan	Percentage of Attacks
Port	28	3.68
ICMP	1	0.13
Vulnerability	296	38.95
Port & ICMP	0	0
Port & vulnerability	42	5.53
ICMP & vulnerability	5	0.66
Port & ICMP & vulnerability	7	0.92
None	381	50.13

This section illustrated how the data collection architecture could be used to address issues associated with the cyber-security threat, like the potential link between scans and attacks.

6 ATTACK TREND ANALYSIS: COMPARISON BETWEEN INTERNAL AND EXTERNAL MALICIOUS TRAFFIC

Often, organizations make security decisions based on the malicious activity observed and recorded on the Internet. Using this method, malicious traffic originating inside of the organization is overlooked. We conducted two studies to compare malicious traffic originating inside of the organization (called *internal traffic*) to malicious traffic originating from outside of the organization (called *external traffic*). We analyzed the correlation between internal and external malicious traffic at two different levels. First, we conducted an analysis at a higher level, correlating the internal and external malicious traffic solely on the basis of the number of connections. We then refined the analysis by considering the number of connections targeted toward specific ports.

The data used for this analysis was collected over a period of 15 weeks from October 4, 2004 to January 16, 2005. The data were subdivided into weeks (week 1–52) with week 1 corresponding to the period October 4–10, 2004. In week 7, data were collected from 6 (out of 7) days and in weeks 5 and 8, data were collected from 5 days. From the malicious traffic on both target computers, we filtered out: (i) ICMP scans that were identified through the protocol and (ii) port scans that were identified based on the number of packets per connections (i.e. connections with a maximum of four packets).

Correlation coefficients were calculated based on the 15 data points obtained during the 15-week long data collection period to form the basis of one aspect of the comparison. We applied Guilford's [20] interpretation of the correlation coefficient (i.e. no correlation = "N", low correlation = "L", moderate correlation = "M", high correlation = "H", and very high correlation = "VH"):

- correlation coefficients lower than 0.2: no correlation (N),
- correlation coefficients between 0.2 and 0.4: low correlation (L),
- correlation coefficients between 0.4 and 0.7: moderate correlation (M),
- correlation coefficients between 0.7 and 0.9: high correlation (H), and
- correlation coefficients higher than 0.9: very high correlation (VH).

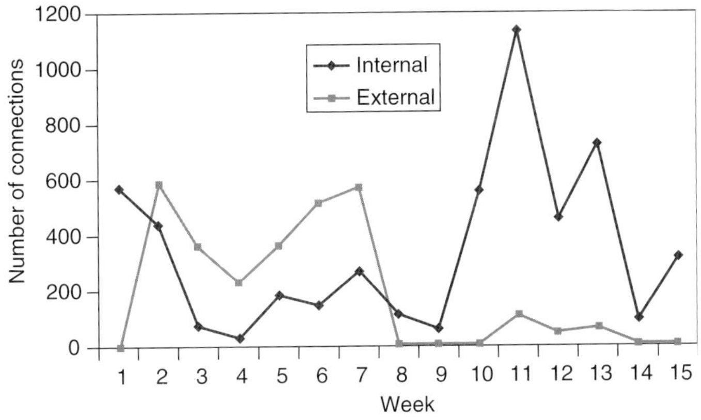

FIGURE 1 Number of internal/external connections per week.

6.1 Correlation 1: Number of Connections

In this section we compare internal and external malicious traffic based on the number of malicious connections. Over the 15-week data collection period, the two target computers received a total of 8029 malicious connections. 5,174 of the connection (64.4%) originated from within the organization and 2855 (35.6%) originated from outside of the organization. Figure 1 shows the number of internal and external connections received each week. No clear relationship existed between internal and external malicious traffic as shown in Figure 1. This is further confirmed by calculating the correlation coefficient, −0.21, which indicated a low correlation between internal and external traffic. The average amount of internal malicious traffic was much higher than the external traffic (345 connections versus 191). As expected, just considering the number of connections was not sufficient for providing a detailed comparison between internal and external malicious traffic.

6.2 Number of Connections per Port

Since the type of malicious activity is linked to the port that is targeted, we focused on the ports targeted most frequently to compare internal and external malicious traffic. The set of ports consists of 21 (ftp), 22 (ssh), 23 (telnet), 80 (www-http), 135 (epmap), 139 (netbios-ssn), and 445 (microsoft-ds). This set of ports was targeted on an average by 87.2% of the internal traffic and 61.4% of the external traffic. Table 4 presents the number of connections associated with internal and external malicious traffic. For most ports, the number of internal connections differed from the number of external ones. To better compare internal and external traffic, the percentage of the number of connections targeting the selected set of ports is also shown in Table 4. For each port, the percentages between internal and external connections varied significantly. These observations are further confirmed when calculating the correlation coefficients between the number of connections per port. The correlation between internal and external malicious traffic was low for ports 21 and 445, and no correlation was observed for ports 22, 23, 80, 135, and 139. Port 445, which was targeted 29% (internal) and 55% (external) of the time, had a correlation coefficient of −0.2. Port 135, targeted 12% (internal) and 25% (external)

TABLE 4 Total Number and Average of the Percentage of Connections per Port and Correlation Coefficients of the Number of Connections per Port

Port	21	22	23	80	135	139	445
Internal	46 (0.6%)	47 (0.6%)	14 (0.2%)	201 (3.2%)	618 (11.7%)	371 (6.3%)	2716 (54.6%)
External	33 (0.4%)	20 (0.3%)	36 (0.8%)	130 (1.9%)	256 (24.7%)	178 (4.0%)	810 (29.3%)
Correlation coefficient	−0.32	−0.11	−0.09	−0.11	−0.14	0.07	−0.2

of the time, had an even lower correlation coefficient of −0.14. These results confirm that internal and external malicious traffic are weakly correlated (i.e. since no correlation coefficient exceeded 0.4, the malicious traffic has a low or no correlation).

6.3 Correlation 3: Correlation Across Ports

To further analyze the malicious traffic per port, we calculated the correlation coefficients between the various ports for the total, internal, and external malicious traffic. Tables 5, 6, and 7, respectively, provide the correlation coefficients for the total, internal, and external malicious traffic.

Table 5 contains the correlation coefficients for the total malicious traffic. These coefficients might lead to some inaccurate conclusions. Indeed, the correlation coefficient between ports 21 and 135 was low for the total traffic but was moderate for the internal and external traffic. Moreover, for ports 21 and 445, the total malicious traffic indicated a moderate correlation, but internal traffic indicated a high correlation and external traffic led to no correlation. These examples show that the total malicious traffic is not sufficient for comparing in detail the correlation across ports and that the origin of the malicious traffic should also be considered.

Based on Tables 6 and 7, we observed that the value of the correlation coefficient between two port numbers often differ between internal and external malicious traffic. The most significant differences between the correlation coefficient values were as follows:

- ports 21 and 445: high correlation (internal) and no correlation (external);
- ports 22 and 445: moderate correlation (internal) and no correlation (external);

TABLE 5 Correlation Coefficients for Total Malicious Traffic

Port	21	22	23	80	135	139	445
21	1						
22	0.71 (H)	1					
23	0.65 (M)	0.47 (M)	1				
80	0.83 (H)	0.88 (H)	0.66 (M)	1			
135	0.24 (L)	0.38 (L)	0.03 (N)	0.39 (L)	1		
139	−0.03 (N)	−0.31 (L)	0.03 (N)	−0.05 (N)	0.3 (L)	1	
445	0.41 (M)	0.58 (M)	−0.2 (L)	0.52 (M)	0.59 (M)	−0.11 (N)	1

TABLE 6 Correlation Coefficients for Internal Malicious Traffic

Port	21	22	23	80	135	139	445
21	1						
22	0.74 (H)	1					
23	0.42 (M)	0.27 (L)	1				
80	0.83 (H)	0.92 (VH)	0.4 (M)	1			
135	0.4 (M)	0.28 (L)	0.04 (N)	0.38 (L)	1		
139	−0.03 (N)	−0.13(N)	0.22 (L)	−0.11 (N)	0.51 (M)	1	
445	0.74 (H)	0.7 (M)	0.08 (N)	0.81 (H)	0.59 (M)	−0.07 (N)	1

TABLE 7 Correlation Coefficients for External Malicious Traffic

Port	21	22	23	80	135	139	445
21	1						
22	0.85 (H)	1					
23	0.85 (H)	0.89 (H)	1				
80	0.91 (VH)	0.87 (H)	0.9 (VH)	1			
135	0.41 (M)	0.48 (M)	0.61 (M)	0.59 (M)	1		
139	0.12 (N)	−0.14 (N)	0.24 (L)	0.28 (L)	0.37 (L)	1	
445	0.15 (N)	−0.08 (N)	−0.12 (N)	0.27 (L)	0.06 (N)	0.29 (L)	1

- ports 23 and 80: moderate correlation (internal) and very high correlation (external);
- ports 23 and 135: no correlation (internal) and moderate correlation (external);
- ports 80 and 445: high correlation (internal) and low correlation (external);
- port 135 and 445: moderate correlation (internal) and no correlation (external).

Attacks are linked to ports in different ways. Different attacks can target the same port (i.e. a DoS as well as a race condition attack may be launched against the same FTP server listening on port 21). Moreover, attacks may target more than one port during the attack (i.e. the Sasser worm targets ports 139 and 445). The relationship between attacks and ports show that attacks can be discriminated at a high level using the ports they target. When comparing internal and external malicious traffic, similar correlation coefficient values between ports indicated that internal and external malicious traffic were similar. Based on Tables 6 and 7, we observed that most of the time the correlation coefficients differ between internal and external traffic. This indicates that, even though some attacks might be similar for internal and external malicious activity, the majority differed. In sum, we found that the amount of malicious traffic differs between internal and external malicious activity and that the type of malicious traffic also differs.

7 APPLICATION OF QUANTITATIVE SECURITY ASSESSMENT

This article shows how security quantification helps to characterize the evolution of the attack threat. We briefly mentioned two applications of quantitative security assessment.

7.1 Critical Infrastructure Protection (CIP)

We mentioned the importance of cyber-security attacks against the critical infrastructure in the introduction. The first step in evaluating this attack threat was to determine the threat profile of the critical information infrastructure. The case study described in this article can be replicated to quantify the attack trend and profiles.

7.2 Intrusion Tolerance

Another solution to deal with evolving attack trends is the application of intrusion tolerance. Verissimo and colleagues [21] define intrusion tolerance as the ability of the system to "address the system faults and attacks in a seamless manner through a common approach to security and dependability". The main premise behind this concept is that because of the pervasiveness of the some form of vulnerability and attacks in the system, it might be beneficial to design the system to withstand attacks as opposed to trying to eliminate all vulnerabilities in the system. Refs. 21 and 22 are two examples where intrusion tolerance is provided at the middleware level.

8 DISCUSSION

Since the threat against critical infrastructure has evolved from a pure physical threat to a fusion of physical and cyber-security threat, particular attention needs to be focused on the cyber-security threat. Risk-based approaches are recommended for measuring and alleviating the cyber-security threat. The core of a risk-based approach lies in the ability to form and evaluate security assumptions. This article illustrates how empirical data can be utilized to quantify such assumptions. In particular, we assessed the assumption that port scans were precursors of an attack. We showed, based on malicious traffic, that the majority of the attacks were not linked to a scan. Another study focused on the assumption that internal and external malicious traffic are similar. We showed that using the number of connections is not sufficient for making a detailed comparison between internal and external malicious traffic. However, when refining the analysis of malicious traffic per port targeted, we showed that through the calculation of the correlation coefficients, internal and external malicious traffic often differs significantly.

Since the cyber-security threat varies greatly over time and external malicious traffic differs from internal traffic, a bias in the risk assessment could come from the use of: (i) expert opinions or (ii) malicious traffic collected on the Internet. This article makes the case for collecting continuous in-house data to accurately assess the cyber-security threat to avoid biasing the overall risk assessment study.

REFERENCES

1. *President Clinton (1996). Executive Order EO 13010 Critical Infrastructure Protection*, http://www.fas.org/irp/offdocs/eo13010.htm.
2. Department of Justice (1998). *The Clinton Administration's Policy on Critical Infrastructure Protection: Presidential Decision Directive 63*, http://www.usdoj.gov/criminal/cybercrime/white_pr.htm.

3. Department of Justice (1998). *Critical Infrastructure Protection*, http://www.cybercrime.gov/critinfr.htm.
4. Department of Justice (1998). Computer Crime and Intellectual Property Section (CCIPS). Presidential Decision Directive 63-protecting the nation's critical infrastructure. *Critical Infrastructure Protection*, Department of Justice, http://www.usdoj.gov/criminal/cybercrime/critinfr.htm#Vb.
5. U.S. General Accounts Office (2003). GAO report number GAO-03-173, Critical Infrastructure Protection, http://www.gao.gov/htext/d03173.html.
6. *President Bush. Executive Order Amendment of Executive Orders, and Other Actions, in Connection with the Transfer of Certain Functions to the Secretary of Homeland Security*. (2003). The White House President George W. Bush, http://www.whitehouse.gov/news/releases/2003/02/print/20030228-8.html.
7. The White House. *Using 21st Century Technology to Defend the Homeland* http://www.whitehouse.gov/homeland/21st-technology.html.
8. Office of Management and Budget. *Circular No. A-130 Revised. Office of Management and Budget*. http://www.whitehouse.gov/omb/circulars/a130/a130trans4.html.
9. U.S. General Accounts Office (1999). GAO/AIMD-00 http://www.gao.gov/special.pubs/ai00033.pdf.
10. *Lew, J. J. (1999). Memorandum For The Heads Of Departments And Agencies* http://www.whitehouse.gov/omb/memoranda/m99-20.html.
11. Bowen, P., Hash, J., and Wilson, M., (2006). *NIST Special Publication 800-100. Information Security Handbook: A Guide for Managers. INFORMATION SECURITY.* csrc.nist.gov/publications/nistpubs/800-100/SP800-100-Mar07-2007.pdf.
12. *National Institute of Standards and Technology (NIST). Special Pub 800-12 –An Introduction to Computer Security: The NIST Handbook*. http://csrc.nist.gov/publications/nistpubs/800-12/.
13. *Federal Information Security Management Act of 2002 (Title III of E-Gov)* (2002). http://csrc.nist.gov/drivers/documents/FISMA-final.pdf.
14. The Honeynet Project (2002). *Know Your Enemy*, Addison-Wesley.
15. Panjwani, S., Tan, S., Jarrin, K., and Cukier, M. (2005). An experimental evaluation to determine if port scans are precursors to an attack. *Proceedings of International Conference on Dependable Systems and Networks (DSN-2005)* Yokohama, Japan. 602–611.
16. http://sourceforge.net/projects/nettime.
17. Socolofsky, T. and Kale, C. A (1991). *TCP/IP Tutorial, RFC 1180*, http://www.ietf.org/rfc/rfc1180.txt.
18. http://www.insecure.org/nmap/.
19. http://www.tenablesecurity.com/products/newt.shtml.
20. Guilford, J. P. (1965). *Fundamental Statistics in Psychology and Education*, 4th ed., McGraw-Hill, New York.
21. Verissimo, P., Neves, N. F., Cachin, C., Poritz, J., Powell, D., Deswarte, Y., Stroud, R., and Welch, I. (2006). Intrusion-tolerant middleware: The road to automatic security. *IEEE Secur. Priv*. **4**(4), 54–62.Jul./Aug.
22. Courtney, T., Lyons, J., Ramasamy, H. V., Sanders, W. H., Seri, M., Atighetchi, M., Rubel, P., Jones, C., Webber, F., Pal, P., Watro, R., Cukier, M. and Gossett J. (2002). Providing Intrusion Tolerance with ITUA, *Supplemental Volume of the 2002, International Conference on Dependable Systems & Networks (DSN-2002)*, Washington, DC, June 23–26, 2002, pp. C-5-1–C-5-3.

DEFEATING SURPRISE THROUGH THREAT ANTICIPATION AND POSSIBILITY MANAGEMENT

WILLIAM L. MCGILL

College of Information Sciences and Technology, Pennsylvania State University, University Park, Pennsylvania

BILAL M. AYYUB

Center for Technology and Systems Management, Department of Civil and Environmental Engineering, University of Maryland, College Park, Maryland

1 RISK ANALYSIS IN AN OPEN WORLD

Expectation is imagination constrained by bounded uncertainty [1]. Rational decisions are based on expectations, and expectations are largely influenced by the bounds a decision maker's imagination places on possible outcomes. When these bounds on expectation are prescribed incorrectly, such as through the illusion of knowledge manifesting from overconfidence, blind sightedness, or faulty reasoning, a failure of imagination could result, which in some contexts could prove harmful to the decision maker. For any decision problem, it is thus important to clearly articulate what is known, what is thought to be known, what is not known, and acknowledge the possibility of unknown unknowns. This last type of uncertainty has been referred to as *ontological uncertainty* [2].

Risk analysis is a tool that informs the decision making process by providing answers to the following three questions for a given future situation (a.k.a. the risk triplet) [3]:

1. What can go wrong?
2. What are the consequences of concern?
3. What is the likeliness of these consequences, all things considered?

In the context of critical infrastructure protection, the first question identifies a set of *plausible initiating events* and *attack profiles*, where a threat scenario an initiating event is the pairing of a threat type with a specific target and an attack profile describes the manner in which the event will occur (e.g. combination of delivery system and intrusion path) [4]. As is mentioned later, failure to identify and consider single plausible initiating event or attack profile increases a defender's vulnerability to surprise. Answers to the latter two questions attempt to make statements about likeliness of each initiating event and the ensuing consequences. Collectively, the triplet of initiating event, consequence dimensions of concern, and likeliness of consequences (to include both likeliness of event and likeliness of consequences given event) define risk [5]. That is, risk is a *multi-dimensional concept*. In practice, probabilistic risk analysis is used to make quantitative statements about risk though quantification is not always necessary to understand risk.

In all decision problems, one must accept either closed-world or open-world assumptions [6]. In the context of risk analysis, *closed-world* thinking assumes a complete set of clearly defined failure events, where any evidence that supports the likeliness of one necessarily supports the unlikeliness of others [7]. For example, if information suggests that an event "A" will occur with a probability p, then closed-world thinking insists that "not A" will occur with a probability $1-p$. Moreover, the closed-world assumption requires that all events contained within the set "not A" possess a clear definition: everything that can possibly happen must be articulated. The *open-world assumption* eases this exhaustiveness requirement, and permits reasoning about the future while accepting the possibility of unanticipated events. However, under open-world thinking, knowledge that supports "A" cannot offer any support to "not A" since not all elements in this set possess a clear definition. That is, the existence of unknown or residual events within a set of possibilities renders statements about likeliness doubtful. To deal with this problem, researchers have proposed maintaining an open mind while entertaining beliefs, and assuming a closed world only for the purposes of decision making by conditioning the set to include only those events that can be articulated [8].

Although the basic risk analysis philosophy is valid for all types of decision making, the mathematical tools available for calculating risk are largely tuned to problems where the probabilities of alternative scenarios and their outcomes can, in principle, be obtained. Notwithstanding the inherent difficulties in assessing these probabilities, extending probabilistic tools to risk analysis when the full spectrum of possibilities is unknown has been met more often with frustration than success. Key examples of this can be found in the domain of security [9], safety [10], international politics [11], high technology [12], project management [13], wastewater treatment [14], civil engineering [15], reliability engineering [16], and just about any other situation where human action plays a key role [1]. Under open-world assumptions, one accepts the possibility of unknown events, which by their very nature lack definition [17]. Without a clear definition of what the residual event is, it is not possible to assign a probability either to the residual event or to a defined event since information cannot offer any weight, directly or indirectly, to an event that lacks definition.

At this point, the distinction must be made between *probable events* and *possible events*. *Probability* is a quantitative measure of likeliness of occurrence of a future event relative to all other events in the same set of possibilities. *Possibility* indicates the degree of membership or belongingness of an event to the set of alternatives, and as a measure provides an upper limit of the actual probability. The connection between probability and possibility is as follows: perfectly possible events can have a probability as high as one, whereas impossible events necessarily have a probability of zero. This notion of possibility is markedly different than probability in that it bounds the potential probability of an event, yet provides no indication of the actual event probability; all it says is that the event is possible. Thus, whereas it is meaningless to assign a probability to events under open-world assumptions, it is perfectly acceptable to construct a possibility distribution over these events.

2 SURPRISE EXPLOITS IGNORANCE

A particularly menacing problem that exploits closed-world thinking is that of surprise. Surprise manifests itself in the unknown, unrecognized, and unrealized, and is a direct by-product of a failure of imagination. According to Grabo [18], surprise occurs when a defender is either unaware of potential hazards or unprepared to defend or respond to unexpected consequences from known, but ignored hazards (i.e., the *counterexpected* event). Each aspect of the prototypical security risk formula (i.e. *risk* = *threat* × *vulnerability* × *consequence* [19]) has elements that contribute to surprise, such as is shown in Figure 1. An event may be unexpected, the probability of its occurrence may be understated, or the resulting consequences may be unanticipated. In general, *surprise exploits defender ignorance*. Figure 2 shows the various types of ignorance, all of which contribute to a defender's vulnerability to surprise. Defeating surprise rests in a decision maker's awareness of possible scenarios and their outcomes, as well as in his preparedness to mitigate a full range of consequences following such events.

Examples of surprise can be found in many areas related to homeland security. In the counterterrorism context, adversaries seek to leverage defender ignorance about adversary intent and capabilities to achieve an asymmetric advantage over their targets. For instance,

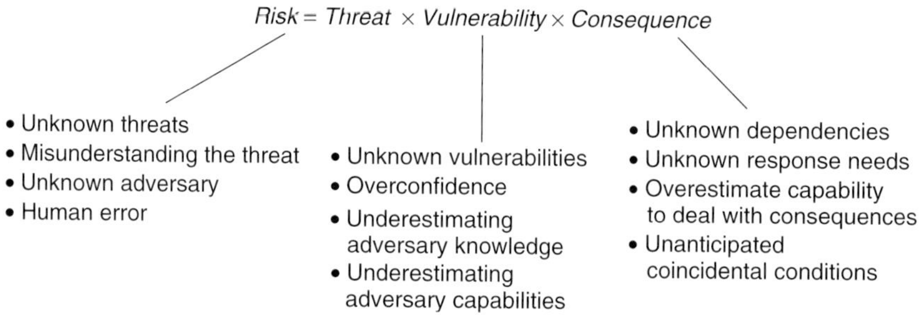

FIGURE 1 Some sources of surprise in risk analysis for critical infrastructure protection.

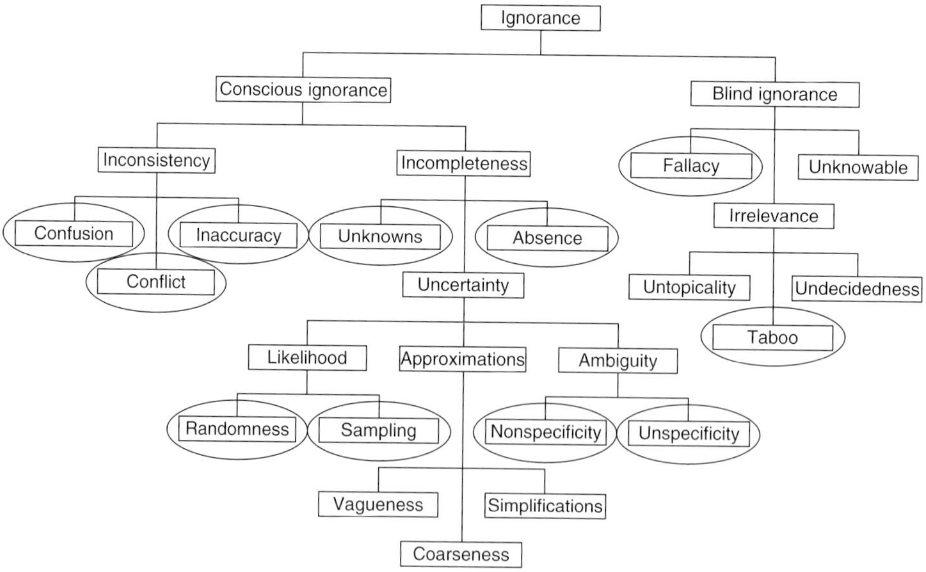

FIGURE 2 Hierarchy of ignorance types highlighting those types that are particularly susceptible to surprise.

the use of airplanes to attack the World Trade Center and Pentagon on 9/11 was arguably a surprise given that a majority of defenders were unaware that such vehicles would be used as projectiles to attack buildings. Manunta [9, 20] highlights the propensity of adversaries to seek opportunities for achieving surprise, and argues that this behavior renders the security problem incompatible with Bayesian probabilities. In the international affairs arena, Cadell [21] notes that potential adversaries engage in deception to deliberately mislead or confuse their opponents to prevent them from learning the deceiver's true intentions or activities. In this sense, deception thrives by manipulating uncertainty to affect a target, and surprise occurs when the actions of the deceived leads to them experiencing unintended and potentially undesirable events and outcomes. For natural-occurring events, Woo [22] notes that "there are many arcane geological hazard phenomena which are beyond the testimony of the living, which would be met with incredulity and awe were they to recur in our own time". Such "black swans" are highly consequential scenarios that are either unknown or have a perceived probability so low as to be considered negligible, yet would result in significant surprise were they to occur [23]. In highly complex technical systems, Johnson [24] suggests that surprise occurs due to unexpected emergent behaviors stemming from the interaction between system components and their environment. Critical infrastructure is among such highly complex technical systems, where unknown interdependencies between infrastructure services may lead to unpredictable cascading consequences [25].

3 IGNORANCE CONTRIBUTES TO VULNERABILITY

Surprise is felt when an event our outcome occurs that was not expected, and can vary in degree according to how far out the realm of possible it was perceived prior to its occurrence [1]. Each possible outcome for a given situation carries with it a degree of *potential surprise* that describes the intensity of this feeling: those outcomes that are deemed possible carry a zero degree of potential surprise, whereas outcomes deemed impossible maximize this measure. Surprise in this sense is experienced by a decision maker whose imagination should ideally reflect the constraints imposed by collective knowledge of his organization. Several authors link these ideas to that of possibility theory [26–28]. Coincidentally, many opponents of probabilistic risk analysis for security suggest that possibility theory provides the mathematical tools needed to support risk and decision analysis, and have made attempts to apply possibilistic techniques to risk analysis problems. Recent work by Karimi and Hüllermeier [29] and Baudrit et al. [30] suggests that progress is being made to propagate both probabilistic and possibilistic information within a quantitative risk analysis framework.

A surprising event is one that is outside the realm of our expectations, and often arises from an inaccurate or insufficient handling of uncertainty. When a situation or decision problem is novel or unique, the challenge is to use the knowledge one has available to set bounds on the scope of imagined future outcomes [31]; more knowledge reduces epistemic uncertainty, whereas lack of knowledge must entertain a wider range of possibilities. According to this point of view, scenarios describing what can happen are considered regardless of their perceived likeliness. Furthermore, the focus is on the outcomes of a scenario, and the goal is to mitigate or constrain the range of possibilities. One example applying this line of thinking has been described in [4] when comparing asset-driven and threat-driven approaches to terrorism risk analysis; the authors asserted that a complete set of threat scenarios that span a complete spectrum of possibilities can be identified based on the physical characteristics of the potential target and without consideration of adversary intent or probability of attack. Thus, only events that are not physically possible are initially ruled out. Unless supported by disconfirming evidence that judge them impossible or consequentially insignificant, all scenarios are considered throughout the analysis [32].

4 DEFEATING SURPRISE THROUGH AWARENESS AND PREPAREDNESS

The key to defeating surprise is awareness and preparedness. Awareness is achieved by acknowledging the possibility of alternative threatening events and preparedness is achieved by taking steps to mitigate the range and scope of potential outcomes that can be realized independent of threat type. That is, awareness decreases a decision maker's vulnerability to surprise from an unexpected event, and preparedness decreases vulnerability to surprise from unanticipated outcomes. This section discusses two approaches for defeating surprise—threat anticipation aimed at increasing awareness of plausible threat scenarios and possibility management aimed at increasing preparedness of unanticipated outcomes following an adverse event.

4.1 Threat Anticipation: Increasing Awareness

The most challenging part of the risk assessment process is identifying an exhaustive set of initiating events. This is especially true in the homeland security context, where adversaries choose from among myriad threat types, targets, and attack profiles to inflict damage, harm, and fear on their targets. To facilitate the scenario identification process, Kaplan et al. [33] developed the theory of scenario structuring (TSS). Beginning with a *success scenario* or *as-planned scenario* that corresponds to nominal performance of a system, TSS seeks to define an exhaustive set of scenarios that negatively deviates from the success state. Initially, this set consists of a single scenario defined generically as the *failure event*. Collectively, the union of the successful scenario with the failure scenario defines a closed universe of possibilities. However, this extreme level of nonspecificity does not facilitate a defensible assessment of event likeliness, but rather only defines the nature of events. As more information is obtained about what can go wrong, the failure event is partitioned into more specific, distinct subsets, where both a clearly defined event (e.g. "*A*") and its complement (e.g. "not *A*") coexist so as to preserve exhaustiveness of the set. The challenge is to partition the set of plausible scenarios in such a way that each partition has a clear definition that facilitates meaningful assessment of probability and consequence, and provides sufficient resolution to support decision making without imposing too much of a cost burden for doing the analysis.

The process of *threat anticipation* seeks to increase awareness by constructing an exhaustive set of plausible initiating events based on the inherent susceptibilities of target elements to a wide range of threat types, independent of demonstrated adversary capabilities and intent. The process for threat anticipation is illustrated in Figure 3 for a notional asset. Moreover, for each identified threat scenario, an exhaustive set of representative attack profiles is constructed based on the compatibility of alternative intrusion paths (i.e. path leading to a target element) with various attack modes, such as a ground vehicle for explosive threats. The outcome of this procedure is a complete list of initiating events and associated attack profiles that provides the basis for follow-on vulnerability

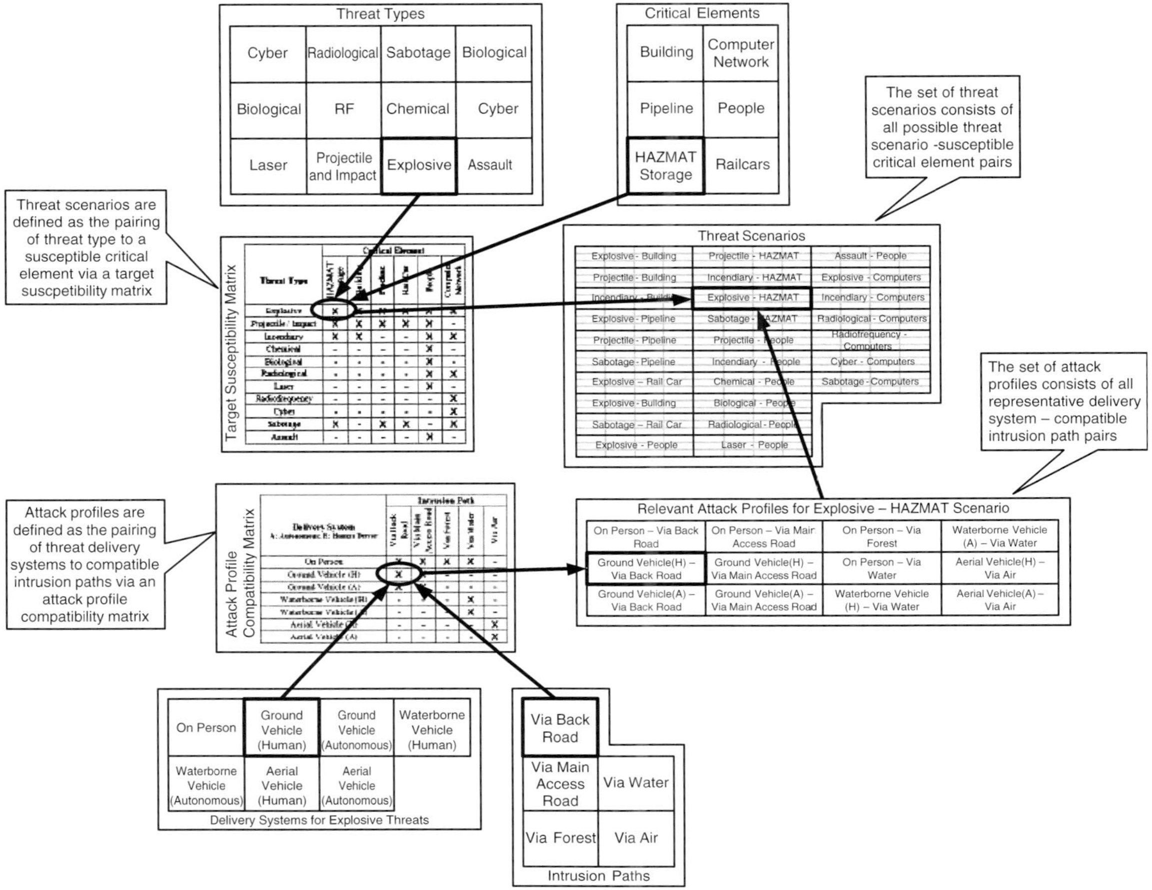

FIGURE 3 Structured approach for threat anticipation.

analysis and risk assessment. More importantly, the mere acknowledgment of possible scenarios, however unlikely they are perceived to be, lessens vulnerability to surprise by bringing them to the decision maker's attention.

4.2 Possibility Management: Improving Preparedness

It is important to note that while an event might come as a surprise, the outcomes following such an event may be manageable. If the capability of an asset or system to deal with the outcomes of a surprising event already exists, then that asset or system has an inherent resilience to surprise due to its ability to recover from known events with similar outcomes. For example, facilities within the oil and gas industry are prepared to deal with explosions due to accidents or system failures, and thus are inherently prepared to mitigate the effects of a malicious explosives attack of similar scale. In contrast, known events may lead to unanticipated outcomes, such as through cascading effects following the collapse of a tree branch on a critical electric power distribution line. In practice, whether an event comes as a surprise is less important than the magnitude of the outcomes following its occurrence, particularly since the range of potential outcomes are what really guide investments in mitigation strategies and preventive measures. In order to defeat surprise, attention should be placed on outcomes that can potentially occur regardless of the initiating event, followed by steps to enhance response and recovery capabilities in light of the possibilities.

This section presents a simple methodology for *possibility management* that seeks to limit the ability of an adversary to achieve surprise through increased knowledge about possible outcomes and improved countermeasures that limit the range of possibilities. A diagram illustrating the steps of the methodology is given in Figure 4. The focus of this methodology is on assessing the possibility of *outcomes* as opposed to the possibility of *events*. Through this approach, a *possibility distribution* over a specified impact measure

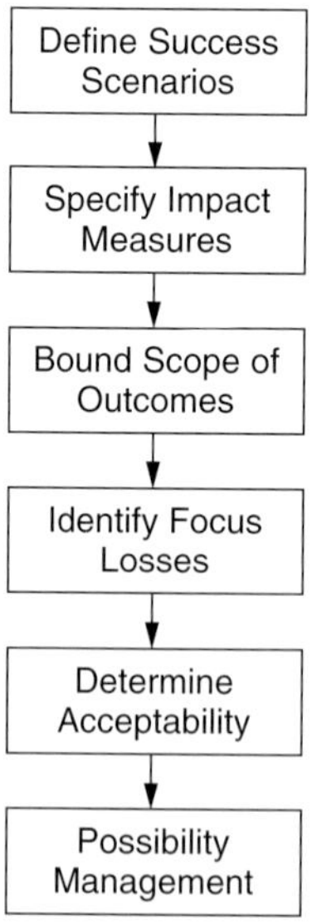

FIGURE 4 Methodology for possibility management.

such as economic loss or recuperation time is constructed by leveraging knowledge of what *cannot* happen so as to constrain the scope of imagined outcomes. Within this range of possibilities, a focus loss can be identified for the purposes of determining whether such a loss can be tolerated. If the focus loss is too great, a precautionary approach [34] can be taken that seeks to decrease or eliminate its possibility. This methodology is broadly applicable to all levels of critical infrastructure and key resource protection, from protection of a single asset to an entire geographical region or infrastructure sector, and can be looked at from an all-hazards perspective.

Step 1: Define success scenarios. This step defines the success scenarios of an asset or system, where the word "success" indicates that the asset or system is functioning as intended [33]. For example, a success scenario of a nuclear power plant might be to provide a specific amount of energy to the energy grid, and a success scenario of the finance and banking infrastructure might be to facilitate the reliable execution of financial transactions between two or more parties. As a suggested rule of thumb, success scenarios for assets are mission focused, and success scenarios for systems are capability focused.

Step 2: Specify impact measures. This step identifies suitable impact measures that quantify the effects of a deviation from each success scenario following the occurrence of a disruptive event. For example, disrupted energy production at a nuclear power plant might be measured in terms of lost revenue or recuperation time. Similarly, the impact of a core damaging event at the same plant might be measured in terms of cost to repair or number of persons exposed to harmful radiation. The appropriate measures and bounds of the ensuing impacts are chosen according to the needs of each individual decision maker.

Step 3: Bound the scope of possible outcomes. This step bounds the scope of possible outcomes by using available knowledge and information to rule out impossible outcomes or discount otherwise perfectly possible outcomes. In particular, this step identifies three points for each impact measure as shown in Figure 5. The first point defines the *best-case scenario* (*BCS*), which by default corresponds to zero consequence or impact. The second point defines the limiting *perfectly possible*

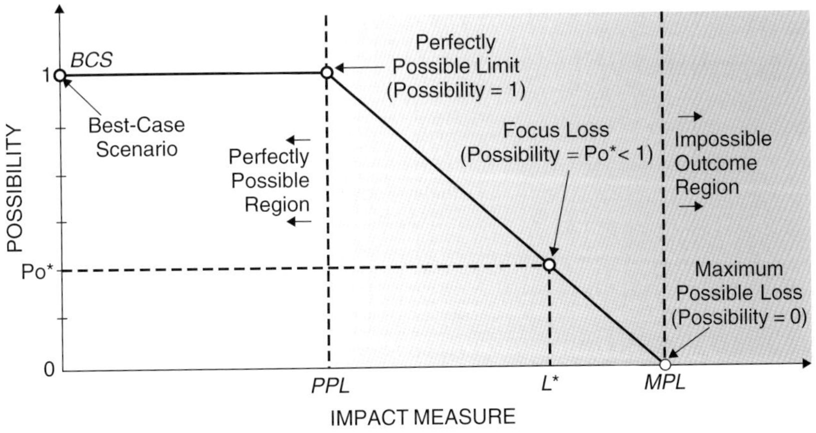

FIGURE 5 Possibility distribution for the outcomes following a disruptive event.

loss (*PPL*) or the point along the impact axis that bounds the set of outcomes that are perfectly possible. The third point defines the *impossibility limit* or *maximum possible loss* (*MPL*) or the point along the impact axis that separates the possible from the impossible. By convention, outcomes deemed perfectly possible carry a possibility of one, whereas outcomes deemed impossible carry a possibility of zero. The possibility of an outcome is a nonincreasing function of impact; thus, the magnitude of *PPL* is always less than or equal to the *MPL*. Under total ignorance, *MPL* and *PPL* coincide, and remain so until knowledge becomes available to judge various outcomes as less than perfectly possible. Given values for *BCS*, *PPL*, and *MPL*, linear piecewise-continuous possibility distribution can be constructed by drawing two line segments: a horizontal line connecting the *BCS* to the *PPL* and another line of decreasing slope connecting *PPL* to *MPL*. Such a possibility distribution is illustrated in Figure 5.

Step 4: Identify focus losses. This step identifies a *focus loss* (L^*) for the purposes of determining whether the current state of the asset or system is tolerable to the decision makers and for constructing scenarios for red teaming or disaster response exercises with the potential to yield this degree of loss. L^* is a single-valued point along the impact axis bounded by *PPL* and *MPL*, that is, $PPL < L^* < MPL$ (Figure 5). For scenario development, Ha-Duong [35] suggests choosing a value for L^* that coincides with a possibility level of about 1/3; however, the exact choice of possibility level is at the discretion of the decision makers and emergency response planners.

Step 5: Determine acceptability of focus loss. Given the focus loss obtained in Step 4, this step assesses whether such a loss is acceptable, tolerable, or manageable. An asset or system can be considered *resilient* if decision maker is prepared to deal with focus loss in such a way that will quickly resume the success scenario irrespective of the exact nature of the disruptive event.

Step 6: Possibility management. This step explores the impact of various proposals to enhance mitigation effectiveness and increase resilience by improving the response and recovery capabilities of the affected asset or system. If the focus loss is unacceptable, precautionary measures such as enhanced response and recovery capabilities may be considered that seek to reduce the focus loss. It is important to note that mathematical possibility theory does not support benefit–cost analysis in a strict sense; however, research has shown that possibility distributions can be transformed into probability distributions [36], and as such statements on the probability of realizing some degree of benefit can be made [5].

In addition, the focus loss can be used to construct scenarios for red team and disaster recovery exercises and emergency response planning [35]. There exist an infinite number of scenarios that could potentially result in the focus loss, and such scenarios can be constructed using methods such as the "lego-block" approach described in [37].

5 FUTURE RESEARCH DIRECTIONS

The discussion and proposed approaches for defeating surprise described in this article are aimed at helping homeland security decision makers cope with an open world by increasing awareness of plausible threat scenarios and attack profiles through threat anticipation and improving preparedness through possibility management. In the technical sense, threat anticipation supports the risk analysis process by developing an exhaustive set of scenarios, and possibility management complements this activity by taking precautionary measures to mitigate loss independent of its cause. Collectively, these techniques serve to decrease a decision maker's overall vulnerability to surprise. Further research along the lines of Pugsley [38] should be pursued to identify a set of factors that contribute to an increased susceptibility to surprise through lack of awareness and preparedness so as to provide guidance to decision makers and organizations on how to improve their defenses against surprise attacks.

REFERENCES

1. Shackle, G. L. S. (1970). *Decision, Order, and Time in Human Affairs*, 2nd ed., Cambridge University Press, Cambridge.
2. Elms, D. G. (2004). Structural safety—issues and progress. *Prog. Struct. Eng. Mater.* **6**(2), 116–126.
3. Kaplan, S., and Garrick, B. J. (1981). On the quantitative definition of risk. *Risk Anal.* **1**(1), 11–27.
4. McGill, W. L., Ayyub, B. M., and Kaminskiy, M. P. (2007). Risk analysis for critical asset protection. *Risk Anal.* **27**(5), 1265–1281.
5. Ayyub, B. M. (2003). *Risk Analysis in Engineering and Economics*. Chapman & Hall/CRC Press, Boca Raton, FL.
6. Walton, D. N. (1990). What is reasoning? what is argument? *J. Philos.* **87**(8), 399–419.
7. Ayyub, B. M. (2004). From dissecting ignorance to solving algebraic problems. *Reliab. Eng. Syst. Saf.* **85**, 223–238.
8. Smets, P., and Kennes, R. (1994). The transferable belief model. *Artif. Intell.* **66**(2), 191–234.
9. Manunta, G. (1999). Security decisionmaking and PRA methodology: does PRA methodology effectively assist security decisionmakers? *J. Sec. Admin.* **22**(2), 1–9.
10. Reason, J. T. (1998). *Managing the Risks of Organizational Accidents*. Ashgate, Hampshire, England.
11. Handel, M. I. (1990). Surprise and change in international politics. *Int. Secur.* **4**(4), 57–85.
12. Perrow, C. (1999). *Normal Accidents: Living with High-Risk Technologies*, Princeton University Press, Princeton, NJ.
13. Pender, S. (2001). Managing incomplete knowledge: why risk management is not sufficient. *Int. J. Proj. Manage.* **19**, 79–87.
14. Demotiér, S., Schön, W., and Denœux, T. (2006). Risk assessment based on weak information using belief functions: a case study using water treatment. *IEEE Trans. Syst. Man Cybern.* **36**(3), 382–396.
15. Beard, A. N. (2004). Risk assessment assumptions. *Civ. Eng. Environ. Syst.* **21**(1), 19–31.
16. Blockley, D. I. (1989). Open world problems in structural reliability. *Proceeding of 5th International Conference on Structural Safety and Reliability*, pp. 1659–1665. San Francisco, CA, August 7–11 1989.
17. Ayyub, B. M., and Klir, G. J. (2006). *Uncertainty Modeling and Analysis in Engineering and the Sciences*. Chapman & Hall/CRC Press, Boca Raton, FL.
18. Grabo, C. M. (2002). *Anticipating Surprise: Analysis for Strategic Warning*. Joint Military Intelligence College, Washington, DC.

19. Broder, J. F. (1984). *Risk Analysis and the Security Survey*, Butterworth Publishers, Boston, MA.

20. Manunta, G. (2002). Risk and security: are they compatible concepts? *Secur. J.* **15**(3), 43–55.

21. Caddell, J. W. (2004). *Deception 101–Primer on Deception*. United States Army War College, Carlisle Barracks, PA.

22. Woo, G. (1999). *The Mathematics of Natural Catastrophes*. Imperial College Press, London.

23. Taleb, N. N. (2007). *The Black Swan: The Impact of the Highly Improbable*. Random House, Allen Lane, UK.

24. Johnson, C. W. (2006). What are emergent properties and how do they affect the engineering of complex systems? *Reliab. Eng. Syst. Saf.* **91**(12), 1475–1481.

25. Rinaldi, S. M., Peerenboom, J. P., and Kelly, T. K. (2001). Complex networks: identifying, understanding, and analyzing critical infrastructure interdependencies. *IEEE Control Syst. Mag.* **21**, 11–25.

26. Klir, G. J. (2002). Uncertainty in economics: the heritage of G.L.S. Shackle. *Fuzzy Econ. Rev.* **VII**(2), 3–21.

27. Prade, H., and Yager, R. R. (1994). Estimations of expectedness and potential surprise in possibility theory. *J. Uncertain. Fuzz. Knowl. Based Syst.* **2**(4), 417–428.

28. Fioretti, G. (2001). A mathematical theory of evidence for G.L.S. Shackle. *Mind Soc.* **2**(3), 77–98.

29. Karimi, I., and Hüllermeier, E. (2007). Risk assessment system of natural hazards: a new approach based on Fuzzy probability. *Fuzzy Sets Syst.* **158**(9), 987–999.

30. Baudrit, C., Couso, I., and Dubois, D. (2007). Joint propagation of probability and possibility in risk analysis: towards a formal framework. *Int. J. Approx. Reas.* **45**(1), 82–105.

31. Chapman, A. (2006). Regulating chemicals–from risks to riskiness. *Risk Anal.* **26**(3), 603–616.

32. Heuer, R. J. (1999). *Psychology of Intelligence Analysis*, Center for the Study of Intelligence, Washington, DC.

33. Kaplan, S., Visnepolschi, S., Zlotin, B., and Zusman, A. (2005). *New Tools for Failure and Risk Analysis: Anticipatory Failure Determination (AFD) and the Theory of Scenario Structuring*. Ideation International, Farmington Hills, MI.

34. Cameron, E., and Peloso, G. F. (2005). Risk management and the precautionary principle: a Fuzzy logic model. *Risk Anal.* **25**(4), 901–911.

35. Ha-Duong, M. (2006). *Scenarios, Probability, and Possible Futures*. Working Paper. URL: http://www.centre-cired.fr/perso/haduong/files/HaDuong-2006-ScenariosProbabilityPossibleFutures.pdf.

36. Dubois, D., Foulloy, L., Mauris, G., and Prade, H. (2004). Probability-possibility transformations, triangular Fuzzy sets, and probabilistic inequalities. *Rel. Comput.* **10**, 273–297.

37. Wideburg, J. (1989). Operational threat assessments for civil defence planning. *Eur. J. Oper. Res.* **43**, 342–349.

38. Pugsley, A. G. (1973). The prediction of proneness to structural accidents. *Struct. Eng.* **51**(6), 195–196.

FURTHER READING

Ayyub, B. M., Klir, G. J. (2006). *Uncertainty Modeling and Analysis in Engineering and the Sciences*. Chapman & Hall/CRC, Boca Raton, FL.

Fioretti, G. (2004). Evidence theory: a mathematical framework for unpredictable hypotheses. *Metroeconomica* **55**(4), 345–366.

Kam, E. (2004). *Surprise Attack: The Victim's Perspective*, 2nd ed., Harvard University Press, Boston, MA.

National Commission on Terrorist Attacks. (2004). *The 9/11 Commission Report: The Final Report of the National Commission on Terrorist Attacks Upon the United States*, Authorized Edition. Norton W. W. & Company, New York, NY.

Shackle, G. L. S. (1979). *Imagination and the Nature of Choice*. Edinburgh University Press, Edinburgh.

MEMETICS FOR THREAT REDUCTION IN RISK MANAGEMENT

ROBERT FINKELSTEIN

Robotic Technology Inc., University of Maryland University College, Adelphi, Maryland

BILAL M. AYYUB

Center for Technology and Systems Management, Department of Civil and Environmental Engineering, University of Maryland, College Park, Maryland

1 PREMISE

Risk, defined as a function of the probability of an event occurring and the consequences given that the event occurs, can be controlled and managed by either reducing the adverse consequences of an event, given that it occurs, or reducing the probability of the event occurring [1, 2]. Memetics can influence both risk components.

Memetics promises to reduce homeland security risks that are a result of human-caused threats by reducing the number of adversaries and thus the probability of an act; increasing the awareness of the risk to the public at large and those responsible for the targeted infrastructure, thus reducing the probability of a successful attack or mitigating its consequences; and enhancing the training of first responders, thus mitigating the consequences of a terrorist act.

2 THE MEME

The word "meme" is a neologism coined by Richard Dawkins [3] in The Selfish Gene (1976), (although it may have had earlier roots) and defined as a self-reproducing and propagating information structure analogous to a gene in biology. Dawkins focused on the meme as a replicator, analogous to the gene, able to affect human evolution through the evolutionary algorithm of variation, replication, and differential fitness. But for military or homeland security applications, the relevant characteristics of the meme are that it consists of information which persists, propagates, and influences human behavior.

2.1 Meme Definitions

There are a plethora of definitions for the meme, with most being variations of Dawkins' original notion of a unit (whatever that means) of cultural transmission, where culture may be defined as the total pattern of behavior (and its products) of a population of agents, embodied in thought, behavior, and artifacts, and dependent upon the capacity for learning and transmitting knowledge to succeeding generations. While none of these definitions is sufficient to allow a meme to be clearly recognized or measured (which is now the focus of research), they do provide an initial grasp of the concept. A few of the many definitions extracted from the literature include [4, 5] the following:

- A self-reproducing and propagating information structure analogous to a gene in biology.
- A unit of cultural transmission (or a unit of imitation) that is a replicator that propagates in the meme pool leaping from brain to brain via (in a broad sense) imitation; examples: "tunes, ideas, catch-phrases, clothes fashions, ways of making pots or of building arches".
- Ideas that program for their own retransmission or propagation.
- Actively contagious ideas or thoughts.
- Shared elements of a culture learned through imitation from others—with culture being defined rather broadly to include ideas, behaviors, and physical objects.
- An element of a culture that may be considered to be passed on by nongenetic means, especially imitation.
- Information patterns infecting human minds.
- While the internal meme is equivalent to the genotype, its expression in behavior (or the way it affects things in its environment) is its phenotype.
- Any information that is copied from person to person or between books, computers, or other storage devices. Most mental contents are not memes because they are not acquired by imitation or copying, including perceptions, visual memories, and emotional feelings. Skills or knowledge acquired by ordinary learning are not memes.
- A (cognitive) information structure able to replicate using human hosts and to influence their behavior to promote replication.
- Cultural information units that are the smallest elements that replicate themselves with reliability and fecundity.
- A rule of behavior, encoded by functional neuronal groups or pathways. (Behavior is action, whether mental or physical. Ideas such as tying shoe-laces or opening a door represent rules of physical action, that is, rules of patterned neural–muscular interaction. Concepts such as apple, seven, or causality, represent rules of mental action, or rules of cognition, that is, rules of patterned neural–neural interaction. Hence, physical movement is governed by memes which represent rules of physical action and thought is governed by memes which represent rules of mental action).

- Any kind, amount, and configuration of information in culture that shows both variation and coherent transmission.
- A pattern of information (a state within a space of possible states).
- A unit of cultural information as it is represented in the brain.
- An observable cultural phenomenon, such as a behavior, artifact, or an objective piece of information, which is copied, imitated, or learned, and thus may replicate within a cultural system. Objective information includes instructions, norms, rules, institutions, and social practices provided they are observable.
- A pattern of information, one that happens to have evolved a form which induces people to repeat that pattern.
- A contagious information pattern that replicates by parasitically infecting human minds and altering their behavior, causing them to propagate the pattern. Individual slogans, catch-phrases, melodies, icons, inventions, and fashions are typical memes. An idea or information pattern is not a meme until it causes someone to replicate it, to repeat it to someone else. All transmitted knowledge is memetic.
- The smallest idea that can copy itself while remaining self-contained and intact—essentially sets of instructions that can be followed to produce behavior.

Our initial effort to provide a pragmatic, functionally useful description of the meme led to the following definition:

A meme is information transmitted by any number of sources to at least an order of magnitude more recipients than sources, and propagated during at least twelve hours.

To distinguish a meme from other sorts of information (e.g. from casual, common daily utterances), we invoke an order of magnitude rule and place an emphasis on the necessity of a threshold for propagation and persistence.

We use Claude Shannon's definition of information as that which reduces uncertainty, as manifested, for example, in the difference between two states of uncertainty before and after a message has been received. In Shannon's formulation, a message carries information inasmuch as it conveys something not already known. Thus, there is a subjective element in that the same message may or may not reduce uncertainty, or can have a different influence or impact, depending on the states of the recipients; a meme, as a subset of information, could have different consequences for different recipients. Memes may be characterized or rated as a function of their ability to propagate (among a few or millions of people) or persist (over days or centuries).

Memes, continuing with Shannon's formulation of information, can also be characterized using entropy as a measure of informational order and disorder, where the entropy of a meme is a function of its size (e.g. number of bits or words). The usefulness of this approach remains to be determined. Additional information on Shannon entropy is provided by Ayyub and Klir (2006) [6].

As a practical approach, we defined metrics and submetrics for evaluating memes, including propagation, persistence, impact, and entropy. The submetrics for the *propagation* metric include the number, type, and dispersion of recipients of the meme. Depending on the problem under consideration, the type of recipients might be characterized or categorized by their economic, social, or educational class; ethnicity or culture; religion; gender; age; tribe; politics; and so on, while the dispersion of recipients might categorized as local, tribal, familial, regional, national, global, and so on. The submetrics for the *persistence* metric distinguish between the duration of transmission of the meme and the duration of the meme in memory or storage. The submetrics for the *entropy* metric distinguish among small, medium, and large memes, which (using an order of magnitude rule) are characterized as less than or equal to 100-K bits, less than or equal to 100-M bits, and greater than 100-M bits. Submetrics for the *impact* metric distinguish between the impact (or potential impact) of the meme on the individual (individual consequence) and its impact (or potential impact) on society as a whole (societal consequence).

2.2 Meme Transmission and Reception

A meme is transmitted after being created in the mind of an individual or retransmitted after being received by an individual from elsewhere. Arriving at a new potential host, the meme is received and decoded. The potential host becomes an actual host if the meme satisfies certain selection and fitness criteria. The new host replicates and transmits the meme (perhaps with a different vector, such as a text message instead of speech). Because the number of memes at any given time exceeds the number of recipients able to absorb them, fitness criteria determine which memes will survive, propagate, persist, and have impact. The selection and fitness criteria include human motivators such as fear (e.g. of going to hell or failing in business) and reward (e.g. of going to heaven or succeeding in business). Alternatively, the meme might be beneficial in a practical way (such as instructions on how to make a hard-boiled egg or an improvised explosive device); entertaining to the recipient, such as a joke ("Why did the terrorist cross the road?") or a song ("Bomb bomb bomb, bomb bomb Iran," as sung by Senator John McCain, as featured on YouTube.com); or consist of a direct appreciative feedback to the recipient (such as providing emotional satisfaction, for example, reinforcement and pride in membership in a nation, tribe, religion, ethnic group, or ideology).

To be readily acceptable to the host, the meme should fit existing constructs or belief systems of the host, or be a paradigm to which the host is receptive. Memes also aggregate and reinforce in complexes (memeplexes) so that a suitable existing framework in the mind of the host is especially susceptible to a new meme which fits the framework (such as a new precept by a religious leader that would be absorbed by a follower of that religion, whereas it would be ignored or escape notice by a nonfollower). Suitable storage capacity, in memory or media, is necessary for the meme to persist, along with enduring vectors (e.g. the meme is literally chiseled in stone or reproduced in many, widely distributed copies of books or electronic media).

New research projects could provide a scientific and quantitative basis for memetics and an exploration of its prospective applicability and value, possibly discovering whether brief memes such as "Death to America" or "Glory to the Martyrs" or "Winston tastes good like a cigarette should" are, in fact, cognitively and functionally different from nonmemes such as "I like your hair," or "Please pass the salt".

2.3 Quantitative Bases

Since Dawkins' revelation about memes, the concept has attracted a coterie of proponents, skeptics, and opponents. In 30 years there has been no significant research of the concept to establish a scientific basis for it—but neither has there been a definitive refutation. To progress as a discipline with useful applications, memetics needs a general theory—a theoretical foundation for the development of a scientific discipline of memetics. It needs a narrowly focused, pragmatically useful definition, and, ultimately, the ability to make testable predictions and falsifiable hypotheses. The discrete meme must be defined, identified, and distinguished in the near-continuum of information, just as the discrete gene can be identified (more or less) in long string of DNA nucleotides (albeit, with current technology a gene may not be clearly identifiable). A quantitative basis for memes must be established, using, for example, tools such as information theory and entropy; genetic, memetic, and evolutionary algorithms; neuroeconomics tools such as functional magnetic resonance imaging (fMRI) and biochemical analyses; and modeling and simulation of social networks and information propagation and impact.

As an example, a transmission probability can be quantified using the following procedure provided herein for illustration purposes using uncertainty measures [6, 7]:

- Assess the information content of memes (or memeplex) using uncertainty measures.
- Assess the internal inconsistency.
- Assess the inconsistency among memes within a host and other memes at potential hosts.
- Assess utilities based on a value structure.
- Assess shaping factors based on meme source, timing, complexity, impact, and so on.
- Aggregate into an overall successful transmission likelihood.

2.4 Military Worth

If memetics can be established as a scientific discipline, its potential military worth includes applications involving information operations (IO) to counter adversarial memes and reduce the number of prospective adversaries while reducing antagonism in the adversary's military and civilian culture, that is, it could have the ability to reduce the probability of war or defeat while increasing the probability of peace or victory.

In the context of asymmetric warfare, IO are of increasing importance for achieving victory (or avoiding defeat). IO consist of the integrated employment of the core capabilities of electronic warfare, computer network operations, psychological operations (PSYOP), military deception (MILDEC), public affairs (PA), and operations security. In concert with specified supporting and related capabilities, IO are deployed to influence, disrupt, corrupt, or usurp adversarial human and automated decision making while protecting one's own. Potentially, memetics can have a major effect on PSYOP, MILDEC, and PA.

PSYOP are intended to induce or reinforce foreign attitudes and behavior favorable to the originator's objectives and to convey selected information and indicators to foreign audiences to influence their emotions, motives, objective reasoning, and ultimately the behavior of foreign governments, organizations, groups, and individuals. PSYOP focuses on the cognitive domain of the battle space and targets the mind of the adversary. It seeks to induce, influence, or reinforce the perceptions, attitudes, reasoning, and behavior of foreign leaders, groups, and organizations in a manner favorable to friendly national and military objectives. It exploits the psychological vulnerabilities of hostile forces to create fear, confusion, and paralysis, thus undermining their morale and fighting spirit.

There are strategic, operational, and tactical PSYOP, as described in Joint Publication 3.53, *Doctrine for Joint Psychological Operations* (5 September 2003) [8].

Strategic PSYOP consists of international activities conducted by US government agencies primarily outside the military arena but which may use Department of Defense (DoD) assets. Operational PSYOP is conducted across the range of military operations, including during peacetime, in a defined operational area to promote the effectiveness of the joint force commander's campaigns and strategies. Tactical PSYOP is conducted in the area assigned to a tactical commander, for a range of military operations, to support the tactical mission.

PSYOP may occur across the spectrum of peace to conflict to war, integral to diplomacy, economic warfare, and military action, ranging from negotiations and humanitarian assistance to counterterrorism. But according to the *Information Operations Roadmap*, DoD (30 October 2003) [9], despite PSYOP being a low-density, high-demand asset which is particularly valued in the war on terrorism, PSYOP capabilities have deteriorated. Well-documented PSYOP limitations include an inability to rapidly generate and immediately disseminate sophisticated, commercial-quality products targeted against diverse audiences, as well as a limited ability to disseminate PSYOP products into denied areas. Remedial action is urgently required—and memetics may be a solution for recent PSYOP difficulties.

MILDEC involves actions executed to deliberately mislead the adversary's military decision makers about friendly military capabilities, intentions, and operations, thereby causing the adversary to take specific actions (or inactions) that will contribute to the accomplishment of the friendly force's mission. According to the *Information Operations Roadmap*, DoD (30 October 2003) [9], MILDEC should be one of the five core capabilities of IO and the value of MILDEC is intuitive. Counterpropaganda includes activities to identify and counter adversary propaganda and expose adversary attempts to influence friendly populations' and military forces' situational understanding. It focuses on efforts to negate, neutralize, diminish the effects of, or gain an advantage from foreign PSYOP or propaganda efforts.

PA operations assess the information environment in areas such as public opinion and attempt to recognize political, social, and cultural shifts. PA is a key component of information-based flexible deterrent options, intended to build the commanders' predictive awareness of the international public information environment and the means to use information to take offensive and preemptive defensive actions. PA is considered to be a lead activity and the first line of defense against adversary propaganda and disinformation (with the caveat that it must never be used to mislead the public, national leaders, or the media). According to the Field Manual FM 46-1, *Public Affairs Operations* (HQ, Department of the Army, Washington, DC, 30 May 1997) [10], PA operations are combat multipliers in that they keep soldiers informed, maintain public support for the soldier in the field, and mitigate the impact of misinformation and propaganda.

Memetics also has potential military worth in supporting the military culture by enhancing recruitment and training. Recruitment may be improved with memetics by influencing the motivation of prospective recruits, enhancing the image of the military, increasing service awareness ("branding"), providing a national perspective and global situational context for serving one's country. Likewise, training can be improved by increasing trainee motivation, providing better explanations for the training, easing comprehension of the training components, enhancing retention of what is learned during training, and solidifying military culture for the trainees (traditions, customs, and mores).

3 HOMELAND SECURITY

In contrast to natural hazards that are indiscriminate and without malicious intent, a unique challenge with assessing risks due to the deliberate actions of intelligent human adversaries is their ability to innovate and adapt to a changing environment. Although one can rely on historical data to estimate annual occurrence rates for natural hazards affecting a region, given that the timescale of geological and meteorological change is much greater than the planning horizon for most homeland security decisions, assets in the same region are always plausible targets for adversaries despite a lack of past incidents [11, 12].

The uncertainty associated with adversary intentions is largely epistemic, and in principle can be reduced given more knowledge about their intentions, motivations, preferences, and capabilities. In general, the threat component of the security risk problem is most uncertain owing to the fact that defenders are often unaware of the adversary's identity and objectives. Less uncertain is the vulnerability component of the risk equation since countermeasures to defeat adversaries are relatively static in the absence of heightened alert. However, since the effectiveness of a security system depends on the capabilities and objectives of the attacker (which is uncertain), the performance of a security system under stress is more uncertain than the consequences following a successful attack. Thus it seems that to build a security risk profile for an asset, it is prudent to start with those aspects of the risk problem that are most certain (i.e. consequence), proceed with the less certain aspects (i.e. vulnerability then threat) as necessary to support resource allocation decisions, and finally proceed to the threat component. Such an approach, however, should recognize that most effective mitigation strategies can be achieved based on threat reduction, changing, or elimination.

In the spectrum of conflict from peace to war, from persuasion to conquest or defeat, from PSYOP to physical destruction, it is far better to mitigate the threat in the beginning by reducing the prospective pool of terrorists. In that early part of the spectrum, there are many problems requiring solutions outside the purview of memetics, such as poverty and the lack of opportunity. But memetics can, potentially, increase the popular demand for realistic solutions to endemic social problems and focus the blame where it belongs. By neutralizing false and incendiary memes that provoke the emergence of terrorists, countermemes can reduce the number of prototerrorists and the probability of consequent terrorist acts.

In the United States, the color-coded terrorist threat warning system has become nearly meaningless in providing specific guidance to the public on responses to terrorist acts. Also, the national critical infrastructure is at terrorist risk from causes outside the purview of memetics, such as insufficient facility security or response systems. But memetics can, potentially, provide the means for increasing the awareness of the public to the terrorist threat in a way that is meaningful as well as memorable. By employing more easily remembered informational techniques, memetics can educate managers of the critical infrastructure about how best to protect their facilities from threat or mitigate its consequences. First responders could also benefit from easily transmitted and absorbed memes to learn the complex tools and techniques needed for countering explosive as well as chemical, biological, and radiological hazards.

4 PROSPECTS

Memetics can ameliorate unfavorable consequences from an adversary's culture and help to deter conflict and reduce animosity through cultural education and generating acceptable solutions to endemic problems. In the case of combat, memetics can enhance tactics, strategy, and doctrine by making an otherwise adversarial situation more acceptable to noncombatants, helping to minimize collateral damage by improving the ability to discern combatants from noncombatants, and encouraging civilians to identify insurgents and terrorists. At the conclusion of combat, memetics can bolster peacekeeping, occupation, and nation-building by making these operations more palatable to civilians and facilitating patience for the "long-haul." During postcombat with continuing insurgence, memetics can help minimize collateral damage by enhancing the ability to discern combatants from civilians and identify insurgents and terrorists.

For homeland security, memetics can reduce the number of prospective adversaries and therefore reduce the probability of an attack. By increasing the awareness of the human-caused risk in the public at large and those who manage the targeted infrastructure, the probability of a successful attack can be reduced, or its consequences mitigated. Likewise, the consequences of an attack can be mitigated by using memes to enhance the training of first responders.

5 FUTURE RESEARCH DIRECTIONS

The discussion provided herein addresses countermeasures relating to the threat component of homeland security risks, and is aimed at aiding decision makers to develop effective strategies for threat reduction and potential elimination. Further research is needed to enhance our understanding of the nature of memes and their attributes. We must develop techniques for identifying memes, such as fMRI, as well as developing simulation methods and simulation environments, that is, sandboxes, to allow analysts to explore the creation of memes and determine their effectiveness using quantifiable metrics and submetrics.

REFERENCES

1. Ayyub, B. M. (2003). *Risk Analysis in Engineering and Economics*, Chapman and Hall/CRC Press, Boca Raton.
2. Kaplan, S., and Garrick, B. J. (1981). On the quantitative definition of risk. *Risk Anal*. **1**(1), 11–27.
3. Dawkins, R. (1976). *The Selfish Gene*, Oxford University Press, Oxford.
4. R. Aunger, Ed., (2000). *Darwinizing Culture: The Status of Memetics as a Science*, Oxford University Press, Oxford.
5. Aunger, R. (2002). *The Electric Meme: A New Theory of How We Think*, The Free Press, New York.
6. Ayyub, B. M. and Klir, G. J. (2006). *Uncertainty Modeling and Analysis in Engineering and the Sciences*, Chapman & Hall/CRC, Boca Raton.
7. Ayyub, B. M. (2004). From dissecting ignorance to solving algebraic problems. *Rel. Eng. Sys. Safety* **85**, 223–238.

8. Joint Chiefs of Staff. Doctrine for Joint Psychological Operations (2003). *Joint Publication 3.53*, Department of Defense, Washington, DC.
9. Department of Defense (2003). *Information Operations Roadmap, 30 Oct. 03*, Washington, DC.
10. Department of the Army (1997). *Public Affairs Operations Army Field Manual FM 46-1, HQ*, Washington, DC.
11. Ayyub, B. M., McGill, W. L., and Kaminskiy, M. (2007). Critical asset and portfolio risk analysis for homeland security: an all-hazards framework. *Risk Anal*. **27**(3), 789–801.
12. McGill, W. L., Ayyub, B. M., and Kaminskiy, M. (2007). A Quantitative Asset-Level Risk Assessment and Management Framework for Critical Asset Protection. *Risk Anal*. **27**(5), 1265–1281.

FURTHER READING

National Commission on Terrorist Attacks (2004). *The 9/11 Commission Report: The Final Report of the National Commission on Terrorist Attacks Upon the United States*, Authorized ed. Norton W. W. and Company New York, NY.

Department of Homeland Security (2006). *National Infrastructure Protection Plan*. URL: http://www.dhs.gov/xlibrary/assets/NIPP_Plan.pdf. Last accessed 27 November 2006.

Hoffman, B. (1998). *Inside Terrorism*, Columbia University Press, New York.

Martz, H. F., and Johnson, M. E. (1987). Risk analysis of terrorist attacks *Risk Anal*., **7**(1), 35–47.

Haimes, Y. Y. (2006). On the Definition of Vulnerabilities in Measuring Risks to Infrastructures *Risk Anal*., **26**(2), 293–296.

HIGH CONSEQUENCE THREATS: ELECTROMAGNETIC PULSE

MICHAEL J. FRANKEL

EMP Commission, Washington, D.C.

1 INTRODUCTION

While the power of a nuclear weapon was seared into the world's psyche during the waning hours of WWII, nuclear scientists and defense specialists have long been aware that the potential for widespread destruction, devastation, and disorder as a consequence of nuclear detonation may exceed even that anticipated by the popular imagination. The locus of such assessment lies in their awareness of the full effects of a nuclear burst, effects not realized in the immediate and particular circumstances of the Japanese wartime experience.

When a nuclear warhead is detonated, the energy bound up in the rest mass of an atom is released. While a large fraction of this energy is originally in the form of radiation—x rays, γ rays, and neutrons—at low burst heights, much of this energy is converted to kinetic and thermal forms, as the atmosphere quickly absorbs the radiated energy, heats up, and launches an extraordinarily powerful blast wave that causes much of the immediate damage to any structures caught in its destructive path. Along with the familiar blast waves comes a veritable witch's brew of destructive insults; a thermal pulse that may ignite fires, along with broken gas lines, overturned stoves, etc., producing secondary fires that can propagate damage beyond the initial blast radius or even coalesce into a highly destructive firestorm, cratering, and ground shocks damaging even well-buried underground structures, an intense but localized ("source region") electromagnetic pulse (EMP), and of course fallout that may extend the effects of the weapon far from the burst site and whose lethal effects may linger long after all other signs of the immediate devastation have disappeared.

But there are also other effects attendant upon a nuclear burst, especially at bursts heights—at relatively high altitude above 30 km or so—which our WWII experience did not adequately presage. These high altitude effects include enhancement of the natural radiation belts that circle the earth (whose existence was unknown during WWII) or the creation of entirely new, albeit temporary, radiation belts that may destroy or degrade artificial satellites in low earth orbits encompassed by their reach [1],[1] ionization effects in the atmosphere, which interfere with communications, direct effects of large X-ray dosages, which can travel unimpeded many thousands of miles through space to affect satellites at great distances from the burst point, and—the subject of this article—by the production of gigantic high altitude EMPs, which may produce unprecedented widespread disruption on the ground [2].[2]

2 WHAT IS EMP?

EMP from weapons detonations in space was discovered as a surprise by-product of early nuclear weapons tests; one of which (STARFISH, a 1.4 Mt device exploded in 1962 at an altitude of 400 km above Johnston Island in the Pacific Ocean), delivered an EMP, which turned out the street lights in downtown Honolulu, a distance of about 800 nautical miles away. Earlier testing in the Soviet Union reportedly damaged buried cables and electric power equipment at a remove of 600 km from detonations of a few hundred kilotons yield at burst heights ranging from about 60 to 300 km [3].

[1] The detonation of Starfish in 1962 was the proximate cause of the almost immediate loss of Telstar, the first commercial telecommunications satellite, launched in 1961. Over the next few months, every single commercial satellite in orbit failed, well before its expected lifetime. No information is publicly available on the fate of a number of satellites performing classified intelligence missions at the time.

[2] We are focused here on the high altitude phenomenon often referred to as high electro magnetic pulse (HEMP) in the national security literature. A ground level burst will produce a localized phenomenon known as *source region electro magnetic pulse (SREMP)*, whose strong electric fields will generally be confined to a radius of a scale with the blast damage radius from a weapon. There are also other forms of EMP related to effects on space systems, such as systems generated electro magnetic pulse (SGEMP), which stem from an interaction between bomb X rays and metallic structures. We will not deal with these, or other EMP exotica, here.

The EMP from a nuclear device comes in two components, a fast pulse and a slow pulse, each emerging from a different physical mechanism. The fast pulse is a result of the initial high energy γ radiation from the bomb traveling toward the earth and in turn producing a stream of electrons heading in the same direction by encounters with the sensible atoms of the atmosphere in a region 20–40 km high, where the atmosphere has thickened enough to intercept the γ rays. The stream of electrons heading toward earth must travel through the magnetic field that surrounds our earth and, as must any charged particle traversing a magnetic field, be forced to "turn", or accelerate, in a direction perpendicular to the field. This near coherent acceleration of the charged electron stream is the source of the fast component, the first radiated EMP field felt on earth. The area coverage of this pulse can be enormous—the phenomenon has been likened to installing a radiating phased ray antenna a thousand miles long up in the sky (Fig. 1)—and its peak electric field may reach tens of kilovolts per meter, a level sufficient to degrade or destroy many commercial electrical components that form the warp and woof of our twenty-first century technology-based society.

The mechanism of the slow pulse is associated with the expansion of the ionized bomb materials in the earth's magnetic field as well as the rise of a bomb heated and ionized patch of atmosphere cutting the earth's geomagnetic field lines. While the fast pulse may last only for nanoseconds and couple unwanted high frequency energy pulses to vulnerable electrical devices, the slow component may last from milliseconds to seconds and couple low frequency, long wavelength pulses to suitable antennae tuned to receive these frequencies. In fact, the US power grid, with conducting runs of wire running from tens to hundreds of kilometers in uninterrupted length, represents such an "antenna" seemingly ideally tuned to the reception and coupling of the frequencies of the slow EMP.

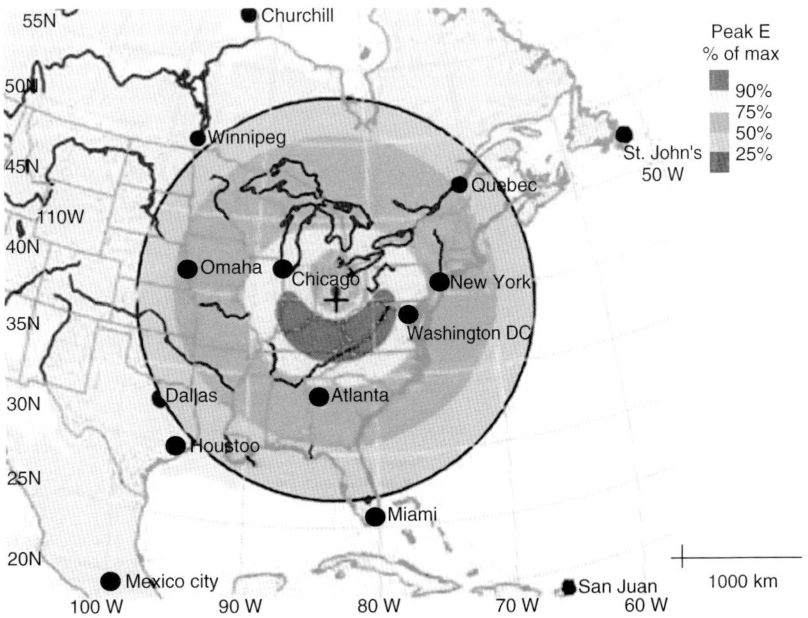

FIGURE 1 Illustrative EMP coverage—fast pulse. Thirty kiloton nominal yield and 100 km burst height.

3 CONSEQUENCES OF AN EMP EVENT ON OUR INFRASTRUCTURES

3.1 Direct Effects

So what can an EMP actually do to you? In terms of directly visible destructive impact on physical structures or biological systems, the answer is nothing. Instead, EMPs are expected to couple unwanted energies into the soft underbelly of our technology-dependent society—the ubiquitous electronic systems, controls, gadgets, means of communication, and power sources that undergird and enable the daily functioning of twenty-first century American civilization. The fast EMP may apply fields of many kilovolts per meter either directly—free field coupling—or through the ubiquitous *ad hoc* antennae, short runs of wire or cable, in the local electrical network to which most devices are connected. As has been repeatedly demonstrated in test programs, levels of insult starting at a few kilovolts per meter are generally sufficient to degrade electronic equipment functioning and higher field levels—easily achievable by a high altitude nuclear source—may even cause permanent physical damage in most commercial-off-the-shelf electronic systems (Fig. 2).

The slow EMP produces long wave pulses capable of coupling electrical energy directly with the long cable runs of the power grid transmission system potentially damaging key components such as hard to replace high voltage transformers. While the fast EMP may be simulated in specialized electric pulse machines, the slow pulse has a natural analog in the EMP produced by geomagnetic storms associated with mass ejection events on the sun and correlated roughly with the 11 year sunspot cycle. Figure 3 demonstrates the physical damage that may result from application of this type of slow pulse. The transistor damage shown in the figure occurred at the same time that the 1989 geomagnetic storm brought down the entire Hydro Quebec system leaving 9 million Canadian customers without power for periods of up to 2 weeks.

FIGURE 2 Electrical arcing damage to a circuit board during current injection EMP simulation testing [4]. [Courtesy of W. Radasky, Metatech Corp.].

FIGURE 3 Permanent damage to high voltage transformer during 1989 geomagnetic storm. [Courtesy of Public Service Electric and Gas Corp.].

High voltage transformers are house-sized, complex, and expensive affairs; no longer manufactured in this country; and require on the order of a year to fill and ship an order. Loss of a significant number of high voltage transformers would undermine any prospect of the power grid to recover in a timely manner and cause almost incalculable economic harm and human misery over a period that might well extend for many months. Figure 4 is a graphical representation of a calculation sponsored by the EMP Commission [4] showing the catastrophic collapse of the power grid in a specific scenario of a high yield detonation over the eastern part of the United States. Red and green circles represent high voltage transformers which are predicted to and have failed, with the colors indicating impressed current directions. The straight lines represent high voltage (>365 kV) distribution lines.

It is not only the slow pulse that couples to the long lines represents a danger to the functioning of the power system but the supervisory control and data systems (SCADAs) and protective relays and breakers—whose earlier electromechanical designs are being increasingly replaced by more versatile but more electrically vulnerable digital electronic systems—are also at risk from the fast pulse. The power grid is an extraordinarily complex network under constant dynamic control, which strives to match generation with the load. It is fair to say that no adequate predictive models exist that fully understand all their potential interactions and—on the order of every decade or so—there is a major power failure, which may stem from one or two precipitating failures which cascaded out of control in some surprising way. Since the coverage of the EMP pulse may extend to a significant fraction of the country, we may envision the result of the failure of potentially hundreds of power grid components over a widely dispersed area, all simultaneously. Worst case estimates anticipate the failure of the power grid over a significant fraction of the country, with many components having suffered permanent physical damage. There is little precedent for any of this, and the time needed to recover from such a state is uncertain. In the worst imaginable, but possible, scenario—the loss of the entire national power system from coast to coast—recovery is particularly uncertain as there is simply no experience available for such a black start condition [5].

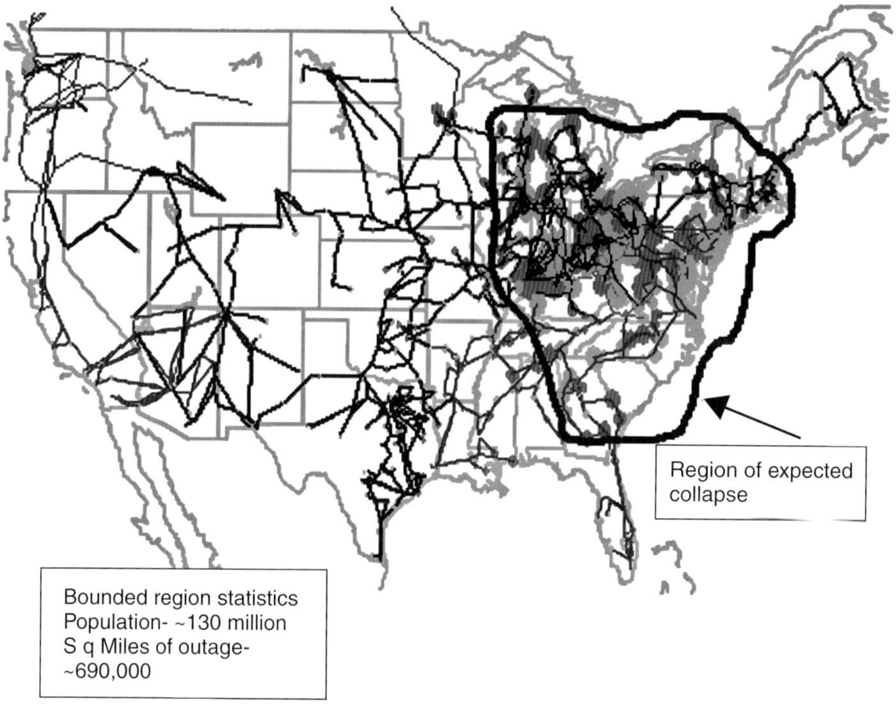

FIGURE 4 Collapse of portion of power grid representing 70% of domestic power generation in high yield burst scenario [4]. [Courtesy of W. Radasky, Metatech Corp.] (See online version for color).

We have discussed the vulnerability of the power grid, but that is not the only critical infrastructure at risk. Significant elements of the computer-controlled national telecommunications network may also be at risk in an EMP attack. Computers, routers, switches, etc., if unshielded, may be damaged by EMPs. The telecommunications system—and thus the nation's financial system—though possessing short-term backup power systems of its own, is in turn ultimately dependent on the functioning of the power grid. The terrorist attack of September 11, 2001, on the World Trade Center exposed telecommunications and concentration of key facilities as serious weaknesses of the financial services industry. Equity markets closed for four days, until September 15, due to failed telecommunications. The New York Stock Exchange could not reopen because key central offices were destroyed or damaged leaving them unable to support operations. According to a senior government economic official, Fedwire (the electronic funds transfer system operated by the Federal Reserve enabling fund transfers between financial institutions), CHIPS (Clearing House Interbank Payments System is an electronic system for interbank transfer and settlement. CHIPS is the primary clearing system for foreign exchange), and SWIFT (Society for Worldwide Interbank Financial Telecommunications provides stock exchanges, banks, brokers, and other institutions with a secure international payment message system) would cease operation if telecommunications are disrupted. He further observed that ACH (Automated Clearing House is an electronic network that processes credit and debit transactions), ATMs (automated teller machines), credit and debit cards all depend on telecommunications. Disruption of these systems would force consumers to revert to a "cash economy".

In the recent report of the Commission to Assess the Threat to the United States from Electromagnetic Pulse (The EMP Commission) [4], the Commission documented the direct effects of EMP irradiation not only on the power and telecommunications infrastructures but also on the transportation, emergency services, banking, oil and gas, water, food, and space infrastructures. Although SCADAs were previously mentioned in conjunction with the power grid, many of the direct vulnerabilities in the other critical infrastructures of this country stem, in turn, from the widespread infiltration of these automated control and monitoring systems into critical functional nodes of most systems that undergird the smooth functioning of our economy and maintain our well being. If the SCADAs that monitor and control the pump pressures in the oil and gas pipelines malfunction or cease working altogether, then fuel will not be delivered to industries and everybody else who needs it; rail systems that rely on sensors to monitor tracks and controls to set rail switches may not function; emergency responder communications networks may be unusable; if electronically controlled, the various pumps that deliver water from reservoirs and underground sources may not function; and stored foods may spoil if dependent on vulnerable refrigeration system.

3.2 Indirect Effects

In an increasingly interdependent world, the power system itself is increasingly dependent on the functioning of the telecommunications system to maintain situational awareness and active remote control. The telecommunications system may maintain itself for a short period by backup, on site power. Uninterruptible power supplies—batteries—may keep operations functioning for a few hours, but generally no longer than a day. Some telecommunications facilities are equipped with backup generators that may keep things going until the fuel runs out. But, these would have to then be replenished by fuel deliveries provided by the transportation system, which in turn require fuels themselves from the oil and gas infrastructures, each of which in turn depends on the availability of power from the grid. Personnel would have to be delivered by the transportation system to repair and maintain the power and they would have to be fed by the food and water infrastructures and paid by the financial system, which in turn depend on telecommunications and power system. A major catastrophe that took down the power system for any extended period of time thus has the potential to produce incalculable misery and ultimately loss of life, with recovery scenarios of uncertain prospects. This mutual interdependence of the critical infrastructures is shown in Figure 5.

Experience demonstrates that such interdependencies and interactions may well be overlooked, until an emergency situation suddenly reveals their existence as they surface to bite us. For example, many of the recovery procedures developed by organizations to deal with emergencies implicitly assume that transportation is available to transport personnel somewhere to go fix something. But, immediately following the precipitating events of 9/11, airplanes (except emergency military transport) were all grounded. In 1991, the accidental severing of a single fiber-optic cable in the New York City region not only blocked 60% of all calls into and out of New York, but it also disabled all air traffic control functions from Washington, DC, to Boston—the busiest flight corridor in the Nation—and crippled the operations of the New York Mercantile Exchange. These key

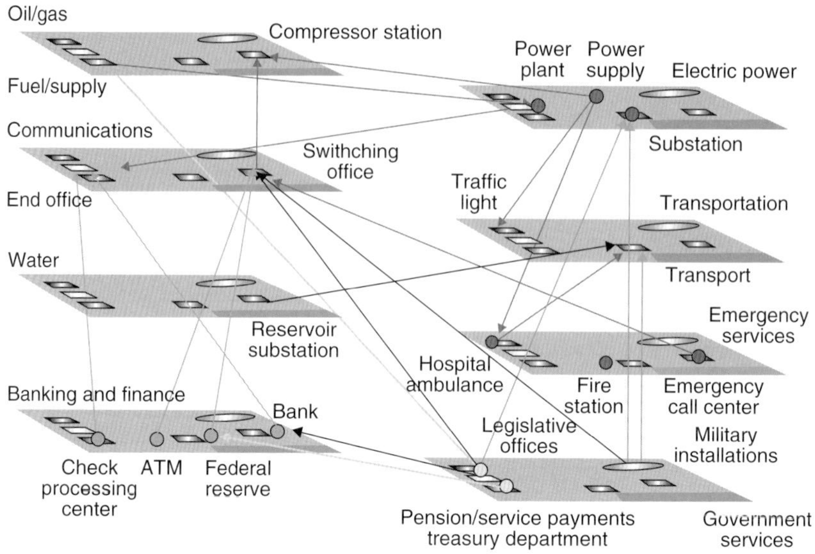

FIGURE 5 Infrastructure interdependencies (diagram courtesy of Sandia National Laboratory). All interdependencies are not shown.

interdependencies are always there, but they are not recognized as warranting advanced contingency planning [6].[3]

Generally, in situations where an infrastructure experiences failure, such as loss of electric power due to storm or some other localized event—the August 2003 electric blackout of the Northeast was ultimately attributed to a localized failure with subsequent unanticipated cascading effects (due to loss of situational awareness through failure of monitoring systems, which prevented timely load shedding that might have been otherwise averted) [7]—many of these infrastructure interdependencies and interactions can be safely ignored. But, in an EMP attack scenario, the energy of EMP is expected to affect the different infrastructures simultaneously through multiple electronic component disruptions and failures over a very wide geographical area. Understanding these cross-cutting interdependencies and interactions is critical to assessing the capability of the full system of systems to recover. The modeling and simulation needed to explore the response of such a complex situation involve a large but finite number of elements and should be amenable to analysis, at least approximately. Efforts to develop a predictive modeling capability are underway, but much progress still needs to be made.

4 THE RISK OF AN EMP ATTACK

Risk is normally accounted as a multiplicative concatenation of threat, vulnerability, and consequences, with each of these factors in turn quantitatively represented as a conditional probability. EMP, however, is also one of a class of very high consequence, low

[3]Nuclear plants require some existing power in a grid and may not be started into a black grid. Coal and gas generating plants also require some existing electricity to run. Presumably, hydroelectric plants might be the first sources brought back on line to nucleate the recovery of other elements. But this would have to be done very carefully and load balancing might be particularly stressing under such circumstances.

probability (or more accurately, unknown) events, which classical risk methodology [8] finds difficult to encompass. EMP is one of a small number of events whose consequences are potentially so large that they hold the possibility of affecting the basic functioning of civil society in our country for an extended period. Although many intuitively believe that the threat may be small, the consequences are so large that risk may not be negligible [9][4] and may result in a finite risk. The EMP Commission, while refusing to assess the likelihood of such an event, nevertheless utilized a capabilities-based threat assessment methodology to conclude that the technical information to accomplish such an attack was widely disseminated amongst potential adversaries. But it is also a truism that vulnerability invites the attention of those who might exploit it, and reducing such vulnerability must inevitably reduce the risk. Given the scale of likely consequences, it is thus remarkable that the Federal government, in its organizational embodiment in the Department of Homeland Security (DHS), has to date given almost no attention at all to this issue. Most critical, and telling, is the lack of any serious study and planning function within DHS, which is the critical first step required to come to grips with the issues reviewed in this article.

It is important to emphasize that such a lack of attention is not uniformly practiced across the Cabinet Departments of this country. The Department of Defense (DOD) has long been aware of the EMP threat to the functioning of a modern military no less dependent on the operation of advanced electronic components than the civilian sector. The DOD has made investments to ensure the continued operation and viability of our fighting forces and military infrastructure in EMP environments. All these are in stark contrast to the present lack of readiness of the civilian infrastructures to withstand, operate through, and recover from any such electromagnetic event.

5 MANAGING THE RISK: RECOMMENDATIONS TO HELP PROTECT OURSELVES

While risk was defined as the product of the likelihoods of threat, vulnerability, and consequence, risk management addresses the steps that may be taken to mitigate, contain, or recover from identified risk [10]. The following general recommendations are offered to address the current demonstrated vulnerabilities of our infrastructures.

- As a mechanism for focusing preparation activities at the DHS, planning is presently organized around consideration of 15 canonical disaster scenarios [11]. These scenarios range from essentially localized disturbances such as a terrorist chemical attack or a natural disaster such as a hurricane to very few scenarios with potentially more widespread effects such as a biological attack. It presently includes one nuclear scenario that envisions a detonation of a small yield device at ground level in a city in the United States. A strong recommendation to DHS is to initiate a planning activity for EMP disasters by adding a 16th scenario of a high altitude nuclear burst. EMP disaster analysis and planning should be augmented by inclusion of EMP scenarios in planning exercises that engage the federal government response with first responders in the state and local government.

[4]As do many others, Schneier makes the point, inter alia, that estimates of risk are greatly conditioned by familiarity. Thus, our intuition about EMP risk may not be worth much. Similarly, many intuitively believed the threat of terrorists crashing commercial airliners into skyscrapers was small.

- As far as possible, attention should be paid to the development of dual benefit solutions to EMP threats. Thus, for example, measures that may be taken to implement protection of the critical elements of the power grid to EMP may also confer other protective benefits against the threat of geomagnetic storms. The latter is a regular natural phenomenon that has demonstrably damaged key elements of the grid in the past (Fig. 3). Recent calculations analyzing the response of the power grid to a so-called 100 year storm (EMP Commission, personal communication) point to an existing known vulnerability to an expected natural phenomenon with the prospect of very severe consequences, with no apparent protection. It is hard not to make analogies to Katrina. As well, measures implemented to monitor and enhance the system's ability to recover from EMP may also enhance the overall reliability and quality of the infrastructure, conferring economic benefit.
- Critical components of each infrastructure should be identified and inventories compiled. Plans should consider the economic and logistical feasibility of storing long lead time replacement items in the vicinity of their prospective use.
- The federal government should enhance its investment in the analysis of infrastructure interactions and the development of validated modeling and simulation tools to support realistic planning simulations and the rapid testing of mitigation and recovery strategies. A good start has already been made here but more effort will be required before a reliable tool capable of exploring far from equilibrium situations with potentially hidden-under-normal-circumstances interaction pathways is available to the planning community. Agent-based approaches may hold particular promise and should be explored.

The Commission to Assess the Threat to the United States from Electromagnetic Pulse has published a number of volumes detailing both the threat and the impact of EMP environments on the military and civilian infrastructures of the country. The volume describing civilian infrastructures was unclassified and contained approximately 70 more specific recommendations addressed in the main volume to the DHS, including recommendations for legislative actions aimed at the Congress. Our final recommendation is that those 70 specific actions be reviewed for action and implemented as rapidly as possible.

REFERENCES

1. Papadopulous, K. D., briefing available at http://www.lightwatcher.com/chemtrails/Papadopoulos-chemtrails.pdf.
2. Defense Threat Reduction Agency, briefing available at www.fas.org/spp/military/program/asset/haleos.pdf.
3. (2004). *Report of the Commission to Assess the Threat to the United States from Electromagnetic Pulse (EMP) Attack*, Vol. 1: Executive Report, p. 4.
4. Radasky, W. *Critical National Infrastructures*, Vol. 3, EMP Commission, calculation, Metatech.
5. Schneider, F. B., Bellovin, S. M., and Inouye, A. S. (1998). Critical infrastructure you can trust: where telecommunications fit, *26th Annual Telecommunications Policy Research Conference*, http://www.tprc.org/abstracts98/schneider.pdf.
6. Report of the Ontario Ministry of Energy. (2003). Following August 2003 blackout. http://www.energy.gov.on.ca/index.cfm?fuseaction=electricity.reports_outage.

7. US-Canada Power System Outage Task Force. (2003). *Interim report: Causes of the August 14 Blackout in the US and Canada*, November, 2003 ftp://www.nerc.com/pub/sys/all_updl/docs/blackout/814BlackoutReport.pdf.
8. Ayyub, B. M. (2003). *Risk Analysis in Engineering and Economics*. Chapman & Hall.
9. Schneier, B. (2003). *Beyond Fear: Thinking Sensibly About Security in an Uncertain World*, Springer.
10. Scouras, J., Cummings, M. C., McGarvey, D. C., Newport, R. A., Vinch, P. M., Weitekamp, M. R., Colletti, B. W., Parnell, G. S., Dillon-Merrill, R. L., Liebe, R. M., Smith, G. R., Ayyub, B. M., and Kaminskiy, M. P.. (2005). *Homeland Security Risk Assessment, Volume I An Illustrative Framework*. RP04-024-01a. Homeland Security Institute, Arlington, VA, November 11.
11. http://media.washingtonpost.com/ wp-srv/nation/nationalsecurity/earlywarning/National PlanningScenariosApril2005.pdf.

FURTHER READING

Glasstone, S., and Dolan, P. J. (1977). *The Effects of Nuclear Weapons*. Department of Defense and the Department of Energy.

Longmire, C. L. (1978). On the electromagnetic pulse produced by nuclear explosions. *IEEE Trans. Electromag. Compat.*, **EMC-20**(1), 3–13.

Messenger, G. C., and Ash, M.S. (1986). *The Effects of radiation on Electronic Systems*, Chapter 8, Van Nostrand Reinhold, New York.

HIGH CONSEQUENCE THREATS: NUCLEAR

CALVIN SHIPBAUGH
Arlington, Virginia

MICHAEL J. FRANKEL
EMP Commission, Washington, D.C.

1 INTRODUCTION

World War II taught the world to measure destructive power by a new metric. Fission weapons, which "split" atoms of uranium or plutonium and release a portion of their binding energy, are measured by their yield in terms of the number of kilotons (kt) of trinitrotoluene (TNT) equivalent to the total energy released. The first fission weapons that were produced in 1945 had yields in the neighborhood of 15–20 kt [1]. This approximate magnitude may crudely be assumed to represent early design attempts at producing nuclear explosions, although the yield of the first French test was reportedly several times greater [2].[1] Lower yields can cause large scale devastation to an urban area. North Korea is the most recent nation to conduct a nuclear test. The yield was very low, and is reported to be less than 1 kt [3]. This yield stands in contrast to prior historical observations of what might have been expected for a first explosion, and illustrates that surprises regarding conventional wisdom on nuclear issues have appeared in recent times.

Thermonuclear weapons, which release excess binding mass energy when hydrogen atoms "fuse", can have yields much higher than those of the most powerful fission weapons. The largest ever tested had a yield of approximately 50 Mt [4], which is more than three thousand times the yield of the fission bomb dropped on Hiroshima. Few nations have demonstrated the ability to manufacture thermonuclear weapons. But, in a volume focused on the risk to homeland security from the activities of well-funded terrorist groups and rogue nations, it would seem prudent to focus our attention on effects of low yield threats, on the order of 1–20 kt. These threat weapons may be relatively unsophisticated devices—improvised nuclear devices (INDs) constructed from stolen or diverted fissile materials—or more sophisticated devices constructed with the aid of experienced weapons designers, and either bought or stolen from a nuclear weapon state whose custodial safeguards are compromised during a period of internal political instability. It has been widely reported that the Department of Homeland Security (DHS) has followed the same track, defining a 10 kt threat for use by federal, state, and local homeland security disaster planning efforts [5].

2 SCOPE OF THE DESTRUCTIVE PHENOMENA

A nuclear explosion in the atmosphere produces a strong blast, much greater than hurricane force winds. This is complicated in urban terrains by combinations of direct and indirect waves that may concentrate the blast flow and enhance intensity. A distinction is made between dynamic pressure, which results from the hammer blow impact of the shock front and its following winds, and overpressure, which uniformly "squeezes" its targets.

A crater will result if a nuclear explosion occurs near a land surface or within the lower levels of a structure, while ground shock will damage some structures beyond the crater. The diameter and depth of the crater depend on the type of material, and varies approximately as the cube root of yield. The apparent crater radius of a 1 kt explosion in dry soil is approximately 20 m [6]. The ground shock effect is enhanced if a nuclear explosive is buried in a few meters of ground [7].[2]

If exploded in or near water, another potential effect is high waves and surge of vapor. The damage caused by waves is, in general, less significant for a low yield explosion than blast or thermal effects, but should be considered for calculating damage to the nearby shore. Vapor clouds that surge outward from the explosion are highly contaminated with radioactive materials.

[1]This test event is known as *Gerboise bleue*. It was said to be 70 kt, and was conducted in 1960.

[2]A 1 kt surface detonation in hard rock generates 100 MPa at a depth of less than 20 m, but a coupled explosion generates the same pressure at approximately 80 m.

The radiant energy of a low yield nuclear explosion can cause severe fire or other thermal damage (e.g. third-degree burns) at distances of several hundred meters to kilometers. The damage depends on transmission through the atmosphere. Factors such as humidity or fog and reflections from the surface affect thc intensity as a function of range. The duration of a pulse depends on yield, so the thermal effect on a target depends on the yield as well as on the energy density at the target. The damage threshold increases with the yield.

The most distinctive feature of a nuclear explosion, apart from the sheer magnitude of energy released, is nuclear radiation [8].[3] An acute whole-body dose from the prompt radiation emerging from the explosion in the range of 150–300 cGy [9] induces nausea and fatigue in most victims in the first hours, and can cause fatalities in a few percent of cases within 6 weeks. The $LD_{50/60}$ dose, at which fatalities reach 50% within 60 days, is often reported to be approximately 450 cGy (without medical treatment). A dose of 600–700 cGy is lethal to most people; that is, $LD_{95/60}$ [9]. Table 1 provides a summary of the known physiological response to radiation dose.

Residual radiation results from the radioactive decay of fission products, radioisotopes created by prompt neutrons, and nuclear materials present in the weapon. The vast majority of this material is carried up by the mushroom cloud and falls back downwind. This phenomenon creates a widespread, lingering hazard. Although fallout presents a long-term risk, it should also be noted that 55% of the theoretical total residual radiation dose released will occur in the first hour, and 75% will occur in the first 12 h. A rule-of-thumb is that a sevenfold passage of time reduces the dose rate by a factor of 10 [6].

Although electromagnetic pulse (EMP) is most often discussed in reference to high altitude explosions where effects can be very widespread, a surface explosion can produce a local or "source region" electromagnetic pulse (SREMP). The intensity of the electric field (i.e. up to a few kilovolts per meter) can be immediately destructive to many kinds of electronics within a deposition region of a few kilometers from the blast point. The effects of such local EMP can also be conducted well beyond the immediate environment of the explosion by coupling electromagnetic energy to wires and cables that conduct the effects to more distant, but electrically connected, sites.

3 CASE STUDY: CONSEQUENCES IN POPULATION CENTERS

We consider now the likely consequence of detonation of a low yield device at zero height of burst—such as might be associated with weapon delivery by a van—in a symbolically and politically important urban center such as Washington, DC.

[3]One Gray (Gy) is the modern term for dose. It is equivalent to 100 rads (radiation absorbed dose). One centi-Gray (cGy) is therefore equivalent to one rad. One rad is defined as 100 ergs absorbed per gram of target material (such as body tissue). The term sievert (Sv) is used to measure the equivalent dose, which is the dose in Gy multiplied by a radiation weighting factor to account for the difference in biological effects among types of radiation. The typical annual background dose, less than one cGy, is three orders of magnitude less than this acute dose. Strictly speaking, biological effects should not be treated in units of cGy (or rads, as in older documents), but for an acute whole-body exposure from a nuclear explosion the radiation weighting factor for both neutrons and gammas is near unity. The average annual effective dose in the United States was estimated (for 1980) to be 3.6 mSv.

TABLE 1 Physiological Response to Radiation Dose [10]

Dose Range (REM Free Air)	Onset and Duration of Initial Symptoms	Performance (Mid-range Dose)	Medical Care and Disposition
0–70	6–12 h: none to slight transient headache, nausea, and vomiting in 5% at upper end of dose range	Combat effective	No medical care, return to duty
70–150	2–20 h: transient mild nausea and vomiting in 5–30%	Combat effective	No medical care, return to duty
150–300	2 h to 3 d: transient to moderate nausea and vomiting in 20–70%; mild to moderate fatigability and weakness in 25–60%	DT:PD 4 h until recovery. UT:PD 6–20 h; DT:PD 6 wk until recovery	3–5 wk: medical care for 10–50%. High end of range death in >10%. Survivors return to duty
300–530	2 h to 3 d: transient nausea and vomiting in 50–90%; moderate fatigability in 50–90%.	DT:PD 3 h until death or recovery. UT:PD 4–40 h and 2 wk until death or recovery	2–5 wk: medical care for 10–80%. Low end of range <10% deaths; high end death >50%. Survivors return to duty
530–830	2 h to 2 d: moderate to severe nausea and vomiting in 80–100%. 2 h to 6 wk: moderate to severe fatigability and weakness in 90–100%	DT:PD 2 h to 3 wk; Cl 3 wk until death. UT:PD 2 h to 2 d, 7 d to 4 wk; Cl 4 wk until death.	10 d to 5 wk: medical care for 50–100%. Low end of range death >50% at 6 wk: High end death for 99%
830–3000	30 min to 2 d: severe nausea, vomiting, fatigability, weakness, dizziness, disorientation; moderate to severe fluid imbalance, headache	DT:PD 45 min to 3 h; Cl 3 h until death. UT:PD 1–7 h; Cl 7 h to 1 d, PD 1–4 d; Cl 4 d until death	1000 REM: 4–6 d medical care for 100%; 100% deaths at 2–3 wk. 3000 REM: 3–4 d medical care for 100%; 100% death at 5–10 d
3000–8000	30 min to 5 d: severe nausea, vomiting, fatigability, weakness, dizziness, disorientation, fluid imbalance, and headache	DT:Cl 3–35 min; PD 35–70 min: Cl 70 min until death. UT:Cl 3–20 min; PD 20–80 min; Cl 80 min until death	4500 REM: 6 h to 1–2 d; medical care for 100%; 100% deaths at 2–3 d
Over 8000	30 min to 1 d: severe prolonged nausea, vomiting, fatigability, weakness, dizziness, disorientation, fluid imbalance, and headache	DT and UT: Cl 3 min until death	8000 REM: medical care immediate to 1 d, 100% deaths at 1 d

3.1 Radiation Consequences

The *Hazard Prediction and Assessment Code* (HPAC) [10] calculates likely consequences of such an explosion scenario. Figures 1 and 2 show the resulting dispersion of radioactive materials (fallout) and estimated casualties following an explosion of a 1 and 20 kt device in downtown Washington, DC.

Not unexpectedly, the resulting radioactive footprint and estimated casualties are more significant in the 20 kt scenario than in the 1 kt case. The total casualties may be expected to decrease if more rapid evacuation is achieved, but it is clear that fallout effects are a significant contributor to the overall hazard. The medical consequences of accumulated dose listed in Table 1 are overlayed with the fallout footprint and population statistics from the Oak Ridge National Laboratory population database [11] to produce the HPAC casualty estimates.

Similarly, potential exposure may be estimated for initial radiation effects. A γ dose of 300 cGy from a low altitude fission explosion is received at a slant range of approximately 850 m for 1 kt and 1500 m for 20 kt (Fig. 3) [6]. The corresponding range for neutrons is approximately 900–1350 m.

3.2 Blast Consequences

The overpressures and dynamic pressures will cause severe blast wave damage within the immediate vicinity of the explosion. Reinforced concrete structures will be damaged at 5 psi peak overpressures and wood frame structures—typical housing construction

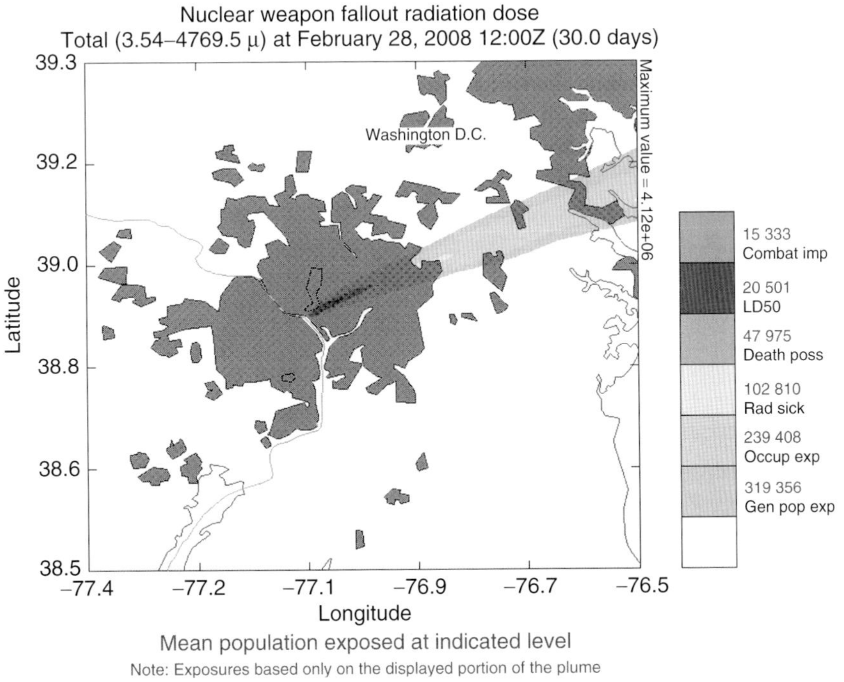

FIGURE 1 One kiloton burst, zero height of burst. 30.0 days accumulated radiation dose (HPAC calculation, historical weather).

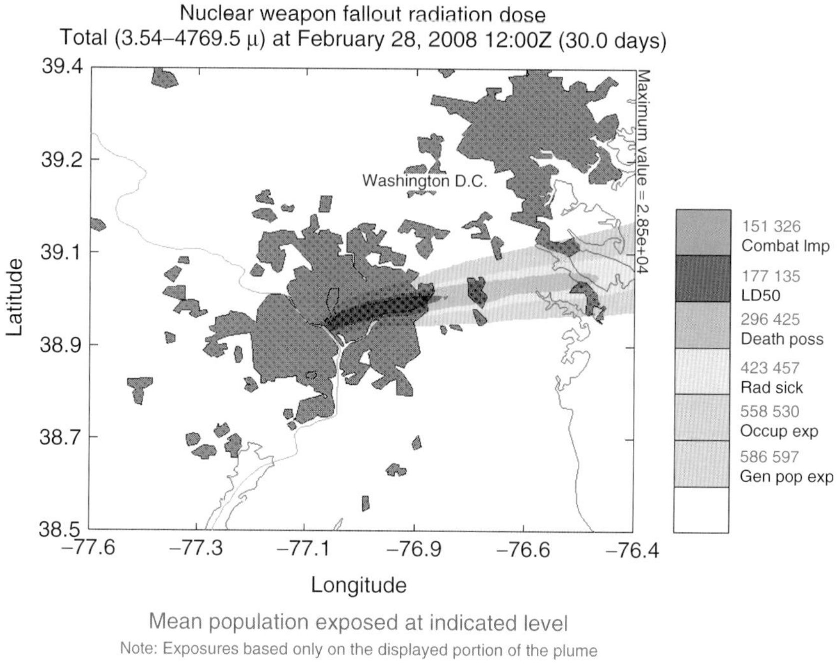

FIGURE 2 Twenty kiloton burst, zero height of burst. Thirty-hour accumulated radiation dose (HPAC calculation, historical weather).

in the United States—will show heavy damage at 2 psi and severe damage at 5 psi. Structures close enough to experience more than 5 psi dynamic pressures may expect to be demolished. At the lower end of the 2–5 psi damage regime, vehicles are likely to be overturned or damaged but may remain operable. Underground components such as pipes and mains will survive the blast, but above-ground structural damage may cause losses at connections. As shown in Figure 3, for a 20 kt surface explosion, 2–5 psi damage range translates into a distance of 2000–1200 m from the explosion point, respectively.

Since the energy of an explosion, which produces the blast is contained in a spherically expanding volume of hot gas, the force of the blast falls off as the cube of the distance from the explosion point and thus gives rise to the "cube-root scaling" typical of explosive phenomena. Thus, at any distance from the detonation site, a 20 kt event will be felt to be *approximately* three times as intense as would a 1 kt event set off at the same point. This is illustrated in Figures 4 and 5, which show the reach of 5 psi curves where severe damage may be expected for the Washington, DC, scenario [12].

3.3 Thermal Consequences

The threshold for a low yield weapon to ignite newspaper is approximately 5 cal/cm^2 [6]. The corresponding prompt effect slant range on a clear day for a near surface *air burst* is 700 and 3200 m, respectively, for 1 and 20 kT devices [6]. Persistent ignition of wood by direct thermal irradiation may be difficult, but combustibles contribute to fires.

Once fires are ignited, there is the possibility that they will spread, although this phenomenon is limited by fire breaks and availability of fuel. Both history and insight

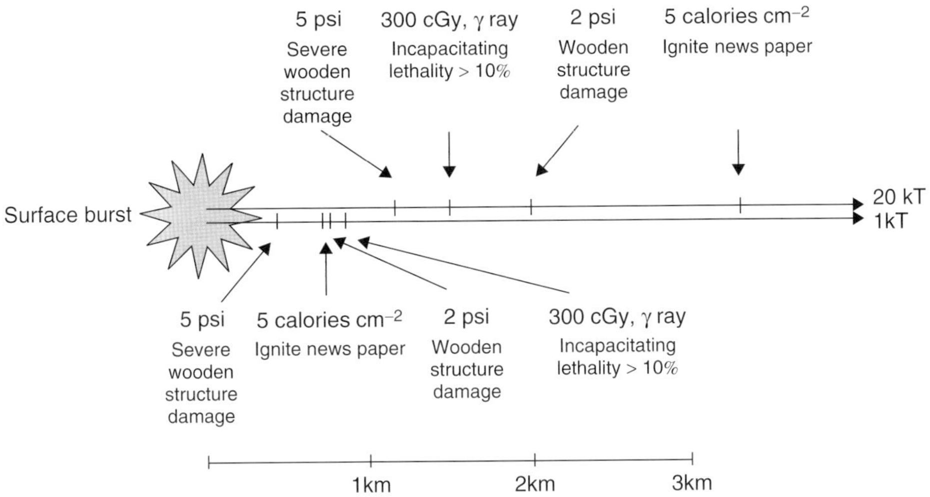

FIGURE 3 Fission explosions—distance versus threshold.

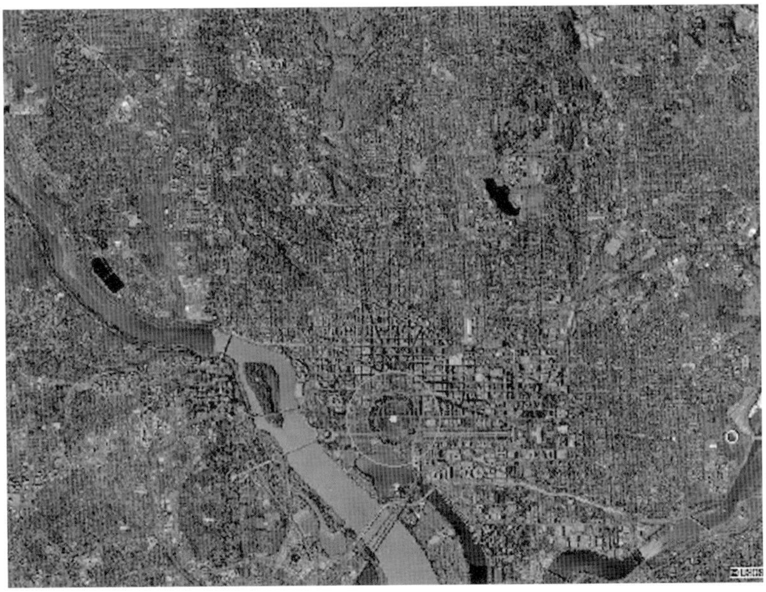

FIGURE 4 Two (blue) and five (yellow) pounds per square inch circles for a 1 kt explosion in downtown Washington, DC. (See online version for color).

gleaned from disruptive events such as earthquakes allow a confident prediction of significant fire activity, as additional damage—such as ruptured gas lines—contributes to ignition and fires. If enough fuel is present, ignition effects might possibly coalesce into a firestorm such as that occurred in Hiroshima, Nagasaki, Hamburg, or Dresden. This could cause more total casualties than the prompt blast effects of the bomb itself. In the

FIGURE 5 Two (blue) and five (yellow) pounds per square inch circles for a 20 kt explosion in downtown Washington, DC. (See online version for color).

two atomic bombings of Japan during World War II, it is impossible to estimate with any confidence the fatalities due to immediate bomb effects and those due to the subsequent firestorms. The likelihood of this extreme fire activity occurring in many modern US population centers may not be high as there is not as much wood as found in Japanese or old European construction.

3.4 Widespread Consequences

In the immediate vicinity of the blast, severe consequences discussed in the preceding sections may be expected to overshadow local concerns about electromagnetic phenomena. The consequences of essentially random equipment failures many kilometers from the blast zone are probably not a significant source of concern.

It is, however, possible to consider the potential for other major effects, secondary to the prompt damage, but which depend on the particular location of the explosion. For example, had the van delivering the nuclear device been rolling down New York's Wall Street rather than Washington DC's Pennsylvania Avenue, it is possible that the financial infrastructure of the United States might be severely compromised. The concern is not merely the shut down of trading markets for an undefined period, but the potential to adversely affect the database systems (ACH, SWIFT, Fedwire, CHIPS[4] [13]) that account for and manage the flow of trillions of dollars per day through the economy. Apart from

[4]Fedwire is the electronic funds transfer system operated by the Federal Reserve enabling fund transfers between financial institutions. CHIPS (clearing house interbank payments system) is an electronic system for interbank transfer and settlement. SWIFT (society for worldwide interbank financial telecommunications) provides stock exchanges, banks, brokers, and other institutions with a secure international payment message system ACH (automated clearing house) is an electronic network that processes credit and debit transactions.

great loss to life and property, the impact of this event would be of very high consequence on the financial and banking system impact [14].[5]

3.5 Uncertainties

The variation of effects given a yield, along with variations of demographics, geography, urban design, and environmental factors, all contribute to uncertainties in consequences. Studies of the effects of very low yield explosions are very rare and have generally not been performed with any scientific rigor. These represent another uncertainty. Understanding the terrorist threat demands an on-going study. Scenario development coupled to scientific and engineering analysis is useful for depiction of potential events, for gaining insight into complex processes and surprises, and for bounding the predictions.

3.6 Risk

A report by the National Commission on Terrorist Attacks Upon the United States finds that "al Qaeda continues to pursue its strategic objective of obtaining a nuclear weapon" [15]. Discussions of the threat presented by nuclear proliferation include recent Congressional concerns regarding networks that were identified years ago as supplying equipment and knowledge that supported the spread of nuclear weapons' potential around the globe [16]. There is a confluence of threats, vulnerabilities, and consequences, and thus a nonzero risk of a high consequence attack.

The quantitative determination of the risk posed by nuclear threats, the evaluation of this risk relative to other high consequence threats, and the prioritized allocation of resources to address such threats are the tasks of the DHS.

4 RISK REDUCTION: RECOMMENDATIONS

Leading the national effort to secure the United States is part of the mission of the DHS [17]. Increased attention is being given to preventing nuclear attacks in the US homeland by terrorist or clandestine operators [18, 19].[6,7] A single nuclear explosion would be a high consequence event that strains or overwhelms regional capability to respond and may inflict debilitating economic and social consequences that affect the entire nation. A scenario involving multiple nuclear explosions is traditionally a problem in the realm of military exchanges, but this may not remain the case. Risk management in this area faces the need for recognizing the large uncertainties, and developing strategies that address widely diverging goals.

[5]Under the assumptions of one study, the consequences for a "typical" low yield nuclear explosion result in a lifetime labor value loss of approximately 200 billion dollars. The losses are amplified by a (crudely estimated) comparable reduction in the gross domestic product and a smaller property loss. The total estimate is hundreds of thousands of casualties and a loss of 1.5 trillion dollars.

[6]The DHS Domestic Nuclear Detection Office (DNDO) was established in 2005 to improve the Nation's capability to detect and report unauthorized attempts to import, possess, store, develop, or transport nuclear or radiological material for use against the Nation, and to further enhance this capability over time.

[7]In 2004, a Task Force of subject-matter experts presented four areas of recommendation to the Department of Defense (DoD). Their findings are as follows: "make immediate operational changes", "improve nuclear intelligence collection capabilities", "start a spiral development program", and "establish joint warfighting requirements and capabilities". These include numerous specific recommendations, such as details regarding spiral development of some types of radiation detection system along with test beds and the S&T base.

The following general recommendations are offered to address potential susceptibilities across a diverse set of conditions.

- As a mechanism for minimizing risk, there is a premium to be gained by emphasizing prevention. During the cold war, this was attained largely due to the skillful use of deterrence. The risk of nuclear terrorism creates a strong need for additional methods, such as intelligence, detection, and interdiction capabilities. The importance of sensing and mitigating attempts to transport weapons and threat materials into the United States has been recognized with the creation of the DHS/DNDO. As was the case with cold war era studies, full systems analyses including political and cultural factors are critical to understanding what is needed to succeed in reducing the likelihood of nuclear explosions throughout the world.
- Given the great uncertainty in the diversity of the nature of the modern nuclear threat, there is compelling reason to examine a range of scenarios to extend existing studies to nontraditional attacks. What are the consequences for attacks against symbolic sites or obscure vulnerable points that may present high leverage for an adversary to inflict extensive damage on civilian infrastructure and key resources? It is advisable to prepare contingencies for strikes that are not well considered in the definitions of what constitutes attractive targets according to conventional wisdom.
- High consequence events are not defined by explosive yield alone. Studies that include a scenario to examine the consequences and response options to subkiloton weapons should be conducted, given the surprisingly low yield of the recent North Korean test. In particular, planning and preparations for responding to the lowest yield events observed in initial testing performed by nations may serve as a set of capabilities to gain experience and develop road maps for further expanding contingencies to better handle more stressing scenarios.
- Further study of the problem of protecting population against radiation effects from a terrorist nuclear attack appears desirable. The issue of remaining shielded indoors (if the structure is sufficiently intact) to await decay of intense radioactivity versus evacuation to minimize exposure time (including inhalation hazards) is important for further analysis under a variety of scenarios.
- This problem is not going to disappear, and the high consequences mean that an investment in analysis will remain an on-going part of our security needs throughout the foreseeable future. Efforts should be supported for identifying, clarifying, and estimating the conditional probabilities that describe the threat. Prioritized investment in response will require further investment in analysis tools to support realistic planning. A single detonation may be a great catastrophe, but it does not end civilization, and the need to continue developing mitigation and recovery strategies in the aftermath remains as a hedge against subsequent events.

The current multipolar nuclear threat environment is widely recognized as more complex and less transparent than the situation faced throughout most part of the cold war. Since 9/11, it has been clear that there are two general categories to address—actions by national actors and terrorist threats which may or may not be completely distinct from state motivations. Mitigation is not just a matter of better technology and traditional strategies. The pursuit of understanding nontechnical factors is very important for risk reduction.

REFERENCES

1. Malik, J. (1985). *The Yields of the Hiroshima and Nagasaki Nuclear Explosions*, Los Alamos National Laboratory, LA-8819, September 1985.
2. http://www.assemblee-nationale.fr/rap-oecst/essais_nucleaires/i3571.asp.
3. Office of the Director of National Intelligence (2006). Public Affairs Office, Washington, DC, October 16, 2006. This announcement was released at: http://www.dni.gov/announcements/20061016_release.pdf.
4. Ministry of the Russian Federation for Atomic Energy (1996). *USSR Nuclear Weapons Tests and Peaceful Nuclear Explosions: 1949 through 1990*, ISBN 5-85165-062-1, Russian Federal Nuclear Center-VNIIEF, 1996.
5. http://www.globalsecurity.org/security/library/report/2004/hsc-planning-scenarios-jul04.htm.
6. Glasstones, S., and Dolan, P. J. (1977). *The Effects of Nuclear Weapons*, U.S. DoD and ERDA, Washington, DC.
7. Buchan, G. C. et al. (2003). *MR-1231, Future Roles of US Nuclear Forces RAND*, Santa Monica.
8. Hall, E. J., and Giaccia, A. J. (2006). *Radiobiology for the Radiologist*, 6th ed., Lippincott Williams & Wilkins, Philadelphia.
9. (1996). *NATO Handbook on the Medical Aspects of NBC Defensive Operations AMedP-6(B), Part I- Nuclear*, FM 8-9, NAVMED P-5059, AFJMAN 44-151, Departments of the Army, The Navy, and the Air Force, Washington, DC, 1 February 1996.
10. Defense Threat Reduction Agency, http://www.ofcm.gov/atd_dir/pdf/hpac.pdf.
11. http://www.ornl.gov/sci/landscan/.
12. Federation of American Scientists *Nuclear Weapons Effects Calculator*, http://www.fas.org/main/content.jsp?formAction=297&contentId=367.
13. http://www.tax.state.ny.us/evta/glossary.htm.
14. Zycher, B. (2003). *A Preliminary Benefit/Cost Framework for Counterterrorism Public Expenditures*. RAND, MR1693.
15. http://govinfo.library.unt.edu/911/staff_statements/staff_statement_15.pdf.
16. (2007). *A.Q. Khan's nuclear Wal-Mart: out of Business or Under New Management?: Joint Hearing before the Subcommittee on the Middle East and South Asia and the Subcommittee on Terrorism, Nonproliferation, and Trade of the Committee on Foreign Affairs, House of Representatives, One Hundred Tenth Congress*, first session, June 27, 2007. http://foreignaffairs.house.gov/110/36424.pdf.
17. http://www.dhs.gov/xabout/strategicplan/.
18. http://www.dhs.gov/xabout/structure/editorial_0766.shtm.
19. Office of the Under Secretary of Defense for Acquisition, Technology, and Logistics. (2004). *Report of the Defense Science Board Task Force on Preventing and Defending Against Clandestine Nuclear Attack*, Office of the Under Secretary of Defense for Acquisition, Technology, and Logistics, Washington DC, June 2004.

FURTHER READING

Glasstone, S. and Dolan, P. J. (1977). *The Effects of Nuclear Weapons*, Department of Defense and the Department of Energy.

Ayyub, B. M. (2003). *Risk Analysis in Engineering and Economics*, Chapman and Hall.

MODELING POPULATION DYNAMICS FOR HOMELAND SECURITY APPLICATIONS

MARK P. KAMINSKIY AND BILAL M. AYYUB
Center for Technology and Systems Management, Department of Civil and Environmental Engineering, University of Maryland, College Park, Maryland

1 INTRODUCTION

Analysis of the risks related to combating terrorism problems is a multidisciplinary subject. It includes practically all the methods of probabilistic risk analysis such as scenario development, event-tree analysis, and precursor analysis. It is worth mentioning that the analysis of real data related to counterterrorist actions is definitely not on this list, which is true for the so-called simulation models [1, 2] as well. The simulation models can add some insights on terrorist population dynamics (TPD), but in contrast to the models considered below, no "quantitative conclusions can be made from the simulation results" [2]. Thus, developing the data analysis methodology related to counterterrorist actions constitutes an important part of the respective methodological tools needed.

International terrorism today is very different compared to what we dealt with in the 1970s and 1980s [3, 4]. The scale of this phenomenon is much larger, so we can apply the notion of *population* to this international community of terrorists to timely investigate its dynamics. The TPD models introduced below are similar (but not identical) to some survival data analysis models, which are, in turn, borrowed from nuclear physics and chemical kinetics theory [5, 6]. One can also find an analogy between the models considered below and the population dynamics models used in Ecology [7, 8], which were introduced independently by Volterra and Lotka in 1920s (the so-called Lotka–Volterra equations). In the suggested population dynamics models, the response function is the number of terrorist cells as a function of time counted from any conveniently chosen origin. It should be noted that *all* the parameters of these nonformal models are physically meaningful, for example, they can include the initial number of the terrorist cells N_0, the disabling rate constant λ or the cell half-life $t_{1/2}$, and the rate of formation of new cells P. The model parameters can be estimated using the data like, say, the annual or monthly numbers of disabled or identified terrorist cells. By estimating these parameters, one can assess other important characteristics, which include, but are not limited to, current number of terrorist cells, the time needed to reduce the number of active cells to a given level, the number of active cells at any given time in the future, and so on. The respective statistical data analysis is reduced to the routine nonlinear least squares estimation (LSE) procedures, which is illustrated by a case study. The article is mainly based on the authors' prior work [9], and also includes some new results on relationship between the Lotka–Volterra model and the suggested TPD model, as well as the above-mentioned case study.

2 CLOSED SYSTEM

For the time being, it is assumed that the system is *closed* (sealed). The closed system is defined in the given context, as such a system, is 100% protected from terrorists penetration from outside, as well as it is 100% protected from creating the terrorist cells inside this system.

Owing to all antiterrorist actions (inside and outside the system), it is reasonable to assume that the number of terrorist cells decreases according to the following differential equation:

$$\frac{dN}{dt} = -\lambda N \tag{1}$$

where λ is a positive rate constant. In nuclear physics, Eq. (1) is known as the radioactive *decay law*, and the constant λ is called *decay constant*. For the considered population of terrorist cells, the $1/\lambda$ can be called the *mean time to capture* or *mean lifetime of terrorist cells*. Another convenient parameter related to λ, which can be borrowed from nuclear physics, is the *half-life* $t_{1/2}$, that is, time needed for disabling of half of the cells, that is, $N(t_{1/2}) = N_0/2$, where N_0 is the initial number of the terrorist cells. Using Eq. (1), one obtains:

$$t_{1/2} = \frac{\ln 2}{\lambda} \tag{2}$$

The number of terrorist cells $N(t)$ at a given instant t is easily obtained by the integration of Eq. (1):

$$N(t) = N_0 \, e^{-\lambda t} \tag{3}$$

It should be noted that from the data analysis standpoint, Eq. (3) is not convenient. In reality, one can observe only the cumulative number of *disabled* cells (denoted below by N_d), which is given by

$$N_d(t) = N_0 - N_0 \, e^{-\lambda t}$$

so that

$$N_d(t) = N_0(1 - e^{-\lambda t}) \tag{4}$$

On the basis of available data, the parameters of Eq. (4), that is, the initial number of the terrorist cells N_0 and the rate constant λ (or the half-life $t_{1/2}$), can be estimated. Eq. (4) is illustrated in Figure 1.

3 SYSTEM WITH FORMATION OF NEW CELLS

A more complicated situation arises, when the system is modeled assuming that it is not 100% protected from terrorist penetration from outside, and it is not 100% protected from formation of the terrorist cells inside the system itself. Similar to the previous case, it is assumed that the terrorist cells are disabled at the rate λ.

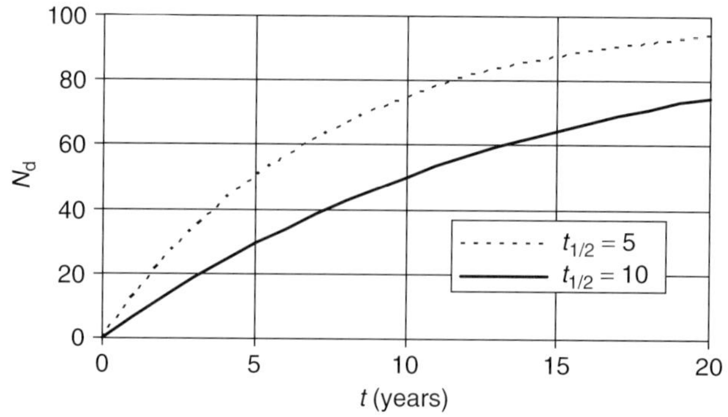

FIGURE 1 Cumulative number of disabled cells N_d as a function of time, $N_0 = 100$.

In this case, Eq. (1) is replaced by

$$\frac{dN}{dt} = -\lambda N + P \tag{5}$$

where P is the rate, at which the new cells are produced due to the penetration from outside the system and/or due to the formation of cells inside the system.

Qualitatively, Eq. (5) shows that one deals with two conflicting processes—the formation of new cells at rate P and their disabling at rate λN. Depending on the sign of the right side of Eq. (5), the number of cells either increases or decreases. The last case is considered below.

The traditional solution of Eq. (5) used in nuclear physics [5] is based on the initial condition $N(0) = N_0 = 0$, which assumes, in the given context, that at the beginning, there are no terrorist cells or their number is negligible compared, say, to the number of cells produced at the rate P during first year. This initial condition does not look realistic for the problems considered, which is why it is not discussed in this paper.

The more realistic initial condition assumes that at the beginning, there is a number of terrorist cells, that is, $N(0) = N_0 > 0$. The solution of Eq. (5) under this initial condition is considered below.

Eq. (5) is a linear differential equation of the first order and of the first degree in N and its derivative. This type of equation may be solved using integrating factors. It can be shown that the solution of Eq. (5) under the initial condition $N(0) = N_0 > 0$ is given by

$$N(t) = \frac{P}{\lambda}\left(1 - e^{-\lambda t}\right) + N_0 e^{-\lambda t} \tag{6}$$

Similar to Eq. (4) from the previous section, Eq. (6) should be rewritten in terms of the observable cumulative number of disabled cells $N_d(t)$ as

$$N_d(t) = \left(1 - e^{-\lambda t}\right)\left(N_0 - \frac{P}{\lambda}\right) \tag{7}$$

where $\frac{P}{\lambda} < N_0$.

On the basis of the available data, the parameters of Eq. (7), that is, the initial number of the terrorist cells N_0, the disabling rate constant λ (or the cell half-life $t_{1/2}$), and the rate of formation of new cells P, can be estimated. Eq. (7) is illustrated in Figure 2.

The figure reveals that the counterterrorist actions (evaluated through the disabling rate constant $\lambda = 0.139$ per year) are less effective when the rate of formation of new cells P is five cells per year compared to the situation when the rate P is only two cells per year. On the basis of Eq. (6), Figure 3 displays the above conclusion in terms of number of active cells.

Using models (6) and (7) one can estimate such important characteristics like the time needed to reduce the number of active cells to a given level, the number of active cells at any given time in the future, and so on.

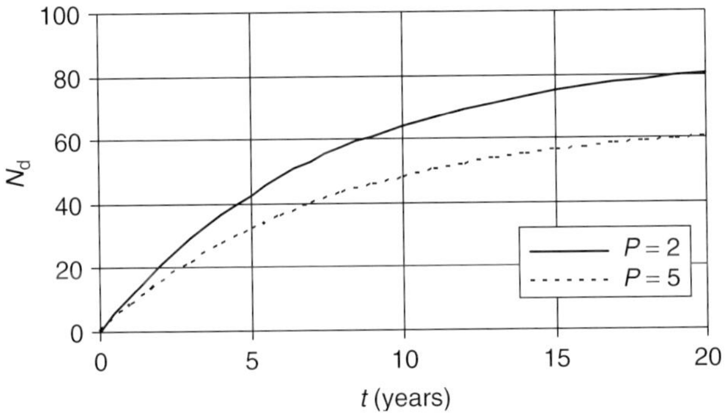

FIGURE 2 Cumulative number of disabled cells N_d as function of time, t: for initial number of cells $N_0 = 100$, disabling rate constant $\lambda = 0.139$ per year ($t_{1/2} = 5$ years), and different rates of formation of new cells $P = 2$ and 5.

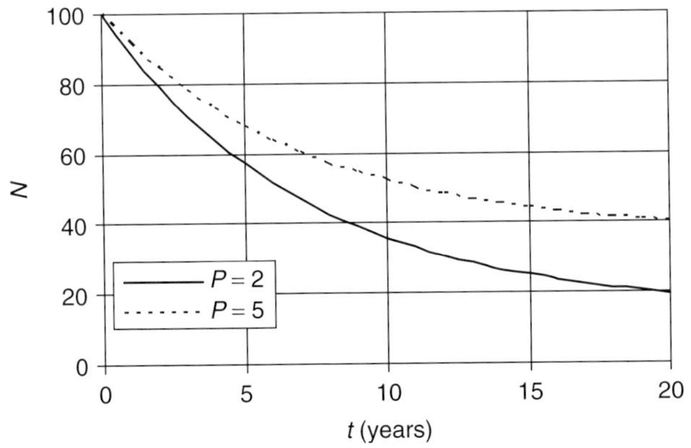

FIGURE 3 Number of active cells N as function of time, t: for initial number of cells $N_0 = 100$, disabling rate constant $\lambda = 0.139$ per year ($t_{1/2} = 5$ years), and different rates of formation of new cells $P = 2$ and 5.

4 TERRORIST POPULATION DYNAMICS MODEL AND LOTKA–VOLTERRA MODEL

As was mentioned above, one can find some analogy between the TPD model and the Lotka–Volterra model. This analogy is considered in more detail below. The Lotka–Volterra model was developed by the founders of Mathematical Ecology [7, 8, 10] as an attempt to understand the dynamics of biological system, in which a predator population interacts with a prey population.

Denote the prey population size by N. It is assumed that the prey population has unlimited food supply, and in the absence of predators, the prey population increases exponentially with a positive rate r, that is,

$$\frac{\mathrm{d}N}{\mathrm{d}t} = rN \tag{8}$$

On the other hand, the prey population decreases due to predator–prey encounters. If the current predator population size is C, then the predator consumption rate of prey is αCN, so that

$$\frac{\mathrm{d}N}{\mathrm{d}t} = rN - \alpha CN \tag{9}$$

where α is the so-called attack rate.

In the absence of food, the size of the predator population C decreases. Thus, in the framework of the Lotka–Volterra model, the predator population is assumed to decline exponentially in the absence of prey:

$$\frac{\mathrm{d}C}{\mathrm{d}t} = -qC \tag{10}$$

where q is the predator mortality rate. This predator population decline is counteracted by predator birth. The predator birth rate is assumed to depend on two factors: the rate, at which the food is consumed, αCN, and the predator's efficiency, f, at turning the food into predator offspring. Thus the following equation for the predator population size can be written:

$$\frac{\mathrm{d}C}{\mathrm{d}t} = f\alpha CN - qC \tag{11}$$

Eqs. (9) and (11) constitute the Lotka–Volterra model.

At this point, our objective is to adjust the Lotka–Volterra model to get the TPD model obtained in Section 3 using some nuclear physics analogy.

To get this model, one has to replace the notion of predators by counterterrorism forces and the notion of prey by a terrorist cell or individual. Using this new terminology, it is reasonable to begin with revising Eq. (9) and the main assumptions on which this equation is based.

In the framework of the Lotka–Volterra model, it is assumed that in the absence of counterterrorism forces (predators), the terrorist (prey) population has unlimited financial and logistic support (food supply) and it increases exponentially according to Eq. (8), which solution is given by

$$N(t) = N_0 \, e^{rt} \tag{12}$$

where N_0 is the initial (at $t = 0$) size of the population.

These assumptions applied to the TPD might be too strong to result in the exponential population growth. Keeping this in mind, the exponential terrorist population growth term in Eq. (9) has to be replaced by a linear growth term, which results in the following equation:

$$\frac{dN}{dt} = r - \alpha C N \tag{13}$$

Denoting αC by λ, one gets the TPD model (5) suggested in [9], but it should be noted that Eq. (13) has one more interpretable parameter α (the attack rate) when the information about factor C (which was introduced above as the antiterrorist manpower, expenses related to antiterrorist operations, etc.) is available.

The solution of Eq. (13) can be written as

$$N(t) = \frac{r}{\alpha C}\left(1 - e^{-\alpha C t}\right) + N_0 \, e^{-\alpha C t} \tag{14}$$

From the data analysis standpoint, Eq. (14) is not convenient. In reality, one can observe only the cumulative number of disabled cells (denoted below by N_d), which is given by

$$N_d(t) = \left(1 - e^{-\alpha C t}\right)\left(N_0 - \frac{r}{\alpha C}\right) \tag{15}$$

If C is a time-dependent factor, Eq. (13) takes on the following form

$$\frac{dN}{dt} = r - \alpha C(t) N \tag{16}$$

The solution of the above equation is out of the scope of this article.

5 COST-EFFECTIVENESS ANALYSIS OF ANTITERRORIST EFFORTS

Let us assume that a steady amount of S dollars is spent annually on the antiterrorist efforts in the framework of the model considered. The cost-effectiveness of the efforts can be evaluated using the following parameter ε:

$$\varepsilon(t) = \frac{S}{\dfrac{dN_d(t)}{dt}} \tag{17}$$

which has the meaning of the cost of disabling one cell during a given year t. Using, for example, Eq. (7), the derivative in the denominator can be written as

$$\frac{dN_d(t)}{dt} = \lambda e^{-\lambda t}\left(N_0 - \frac{P}{\lambda}\right) \tag{18}$$

so that the cost-effectiveness equation (Eq. 17) takes on the following form:

$$\varepsilon(t) = S\frac{e^{\lambda t}}{\lambda\left(N_0 - \dfrac{P}{\lambda}\right)} \tag{19}$$

Figure 4 illustrates the cost-effectiveness ε time dependence for the same model parameters as it is shown in the previous figures. The figure shows that the cost per terrorist cell disabled is exponentially increasing as time goes on. The cost-effectiveness parameter $\varepsilon(t)$ can be applied to assess a time dependence of the risk associated with potential terrorist actions. The respective risk assessment is out of the scope of this article. Nevertheless, assuming that an acceptable risk level [11] is established, it is clear that the effectiveness parameter $\varepsilon(t)$ can be used for making a timely decision about the necessity of revising current antiterrorist policy.

6 STATISTICAL DATA ANALYSIS

The format of the data needed to fit the above discussed models can be given by a simple table with the following three columns:

1. Time (year or month)
2. Number of disabled cells
3. Optional column—cell origin (foreign or domestic)

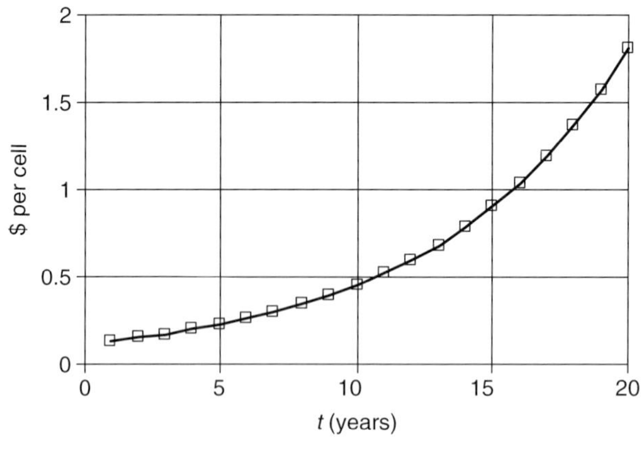

FIGURE 4 Cost of disabling one cell during a given year t per 1$ invested annually. Initial number of cells $N_0 = 100$, disabling rate constant $\lambda = 0.139$ per year ($t_{1/2} = 5$ years), and rate of formation of new cells $P = 5$.

Applying Eq. (7) to fit the respective data is far from being a trivial statistical problem, but it is doable. Moreover, applying Eq. (7) to real data can provide some extra benefits. For example, if the respective data can be divided into two groups—one for the terrorist cells of domestic origin and the other for the cells associated with terrorist penetration from outside [12], Eq. (7) can be applied separately to each group providing some more specific information, for example, one can get the following two equations with more informative parameters.

In the case of the terrorist cells of domestic origin, Eq. (7) takes on the following form:

$$N_{\mathrm{dd}}(t) = (1 - \exp(-\lambda_{\mathrm{d}}t))\left(N_{0\mathrm{d}} - \frac{P_{\mathrm{d}}}{\lambda}\right) \tag{20}$$

where the parameters $N_{0\mathrm{d}}$, P_{d}, and λ_{d} are associated with the cell of a domestic origin only. Compared to model (4), model (20) does not assume that the system of interest is 100% protected from the formation of the terrorist cells inside this system.

Correspondingly, in the case of the cells associated with (foreign) terrorist penetration only from outside is given by

$$N_{\mathrm{df}}(t) = (1 - \exp(-\lambda_{\mathrm{f}}t))\left(N_{0\mathrm{f}} - \frac{P_{\mathrm{f}}}{\lambda}\right) \tag{21}$$

where the parameters $N_{0\mathrm{f}}$, P_{f}, and λ are related to the terrorist cells of a foreign descent.

Obviously, models (20) and (21), when their parameters estimated, provide the information that can be used for balancing the efforts related to counterterrorism actions inside the system and the efforts protecting its borders.

6.1 Case Study

The *Sourcebook of Criminal Justice Statistics 2003* [13] provides the data on terrorist incidents and preventions in the United States from 1980 up to 2001. In this source, terrorism prevention is defined as "a documented instance in which a violent act by a known or suspected terrorist group or individual with the means and a proven propensity for violence is successfully interdicted through investigative activity". For lack of other available data, it is excusable to assume that the annual number of preventions is close to the annual number of disabled cells (note that, strictly speaking, the number of terrorism prevention is equal or less than the number of the respective disabled cells).

Based on this assumption, one can use model (7) to fit the data on the terrorism preventions. In this case, model (7) takes on the following form:

$$N_{\mathrm{p}}(t) = \left(1 - \mathrm{e}^{-\lambda t}\right)\left(N_0 - \frac{P}{\lambda}\right) \tag{22}$$

where N_{p} is the cumulative number of preventions (disabled cells) and $\frac{P}{\lambda} < N_0$.

For the given case study, the selected data related to the time interval 1984–1994 were used. The data are displayed in the first two columns of Table 1.

In order to better interpret these data, the reader is referred to the Naftali's monograph on the history of modern terrorism [4].

Based on the data given in Table 1, the nonlinear least square estimates of the parameters of model (20) were obtained as follows:

TABLE 1 Number of Terrorism Preventions

Year	Number of Terrorism Preventions	Cumulative Number of Preventions (Observed)	Cumulative Number of Preventions (Fitted Using Eq. (22))
1984	9	9	15.1
1985	23	32	27.4
1986	9	41	37.4
1987	5	46	45.5
1988	3	49	52.1
1989	7	56	57.5
1990	5	61	61.8
1991	5	66	65.3
1992	0	66	68.2
1993	7	73	70.5
1994	0	73	72.4

disabling rate $\lambda = 0.208$ per year ($t_{1/2} = 3.339$ years),
initial (related to 1983) number of cells $N_0 = 131$, and
rate of formation of new cells $P = 10.534$ per year.

Cumulative numbers of preventions fitted using Eq. (20) are given in the right column of Table 1, and illustrated in Figure 5. The fraction of variation of N_p explained by the fitted model is 0.974.

7 CONCLUSIONS

Among the numerous methods of risk analysis related to the problems of combating terrorism, the data analysis does not always receive adequate attention mainly owing

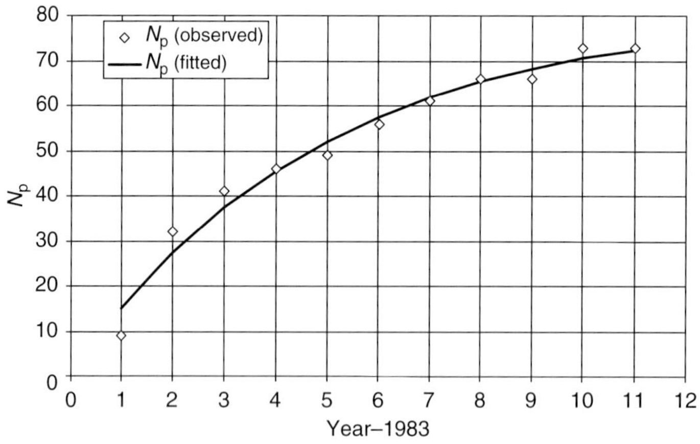

FIGURE 5 Observed and fitted cumulative number of terrorism preventions N_p.

to the classified status of most relevant data. Nevertheless, developing methodology of the data analysis related to counterterrorist actions constitutes an important part of the respective methodological tools needed.

As an example of these tools, the TPD model is introduced. The model describes the evolution of the system in terms of time dependence of the number of terrorist cells $N(t)$ acting inside the considered system. The model includes only interpretable parameters, such as the initial number of terrorist cells N_0, the disabling rate constant λ (or the cell half-life $t_{1/2}$), and the rate of formation of new cells P.

The suggested cost-effectiveness parameter $\varepsilon(t)$ is based on the cost per terrorist cell disabled as a function of time. If combined with a given acceptable risk level related to terrorist actions, the parameter can be used for making a timely decision regarding the necessity of revising current antiterrorist policy.

Another important issue raised concerns balancing the efforts related to counterterrorism actions inside the system and the efforts needed to protect its borders.

From the standpoint of data collection and analysis, it is important that the only observable variables needed are the annual (or monthly) numbers of disabled (or identified) terrorist cells. On the basis of these data, the suggested model can evaluate the number of terrorist cells $N(t)$ acting inside the system at any given time (including current and predicted numbers), the initial number of the terrorist cells N_0, the disabling rate constant λ (or the cell half-life $t_{1/2}$), and the rate of formation of new cells P.

On the basis of the results of model parameter estimation, one can estimate such important characteristics like the time needed to reduce the number of active cells to a given level, the number of active cells at any given time in the future, and so on.

Using a case study, it is shown that the model parameters can be successfully estimated applying the nonlinear least squares fitting. It should be noted that the suggested models can be applied to other criminal populations, for example, illegal immigrants and drug offenders.

REFERENCES

1. Smith, R. Armed Forces Communications and Electronics Association (AFCEA). (2001). Modeling and simulation adds insight on terrorism. *Signal Mag* December, 31–35.
2. Raczynski, S. (2004). Simulation of the dynamic interactions between terror and anti-terror organizational structures. *JASSS*. **7**(2), http://jasss.soc.surrey.ac.uk/7/2/8.html.
3. Linden, E. V. Ed. (2004). *Foreign Terrorist Organizations: History, Tactics and Connections*. Nova Science Publishers, Inc., New York.
4. Naftali, T. (2005). *Blind Spot. The History of American Counterterrorism*, Basic Books, New York.
5. Bertulani, C. A., and Schechter, H. (2002). *Introduction to Nuclear Physics*, Nova Science Publishers, Inc., New York.
6. Draper, N. R., and Smith, H. (1998). *Applied Regression Analysis*, 3rd ed., Wiley, New York.
7. Begon, M., Harper, J. L., and Townsend, C. R. (1996). *Ecology: Individuals, Populations, and Communities*, 3rd ed., Blackwell Science, Cambridge, MA.
8. Gotelli, N. J. (1998). *A Primer of Ecology*, 2nd ed., Sinauer Associates, Sunderland, MA.
9. Kaminskiy, M., and Ayyub, B. (2006). Terrorist population dynamics model. *Risk Anal. Int. J.* **26**(3), 747–752.

10. Volterra, V. (1931). Variations and fluctuations of the number of individuals in animal species living together. In *Animal Ecology*, Translated from 1928 edition by R. N. Chapman, Ed. McGraw-Hill, New York.
11. Ayyub, B. M. (2003). *Risk Analysis in Engineering and Economics*, Chapman & Hall/CRC Press, Boca Raton, FL.
12. Pate-Cornell, E. (2004). On signals, response, and risk mitigation. In *Accident Precursor Analysis and Management*, J. R. Phimister, V. M. Bier, H. C. Kunreuther, Eds. The National Academic Press, Washington, DC.
13. U.S. Department of Justice, Federal Bureau of Investigation. (2006). *Terrorism 2000/2001*, http://www.albany.edu/sourcebook/pdf/t3173.pdf.

FURTHER READING

Harris, B. (2004). *Mathematical methods in combating terrorism*. *Risk Anal. Int. J.*. **24**(4), 985–989.

Farley, J. D. (2003) Breaking Al Qaeda cells: A mathematical analysis of counterterrorism operations (A guide for risk assessment and decision making). *Studies in Conflict & Terrorism*. **26**, 399–411.

Wilson, A. G., Wilson, G. D., Olwell, D. H., Eds. (2006). *Statistical Methods in Counterterrorism: Game Theory, Modeling, Syndromic Surveillance, and Biometric Authentication*, Springer, New York.

SENSING AND DETECTION

PROTECTING SECURITY SENSORS AND SYSTEMS

TODD P. CARPENTER
Adventium Labs, Minneapolis, Minnesota

1 INTRODUCTION

Security systems are key protection elements which prevent, detect, mitigate, minimize, and aid recovery from natural, accidental, and intentional threats ranging from weather effects, illness and disease, to misguided youth, hardened criminals, and terrorists. The security systems can be focused on the outside, perimeter, or be inward facing. The systems can be self-contained, such as common smoke detectors, or they can include multiple sensors of different types, integrated together over a distributed sensing and power network with storage of events, sensor fusion and event correlation, and automatic escalation of event notification based on severity. These systems can be purely physical (e.g. military trip wire and flash bang), or heavily supported by electronic sensing and analysis. Security systems considered in this article integrate some type of sensor, reasoning (is it an event or not?), and alarm or state annunciation. Simpler physical security systems, such as a door lock or a Sargent and Greenleaf [1] lock on a file cabinet, are not explicitly addressed. Locks have their own fascinating and evolving issues, which can be explored in [2–5]. Systems considered here share the potential for physical, cyber, and social engineering vulnerabilities that attackers can exploit to overcome the security system. This article provides an overview of threats, potential weaknesses, and some risk-reduction approaches and resources to consider when designing, developing, or applying a security system.

2 BACKGROUND

Prevention or deterrence, confidence bolstering, detection, mitigation, minimization of effects, and recovery, including providing forensic evidence, are potential benefits of security systems. The overt presence of a security system, such as visible cameras or biometric sensors, might deter some class of criminals or other adversaries from attempting to enter the premises. The same security system in a bank might inspire confidence in potential customers—if the bank looks safe, they might be more inclined to do business there. Security event detection and subsequent notification of guards or other authorities are perhaps what people normally think of as security systems. Security systems support mitigation when they provide situation awareness that the responders can leverage to address ongoing events, such as cameras or motion detectors that indicate conditions or attacker locations. Minimizing the effects or reducing the potential impact of a successful attack, might be considered a side effect of a security system. (For instance, installing a real security system on a high-value asset, such as critical infrastructure, is rarely a trivial exercise. In the process, one might presume that subject matter experts analyze what is being protected. If toxic chemicals in storage tanks are the asset, one might reassess how much material needs to be on hand and reduce the amount if possible.) Security systems might also help recovery and forensics efforts by providing causal information and information used to identify the attackers.

Understanding why a security system is installed can give attackers clues about how to defeat aspects of the system about which the attacker cares, or it might completely deter the attacker. During reconnaissance, a potential attacker might size up security systems and select the least protected target. There's an old adage that says you only have to make your security better than the neighbor. That does not really apply when protecting our intertwined infrastructures. Seasoned professionals might note model numbers and designs of the security units (Decoy cameras might not provide much value in this case. Unfortunately, deterrence does not always work. Despite fences, locks, visible cameras, energized high-voltage lines, and fatalities from earlier attempts, thieves are attacking the US electric grid for copper to sell on the scrap market [6]), and devise specific attacks, or choose to accept the exposure that would go along with an attack.

The following subsections will describe the broad classes of attacks and provide an abstract model of a security system. The next section will discuss specific threats, attacks on the sensor and security systems based on attacker goals and capabilities, and possible solutions. The remaining sections will cover future directions and recommended approaches to designing and acquiring security systems.

2.1 Classes of Attacks

Social engineering is the practice of discovering and exploiting humans to gain information or access. It can be as simple as ignoring the security system and merely joining in with other people entering work in the morning, from their smoke break on the loading dock, or mingling with a crowd reentering a building after a fire alarm drill (which are common during fire-safety week in the United States), or as complex as wooing a bank staff personnel with chocolates, and walking away with €21m of diamonds [7]. Unfortunately, a worthy treatment of this topic is beyond the scope of this article. The Art of Deception [8] is a highly recommended read, because a fool with a tool is still a fool: If the people specifying, researching, designing, building, operating, and managing the security system are not appropriately trained, technology might not impede the determined attackers.

Cyber attacks are a broad class of electronic-based attacks, which affect code or data in the security system. For this treatment, physical attacks are those that employ physical effects or tangible material to cause desired sensor behavior (e.g. use of fingerprint duplicate to gain access through a reader) or physically change the state of the security system (e.g. sever an alarm wire so that the alarm cannot be activated). We are not considering physical attacks such as armed attackers overcoming the guard, since you probably want your security system to provide enough notification that your guards would be prepared for and fend off any credible attack of that sort. Many attacks combine multiple aspects of physical and cyber attacks; for example, a denial-of-service cyber attack might create an opportunity for the compromised maintenance personnel to add access credentials enabling future attacks. Another attack could create specific false security events through some physical or cyber interactions. Eventually, the guards could get used to ignoring that particular alarm, or disable it altogether, which would be an example of leveraging cyber or physical attacks into social engineering. Incidentally, avoiding alarm saturation leading to guard conditioning of this sort is one of the reasons why low "false-positive" rates are so important.

2.2 Security Sensors and Systems

The abstract security system shown in Figure 1 identifies the external interfaces to a security system and major internal components. Examples of classes of security sensors and how they are used and implemented are shown in Table 1. Information from the sensors is processed according to configuration settings, optionally stored in a historical database along with timestamp and other identifying information, and state and alarms are presented to the operator. The security system might also actuate some mechanical device, such as unlocking a door. The primary inputs to the security system are power, environmental conditions (e.g. ambient light, temperature, motion, the supporting structure, humidity, or precipitation), control inputs from the operator, and the actual sensed security events. Processing can be performed on custom hardware (e.g. smoke

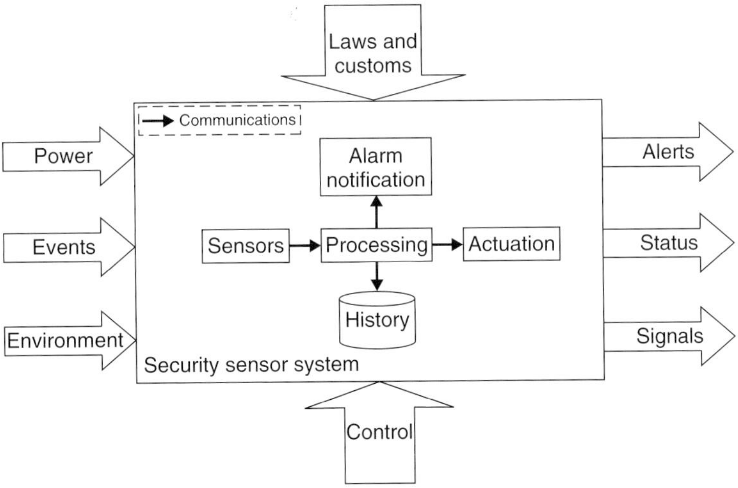

FIGURE 1 Abstract security system: showing both the external environment with inputs, constraints, outputs, and control, and internal system components of sensor(s), communications, processing, history, and notification.

TABLE 1 Security Sensor Examples: Broad Classes of Sensors Employed in Security Systems, Including Example Applications and Implementations

Sensor	Example Uses	Example Sensor Implementations	See Also
Contact	Door open/close/event	Simple mechanical or magnetic reed switches, photoelectric or hall effect sensors	Perimeter Security Contact and Proximity Sensors
Motion	Identify motion in area	Passive infrared, active IR, visible light camera, microwave, buried cable capacitance, accelerometers	Video Surveillance Sensors: Cameras and Digital Video Recorders; Video Sensor Systems and Integrated Cameras
Fingerprint	Identify individuals	Capacitance, thermal, visual, pressure	Checkpoint Access Control Sensors for Identity Management; Automated Fingerprint Indentification Systems
Hand geometry	Identify individuals	Capacitance, thermal, visual, pressure	Checkpoint Access Control Sensors for Identity Management
Iris	Identify individuals	Visible, IR camera	Checkpoint Access Control Sensors for Identity Management; Video Surveillance Sensors: Cameras and Digital Video Recorders; Video Sensor Systems and Integrated Cameras; Eye and Iris Sensors
Voice	Identify individuals	Microphone	Voice Recognition and Speaker Identification Sensors
Facial	Detect and identify individuals	Visible, IR camera, light detecting and ranging (LIDAR) laser	RADAR and LIDAR Perimeter Protection Sensors; Infrared Sensors and Systems; Face Recognition Sensors 2D Face Recognition Sensors and Their FR Algorithms; 3D Face Recognition Sensors-Technological Challenges and Potential Benefits; Normalization and Morphing of Face Images for Improved Face Recognition Sensors
Pressure plate	Detect occupation	Mechanical, piezo, and accelerometers	Perimeter Security Contact and Proximity Sensors

TABLE 1 ***(Continued)***

Sensor	Example Uses	Example Sensor Implementations	See Also
Particulate	Detect smoke or other particulates	Ionization, optical detection	Knowledge Extraction from Surveillance Sensors; Checkpoint Access Control Sensors for Identity Management; Perimeter Security Contact and Proximity Sensors; RADAR and LIDAR Perimeter Protection Sensors; Video Surveillance Sensors: Cameras and Digital Video Recorders; Video Sensor Systems and Integrated Cameras; Infrared Sensors and Systems; Remote Un-Manned Surveillance Sensors; Sensors for Airborne Video Surveillance Applications; Physical Forms of CB Threats; Design Considerations in Development and Application of Chemical and Biological Agent Detectors; Sensing Dispersal of Chemical and Biological Agents in Urban Environments; Technologies for Sensing Terrorist use of Chemical and Nerve Agents Threats; Use of Biological Agent Sensors in Homeland Security; Sensing Releases of Highly Toxic and Extremely Toxic Compounds
Chemical	Detect presence of chemical agent	Capacitive, reactive, micro-electro mechanical systems (MEMS) vibration	Knowledge Extraction from Surveillance Sensors; Checkpoint Access Control Sensors for Identity Management; Perimeter Security Contact and Proximity Sensors; RADAR and LIDAR Perimeter Protection Sensors; Video Surveillance Sensors: Cameras and Digital Video Recorders; Video Sensor Systems and Integrated Cameras; Infrared Sensors and Systems; Remote Un-Manned Surveillance Sensors; Sensors for Airborne Video Surveillance Applications; Physical Forms of CB Threats;

(continued overleaf)

TABLE 1 ***(Continued)***

Sensor	Example Uses	Example Sensor Implementations	See Also
			Design Considerations in Development and Application of Chemical and Biological Agent Detectors; Sensing Dispersal of Chemical and Biological Agents in Urban Environments; Technologies for Sensing Terrorist use of Chemical and Nerve Agents Threats; Use of Biological Agent Sensors in Homeland Security; Sensing Releases of Highly Toxic and Extremely Toxic Compounds
Biological	Detect presence of biological agent	Reactive, fluorescence	Knowledge Extraction from Surveillance Sensors; Checkpoint Access Control Sensors for Identity Management; Perimeter Security Contact and Proximity Sensors; RADAR and LIDAR Perimeter Protection Sensors; Video Surveillance Sensors: Cameras and Digital Video Recorders; Video Sensor Systems and Integrated Cameras; Infrared Sensors and Systems; Remote Un-Manned Surveillance Sensors; Sensors for Airborne Video Surveillance Applications; Physical Forms of CB Threats; Design Considerations in Development and Application of Chemical and Biological Agent Detectors; Sensing Dispersal of Chemical and Biological Agents in Urban Environments; Technologies for Sensing Terrorist use of Chemical and Nerve Agents Threats; Use of Biological Agent Sensors in Homeland Security; Sensing Releases of Highly Toxic and Extremely Toxic Compounds
Radiation	Detect presence of ionizing radiation	α, β, γ, neutron, X-ray radiation detectors using photomultipliers, charge-coupled devices, gaseous detectors	Radioactive Materials Sensors; Sensing Dirty Bombs

TABLE 1 (*Continued*)

Sensor	Example Uses	Example Sensor Implementations	See Also
Temperature	Detect temperature variations	IR sensor, thermistor, thermocouple	Perimeter Security Contact and Proximity Sensors
Card reader	Identify bearer	Magnetic, RF	Checkpoint Access Control Sensors for Identity Management
RFID	Identify bearer	Passive RF, active RF	Checkpoint Access Control Sensors for Identity Management
Explosives	Detect explosives	Chemical vapor detectors, neutron detectors	Sensors for Weapons Concealment in Shipments; Sensing Body Worn Concealed Weapons and Explosives; Detection of Metallic and Ceramic Concealed Weapons in Hand Carried Articles; Detection of Concealed Explosives and Chemical Weapons in Hand carried Baggage

detectors, embedded motion sensors) or on general purpose processors running standard operating systems (e.g. web-enabled cameras running on Linux or integrated security systems built on top of Microsoft Windows). The communications connecting all these elements might be copper, fiber, wireless (radio frequency (RF) or microwave), or any combination thereof. The sensor communications network might be separate from the rest of the security network, or everything might be hosted on the corporate local area network (LAN). The next section describes how all of these elements can be exploited by attackers to gain some desired advantage.

3 THREATS, VULNERABILITIES, AND RISK MITIGATION

We begin here with an overview discussion of specific threats, vulnerabilities, and risk mitigation without going into exhaustive detail. More detail will be provided in the various technical articles and case studies in the rest of the handbook. Other sources of vulnerability information can be found at popular conferences such as DEFCON [9] and Black Hat [10] and online security tracking and reporting such as 'SysAdmin, Audit, Network, Security (SANS) Institute' [11].

3.1 Threats

Considered generically, attackers might be grouped into different categories, including petty thieves, disgruntled employees, and criminal organizations. If we consider security systems protecting US critical infrastructure, we might also include joy riders (teenagers in the countryside with a new rifle shooting insulators on high-voltage power lines), environmental terrorists (ranchers toppling high-voltage power towers), state-sponsored terrorists, strategic threats, and naturalized radical elements. Of these, the deterrence effects of strong security systems might eliminate many of the nuisance attacks and petty theft; though we have seen it does not deter copper theft. The other threats are interesting because they possess technical and monetary resources, and indirectly affecting US critical infrastructure is plausibly within their goals. See also Part I, Organizational, Structural, and Policy Issues for a more detailed treatment of possible threats and motivations.

Of particular note are the capabilities that potential attackers bring to bear, whether it is cash to buy off-the-shelf tools, fabricate new tools, or invest in research and evaluation of sensors and systems to identify vulnerabilities. Such attackers can also expend human capital—people willing or coerced into probing security systems.

3.2 Individual Vulnerabilities

Considered independently, the ways by which physical phenomenology is sensed by each of the security sensors, shown in Table 1, have characteristics that can be exploited. Switches can be shorted or mechanically altered. Motion sensors can be overwhelmed by intense light or fooled by slow movement. Voiceprints can be fooled by recordings.

Security sensors that interact with cards and radio frequency identification (RFID) also have issues. Barcode and magnetic strip cards can be duplicated. Smart cards, such as in the new contactless e-passports, can be cloned [12–15]. RFID and card readers can be spoofed and jammed [16, 17] or communicated with from a distance far exceeding their design specifications [18]. Industry advocates claim that this is not a problem [19], perhaps ignoring the fact that one difference with contactless cards is that the bearers have no way of detecting when their personal information have been compromised.

Additional problems might arise when vulnerable biometrics on the smart cards are solely relied upon, especially if the biometric sensing and matching is automated. Vulnerabilities of biometric sensors is a whole topic by itself [20–23] and the handbook section Terrorist Identification and Recognition, and it is recognized that "there is no one perfect biometric that fits all needs" [24]. Hollywood is full of tales of biometric compromises, some of which even work, such as disguises, lifting fingerprint from surfaces, reactivating latent fingerprints on readers, and gelatin fingerprints made with molds or inkjet printers [25]. To combat latent fingerprint reactivation, fingerprint "swipe" readers require a finger to be drawn across a linear imager, as opposed to pressed against a 2D plate. However, such systems do not inherently address the problem of prosthetics made from fingerprints left elsewhere or photographed.

Iris recognition systems are attractive due to the reliability and ease of biometrics capture without having to touch sensors that other people have touched or submit to possibly intrusive or damaging scans. This is an issue with fingerprint and handprint devices, and a concern with retina scans. However, some iris systems have been compromised with paper copies of irises [26] and printed contact lenses, and have had to develop countermeasures [27].

Facial recognition systems are similarly attractive since faces are generally considered public information and do not require contact for acquiring images. This is not universal—some cultures claim that it is their right to keep their faces covered. The 2D facial recognition systems have been spoofed with photocopied images and disguises. Researchers have even regenerated (false) images from stored templates (a template is an abstraction of biometric characteristics, so the original image does not need to be stored) that are sufficient enough to fool biometric systems [28].

Chemical and biological sensors can be vulnerable to false positives (getting swamped with false readings), suffer time lag for individual measurements, and can require significant maintenance and upkeep. Radiological and nuclear sensors are vulnerable to shielding, though detection technology continues to advance (*see* Radioactive Materials Sensors).

3.3 Combined Sensor Vulnerabilities

Since all these sensors are individually vulnerable, the careful user will apply sensors in combination to secure an area. For instance, "liveness" detectors coupled with a biometric sensor can rely on different physical attributes, such as conductivity, moisture content, temperature or thermal emissions, and movement. However, when the specific measured characteristics are understood, prosthetics can be developed to mimic "liveness" [26].

Movement can be simulated with motion video (e.g. playback on a monitor held in front of the video capture device), eyeholes cut though a face picture and held over the attacker's face, or even clipping a pupil hole in a picture of an iris and holding it over the attacker's eye. Video systems that capture multiple bands (e.g. visible, near, and far infrared (IR)) make spoofing more complex, but not impossible. Of course, such trivial attacks are unlikely to work when a security guard is present and attentive, but makeup printing systems [29], prosthetics, and printed contact lenses can potentially help the attacker blend appropriately.

However, as the complexity of spoofing the sensor(s) increases, at some point the attacker will start to look at other parts of the security system. This might also be driven by the attacker's risk sensitivities. For instance, suicide terrorists might only care about premature detection, but once the attack is underway, detection and identification might even serve the cause. Such attackers might be satisfied with a quick-and-dirty approach that gets them through the obvious security sensors, even though the whole attack might not stand up to forensics or even close scrutiny. In contrast, state-sponsored strategic threats, such as long-term economic warfare, might rely on clandestine operations, so detection and identification are highly undesirable. Overt sensor jamming might prevent identification, but unless it looks like a normal system failure, it might be enough to detect an attack is underway, and other system evidence could lead to threat mitigation and exposure.

3.4 Security System Vulnerabilities

When overcoming the sensors themselves becomes too complex or costly, the attacker can turn to vulnerabilities in the supporting system. The first thing that the attacker might do is avoid sensors altogether, for example, by entering through an unprotected window or at the loading dock. We will rashly assume that a security system installation is better designed than that, and focus on the elements described in Figure 1. One method to attack the sensor system is a simple denial of service: disable the power, for instance, by cutting cables or shorting lines. One would expect critical installations to have backup power, so the thoughtful attacker might also foul the diesel backup fuel. A remote site might have internal battery backups, so attacks on the backup power might not be easy to carry out. However, depending on the responsiveness of repair crews, the batteries might run out before crews can arrive. A good system's design would ensure that the repair service-level agreement is within the worst-case battery lifetime. It would also require a hard-to-masquerade heartbeat from the security system, so losing either power or communications would invoke an immediate response.

A more extreme power-based attack is to invoke a high-voltage transient on the power or communications lines to destroy security sensors and systems. Well-designed equipment will have surge protection. However, surge protection has voltage, energy, and frequency limits, which mean the attack can escalate. A disadvantage of this type of attack is that it is difficult to remain clandestine, especially if there is collateral damage on the surrounding power grid.

The attacker might also leverage environmental conditions to exploit weaknesses in the security system. For instance, heavy precipitation storms might temporarily reduce the effectiveness of exterior sensors, such as motion detectors, visible light cameras, acoustic sensors, and even buried RF and microwave loops. Extreme temperatures and humidity might affect a variety of systems, including visible and IR cameras and even fingerprint readers. Such systems should be carefully evaluated under worse case situations, and if the sensors fail, at least the guards should be warned to be particularly vigilant under those circumstances.

Environmental effects might also be induced by the attacker. For example, IR lasers can cause localized heating that could temporarily disable or destroy sensors or communications. This could be done from a distance, over a long period of time, during warm days so that it looks like a normal, heat-induced failure. Heating, ventilation, and air conditioning (HVAC) systems and controls could be tampered with; causing control or server-room conditions that can result in downtime for the security system or supporting infrastructure. The author has been on security audits in critical infrastructure where there was no access control beyond the front door of the installation, where the HVAC control workstation sat in a dusty corner without the benefit of even a screen blanker with password protection. It would have been trivial to update HVAC set points for some desired future time.

Inputs can also be forged, such as submitting false enrollment documents, then showing up for badging. Hopefully this requires significant social engineering, on-site presence, and explicit signature from an authority, rather than just a couple of forged e-mails to security staff. It would be unfortunate if industry-wide cost-cutting approaches (e.g. six sigma and lean manufacturing) optimized away necessary security cross-checks due to overemphasis on short-term cost.

Everything within the security system in Figure 1 can itself be a target to the attacker. Communications can be physically interrupted, corrupted, or altered, especially if cables are insufficiently protected. Wireless networks are growing in popularity, yet common standards like 802.11b/g have serious weaknesses [30]. Wired equivalency privacy (WEP) and WiFi protected access (WPA), security protocols for Wi-Fi networks attack toolkits are available on the Internet (*see* Wireless Security), so industrial grade security must be layered on top of whatever is provided by the underlying protocol. The wired protocols are not inherently secure anymore; it is just that the attacker needs a closer physical proximity than required by wireless. Therefore, security communications should be secured even on internal wired networks, lest a daytime visitor or nighttime cleaning crew inserts a router between a legitimate host and the network jack on the wall and siphons away all sorts of useful information.

The security processing itself can be attacked. Unless the operating systems that host the security software are locked down (all unnecessary services disabled, strict access policies, etc.) and operating system (OS) patches are applied as soon as they are available, the underlying platform can be corrupted. Once that platform is owned by an attacker, software on top of it can no longer be trusted. Attackers could potentially have arbitrary control authority, including remote control of the display console using popular tools such as VNC or PCAnywhere. Even without an operating system level attack, access to the security system software might be gained by something as basic as default passwords or shared passwords. Installation policies should require changing all default passwords, but this can be tricky if the vendor buries features. For example, I once had a web-enabled security camera with a 10/100 Ethernet port on it, so it could sit right on the LAN, and it could serve FTP images, push images over e-mail or FTP, stream video, and so on, and included quite capable video compression and motion detection software. Unfortunately, I lost the administrator password and could not find a hardware reset pin; so I checked the manufacturer's website. Someone else had asked about recovering the administrator password, and the response was to provide the IP address of the camera, and the manufacturer would reset the camera remotely. Note that the manufacturer was in Korea.

The security system's history and configuration information are also target rich for an attacker. Such information might be contained in application files or in a commercial database that has its own cyber vulnerabilities. Attackers can change configuration information to disable features or add themselves to access control lists or insiders who could escalate their existing privileges. They can also delete incriminating events from the historical logs, or add/modify events to incriminate others or obfuscate causality to impede forensic investigations.

The actuators and alarms, including displays, can also be attacked. Examples of actuators controlled by the security system include access control portals (doors) and illuminators. Promiscuous eavesdropping on communications might enable capture of commands that unlock the desired doors. Security systems should have strong cryptographically secured communications that prevent such playback attacks and spoofing attacks. Displays can be altered to misdirect or suppress alerts. For instance, HTML-injection attacks can intercept and modify display information on web-based security interfaces.

In summary, all of the elements of a security system have vulnerabilities which, if exploited, can subvert the intent of the security system. The following section describes possible techniques to help assess the risk in the system and make decisions about mitigation techniques.

3.5 Risk Mitigation

Risk is a key security concept and is often considered as a product of the threat, vulnerability, and consequences. Therefore, a big threat with small vulnerability and consequences might be less important to address than something that is medium in all those categories. Humans do not always assess risk uniformly [31], so transparent methods that capture incremental results and rationale, and track the reasoning process are important for making reasoned and defensible decisions. There are, of course, multiple ways to design, select, install, staff, train, and maintain security systems to minimize risk. Presumably there are even multiple correct methods, but there are some clear-cut bad ideas.

One example of what not to do is blindly trust vendor claims. Consider a USB memory stick that claimed a self-destruct system activated if users do not know the correct password. This is conceivably the type of device that well-intentioned people could use to store passwords, private keys, or personal information. One group purchased several such devices and discovered multiple hardware and software hacks to defeat the protections. Perhaps even worse, no evidence of any "self-destruct" capability was discovered. [32]

However instructive such tales are, there are probably infinitely many ways to do things wrong. The next subsections describe some process steps that one can employ to better design, specify, implement, and maintain security systems. These, or equivalent processes, can be (should be) addressed by security sensor developers, security system providers, and the end users.

If trained staff is unavailable for the necessary analysis, design, and assessment, appropriately credentialed third-party services should be acquired. Credentials to consider include, but are certainly not limited to, Physical Security Professional (PSP), Certified Information Systems Security Professional (CISSP), Certified Information Privacy Professional (CIPP), Information Systems Security Architecture Professional (ISSAP), and appropriate training provided to US military, such as force protection and the full spectrum integrated vulnerability assessment (FSIVA). If you intend to rely on an individual, do not take his/her credential claim at face value—check up on it to see if it is valid. In addition, do not rely too much on single individuals. As mentioned before, security systems face social engineering, physical, and cyber threats. So, build a team versed in all these aspects, as well as understand the objectives of the business or infrastructure applying the security system, and has someone who understands the threat environment.

Credentials can also apply to the security systems and individual components in the system. Listings of Underwriters Laboratories provide an indication that professionals have looked at some aspects of the system. A common criteria certification is another good indicator that the system or component has undergone some level of structured design and evaluation by security experts.

Next, pick an assessment process, rather than creating your own. There are multiple processes out there, and you can select one to suit your needs and adapt as appropriate. For example, a long-time-available approach is NIST SP-800-30, "Risk Management Guide for Information Technology Systems" [33]. The security risk assessment methodologies (RAM) of Sandia National Laboratories [34] are possibly already tailored for your critical infrastructure. American Society of Mechanical Engineers (ASME) offers risk analysis and management for critical asset protection (RAMCAP), a risk-based methodology to support resource allocation for critical asset protection [35]. FSIVA and the related Joint Staff Integrated Vulnerability Assessment (JSIVA) are supported by the National Guard and the Defense Threat Reduction Agency (DTRA). In the author's opinion, these methods are all excellent and share many similarities. You are encouraged to pick one—when multiple good choices are available, it is more productive to make progress following a good process rather than spending a whole lot of time deciding which one to follow.

The rest of this article uses NIST SP-800-30 as the baseline for an extremely brief overview of an assessment process, identifying several critical steps. NIST SP-800-30 is convenient since it is a freely available public standard. It is adaptable to multiple domains—it works for evaluating the security sensors, the system in which the sensors reside, and the installation at the end user. One minor adaptation we make is to explicitly consider it a cyclic, repetitive process, since security is not a finished product. Good security must continually adapt to evolving threats, so assessments and changes to the system are periodically necessary.

The first step of the risk assessment is to define the scope of the effort. The boundaries of the system, area, region, base, or critical are identified, along with the resources and the information that constitute the system under evaluation. Characterization establishes the scope of the risk assessment effort, delineates the operational authorization (or accreditation) boundaries, and provides information (e.g. infrastructure dependencies, personnel, communications, and responsible division or support personnel) essential to defining the risk.

Next, we consider the various threat actors or adversaries that might be motivated to attack the system. One approach is to have the stakeholder decision makers (not necessarily the analysts in subsequent steps) rank the relative importance of the various adversary organizations, based on business or political realities. For instance, it might not be reasonable to expect a rural infrastructure operator to withstand a focused attack from a state-sponsored strategic threat. Early knowledge of the threat allows the remaining analysis resources to focus on more realistic threats. One approach to ranking the adversaries once they are identified is to use Dr Thomas Saaty's Analytic Hierarchy Process [36, 37] for implementing this ranking, since a potentially a large number of items need to be ranked relative to each other. Note that this mechanism does not say these are the biggest threats; but will just show where the assessment is going to focus resources. This is already common practice; for instance, studies document the significance of the insider threat [38, 39], yet many organizations still focus their resources outwardly. Motivation for this could be political and marketability reasons as well as technical reasons.

After the attackers are identified and ranked, experts familiar with the adversary groups capture abstract attackers' goals, methods, and capabilities. Do they want to steal things or destroy them? Do they plan to get away or are they suicides? Are security systems deterrents for them? Can your system provide early enough detection of an imminent attack to provide opportunity for mitigation? The point is not to try to predict what the attacker will do, but rather to develop guidance on what security measures might provide utmost benefit for the resources you have available. These can be combined and ranked, using the earlier adversary ranking as weighting factors to the individual goals and capabilities, using matrix-based decision support techniques such as quality function deployment (QFD). It is useful to separate goals and ways of achieving those goals from capabilities and resources.

Now we focus on the system which is being secured, and assume the attacker's perspective and consider what within that system can be used to achieve the attacker's goals. Next we develop high-level scenarios that might achieve these goals and capture associated attack costs and results. These scenarios can then be ranked against the attacker's goals and capabilities to give a relative ranking of attractiveness from the attacker's perspective. Then the scenarios are mapped according to their attractiveness against their significance or impact from the defender's perspective. This is an analog to the typical likelihood versus impact risk mapping, and provides an overall ranking of scenarios that fuses both attacker and defender perspectives.

Since protecting against individual scenarios is a losing proposition, the next step is to disassemble and rank the vulnerabilities that enable these scenarios. Mitigating controls (e.g. specific security sensors, protections on communications, redundant power supplies, and maintenance and backup schedules) are then identified and mapped against these common mode enablers according to potential efficacy. This provides a ranking of which mitigations are most valuable, as well as a long-term plan for what to address as new resources become available.

Finer grained analysis can help determine the residual risk. An attack tree process [40] allows one to model the system under assessment and include specification of defensive mitigations (e.g. layered defenses such as cameras, locks, multiple barriers, and guards), and costs of overcoming these mitigations. Attackers are modeled with abilities and the attack tree tool maps the attackers' capabilities to the attack graph, and prunes infeasible attacks. It is particularly useful for prevention/detection/mitigation trade-offs, and can help to quantify the difference in residual risk between adding, for instance, cameras versus additional security staff, and where the "sweet spot" is for a particular site.

Another approach to consider is penetration testing, although this should be viewed as additional information—penetration tests are insufficient by themselves. It is worth the effort to consider standard steps in a penetration test framework [41] and make sure that the low hanging fruit documented there does not apply to you.

4 FUTURE RESEARCH DIRECTIONS

For the security sensor or system designer, much work remains in standards to specific security requirements, especially to cover multidisciplinary attacks that cover social engineering, physical, and cyber attacks. There is a lot of security art and lore in the industry (e.g. security checklists), but the underlying science and ability to quantitatively rate all aspects of security is lacking. Basic academic research is necessary to advance the science, followed by investment in engineering tools that can bring the science into the state of the practice.

Antitamper capabilities are an additional aspect to consider for system components. However, the state-of-the-art in commercial systems is still evolving, and one can search the Internet for Xbox hacks to see how effective tenacious, organized attackers are. New "trusted platform modules" (TPM), if properly integrated with the operating system (either of smart sensors or the security integration system), can provide a foundation for a chain of trust, but the effort required to achieve the necessary integration is still significant, and attacks on TPM-enabled machines have been reported. Beyond the TPM, IBM's secure processor architecture, which brings the protection boundary onboard the main CPU, is another step up, and is something to watch in the future; especially as such systems gain traction. Clearly, pure software-based approaches are insufficient for protection, as is any single-mode defense. Aspects of an antitamper defense-in-depth strategy might include device level protections, such as those provided by Altera Stratix II FPGA or Xilinx CoolRunner II CPLDs. Coatings, such as Foster-Miller Inc. provide, offer additional protections that raise the cost of getting into chips. Enclosure protections, such as those provided by Mektron Systems Limited and WL Gore, can wrap the alarm devices with tamper resistant coatings. Several software packages are available, which can help resist attack, such as Accord Solutions—BruteSafe, Arxan Defense Systems—EnforcIT, and Architecture Technology Corporation—Tornado, Twister, and Cloakware. However, all of these techniques, even when applied together, might only delay certain classes of attackers.

5 CONCLUSIONS

Security systems are an important aspect of protecting our critical infrastructure, but automation, sensors, and reliance on vendor claims can only go so far. Careful design, attention to policies, training of staff, and continual reassessment are necessary to maintain a strong security posture.

REFERENCES

1. Sargent and Greenleaf. http://www.sglocks.com/, 2007.
2. Locksport International. http://www.locksport.com/, 2007.
3. Lock Picking 101. http://www.lockpicking101.com/, 2007.
4. The document formerly known as *The MIT Lockpicking Guide*, http://people.csail.mit.edu/custo/MITLockGuide.pdf, 2007.
5. Blaze, M., Safecracking for The Computer Scientist, Department of Computer and Information Science University of Pennsylvania. (2004). http://www.crypto.com/papers/safelocks.pdf.
6. Central Electric Cooperative Connections. (2007). *Copper Wire Theft Can be Deadly*, http://www.cme.coop/miscellaneous/pdf/May07/May12–13.pdf, 2008.
7. The Independent. (2007). *Thief Woos Bank Staff with Chocolates—then Steals Diamonds Worth £14m*, http://news.independent.co.uk/europe/article2369019.ece.
8. Mitnick, K. D. (2001). *The Art of Deception*, John Wiley & Sons, New York, ISBN: 978-0471237129.
9. DEFCON, http://www.defcon.org/, 2007.
10. Black Hat Conference, http://www.blackhat.com/, 2007.
11. SANS Institute, http://www.sans.org/, 2007.
12. Grunwald, L., DN-Systems GmbH Germany. (2004). *New Attacks against RFID-Systems*, Black Hat, https://www.blackhat.com/presentations/bh-usa-06/BH-US-06-Grunwald.pdf.
13. Jacobs, B. and Wichers Schreur, R. (2005). *"Biometric Passport", Security: Applications, Formal aspects and Environments, in the Netherlands Workshop*, Radboud University, Nijmegen, http://wwwes.cs.utwente.nl/safe-nl/meetings/24-6-2005/bart.pdf.
14. The Register (2007). *How to Clone a Biometric Passport While it's Still in the Bag*, http://www.theregister.com/2007/03/06/daily_mail_passport_clone/.
15. Wired (2006). *Hackers Clone E-Passports*, http://www.wired.com/science/discoveries/news/2006/08/71521.
16. Rieback, M. (2006). *A Hacker's Guide to RFID Spoofing and Jamming*, DEFCON 14, Vrije Universiteit Amsterdam, Amsterdam.
17. Mahaffey, K. (2005). *Flexilis "Passive RFID Security"*, Black Hat, Las Vegas, United States, https://www.blackhat.com/presentations/bh-usa-05/bh-us-05-mahaffey.pdf.
18. Kfir, Z. and Wool, A. (2005). *Picking Virtual Pockets using Relay Attacks on Contactless Smartcard Systems*, Tel Aviv University, http://eprint.iacr.org/2005/052.pdf.
19. RFID Journal Industry Group. (2007). *Says E-Passport Clone Poses Little Risk*, http://www.rfidjournal.com/article/view/2559/.
20. NIST Biometrics http://www.itl.nist.gov/div893/biometrics/, 2007.
21. International Biometric Group IBG. http://www.biometricgroup.com/, 2007.
22. Biometrics Consortium. http://www.biometrics.org/, 2007.

23. InterNational Committee for Information Technology Standards (2006). *Study Report on Biometrics in E-Authentication*, INCITS M1/06-0112, http://www.incits.org/tc_home/m1htm/2006docs/m1060112.pdf.
24. Podio, F. L. and Dunn, J. S. (2001). *Biometric Authentication Technology: from the Movies to Your Desktop*, NIST, http://www.itl.nist.gov/div893/biometrics/Biometricsfromthemovies.pdf.
25. Matsumoto, T. Matsumoto, H., Yamada, K., Hoshino, S. (2002) Impact of artificial gummy fingers on fingerprint systems. *Proceedings of SPIE Volume No. 4677*, Optical Security and Counterfeit Deterrence Techniques IV http://www.imaging.org/store/physpub.cfm?seriesid=24&pubid=562, San Jose, CA.
26. Matsumoto, T. (2004). *Gummy Finger and Paper Iris: An Update*, Yokohama National University, Workshop on Information Security Research, Fukuoka, http://www-kairo.csce.kyushu-u.ac.jp/WISR2004/presentation12.pdf.
27. International Biometric Industry Association. (2005). *Iridian Technologies Announces Enhanced Countermeasures to Detect Printed Contact Lenses*, http://www.ibia.org/biometrics/industrynews_view.asp?id=148.
28. Adler, A. (2003). *Sample Images can be Independently Restored from Face Recognition Templates*, University of Ottawa, Ontario, http://www.site.uottawa.ca/~adler/publications/2003/adler-2003-fr-templates.pdf.
29. Liang, J. J. (2006). Desktop Personal Digital Cosmetics Make Up Printer, US Patent Application 20060098076.
30. Wagner, D. (2004). Security in wireless sensor networks. *Commun. ACM* **47**(6), 53–57, ISSN:0001-0782.
31. Slovic, P. (2000). *The Perception of Risk*, Earthscan Publications, London, ISBN: 978-1853835285.
32. Tweakers.Net. (2007). *Secustick Gives False Sense of Security*, http://tweakers.net/reviews/683/1.
33. NIST SP 800 30. (2007). *Risk Management Guide for Information Technology Systems*, http://csrc.nist.gov/publications/nistpubs/800-30/sp800-30.pdf.
34. Sandia National Laboratories. *Security Risk Assessment Methodologies*, http://www.sandia.gov/ram/, 2007.
35. ASME. (2007). *Risk Analysis and Management for Critical Asset Protection (RAMCAP)*, http://www.asme-iti.org/RAMCAP/.
36. Saaty, T. L. (1980). *The Analytic Hierarchy Process*, McGraw-Hill, New York.
37. Using the analytic hierarchy process for decision making in engineering applications: some challenges. *Int. J. Ind. Eng. Appl. Pract.* **2**(1), 35–44, 1995.
38. National Security Telecommunications And Information Systems Security Committee. (1999). *The Insider Threat to U.S. Government Information Systems*, http://www.cnss.gov/Assets/pdf/nstissam_infosec_1-99.pdf.
39. Reconnex. (2005). Insider Threat Index—Year to Date Findings, http://akamai.infoworld.com/weblog/zeroday/archives/files/Reconnex%202005%20Insider%20Threat%20Findings.pdf.
40. Amenaza. *SecurITree: The Hostile Risk Modeling Software*, http://www.amenaza.com/, 2007.
41. Orrey, K. and Lawson, L. (2007). *Penetration Testing Framework*, http://www.vulnerability assessment.co.uk/Penetration%20Test.html.

FURTHER READING

A highly recommended book is Ross Anderson's "Security Engineering," especially for system's designers. He drives home the point that the little details really do matter, and good security is an engineering process, not just cobbling together functionality. Anderson, R. J. (2001). *Security Engineering: A Guide to Building Dependable Distributed Systems*, John Wiley & Sons, ISBN: 978-0471389224.

The TRADOC DCSINT Handbooks on terrorism are also recommended to put the threat in perspective. *A Military Guide to Terrorism in the Twenty-First Century*, US Army Training and Doctrine Command, August 2005, http://handle.dtic.mil/100.2/ADA439876.

If you are a security system developer who thinks they are smart enough to develop or use anything other than standard, approved encryption packages, you must first read the basics such as Bruce Schneier's "Applied Cryptography" and "Secrets and Lies". Then read the crypto mini-FAQ and follow sci.crypt for a couple years. If you are a purchaser of security systems, I do not encourage you to use security systems (or any other systems) that claim proprietary encryption algorithms.

Schneier, B. (1995). *Applied Cryptography: Protocols, Algorithms, and Source Code in C*, 2nd Edition, John Wiley & Sons, New York, ISBN: 978-0471128458.

Schneier, B. (2004). *Secrets and Lies: Digital Security in a Networked World*, John Wiley & Sons, ISBN: 978-0471453802.

THREAT SIGNATURES OF EXPLOSIVE MATERIALS

LISA THEISEN

Contraband Detection, Sandia National Laboratories, Albuquerque, New Mexico

1 INTRODUCTION

Explosives detection systems can be categorized by application (e.g. detect explosives concealed on people, in packages, or in vehicles) as well as by technology (e.g. imaging or anomaly detection vs. trace detection). The unique characteristics of explosives have enabled researchers to develop a variety of explosives detection systems that capitalize on those properties that separate explosives from other materials. Properties such as vapor pressure, density, dielectric number, and effective atomic number are exploited in the detection of explosives.

Explosives detection methods are divided into two main technologies: trace detection and bulk detection. Trace detection technology searches for residue of explosives and bulk detection technology for larger amounts of explosives. Trace explosives techniques collect and analyze a sample of air or residue to detect explosive vapor and/or particles. The sample is collected and analyzed, and on the basis of a chemical analysis the material is detected. To generate a sample to collect and analyze, trace detection depends on vapor pressure. Sometimes taggants, which are additions to commercially manufactured explosives, can be the target for a successful trace detection. Bulk explosives techniques measure characteristics of the materials in question in an attempt to detect the possible presence of large amounts of explosives. Bulk techniques use a probing radiation source to detect variations in density, dielectric number, and effective atomic number of explosives. While none of these characteristics are unique to explosives, they can indicate a high probability of the presence of explosives.

2 SCIENTIFIC OVERVIEW

2.1 Explosive Terms Defined

An *explosive* is defined as a chemically unstable solid or liquid material that undergoes an extremely rapid conversion to form other stable gaseous materials (or products). The transformation is self-propagating, and results in explosion with the liberation of heat and the production of a shock wave from the rapid gas formation. This transformation is called *combustion* and requires sufficient oxygen and fuel to be present to maintain the reaction.

The three main types of chemical explosives are propellants, primary, and secondary. Propellants are combustible materials that burn violently and produce large volumes of gas, but do not detonate. Propellants are used to propel projectiles such as bullets or rockets. Primary explosives are characterized by their sensitivity to heat or shock and tend to detonate readily and rapidly and are used mainly to detonate secondary explosives and thereby initiate an explosive chain. Secondary explosives are generally more powerful (larger detonation velocity) than primary explosives. Secondary explosives are designed to be insensitive to heat and shock to ensure safe handling and storage. Therefore, explosive devices typically utilize a small amount of primary explosives to detonate a larger amount of secondary explosives. This article focuses on secondary explosives (also known as *high explosives*) and there will be no further discussion of propellants and primary explosives.

3 ELEMENTS THAT COMPRISE EXPLOSIVE MATERIALS

Generally, secondary explosives contain oxygen (O), nitrogen (N), carbon (C), and hydrogen (H), where the fuel and the oxidizer bond together in the same molecule. In an explosive, carbon and hydrogen are the fuel and oxygen is the oxidizer. Figure 1 shows a selection of secondary explosives that have a wide variety of structures. Explosive compounds can be divided into classes, which are characterized by their functional groups, as follows:

- nitroaromatic compounds contain $C{-}NO_2$ groups (e.g. TNT, 2,4,6-trinitrotoluene);
- nitramines contain $N{-}NO_2$ groups (e.g. RDX, also known as *hexogen* or *cyclonite*);
- nitrated esters contain $C{-}O{-}NO_2$ groups (e.g. PETN, pentaerythritol tetranitrate);
- peroxide compounds contain O–O bonds (e.g. TATP, triacetone triperoxide).

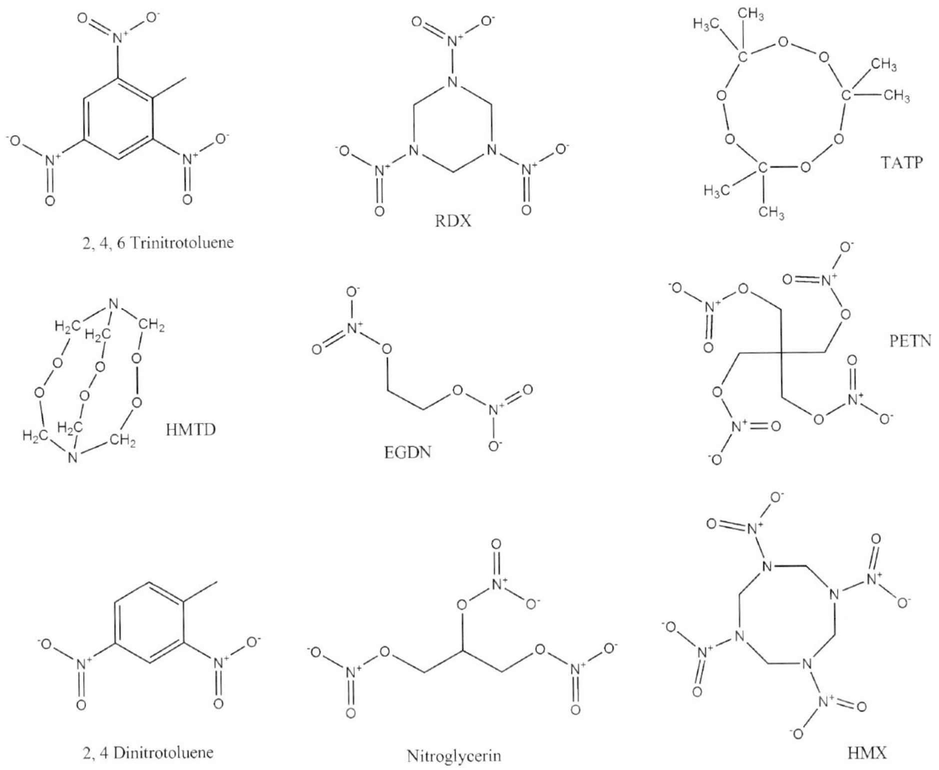

FIGURE 1 Structures of secondary explosives including nitramines (RDX and HMX), nitroaromatics (TNT and DNT), and nitrated esters (PETN, EGDN, and NG) and of peroxide-class compounds (TATP and HMTD).

4 VAPOR PRESSURE OF EXPLOSIVE MATERIALS

All solids and liquids emit some amount of vapor into their surroundings (the environment). The released vapor is in equilibrium with the remaining solid or liquid and this unique property is called *vapor pressure*. Some explosive materials do not readily form a vapor (low vapor pressure explosives) and therefore have a low amount of vapor in the air. Some explosive materials exhibit a high vapor pressure and thus have a substantial amount of vapor in the air. The quantity of available vapor can affect opportunities for detection.

Figure 2 shows the maximum vapor concentrations of different explosives in air at room temperature. Note that the vertical axis is logarithmic; each hash mark corresponds to a factor-of-ten increase in vapor concentration. The vapor pressures of explosive materials (as illustrated in Fig. 2) span more than 12 orders of magnitude. This variation means that, depending on the vapor pressure of the explosive of interest, one would or would not anticipate the presence of explosive vapor for detection purposes.

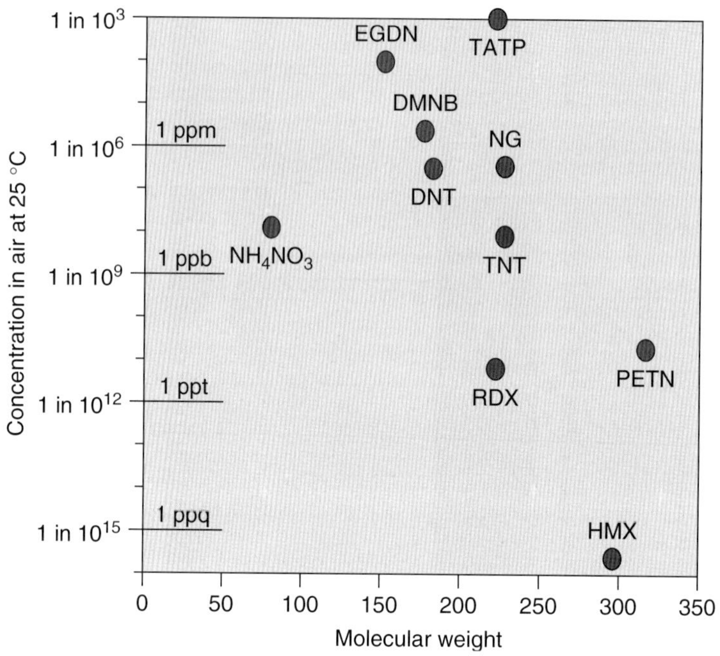

FIGURE 2 Vapor concentration of high explosives in saturated air at 25°C (vapor concentrations are approximate). Top—high vapor pressure compounds. Bottom—low vapor pressure compounds.

In general, explosives can be categorized by their vapor pressures and vapor concentrations, as follows:

- *High* vapor pressure explosives include ethylene glycol dinitrate (EGDN), TATP, nitroglycerin (NG), and 2,4-dinitrotoluene (DNT). These explosives have equilibrium vapor concentrations in air on the order of about 1 ppm or greater, which means that there will be roughly one molecule of explosive vapor for every million molecules in the air.
- *Medium* vapor pressure explosives have equilibrium vapor concentrations in air near 1 ppb. The medium vapor pressure group includes TNT and ammonium nitrate (NH_4NO_3).
- *Low* vapor pressure explosives have equilibrium vapor concentrations in air near or below 1 ppt, approximately 1000 times lower than the medium vapor pressure explosives. The low vapor pressure group includes HMX (octogen), RDX, and PETN. These vapor pressures are for the pure materials.

5 EXPLOSIVES DETECTION

5.1 Overview

Explosives detection methodologies are divided into two major categories: trace detection and bulk detection methods (Fig. 3.) Trace and bulk explosives detection methods are very complementary and exhibit different strengths.

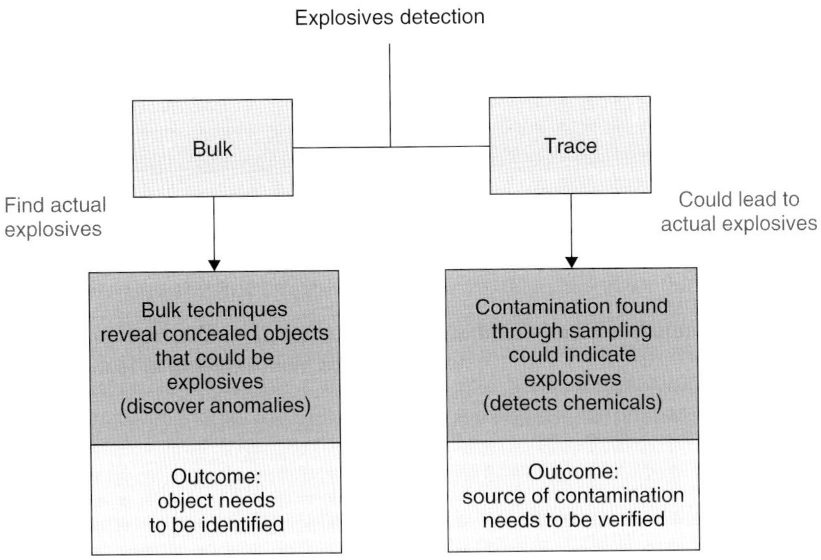

FIGURE 3 The two types of explosives detection methodologies: trace and bulk.

- *Trace explosives detection* looks for *residue or contamination* from handling or being in proximity to explosive materials. Trace detection involves the chemical detection of explosives by collecting and analyzing a sample that contains tiny amounts of explosive vapor or particles. Canines are considered a biological trace detection method because they sample odors.
- *Bulk explosives detection* seeks the *actual explosive material* (a visible amount of explosives). The detectable mass is much larger for bulk detection than the quantity of material for trace detection. Bulk detection is usually imaging based or exploiting other molecular properties (i.e. density) of the explosive. Bulk detection methods are less dependent than trace detection methods on sample collection due to the anticipated larger amount of material present. Bulk detection is not affected by an explosive contamination background.

5.2 Trace Explosives Detection

Trace explosives detection is the acquisition and analysis of small, physically acquired samples in search of explosive residue (or contamination) in very low quantities, either in the form of vapor or particles (Fig. 3).

With trace detection, not all explosives alarms indicate a real bomb threat. A valid explosives detection alarm can occur if the object under inspection has been exposed to trace amounts of explosive material for legitimate reasons. For example, when screening people, it is possible to generate a valid alarm with a frequently prescribed heart medication, even though no bomb is present. Additionally, some regions around the world may have an environment contaminated with a background signature of explosive residues.

Even though trace explosives detection seeks out only the residue present on an item, person, or vehicle, an alarm may ultimately reveal a large amount of explosives. For example, a shipping box is examined by a trace technology; the sensor may detect trace residue on the box exterior. The magnitude of the detection from the trace technology does not necessarily relate to the quantity of explosive that may be present.

Alarm resolution is a very important issue when using trace explosives detection technologies, because the explosive materials could be legitimately used.

The following are two sample types for trace detection analysis:

- *Vapor.* Gas-phase molecules that are emitted from a solid or liquid explosive. The concentration of explosives in the air is related to the vapor pressure of the explosive material and to other factors such as temperature, humidity, and air circulation.
- *Particle.* Microscopic particles of the solid explosive materials that adhere to surfaces (i.e. by direct contact with the explosive or, indirectly, through contact with someone's hands who has been handling explosives).

5.3 Explosive Vapor Detection

Explosive vapor detection is the sampling and analysis of airborne, vapor-phase explosive materials. The sample is collected without contacting the surface of the sampled item. Sampling strategies are critical because of the broad range of vapor pressures that could be encountered. The large variation in the range of all explosive materials' vapor pressures presents a huge challenge in the successful detection of trace amounts of explosives via a vapor sample.

High vapor pressure explosive materials have a significant amount of vapor present at any time, and any small particles that may be present tend to evaporate. Usually, high vapor pressure explosive materials are best detected with vapor detection. Dynamites, which usually contain EGDN and/or NG as an explosive ingredient, are examples of high vapor pressure materials and these can usually be detected from their vapor. Detecting these compounds by swiping surfaces (i.e. particle collection) is also possible, but may be less effective.

5.4 Explosive Particle Detection

Particle detection is the acquisition and analysis of microscopic solid explosive materials. The sample is collected by swiping the surface of the item. This sampling method requires direct contact to remove explosive material particles from a surface with a swipe pad provided by the manufacturer. The swipe pad is then inserted into a sampling port on the explosives detection system, and within seconds, it is analyzed for the presence of explosives. Personnel screening is not usually performed with surface swiping because many individuals would find the method invasive.

Low vapor pressure explosive materials have a significant amount of solid material present and produce very little vapor. Efforts to detect these compounds using trace technology must focus on swipe collection of particles.

Low vapor pressure explosives tend to be sticky, and a person handling a piece of solid explosive material will quickly transfer large amounts of contamination (explosive particles) to the hands. Explosive material contamination will be transferred to any additional surfaces touched by the hands, which likely will include the person's clothing as well as doorknobs, tabletops, and other objects he or she touches.

Particle contamination consists of microscopic solid particles, often on the order of a few micrograms. Most bomb builders and carriers will not be meticulous, which would result in particle contamination.

Although it is difficult to generalize how much explosives contamination is in a fingerprint, a typical fingerprint will contain numerous particles. When working with low and medium vapor pressure explosives at room temperature, more explosive materials are found to be contained in the fingerprint than would be present in a liter of air saturated with vapor (by a factor of 1000–1,000,000). Thus, for low and medium vapor pressure explosives, explosives detection is usually based on particle detection. The pure explosive materials, RDX and PETN, have extremely low vapor pressures, and the vapor pressures of plastic explosives such as C–4, Semtex, and Detasheet (which contain RDX and/or PETN as the explosive material) are even lower due to the presence of oils and plasticizing agents that give the material its form. Swipe collection is the preferred method to obtain a sample for these explosives.

Surface swiping (also called *particle sampling*) works best with small packages, briefcases, and purses (most notably, at many airports), but can be adapted to sampling larger suspect items, such as vehicles.

The choice of obtaining a vapor or particle sample is dependent on the explosive's vapor pressure. Vapor pressure is highly temperature dependent and has a dramatic effect on the amount of explosive vapor present. On a cold day, little explosive vapor is available and more vapor is available on hot days. Therefore, the most appropriate trace detection method depends on the explosive material and environmental conditions. Sometimes the situation will determine the most appropriate trace detection method. For an unknown or unattended item that may contain a detonable explosive, vapor collection is preferred in spite of the advantage of better sample collection through swipe methods.

5.5 Taggants

When low vapor pressure explosives (such as plastic explosives) are manufactured by legitimate suppliers, they are spiked with a high vapor pressure, nitrogen-containing compound called a *taggant* to make them more easily detectable. In 1989, the United Nations' International Civil Aviation Organization (ICAO) adopted a standard on the addition of taggants to manufactured plastic explosives. These taggants have high vapor pressures (similar to NG or EGDN) that make vapor detection of plastic explosives possible. However, relying on the presence of the taggant for vapor detection of plastic explosives is risky. Homemade (and some foreign-made) plastic explosives do not contain taggants. Also, as plastic explosives age, taggant material is lost to the environment because of its high vapor pressure. Nevertheless, detection of one of the taggants using vapor sampling with a trace explosives detection system should be interpreted as possibly indicating the presence of a plastic explosive.

5.6 Trace Detection Technologies Equipment

The explosives and security sector is evolving and the list of new commercially available detector technologies is expanding. A critical part of any trace explosives detection system will be the ability to separate different materials in the sample to individually analyze them. The individual component's separation is accomplished typically by the addition of a gas chromatographic (GC) column on the inlet portion of the detector. A GC column is a very adaptable hardware and can be used in combination with almost any detector that uses gas flow.

5.7 Canine Detection of Explosives

Trained canines (dogs) are used more than any other technology for the detection of explosives under real-world conditions. A dog's nose is the best vapor sensor that evolution has offered and it competes favorably with man-made detection technologies under many circumstances. A dog's nose is orders of magnitude more sensitive than a human's nose for detecting airborne odors.

Dogs are the detection method of choice for applications that involve any search component. Dogs are highly mobile and can search a building significantly faster than any other technology. Dogs can also rapidly follow a scent to its source.

In principle, dogs can be trained to detect any explosive material. Training is crucial in the development of an effective explosive-detecting dog. Dogs undergo a regular retraining program to maintain optimal performance. A dog will perform best when work conditions closely match the training conditions.

Dogs have a low purchase cost but have high continuing costs, including the handler's labor and refresher training for the handler and dog team. Dogs require a dedicated handler throughout their working lifetime. A handler–dog team's schedule might be 8- to 10-h days, 5 days a week, with rest-times every 2 h. If a dog has no major health problems, it can work from 8 to 10 years.

5.8 Bulk Detection

Bulk explosives detection is the second major category of explosives detection methodologies (as shown in Fig. 3). Bulk detection is characterized by the inspection of a sample *in situ* (where no attempt is made to obtain a sample for analysis) for visible, or macroscopic, amounts of explosives. Bulk explosives detection techniques use a probing radiation source (i.e. electromagnetic (EM) radiation or neutrons) to irradiate and interrogate an object, such as a suitcase full of clothing and toiletries. Upon irradiation, the response is measured from all the materials present in the object and a decision is made whether any material in the object is an explosive or not.

Bulk explosives techniques measure some bulk properties of the material in question in an attempt to detect the possible presence of visible quantities of explosives. Some of the properties of interest are density, dielectric number, and effective atomic number (Z number) of the material.

Bulk explosives techniques are divided into two broad technology categories called *imaging* and *nuclear based* (Fig. 4). Imaging techniques are typically X-ray, microwave, or millimeter-wave interrogation technologies. Nuclear-based techniques probe the atomic nuclei of the material under inspection. Some nuclear-based techniques interrogate a sample with high-energy radiation and monitor the detection of γ rays. Another nuclear-based technique uses radio waves for determining the presence of quadrupolar nuclei.

5.9 Electromagnetic Radiation

EM radiation can be defined as the energy of a specific wavelength traveling through space, such as solar rays. Figure 5 shows the EM spectrum with wavelengths ranging from short (10^{-12} m) to long (10^2 m). The long-wavelength radiation on the left-hand side of Figure 5 has low energies and the short-wavelength radiation has high energies.

EM radiation can be categorized as ionizing or nonionizing radiation. The terms *ionizing* and *nonionizing* radiation indicate whether the EM radiation possesses enough

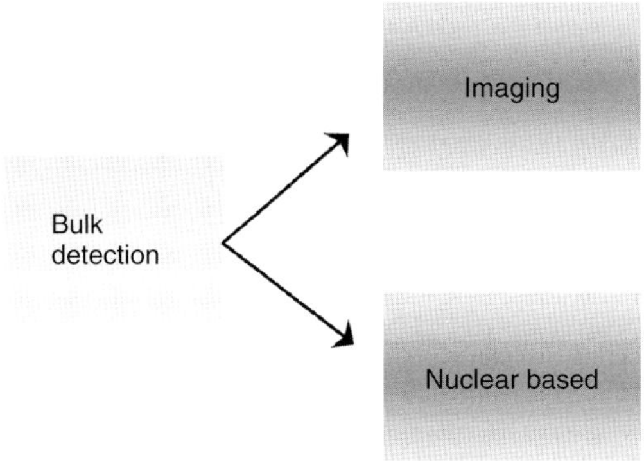

FIGURE 4 Two broad categories that describe bulk explosives detection. Imaging-based techniques examine a material's bulk property. Nuclear-based techniques probe the nucleus of the material under inspection.

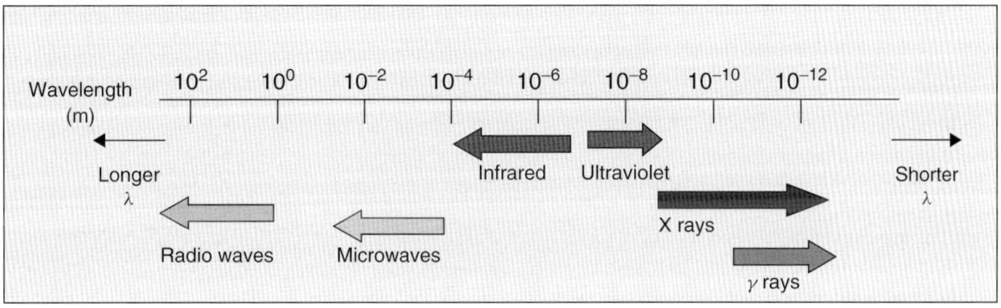

FIGURE 5 Electromagnetic spectrum illustrates the broad range of radiation and how the different wavelength regions are designated. Only a very tiny portion of the electromagnetic spectrum is visible to the human naked eye and it encompasses a range of approximately $4–7 \times 10^{-7}$ m, which is in between infrared and ultraviolet radiation.

energy to cause ionization in the atoms with which it interacts. Ionization is the process of removing electrons from atoms in molecules (such as water, protein, and DNA) with EM radiation. Nonionizing radiation is lower in energy than ionizing radiation and will not remove electrons from atoms. Examples of nonionizing radiation are radio waves, microwaves, and visible light.

Energy of ionizing radiation can be high, and it can originate from radioactive materials, high-voltage equipment, and stars. There are four types of ionizing radiation: α particle, β particle, γ ray (X rays), and neutron particle. The main difference between X rays and γ rays is their different origins. X rays originate from energetic electrons and γ rays from an atom's nucleus.

5.10 Imaging Technologies

The imaging category encompasses bulk techniques that use EM radiation of different wavelengths to interrogate items of interest to produce an image of the object.

Commercially available imaging technologies utilize X rays (including single- and dual-energy X rays, and backscatter X rays), microwaves, and millimeter waves.

Microwaves and millimeter waves use nonionizing radiation. X rays are ionizing radiation and can potentially affect the normal function of cells in the human body, and so should be used with caution. A cumulative radiation dose could have adverse health effects.

Figure 6 pictorially represents three common ways that EM radiation interacts with matter, composed of atoms and molecules. The three interaction outcomes are

- passes through the material with no interaction (transmission);
- interacts with the material (absorption and/or emission process);
- strikes the material and deflects off its original course (scattered or backscattered).

The three outcomes occur in distinct percentages determined by the energy of the probing radiation and the bulk characteristics of the materials.

X-ray techniques exploit the interaction characteristics of explosives as compared to other common materials. The properties such as density and effective atomic number (Z) are keys for distinguishing high-Z materials (metals) from low-Z materials (including H, O, and C elements).

Imaging technologies that are not X-ray-based are sometimes called *anomaly detectors*. These techniques seek changes in a material's property (other than elemental composition) as it scans the material under inspection. For example, dielectrometry measures the dielectric constant differences in an item with microwaves.

5.11 Nuclear-Based Technologies

Nuclear-based technologies interrogate the nucleus of the material under inspection with EM radiation and monitor for its characteristic emission. Current commercially available nuclear-based technologies utilize neutrons or radio waves as the interrogating radiation.

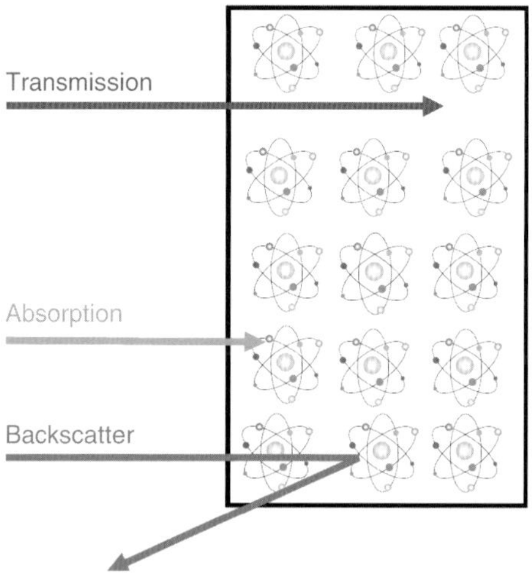

FIGURE 6 Interaction of electromagnetic radiation with matter. Matter (composed of atoms and molecules) is represented by the rectangle. The top line shows transmission, where there is no interaction between the matter and radiation. The middle line shows absorption, where there is an interaction between the atoms (molecules) of the matter and the radiation. The amount of absorption varies with the wavelength of the original radiation. The bottom line shows backscatter where the radiation is no longer on its original trajectory.

Neutron technologies utilize thermal or high-energy neutrons to interrogate the material under investigation. The neutrons are absorbed by atoms in the material and the atoms with incorporated neutrons are energetically excited. The excited atoms release their excess energy by emitting γ ray with characteristic energy. Neutrons and γ rays are ionizing radiations and so care should be exercised with this radiation hazard.

A radio frequency (RF) field is used to interrogate quadrupolar nuclei. The exposed quadrupolar nuclei are excited to a higher-energy state. Upon removal of the field, the nuclei relax back to their original lower-energy state along with the emission of radiation of characteristic energy. The energy depends on many factors, including the type of atom in the material. RF radiation is nonionizing radiation, but care should still be exercised not to unnecessarily expose humans.

Nuclear-based technologies can provide greater specificity and are more material specific for explosives than imaging technologies. To positively identify a material uniquely as an explosive, a chemical analysis technology such as mass spectrometry is required. All of the bulk detection technologies have strengths and weaknesses. If enough information is gathered on a suspect material, a determination of the presence of explosives may be made.

Although not all properties measured with a bulk technique are unique to explosives, they can indicate a high probability of the presence of explosives. The false alarm rate for bulk detection devices can be low enough in general to allow for automatic detection of explosives-like materials (e.g. in luggage screening). Alarm resolution is still an important issue when using bulk detection technologies.

5.12 Summary and Future Research Directions

Trace and bulk explosives detection techniques are complementary and have different strengths. The feasibility of having multiple pieces of purchased equipment on hand is usually constrained by equipment cost, throughput, and operational needs of a facility. Future equipment design should incorporate multisensor platforms to increase the detection functionality.

Until recently, low vapor pressure explosives, especially military explosive materials, have been the main materials of interest for explosives detection. The recent emergence of nontraditional explosive materials is challenging and the following are areas of recent interest:

- standoff explosives detection
- improvised explosive devices (IEDs)
- homemade explosives

6 RESEARCH AND FUNDING DATA

In December 2006, the US government still did not have a FY2007 budget in place and was operating in a continuing resolution mode. As the budget progressed through Congress, there was some fluctuation in the numbers.

The two largest agencies receiving explosives detection funding are the Department of Homeland Security (DHS) and the National Science Foundation (NSF). DHS earmarked \$87 million for explosives countermeasures research, which is an increase over previous years [1, 2]. The overall budget dollars for all DHS R&D activities is expected to drop approximately 10%. NSF's total budget is expected to grow about 8% from the previous year. A $20 million request was earmarked within NSF for fundamental research for new explosives sensor systems and technologies [3–5].

The European Union is spending approximately €15 million for security research projects. One of the new projects is for improving the detection of explosives, including liquids at airports [6].

The outlook for explosives detection funding in FY2008 looks promising. The Bush administration has placed emphasis on efforts that support standoff detection of conventional explosives [7].

7 CRITICAL NEEDS ANALYSIS AND RESEARCH DIRECTIONS

The recent emergence of nontraditional, homemade explosive materials has forced the broadening of the term *explosives threat*. The adversary is not constrained by politics or layers of bureaucracy and the explosives research community should not be either.

The explosives research community (which includes government, industry, and academia) needs to become more agile and nimble to meet this evolving threat by thinking outside of the box and streamlining proposals to minimize politics and bureaucratic red tape.

A knowledge-base center comprising researchers with broad, cohesive, diversified backgrounds needs to be set up to provide a network for the explosives research community. The network would have innovative ways to share information between collaborators. The enhanced communication would minimize the duplication of research efforts and pass along research dead ends. The network would allow the community to be more nimble and responsive when a new threat arrives.

ACKNOWLEDGMENT

This technical article has been authored by a contractor of the US government. Accordingly, the US government retains a nonexclusive, royalty-free license to publish or reproduce the published form of this article, or to allow others to do so, for US government's purposes. Sandia National Laboratories is a multiprogram laboratory operated by Sandia Corporation, a Lockheed Martin Company, for the US Department of Energy's National Nuclear Security Administration under Contract DE-AC04-94AL8500.

REFERENCES

1. American Association for the Advancement of Science *DHS R&D Falls in 2007 Budget*, http://www.aaas.org/spp/rd/dhs07p.htm, accessed November 2006.
2. Knezo, G. J. *Homeland Security Research and Development Funding, Organization, and Oversight*, CRS Report for Congress, Congressional Research Service, the Library of Congress, Order Code RS21270, accessed November 2006, http://www.fas.org/sgp/crs/homesec/RS21270.pdf.

3. Office of Science and Technology Policy, Executive Office of the President, Homeland Security *Research and Development in the President's 2007 Budget*, http://www.ostp.gov/html/budget/2007/1pger_HomelandSec.pdf, accessed November 2006.
4. Davey, M. E., Matthews, C. M., Moteff, J. D., Morgan, D., Schact, W. H., Smith, P. W., and Morrissey, W. A. *Federal Research and Funding: FY2007*, CRS Report for Congress, Congressional Research Service, the Library of Congress, Order Code RL33345, accessed November 2006. http://www.fas.org/sgp/crs/miisc/RL33345.pdf.
5. Ember, L. R., Janson, D. J., Hess, G., Hileman, B., Johhson, J., and Morrissey, S. (2007). R&D Budget Lacks Balance. *Chem. Eng. News* **84**(8), 27–32. February 20, 2006.
6. *Europa Press Releases, 15 Million Funding for Security Research to Combat Terrorism*, (2006). Reference: IP/06/1390 Date: 13/10/2006, http://europa.eu/rapid/pressReleasesAction.do?reference=IP/06/1390&format=HTML&aged=0&language=EN&guiLanguage=en, accessed November 2006.
7. Marburger, J. H. III, and Portman, R. *Memorandum for the Heads of Executive Departments and Agencies*, http://www.ostp.gov/html/m06-17.pdf, accessed November 2006.

FURTHER READING

Akhavan, J. (2004). *The Chemistry of Explosives*, Royal Society of Chemistry, Cambridge.

Theisen, L., Hannum, D. H., Murray, D. W., and Parmeter, J. E. *Survey of Commercially Available Explosives Detection Technologies and Equipment 2004*, National Institute of Justice, http://www.ncjrs.gov/pdffiles1/nij/grants/208861.pdf, accessed November 2006.

Yinon, J. (1999). *Forensic and Environmental Detection of Explosives*, John Wiley and Sons, Chichester.

RADIOACTIVE MATERIALS SENSORS

R. M. Mayo[1]

U.S. Department of Energy, National Nuclear Security Administration, Washington, D.C.

D. L. Stephens[2]

Pacific Northwest National Laboratory, Richland, Washington

[1]Present Address: Johns Hopkins University Applied Physics Laboratory, Baltimore, Maryland
[2]Present Address: Battelle Boulevard, P.O. BOX 999, MSIN K8-34, Richland, Washington

1 INTRODUCTION

The likelihood of a radiological or nuclear (RN) attack on the United States or its allies is a subject of contemporary debate and concern [1, 2]. What is undeniable, however, is the potential consequence of such an incident, making the risk unacceptable to ignore. Means of preventing such events are many and diverse, and constitute much of the federal government's efforts in nuclear nonproliferation, counter-proliferation, and counter-terrorism. Such efforts include an array of activities like the establishment of treaties and agreements that limit the use of nuclear technologies and materials, and provide for enforcement measures like MPC&A (materials protection, control, and accountability) designed to secure nuclear materials in place at the point of origin or during transit. As well, safeguards are established by international agreement to secure nuclear facilities and processes against diversion of material [3]. Dismantlement of nuclear devices and facilities is also an integral part of establishing international nuclear security, as are capabilities to interdict stolen, diverted, or smuggled material in such programs as the Second Line of Defense (SLD) and Mega-Ports Initiative (MI) that monitor land border crossings and seaports, the Proliferation Security Initiative (PSI) that provides for interdiction at sea, and cargo and vehicle screening at US ports of entry. The discovery of undeclared material or devices triggers an incident response architecture with the end goal of material disposition or consequence management in the event of a detonation or dispersal.

The common thread in the aforementioned is an implied ability to sense or detect material as it is being diverted, smuggled, moved, or stored for an illicit purpose [4, 5]. All RN materials emit nuclear radiation, making radiation sensing among the most common and effective means of detecting, locating, identifying, and characterizing the form, function, and threat associated with RN materials. The consequent nuclear or radio-activity of these materials emits sub-atomic particles of radiation: alpha(α), beta(β), gamma(γ), and neutrons(n). Among these, the most important for the search and screening applications that will be the subject of this discussion are γ and n as they are charge neutral and, hence, much more highly penetrating of intervening or shielding material. These particles are typically emitted by RN materials with energies in the range ~0.06 to 10 MeV for γs and ~0 to 14 MeV for neutrons, and carry information in their energy content that is useful in identifying the source material. Indeed, the emission spectrum is characteristic of the parent isotope (decay) or interaction process (induced reactions), making spectroscopy a useful tool in identification.

Radiological threats might take the form of a radiological dispersal device (RDD) or a radiological exposure device (RED). In the former, a conventional explosive detonation, aerosolization, or other means of dispersement might be used to spread radioactive material, while in the latter a large quantity of material may simply be accumulated and placed in a common meeting location or other venue where people congregate or pass-by thereby exposing them to high doses of radiation. Common radioisotopes used in medicine and industry such as ^{60}Co (1.17 and 1.33 MeV γ) or ^{137}Cs (0.662 MeV γ) are among those considered likely to be misused in this way for their availability and relatively high specific activity. Nuclear threat materials, on the other hand, are much more weakly radioactive and, therefore, more difficult to detect with passive radiation instruments. Materials in this class include highly enriched uranium (HEU, comprising greater than 20% ^{235}U) and weapons grade plutonium (WGPu, comprising ~90% or greater ^{239}Pu). These and other materials are classified as special nuclear materials (SNM) for their ability to sustain fission chain reactions and constitute the fuel for nuclear explosive devices. SNMs, and especially HEU, have difficult detection problems by virtue of lower radiation emission rates per unit mass and weaker radiation energy. In addition, threat materials containing uranium are more difficult to distinguish from background and normally occurring radioactive materials (NORM) since uranium and its decay products (also radioactive) are naturally occurring in the environment. SNM is also an emitter of neutrons through the spontaneous fission decay branch, and many plutonium detection and assay techniques rely upon the detection of neutrons by the minor isotope ^{240}Pu. As there are few natural sources of neutrons in the environment, sensing these particles is considered a strong indicator of threat, though not a definitive one. Spallation reactions in high Z materials induced by cosmic ray shower products also produce neutrons. This source needs careful characterization to distinguish from threat sources. In addition, HEU is a particularly weak emitter of neutrons. Passive detection is often not practical.

For much of what follows it will prove beneficial to consider two general classes of detection scenarios, search and screening, as a framework to guide discussion. Many of the nonproliferation, counter-proliferation, and counter-terrorism problems mentioned above take advantage of these operational schemes to conduct effective material control and interdiction operations. Screening includes procedures such as the establishment of radiation monitoring portals or other choke points at ports of entry, ports of departure, or facilities entrances. The objective is to survey persons, vehicles, and cargo for sources of radioactivity. Once radiation detection is confirmed, it is required to divert suspect conveyances to secondary inspection for identification or adjudication of the radiation alarm, so that both detection and identification equipment are required. In search, the scenario is quite different, especially in the early stages of the process in which a search team may be required to scan a limited region or building for threat sources of radiation. Proximity detection through gross counting detection and operational procedures, direction finding sensors, or imagers may prove useful to locate a source. Identification instruments employed as high confidence ID are almost always required.

In the following sections, we discuss the state-of-art and ongoing research and development for these and other radiation sensing technologies that have application to the above scenarios. A brief mention has already been made regarding the importance of spectroscopic (or energy information) especially in γ-ray sensing for threat identification. The next section elaborates on this substantially. This will be followed by a discussion of radiation imaging and its particular importance to the search mission. Interrogation of nuclear materials by active means (external stimulation of nuclear signatures by energetic particles of radiation) will be covered in a subsequent section followed by a discussion on advancement on detector materials. Finally, a discussion is provided on the integration of sensor component technologies into detection systems and associated electronics, before a brief closing discussion on Federal involvement and resources.

2 GAMMA RAY AND NEUTRON SENSING AND SPECTROSCOPY

The detection of γ rays and neutrons is a mature field with significant and continuing innovation over the last several decades. Realizing the potential for nuclear/radiological terrorism, these detectors have received renewed interest in the first decade of the twentifirst century for their role in the detection of SNM and other radioactive threat materials. This renewed effort has focused on improving energy resolution, efficiency, and in some cases the mechanical properties of detectors. The national and homeland security application of γ and neutron detectors has focused on the detection of characteristic radiation that can uniquely identify threat materials from other benign radioactive material. This diverse set of missions employing γ and neutron detectors results in a large variety of different sensor systems when optimized for the particular operational environments and applications. These may vary from the screening of material at a fixed point of entry, often called portal monitoring, to the search for materials where the detector is moving through an environment to find nuclear materials. There are many operational and technical differences between these two scenarios, yet the single factor that dominates detection sensitivity differences is the way radiation background is experienced in each scenario. In the case of a fixed system, the detector operates in a predictable background that typically varies slowly when compared with the frequency and duration of screened items. It can be carefully recorded and characterized, sometimes over long times, before the sensor witnesses a radioactive material event passage. In the search scenario, however, variations in background are often substantially larger in magnitude and of significantly increased frequency. When combined, these effects generally reduce the overall system sensitivity for a desired rate of false positive detection [6]. The false positive detection rate is of paramount importance to the discussion of operational practicality since the total cost of operating a sensor system is proportional to the total number of detections or alarms (including false alarms) that need to be resolved. For most real world situations, it is reasonable to expect that the number of real events where an SNM anomaly is encountered is diminishingly small, yet because of the relative abundance of NORM materials and other nuisance alarms, the number of false positives will dominate the operational cost in the deployed system. Research in radiation detection systems must be viewed in this context, placing a premium in high confidence performance. In an idealized system the false detection probability must approach zero while the detection of defined threats, especially SNM, must simultaneously approach unity. Advancing deployed detection systems toward this ideal operational condition is driving much of the current research and development for national and homeland security applications of radiation detection. Specific R&D focus areas resulting from this operational need include the improvements in unique identification, and in some cases quantification, of radioisotopes present. This can be accomplished by achieving better energy resolution, better timing resolution, and higher detection efficiency.

The detection of γ rays and neutrons is possible because of their ability to produce ionization in materials. This *Ionizing Radiation* deposits energy in the detector material producing an electrical signal that can be recorded. Detector development, therefore, focuses on the optimization of detector materials and electrical signal readout technologies. For a more complete treatment on the fundamentals of radiation detection see Ref. [7].

There are two basic γ-ray detector classes common in national and homeland security applications, semiconductor and scintillation detectors. In semiconductor detectors electron–hole pairs are produced directly in incident γ-ray interactions with the detector material. These electrically charged particles are subsequently drifted to oppositely biased electrodes and collected. The highest performing among the semiconductor detectors is the high-purity germanium (HPGe) single crystal detector. Semiconductor detectors are capable of very good γ-ray energy resolution and have moderate efficiency. The HPGe detector, while unsurpassed in overall performance, has a significant operational drawback; it must be cooled to liquid nitrogen temperatures for effective operation. Initial cooling often requires many hours to days, and this low temperature must be maintained during operation. In field situations, the logistics of providing the consumable liquid nitrogen can be operationally limiting or at least cumbersome. In the past decade there have been advances, including the introduction of mechanical coolers for these devices. To date, these systems are not capable of supporting large efficient crystals in person-portable devices with operational lifetime longer than a few hours. This has led to renewed interest in identifying new semiconductor materials for radiation detection that can optimally operate at room temperature. Additional information is provided in the section Radiation Detection Materials. A notable example of recent success in the search to replace HPGe with a room temperature material is the development of cadmium zinc telluride (CdZnTe, or CZT) detectors. These detectors have been the subject of much research over the past 15 years and are just now penetrating the market as radiation detectors for operational use. While this new detector material is capable of operating at room temperature, it has yet to demonstrate as high an energy resolution as HPGe and is only available in volumes of a few cubic centimeter, whereas HPGe is available in hundreds of cubic centimeter volume. Recent developmental progress, however, is continuing to push the performance of detectors based on this material so that energy resolution is now within a factor of three of HPGe and larger volume crystals are becoming available. In light of recent advancement CZT promises to be an important new detector material for national and homeland security applications because of its room temperature operation, low relative power draw, and relatively high energy resolution resulting in a potentially more practical isotope identification capability. The first ever high efficiency (~20%), high resolution (~1%) handheld detector with source directional indication based on an array of 18 pixilated CZT detectors is the "Gamma Tracker" system [8] shown in Figure 1.

The second class of γ-ray detector in common use in the national and homeland security arena is the scintillator detector. The scintillating materials in these detectors

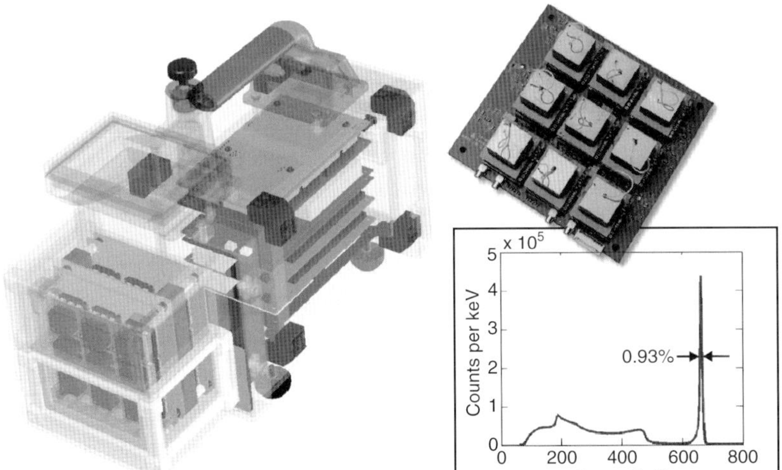

FIGURE 1 Gamma-Tracker [8], a high efficiency, high resolution handheld radioisotope identifier with directional source indication based on 18 pixilated CZT detectors. Breakaway design drawing as well as insets of CZT detector array mounted on its front end electronics board, and a sample composite spectrum of ^{137}Cs demonstrating ~1% resolution at 662 keV are shown.

produce visible and near UV light photons through scintillation in an often complex cascade of events that result from the γ-ray ionization in the detector material. The most common medium energy resolution scintillator is the inorganic material thallium-doped sodium iodide (NaI(Tl)). This detector has been the mainstay of γ-ray detection and spectroscopy for laboratory and field measurement since its discovery in the 1940s. NaI(Tl) and its other doped variants have moderate energy resolution, high efficiency, operate at room temperature, and can be grown in volumes as large as thousands of cubic centimeters. Other notable scintillators utilized in security applications are the organic scintillators. There are many organic scintillator materials for γ-ray detection, but the most common for security applications is poly(vinyl toluene) (PVT) solidified in a polymer matrix. These so called *plastic* scintillators are widely used in large area detectors deployed for portal monitoring as they are relatively inexpensive and easy to manufacture on very large scales. They are of rather poor energy resolution, however, as a result of the predominance of Compton scattering events induced by incident γ rays due to the low atomic number of PVT constituents. The energy resolution of these scintillators is not sufficient to uniquely identify specific radioisotopes in real world operations [9]. In these situations, the low energy resolving power may, at best, only be used to broadly classify radioactive sources as possibly containing threat materials.

Recent development progress has culminated in the commercialization of higher performing scintillation detectors such as those based on the inorganic crystal lanthanum tribromide ($LaBr_3$). They have detector characteristics similar to NaI(Tl) but with much better energy resolution [10]. An additional challenge unique to scintillators and one that continues to be the subject of vigorous investigation is the need to efficiently collect the visible or near-UV light produced by radiation energy deposition. This function is typically relegated to the Photo-Multiplier Tube (PMT), a vacuum tube devices that convert incident photons to an electrical signal. Improvements in the efficiency and robustness of PMTs, as well as improvements upon recently introduced photo-diode photon detectors will improve the overall performance of scintillator detectors.

In addition to detectors in the semiconductor and scintillator detector categories resides a class of calorimetric techniques that have recently achieved record [11] γ-ray energy spectroscopic performance. These detectors operate at superconducting transition temperatures (typically in the range of 0.1 K, below the liquid helium boiling point) and resolve γ-ray energy content by a very high precision temperature measurement of the heat deposited in a metal absorber utilizing a superconducting transition edge sensor (typically Cu/Mo). These bolometric techniques are achieving γ-ray energy resolution an order of magnitude better than the best HPGe. The most significant drawback is their inherently small size ($<1\ mm^3$) required to have effective thermal absorption and readout in reasonable time. This limits effective operation to low energy γ rays and hard X rays (<200 keV), and for specialized long counting laboratory applications. Among the technical challenges being addressed in current R&D are increasing efficiency with multiplexed large arrays, ruggedization for potential field or mobile laboratory use, and applicability to α particle spectroscopy. Their ultimate application to homeland and national security missions is currently being evaluated.

In contrast to the atomic processes involved in the predominance of γ-ray interactions, indirect ionization of detector materials by neutrons results from nuclear scattering or absorption processes. These neutron-induced processes produce fast primary charged particles (usually protons, alpha particles, or other light ions) that then slow in the material and ultimately produce ionization. The net result is once again ionization of the detector material. This allows similar detection techniques to be used as those described earlier for γ rays albeit with different materials that have an affinity for neutrons.

Neutron detectors are typically divided into two distinct categories associated with the most likely interaction mechanism at a given range of incident neutron energy. In the low energy, or *thermal*, range (where neutrons are in ambient thermal equilibrium), detection usually requires nuclear absorption reactions. Absorption reaction based detectors contain materials with isotopes like ^{3}He, ^{6}Li, ^{10}B, and ^{157}Gd that have high neutron absorption probability. These reaction based neutron detectors preserve no neutron energy spectral information for incident thermal neutrons. In applications typical of national and homeland security, these detectors either operate as scintillators or semiconductors, having the same basic operation as described for γ rays, or as gas ionization detectors. (The gas ionization detector operates by directly measuring the ionization products in the gas created by the neutron-induced reaction products.) The dominant thermal neutron detector in national and homeland security applications is the ^{3}He gas ionization detector [7]. This detector is particularly favored for thermal neutron detection because it provides excellent discriminating ability between neutrons and γ rays, being perhaps as much as 10^5 times more sensitive to neutrons than to γ. In most operational situations, mixed fields of γ rays and neutrons are encountered. Since confident detection of neutrons is often considered a stronger indicator of the presence of an SNM threat, it is important to have γ-ray discrimination as high as possible to avoid false positive detection.

Neutrons with energies typical of those emitted in nuclear decay, including fission in SNM, are initially of high neutron energy and are referred to as *fast neutrons*. Fast-neutron detectors can be built to exploit either nuclear absorption reactions or scattering. Common reaction based detectors for fast neutrons utilize absorption reactions in ^{3}He or ^{6}Li. The probability of these reactions occurring at fast-neutron energies is much lower than for thermal energies. Nevertheless, they have been exploited for fast-neutron detection. Fast-neutron detectors that exploit nuclear scattering, however, rely on the energetic recoil of the struck target nucleus to produce ionization in the detector medium. These detectors are commonly designed with high hydrogen content since the kinematics of the scattering collision favors energy transfer to a target proton [7]. These detectors are most usually scintillators and only measure the energy transferred in the elastic scattering collision. In general a fast neutron can undergo many such scattering events before it reaches thermal energy and is absorbed in a nuclear reaction. In practice, fast-neutron detection may also be accomplished by taking advantage of the energy loss from neutron scattering in hydrogen-rich materials (a process referred to as *moderation*) to reduce the energy of an incident fast neutron to thermal energies so that it can then be detected by a ^{3}He based thermal neutron detector. This type of fast-neutron detector typically requires large masses of passive moderation material, typically polyethylene. In many operational settings, because of strict requirements on timing and/or additional energy content information, it is desirable to field direct fast-neutron detectors rather than those that moderate first before detecting.

Incident neutron energy spectrum of fast neutrons is, many times, an important quantity for national and homeland security applications since spectral details can reveal threat source characteristics and, perhaps, help distinguish among threat sources. Neutron spectroscopic techniques for thermal energy neutrons, on the other hand, are rather complex and not likely to provide additional information important in these applications. Spectroscopy for fast neutrons is typically achieved by a convolution technique whereby neutrons are moderated by varying amounts of hydrogenous material followed by detection of the slowed neutron component, typically in a ^{3}He or ^{6}Li based detector. The relative populations of thermal neutrons in each moderation group allows the unfolding of a low energy resolution neutron spectra, devoid of fine spectral details yet representative of general spectral shapes that are often information rich. Fast-neutron spectroscopy may also be based on detector systems that record multiple scattering events in successive detector planes. Measuring event-wise energy deposition, interaction position, and time-of-flight between interactions allows for full kinematic reconstruction [12], as well as neutron source imaging as discussed in the section below. Detection challenges arising from inefficiencies in fast-neutron spectroscopy (since it requires multiple events or successive moderation) are somewhat mitigated in homeland and national security application by the very low neutron background rates. A notable exception is the increased neutron background found aboard large seagoing vessels produced by spallation interactions of cosmic rays with the large quantities of modest to high Z materials (especially iron) present. In these particular cases, neutron spectroscopic or temporal correlation information may provide a means to distinguish SNM threats from this background source.

Recent R&D in neutron detection has focused on identification of replacements for ^{3}He gas ionization chambers, and the development of direct high energy neutron detectors. Gas ionization neutron detectors are not desirable for many operational scenarios. Alternatives such as intrinsic and convertor layer semiconductor neutron detectors [13] have been the focus of recent investigation. These alternatives, however, have thus far failed to achieve the same high detection efficiency and γ-ray discrimination of ^{3}He detectors. Much of the current R&D for fast-neutron detectors is focused on the discovery and development of new organic scintillators [14] in either liquid or solid form, and in the development of faster and more sensitive neutron scatter spectrometer/cameras. The more in-depth exploitation of neutron correlated signatures from fission is a potentially productive area for further R&D as well.

3 RADIATION IMAGING

Radiation imaging is a technique that promises two potential benefits to homeland and national security missions. Imaging methods can reduce the contribution of background radiation to the observed signal, thereby raising the probability of detection by increasing the signal-to-noise (S/N) ratio in a narrow field of view, and by providing a visual image of the radiation source. Radiation imaging techniques have been and are continuing to be developed for both γ rays and neutrons. For relatively low energy γ rays, physical imagers are optimal. These imagers use physical optics to create images of γ rays on position sensitive detectors and have been demonstrated [15] for stand-off detection of γ-emitting materials. The optics are made from high atomic number materials such as lead, and are ordered into patterns that occult a portion of the incident γ flux, thereby producing a unique image at the detector plane. Techniques vary from the more complex coded aperture methods to the relatively simple pinhole camera. In practice, the exploitation of these techniques often results in massive detection systems because of the large amount of lead or other high Z shielding required, especially as the area of the detector increases in an effort to increase detection sensitivity. Systems of this type pose design challenges that trade detection efficiency for simplicity and portability. Pinhole cameras, being the simplest possible encoding, suffer the largest efficiency penalty of all physical optics imagers since a sharp image requires a small aperture. On the other end of the scale, coded aperture data encoding allows the maximal quantity of photons to reach the active detector (up to 50% of the incident flux) at the expense of a more complicated image reconstruction process. Physical optics systems additionally suffer from reduced efficiency as γ-ray energy increases for fixed aperture thickness.

The Compton Imager, on the other hand, performs more efficiently at higher γ-ray energies. As the name implies, these systems exploit the Compton Scattering mechanism whereby partial energy loss from an incident γ ray is measured in multiple detector volumes on successive scattering events. The technique does require multiple interactions in the detector, thus reducing the overall detection efficiently. However, much more imaging information is afforded per detected photon both because there is no coding (or aperture) penalty and because the sequence of scattering events from the same incident γ ray provides nearly unique information regarding its incident direction. This technique has long been implemented in astrophysical imaging systems and is now beginning to be developed for national and homeland security applications in larger fixed, portable, and small handheld detectors [8] for direction finding in search and screening.

Neutron imaging techniques are functionally very similar to those applied in γ-ray imaging, though employing different detection materials as described above. In the case of thermal neutron imaging, physical methods such as coded aperture techniques work well. Materials with high thermal neutron absorption affinity (B, Li, Cd) usually comprise or are constituents of the aperture. In the case of high energy neutrons, scattering techniques analogous to Compton imaging are an area of current investigation and development. Figure 2 illustrates a recently developed instrument [12] for imaging fast neutrons ($\sim$1–10 MeV) referred to as a neutron double scatter camera. These techniques have recently demonstrated [12, 16] the additional capability to directly measure neutron energy spectra. The challenge with all imaging approaches is in justifying the penalty paid in loss of overall efficiency for encoding position information. Operational requirements must be carefully considered when selecting imaging techniques over other detection methods.

4 ACTIVE INTERROGATION

Although the passive detection technologies described above have proven quite effective in many situations and means of improving upon these continue through aggressive research and development campaigns, relying exclusively on passive means is likely to be ineffectual in some circumstances. Continually increasing requirements to detect ever smaller quantities of weakly emitting materials like HEU through material barriers and shielding raises detection and identification challenges beyond that practical of passive sensors. In these cases, only active stimulation of nuclear signatures can provide confident and timely detection, identification, and characterization.

Active interrogation refers to techniques that apply external sources of radiation, which induce reactions (usually nuclear) in SNM, thereby greatly improving detectability and/or reducing detection time. This improvement is possible because rates for induced reactions can be controlled through the intensity of the external source of radiation and are often many orders of magnitude greater than those of spontaneous decay. In this manner, active interrogation results in a greater rate of reaction product emission, a different variety of product particle, and/or unique energy spectral features of the reaction products, each of which can enhance our ability to detect.

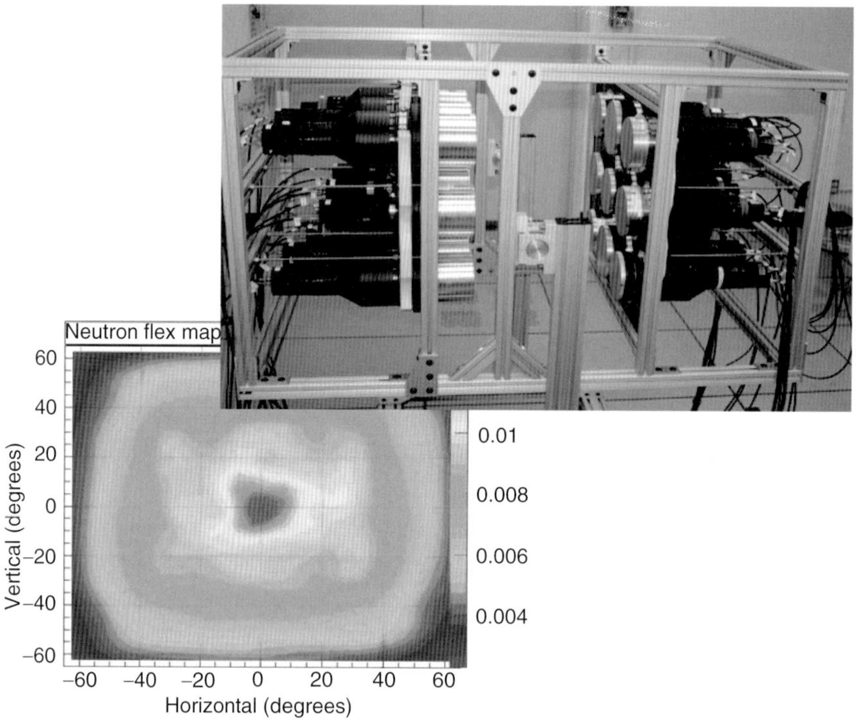

FIGURE 2 Neutron scatter camera [12], a liquid scintillator based technique to witness fast-neutron scattering events in two successive interaction planes. Emission images and neutron energy spectra can be reconstructed from these data. The inset shows an image of a small ^{252}Cf spontaneous fission neutron source imaged at ~10 m through the steel hull of a sea vessel.

A wide variety of induced reactions is possible with fission, inelastic scattering, capture, and resonant absorption being the most prevalent. Only SNM fissions upon absorbing low energy neutrons, making thermal neutron-induced fission a unique signature. Unfortunately, low energy neutrons only weakly penetrate shielding material (especially low Z materials). Higher energy neutrons and γ rays can be made to induce fission, however, with some loss of specificity since these particles (given sufficient energy) can eject neutrons from many other materials. Neutrons may also be captured or inelastically scattered by SNM resulting in the emission of characteristic γ rays, though these reactions proceed at somewhat lower rates. Newly emerging on the active interrogation scene for SNM detection is resonant absorption and re-emission of characteristic γ [17] referred to as *Nuclear Resonance Fluorescence* (NRF). Figure 3 depicts the experimental setup by which recent NRF signature measurements of HEU were recently conducted. This technique is in early stages of development and exploitation as a means of detection, yet shows great promise by virtue of the characteristic nature of emission and penetrability of energetic γ in the range of 1–3 MeV for NRF in SNM [18]. Further exploitation of these and other signatures and detection modalities (like fission multiplicity and time correlation [19]) are ongoing subjects of investigation.

A wide array of sources is being developed to support active interrogation detection applications. Continual improvements and novel means are in high demand. Small accelerators, especially RF linacs, are the workhorses in most energetic photon interrogation

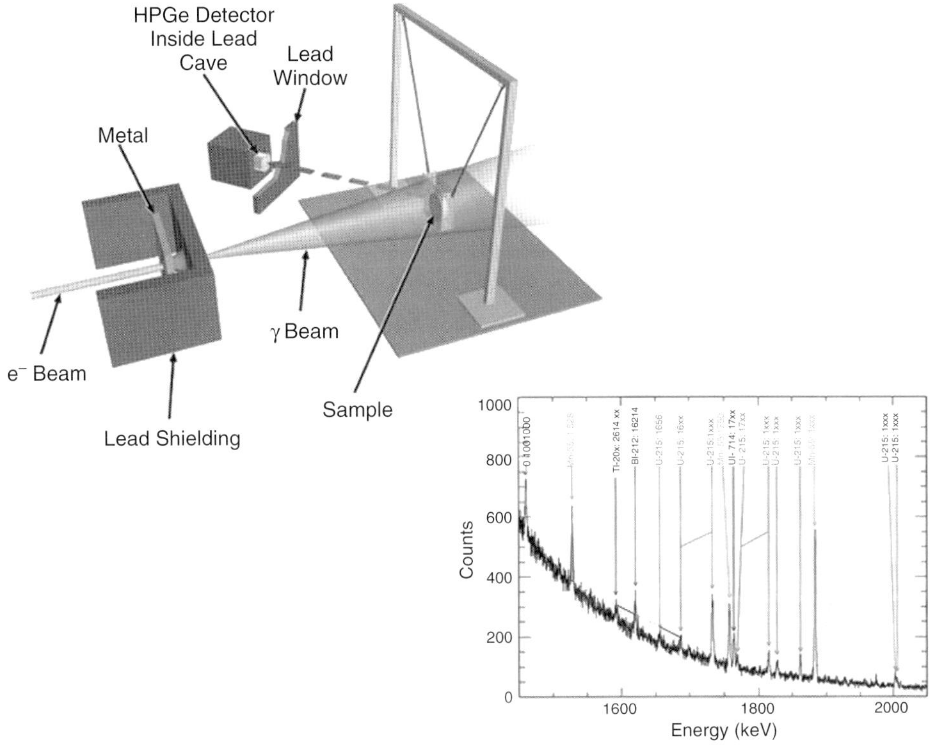

FIGURE 3 Nuclear Resonance Fluorescence (NRF) [18] measurement setup. Inset γ energy spectrum shows first ever direct measurements of ^{235}U NRF lines in this region of interest (1450–2050 keV). Additional lines from background radiation and NRF in surrounding materials are also present.

schemes where accelerated electrons (usually to tens of MeV) are stopped on high Z targets to produce bremsstrahlung [20] radiation. Such sources are being pursued for both photo-fission and NRF interrogation. Bremsstrahlung beams, however, yield γ energy distributions overwhelmingly dominated by low energy particles. Since the desired induced reactions require only those particles in the high energy tail, these sources produce much more radiation than necessary. This may result in more than necessary radiation exposure to operators and bystanders making these sources far from ideal. Narrow energy band sources are desired, which are either energy selectable or tunable for both photo-fission or NRF applications. γ producing nuclear reaction based sources (e.g. p + ^{19}F) and laser compton backscattering (LCB) sources [21] are being pursued. As compactness is a critical driver for many operations, laser wake field accelerators (LWFAs) are also being investigated as LCB drivers [22].

Fusion reaction (D–D or D–T) based high energy neutron sources are available commercially. Though some are compact, they suffer from lack of robustness and short lifetimes as well as modest output intensity. Addressing these issues, in addition to electronic collimation and high repetition rates and pulse shaping for pulsed sources, is the subject of substantial research investment [23]. Endoergic reaction based sources (e.g. p-Li) that produce moderate energy neutrons (~60 keV) are of interest for their modest penetrating ability while distinctly different in energy from the resultant fission neutrons from SNM [24]. A portion of the development community is pursuing more exotic charged particle sources (especially protons and muons) for very long stand-off applications (up to kilometers). While practical applications may be far off, further development of source technologies and more complete understanding of beam particle interactions in air and target materials will surely advance state-of-the-art in active interrogation. Finally, detector development especially for the harsh environment, fast timing requirements, and particle discrimination requirements imposed by active interrogation applications is yet another very active investment area.

5 RADIATION DETECTION MATERIALS

Detecting particles of nuclear radiation relies on sensing the interaction of those particles with the material media of the detector. Since nuclear radiation is ionizing (either directly or indirectly), charged pairs are produced in such interactions and can be either drifted and directly collected in semiconductor detectors, or migrate to activator sites and produce scintillation light which can then be detected by a photo-detector (PMT or photo-diode) in scintillator detectors. Examples of both semiconductor and scintillator type detectors can be found in the solid, liquid, or gaseous states, though the bulk of contemporary research and development is focused on solid state detectors. Transportation regulation accommodation, avoidance of toxic and volatile material forms, and maximizing radiation stopping power are all strong motivators driving the technology in this direction.

Recent motivation to increase National and Homeland Security detection capability is beginning to provide impetus through more aggressive funding support for new radiation detection materials discovery, a research area that has been investment starved for decades [25]. Materials with long detection legacy like NaI (scintillator) and HPGe (semiconductor) continue to bear the brunt of the detection workload. Although well characterized and carefully outfitted for operational requirements, each has severe shortcomings. NaI is hydroscopic requiring enclosure and is sensitive to temperature fluctuations requiring gain control, but of more concern is its only modest ability to distinguish among radiation particles of differing energy, having resolving ability in the range of 6–10% (full width at half maximum in the photopeak). Although adequate for some purposes, for others this is insufficient discriminating ability. The search for better performing scintillator materials is ongoing. Recent progress on emerging materials like $LaBr_3$ [10] has been encouraging, though this material is not without its own peculiar problems especially fracture during growth. Still more recently SrI_2 and several candidates in the Elpasolite class have emerged as promising new scintillators.

Semiconductor materials are generally performing better with respect to energy resolution (HPGe resolution can be as low as $\sim$0.25%) though these materials are not without their own shortcomings. HPGe, for example, must be cooled to liquid nitrogen temperatures before such performance can be realized, a serious maintenance burden for field operations. Candidate replacement materials have shown neither the performance nor the ability to be fabricated into detectors of comparable sizes. By far the most promising replacement for HPGe is CZT. Though not in widespread use, yet and still requiring additional improvement [26], promising new results from CZT are emerging. Less than 1% resolution at room temperature is consistently reported (Figure 1) with results as low as 0.7% not uncommon. Improvements continue and 0.5% resolution is expected imminently as a more thorough and mechanistic understanding of materials properties

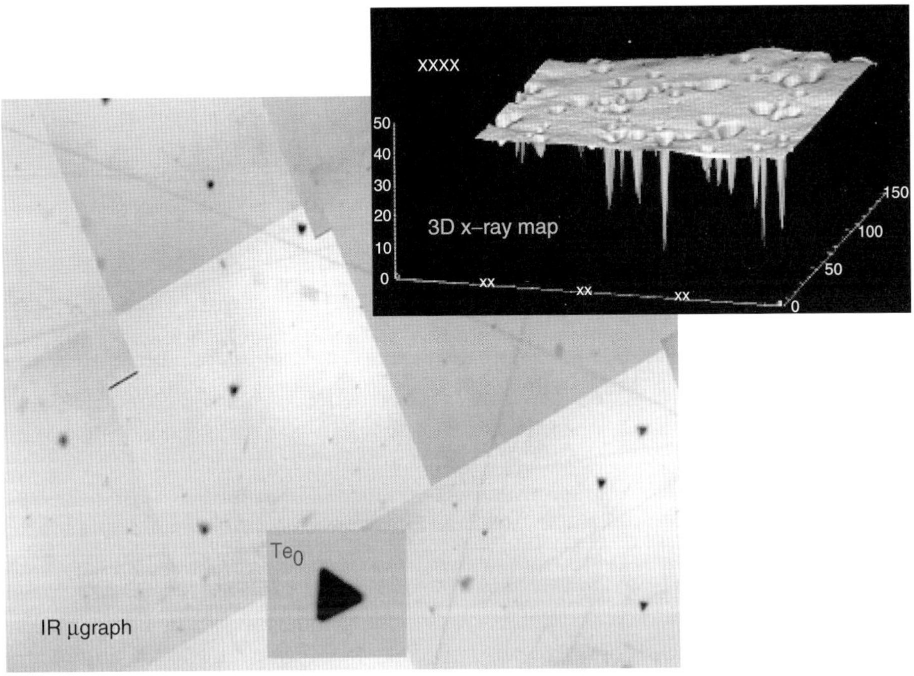

FIGURE 4 IR transmission micrograph of a CZT single crystal showing presence of Te secondary phases. These material defects are well correlated with charge trapping as demonstrated by X-ray excitation from a 10 μm diameter beam line at the National Synchrotron Light Source (inset).

and charge transport develops. Figure 4 depicts recent microscopic examination in CZT where the appearance of Te secondary phases (as seen in IR micrograph) correlates well with reduction in charge collection as demonstrated by illumination with a 10 μm diameter X-ray beam line at the National Synchrotron Light Source (inset).

In addition to recent improvements made in the material properties, novel electronic readout provides a means to correct for material nonuniformities [27] so that better than intrinsic performance can be realized. The search for comparable or better performing materials that are easier and cheaper to grow in larger sizes continues. Recent indications suggest that nano-structured materials may hold promise as high performing detector materials or imbedded in a host matrix as scintillators or semiconductors [28].

There continues to be high demand for materials and structures to detect both fast and thermal neutrons. Lithium or boron doped (sometimes as nanoparticles) glasses and plastics are actively being pursued as efficient thermal neutron scintillator materials [29], while those materials either embedded or layered on Si devices are currently under investigation as thermal neutron semiconductors. The detection of fast neutrons is most often accomplished via liquid or plastic scintillators, with stilbene being considered the most effective for national and homeland security applications owing to the ability to readily distinguish among neutron and γ interactions in the material. Having highly toxic starting materials, however, renders this material somewhat unattractive in providing a mass produced detection infrastructure. The search for new materials continues. Recently, salicylic acid derivatives have been investigated as promising replacements.

6 SYSTEM INTEGRATION

Radiation detection systems for homeland and national security require solutions that can meet the needs of the often harsh and complex operational environments. They must also be used effectively by a wide variety of users whose training and familiarity often varies considerably. The challenges associated with rapid transition of radiation detection technologies from the laboratory to these new venues requires the continued development and refinement of methods that result in high confidence operation. Areas of active research in automated isotope identification, low power computing, high efficiency cooling methods for detectors, improved mechanical properties of radiation detection materials, and imaging algorithms are a few specific examples. Algorithms for isotope identification in γ-ray spectra have been an area of particular emphasis in recent years. Expertise in γ spectra analysis often requires years for even a highly technical individual to master, yet operational end users must rely on spectroscopic systems to routinely determine the composition of suspect items. The current state-of-the-art in this field necessarily focuses on both improved automated extraction of spectral features of interest and spectral deconvolution. Optimizing electronic components is yet another area of vigorous investigation and investment. As a general rule, the detection community attempts to utilize the latest in electronics technology by adapting it for use by custom integration. This is made apparent in the custom application of the latest communications hardware, custom ASICs, and field programmable gate arrays in radiation detection systems. On a final note, for deployed systems to be effective they must be optimized as systems rather than as stand-alone radiation sensors. Optimization necessarily includes test and evaluation specific to the ultimate system use, and the need for good systems engineering is sometimes overlooked in the push for better subsystem performance.

7 CLOSING REMARKS

In this short article we have described our understanding of the state-of-the-art in nuclear and radiological detection technologies. This is by no means a complete discussion, but should provide a reasonable introduction to the interested reader and new researcher. We have made every attempt to identify references that highlight contemporary work that has the most impact, from the leaders in their respective fields to provide a starting point for further investigation.

It was our further motivation to identify technical gaps or shortfalls wherever possible as a means of motivation. Many of these, including the development of new materials for higher performance radioisotope ID, compact intense sources for active interrogation with low unintended dose impact, and stand-off detection technologies rise to the level of *grand challenge*. For the eager researcher there is no shortage of opportunities to make significant impact to national and homeland security.

The technical disciplines described herein are rich and dynamic with many and diverse issues to be addressed. Indeed, RN detection technologies are interdisciplinary requiring expertise not only in radiation and detector physics, but also in microelectronics, shock and vibration isolation, thermal management, algorithms development, data reduction, and information management to name a few. The field has historically benefited from related disciplines in nuclear and high energy physics, astrophysics, and weapons physics and is now leveraging clever solutions from the telecommunications, pharmaceutical, and biotechnology fields.

Support for research in all areas discussed in this article is provided across the federal government but development for security applications is concentrated within the Departments of Energy (DOE), Department of Defense (DoD), and Department of Homeland Security (DHS). The most directly relevant organizations within these departments are the Office of Nonproliferation Research and Development in the National Nuclear Security Administration of DOE, the Nuclear Technologies Directorate in the Defense Threat Reduction Agency of DoD, and the Transformational and Applied Research and Development Office in the Domestic Nuclear Detection Office of DHS. All three organizations support vigorous research and development programs conducting R&D in the national laboratories, universities, and industry.

REFERENCES

1. Ferguson, C. and Potter, W. (2004). *The Four Faces of Nuclear Terrorism*, Center for Nonproliferation Studies, Monterey, CA.
2. Levi, M. (2007). *On Nuclear Terrorism*, Harvard University Press, Cambridge, MA.
3. Hecker, S. S. (2008). Preventing nuclear weapon proliferation as nuclear power expands. *MRS Bulletin* **33**, 340–342.
4. Byrd, R. C., Moss, J. M., Priedhorsky, W. C., Pura, C. A., Richter, G. W., Saeger, K. J., Scarlett, W. R., Scott, S. C., and Wagner, R. L. (2005). Nuclear detection to prevent or defeat clandestine nuclear attack. *IEEE Sensors J*. **5**(4), 593–609.
5. The Royal Society (2008). *Detecting Nuclear and Radiological Materials*, RS policy document 07/08. Available at http://royalsociety.org/displaypagedoc.asp?id=29187.
6. Stephens, D. L., Runkle, R. C., Carlson, D. K., Peurrung, A. J., Seifert, A., and Wyatt, C. (2005). Induced temporal signatures for point-source detection. *IEEE Trans. Nucl. Sci*. **52**(5), 1712–1715.
7. Knoll, G. F. (2000). *Radiation Detection and Measurements*, John Wiley & Sons, New York.
8. Myjak, M. J. and Seifert, C. E. (2008). Real-Time compton imaging for the GammaTracker handheld CdZnTe detector. *IEEE Trans. Nuc. Sci*. **55**(2), 769–777.
9. Ely, J., Kouzes, R. T., Schweppe, J. E., Siciliano, E. R., Strachan, D. M., and Weier, D. R. (2006). The use of energy windowing to discriminate SNM from NORM in radiation portal monitors. *Nucl. Instrum. Methods Phys. Res., Sect. A* **560**(2), 373–387.
10. Shah, K. S., Glodo, J., Klugerman, M., Moses, W. W., Derenzo, S. E., and Weber, M. J. (2003). $LaBr_3$:Ce scintillators for gamma-ray spectroscopy. *IEEE Trans. Nucl. Sci*. **50**(6), 2410–2413.
11. Horansky, R. D., Ullom, J. N., Beall, J. A., Doriese, W. B., Durican, W. D., Ferreira, L., Hilton, G. C., Irwin, K. D., Reintsema, C. D., Vale, L. R., Zink, B. L., Hoover, A., Rudy, C. R., Tournear, D. M., Vo, D. T., and Rabin, M. W. (2007). Superconducting absorbers for use in ultra-high resolution gamma-ray spectrometers based on low temperature microcalorimeter arrays. *Nucl. Instrum. Methods Phys. Res., Sect. A* **579**(1), 169–172.
12. Mascarenhas, N., Brennan, J., Krenz, K., Lund, J., Marleau, P., Rasmussen, J., Ryan, J., and Macri, J. (2006). Development of a neutron scatter camera for fission neutrons. *Nuclear Science Symposium Conference Record, 2006. IEEE*. 2006 Nov; San Diego, CA, 1, pp. 185–188.
13. Almaviva, S., Marinelli, M., Milani, E., Prestopino, G., Tucciarone, V. C., Verona-Rinati, G., Angelone, M., Lattanzi, D., Pillon, M., Montereali, R. M., and Vincenti, M. A. (2008).

Thermal and fast neutron detection in chemical vapor deposition single-crystal diamond detectors. *J. Appl. Phys.* **103**(5), 054501.

14. Budakovsky, S. V., Galunov, N. Z., Karavaeva, N. L., Kim, J. K., Kim, Y. K., Tarasenko, O. A., and Martynenko, E. V. (2007). New effective organic scintillators for fast neutron and short-range radiation detection. *IEEE Trans. Nucl. Sci.* **54**(6), 2734–2740.
15. Ziock, K. P., Collins, J. W., Fabris, L., Gallagher, S., Horn, B. K. P., Lanza, R. C., and Madden, N. W. (2006). Source-search sensitivity of a large-area, coded-aperture, gamma-ray imager. *IEEE Trans. Nucl. Sci.* **53**(3), 1614–1621.
16. Vanier, P. E., Forman, L., Dioszegi, I., Salwen, C., and Ghosh, V. J. (2007). Calibration and testing of a large-area fast-neutron directional detector. *Nuclear Science Symposium Conference Record, 2007. NSS '07. IEEE*. 2007 Nov; Honolulu, HI, 1, pp. 179–184.
17. Bertozzi, W., Korbly, S. E., Ledoux, R. J., and Park, W. (2007). Nuclear resonance fluorescence and effective Z determination applied to detection and imaging of special nuclear material, explosives, toxic substances and contraband. *Nucl. Instrum. Methods Phys. Res., Sect. B* **261**(1), 331–336.
18. Warren, G. A., Caggiano, J. A., Hensley, W. K., Lepel, E., Pratt, S., Bertozzi, W., Korbly, S. E., Ledoux, R. J., and Park, W. H. (2006). Nuclear Resonance Fluorescence of 235U. *IEEE Nuclear Science Symposium Conference Record*. 2006 Nov; San Diego, CA, 2(1), pp. 914–917.
19. Padovani, E., Clarke, S. D., and Pozzi, S. A. (2007). Feasibility of prompt correlated counting from photon interrogation of concealed nuclear materials. *Nucl. Instrum. Methods Phys. Res., Sect. A* **583**(2), 412–420.
20. Jones, J. L., Blackburn, B. W., Watson, S. M., Norman, D. R., and Hunt, A. W. (2007). High-energy photon interrogation for nonproliferation applications. *Nucl. Instrum. Methods Phys. Res., Sect. B* **261**(1), 326–330.
21. Hartemann, F. V., Brown, W. J., Gibson, D. J., Anderson, S. G., Tremaine, A. M., Springer, P. T., Wootton, A. J., Hartouni, E. P., and Barty, C. P. J. (2005). High-energy scaling of Compton scattering light sources. *Phys. Rev*. **8**(10), 100702.
22. Nakamura, K., Nagler, B., Toth, C., Geddes, C. G. R., Schroeder, C. B., Esarey, E., Leemans, W. P., Gonsalves, A. J., and Hooker, S. M. (2007). GeV electron beams from a centimeter-scale channel guided laser wakefield accelerator. *Phys. Plasmas* **14**(5), 056708.
23. Reijonen, J., Gicquel, F., Hahto, S. K., King, M., Lou, T. P., and Leung, K. N. (2005). D-D neutron generator development at LBNL. *Appl. Radiat. Isot*. **63**(5), 757–763.
24. Dietrich, D., Hagmann, C., Kerr, P., Nakae, L., Rowland, M., Snyderman, N., Stoeffl, W., and Hamm, R. (2005). A kinematically beamed, low energy pulsed neutron source for active interrogation. *Nucl. Instrum. Methods Phys. Res., Sect. B* **241**(1), 826–830.
25. Peurrung, A. J. (2008). Material science for nuclear detection. *Mater. Today* **11**(3), 50–54.
26. Bolotnikov, A. E., Camarda, G. S., Carini, G. A., Cui, Y., Li, L., and James, R. B. (2007). Cumulative effects of Te precipitates in CdZnTe radiation detectors. *Nucl. Instrum. Methods Phys. Res., Sect. A* **571**(3), 687–698.
27. He, Z., and Sturm, B. W. (2005). Characteristics of depth-sensing coplanar grid CdZnTe detectors. *Nucl. Instrum. Methods Phys. Res., Sect. A* **554**(1), 291–299.
28. McKigney, E. A., Del Sesto, R. E., Jacobsohn, L. G., Santi, P. A., Muenchausen, R. E., Ott, K. C., McCleskey, T. M., Bennett, B. L., Smith, J. F., and Cooke, D. W. (2007). Nanocomposite scintillators for radiation detection and nuclear spectroscopy. *Nucl. Instrum. Methods Phys. Res., Sect. A* **579**(1), 15–18.
29. Neal, J. S., Boatner, L. A., Spurrier, M., Szupryczynski, P., and Melcher, C. L. (2007). Cerium-doped mixed-alkali rare-earth double-phosphate scintillators for thermal neutron detection. *Nucl. Instrum. Methods Phys. Res., Sect. A* **579**(1), 19–22.

KNOWLEDGE EXTRACTION FROM SURVEILLANCE SENSORS

RAMA CHELLAPPA, ASHOK VEERARAGHAVAN, AND ASWIN C. SANKARANARAYANAN
Center for Automation Research and Department of Electrical and Computer Engineering, University of Maryland, College Park, Maryland

1 INTRODUCTION

In the last decade, surveillance and monitoring has become critical for homeland security. With this increasing focus on surveillance of large public areas, a traditional human-centric surveillance system where a human operator watches a bank of cameras is largely being supported with automated surveillance suites. In a typical public area such as an airport or a train station there could be anywhere between 50 to a few hundred cameras deployed all over the area. It is almost impossible for human operators to keep a close watch on all of these cameras and continuously and robustly identify subjects and events of interest. Therefore, there is a greater need for automated analysis of the data obtained from multiple video cameras and other sensors that are distributed all around the area of interest.

Several current state-of-the-art surveillance systems work in aid of human operators. These surveillance systems have a host of sensors [visual, audio, infrared (IR) etc.] that are distributed in the area of interest. These sensors are in turn networked and connected to a central command center, where sophisticated algorithms for varied tasks such as person detection, tracking, and recognition; vehicle detection and classification; detection of restricted zone incursions, activity analysis; anomalous activity detection and so on, can be performed in real-time. One outcome of this automated analysis is to select a subset of the video cameras that are of interest so that these can then be monitored closely by human operators. This may significantly reduce the burden on the human operator. Moreover, when these sophisticated data mining techniques detect threats or anomalous behaviors they immediately alert the security personnel in the command center thereby reducing the time between an event and its detection.

We first discuss the various sensor modalities that are used in typical surveillance systems and discuss their advantages and disadvantages. We will then discuss in detail algorithmic paradigms for recovering important information from the data gathered by these sensors. Finally, we discuss the current challenges in fusion of information from a distributed web of sensors and discuss future trends.

2 SENSOR TYPES AND KNOWLEDGE EXTRACTION

There exist a wide range of sensors that find use in surveillance applications. These sensors are drawn from the families of line-of-sight and non-line-of-sight sensors. We discuss them here in the order of increasing complexity of acquired data and the subsequent processing required for extracting the information of interest. In many cases, this correlates directly to the intrinsic capability of the sensor type.

2.1 Motion Sensors

Motion sensors register the presence/absence of humans in a small region by actively using laser to sense or passively noting the presence of IR to detect motion [1–3]. In this way, at each time instant, motion sensors report sparse information without much capability to characterize the identity of the target, and as a consequence disambiguate between multiple targets or the action performed by the target(s). However, a dense deployment of motion sensors along with sparse deployment of complementary sensors such as cameras can form a very powerful sensor suite. Motion sensors, typically consume much less power compared to a camera and it is easier to process the information that it generates. Further, it is possible to devise clever fusion strategies that allow for intelligent steering of cameras, so that the cameras' fields of view (FoVs) cover areas in which there is significant activity. Recently, deployments of such motion sensors have gained immense popularity. There are publicly available data sets which allow exploring such data [4].

2.2 Acoustic, Seismic, and Radar Sensors

Acoustic and seismic sensors detect pressure waves and record a profile of the time-varying strength of the wave. This information, especially its description in the time–frequency space encodes a rich description of target properties [5]. The amplitude of the signal is a function of the source power and the range of the source from the sensor. Fourier analysis of the sensed signal reveals the harmonic content of the source that could possibly be used for tracking and identification [6]. When an array of microphones is used, it is possible to estimate the direction of arrival (DoA) of the source [7]. The data collected using an array can then be used to compute a time delay between the signals received between a pair of microphones. This time delay is proportional to the difference of the distance between the source and the individual microphones. Given sufficient number of microphones (or equivalently, time-delays) it is possible to estimate a source location that could result in such time delay measurements. However, in the case of collocated microphone arrays, the narrow baseline only allows for the reliable estimation of the DoA of the source. Given multiple DoA estimates from different arrays, it is possible to estimate source location by triangulation. Acoustic microphones provide omnidirectional sensing at cheaper power costs, while losing the ability to robustly determine the identity of the target. Further, acoustic microphones do not scale well with the number of targets in a scene and have poorer sensing range than video cameras. Similar algorithms can be used for range-only tracking using radar sensors [8].

2.3 Visible and Infrared Imaging

Video sensors are among the most prominent sensors used for surveillance [9, 10]. Visual sensors allow for the estimation of target parameters such as its location in a world coordinate reference, its texture, motion, and specifics the nature of its interaction with the scene and other targets. The information captured in the visual modality essentially allows for the extraction of significant amount of knowledge about the scene and events that occur in it. Further, the physics of image formation from the 3D world to the video provides constraints unique to the visual modality. These constraints are of immense use in well-conditioning of estimation algorithms that infer quantities of interest.

2.4 Multimodal Sensor Fusion

While video sensors provide a rich characterization of the scene, they suffer from a problems such as limited FoV, high operating costs in terms of power and computational complexity. In contrast, sensors such as motion detectors and acoustic microphones provide complementary capabilities that can be used in conjunction with visual sensors. As an example, motion sensors have been used to steer pan-tilt-zoom (PTZ) cameras to capture targets of interest [11]. This allows the PTZ cameras to be used only when there are targets in the scene and further, allows a large area to be sensed by the camera. In this sense, the two sensor types form a complementary pair that is useful for indoor surveillance. Acoustic microphone arrays provide similar support in the case of outdoor surveillance. Acoustic video fusion helps in providing omnidirectional sensing capabilities and providing robust tracking of objects even when one of the modality fails [12]. Similarly, acoustic-radar nodes provide position tracking for outdoor surveillance, wherein the DoA comes from the acoustic modality and the range information is provided by the radar [13].

3 SURVEILLANCE TASKS

A typical surveillance system must be capable of performing the following tasks—detection and tracking of humans and vehicles in the scene, person identification, vehicle classification, activity analysis, and anomalous activity detection.

3.1 Detection

The first and foremost task in typical surveillance systems is to detect objects of interest in the scene. Since these objects may have very different appearance characteristics it becomes extremely challenging to detect general objects. Hence, most surveillance systems resort to detection of moving objects. This is reasonable since in surveillance we are typically interested in humans and vehicles and their interaction with other humans and the environment. Motion-based object detection can be achieved both via passive sensors such as IR sensors or video cameras and via active sensors which are based on laser (similar to the ones used in elevator doors). Among these most surveillance systems use an intelligent mixture of passive IR sensors and video cameras [3, 4]. The advantage of using passive IR sensors is that they are very cheap to install, and they can therefore be densely distributed over the surveillance area. Moreover, these sensors provide just 1 bit of information, indicating the presence or absence of a human within its sensing range. Therefore analysis and integration of the information provided by a distributed web of these sensors is a rather simple task. However, these sensors do not provide any information about the appearance and therefore, the identity of the targets. This identity information is extremely important for several surveillance tasks. Therefore, these motion detection sensors are usually used in conjunction with several PTZ cameras. These cameras can take the initial target detection output of these sensors and then zoom into the targets of interest thus enabling high resolution capture of target appearance so that other tasks such as recognition, classification, and action analysis may be performed [11, 14].

3.1.1 Detection via Video Cameras. Background subtraction [15, 16] is the simplest and the most common computational algorithm that is used in order to detect moving objects from videos. The basic idea behind background subtraction is that an observed image is identical to the background model at all pixels that are static while the pixels that correspond to the moving object affect the appearance of those pixels that are on these moving objects. The background model may itself be a single static template (indoor single static camera with known scene environment) or a dynamic model (e.g. mixture of Gaussian) whose parameters are updated with time. In typical surveillance scenarios, the background model is dynamic, changes with the changing illumination conditions in the scene, and is updated on-the-fly. At each frame the background model is subtracted from the acquired video frame and the difference image indicates the regions in the image where there are potential moving objects (see Figure 1). Simple analysis of these pixels such as connected component analysis is performed in order to reliably detect objects of interest in the video.

3.2 Tracking

Once the targets of interest have been detected, the next task at hand is to track each of these targets using the multiple cameras surveying the scene [18]. Most algorithms maintain an appearance model for the detected targets, and use this appearance model in conjunction with a motion model for the target to estimate the target position at each individual camera. Such tracking can be achieved using deterministic approaches that solve an optimization problem [19] or using stochastic approaches that estimate the posterior distribution of the target location using Kalman filters [20] or more commonly particle filters [21, 22].

3.2.1 Appearance Model Based Monocular Tracking. At each individual camera, the detector outputs a set of pixels that correspond to a single moving object. Using some simple cues such as the size and shape of the set of detected pixels a bounding box enclosing the entire object is then drawn. The color/intensity values denoting the appearance of

FIGURE 1 In surveillance, it is common to use object motion for detection. In this example, we have available a target free image (a) of the scene constructed using the algorithm described in [17]. Given an image containing moving objects (b), we can compare the two to obtain possible target locations (c). The target free image models a static world and in this instance, the motion of the target forms an important cue for detection.

the object within this bounding box is then used in order to build/update the appearance model of the object. The simplest possible tracking algorithm is to naively search for bounding boxes with similar appearance in the next frame of the video. But such an approach will fail in several scenarios such as occlusions, change in illumination and pose of the targets. In order to account for these slow changes in the appearance of the target, most tracking algorithms build an on-line dynamic appearance model. For example, in [22], an on-line color appearance model is built using a mixture of Gaussian for each pixel within the object. The parameters of this dynamic appearance model are then updated using the current tracked output. In the case of the mixture of Gaussian model, the model parameters are comprised of the mixture weights and the mean and variances of each Gaussian. Moreover, in most cases some information about the motion characteristics of the object is also typically available. For example, in several surveillance scenarios, for tracking humans and vehicles it is reasonable to assume a simple constant velocity motion model. Both the motion model and the appearance model are used together in order to formulate a filtering problem in which the video sequence forms the observations. Traditional filtering methods such as a Kalman filter or more recent Monte Carlo methods such as a particle filter [21] are then used to recursively estimate the location and the appearance of the target in each subsequent frame of the video. Particle filtering [23] is an estimation technique that relies on approximating the posterior probability density (of the filtering problem) with a set of particles/samples and propagating these samples to recursively estimate the posterior density function. In particular, particle filtering techniques find applicability for a wide range of computer applications given the inherent nonlinearity in their formulations.

3.2.2 Multicamera Tracking. In the case of multicamera networks, another important issue to address is the association of targets across camera views. In the case of cameras with overlapping fields of view, the geometric relationship between the fields of view of the cameras along with the appearance information of the targets may be used in order to perform target association across views [24] (see Figure 2). For cameras that have

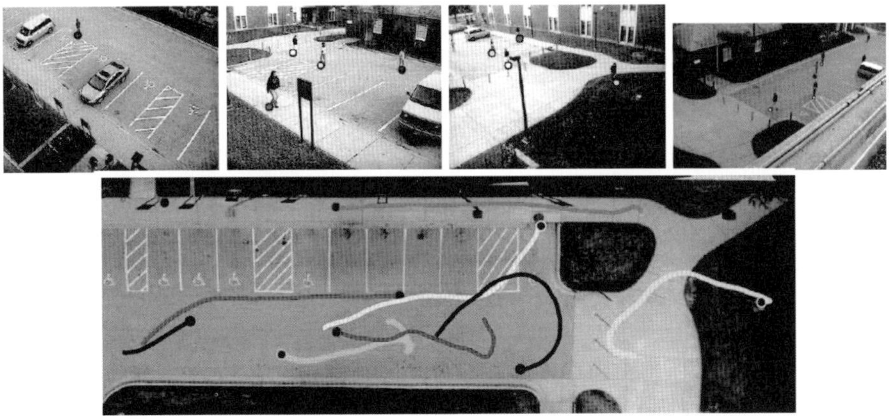

FIGURE 2 Inputs form six cameras are used to track object locations on the ground plane using the algorithm described in [24]. (Top row) Shown are target locations on the respective image plane in four different cameras. (Bottom row) The collection of tracks overlaid on a top view of the world is also shown. Tracking using multiview data is robust to occlusions and low signal to noise rations (SNR).

nonoverlapping fields of view, target association must be achieved using either a 3D site model or by learning the relationship between patterns of entry and exit in the various camera views [25]. In several restricted surveillance scenarios, 3D site models may be available along with the location and the orientation of the cameras. If such 3D site models are available, then they can be used along with the camera parameters in order to reliably and accurately associate targets across camera views. We refer the interested reader to a recent survey of multicamera-based algorithms for details pertaining to target detection, tracking, and classification using multiple cameras [26].

3.3 Recognition and Classification

Having detected and tracked targets the next task is to recognize the objects that we have tracked. The simplest method for identifying human subjects is to use appearance models for the face and perform image and video-based face recognition. In far-field surveillance scenarios where the number of pixels on the target's face may be very small, other cues such as gait, which is related to the manner of walking may be used in order to identify people. In the simplest such scenario, this recognition may be performed using just the shape or the silhouettes of the objects being tracked. This will ensure that the target classification is robust to variations in lighting, color, clothing and so on, that will affect the color image significantly more than the binary silhouette. In calibrated multicamera settings, one may also fuse the 2D appearance models obtained at individual camera views in order to obtain a 3D texture mapped model and then perform recognition using these 3D models.

3.3.1 2D Appearance Models for Recognition. For the sake of simplicity, let us assume that the objects of interest are faces, though the approach can be extended to generic objects. A simple 2D appearance template is first extracted and stored. This template is usually an image of the person's face under uniform illumination conditions. In [27], a particle filter is used to simultaneously estimate both the position of the target's face and the identity of the individual being tracked. The 2D appearance of the individual is modeled as a mixture of Gaussian and the parameters of the mixture density are estimated from the images in the gallery. The top row of Figure 3 shows the stored 2D appearance templates for the individuals in the gallery. In the bottom row are two images from a test sequence with the bounding box showing the location of the target's face. The image within the bounding box is matched with the stored 2D appearance models in the gallery in order to perform classification. Such a completely 2D-based recognition scheme suffers from limitations when the 3D pose of the face varies significantly. In order to tackle the pose problem effectively, multiple cameras along with 3D face models need to be used.

3.3.2 Silhouettes: Gait-based Person Identification. Gait is defined as the style or manner of walking. Studies in psychophysics suggest that people can identify familiar individuals using just their gait. This has led to a number of automated vision-based algorithms that use gait as biometric. Gait as a biometric is nonintrusive, does not require cooperation from the subjects, and performs reliably at moderate to large distances from the subjects. Typical algorithms for gait-based identification first perform background

FIGURE 3 (Top row) 2D appearance models for the individuals in the gallery. (Bottom Row) Two images from a video sequence in which a person is walking. The target's face is being tracked and the image within the bounding box of the tracked face is matched with the 2D appearance models in the gallery in order to perform recognition. (Image courtesy of [27]).

subtraction to obtain a binary image indicating the silhouette of the human. The first row in Figure 4 shows the tracked bounding box around the target in each frame of the video. Note that since the subject is moving, tracking needs to be performed before such a bounding box can be extracted. Background subtraction within the bounding box provides the binary image shown in the second row. The third row shows the shape of the extracted shape feature which is used in order to perform recognition. Note that the characteristics of the shape and identity of the individual and the subject's unique mannerisms during walking are captured in the sequence of shapes shown in the third row of Figure 4. Gait-based person identification differs from traditional biometrics in the sense that while traditional biometrics work on static features, gait as a biometric takes a sequence of deforming shapes in order to perform recognition. This means that dynamical models that capture both the shape and the kinematics of the feature need to be developed. Typical algorithms perform this shape sequence matching either using dynamic programming (dynamic time warping) or using state space methods (hidden Markov model (HMM)—[28, 29]) or by performing linear system identification (auto-regressive moving average (ARMA) model—[30]). In all these approaches the corresponding shape features or the model parameters for each of the subjects in the gallery is stored and is then used for comparison during the test/verification phase.

3.4 Event Analysis and Action Recognition

Event analysis in the context of surveillance systems can be broadly divided into those that model actions of single objects and those that handle multiobject interactions. In the case of single objects, we are interested in understanding the activity being performed by the objects. This is important in several scenarios so that vision-based surveillance systems may be able to understand and interpret the actions of humans within their field of view so

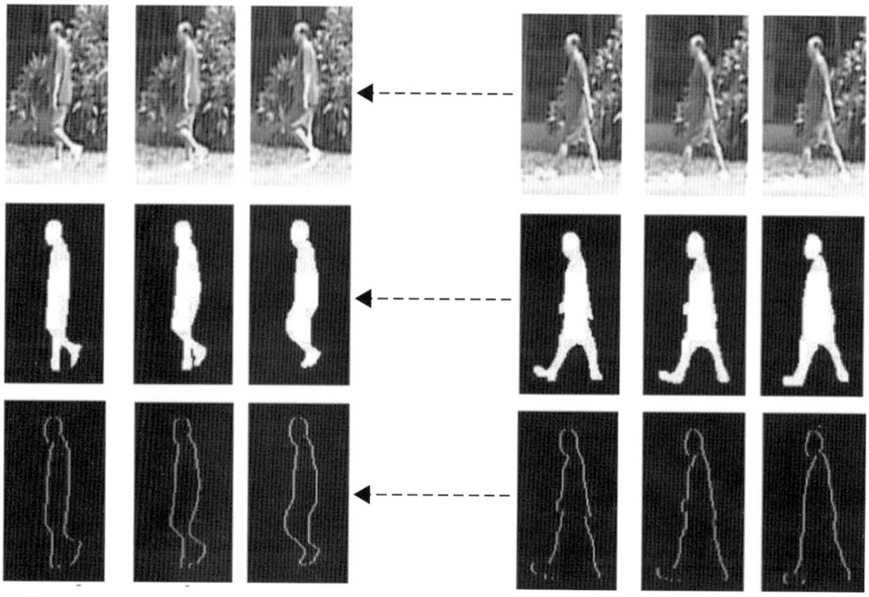

FIGURE 4 (Top row) Image within the tracked bounding box as a subject is walking. (Middle row) Binary Background subtracted image sequence. (Bottom row) Extracted Shape Sequence containing the discriminative information for classification. (Image courtesy of [30]).

that they can effectively react to those actions. This kind of action recognition is usually performed in indoor surveillance when the number of pixels occupied by human subjects in the video is significantly large (of the order of 100×100 pixels). First, background subtraction is performed and a shape/silhouette feature vector [30] is computed. This shape representation is suitable to identify the activities while marginalizing nuisance parameters such as the identity of the object or view and illumination. Stochastic models such as HMMs [31] and linear dynamical systems [32] have been shown to be efficient in modeling activities. In these, the temporal dynamics of the activity are captured using state—space models, which form a generative model for the activity. Given a test activity, it is possible to evaluate the likelihood of the test sequence arising from the learnt model.

Modeling interactions between multiple objects are of immense importance in many outdoor and far-field surveillance scenarios. Consider a parking lot where we may be interested in interactions between humans and vehicles. Examples of such interactions include an individual exiting a building and driving a car, or an individual casing vehicles. Several other scenarios, such as abandoned vehicles and dropped objects also fit under this category. Such interactions can be modeled using context-free grammars [33, 34] (see Figure 5). Detection and tracking data are typically parsed by the rules describing the grammar and a likelihood of the particular sequence of tracking information conforming to the grammar is estimated. Other approaches rely on motion analysis of humans accompanying the abandoned objects.

In activity analysis, the challenges are in making algorithms robust to variations in pose, illumination, and identity. In this regard, the choice of feature vector chosen to describe objects is very important. Further, there is a need to bridge the gap between

FIGURE 5 Example frames from a detected casing incident in a parking lot [34]. The algorithm described in [34] was used to detect the casing incident.

the semantics of interactions described using grammars and the feature vectors used to identify individual actors and their activities.

4 CONCLUSIONS AND FUTURE WORK

In this article, we presented several issues related to the process of knowledge extraction from sensors for typical surveillance applications. Specifically, we discussed the relationship between sensor types and their sensing capabilities and stressed the importance of multimodal surveillance for practical systems. Further, we discussed the kind of information that is of typical interest to an end user and provided an overview of the algorithmic procedure that extracts such information.

There are several challenges and problems in information extraction from surveillance sensors especially in the context of distributed sensing and distributed computing. In particular, with the increasing use of unmanned air vehicles (UAVs) connected with static sensors using wireless networks, it is important to account for bandwidth and power constraints in the design. This is particularly important for robust information extraction as the connectivity of network depends on the lifetime of the individual nodes especially when inputs from multiple sensors are required. Therefore, a systematic and detailed study of both power and energy optimization versus algorithm performance is necessary. Recently, several research groups have started looking at this problem especially for visual sensor networks [35, 36].

Another key area of future research is in the visualization of the sensed information. The need to present surveillance information in a holistic framework to the end user is paramount in large scale sensor networks. With possibly hundreds of nodes, the correlations between events and information coming from the various parts of the network need to be presented in an intuitive way so that their semantics are easily perceived. We are currently building a test bed for novel visualization schemes in order to provide an end user the freedom in viewing the scene and the activities being performed from *arbitrary points of view*. The end user is presented with a virtual depiction of scene with synthetic virtual actors depicting the events of the scene. The information extraction using the sensors are used to locate, clothe, and for animating the actions of the virtual actors. One potential application of this technology is in scene monitoring where the security personnel can freely move around the scene without having to watch a fixed set of Closed Circuit Television (CCTV) screens (where the spatial coherence between views and the activities are lost). Further, this can be combined with algorithms that alert the personnel when events of interest occur. Another area of potential use for such technology is in *privacy respecting* surveillance, wherein information that reveals the identity of the individual(s) is optionally suppressed by the visualization interface.

REFERENCES

1. Aipperspach, R., Cohen, E., and Canny, J. (2006). Modeling human behavior from simple sensors in the home. *Lect. Notes Comput. Sc*. **3968**, 337.
2. Ivanov, Y. A., Wren, C. R., Sorokin, A., and Kaur, I. (2007). Visualizing the history of living spaces. *IEEE Transactions On Visualization and Computer Graphics* **13**(6) 1153–1160.
3. Wilson, D. H., and Atkeson, C. (2005). Simultaneous tracking and activity recognition (STAR) using many anonymous, binary sensors. In *The Third International Conference on Pervasive Computing*, Springer, Heidelberg, pp. 62–79.
4. Wren, C., Ivanov, Y., Leigh, D., and Westhues, J. (2007). *The MERL Motion Detector Dataset: 2007 Workshop on Massive Datasets*, Tech. Rep., Technical Report TR2007-069, Mitsubishi Electric Research Laboratories, Cambridge, MA. August.
5. Johnson, D. H., and Dudgeon, D. E. (1993). *Array Signal Processing*, Prentice-Hall, Englewood Cliffs, NJ.
6. Chen, V. C., and Ling, H. (2002). *Time-Frequency Transforms for Radar Imaging and Signal Analysis*, Artech House, Boston, MA.
7. Cevher, V., Velmurugan, R., and McClellan, J. H. (2006). Acoustic multi target tracking using direction-of-arrival batches. *IEEE Tran. Signal Processing*. **55**(6), 2810–2825.
8. Cevher, V., Velmurugan, R., and McClellan, J. H. (2006). A range-only multiple target particle filter tracker. *In Proc. IEEE Intl. Conf. Acoustics, Speech, and Signal Proc. ICASSP* **4** (14–19), IV.
9. Regazzoni, C. S., Fabri, G., and Vernazza, G. (1999). *Advanced Video-Based Surveillance Systems*, Kluwer Academic Publishers.
10. Remagnino, P., Jones, G. A., Paragios, N., and Regazzoni, C. S. (2001). *Video-Based Surveillance Systems: Computer Vision and Distributed Processing*, Kluwer Academic Publishers.
11. Wren, C. R., Erdem, U. M., and Azarbayejani, A. J. (2005). Automatic pan-tilt-zoom calibration in the presence of hybrid sensor networks. In *Proceedings of the Third ACM International Workshop on Video Surveillance & Sensor Networks*, ACM, New York, pp. 113–120.
12. Cevher, V., Sankaranarayanan, A. C., McClellan, J. H., and Chellappa, R. (2007). Target tracking using a joint acoustic video system. *IEEE Transactions on Multimedia* **9**(4), 715–727.
13. Cevher, V., Borkar, M., and McClellan, J. H. (2006). A joint radar-acoustic particle filter tracker with acoustic propagation delay compensation. *In Proc. 14th EUSIPCO*. Florence, Italy.
14. Hampapur, A., Brown, L., Cornell, J., Ekin, A., Haas, N., Lu, M., Merkl, H., Pankanti, S., and I. B. M. T. J. W. R. Center, Hawthorne, NY (2005). Smart video surveillance: exploring the concept of multiscale spatiotemporal tracking. *IEEE Signal Proc. Mag*. **22**(2), 38–51.
15. Piccardi, M. (2004). Background subtraction techniques: a review. *Systems, Man. and Cybernetics, IEEE International Conference on* **4**, 3099–3104.
16. Elgammal, A., Duraiswami, R., Harwood, D., and Davis, L. S. (2002). Background and foreground modeling using nonparametric kernel density estimation for visual surveillance. *P. IEEE* **90**(7), 1151–1163.

17. Joo, S., and Zheng, Q. (2005). A temporal variance-based moving target detector. *IEEE International Workshop on Performance Evaluation of Tracking and Surveillance (PETS)*. Breckenridge, Colorado, January.
18. Yilmaz, A., Javed, O., and Shah, M. (2006). Object tracking: a survey. *ACM Comput. Surv. (CSUR)* **38**(4).
19. Comaniciu, D., Ramesh, V., and Meer, P. (2000). Real-time tracking of non-rigid objects using mean-shift. *CVPR*, Hilton Head Island, South Carolina, **2**, 142–149.
20. Broida, T. J., Chandrashekhar, S., and Chellappa, R. (1990). Recursive techniques for the estimation of 3-d translation and rotation parameters from noisy image sequences. *IEEE Trans. Aerosp. Electron. Syst*. **AES 26**, 639–656.
21. Isard, M., and Blake, A. (1998). Icondensation: Unifying low-level and high-level tracking in a stochastic framework. *European Conference on Computer Vision* **1**, 767–781.
22. Zhou, S. K., Chellappa, R., and Moghaddam, B. (2004). Visual tracking and recognition using appearance-adaptive models in particle filters. *IEEE T. Image Process*. **11**, 1434–1456.
23. Doucet, A., de Freitas, N., and Gordon, N. (2001). An introduction to sequential Monte Carlo methods. *Sequential Monte Carlo Methods in Practice*, Springer, pp. 4–11.
24. Sankaranarayanan, A. C., and Chellappa, R. (2008). Optimal multi-view fusion of object locations. *IEEE Workshop on Motion and Video Computing (WMVC)*. Copper Mountain, CO, pp. 1–8.
25. Makris, D., Ellis, T., and Black, J. (2004). Bridging the gaps between cameras. *IEEE Conference on Computer Vision and Pattern Recognition*, Washington, DC, **2**, pp. 1–8.
26. Sankaranarayanan, A. C., Veeraraghavan, A., and Chellappa, R. (2008). Object detection, tracking and recognition using multiple smart cameras. *P. IEEE* **96**(10), 1606–1624.
27. Zhou, S., Krueger, V., and Chellappa, R. (2003). Probabilistic recognition of human faces from video. *Comput. Vis. Image Underst*. **91**, 214–245.
28. Kale, A., Rajagopalan, A. N., Sundaresan, A., Cuntoor, N., Roy Cowdhury, A., Krueger, V., and Chellappa, R. (2004). Identification of humans using gait. *IEEE T. Image Process*. **13**, 1163–1173.
29. Liu, Z., and Sarkar, S. (2006). Improved gait recognition by gait dynamics normalization. *IEEE Trans. Pattern Anal. Mach. Intell*. **28**, 863–876.
30. Veeraraghavan, A., Roy-Chowdhury, A. K., and Chellappa, R. (2005). Matching shape sequences in video with applications in human movement analysis. *IEEE Trans. Pattern Anal. Mach. Intell*. **27**, 1896–1909.
31. Rabiner, L. R. (1989). A tutorial on hidden Markov models and selected applications inspeech recognition. *P. IEEE* **77**(2), 257–286.
32. Turaga, P. K., Veeraraghavan, A., and Chellappa, R. (2007). From videos to verbs: mining videos for activities using a cascade of dynamical systems. *In IEEE Computer Vision and Pattern Recognition, 2007*. Minneapolis, MN, pp. 1–8.
33. Moore, D., and Essa, I. (2001). Recognizing multitasked activities from video using stochastic context-free grammar. *Workshop on Models versus Exemplars in Computer Vision*. Kauai, HI.
34. Joo, S. W., and Chellappa, R. (2006). Recognition of multi-object events using attribute grammars. *IEEE International Conference on Image Processing*. Atlanta, GA, pp. 2897–2900.
35. Shen, C., Plishker, W., Bhattacharyya, S. S., and Goldsman, N. (2007). An energy-driven design methodology for distributing dsp applications across wireless sensor networks. *IEEE Real-Time Systems Symposium*. Tucson, Arizona, December.
36. Schlessman, J., and Wolf, W. (2004). Leakage power considerations for processor array-based vision systems. *Workshop on Synthesis And System Integration of Mixed Information Technologies*. Kanazawa, Japan.

RADAR AND LiDAR PERIMETER PROTECTION SENSORS

Frank W. Griffin

Sandia National Laboratories, NWS DoD Program Design & Implementation, Albuquerque, New Mexico

Rick Hurley

Sandia National Laboratories, RF and Optics Microsystem Applications, Albuquerque, New Mexico

Dan Keller

Sandia National Laboratories, Strategic Business Enterprise Services, Albuquerque, New Mexico

1 INTRODUCTION

With the goal of improving the adversary interdiction time line, ground-based radar and light detection and ranging (LiDAR) technologies offer options for intrusion detection and situational awareness, beyond facility perimeters. However, failure to understand and address the variables that affect successful employment of these technologies can result in costly installation of an inappropriate technology for the given security application. Far too often, security system designers work backwards by allowing advertised performance or cost to determine technology selection rather than the objectives and requirements of the security application.

Sandia National Laboratories (SNL) conducts evaluations of extended detection technologies for US government customers and their specific applications. On the basis of this experience, the authors assert that installation mistakes can be avoided using an approach that first assesses and defines the objectives of the application (what the application must detect, where, and for what reason), and then matches an appropriate technology to meet these objectives. This process is not always linear and may require an iterative approach.

The unique operating environment (terrain, weather, etc.) also plays a key role in appropriate technology selection. Other variables include system-specific capabilities and limitations, user-defined settings, and how a given technology is installed and employed.

2 TECHNOLOGY OVERVIEW

The following is an advanced discussion of radar and LiDAR operational methods and design variables. Each technology's operational methods have inherent strengths and limitations, and some systems employ or combine more than one operational method.

For information regarding specific systems, the reader is encouraged to contact product vendors. Reference herein to any specific commercial product, process, or service by trade name, trademark, manufacturer, or otherwise, does not necessarily constitute or imply its endorsement, recommendation, or favoring by SNL, Lockheed Martin, the US government, any agency thereof, or any of their contractors or subcontractors.

2.1 Light Detection and Ranging Technology (LiDAR)

LiDAR systems are active infrared technologies that use the near infrared (NIR) portion of the electromagnetic spectrum (750–3000 nm) for transmission of energy to illuminate targets. For target identification and tracking, target position and velocity changes are noted between each scan of the area against the known background. Because of the short wavelength of NIR energy, these systems are very sensitive to detecting small scene changes; however, increased sensitivity also tends to produce a relatively high nuisance alarm rate (NAR) from alarms attributed to animals and other nuisance alarm sources (a high NAR, which is the number of nuisance alarms averaged over a specified time period, may lead to operator overload and decrease confidence in the effectiveness of the system). Issues with these systems include laser safety concerns, short detection ranges for available systems (150–250 m maximum), and performance limitations in poor visibility or foggy conditions.

2.2 Radar Technology

Tactically deployable radar systems have been in use since the 1970s, and were originally designed to allow an operator to "listen" for troop or vehicle movement within an area of interest [1, 2]. Depending on the radar's horizontal and vertical beam width and the distance to the target, a large area several kilometers away may be monitored. Operating in the X-band (8.5–10.68 GHz) or Ku-band (12–18 GHz), these systems use a pulse repetition rate of several kilohertz. The frequency shift of return signals due to the Doppler effect is analyzed for target detection and identification [3]. These systems were not originally designed to operate continuously from a permanent position, but only as needed for the given tactical situation. Today, interest in using these systems for static security applications has gained momentum due to technology advances.

2.2.1 Pulsed-Doppler Scanning Radar. For basic pulsed-Doppler radar systems, target range is calculated from the time it takes an emitted signal pulse to travel from the radar emitter, bounce off the target, and then return to the system. A Doppler shift occurs when a transmitted carrier frequency bounces off a moving target. This shift is a measure of relative radial velocity and direction, and is used to distinguish a moving target from stationary objects [3]. For early ground-based radars, this information was presented as audible signal changes, and system effectiveness was dependent on the skill and training of the operator.

Today, many pulsed-Doppler systems employ a graphical user interface to display target information and use lower power or pulse compression (less power over a longer pulse duration), improved processing and filtering techniques, variations of modulation methods, and various antenna configurations to improve performance.

The primary advantage of a pulsed-Doppler system is the ability to identify and isolate target movement in an area with relatively high energy returns from background clutter. A disadvantage is that a high NAR is often generated by reflections from rainfall and vegetation moving in the wind, thus creating multiple signal returns (Fig. 1).

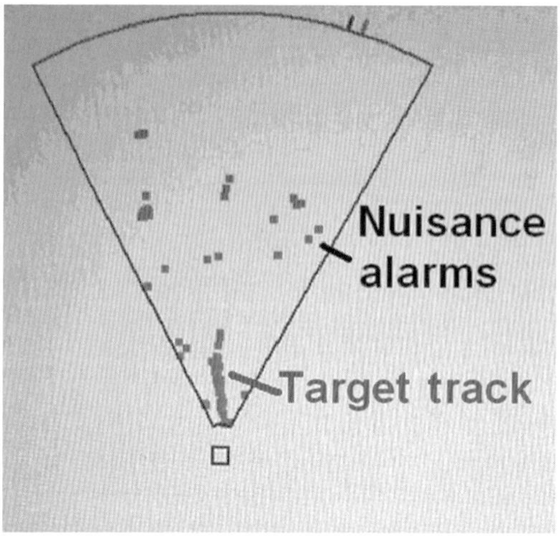

FIGURE 1 Display from pulsed-Doppler radar: reflections from moving vegetation as nuisance alarms.

2.2.2 Frequency Modulated Continuous Wave Scanning Radar. Advancements in computer processing speed, low cost techniques for generating linear frequency sweeps, and improved filtering have led to development of ground-based versions of frequency modulated continuous wave (FMCW) systems. Used for years by aircrafts, FMCW radars generally employ simultaneously operating transmitter and receiver antennas [3] and operate in the Ku or the Ka (27–40 GHz) frequency band.

There are significant differences between FMCW short-range and long-range systems. Long-range systems use Doppler discrimination to identify moving targets; short-range systems use high resolution range bin cell sizes (e.g. 0.5 m) and track target movement across multiple range bins to determine direction and velocity. By using higher resolution range bins, short-range systems use background clutter and an adaptive threshold to account for environmental changes occurring over time. This allows for filtering of energy received from stationary nuisance alarm sources.

2.2.3 Moving Target Indication (MTI) versus Pulse Doppler. Moving target indication (MTI) radars are based on the same principles as pulsed-Doppler systems. The difference is that an MTI system typically operates with an ambiguous Doppler measurement and unambiguous range measurement. This means that the system has a blind speed where detection will not occur and does not address second-time-around echoes from far targets. Pulsed-Doppler systems have no blind speed, but do experience range ambiguities and limitations [3].

2.2.4 Electronic Scanning or Phased Array Radar. For an electronic scanning or phased array radar, the system scans by electronically varying the electrical current phase across the antenna aperture [3]. These systems have scan area limitations. Depending on the system, scan angle is usually limited to less than 90°. This problem can be offset by using multiple antennas to cover large areas of interest. Electronic scanning offers a potential maintenance cost savings (no moving parts requiring periodic replacement), but emerging systems require additional design work to improve processing and scanning capabilities.

2.3 Design Variables

Modern radar and LiDAR systems vary considerably in design and operational characteristics. Variables include, but are not limited to, the number and size of range bin cells, horizontal and vertical beam width, and scan angle and rate.

2.3.1 Range Bin Cells and Resolution Cell Size. All radars use range and angular range bins for processing. In pulsed-Doppler systems, these cells are used to integrate (average) multiple returns from a single target to minimize the effects of noise, improve the probability of detection, and reduce nuisance alarms [3]. For long-range pulsed-Doppler systems, these range bins can be 50–100 m in length, and the strongest signal can be processed to approximate the range from the radar to the target within that cell.

For FMCW radars, smaller resolution cell sizes offer higher fidelity for target detection, but reduce the maximum detection distance of the radar. In many cases, when vendors offer multiple ranges for FMCW radar, they are increasing the cell size to effectively double the range capability of the system. While this adds distance, it reduces radar sensitivity because the cell size is now larger and more energy is returned from the background within that cell (i.e. a higher clutter level).

2.3.2 Horizontal and Vertical Beam Width. For radars, a narrow vertical beam width is desirable for a system that tracks targets several kilometers away (e.g. 3° or less), but this limits effectiveness at ranges less than 1 km if there are undulations in the terrain. Generally, a larger vertical beam width will cover variations in terrain easier than a narrow beam width. The downside is that target location is generally reported as two-dimensional information (e.g. latitude and longitude), and target elevation information relative to the system is usually not provided.

Typically, the horizontal beam width determines the angular resolution and angular cell size of the radar. When a system averages returns from the environment to establish a background clutter level, the width of the angular cell size will have a significant impact on the amount of energy received from background clutter in that cell. For pulsed-Doppler systems, horizontal beam width also determines how many times an object will be interrogated during a single scan.

2.3.3 Scan Angle, Scan Rate, and Processing Time. The scan angle, scan rate, and processing time required for a system to identify a target are important variables for system selection. Long-range systems typically have a longer scan rate than short-range systems. Problems arise if the selected system scan rate is slower than the time it would take a target to traverse the area the radar is expected to cover. For example, consider a system with a fixed scan angle of 360°, a 15-s scan rate, and a target acquisition parameter that requires three consecutive target interrogations, or "hits," before generating an alarm. For this system, processing time is greater than 45 s, and a fast-moving target (such as a vehicle) may travel a considerable distance before being detected.

2.3.4 User Interface and Clutter Maps. Most modern systems employ clutter or background return maps presented through the user interface. Essential during system operation, clutter maps also allow the security engineer to easily identify areas of concern, as illustrated in Figure 2, which shows an area containing a known reflection source (a fence) as presented by two different radar systems' user interfaces. Using the clutter map as a reference, the security system engineer can make adjustments to the system or the installation configuration to mitigate application-specific concerns.

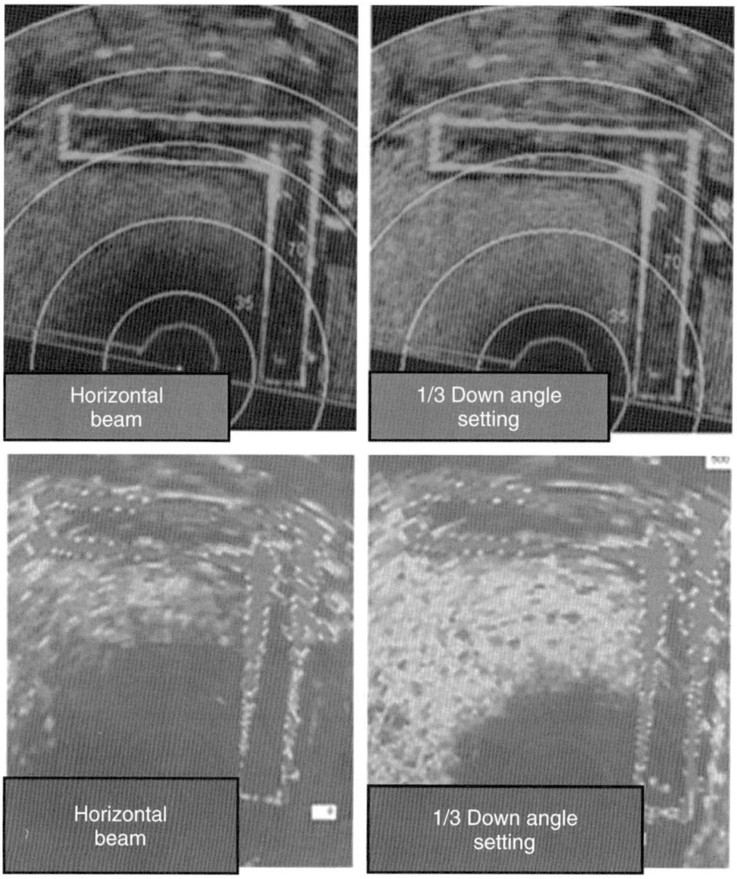

FIGURE 2 Clutter map and increased clutter due to angle of incidence variation for two radar systems.

2.4 Operational and Performance Variables

When selecting a radar or LiDAR system, the security system designer should also understand and address technology and system-specific operational and performance variables.

2.4.1 ***Clutter.*** Obstructions produce high clutter returns and reduce the operator's ability to distinguish target returns relative to the obstruction (Fig. 3). These returns are scene-dependant (i.e. dependant on the angle and geometry of the obstructions within the scene). Changing weather conditions, terrain undulations, and the presence of tall vegetation also cause clutter problems.

For short-range LiDAR systems (150–250 m), clutter sources are generally less of a problem than for radar systems because emitted (light) energy does not behave in the same way as radiated (radio frequency) energy, to background clutter. For LiDAR systems, the problem posed by a high clutter environment is shadowing from line-of-sight obstructions.

Redundancy (multiple radars overlapping the same area of observation) or use of other detection technologies may be needed within an application to mitigate high clutter concerns.

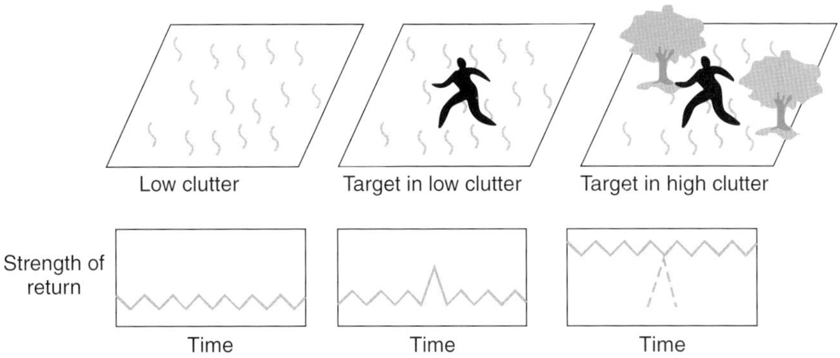

FIGURE 3 Target in low and high clutter resolution cells.

2.4.2 ***Dead Spot.*** Because of their operating characteristics and beam width, radar and LiDAR systems are surrounded by an area of no detection (a dead spot) that varies in range and area, dependent on the radar system. If installation height is increased, the area of the dead spot will also increase. The size of the dead spot can be reduced by adjusting the radar's angle of incidence; however, testing has shown that this reduces far-range performance and increases near-range returns, which may lead to receiver saturation (high clutter). Theoretically, LiDAR systems do not have the same saturation issue as radar systems, but this has not been verified through SNL testing.

2.4.3 ***Environmental Effects.*** Rain, snow, and fog can increase clutter within a resolution cell and attenuate the transmitted signal at its operating frequency, diminishing effective radar range. The effects of precipitation are less significant for lower operational frequencies. However, most lower-frequency radars are pulsed-Doppler systems that have a greater sensitivity to nuisance alarms from rainfall, due to the movement and density of the rain. (Circular polarization of a pulsed-Doppler system has been advertised to help minimize returns received from rainfall, but this has not been verified by SNL). Figure 4 provides an example of target returns received during testing from a relatively light but fast-moving rainstorm.

LiDAR systems have issues relating to dispersion and reflection of light energy due to precipitation. Fog creates a diffuse reflectance and can be the source of nuisance alarms (Fig. 5).

FMCW systems that use adaptive threshold processing have been demonstrated to function without generating nuisance alarms in light and medium rainstorms (heavy downpour effects have not been evaluated by SNL), as illustrated in Figure 6 which shows an increase in background clutter but not in nuisance alarms. However, rain degrades the detection capability through a decrease in the maximum detection distance.

While many systems provide manual rain filters or can be configured to minimize rain effects, the security system designer should note that the more an operator is required to make system adjustments, the higher the chance that the system will be adjusted incorrectly or not be readjusted when conditions return to normal.

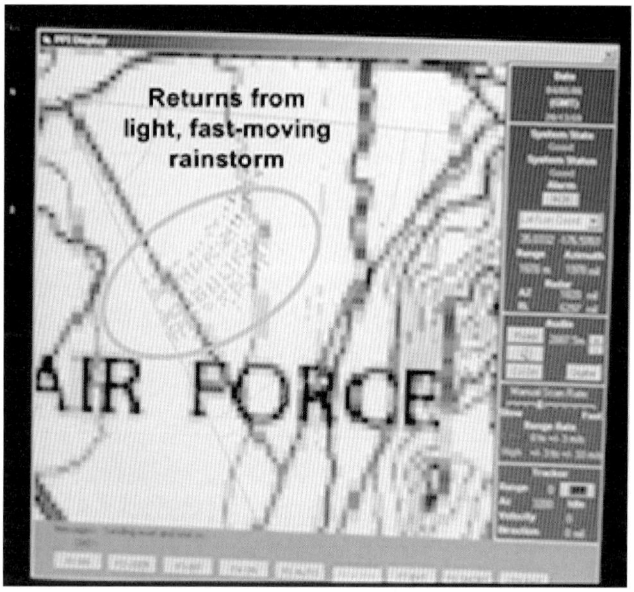

FIGURE 4 Operator display (rainstorm) with pulsed-Doppler system.

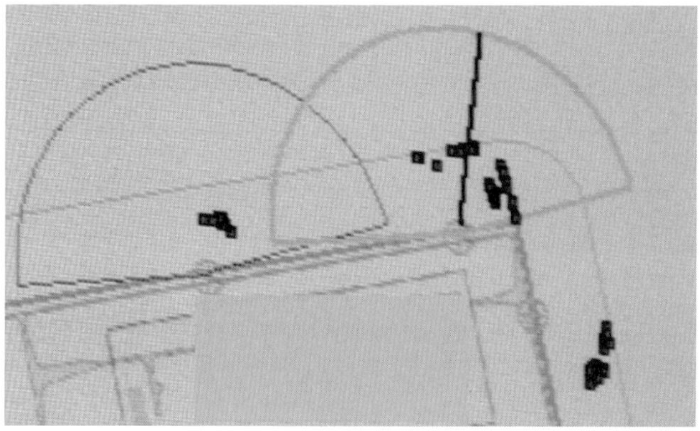

FIGURE 5 Fog generated nuisance alarms for LiDAR system—example application.

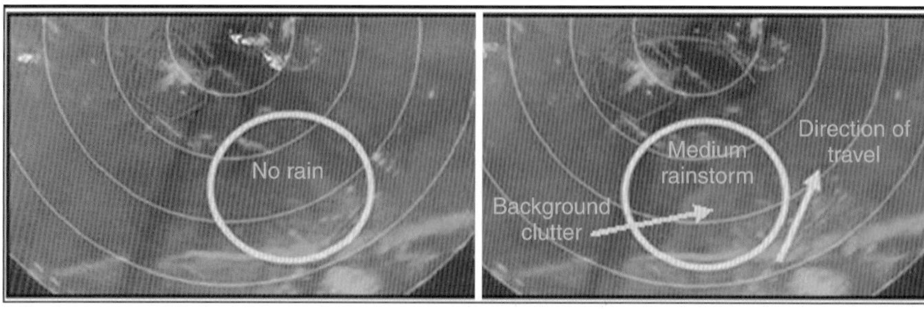

FIGURE 6 Rainstorm affecting background clutter level for FMCW radar.

3 DEFINE AND CHARACTERIZE THE SECURITY APPLICATION

Radar and LiDAR technologies appear to offer a cost-effective alternative to the traditional perimeter security applications. In some circles, the concept of using these technologies to maintain a "virtual fence" has taken root. Because of the unique issues encountered by radar and LiDAR applications, the authors assert that these systems should be employed and characterized as situational awareness or extended detection *enhancements* to traditional perimeter security systems, not as a replacement.

Further, technology selection and installation considerations generally applied to most perimeter detection applications cannot be as easily applied to extended detection applications. What works well for one application (technology selection and installation configuration) may not work for another application with different objectives, environmental conditions, and operational use variables. To match the best technology to a given application, the security system designer must first characterize the application.

3.1 Application Objectives

At the highest level, the application objective consists of what is to be detected, where, and for what reason. For traditional perimeter security systems, the objective is usually to detect a target as it crosses a defined line of demarcation (perimeter), to provide alarm assessment (often using video cameras), and to delay the target long enough for an appropriate response [2]. Target intent is easily determined (e.g. someone climbing the perimeter fence is not supposed to be there).

Defining objectives for an extended detection application is more involved. There is not always a defined line of demarcation, and some targets detected within the area of interest may be authorized to be there, such as traffic on a roadway. Alarm assessment is more difficult at extended ranges (as is determining intent), and initiating a response beyond a perimeter may not be practical.

To define application objectives, the security system designer must first determine what target types represent the primary concern for the site (such as personnel on foot or fast-moving vehicles, etc.) to select the technology best suited to detect these target types for the given area of interest. Target types are generally identified through site-specific threat analyses and historical precedence.

3.2 Application Requirements

Application requirements define how the application must perform to achieve its objectives against the defined target set.

3.2.1 Quantifying Performance. To quantify performance, security systems are evaluated using performance metrics to verify compliance with established performance requirements. Metrics include such performance thresholds as a minimum NAR and probability of detection, which is a measure of system effectiveness based on target sensing, assessment, and alarm communication probabilities. Because of differing system and environmental variables, radar and LiDAR systems will generally have a higher NAR and lower probability of detection than is acceptable for perimeter sensors, and the security system designer and system effectiveness analyst should understand that detection probability will decrease and the NAR may increase as detection range increases.

3.2.2 Target Tracking and Target Intent. For long-range applications, it may be impractical to initiate an immediate response at the point of detection because target intent may be difficult to determine; therefore, accurately tracking target movement is as important as detecting the target. For example, a target moving rapidly toward a facility or line of demarcation may warrant a more heightened state of alert than a nonthreatening target passing through the area.

3.2.3 Operational Mode. Does the application require the system to remain in continual operation, or will it be turned on only as needed for enhanced situational awareness? Systems operating continually will likely require more frequent maintenance and repair than those operated less frequently, so a system's mean time between failure (MTBF) rate, operational mode, and the availability and cost of spare components should be factored into its selection. For example, electronic scanning systems have fewer movable parts, which may translate into lower maintenance costs for these systems.

3.2.4 Area of Interest and Range. Radar and LiDAR systems are line-of-sight technologies that work best for applications relatively clear of objects and terrain undulations that create areas of shadowing within the area of interest. The authors recommend characterizing the potential area of interest using three-dimensional terrain modeling to identify these areas of concern. This assists with determining system placement location for maximum coverage. Other measures may include raising the installation height of the system, deploying more than one system, or employing additional sensor technologies.

The security system designer should not simply accept the advertised maximum range of a technology to define the range of the application. Generally, the farther the range, the more difficult it is to assess the target due to distance and more line-of-sight obstructions. Effective range may also decrease for some systems during inclement weather. For short-range applications, it is also a mistake to assume that a system designed to detect at longer ranges will exhibit the same performance at shorter ranges.

4 TECHNOLOGY SELECTION

4.1 Systems Approach and Additional Deployment Considerations

For technology selection, the security system designer, using a systems approach, must address deployment considerations and adopt methods for dealing with alarm assessment issues, the effects of excessive numbers of actual and nuisance alarms, and must implement an effective Concept of Operations plan. The selected product must also be appropriately installed to take full advantage of its positive strengths while minimizing the negative effects of its limitations.

4.1.1 Installation Height and Antenna Tilt Angle. Radar installation height and antenna tilt angle are key factors affecting the signal-to-clutter ratio. Background energy (clutter) increases with a system's angle of incidence. Increasing installation height reduces shadowing effects from obstructions; however, increased height decreases sensing and tracking performance since the clutter return is higher due to an increased angle of incidence with the environment.

Radar and LiDAR systems should be mounted as low as possible, dependent on intervening obstructions within the application. Figure 2 illustrates the changes seen on clutter maps caused by changing the angle of incidence. When energy returns from the ground are strong (dark areas), sensitivity is decreased for detecting new targets because there is already a significant amount of energy returning in those range and angular cells.

4.1.2 Assessment of Alarms and Tracks. For the standard perimeter, fixed cameras provide assessment of both legitimate targets and nuisance alarm sources. In general, the operational environment for perimeter security systems is well maintained, is usually clear of line-of-sight obstructions, and is well illuminated at night [2].

The extended detection environment, however, is full of visual obstructions that may limit effective assessment of alarms, and may require assessment under low or no-light conditions. Distance is also an important factor, as the probability of (correct) assessment of alarms decreases as detection distance increases.

If visual assessment is a requirement, the desired effective range of a radar or LiDAR system should be matched to an equally effective assessment system. An ineffective assessment system may decrease or cancel out the interdiction time line advantage afforded by an effective detection system.

4.1.3 Nuisance Alarms and Actual Alarms. Even if a system has been tested and the results indicate a low NAR, it is important to understand that those values will increase with each system added to the design, and as a function of desired detection range. The capability to detect targets at the maximum range of the system is not always the most effective use of radar and LiDAR technologies, particularly if monitoring a large area of interest becomes unmanageable due to a high NAR.

Most radar and LiDAR systems have the ability to mask off areas of noninterest. Masking features could be employed during periods of high traffic along known pathways. In this application, it is recommended to define a line of detection and designate routes for restricted passage of authorized personnel within the detection envelope. Without these measures, system effectiveness is decreased during peak traffic periods.

4.1.4 Technology Integration. The security system designer must also address how the radar or LiDAR system will integrate with the existing site physical security system. Factors to consider are alarm message communications using eXtensible Markup Language (XML) protocols, visual assessment technologies (if used), and communications links.

In the authors' opinion, a site's perimeter security system alarm console should not be linked with radar/LiDAR operator interfaces as the radar/LiDAR operator may deal with nuisance alarms and actual alarms at a rate much higher than a perimeter security system operator. Combining the two functions may result in a reduction in effectiveness for the site's security system as a whole.

4.1.5 Life Cycle Costs. These include initial installation, maintenance, and costs for repairs and replacement parts throughout the operational life of the system. Variables affecting costs include system MTBF (and time to repair), availability of spare parts, and how the system is employed (i.e. continual versus occasional operation). Not considering long-term costs can lead to long periods of down time due to unforeseen maintenance costs.

4.1.6 Response—Concept of Operations. Concept of Operations for a realistic response is critical to the effectiveness of a radar or LiDAR application. There are several issues to be addressed, such as:

- definition of what situations to respond to and what level of response is appropriate for each situation;
- identification of a challenge line or challenge distance(s) from the facility;
- determination of a response location or range (i.e. where the target must be challenged).

4.2 System Selection

In general, when reviewing potential radar or LiDAR systems for a given application, it is important to consider the following:

- *Existing integration capabilities (XML protocols, assessment technologies, etc.).* Does the system integrate easily with the selected system platform and/or assessment system?
- *Range and angular resolution requirements.* At what range are you trying to detect a target and why—can the response force effectively use the information given for a target several kilometers away? How important is it to accurately know the target's location, and is it sufficient to accurately slue or position an integrated assessment system?
- *Scan angle limitations.* Is the system installed on a fence directly adjacent to the area of interest? (Installation may require a scan angle of 180° or 270°.)
- *Scan rate limitations.* Is the system scanning at the correct speed to pick up its intended target in time to support timely response? (If a system is overlooking an area of interest from a distance of 1 km away, a slow scan speed may be acceptable for the smaller field of view relative to the system; however, a system with a shorter range may require 1 rps, for example.)
- *Processing delays, limitations, or strengths (system alarm criteria, level of filtering, tracking capabilities, etc.).* Do filtering techniques impede or slow down the alarm reporting time line? Does the system require a large number of operator adjustments?
- *Tested or advertised MTBF, cost of system, cost of maintenance.* All are equally important for calculating life cycle costs.
- *Down time associated with repairs and support.* Is the manufacturer or one of their subsidiaries located in your country? What effects will customs considerations have on emergency repairs, and how does this affect security?
- *System-specific limitations.* Can system limitations be compensated for by other security measures? All sensor technologies have inherent limitations that may be exploited, requiring other measures to compensate for these limitations.

4.3 Testing

One important point that should be adopted from this article is the need for on-site performance testing. For this, a concise but comprehensive test plan must be developed to address application-specific security compliance requirements. This test plan should also include tests designed to identify and define technology strengths and limitations while operating within the unique operational environment. Testing should be designed to ensure the system is installed and calibrated to provide optimal performance.

Radar and LiDAR systems should also be tested during degraded environmental conditions to ensure adequate performance. Security system designers should not rely solely on advertised performance specifications provided by product vendors as these metrics are usually collected under "best-case" evaluation conditions.

5 TECHNOLOGY ADVANCEMENTS AND CURRENT ISSUES

With an increased interest in radar and LiDAR technology, increased funding is being allocated to improve and further develop these products. Below is a brief description of some recent improvements, as well as areas that still require attention.

5.1 Signal Processing Changes

Improvements in signal processing allow design engineers to further filter or process the information returned to the system. This is seen in tracking algorithms that take multiple target returns and look for target speed, direction, signal strength, and general characteristics to provide the operator with a single labeled track of a target as opposed to a group of individual dots on the user interface. By establishing target tracks, software developers can then create sensor fusion engines that process returns from multiple units to show a single target on a user interface. As more platforms begin to offer this capability, the focus may change to improved tracking using integrated assessment and surveillance video systems that will consider target elevation.

5.2 Elevation Tracking and Multipoint Calibration for Integrated Assessment Systems

Most ground-based radar and LiDAR systems only provide two-dimensional location data for targets detected within their vertical and horizontal beam pattern. Although these data are useful for sluing a camera system to the proper latitude/longitude coordinates, there is a need for collection and interpretation of elevation data for proper vertical orientation of the camera system. This becomes more apparent as the range from the radar to the target increases (few environments are perfectly level).

Inclusion of a three-dimensional mapping capability into the control software would improve tracking capabilities for integrated assessment systems. Another possible solution would be to allow the operator to create an initial multipoint calibration for the camera, allowing the camera to slue to fixed points at different elevations. The elevation values for camera travel would be linearly interpreted between elevation points. Thus, the more points included in the calibration, the more accurately the camera would follow the terrain, as illustrated in Figure 7.

5.3 Integration Issues and Needs

Radar systems for most installations within the United States deal with everyday high traffic areas and with segregating potential target movement from normal traffic (assessment

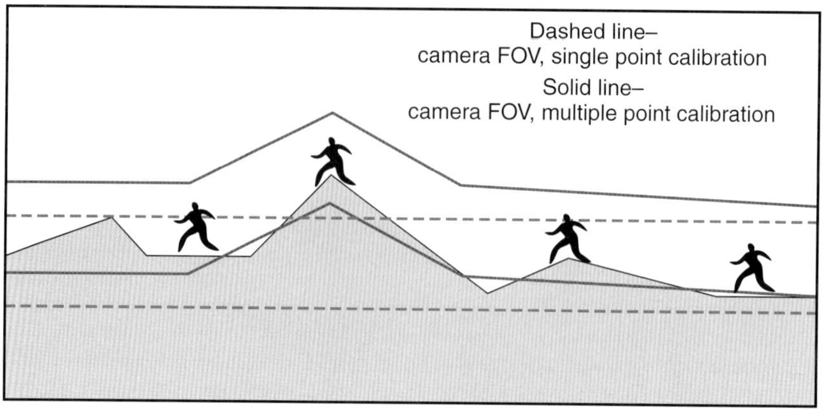

FIGURE 7 Camera multiple point versus single point calibration.

and intent). The human operator may experience complacency over time, and a system that returns too many nuisance alarms for the operator to manage creates operator overload, degrading the effectiveness of the system.

Ideally, security system designers should work toward developing a nearly autonomous system that alerts the operator only when a "real target" is identified. This is a challenging task, and may introduce other security concerns with regards to vulnerabilities. Therefore, making the system more accurate and usable should be a primary design focus. When an alarm or target track is displayed for an operator, accurate assessment (i.e. camera tracking with sufficient imaging resolution) is a must.

5.4 Identification of Friends and Foes

For many applications, site personnel are authorized within an area of interest. Having a means to identify these individuals would assist system operators with distinguishing between authorized and unauthorized targets (friend or foe). What is needed is a means to identify authorized individuals directly on the radar or LiDAR user interface, and some effort has been made toward realizing this capability.

6 SUMMARY OF MAIN POINTS

In summary, the following are the main points security system designers should take from this overview:

- Failure to address radar or LiDAR system design, performance, operational, and installation variables can lead to costly installation of an inappropriate technology for a given security application.
- Successful employment of radar or LiDAR technologies first requires characterization of the security application in order to effectively match an appropriate technology to meet application objectives.
- Cost and advertised performance (e.g. advertised detection range) should not drive technology selection or define application objectives.

- Radar and LiDAR technologies should be employed as situational awareness or extended detection enhancements to traditional facility security systems, not as a replacement.
- On-site performance testing under all environmental conditions is crucial to ensuring optimal performance.

REFERENCES

1. Tryniski, M. (2004). *AN/PPS-5D Radar Overview Presentation*. Syracuse Research Corporation, North Syracuse, NY.
2. Garcia, M. L. (2008). *The Design and Evaluation of Physical Protection Systems*, 2nd ed. Butterworth-Heinemann.
3. Skolnik, M. I. (1980). *Introduction to Radar Systems*. McGraw-Hill Book Company, New York, NY.

DESIGN CONSIDERATIONS IN DEVELOPMENT AND APPLICATION OF CHEMICAL AND BIOLOGICAL AGENT DETECTORS

DONNA C. SHANDLE
Nuclear, Chemical, and Biological Contamination Avoidance, Aberdeen Proving Ground, Maryland

1 BACKGROUND

This study will address the differences and similarities in requirements that drive design features and product development of a handheld chemical vapor detector and a man-portable, vehicle-mounted standoff chemical detector for use by the uniformed services, and what changes or additional design/test work is required to ensure that the systems meet the needs of a Homeland Security application. It is apparent that the missions of the uniformed services likely differ in some respects from those of Homeland Security. However, the application of many systems developed for the services to that of Homeland Security involves minor, if any, adjustments resulting in significant capability for this governmental department. The issues to be answered relate to where and how the systems would be employed. Once these questions are addressed, immediate progress toward providing capability can occur.

1.1 How Requirements are Derived

It is worth noting that the Department of Defense follows a disciplined requirements development process coupled with standard systems engineering processes. From the definition of the customer's use concept, a developer derives performance specifications. Categories and related questions to be answered include, but are not limited to the following:

1. Intended users
   - Who will operate the system? This question drives human factors requirements such as size, weight, ease of maintenance, modular design, use of common parts, and location and size of any switches, knobs, buttons, or indicators on the system.
2. Operational environment
   - Under what climatic conditions will the system be used? Specifically, what are the temperature extremes for operation as well as for storage?
   - What are the relative humidity conditions in which the system must operate?
   - Must the system operate in rain, high wind conditions, and maritime atmospheres?
   - What are the intended environments for system operation? This question addresses interferents that may be naturally occurring or specific to an area of operation such as one where there is a heavy concentration of volatiles, for example. These environments affect the ability of the system to detect the target materials of interest. This question also addresses the safety aspects of the system and its operating conditions.
   - Will the system be used in a stationary or mobile scenario?
3. Operational use cases
   - What is the purpose of the detector in the mission? That is, what protection is the system required to provide?
   - What materials must the system detect, at what concentrations, and what is the required response time at each concentration? If the detector is used by personnel in protective clothing, the minimum level of detection may differ from that if the operator were wearing protective clothing/masks, for example.
4. Mandated integration/interfaces
   - With what other systems must this detector interface or communicate?
   - What other systems will be operating in close proximity to this detector?
5. Transport and storage
   - How will the system be transported to the area where it will be employed? This question is significant since it determines the maximum weight of the system as well as the means of carry/transport.
   - Will it be transported by vehicle, can it include a carrying case, what straps or attachments are needed, or is it carried by one or multiple persons?
6. Size/weight/power
   - What is the mission duration? This question establishes requirements that impact selection of power sources (battery or line), definition of scheduled maintenance

cycles, reliability, all logistics aspects such as types and locations of repair facilities, spare parts provisioning, and the need for specialized maintainer skills, as examples.

7. Cost
   - What is the target cost for the system on initial purchase?
   - What is the sustainment cost over the life cycle of the system?
   - What is the anticipated density for issue to users?

1.2 Comparison of Homeland Security—Specific Requirements to those of Uniformed Services

1. Intended user
   - The users of the systems for a Homeland Security mission are likely to have a higher level of technical knowledge/understanding regarding detectors.
   - The Homeland Security users could likely accommodate less-detailed training and premission operational trials.
2. Operational environment and use cases
   - That the mission of Homeland Security is within the continental United States reduces the extreme environments in which vapor detector systems must be stored and operated.
   - The lower detection limits are considered comparable for both missions. However, it is likely that the interferents in the environments may differ particularly if the Homeland Security mission extends into industrial settings where vapor atmospheres are prevalent, thus presenting a significant challenge to detect and differentiate between chemical warfare agents, toxic industrial compounds, and nonhazardous industrial compounds. This is specifically important if the materials of interest are in lower concentrations than the vapor interferents in the atmosphere.
   - Maintenance requirements for the Homeland Security scenarios differ in nature from the needs of the uniformed services and can be assumed to be more easily accommodated using local maintenance facilities vice having the systems repaired or maintained in the field by the operator of the detector.
3. Size/weight/power/integration/interfaces
   - Although the Homeland Security scenarios will entail outdoor, stand-alone missions, there is also the high likelihood that some mission scenarios will employ detector systems with sources other than battery power. For example, in an airport security scenario, detectors could easily be installed using house power, thereby eliminating a large logistics burden, which translates into cost avoidance per operating hour. Likewise, detector systems that are installed versus carried, lessen the concern for power since these can be hard-wired using commercial power in lieu of batteries that have a specific life.
   - The size requirements for both unformed services and Homeland Security are considered comparable, particularly for the counterterrorism missions where mobility and portability are of primary importance. For point detectors in

particular, small, lightweight, and easy to operate are essential characteristics for both customer bases.

4. Cost
   - The structure and facilities available to the Homeland Security teams also reduce the design features that provide for ruggedness and hardness in detector systems.
   - Owing to the complexity of some of the technologies in the current detectors, particularly passive infrared technology and the detection components within these systems, operator maintenance is not feasible; therefore a skilled maintenance facility is essential in the case of both the uniformed services and Homeland Security.

2 DESIGN CONSIDERATIONS

The design of a chemical vapor detector revolves around the technology employed that addresses the required performance. In the case of vapor detectors, the technology must address the characteristics of the chemical agent vapor that make them distinguishable from other nonthreat vapors in the atmosphere. The molecular structure of the materials of interest, the pattern of chemical bonds present in the substance, or unique spectra of the vapor are examples of characteristics that technologies can target. There are two mature technologies that have been incorporated into current chemical vapor detectors that each detect and identify agent vapors using different technologies, which impact the design considerations of each system. The first technology, ion mobility spectrometry, is utilized in vapor detectors classified for use as "point", since the detection is made within several inches to a foot of the agent, while passive infrared detection is incorporated in standoff detectors, for the purpose of enabling detection without exposure to the agent itself. Both of these technologies and systems are applicable to Homeland Security and counterterrorism missions.

2.1 Ion Mobility Spectrometry in Point Chemical Vapor Detectors

Ion mobility spectrometry technology is applied in agent vapor detectors to measure time of flight of ions of the materials of interest. The application of this technology requires that a vapor be ingested into the opening of the detector and pass through an ionizing process. The ions then drift down a tube owing to the force created by an applied voltage. The speed with which the ions travel are a fixed characteristic for each material at a given voltage. The arrival of the ions at the end of the drift tube creates a spectrum suitable for analysis through detection algorithms [1]. Detection of a material of interest is made when the result of the application of a set of algorithms to the spectrum of each material meets predetermined criteria. The technology lends itself to packaging into a small, lightweight, handheld device and boasts added benefits of low power consumption and minimal maintenance. Advances in the technology have successfully eliminated the radioactive ionizing source, thereby increasing the safety of the user and the maintainer, while reducing the environmental considerations upon disposal of the system at the end of its useful life. This improvement also reduces the ownership and accountability burden imposed by radioactive source licensure [2].

The incorporation of ion mobility spectrometry technology into a new detector has enabled the Department of Defense to generate test data that demonstrates detection performance that meets system requirements for the Joint Chemical Agent Detector, a handheld, point detection system for use by the uniformed services [3]. As stated above, the advancement of the Joint Chemical Agent Detector over existing detection systems that employ ion mobility spectrometry is that it accomplished the ionization by a source other than a radioactive one. This favorable feature enhances the safety aspects of use, handling, and maintenance and impacts the design considerations as evidenced by the system's modularity, the reduced power allocation, and the overall cube of the final package [4]. The Department of Defense has also shown that this technology lends itself to detection of additional materials, such as toxic industrial chemicals, upon development of the appropriate algorithms and detection windows [5].

2.2 Passive Infrared Detection in Standoff Vapor Detection Systems

Development of a standoff detector for chemical agent vapor has effectively employed passive infrared technology to detect a vapor cloud at ranges of at least half a kilometer from the detector [6]. The application of this technology in a battlefield scenario assumes that the target material appears as a cloud with sufficient dimensions to create a change in the optical signature, due to a small temperature difference from the background, of the order of a few degrees, when viewed through the system optics after spectra were analyzed through the algorithm. Therefore, use of this technology incorporated a sophisticated optics package that presented design challenges such as packaging the scanner, its cooling system, maintaining module pressurization, and measuring signal-to-noise ratios. The output from the optics is a signal distinguishable from the background at sufficient amplitude to produce a spectrum that is then processed through algorithms to discriminate between a scene of nonagent-containing atmosphere and one that includes a cloud with the features of a threat agent. The packaging of the modules that contain the optics and associated coolers, elements for system orientation, and computer for signal processing must be optimized to meet system requirements of environmental conditions for on-the-move operations, weight, and power considerations. The determination of probability of detection at a given range is directly related to the optics of the system. The successful operation of the system depends on maximizing the opportunities for the detector to view the cloud of interest. More opportunities are a function of the size of the field of view and field of regard, which are optics design parameters. The relationship is that the smaller the field of regard, the higher the number of opportunities for the optics to view the cloud within a given time. Additionally, the smaller the instantaneous field of view, the greater the signal-to-noise ratio, which enables better resolution, and thus more accurate detections as well as enabling an increased range at which a cloud can be detected. An additional requirement for the Joint Service Lightweight Standoff Chemical Agent Detector was that the system detect in a 360° arc across an elevation range of -3 to $+20^\circ$ while on the move [6]. Its predecessor, the M21 Remote Standoff Chemical Agent Alarm, provided standoff detection at seven distinct snapshots at the horizon across a 60° arc from a stationary position. This additional requirement for on-the-move operations presented an algorithm challenge in that there needed to be a reference to the "before" scene in order to make the determination of the temperature differential due to the presence of an additional atmospheric feature. The requirement for on-the-move detection eliminated the use of the background subtraction algorithm that was employed in the M21 [7]. The development of an analytical technique for the spectra in light of the sizable data sets generated by the Joint Service Lightweight

Standoff Chemical Agent Detector was perhaps the biggest challenge for the developers. Improving detection sensitivity required trade-offs with false alarm rates; therefore, the algorithm designers conducted many iterations of algorithm adjustments to optimize the parameters that affected the probability of detection and false alarm rate.

2.3 General Design Considerations for Agent Detectors

Today's detection systems are more complex than the prior generation of detectors, but in spite of the increased complexity of current technologies, there is an even stronger requirement for a high degree of simplicity in the user interface. Thus, equipment designers tackle the challenge of presentation of comprehensible information on a small system display. The degree and detail of information, time available to act once information is received, and questions that must be answered by the information presented are all specific to the mission of the customer, whether it is Homeland Security or the uniformed services. Additionally, the level of user training and education impacts the grade level that the designer uses to ensure the user's total comprehension of the information. As an example, the requirements for the Joint Chemical Agent Detector are stated in terms of the product of concentration and time. Ion mobility spectrometry provides information regarding concentration of a given agent. However, a given concentration of a nerve agent produces a much different biological response than an equal concentration of a vesicant [8]. Thus, if the display were to show a response strictly based on concentration detected, the user would need knowledge of agent toxicity in order to understand the significance of the system display. To overcome this issue, the designer related the relative hazard of each agent to the detected concentration and displayed that data in terms of a bar response [4]. As the bar response incorporated agent-specific toxicity values in the algorithm, the display output was normalized, enabling the user to understand his/her level of danger no matter which agent was detected.

2.4 Design Trade-Off Process

Because users defining system requirements often ask for performance that exceeds the current state of the art, equipment designers conduct a trade-off analysis during the design and systems engineering phase of the effort. In that analysis, they document the capabilities of the technology based on work accomplished during the research phase of the effort. When a basic technology in research is proposed as a candidate to answer a requirement, the capability of the candidate technology is often reduced when applied in a nonlaboratory environment. The engineering work that continues during the development of the technology into a usable piece of hardware considers the hardness, ruggedness, human interface, and environmental conditions in which the system must operate. The incorporation of these factors into the development results in the necessity for trade-off studies to prioritize the requirements, assign a risk of meeting required levels of performance, and perform an optimization study that presents an estimate of the best performance that the system could achieve [9]. The use of a science-based system model provides an effective, cost-saving tool to exercise critical system parameters during the optimization process.

As an example, the physical characteristics of the optics incorporated in standoff detection systems influenced the range and probability of detection of the Joint Service Lightweight Standoff Chemical Agent Detector. The users desired maximum probability of detection and range of detection with minimum false alarms. There were at least two parameters that were nonnegotiable in the early prototype systems, these being the window material and the field of view dimensions. These were fixed based on the required area of interrogation (originally 360°s horizontally and -10 to $+50^{\circ}$ vertically) and the time to detect. The designer traded off the probability of detection at a given range to maintain an acceptable false alarm rate. The design parameters that impacted detection performance pertained to the optics characteristics—field of regard and instantaneous field of view—since these directly resulted in the signal-to-noise ratio that was processed into spectra for analysis and time to detect. Additional tradeoffs were made in resolution of the wave numbers as well as in the algorithms. The analysis of test data from field trials enabled the designers to estimate and then optimize the relationship of field of view, field of regard, and signal-to-noise ratios on the range and detection probability of the system to simulant clouds [10]. The analysis also identified the impact on increase in false alarm rates should the algorithms be altered [10].

2.5 Application of Existing Chemical Vapor Detectors to Homeland Security Mission

To provide flexibility to the user in a uniformed services or Homeland Security mission, a solution that consists of a suite of systems best addresses detection levels between submiosis and large, overwhelming concentrations. In the Homeland Security mission, the desire to detect materials ranges from precursors to the actual finished product. To date, there is no single technology that has the capability to detect across this wide range. Therefore, the development of packages of detection equipment for use in these scenarios must consider a suite of detectors that employ several technologies and offer a wide variety of functions. The toolbox of detection equipment may also utilize commercially available products, which in combination can provide an extensive capability for the teams. The Department of Defense has successfully leveraged commercial products in both an as-purchased configuration as well as with some modification to meet more specific requirements. The Joint Chemical Agent Detector is an example of a commercially available system that the Department of Defense, in conjunction with the manufacturer, conducted system modifications to meet service-unique requirements [4].

3 DEFINING SYSTEM PERFORMANCE

This study opened with a discussion of the need to determine system requirements. Once defined, there must be a process to determine actual system performance against the requirements. Without performing an evaluation with actual test data, any system that is fielded carries a level of risk of having lesser capability than required or being inappropriate for the intended mission. Therefore, the step in the development that closely follows requirements definition is the development of the performance evaluation strategy.

3.1 System Evaluation

The process for conducting a system analysis is initiated with a review of the threat against which there is no defense. The Science and Technology community offers a candidate technology that forms the foundation for systems development. This community also identifies the critical parameters that make the technology effective against the threat, and having performed a sensitivity analysis, suggests a required range of measurements for each parameter. This information forms the basis for the statistical design of experiments, to include accuracy and precision of each measurement that the Science and Technology community establishes in concert with the development community to ensure proper application and evaluation of the technology. In addition to reliability and confidence in terms of numbers and repetitions of trials, the design of experiments also considers the questions identified in the Section 1, so that all tests address the full gamut of conditions to which a system will be exposed in its life cycle to include disposal. To be efficient in providing equipment to the users, the evaluation strategy for any given system must be documented ideally before design and hardware build is initiated. If the evaluation strategy is developed based on the requirements and is finalized before the hardware is readied for testing, the parties who are developing the strategy do so without preconceived ideas of what the hardware will be or its performance characteristics. Likewise, when the developer understands how the system will be evaluated, he/she will be able to focus on the performance traits that are of highest priority to the user. This enables a situation whereby the provider offers a solution that most closely meets the customer's needs. The Department of Defense application of these concepts is illustrated in Figure 1.

To adequately evaluate a detection system, one must establish a series of tests that generate data to answer the questions that are pertinent to the use and purpose of the system. For example, the questions posed in Section 1.1 ask "how and to what degree will this equipment protect the intended audience?" The two modes of use for the held Joint Chemical Agent Detector are monitor and survey [4]. Each mode has a different purpose, which creates different test conditions and requirements. In a monitor mode, the employment concept is for use in a preexposure/preattack clean environment; therefore, the system must provide an alert at first indication of the presence of a chemical agent. In the survey mode, the assumption is that the attack has occurred, and the user is surveying to define areas of contamination as well as checking for any possible contamination after decontamination operations. These two conditions present different levels of acceptable performance.

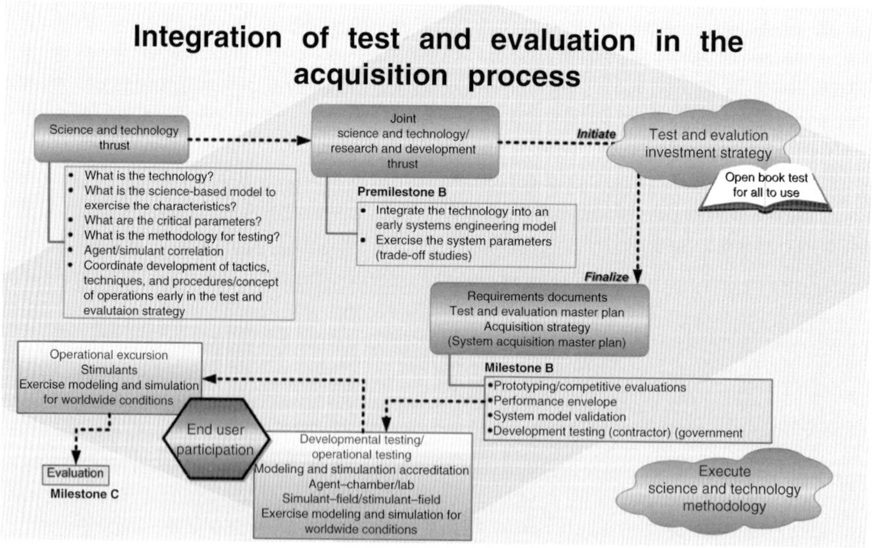

FIGURE 1 Test process development.

The Joint Chemical Agent Detector testing for detection of agent in the monitor mode required the introduction of an agent stream that replicated a concentration profile representing the onset through full development of an agent vapor cloud. To determine whether the detection performance was acceptable, the data collected during the test trials included concentration as a function of time, time for the system to detect, concentration at which a detection occurs, temperature and relative humidity of the agent stream, temperature and relative humidity of the system under test, occurrence of any false alarms, and detection and identification of the agent by the system under test as depicted by its display [11]. In some cases, actual spectra were collected during agent trials to provide information regarding the accuracy of the drift tube functioning [12]. This data was essential to understand how the system was performing and if its performance would provide utility for the user. Additionally, in the monitor mode, the time to alarm when exposed to agent vapor must allow the user sufficient time to alert others to don protective gear or initiate collective protection in vehicles or facilities. Therefore, the product of the concentration level at detection and the time to alarm is an indication of the dose that a user would experience. This factor and the toxicity of the detected substance translate into a relative hazard level to the user as discussed above. Similarly, in the survey mode, the Joint Chemical Agent Detector is used in both postattack and post-decontamination scenarios to determine if the equipment is contaminated, and once decontaminated, if the decontamination process was sufficient to declare the equipment safe for use by personnel in an unprotected posture. These evaluation questions were answered from an agent performance data set [5] with two sampling intervals that represented the mission scenarios. The efficiency of the design of experiments generated a data set that could determine the degree to which the system enabled the user to meet both missions of detect and alert to protect as well as sort for and validate after the decontamination process.

When the Joint Chemical Agent Detector is used in its intended environment, and since it is a handheld device, the designer paid strict attention to the human factor aspects. The Joint Chemical Agent Detector was required to operate in "basic" operational environmental [3] conditions, which translated to a minimum temperature of -25°F [13]. The human factors implication of this requirement meant that the size of the system operator could range from the 5th percentile female to 95th percentile male. Additionally, since the range of operations included arctic conditions, operators would be wearing arctic mittens, which increase the size of the gloved hands while decreasing dexterity. Design considerations included the location and separation of switches and buttons as well as the size and readability of the display. These considerations influenced the operation of the instrument where size, weight, and ease of use were traded in favor of optimized performance.

3.2 Parametric Determination

When designing a test program for a given technology, it is essential to understand what parameters of the technology have the greatest influence on whether the technology performs appropriately. These are then the parameters that must be tested thoroughly and with greater accuracy and precision, than a parameter that may be relevant to the overall system performance, such as weight or cube, but which has little impact on whether the applied science will be effective.

The Science and Technology community typically proposes a candidate technology as the foundation for a development program. It also identifies the key parameters of the technology and the scientific explanation of how the technology addresses the threat characteristics. When a new technology is introduced or applied in a new use concept, the need to develop an appropriate test methodology arises. It is essential to test and collect data for a technology in a scenario appropriate to the application for which it will be used. In ion mobility spectrometry, the mobility peaks of samples, be they chemical agent vapors or naturally occurring vapors such as water, are affected by the environmental conditions of temperature, humidity, and pressure. Since all the peaks shift an equal amount under the same environmental conditions, the impact of the environmental conditions can be neutralized if a reference reactant ion peak is incorporated in the design. Alternatively, the design of a system that did not require a reference reactant ion peak would have required internally controlled temperature, humidity, and pressure to ensure that the system saw the absolute mobility of the sample in the drift tube under all environmental conditions. This design would have added serious size, weight, and power demands, which were unacceptable to the user, because, in the Joint Chemical Agent Detector, the minimum size and weight requirements were extremely high priority to the user. Therefore, developing a methodology to determine the applicability of the ion mobility spectrometry technology for use in the Joint Chemical Agent Detector by measuring the ion peak of a substance without reference to a reactant ion peak is of no value in determining what material is present. Similarly, in an evaluation strategy, measuring the weight of a system to five significant decimal places when the impact of this precision has no bearing on the effectiveness of the ion mobility spectrometry is irrelevant and wasteful in terms of value to the overall system assessment. Determining the correct parameter to measure, establishing the relevant accuracy and precision of the measurement, identifying appropriate instrumentation to attain the required measures, and designing an experiment, which captures the use conditions of the system, form the basis of the methodology development effort. The final result must be a data set that defines the performance of the system void of experimental artifacts. The Department of Defense's approach to test process development is pictured in Figure 2.

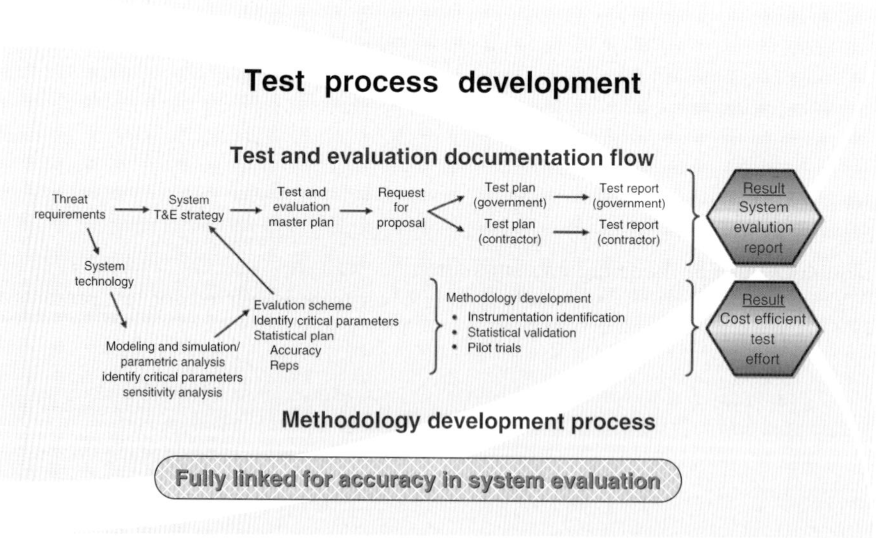

FIGURE 2 Integration of test and evaluation in the acquisition process.

3.3 Use of Modeling and Simulation as an Element of the Evaluation Strategy

The systems engineering approach to development includes requirements development, parametric determination, design of experiments, trade-off analysis, modeling and simulation, and experimentation. In the development of detectors for chemical agent vapors, the incorporation of modeling and simulation has played an important role. In all developments, the evaluators require data regarding performance against live agents. The single most important difference between detector developments with ion mobility spectrometry technology and passive infrared technology was the ability to conduct qualification tests that represented the actual mode of employment. For the small handheld point detector such as the Joint Chemical Agent Detector, tests were conducted in a chamber in which the systems were exposed to an agent environment as they would experience in the open air [14]. Responses were directly related to statements of performance. The evaluation of the Joint Service Lightweight Standoff Agent Detector precluded testing in the system's intended environment due to the prohibition of use of agent in the open environment [15]. Therefore, the evaluation strategy determined performance through a series of chamber tests with a live agent that had a truncated path length from the instrument to the target material, similarly truncated path length chamber tests with simulants, field-tests with simulants, and a physics-based model that predicted performance based on the test data and scientific first principles [16]. The development and evaluation community made performance statements based on a strategy consisting of empirical data and modeling results. The models were verified and validated by independent sources and then accredited by the organization using the models to perform the system evaluation. The process of model development, verification, and validation is a standardized, documented procedure used in industry as well as the federal government [17]. The advantages of using modeling include creating the ability to gain system knowledge when testing is unavailable, as well as asset conservation.

4 SUMMARY AND CONCLUSIONS

The development of chemical warfare agent vapor detectors by the Department of Defense has been focused on the use scenarios specific to the uniformed services. Using a systems engineering approach, hardware developers respond to the user's requirements, which describe the need for the systems based on new or increased capability, to protect against known or changing threats. The identification of the critical parameters of a given technology by the Science and Technology community aids in developing an appropriate evaluation strategy that must include the design of experiments, development of new methodology, use of modeling and simulation, as well as acquisition of empirical data. Developing the evaluation strategy before any hardware is offered for test presents the most unbiased and unconstrained opportunity for the community to define and prioritize the data requirements.

Use of modeling and simulation in the qualification of systems was essential in the acceptance of a system for which outdoor testing with prohibited materials was not permitted. The development of a physics-based system model was a foundation block for the modeling and simulation effort for the standoff detection system that demonstrated the performance of the system against the threat materials. In the development of agent detector systems, challenges against actual threat material are essential to assure confidence in the system. Data exists for currently available systems that demonstrate effective chemical agent vapor detection performance by both ion mobility spectrometry point and passive infrared standoff systems [16]. The application of this equipment to Homeland Security scenarios requires early analysis of employment concepts to determine similarities in use to enable leveraging of performance data and development of maintenance concepts and facilities to support the Homeland Security mission. The demonstrated performance of the Joint Chemical Agent Detector, a point detector, and the Joint Lightweight Standoff Chemical Agent Detector, a standoff detector, against chemical warfare agents in the challenging environments typical of military applications is a prime example of the opportunity to leverage Department of Defense development to a Department of Homeland Security application. That the use concepts and environments of the Department of Homeland Security are less harsh ensure favorable performance while enhancing commonality and standardization of equipment across federal agencies.

REFERENCES

1. Eiceman, G. A., and Karpas, Z. (2005). *Ion Mobility Spectrometry*, 2nd ed., CRC Press, Taylor & Francis Group, 6000 Broken Sound Parkway NW, Suite 300, Boca Raton, FL, 33487-2742, pp. 3–5.
2. Title 10 Code of Federal Regulations Part 20, Standards for Protection Against Radiation. (2007).
3. Capability Production Document for Joint Chemical Agent Detector (JCAD). (2008). *Increment: I, ACAT III, Validation Authority: Joint Requirements Office Committee (JROC), Milestone Decision Authority: Joint Program Executive Officer for Chemical and Biological Defense, Designation: JROC Interest prepared for Milestone (MS) C Decision*, Version 11.
4. Performance Specification Detector, Joint Chemical Agent (JCAD). (2008). *Increment 1, EA-D-10009, Prepared for: JPM, NBC Contamination Avoidance, SFAE-CBD-NBC, Aberdeen Proving Ground, MD 21010-5424*, Submitted by Kim-Tuyen Le, Chairman, Configuration Control Board, JCAD Team.
5. System Evaluation Report for Joint Chemical Agent Detector (JCAD) Increment 1. (2008). *Assessment produced by U.S. Army Evaluation Center*.
6. Joint Service Lightweight Standoff Chemical Agent Detector (JSLSCAD) Capability Production Document (CPD) Milestone C. (2005). *Increment 1, Acquisition Category (ACAT) III, version 1*.
7. Lyons, R. C., and Milchling, S. S. (1997). *ERDEC Technical Report TR-346, XM21 Remote Sensing Chemical Agent Alarm (RSCAAL) Risk Reduction Program*.
8. USACHPPM. (2008). *Health-based Chemical Vapor Concentration Levels for Future Systems Acquisition and Development*, USACHPPM Technical Report No. 64-FF-0722-07, U.S. Army Center for Health Promotion and Preventative Medicine.
9. Tauras, D. G., and Kichula, J. (1995). *Contributor, "Conduct Systems Engineering Trade Studies Process", Rev 3*, Defense Acquisition University.
10. Flanagan, M. J., and Hammond, B. (2006). *Final Technical Report for the Joint Service Lightweight Standoff Chemical Agent Detector Engineering and Manufacturing Development Program, prepared by General Dynamics Armament and Technical Products, Charlotte, N.C. under Contract No DAAM01-97-C-0030*.
11. Joint Chemical Agent Detector (JCAD) ACAT II (with OSD T&E oversight) Test and Evaluation Master Plan (TEMP) Increment 1 Supporting Multi-Service Operational Test and Evaluations, and Full Rate Production, Version 2.2, 13 June 2007, submitted by Kyle T. Burke, COL USA, Joint Project Manager NBC Contamination Avoidance, approved by Charles, E. McQueary, Ph.D., Director, Operational Test and Evaluation, 10 August 2007.

12. TRIMSCAN. (2009). *Time Retentive Ion Mobility Scan Analysis (TRIMSCAN) Manual/User's Guide 8X/699-399 Issue AJ (Version 0.5)*. Smiths Detection Limited, Watford.
13. Department of Defense Test Method Standard. (2008). *MIL-STD 810G, Environmental Engineering Considerations and Laboratory Tests*, 31 Oct 2008.
14. Detailed Test Plan for the Production Qualification Test (PQT) for the Chemical Warfare Detection and Identification Testing for the Joint Chemical Agent Detector (JCAD) Increment 1 Gate 2, Test Project No 2006-DT-DPG-JCADX-C8866, WDTC Document No. WDTC-TP-05-135, December 2005, prepared by Louis Anderson, R. James Berry, Brad Rowland, and George Law, West Desert Test Center, U.S. Army Dugway Proving Ground, Dugway, UT 84022-5000.
15. *US Code Title 50:War and National Defense-50 USC 1512 Sec 1512*, January 2003.
16. Larsen, E. C., Bankman, I. N., Lazarevich, A. K., and Rogala, E. W. (2006). *The Johns Hopkins University Applied Physics Laboratory, Young, Randy S., Althenbaugh, Ryan, Reherman, Jason, Joint Project Manager Nuclear Biological Chemical Contamination Avoidance, "Verification and Validation (V&V) Report for the Joint Service Lightweight Standoff Chemical Agent Detector (JSLSCAD) Increment 1 Model and Simulation (M&S).*
17. Testing and Evaluation of Standoff Chemical Agent Detectors. Committee on Test and Evaluation of Standoff Chemical Agent Detctors, Board on Chemical Sciences and Technology, Division on Earth and Life Science, National Research Council of National Academies. (2003). *"Testing and Evaluation of Standoff Chemical Agent Detectors" report by the National Academy of Sciences (NAS) Committee on Testing and Evaluation of Standoff Chemical Detectors.* The National Academies Press, Washington, D.C., www.nap.edu.

SENSING DISPERSAL OF CHEMICAL AND BIOLOGICAL AGENTS IN URBAN ENVIRONMENTS

ANGELA M. ERVIN, ANNE E. HULTGREN, EDWARD P. RHYNE, AND KEITH B. WARD
Department of Homeland Security Science and Technology Directorate, Washington, D.C.

1 RATIONALE FOR CHEMICAL AND BIOLOGICAL SENSORS IN URBAN ENVIRONMENTS

The urban environment consists of a variety of unique venues, including large indoor and outdoor gathering spaces, office and residential buildings, and regional and national transportation systems, all containing large populations to be protected. The potentially devastating effects of the release of a chemical or biological agent in an urban environment, including massive loss of life and property, have been documented in several recent reports [1–4]. The Department of Homeland Security (DHS) has the task of protecting these urban environments against the threat of a chemical or biological agent attack from international or domestic terrorists, natural disasters, and industrial accidents [5–7]. DHS Science and Technology (S&T) Directorate is addressing this challenge by developing appropriate detection technologies, performing assessment and impact studies, as well as validating decontamination and restoration procedures and protocols to ensure that we are prepared to respond to and mitigate the damage from such a chemical or biological attack or accident.

A DHS preparedness plan to counter chemical and biological threats begins with deploying an extensive system of chemical and biological sensors throughout the urban environment to provide early warning of an attack or accident [8, 9]. There are significant challenges with deploying sensors in an urban environment, such as operating in a densely populated civilian population where interpretation and response to a sensor alert must be understood and managed in a dynamic environment, operating in the presence of sensor interferent materials, and operating in the midst of legitimate activities which involve near neighbor or simulant materials. All of these challenges can degrade sensor performance by creating false positive or false negative detections. DHS S&T is also sensitive to the commercial market drivers which demand that the technology solutions are low cost to purchase and maintain as the operation of these sensors will be an additional burden on facility operations. There are systems currently in place to provide detection and response to a chemical or biological incident; however there are still significant shortcomings with these technologies. The current first responder tools are plagued with false alarms and difficultly in interpretation of the results, while the installed systems have long detection times and are prohibitively expensive for general deployments. The technologies and sensor network architectures to fulfill the DHS preparedness plan and improve upon current sensors are described in detail in this article.

2 CHEMICAL AND BIOLOGICAL DETECTION ARCHITECTURE

The chemical agent detection architecture concept for homeland security considers the very different requirements for response to an indoor attack versus the response to an outdoor attack. Indoor protection of high-asset buildings and facilities (e.g. subway stations, airports, etc.) are oriented toward a "detect-to-warn" scenario. In this scenario, autonomous sensors with a rapid response time are widely deployed in the indoor environment. These rapid sensors will issue an alert of an attack in time to allow effective countermeasures (e.g. shutting down or redirection of heating, ventilation and air conditioning (HVAC) units) to contain the threat and reduce exposure to the building occupants. This protection architecture will save the greatest number of lives via sufficient warning time, and also provide critical information to first responders prior to entering the scene of an incident. The response plan to an outdoor attack constitutes a "detect-to-protect" scenario, which are aimed at providing first responders information to quickly assess the type and extent of the chemical release and the type of protective protection equipment (PPE) to don. The lightweight, portable nature of the devices under development for outdoor environments makes them very mobile, and hence can be used to track chemical agent plumes, either by hand or secured to a fire truck or response vehicle, to assist emergency personnel in determining evacuation areas and decontamination procedures when necessary. The detection of low vapor pressure chemical threat compounds (e.g. $<10^{-4}$ Torr) is of special concern for homeland security because of their persistence and long-term contamination in the environment. Although chemical attacks are not likely to cause the widespread casualties that could be observed in the case of a biological attack, chemicals are easier to obtain and package for dissemination amongst a large crowd of people (e.g. the 1995 release of Sarin in the Tokyo subway) and thus continue to represent a serious threat to homeland security.

The goal of the biological agent detection architecture concept for homeland security protection of urban areas is to enable the rapid detection of the presence of disease causing agents in advance of the presentation of medical symptoms in the population. The architecture must also take into consideration the diversity of biological threat agents as well as the variety of release scenarios for the protection of both indoor and outdoor environments. This goal is accomplished through the wide deployment of three classes of biological agent detection sensors, supported by the national network of public health laboratories. Two of these classes are low confidence sensors intended for use in indoor environments. These sensors would initiate low consequence responses, such as shutting down or redirecting HVAC units, closing portions of buildings, and initiating analysis from the high confidence detection sensors. The third class of sensor is a high confidence sensor intended for use in a "detect-to-treat" mode, which would initiate high consequence responses, such as the quarantine of people, evacuation and restriction of affected areas, and the distribution of medical countermeasures. In order to maximize the benefit from national investment in these high confidence biological detectors, the locations for their permanent installation are determined through studies to protect the maximum percentage of the US population as well as the occupants and visitors of high-profile indoor spaces.

Lastly, the feasibility of an integrated architecture to include the detection of chemical, biological, radiological, nuclear, and explosive (CBRNE) threats within regional areas is being evaluated. This architecture would enable state and city agencies and private facilities to rapidly share information and decisions made in detection, crisis management, and response. By providing a complete view of an incident, the decision support needed for an effective response in the event of an attack will be available and easily transferable among the relevant agencies.

2.1 Low Consequence Sensor Performance Requirements

A low consequence sensor is one that initiates an action that is invisible to the general public as a result of an alert from the sensor. In the chemical detection arena, two projects fall into this category. Both projects are developing portable technologies for use by first responders to help with the assessment and characterization of an incident. The goal of the Lightweight Autonomous Chemical Agent Identification System (LACIS) Project is to develop an autonomous, hand-portable detection system with the ability to detect and identify a broad range of chemical agents. Table 1 lists the detailed performance metrics for this project. The lower limit of detection (LOD) and response time of detectors

TABLE 1 DHS S&T Performance Goals for Low- and High Consequence Chemical Sensors

	Low Consequence		High Consequence
Parameter	LACIS	LVPCD	ARFCAM
Agents detected	17 chemical hazards: chemical warfare blood, vesicant, nerve, choking, and blister agents; TICs	Persistent chemical compounds (vapor pressures 10^{-4} Torr) on a variety of surfaces	17 chemical hazards: chemical warfare blood, vesicant, nerve, choking, and blister agents; TICs
Response time	2 min (IDLH and LOD)	2 min	1 min (IDLH); 15 min (PEL)
False negative rate	0.1%	1%	5%
False positive rate	1%	1%	1 per yr
Unit size	0.5 cu ft; 5.0 lbs	20 kg	
Unit acquisition cost	$2000 (quantities of 1000)	TBD	$1000 (quantities of 10,000)
Maintenance interval	6 mo	1 h battery life	6 mo
Environmental conditions	10–60°C and 0–90% RH	0–40°C and 0–90% RH	10–60°C and 0–90% RH
Communications	Wireless	TBD	Building network
Operations	Handheld	3 m standoff, man portable	Installed, continuous and automated

IDLH, immediate danger to life and health; LOD, limit of detection; PEL, permissible exposure level; TBD, to be determined; TIC, toxic industrial chemical; RH, relative humidity.

developed in LACIS will provide first responders with a tool to determine areas having dangerous chemical concentration levels and to determine if protective garments will be required for their activities. The second chemical detector technology that is considered a part of the low consequence scenario is the Low Vapor Pressure Chemical Detector System (LVPCDS) Project. The goal of this project is to develop and field-test a system focused on detecting and identifying toxic, low vapor pressure chemicals without contacting the contaminated surface. The LVPCDS performance metrics are also included in Table 1. Detectors developed in the LVPCDS Project will provide emergency personnel with a capability to detect and identify persistent chemical threats *in situ*, thereby eliminating the need for surface sampling (via swipes or swabs) and subsequent laboratory analysis.

The low consequence biological detection sensors are being developed for long-term unattended operation in indoor environments through a sensor network system. The performance requirements for these sensors assume that a high concentration release occurs in an indoor environment, and therefore the sensors will be exposed to a large amount of the biological agent due to the containment of the agent within the space. For operational use in occupied buildings, the sensor network must be affordable and have an overall false positive rate of once per many years depending on the operational requirements of the facility. In addition, the maximum benefit from these sensors will be in providing the earliest possible warning, less than 20 min after the release, to building occupants of an attack. This warning will limit the number of people who enter a potentially contaminated area and reduce the spread of the contaminant into other areas by shutting

TABLE 2 DHS S&T Performance Goals for Low and High Consequence Biological Sensors

	Low Consequence		High Consequence
Parameter	Triggers	Confirmers	BAND
Agents detected	Biological vs. Nonbiological Discrimination	20 agents: CDC A&Blist	20 agents: CDC A&B list
Limit of detection	1000 CFU/L air (spores) 5 ng (toxin)	100 CFU/L air (spores) 0.5 ng (toxin)	100 organism/collection (spores, cells and viruses) 10 ng (toxin)
Response time	2 min	15 min	3 h sample, 1 h analysis
False positive rate	1 per wk	0.001%	0.00001%
Unit size	2 cu ft	5 cu ft	Modest packaging
Unit acquisition cost	$1000–10,000 (quantities of 1000)	$50,000 (quantities of 1000)	$25,000 (quantities of 1000)
Yearly O&M cost	$50–2000/sensor	$5000/sensor	$10,000/instrument
Maintenance interval	1 mo	1 mo	1 mo
Sample archive	None	1 mo (nonviable)	5 d
Environmental conditions	Indoor	Indoor	Indoor and outdoor
Communications	Building network	Building network	Wireless
Operations	Installed in building environment	Installed in building environment	Installed, continuous and automated

CDC A&B, Centers for Disease Control A&B agent list; O&M, operation and maintenance; CFU/L air, colony forming unit per liter of air.

down or redirecting building airflows. To meet these aggressive sensor network requirements, two classes of sensors (triggers and confirmers) are under development in the DHS Detect-to-Warn Project as components of an overall sensor network. The detailed performance requirements for the triggers and confirmers are listed in Table 2. Trigger sensors continuously test the environment for an increase in respirable biological material, above levels either anticipated or typically observed. Trigger sensors can accomplish this nonspecific biological detection within 1 min, although they are expected to have a high false positive rate due to naturally occurring biological materials in the environment. Once this elevation of biological material is detected, a confirmation sensor is initiated to automatically collect a sample and determine if the material is a biological threat. The confirmation sensors are very specific to biological threat materials, have automated sample collection, purification, and proessing ability, and also perform very rapid detection assays to meet the 10–20 min network detection goal. This triggered confirmer architecture lowers the false positive rate of the overall detection network to meet the stringent requirements of facility owners, since the confirmer sensors are run only a fraction of the time.

2.2 High Consequence Sensor Performance Requirements

A high consequence sensor is one that initiates an action that would be obvious and inconvenient to the public (e.g. building evacuation) from an alert by the sensor. In the chemical detection arena, the indoor facility chemical threat monitor project, the Autonomous Rapid Facility Chemical Agent Monitor (ARFCAM) Project is focused on high consequence detection. The ARFCAM Project goal is to develop a monitor to detect chemical threats in closed or partially enclosed facilities. The performance specifications for this project are shown in Table 1. This rapid and ultrasensitive detection capability will provide facility protection against both intentional releases and slow leaks of dangerous chemical materials.

In the biological detection arena, the Bioagent Autonomous Networked Detector (BAND) Project is focused on development of the high consequence detection systems. Deployed outdoors in an urban area in a "detect-to-treat" mode, these detection systems are anticipated to continuously sample the air, extract aerosol particulates, and analyze them at least once every three hours, providing an integrated detection of any airborne bioagent released within the preceding 3 h window. The DHS S&T performance goals of the BAND systems are given in Table 2. The key aspect of these systems is that they provide high confidence results to the public health community, which can be utilized as part of their decision process in response to a biological threat. To ensure that the BAND systems are delivering high confidence results, multiple agencies and the public health community have developed inclusivity (representative samples of potential pathogens) and exclusivity (background materials and near-neighbors) panels to use in testing the efficacy of the systems prior to their operational use. The tests of the sensors using these panels, as well as other system level tests such as long-term field tests, are designed to illustrate the system performance and establish that the BAND systems, or any other developmental systems, meet the minimum performance standards to be considered valid by the public health community.

In addition, two projects executed by DHS S&T are implementing integrated architectures to include the detection of CBRNE threats within high-profile regional areas. The Integrated CBRNE (iCBRNE) Project goal is to develop and deploy citywide detection, alert, and emergency response communication systems using mature technologies and tools. This would enable a network of city agencies to rapidly share critical information to make informed decisions with respect to detection, crisis management, and response. The Regional Technology Integration Initiative (RTII) Project intends to provide a complete networked chemical, biological, and explosive (CBE) sensor system in a model facility, using mature and validated detectors. In addition, the RTII Project will assist with training and conducting local exercises with the facility and first responder community on the operation and response to the CBE sensor network. This model CBE protection system will serve as an example for how to deploy an effective protection system for similar facilities throughout the country. In executing these two architecture focused projects, DHS S&T is facilitating the successful transition of these new technology solutions to the public and private community as well as offer a venue for testing and transition to commercial-off-the-shelf (COTS) of other DHS S&T developed sensors.

3 CHEMICAL AND BIOLOGICAL SENSOR TECHNICAL APPROACH

This section contains a description of the technologies currently under development to address the need for chemical and biological detection sensors for homeland defense. Note that neither the discussion nor the ordering of the technologies indicates a DHS S&T preference.

3.1 Low Consequence Chemical Sensor Technologies

The current technology development for low consequence chemical sensors in the LACIS Project is focused on three different analytical approaches described in this section. The first approach is a microelectromechanical system (MEMS)-based platform using an array of semiconducting metal oxide (SMO) materials to detect and identify a select series of chemical analytes. The SMOs are processed in such a way so as to induce control of target analyte reactions with a given sensing material imparting partial differential selectivity that, when coupled with signal processing and pattern recognition algorithms, affords detection and identification of chemical analytes with consistent accuracy and reproducibility [10].

The second approach is based on a hybridization of ion mobility spectrometry (IMS) and a polymer nanocomposite array (NCA). The IMS engine is based on the DoD-developed lightweight chemical detector (LCD) [11]. In IMS, gas-phase ions are separated by difference in mobility of the ions under the influence of an electric field gradient [12]. The IMS technology is a mature analytical approach with fundamental studies dating back to the 1950s. IMS parameters include high-speed analysis, portability, high reliability, and low cost. The NCA is a resistance-based detection modality where the interaction of a chemical analyte with a selective polymer matrix disturbs the inherent conductivity and the resulting change in resistance can be related to concentration of the analyte. The combination of the NCA and IMS data streams and processing with appropriate data fusion algorithms increases the detection space of this handheld chemical detection device.

The third approach under investigation in the LACIS Project is based on the laboratory gold-standard, mass spectrometry (MS). During a typical MS analysis, gas-phase ions are produced and undergo fragmentation via electron impact ionization [13]. These fragmented ions are then separated according to their mass-to-charge ratio (m/z) and their abundance. Due to the presence of multiple fragmented species, information rich spectra are obtained which can then be compared to a spectral library for identification. The technological challenges for these "detect-to-protect" devices are being able to detect a much broader range of chemical analytes than current COTS items with minimal (to no) false alarms, particularly false negative responses, in a rugged, form-fit low cost ($<$\$2000) package.

The current technology development for low vapor pressure chemical hazards in the LVPCDS Project is based on a technology known as laser interrogation of surface agents (LISA). The LISA platform is based on UV–Raman spectroscopy, and leverages Department of Defense (DoD) investments. UV–Raman spectroscopy uses ultraviolet (UV) light (typically 200–300 nm) to interrogate samples, which results in spectra with enhanced signal intensities versus longer wavelength Raman spectroscopy and thus improved signal to noise [14]. The current LISA technology in the man-portable configuration has several weaknesses, foremost being the high acquisition cost. Other technological challenges include high fluorescence on certain backgrounds and high shot noise, a short scan range ($<$3 m), and a small spot size for detection leading to long scan times to cover larger areas.

3.2 Low Consequence Biological Sensor Technologies

The current technology development for low consequence biological sensors focuses on triggers (1–3 min detection) and rapid confirmers (6–15 min detection). To meet the time goals for trigger detection of biological aerosols, the "detect-to-warn" sensor development has focused on optical techniques that offer very rapid, often single-particle detection schemes. However, these optical techniques have limited discrimination capability and are prone to false alarms from naturally occurring biological and nonbiological materials in the environment. The most mature optical detection technique is based on detection of the three amino acids that have fluorescent properties in the UV region (tyrosine, tryptophan, and phenylalanine [15]). These fluorescent properties are measured by the sensors to detect the presence of aerosol particles with similar excitation and emission properties as those of the amino acids, indicating the presence of particles of potential biological origin [16, 17]. Additional optical techniques are being investigated to exploit characteristics of biological materials in other spectral regions, such as infrared detection [18], surface-enhanced Raman spectroscopy [19], laser-induced breakdown spectroscopy [20], and spark-induced breakdown spectroscopy [21]. Finally, another trigger sensor technique is to add a biologically specific optical tag to collected aerosol particles to enhance the detection of DNA or protein containing biological material from nonbiological environmental interferents [22]. These developmental trigger techniques are currently in the evaluation process of the performance characteristics of the technology to ensure confidence in the deployment of the sensors, and the engineering analysis for adequate fieldability of the resulting sensor designs.

"Detect-to-warn" confirmer sensors will provide a more specific analysis of collected material for rapid field presumptive identification. The indoor deployment of these sensors assumes a relatively high concentration of biological material to allow for the rapid detection timeline necessary for early warning of a biological attack. The confirmation sensor development has focused on moving standard laboratory techniques into autonomous, rapid, field deployable devices. Improvements have been made in the engineering of real-time polymerase chain reaction (rt-PCR) devices to reduce the temperature ramping times through reduction in the volume of the PCR reaction and shuttling of the reaction between fixed temperature zones. These improvements have resulted in devices capable of 6–8 min complete rt-PCR reactions using laboratory standard reagents [23, 24]. Advances in the design and production of microfluidic devices have utility in the creation of high-surface area antibody or nucleic acid structures with small analysis volumes, and for rapid mixing and binding of reagents leading to rapid detection assays [25]. Another emerging detection technology exploits cell-based sensing, where the response of simple living systems to the environment is monitored to rapidly detect the presence of biological agents [26]. Finally, new MS techniques are also under development to incorporate sample preparation and protein detection in cluttered environmental samples on automated MS platforms [27]. These confirmer technologies face several significant challenges before they will be transitioned to operation, including engineering for completely unattended operation and maintenance by facility personnel, and validation testing to demonstrate specificity and false positive characteristics of the sensors.

3.3 High Consequence Chemical Sensor Technologies

The current technology development for high consequence chemical sensors in the ARFCAM Project is focused on three different analytical approaches described in this section. The first approach is based on IMS. As described in Section 3.1, gas-phase ions are separated by difference in mobility of the ions under the influence of an electric field gradient [12]. IMS parameters include high-speed analysis, portability, high reliability, and low cost.

The second approach is a coupling of the IMS with a miniature gas chromatograph (GC). In GC, constituents within a complex gas sample can be separated by sweeping the sample into a column with an inert gas (in this application the inert gas is ambient air). The chemical constituents that make up the gas sample interact with a stationary phase, which either fills or coats the column at different rates depending on their chemical or physical properties. Hence, the sample constituents will exit the end of the column into the IMS in a sequential manner and undergo further separation and detection. By placing the GC on the front end of the IMS, the GC acts to "pre-filter" the gas sample to increase selectivity of the IMS detection.

The third approach is a coupling of the IMS with differential mobility spectrometry (DMS). As is the GC in the GC–IMS approach, the DMS is positioned as a pre-filter to the IMS. The DMS technology was discovered in the former Soviet Union and is proving to be a very promising technology [12, 28]. DMS separates ionized compounds based on their differential mobilities that are a function of their charge, mass, and cross-sectional area. By applying both an RF and DC field to the DMS sensor, the sensor selects a chosen ion or collection of ions, which are then swept into the IMS for further separation and detection. The technological challenges for these "detect-to-warn" systems are being able to detect a much broader range of chemical analytes than current facility monitor systems, having minimal false alarms, particularly false positive responses, and being configurable and interoperable with current facility security measures, all at a low cost (~\$10,000).

3.4 High Consequence Biological Sensor Technologies

During the BAND Project, several technologies have been investigated for both toxin and nucleic acid detection to meet the aggressive goal of identifying 20 agents in one system. To meet the BAND requirements, each system must be able to detect several diverse classes of agents, including toxins, viruses (RNA and DNA), and bacteria, within the analysis cycle time and at the LOD. The toxin detection methodologies exploit specific antibody reactions with various detection methodologies. The detection methods range from immuno-PCR [29] to traditional sandwich assays on beads or on long DNA constructs.

The nucleic acid detection methodologies employ both amplifying techniques and nonamplifying techniques. Each of these methodologies present opportunities for robust nucleic acid detection; however, they also face important challenges for creating a system that is capable of autonomous operation in extreme environments for 30 d without maintenance. The amplifying techniques utilize traditional PCR with various signal transduction methods, including microarrays, fluorescent probes, and padlock probes [30] with molecular beacons [31]. The amplifying technologies have the challenge of reducing reagent use and maintaining reagent activity for the month long maintenance cycles, as well as enabling high-multiplexed capability to identify 20 or more threat agents simultaneously. The nonamplifying technology under investigation in the BAND Project, direct linear analysis (DLA), creates a unique bar code of large DNA molecules [32]. By using long fragments of DNA and nonspecific fluorescent tags, DLA utilizes a universal reagent set reducing operating costs while maintaining the high confident identification. DLA has the challenge that each DNA fragment of the appropriate length is investigated to determine the content of the sample, and therefore throughput challenges exist with this technique.

4 FUTURE TECHNOLOGY NEEDS AND IMPROVEMENTS

Although much progress has been made in the area of detection of chemical warfare agents, toxic industrial chemical vapors, and known pathogenic biological agents, there is still considerable room for improvement. These improvements are necessary for all of the chemical and biological detection technologies to allow for confident and widespread use of these sensors. Further reduction of false alarms is of key importance because decisions on whether to perform low-regret actions (such as shutting off or redirecting HVAC units) versus high-regret actions (such as evacuating buildings) will need to be made by facility operators based on the output from these sensor systems. The financial loss from evacuating a building after a false alarm can be very costly, so it is critical that the results from the sensors be extremely reliable. The continued reduction in sensor cost, while maintaining desired performance, is a major consideration for all potential customers of these technologies. These cost reductions will be essential so that more buildings, transit systems, and emergency personnel can afford the protection these sensors will provide. Finally, for both chemical and biological detection sensors, interoperability with existing incident commander procedures and protocols is an important future goal and should be considered early on in the technology development phase.

Major challenges facing chemical detection sensors include decreasing detection speed, while increasing the sensitivity and reliability of the sensors. One single technology will most likely not be able to detect such a broad range of threats, hence the future will likely see the integration of multiple sensors with data fusion functionality. Specific improvements for the portable chemical sensors will be continued reduction in sensor weight and size, as well as increased single battery operation time. These improvements are critical to meet the operational requirements of emergency responders.

For biological confirmation sensors, a major challenge for future development will be in moving beyond the paradigm of sensing specific biological agents to sensing the fundamental elements of what causes illness. Also desired in future sensors is the ability to keep field collected organisms viable for further investigation such as antibiotic susceptibility, and the ability to quantify the amount of organisms that are collected. A portable biological detection sensor will allow emergency personnel to add another layer of detection capability to their protective architecture. Finally, continued improvements in the long-term stability of reagents, the reduction in reagents needed for detection assays, and automated sample preparation platforms are required to make the distributed deployment of these sensors practical.

Looking beyond the next few years, the rapidly advancing field of nanotechnology is expected to positively impact the detection technology community. Highly sensitive and selective sensors are expected to result from the ongoing research that is exploiting the unusual surface area and thereby the increased reactivity and catalytic properties of nanomaterials. Additionally, improved fabrication processes are revealing reproducible and low cost production of these novel materials.

Significant progress has been made in the past few years in moving laboratory detection techniques into the first generation of fieldable chemical and biological sensors. More work remains to be done to engineer sensor solutions to address the challenges and needs of the diverse set of operators and users of these sensors in the urban environment. Protection of the urban infrastructure and population against a release of a chemical or biological threat will be realized as these sensors are widely deployed and operated.

REFERENCES

1. Danzig, R. (2003). *Catastrophic Bioterrorism—What is to be done?* Center for Technology and National Security Policy, Washington, DC.
2. Romano, J. A. Jr., Lukey, B. J., and Salem, H. (Eds.) (2008). *Chemical Warfare Agents: Chemistry, Pharmacology, Toxicology, and Therapeutics*, 2nd ed., CRC Press, Boca Raton, FL.
3. Tucker, J. B. (2006). *War of Nerves*, Pantheon Books, NY.
4. Langford, R. E. (2004). *Introduction to Weapons of Mass Destruction*, John Wiley & Sons, Inc., Hoboken, NJ.
5. Bush, G. W. (2002). *National Strategy to Combat Weapons of Mass Destruction. National Security Presidential Directive 17/Homeland Security Presidential Directive 4*, Office of the Press Secretary, Washington, DC.
6. Bush, G. W. (2004). *Biodefense for the 21st Century. National Security Presidential Directive 33/Homeland Security Presidential Directive 10*, Office of the Press Secretary, Washington, DC.
7. Bush, G. W. (2007). *Homeland Security Presidential Directive 22 (classified)*.
8. Vitko, J. Jr., Franz, D. R., Alper, M., Biggins, P. D. E., Brandt, L. D., Bruckner-Lea, C., Burge, H. A., Ediger, R., Hollis, M. A., Laughlin, L. L., Mariella Jr, R. P., McFarland, A. R., and Schaudies, R. P. (2004). *Sensor Systems for Biological Agent Attacks: Protecting Buildings and Military Bases*, National Academies Press, Washington, DC.
9. Franz, D. R., Johnson, N. L., Bahnfleth, W. P., Bruckner-Lea, C., Buchsbaum, S. P., Friedlander, S. K., Hamlet, M., Knoop, S. L., Maier, A., Schaudies, R. P., Sextro, R. G., Stetzenbach, L. D., Thomas-Mobley, L. M., and Walt, D. R. (2007). *Protecting Building Occupants and Operations from Biological and Chemical Airborne Threats: A Framework for Decision Making*, National Academies Press, Washington, DC.
10. Derringer, T. *Unpublished Results from DHS S&T Phase II Independent Testing & Evaluation*, Battelle Memorial Institute, Columbus, OH (contract HSHQDC-06-F-00025).
11. Press Release (2007). *Smiths Detection Awarded Initial Contract to Supply M4 JCAD, the US Military's New generation Chemical Agent Detector*, 24 July 07, http://www.smiths-group.com/press_release_details.aspx?releaseID=264.
12. Eiceman, G. A., and Karpas, Z. (2005). *Ion Mobility Spectrometry*, 2nd ed., Taylor & Francis Ground, Boca Raton, FL.
13. de Hoffmann, E., and Stroobant, V. (2007). *Mass Spectrometry: Principles and Applications*, 3rd ed., John Wiley & Sons, Ltd, West Sussex.
14. Skoog, D., Holler, F. J., and Nieman, T. (1998). *Principles of Instrumental Analysis*, 5th ed., Brooks/Cole, United States.
15. Teale, F. W. J., and Weber, G. (1957). Ultraviolet fluorescence of the aromatic amino acids. *Biochem. J*. **65**(3), 476–482.

16. Kierdaszuk, B., Gryczynski, I., Modrak-Wojcik, A., Bzowska, A., Shugas, D., and Lakowicz, J. R. (1995). Fluorescence of tyrosine and tryptophan in proteins using one- and two-photon excitation. *Photochem. Photobiol*. **61**(4), 319–324.

17. Jeys, T. H., Herzog, W. D., Hybl, J. D., Czerwinski, R. N., and Sanchez, A. (2007). Advanced trigger development. *Linc. Lab. J*. **17**(1), 29–62.

18. Stuart, B. H., and Ando, D. J. (Eds.) (1997). *Biological Application of Infrared Spectroscopy*, John Wiley & Sons, West Sussex.

19. Jarvis, R. M., and Goodacre, R. (2004). Discrimination of bacteria using surface-enhanced Raman spectroscopy. *Anal. Chem*. **76**(1), 40–47.

20. Panne, U., and Hahn, D. (2006). Analysis of aerosols by LIBS. In *Laser Induced Breakdown Spectroscopy*, A. W. Miziolek, V. Palleschi, and I. Schechter, Eds. Cambridge University Press, Cambridge, pp 194–254.

21. Hunter, A. J. B., and Piper, L. G. (2006). Spark-induced breakdown spectroscopy: a description of an electrically generated LIBS-like process for elemental analysis of airborne particulates and solid samples. In *Laser Induced Breakdown Spectroscopy*, A. W. Miziolek, V. Palleschi, and I. Schechter, Eds. Cambridge University Press, Cambridge, pp. 585–614.

22. Reichardt, T. A., Bisson, S. E., Crocker, R., and Kulp, T. J. (2008). Analysis of flow-cytometer scattering and fluorescence data to identify particle mixtures (conference proceedings paper). In *Optics and Photonics in Global Homeland Security IV (Proceedings Volume 6945)*, C. S. Halvorson, D. Lehrfeld, and T. T. Saito, Eds. SPIE Publications, France.

23. Neuzil, P., Zhang, C., Pipper, J., Oh, S., and Zhou, L. (2006). Ultra fast miniaturized real-time PCR: 40 cycles in less than six min. *Nucleic Acids Res*. **34**(11):e77 (9 pages).

24. Kopp, M. U., de Mello, A. J., and Manz, A. (1998). Chemical amplification: continuous-flow PCR on a chip. *Science* **280**(5366), 1046–1048.

25. Northrup, M. A. (2004). Microfluidics: a few good tricks. *Nat. Mater*. **3**(5), 282–283.

26. Rider, T. H., Petrovick, M. S., Nargi, F. E., Harper, J. D., Schwoebel, E. D., Mathews, R. H., Blanchard, D. J., Bortolin, L. T., Young, A. M., Chen, J., and Hollis, M. A. (2003). A B cell-based sensor for rapid identification of pathogens. *Science* **301**(5630), 213–215.

27. Bunger, M. K., Cargile, B. J., Ngunjiri, A., Bundy, J. L., and Stephenson, J. L. Jr. (2008). Automated proteomics of E. coli via top-down electron-transfer dissociation mass spectrometry. *Anal Chem*. **80**(5), 1459–1467.

28. Buryakov, I. A., Krylov, E. B., Makas, A. L., Nazarov, E. G., Pervukhin, V. V., and Rasulev, K. (1991). Separation of ions according to mobility in strong AC electric fields. *Sov. Tech. Phys. Lett*. **17**(6), 446–447.

29. Sano, Takeshi., and Smith, Cassandra. L. (1992). Cantor, Immuno-PCR: very sensitive antigen detection by means of specific antibody-DNA conjugates. *Science* **258**, 120–122.

30. Marras, S. A. E., Kramer, F. R., and Tyagi, S. Genotyping single nucleotide polymorphisms with molecular beacons. In *Genotyping Single Nucleotide Polymorphisms: Methods and Protocols*, P. Y. Kwok, Ed. The Humana Press, Totowa, NJ, Vol. 212, pp. 111–128.

31. Hardenbol, P., Baner, J., Jain, M., Nilsson, M., NAmsaraev, E. A., Karlin-Neumann, G. A., Fakhrai-Rad, H., Ranoghi, M., Willis, T. D., Landegren, U., and Davis, R. W. (2003). Multiplexed genotyping with sequence-tagged molecular inversion probes. *Nat. Biotechnol*. **21**(6), 673–678.

32. Chan, E. Y., Goncalves, N. M., Haeusler, R. A., Hatch, A. J., Larson, J. W., Maletta, A. M., Yantz, G. R., Carstea, E. D., Fuchs, M., Wong, G. G., Gullans, S. T., and Gilmanshin, R. (2004). DNA mapping using microfluidic stretching and single—molecule detection of fluorescent site-specific tags. *Genome Res*. **14**, 1137–1146.

SENSING RELEASES OF HIGHLY TOXIC AND EXTREMELY TOXIC COMPOUNDS

D. ANTHONY GRAY
Syracuse Research Corporation, North Syracuse, New York

1 INTRODUCTION

Chemical terrorism (i.e. using or threatening to use chemicals as weapons of terror) is a serious threat in modern society, which is likely to continue for many years. The level of the concern about this threat can be seen in many of the articles in this volume, and the explosion of information on the subject is available in scholarly documents, journal articles, Internet (e.g. [1, 2]), symposia, and short courses dedicated to the subject. In addition, the currency of at least one aspect of this threat was underscored when Iraqi insurgents successfully dispersed chlorine for the first time by exploding an improvised explosive device (IED)/chlorine tank combination in early 2007 as a means of dispersing an industrial chemical agent (see [3, 4]). Because of the magnitude of the threat both to civilians and military personnel and the potential consequences, it is incumbent on both governments and industry to prepare for such an event to the extent possible.

One of the ways to prepare for chemical terrorism events is to develop sensors capable of identifying and quantifying the chemical in use so that when an event occurs, it can be rapidly and accurately identified and its effects can be mitigated. A second way to prepare is to identify a list of candidate chemicals that have toxicological and physicochemical properties that are consistent with use as a chemical agent (e.g. a chemical that is both toxic and dispersible). Both of these activities are important to preparing for such an event for a number of reasons including the two major ones important to this discussion. Firstly, there are more than 100,000 chemicals currently in commerce, many of which have low toxicities, limited availabilities, or physical properties that are not consistent with those needed to be a serious threat to public health. It is therefore both impractical and imprudent to develop the large array of expensive sensors that would be needed to support the capability of identifying and quantifying all commercial chemicals and especially for those having a low likelihood of being considered as candidates for terrorist activity. Secondly, because there are so many commercially available chemicals that cannot be selectively identified with current sensor technology, there is a high likelihood that the identity of the toxic industrial chemical used in a chemical attack will be unknown to responders and health care providers throughout at least much of the initial portion of the event. This lack of information, however, may both lead to additional exposures to, for example, emergency responders, others at the scene, and health care providers at hospitals treating the victims (especially if the chemical has few or no organoleptic warning properties) and hamper or prevent delivery of life saving procedures to the exposed population.

For example, methyl sulfate has few warning properties at lethal concentrations and causes delayed pulmonary edema so that immediate symptoms may not be present in the exposed population. However, when symptoms appear 8–12 h after exposure, victims presenting respiratory distress may quickly overwhelm the capacity of the health care system to provide the needed respirators. This potentially could lead to significant loss of life that might be prevented if respirators were widely available. However, if the chemical substance were known at the outset of the event, valuable time would be gained to locate additional respirators that might be made available to the affected community from relatively distant points. This article is therefore devoted to describing a methodology for identifying potential chemical threat agents, identifying a set of toxic industrial chemicals that fit the criteria, and describing example sensor availabilities for selected highly and extremely toxic chemicals.

Any discussion of the type presented below, which lists a set of highly toxic industrial chemicals that could be used in a terrorist event and thereby potentially aiding terrorists, is not complete without a discussion of the propriety making such a list public. The types of issues associated with this article therefore parallel those discussed by Phillip A. Sharp in his editorial on the conduct of responsible science and the prudence of publishing the DNA sequence and reconstruction of the 1918 Spanish influenza virus [5]. His conclusion and that of a committee of the US National Academies charged with considering these issues were that the publication was the correct action on the basis of both national security and public health. While the consequences of publishing the DNA sequence and reconstruction of the 1918 influenza virus, which killed an estimated 50 million people worldwide, are much higher than a list of highly toxic industrial chemicals (release of a reconstructed version of this virus could potentially result in many deaths and therefore justified review at the national level), these same issues are relevant for publishing a list of highly toxic chemicals. It is the sincere hope of the author that the information in this article will spur new and innovative technologies including advances in the sensor development and manufacturing community, advances in the development of medical countermeasures that can be used to better treat victims, and advances in the development of replacement chemicals that will eliminate the need for the production of commercial quantities of these toxic chemicals. It is therefore the goal of this article that the information contained in it will result in greater security and safety for the people of all countries that suffer from a terrorist threat.

Finally, the list presented here is designed to be neither an exhaustive list of all possible threat agents nor a compilation of chemicals developed from highly authoritative and critically reviewed information sources (as is the case with, for example, the National Advisory Committee for Acute Exposure Guideline Levels for Hazardous Substances). Rather it is designed to be an initial survey of industrial chemical threat agents that can be used as a starting point for further discussion and development.

2 IDENTIFICATION OF EXTREMELY AND HIGHLY TOXIC CHEMICALS AND IMPACT TO SENSOR DEVELOPMENT

The US and other governments, in both recognition and response to the need described above, have sought to identify a set of toxic industrial chemicals that if used by terrorists would cause significant harm (see, for example, Hauschild and Bratt [6] and Federal Register Vol. 72 17687–17745 [2007]), although no list specifically dedicated to extreme toxicity could be found in the available literature. In addition to identifying candidate threat chemicals, many of these efforts have also sought to quantify and prioritize the threats so that, for example, research and development work can be directed to those chemicals that will likely cause the highest impact if used. The criteria used to identify and prioritize these chemicals include acute lethality, availability, flammability, and reactivity and acknowledge that many factors will influence the degree of harm an event inflicts. A terrorist event involving industrial chemicals, of course, could take many forms including, for example, attacks on the infrastructure of a chemical manufacturing or storage facility (and thereby releasing toxic chemicals to the surrounding area), attacks on the chemical transportation system (e.g. destroying a rail tank car or cars carrying toxic industrial chemicals in a populated area), or attacks on the structures in populated or sensitive areas (e.g. a chemical release in a shopping mall, office building, or military installation). Factors that will influence the magnitude of the effect of these types of scenarios include the exact location of the attack (e.g. a major densely populated city vs. a smaller less densely populated area), weather (e.g. temperature, wind speed and direction, and precipitation), time of day and year (e.g. a mall at the height of the Christmas shopping season), dispersal method, and chemical used. While specific scenarios vary, it is clear that, if well executed, a terrorist attack using a highly toxic industrial chemical or chemicals can result in loss of life, significant injury, and potentially substantial economic harm. Other preparedness activities, such as determining the likely health and environmental effects from chemical releases/exposures, developing medical countermeasures, and developing and implementing training for the medical and emergency community to respond adequately to an event, are also important activities that must be undertaken before we can adequately respond to an event.

The first step in identifying a set of highly toxic chemicals is to determine what constitutes an extremely or highly toxic industrial chemical. As with previous efforts, the list of chemicals presented below is based on an assessment of acute lethality information by the inhalation and dermal routes and reported as an LC_{50} or LD_{50} (i.e. the air concentration or dose to the skin, respectively, that causes 50% of the population of test animals to die). The inhalation and dermal routes were considered to be the most important routes of exposure during a terrorist event involving industrial chemicals (oral exposure was considered to be less important) because the scenarios considered most relevant for toxic industrial chemicals involve dispersal of the chemical in the air in a building or at an event. During these releases, people will be exposed by both breathing the air that contains the chemical and potentially by touching surfaces contaminated with the chemical. Acute toxicity was considered to be the most relevant and most available measure since the types of events considered most likely for industrial chemicals are short-term events. In addition, lethality was considered to be the most relevant measure of acute toxicity because this endpoint will likely be a major purpose of a terrorist event and compilations of acute lethality are available (see below for further discussion). Although dated, one of the most commonly used (and thereby accepted) metrics for categorizing chemicals into acute lethality classes is that compiled by Hodge and Sterner [7, 8] and Hine and Jacobsen [9], and are summarized in Table 1. Chemicals under consideration

TABLE 1 Acute Toxicity Classes

		Routes of Administration		
Toxicity Rating	Commonly Used Term	Oral LD_{50} (Single Dose to Rats) (mg/kg)	Inhalation LC_{50} (4-h Exposure to Rats) ppm[a]	Dermal LD_{50} (Single Dose to Rabbit Skin) (mg/kg)
1	Extremely toxic	≤ 1	<10	≤ 5
2	Highly toxic	1–50	10–100	5–43
3	Moderately toxic	50–500	100–1000	44–340
4	Slightly toxic	500–5000	1000–10,000	350–2810
5	Practically non-toxic	5000–15,000	10,000–100,000	2820–22,590
6	Relatively harmless	$\geq 15{,}000$	$>100{,}000$	$\geq 22{,}600$

[a]Neither Hodge and Sterner [8] nor Hine and Jacobsen [9] use the term "LC_{50}" to describe the criteria for inhalation toxicity, rather, they use "vapor exposure mortality of 2/6–4/6 rats" as the measure. This has been interpreted to be equivalent to LC_{50} for the purposes of this discussion.

here, therefore, have either inhalation LC_{50}s that are less than or equal to 100 ppm for a 4-h exposure to rats or dermal LD_{50}s less than or equal to 43 mg/kg for exposure to rabbits.

Acute lethality is not the only measure of acute toxicity or necessarily the only one that should be considered when identifying chemicals of concern for use in a terrorist event. Rather, other endpoints may be equally important or at least significant in determining the final outcome of an event. For example, a moderately toxic chemical that is a lachrymator may cause the same effects as a highly toxic chemical that is not a lachrymator, and those exposed to the lachrymator cannot escape because of blurred vision and experience longer exposure times. Similarly, chemicals with lower toxicities, but without organoleptic warning properties such as smell or taste, may also have the same outcome if used in an event since air concentrations will go undetected by the exposed population and exposure concentrations and durations can be higher and longer, respectively. Other endpoints will also determine the outcome of an event and include, for example, long-term effects from short-term exposures (e.g. effects to a developing fetus and neurological effects) and delayed toxicity (e.g. pulmonary edema). Although metrics other than lethality, therefore, can be used to identify a set of candidate chemicals, they are typically much more difficult to consistently ascribe, have small data sets, or require significant assessment of the data (involving professional judgment) before conclusions can be made about toxicity and applied to a set of chemicals.

After determining the metrics for identifying extremely and highly toxic chemicals, sources of the needed information were located. One of the most comprehensive (albeit not exhaustive or peer reviewed) compilations of acute lethality information is contained within the Registry of Toxic Effects of Chemical Substances RTECS, available from Elsevier MDL; much of the acute lethality information in RTECS is also available on the ChemIDplus, an advanced website maintained by the National Library of Medicine NLM, http://chem.sis.nlm.nih.gov/chemidplus/). Information from RTECS was supplemented with acute toxicity information from the Syracuse Research Corporation implementation of the Toxic Substances Control Act Test Submissions TSCATS database, which contains unpublished data submitted by the chemical industry to the Environmental Protection Agency (EPA) under Toxic Substances Control Act (TSCA) and provides abstracts with endpoint data for some of the submissions in the database (see http://www.syrres.com/||esc/||tscats.htm). There were approximately 7100 lethality records in the combined data sets for inhalation and dermal exposure, but only slightly more than 4200 records had information for the selected species.

The list was further edited by (i) removing those chemicals with inexact LD_{50} or LC_{50} values (e.g. an LD_{50} value identified as "$>$10 mg/kg" was removed because it does not exactly identify a level), (ii) removing chemicals marketed as variable mixtures (e.g. ligroine), and then (iii) selecting only those that were commercially available. Commercial availability was determined by identifying a set of over 50,000 chemicals that are in EPA's Inventory Update Rule(IUR—this source lists organic chemicals produced or imported into the United States in volumes greater than or equal to 10,000 lb/year), the Directory of Chemical Producers (worldwide) published by SRI Consulting, or identified as commercially available on the ChemChannels website (http://www.chemchannels.com). Inclusion on one of these lists was considered to be sufficient to mean that the chemical was readily available and could be purchased somewhere in the world without significant difficulty.

Finally, units were converted when necessary to ensure that the comparisons with the selection criteria were consistent. In both RTECS and TSCATS, the units for inhalation LC_{50} values can be reported either in terms of parts (e.g. ppm or ppb) or mass per unit volume (e.g. mg/m^3). The units for dermal LD_{50} values can also be expressed in terms of either mass or volume per unit body mass (e.g. milligram per kilogram body mass or microliter per kilogram body mass). Inhalation LC_{50} units in terms of mass per volume were converted to ppm then adjusted to a 4-h exposure assuming linearity (Haber's rule). Whenever converting from units of mass per volume, care was taken to ensure that the theoretical saturation concentration in ppm (i.e. [vapor pressure in mmHg/760 mmHg] $\times$ 10^6) was at least equal to the calculated LC_{50} in ppm. The needed vapor pressures for this calculation were obtained from Syracuse Research Corporation's PHYSPROP database, if available, or estimated using the EPISuite estimation programs (see http://www.epa.gov/opptintr/exposure/pubs/episuite.htm). If the LC_{50} was reported in ppm or ppb, it was assumed that the reported concentration was achieved as a vapor independent of the calculated saturation concentration. Since animal exposures to aerosol concentrations in terms of mass per unit volume cannot be expressed in terms of ppm, this process eliminates those chemicals with vapor pressures too low to achieve a vapor concentration equal to their experimental LC_{50} (the selection criteria assume vapor exposures). Dermal LD_{50} values in terms of volume per unit body mass were converted to milligram per kilogram units assuming a density of 1 g/ml, thus the maximum dermal LD_{50} value consistent with the highly toxic criteria in Table 1 would be 0.043 ml/kg or 43 μl/kg. The final set of chemicals matching the criteria included 126 and 31 chemicals that matched the inhalation (Table 2) and dermal criteria (Table 3, respectively. Tables 2 and 3 also contain

TABLE 2 Industrial Chemicals Extremely or Highly Toxic by the Inhalation Route of Exposure

CAS No.	RTECS No.	Chemical Name	Molecular Formula	Molecular Mass	LC_{50} (ppm)	Toxic Effects		
1752-30-3	AL2338000	Acetone thiosemicarbazide	$C_4H_9N_3S$	131.22	6.71	Tremor	Convulsions or effect on seizure threshold	Changes in the structure or function of the salivary glands
107-02-8	AS9660000	Acrolein	C_3H_4O	56.07	7.85			
2937-50-0	LV7168000	Allyl chloroformate	$C_4H_5ClO_2$	120.54	6.57	Dyspnea (labored breathing)		
26842-43-3	SF1435000	2-Amino-2,4-dimethyl pentanenitrile	$C_7H_{14}N_2$	126.23	25	Changes relating to olfaction other than a deviated or ulcerated nasal septum, change of the olfactory nerve, or change in the sense of smell	Somnolence or a generally depressed activity level	Tremor
19355-69-2	UD5155000	2-Aminoisobutyronitrile	$C_4H_8N_2$	84.14	27.8	Changes relating to olfaction other than a deviated or ulcerated nasal septum, change of the olfactory nerve, or change in the sense of smell	Somnolence or a generally depressed activity level	
7647-18-9	CA1722000	Antimony pentachloride	Cl_5Sb	299	29.4			
7784-42-1	CC9968000	Arsine	AsH_3	77.95	5.1			
98-87-3	CW9716000	Benzal chloride	$C_7H_6Cl_2$	161.03	30.5	Changes in recordings from specific areas of the brain and/or membranous coverings	Excitement	Respiratory depression
100-44-7	XW9324000	Benzyl chloride	C_7H_7Cl	126.59	75	Respiratory depression		

111-44-4	KX1708000	Bis(2-chloroethyl)ether	$C_4H_8Cl_2O$	143.02	56.4			
542-88-1	KX1750000	Bis(chloromethyl)ether	$C_2H_4Cl_2O$	114.96	12.3	Chronic pulmonary edema (excess fluid accumulation and retention over time)	Hemorrhage (copious bleeding)	
1461-23-0	WR3458000	Bromotributyltin	$C_{12}H_{27}SnBr$	369.99	0.859	Somnolence or a generally depressed activity level	Structural or functional change in the trachea or bronchi	Hemorrhage (copious bleeding)
78-94-4	EX4312000	3-Buten-2-one	C_4H_6O	70.1	2.44			
4262-43-5	CD1127000	*t*-Butylarsine	$C_4H_{11}As$	134.07	73.5	Unspecified liver changes	Unspecified changes in the kidneys, ureter, and/or bladder	Unspecified hair changes
75-87-6	FW7714000	Chloral	C_2HCl_3O	147.38	73	Ptosis (drooping of the eyelid)	Somnolence or a generally depressed activity level	Dyspnea (labored breathing)
7782-50-5	FX8204000	Chlorine	Cl_2	70.9	73.3			
7790-91-2	FX8372000	Chlorine fluoride	ClF_3	92.45	74.8			
13637-63-3	FX8386000	Chlorine pentafluoride	ClF_5	30.45	30.5	Lachrymation (excessive tearing)	Changes in the structure or function of the salivary glands	Corneal damage

(*continued overleaf*)

TABLE 2 ***(Continued)***

CAS No.	RTECS No.	Chemical Name	Molecular Formula	Molecular Mass	LC_{50} (ppm)	Toxic Effects		
107-20-0	AB4928000	Chloroacetaldehyde	C_2H_3ClO	78.5	50.6	BP elevation not originating from effects on the autonomic nervous system	Respiratory obstruction	Unspecified respiratory changes
79-11-8	AH6664000	Chloroacetic acid	$C_2H_3ClO_2$	94.5	46.6			
78-95-5	UG8134000	Chloroacetone	C_3H_5ClO	92.53	65.5	Somnolence or a generally depressed activity level	Ataxia (incoordination)	Unspecified hair changes
2315-36-8	AC3934000	2-Chloro-*N*,*N*-diethyl acetamide	$C_6H_{12}ClNO$	149.64	51.5	Excitement	Muscle weakness	Unspecified respiratory changes
1622-32-8	KT2954000	2-Chloroethanesulfonyl chloride	$C_2H_4Cl_2O_2S$	163.02	63			
107-07-3	KU7910000	2-Chloroethanol	C_2H_5ClO	80.52	88.1	Somnolence or a generally depressed activity level	Dyspnea (labored breathing)	
107-27-7	PA7966000	Chloroethyl mercury	C_2H_5ClHg	265.11	63.5			
107-30-2	KX3066000	Chloromethyl methyl ether	C_2H_5ClO	80.52	96.3	Chronic pulmonary edema (excess fluid accumulation and retention over time)	Hemorrhage (copious bleeding)	
106-48-9	SM5446000	*p*-Chlorophenol	C_6H_5ClO	128.56	2.09			

104-12-1	NZ5348000	4-Chlorophenyl isocyanate	C_7H_4ClNO	153.57	18	Changes relating to olfaction other than a deviated or ulcerated nasal septum, change of the olfactory nerve, or change in the sense of smell	Dyspnea (labored breathing)	Unspecified gastrointestinal changes
2909-38-8	NZ5334000	3-Chlorophenyl isocyanate	C_7H_4ClNO	153.57	6.69	Somnolence or a generally depressed activity level	Cyanosis (bluish or purplish cast to skin and mucous membranes due to lack of oxygen	Respiratory stimulation
76-06-2	PD3906000	Chloropicrin	CCl_3NO_2	164.37	14.4	Unspecified respiratory changes		
26447-14-3	UD4312000	Cresyl glycidyl ether	$C_{10}H_{12}O_2$	164.22	42	Somnolence or a generally depressed activity level	Excitement	Muscle weakness
4170-30-3	HB4606000	Crotonaldehyde	C_4H_6O	70.1	34.9			
460-19-5	HD6146000	Cyanogen	C_2N_2	52.04	87.5	Respiratory obstruction	Lachrymation (excessive tearing)	Somnolence or a generally depressed activity level

(continued overleaf)

TABLE 2 ***(Continued)***

CAS No.	RTECS No.	Chemical Name	Molecular Formula	Molecular Mass	LC_{50} (ppm)	Toxic Effects		
675-14-9	YC3052000	Cyanuric fluoride	$C_3F_3N_3$	135.06	3.1	Excitement	Chronic pulmonary edema (excess fluid accumulation and retention over time)	Dyspnea (labored breathing)
17702-41-9	HP5880000	Decaborane	$B_{10}H_{14}$	122.24	46	Convulsions or effect on seizure threshold	Unspecified liver changes	Unspecified changes in the kidneys, ureter, and/or bladder
56-18-8	KC1302000	3,3′-Diaminodipropyl-amine	$C_6H_{17}N_3$	131.26	5.59	Dyspnea (labored breathing)	Cyanosis (bluish or purplish cast to skin and mucous membranes due to lack of oxygen	Unspecified hair changes
105-83-9	KC1316000	3,3′-Diamino-*N*-methyl dipropylamine	$C_7H_{19}N_3$	145.29	11.8	Acute pulmonary edema (rapid onset of excess fluid accumulation in the lungs)	Dyspnea (labored breathing)	Unspecified hair changes
19287-45-7	IB2688000	Diborane	B_2H_6	27.68	40			
534-07-6	UG8862000	1,3-Dichloroacetone	$C_3H_4Cl_2O$	126.97	2.79			

110-57-6	EW9534000	*trans*-1,4-Dichloro-2-butene	$C_4H_6Cl_2$	125	86	Lachrymation (excessive tearing)	Unspecified respiratory changes	Changes in the structure or function of the salivary glands
79-35-6	LD6370000	1,1-Dichloro-2,2-difluoroethylene	$C_2Cl_2F_2$	132.92	92.9	Somnolence or a generally depressed activity level	Unspecified liver changes	Unspecified changes in the kidneys, ureter, and/or bladder
22591-21-5	EW3934000	1,1-Dichloro-3,3-dimethyl-2-butanone	$C_6H_{10}Cl_2O$	169.06	93	Unspecified changes associated with the eye	Unspecified respiratory changes	
62-73-7	TB3360000	Dichlorvos	$C_4H_7Cl_2O_4P$	220.98	1.66	Lachrymation (excessive tearing)	Tremor	Changes in the structure or function of the salivary glands
1464-53-5	EU5824000	1,2 : 3,4-Diepoxybutane	$C_4H_6O_2$	86.1	90			
2524-04-1	TB8148000	Diethyl phosphorochlori-dothionate	$C_4H_{10}ClO_2PS$	188.62	20			
627-44-1	PA8694000	Diethylmercury	$C_4H_{10}Hg$	258.73	24.4	Excitement	Ataxia (incoordination)	Dyspnea (labored breathing)
627-54-3	KT2996000	Diethyltelluride	$C_4H_{10}Te$	185.74	3.16			

(*continued overleaf*)

TABLE 2 ***(Continued)***

CAS No.	RTECS No.	Chemical Name	Molecular Formula	Molecular Mass	LC_{50} (ppm)	Toxic Effects		
28178-42-9	CV6664000	2,6-Diisopropylphenyl isocyanate	$C_{13}H_{17}NO$	203.31	5.65			
624-92-0	KE2380000	Dimethyl disulfide	$C_2H_6S_2$	94.2	2.06			
2524-03-0	TB8176000	Dimethyl phosphorochlo-ridothionate	$C_2H_6ClO_2PS$	160.56	51.8			
77-78-1	WX4466000	Dimethyl sulfate	$C_2H_6O_4S$	126.14	8.72	Dyspnea (labored breathing)	Cyanosis (bluish or purplish cast to skin and mucous membranes due to lack of oxygen	Hemorrhage (copious bleeding)
2867-47-2	PC0994000	2-(Dimethylamino)ethyl methacrylate	$C_8H_{15}NO_2$	157.24	96.4			
2439-35-2	AT2044000	2-(Dimethylamino)ethyl acrylate	$C_7H_{13}NO_2$	143.21	11.3	Ptosis (drooping of the eyelid)	Somnolence or a generally depressed activity level	Dyspnea (labored breathing)
24424-99-5	ID4235000	Di-*t*-butyldicarbonate	$C_{10}H_{18}O_5$	218.28	11.2	Unspecified changes associated with the eye	Dyspnea (labored breathing)	Weight loss or decreased weight gain
106-91-2	PC1022000	2,3-Epoxypropyl methacrylate	$C_7H_{10}O_3$	142.17	45	Dyspnea (labored breathing)	Respiratory stimulation	Unspecified changes in the blood

541-41-3	LV7252000	Ethyl chloroformate	$C_3H_5ClO_2$	108.53	47.3	Changes in lung weight	Weight loss or decreased weight gain	
2941-64-2	FQ5460000	Ethyl chlorothioformate	C_3H_5ClOS	124.59	41.2	Cardiomyopathy, including infarction (disease of the heart muscle, including tissue death resulting from local blood supply loss)	Emphysema (a condition of the lungs marked by enlargement of the alveoli either by dilatation or by breakdown of their walls, resulting in labored breathing and increased susceptibility to infection)	Diffuse hepatitis (liver cells are killed in a random manner, not relating to any particular lobe or zone of the liver)
1498-64-2	TB9142000	O-Ethyl-dichlorothiophosphate	$C_2H_5Cl_2OPS$	179	32			
1498-51-7	TB9114000	Ethyl phosphorodichloridate	$C_2H_5Cl_2O_2P$	162.94	84.6			
151-56-4	LE0924000	Ethylenimine	C_2H_5N	43.08	28.4			
24468-13-1	FQ5330000	2-Ethylhexyl chloroformate	$C_9H_{17}ClO_2$	192.71	34.3			

(continued overleaf)

TABLE 2 *(Continued)*

CAS No.	RTECS No.	Chemical Name	Molecular Formula	Molecular Mass	LC_{50} (ppm)	Toxic Effects		
7782-41-4	LR2170000	Fluorine	F_2	38	46.3	Conjunctive irritation	Dyspnea (labored breathing)	Weight loss or decreased weight gain
371-62-0	KV2324000	2-Fluoroethanol	C_2H_5FO	64.07	3.18			
79-14-1	MG6160000	Glycolic acid	$C_2H_4O_3$	76.06	2.28	Changes relating to olfaction other than a deviated or ulcerated nasal septum, change of the olfactory nerve, or change in the sense of smell	Dyspnea (labored breathing)	Weight loss or decreased weight gain
77-47-4	HK1106000	1,2,3,4,5,5-Hexachloro-1,3-cyclopentadiene	C_5Cl_6	272.75	1.6	Somnolence or a generally depressed activity level	Unspecified respiratory changes	
18406-41-2	JW7266000	Hexamethoxydisilylethane	$C_8H_{22}O_6Si_2$	270.48	2.4	Unspecified respiratory changes		
74-90-8	MY6300000	Hydrogen cyanide	CHN	27.03	20			
7783-66-6	NW2408000	Iodine pentafluoride	F_5I	221.9	98.1	Convulsions or effect on seizure threshold	Acute pulmonary edema (rapid onset of excess fluid accumulation in the lungs)	Fatty liver degeneration
13463-40-6	NX6062000	Iron carbonyl	C_5FeO_5	195.9	10	Somnolence or a generally depressed activity level	Dyspnea (labored breathing)	Weight loss or decreased weight gain

30674-80-7	PC1218000	2-Isocyanatoethyl methacrylate	$C_7H_9NO_3$	155.17	6			
55-91-4	TC3948000	Isoflurophate	$C_6H_{14}FO_3P$	184.17	1.99	Muscle weakness	Dyspnea (labored breathing)	Changes in the structure or function of the salivary glands
10471-78-0		2-Isopropenyl-2-oxazoline	C_6H_9NO	230.88	7.7			
68-11-1	AJ4690000	Mercaptoacetic acid	$C_2H_4O_2S$	92.12	55.7	Food intake in animals	Dyspnea (labored breathing)	Changes in lung weight
920-46-7	PC1750000	Methacryloyl chloride	C_4H_5ClO	104.54	14			
558-25-8	PD2016000	Methanesulfonyl fluoride	CH_3FO_2S	98.1	1.75	Inhibition, induction or change in blood or tissue levels of true cholinesterase		
79-22-1	FQ5362000	Methyl carbonochloridate	$C_2H_3ClO_2$	94.5	22			
453-18-9	AJ2058000	Methyl fluoroacetate	$C_3H_5FO_2$	92.08	3.32			
421-20-5	LU2884000	Methyl fluorosulfate	CH_3FO_3S	114.1	1.25	Lachrymation (excessive tearing)	Changes in motor activity (specific assay)	Dyspnea (labored breathing)
60-34-4	MX5208000	Methyl hydrazine	CH_6N_2	46.09	34			
624-83-9	NZ5754000	Methyl isocyanate	C_2H_3NO	57.06	9.15	Lachrymation (excessive tearing)	Dyspnea (labored breathing)	

(continued overleaf)

TABLE 2 (*Continued*)

CAS No.	RTECS No.	Chemical Name	Molecular Formula	Molecular Mass	LC_{50} (ppm)	Toxic Effects		
676-97-1	SZ9660000	Methyl phosphonic dichloride	CH_3Cl_2OP	132.91	26	Excitement	Dyspnea (labored breathing)	Unspecified dermatitis after systemic exposure
12108-13-3	OV9576000	2-Methylcyclopentadienyl manganese tricarbonyl (MMT)	$C_9H_7MnO_3$	218.1	8.52	Conjunctive irritation	Somnolence or a generally depressed activity level	Dyspnea (labored breathing)
35203-06-6	CT7938000	*N*-Methylene-6-ethyl-2-methylaniline	$C_{10}H_{13}N$	147.24	93	Unspecified changes associated with the eye	Somnolence or a generally depressed activity level	Dyspnea (labored breathing)
140-76-1	UZ4004000	2-Methyl-5-vinylpyridine	C_8H_9N	119.18	19.4			
7786-34-7	HB6104000	Mevinphos	$C_7H_{13}O_6P$	224.17	3.5	Tremor	Acute pulmonary edema (rapid onset of excess fluid accumulation in the lungs)	Changes in the structure or function of the salivary glands
7783-77-9	QK5418000	Molybdenum hexafluoride	F_6Mo	209.94	38.8	Structural or functional change in the trachea or bronchi	Respiratory obstruction	Hemorrhage (copious bleeding)
13463-39-3	RB0042000	Nickel carbonyl	C_4NiO_4	170.75	4.38			
10102-44-0	RE9912000	Nitrogen dioxide	NO_2	46.01	88			

10544-72-6	RF0322000	Nitrogen tetroxide	N_2O_4	92.02	27.9	Convulsions or effect on seizure threshold	Unspecified cardiac changes	Acute pulmonary edema (rapid onset of excess fluid accumulation in the lungs)
62-75-9	JF9856000	*N*-Nitrosodimethylamine	$C_2H_6N_2O$	74.1	78	Unspecified changes associated with the eye	Somnolence or a generally depressed activity level	Hemorrhage (copious bleeding)
57-57-8	RX2352000	2-Oxetanone	$C_3H_4O_2$	72.07	37.5			
10028-15-6	SA2240000	Ozone	O_3	48	4.8	Acute pulmonary edema (rapid onset of excess fluid accumulation in the lungs)	Hemorrhage (copious bleeding)	
376-53-4	AV2842000	Perfluoroadiponitrile	$C_6F_8N_2$	252.08	6.01	Somnolence or a generally depressed activity level	Unspecified respiratory changes	Unspecified changes in the kidneys, ureter, and/or bladder
376-89-6	ME8988000	Perfluoroglutaronitrile	$C_5F_6N_2$	202.07	8.11	Somnolence or a generally depressed activity level	Unspecified respiratory changes	Unspecified changes in the kidneys, ureter, and/or bladder
108-95-2	SL9170000	Phenol	C_6H_6O	94.12	82.1			

(continued overleaf)

TABLE 2 (*Continued*)

CAS No.	RTECS No.	Chemical Name	Molecular Formula	Molecular Mass	LC_{50} (ppm)	Toxic Effects		
103-71-9	CY2520000	Phenyl isocyanate	C_7H_5NO	119.13	4.52	Somnolence or a generally depressed activity level	Acute pulmonary edema (rapid onset of excess fluid accumulation in the lungs)	Cyanosis (bluish or purplish cast to skin and mucous membranes due to lack of oxygen
140-29-4	AL4130000	Phenylacetonitrile	C_8H_7N	117.16	44.9	Altered sleep time, including changes in the righting reflex	Muscle contraction or spasticity	Dyspnea (labored breathing)
2687-12-9	CW3633000	1-Phenyl-3-chloro-1-propene	C_9H_9Cl	152.63	4.65	Somnolence or a generally depressed activity level	Respiratory depression	Decrease in body temperature
638-21-1	SY6930000	Phenylphosphine	C_6H_7P	110.1	38	Dyspnea (labored breathing)	Lachrymation (excessive tearing)	
298-02-2	TC1092000	Phorate	$C_7H_{17}O_2PS_3$	260.39	0.258			
7803-51-2	SY4830000	Phosphine	H_3P	34	11	Dyspnea (labored breathing)		
10025-87-3	TD6832000	Phosphorus oxychloride	Cl_3OP	153.32	32			
1185-09-7		1,1,2,2-Tetrachloroethyl sulfenyl chloride	C_2HCl_5S	174.16	6.2			

5216-25-1	XX3766000	*α*,*α*,*α*-*p*-Tetrachloro toluene	$C_7H_4Cl_4$	229.91	13.3	Somnolence or a generally depressed activity level	Convulsions or effect on seizure threshold	Structural or functional change in the trachea or bronchi
78-00-2	TU5040000	Tetraethyl lead	$C_8H_{20}Pb$	323.47	16.1	Excitement	Cyanosis (bluish or purplish cast to skin and mucous membranes due to lack of oxygen	
597-64-8	WR6118000	Tetraethyltin	$C_8H_{20}Sn$	234.97	11.9	Muscle weakness	Dyspnea (labored breathing)	Hemorrhage (copious bleeding)
2778-42-9	CV6454000	Tetramethyl-*m*-xylylene diisocyanate	$C_{14}H_{16}N_2O_2$	244.32	2.3	Dyspnea (labored breathing)	Respiratory stimulation	Decrease in body temperature
509-14-8	PD2464000	Tetranitromethane	CN_4O_8	196.05	18	Methemoglobinemia (a condition in which the iron in a hemoglobin molecule is unable to transport oxygen effectively to the tissues)-carboxyhemoglobin (carbon monoxide bound to hemoglobin preventing oxygen exchange)		

(*continued overleaf*)

TABLE 2 (*Continued*)

CAS No.	RTECS No.	Chemical Name	Molecular Formula	Molecular Mass	LC_{50} (ppm)	Toxic Effects		
294-93-9	XJ3220000	1,4,7,10-Tetraoxacyclo dodecane	$C_8H_{16}O_4$	176.24	27	Sensory change involving peripheral nerves	Unspecified liver changes	Weight loss or decreased weight gain
108-98-5	DB3304000	Thiophenol	C_6H_6S	110.18	33	Respiratory stimulation	Somnolence or a generally depressed activity level	Lachrymation (excessive tearing)
3982-91-0	XS4662000	Thiophosphoryl chloride	Cl_3PS	169.38	20			
7646-78-8	XU7588000	Tin tetrachloride	Cl_4Sn	260.49	9			
7550-45-0	XV3248000	Titanium chloride	Cl_4Ti	189.7	51.6			
584-84-9	CX0518000	2,4-Toluene diisocyanate	$C_9H_6N_2O_2$	174.17	14	Lachrymation (excessive tearing)	Excitement	Dyspnea (labored breathing)
1321-38-6		Toluene diisocyanate (mixture of 2,4- and 2,6-isomers)	$C_9H_6N_2O_2$	196.24	6			
921-03-9	UH4522000	Trichloroacetone	$C_3H_3Cl_3O$	161.41	29.5			
76-02-8	AO0392000	Trichloroacetyl chloride	C_2Cl_4O	181.82	63.9			
594-42-3	PC6468000	Trichloromethanesulfenyl chloride	CCl_4S	185.87	2.75			
98-07-7	XX3976000	α,α,α-Trichlorotoluene	$C_7H_5Cl_3$	195.47	9.5	Muscle contraction or spasticity	Respiratory depression	Weight loss or decreased weight gain

998-30-1	WI6062000	Triethoxysilane	$C_6H_{16}O_3Si$	164.31	37.2	Ataxia (incoordination)	Acute pulmonary edema (rapid onset of excess fluid accumulation in the lungs)	Changes in kidney tubules, including acute renal failure and acute tubular necrosis
35037-73-1		Trifluoromethoxyphenyl isocyanate	$C_8H_4F_3NO_2$	203.12	6.3			
98-16-8	XX8456000	*α,α,α*-Trifluoro-*m*-toluidine	$C_7H_6F_3N$	161.14	66.8	Muscle weakness	Cyanosis (bluish or purplish cast to skin and mucous membranes due to lack of oxygen	Respiratory depression
2487-90-3	WI6188000	Trimethoxysilane	$C_3H_{10}O_3Si$	122.22	42			
141-59-3	SF2044000	2,4,4-Trimethyl-2-pentane thiol	$C_8H_{18}S$	146.32	50.1	Convulsions or effect on seizure threshold	Unspecified respiratory changes	Weight loss or decreased weight gain
15112-89-7	WI5754000	Tris(dimethylamino)silane	$C_6H_{19}N_3Si$	161.37	57	Lachrymation (excessive tearing)	Ataxia (incoordination)	Changes in the structure or function of the salivary glands
100-43-6	VA3262000	4-Vinylpyridine	C_7H_7N	105.15	39.5			

TABLE 3 Industrial Chemicals Extremely or Highly Toxic by the Dermal Route of Exposure

CAS No.	RTECS No.	Chemical Name	Molecular Formula	Molecular Mass	LD_{50}	Toxic Effects		
75-86-5	ON3024000	Acetone cyanohydrin	C_4H_7NO	85.12	17 μl/kg			
309-00-2	JD4970000	Aldrin	$C_{12}H_8Cl_6$	364.9	15 mg/kg	Changes relating to olfaction other than a deviated or ulcerated nasal septum, change of the olfactory nerve, or change in the sense of smell	Convulsions or effect on seizure threshold	Excitement
107-11-9	BA4858000	Allylamine	C_3H_7N	57.11	35 mg/kg	Primary irritation after topical exposure (causes redness, swelling, and/or pain, and other evidence of irritation upon first contact)	Chronic pulmonary edema (excess fluid accumulation and retention over time)	
28772-56-7	GZ7154000	Bromadiolone	$C_{30}H_{23}BrO4$	527.44	2.1 mg/kg			
95465-99-9	TC3388000	Cadusafos	$C_{10}H_{23}O_2PS_2$	270.42	24.4 mg/kg			
8065-48-3	TC8792000	Demeton	$C_8H_{19}O_3PS_2$. $C_8H_{19}O_3PS_2$	516.72	24 mg/kg			
814-49-3	TB7994000	Diethyl chlorophosphate	$C_4H_{10}ClO_3P$	172.56	8 μl/kg			
926-64-7	AL3304000	Dimethylamino acetonitrile	$C_4H_8N_2$	84.14	32 mg/kg			
1122-58-3	UZ0910000	4-Dimethylamino pyridine	$C_7H_{10}N_2$	122.19	13 mg/kg			
77-77-0	KW9058000	Divinyl sulfone	$C_4H_6O_2S$	118.16	22 μl/kg	Primary irritation after topical exposure (causes redness, swelling, and/or pain, and other evidence of irritation upon first contact)		

2104-64-5	TA6020000	EPN (Ethoxy-4-nitrophenoxyphenylphosphine sulfide)	$C_{14}H_{14}NO_4PS$	323.32	30 mg/kg	Somnolence or a generally depressed activity level	Unspecified gastrointestinal changes
13194-48-4	TC3416000	Ethoprop	$C_8H_{19}O_2PS_2$	242.36	2.4 mg/kg		
107-16-4	AL3556000	Glycolonitrile	C_2H_3NO	57.06	5 mg/kg	Primary irritation after topical exposure (causes redness, swelling, and/or pain, and other evidence of irritation upon first contact)	
105-31-7	MT5768000	1-Hexyn-3-ol	$C_6H_{10}O$	98.16	15.8 mg/kg		
78-97-7	ON2940000	Lactonitrile	C_3H_5NO	71.09	20 μl/kg		
950-10-7	KF13860-00	Mephosfolan	$C_8H_{16}NO_3PS_2$	269.34	28.7 mg/kg		
126-98-7	UI2520000	Methacrylonitrile	C_4H_5N	67.1	12.5 mg/kg		
453-18-9	AJ2058000	Methyl fluoroacetate	$C_3H_5FO_2$	92.08	20 mg/kg	Convulsions or effect on seizure threshold	
556-61-6	PC6272000	Methyl isocyanate	C_2H_3NS	73.12	33 mg/kg		
3680-02-2	WW8330000	Methyl vinyl sulfone	$C_3H_6O_2S$	106.15	32 μl/kg		
7786-34-7	HB6104000	Mevinphos	$C_7H_{13}O_6P$	224.17	4.7 mg/kg		
5903-13-9	AE0938000	Nissol	$C_{13}H_{12}FNO$	217.26	1.75 mg/kg		
311-45-5	TB4298000	Paraoxon	$C_{10}H_{14}NO_6P$	275.22	5 mg/kg	Direct parasympathomimetic (mimicking the "rest and relaxation" action of the parasympathetic nervous system, for example, reduces heart rate, dilates blood vessels, and reduces blood pressure, etc.)	

(*continued overleaf*)

TABLE 3 ***(Continued)***

CAS No.	RTECS No.	Chemical Name	Molecular Formula	Molecular Mass	LD_{50}	Toxic Effects		
56-38-2	TC9772000	Parathion	$C_{10}H_{14}NO_5PS$	291.28	15 mg/kg			
298-02-2	TC1092000	Phorate	$C_7H_{17}O_2PS_3$	260.39	3.1 mg/kg			
947-02-4	NP4102000	Phosfolan	$C_7H_{14}NO_3PS_2$	255.31	23 mg/kg			
107-19-7	UO7966000	Propargyl alcohol	C_3H_4O	56.07	16 mg/kg	Unspecified dermatitis after systemic exposure		
3689-24-5	XS8050000	Sulfotep	$C_8H_{20}O5P_2S_2$	322.34	20 mg/kg			
13071-79-9	TC0350000	Terbufos	$C_9H_{21}O_2PS_3$	288.45	1 mg/kg			
2001-95-8	YZ0098000	Valinomycin	$C_{54}H_{90}N_6O_{18}$	1111.5	5 mg/kg	Tremor	Convulsions or effect on seizure threshold	Weight loss or decreased weight gain
3984-22-3	JX5551000	2-Vinyl-1,3-dioxolane	$C_5H_8O_2$	100.13	25 mg/kg			

toxic effects information when available from RTECS or ChemIDplus Advanced at NLM.

Tables 4 and 5 contain membership in regulatory lists and toxicity guideline values for chemicals toxic by the inhalation and dermal routes, respectively. Regulatory lists included in the tables are the Department of Homeland Security's interim final rule on Chemical Facility AntiTerrorism Standards (DHS, see Federal Register Vol. 72 17687–17745 [10]), Chemical Weapons Convention (Organisation for the Prohibition of Chemical Weapons, OPCW), the US High Production Volume list (US HPV), and European Union High Production Volume Chemicals list (EU HPVC). Tables 4 and 5 also contain (i) Acute Exposure Guideline Levels (AEGL-3—the airborne concentration of a substance above which it is predicted that the general population, including susceptible individuals, could experience life-threatening health effects or death), (ii) Emergency Response Planning Guideline (ERPG-3—the maximum airborne concentration below which it is believed that nearly all individuals could be exposed for up to 1 h without experiencing or developing life-threatening health effects), and (iii) Temporary Emergency Exposure Limit (TEEL-3—the maximum concentration in air below which it is believed nearly all individuals could be exposed without experiencing or developing life-threatening health effects).

Of the 126 industrial chemicals listed in Table 2 (toxic by inhalation), 44 (35%) have LC_{50} values that are less than or equal to 10 ppm, which is extreme toxicity by the Hodge and Sterner/Hine and Jacobsen scale. Similarly, 9 of the 31 chemicals (29%) listed in Table 3 (toxic by dermal exposure) have toxicities at or below 5 mg/kg or 5 μl/kg (or extreme toxicity). More than half (16 of 31) of the dermally toxic chemicals have pesticidal uses, while those identified with inhalation toxicity typically are used exclusively in industrial/commercial applications. There are nine isocyanates (e.g. trifluoromethoxyphenylisocyanate, 4-chlorophenyl isocyanate, and toluene diisocyanate), eight halophosphorus compounds (e.g. isoflurophate, diethyl phosphorochloridothionate, and thiophosphoryl chloride), and seven acid chlorides (e.g. trichloroacetyl chloride, thiophosphoryl chloride, and 1,1,2,2-tetrachloroethylsulfenyl chloride) in Table 2. The chemicals in the two tables range from reactive inorganic (e.g. chlorine, tin tetrachloride, and diborane) and organic (e.g. methyl fluorosulfate, methyl hydrazine, and acrolein) chemicals to organometallics (e.g. methylcyclopentadienyl manganese tricarbonyl (MMT), bromotributyltin, and diethyl telluride), hydrides (e.g. arsine, phenyl phosphine, and phosphine), and α-halo compounds (e.g. chloroacetaldehyde, 1,3-dichloroacetone, and methyl fluoroacetate).

This wide range of chemical structures presents a problem for both industry and government to develop sensors that are sufficiently specific and sensitive to detect this range of chemicals at concentrations that will be protective of the potentially exposed populations. Specificity is important because, as noted above, one of the problems facing emergency response and medical teams is if the identity of the chemical used in a terrorist event is unknown, medical countermeasures appropriate to the toxicant used may not be available or even considered by the teams. Therefore, developing sensors specific to a chemical or class of chemicals (e.g. isocyanates) or capable of identifying the sensed chemical (e.g. gas chromatograph/mass spectrometers) is a critical

TABLE 4 Membership on Regulatory Lists and Guideline Values for Chemicals Toxic by the Inhalation Route

CAS No.	RTECS No.	Chemical Name	DHS	OPCW	US HPV	EU HPVC	AEGL-3 10 min (ppm)	ERPG-3 (ppm)	TEEL-3
1752-30-3	AL2338000	Acetone thiosemicarbazide							100 mg/m^3
107-02-8	AS9660000	Acrolein	•		•	•	6.2		
2937-50-0	LV7168000	Allyl chloroformate							
26842-43-3	SF1435000	2-Amino-2,4-dimethylpentanenitrile							
19355-69-2	UD5155000	2-Aminoisobutyronitrile			•				
7647-18-9	CA1722000	Antimony pentachloride							125 mg/m^3
7784-42-1	CC9968000	Arsine	•				0.91		
98-87-3	CW9716000	Benzal chloride			•	•			500 mg/m^3
100-44-7	XW9324000	Benzyl chloride			•	•		25	
55-91-4	TC3948000	Bis(1-methylethyl) phosphorofluoridic acid ester							3.6 mg/m^3
111-44-4	KX1708000	Bis(2-chloroethyl) ether			•				100 ppm
542-88-1	KX1750000	Bis(chloromethyl) ether	•					0.5	
1461-23-0	WR3458000	Bromotributyltin							
78-94-4	EX4312000	3-Buten-2-one							0.25 ppm
4262-43-5	CD1127000	*t*-Butylarsine							
75-87-6	FW7714000	Chloral			•	•			250 mg/m^3
7782-50-5	FX8204000	Chlorine	•		•	•	50		
7790-91-2	FX8372000	Chlorine trifluoride	•				84		
13637-63-3	FX8386000	Chlorine pentafluoride	•						60 ppm
107-20-0	AB4928000	Chloroacetaldehyde							45 ppm
79-11-8	AH6664000	Chloroacetic acid			•	•			20 ppm
78-95-5	UG8134000	Chloroacetone			•				7.5 ppm
2315-36-8	AC3934000	2-Chloro-*N*,*N*-diethylacetamide							
1622-32-8	KT2954000	2-Chloroethanesulfonyl chloride							150 mg/m^3
107-07-3	KU7910000	2-Chloroethanol				•			7 ppm

107-27-7	PA7966000	Chloroethyl mercury							2.5 mg/m^3
107-30-2	KX3066000	Chloromethyl methyl ether	•		•		2.6		
106-48-9	SM5446000	*p*-Chlorophenol				•			400 mg/m^3
104-12-1	NZ5348000	4-Chlorophenyl isocyanate							
2909-38-8	NZ5334000	3-Chlorophenyl isocyanate							
76-06-2	PD3906000	Chloropicrin	•	•				1.5	
26447-14-3	UD4312000	Cresyl glycidyl ether				•			
4170-30-3	HB4606000	Crotonaldehyde	•		•	•	44		
460-19-5	HD6146000	Cyanogen	•						15 ppm
675-14-9	YC3052000	Cyanuric fluoride							75 mg/m^3
17702-41-9	HP5880000	Decaborane							15 mg/m^3
56-18-8	KC1302000	3,3′-Diaminodipropylamine							
105-83-9	KC1316000	3,3′-Diamino-*N*-methyldipropylamine							
19287-45-7	IB2688000	Diborane	•				7.3		
534-07-6	UG8862000	1,3-Dichloroacetone							2 mg/m^3
110-57-6	EW9534000	*trans*-1,4-Dichloro-2-butene			•				7.5 ppm
79-35-6	LD6370000	1,1-Dichloro-2,2-difluoroethylene							
22591-21-5	EW3934000	1,1-Dichloro-3,3-dimethyl-2-butanone			•				
62-73-7	TB3360000	Dichlorvos			•				100 ppm
1464-53-5	EU5824000	1,2:3,4-Diepoxybutane							10 ppm
2524-04-1	TB8148000	Diethyl phosphorochloridothionate			•	•			
627-44-1	PA8694000	Diethylmercury							2.5 mg/m^3
627-54-3	KT2996000	Diethyltelluride							
28178-42-9	CV6664000	2,6-Diisopropylphenyl isocyanate							
624-92-0	KE2380000	Dimethyl disulfide			•	•		250	

(*continued overleaf*)

TABLE 4 (*Continued*)

CAS No.	RTECS No.	Chemical Name	DHS	OPCW	US HPV	EU HPVC	AEGL-3 10 min (ppm)	ERPG-3 (ppm)	TEEL-3
2524-03-0	TB8176000	Dimethyl phosphorochloridothionate			•	•			150 mg/m^3
77-78-1	WX4466000	Dimethyl sulfate			•	•			7 ppm
2867-47-2	PC0994000	2-(Dimethylamino)ethyl methacrylate			•	•			
2439-35-2	AT2044000	2-(Dimethylamino)ethyl acrylate			•	•			
24424-99-5	ID4235000	Di-*tert*-butyldicarbonate							
106-91-2	PC1022000	2,3-Epoxypropyl methacrylate			•				
541-41-3	LV7252000	Ethyl chloroformate			•	•			10 ppm
2941-64-2	FQ5460000	Ethyl chlorothioformate			•				
1498-64-2	TB9142000	*O*-Ethyl dichlorothiophosphate							
1498-51-7	TB9114000	Ethyl phosphorodichloridate			•				
151-56-4	LE0924000	Ethylenimine	•		•	•	51		
24468-13-1	FQ5330000	2-Ethylhexyl chloroformate			•	•			
7782-41-4	LR2170000	Fluorine	•				39		
371-62-0	KV2324000	2-Fluoroethanol							1.25 ppm
79-14-1	MG6160000	Glycolic acid			•	•			0.75 mg/m^3
77-47-4	HK1106000	1,2,3,4,5,5-Hexachloro-1,3-cyclopentadiene			•	•			0.0179 ppm
18406-41-2	JW7266000	Hexamethoxydisilylethane							
74-90-8	MY6300000	Hydrogen cyanide	•	•	•	•	27		
7783-66-6	NW2408000	Iodine pentafluoride	•						
13463-40-6	NX6062000	Iron carbonyl	•		•	•	0.23		
30674-80-7	PC1218000	2-Isocyanatoethyl methacrylate						1	
10471-78-0		2-Isopropenyl-2-oxazoline							
68-11-1	AJ4690000	Mercaptoacetic acid			•	•			
920-46-7	PC1750000	Methacryloyl chloride							6 ppm
558-25-8	PD2016000	Methanesulfonyl fluoride							3.49 ppm

79-22-1	FQ5362000	Methyl carbonochloridate	•		•	•			4 ppm
453-18-9	AJ2058000	Methyl fluoroacetate							5 mg/m^3
421-20-5	LU2884000	Methyl fluorosulfate							0.25 ppm
60-34-4	MX5208000	Methyl hydrazine	•				16		
624-83-9	NZ5754000	Methyl isocyanate	•		•		1.2		
676-97-1	SZ9660000	Methyl phosphonic dichloride	•	•					15 mg/m^3
12108-13-3	OV9576000	2-Methylcyclopentadienyl manganese tricarbonyl (MMT)			•				7.5 mg/m^3
35203-06-6	CT7938000	*N*-Methylene-6-ethyl-2-methylaniline			•				
140-76-1	UZ4004000	2-Methyl-5-vinylpyridine							40 mg/m^3
7786-34-7	HB6104000	Mevinphos							
7783-77-9	QK5418000	Molybdenum hexafluoride							
13463-39-3	RB0042000	Nickel carbonyl	•				0.46		
10102-44-0	RE9912000	Nitrogen dioxide					34		
10544-72-6	RF0322000	Nitrogen tetroxide	•			•			20 ppm
62-75-9	JF9856000	*N*-Nitrosodimethylamine							19 mg/m^3
57-57-8	RX2352000	2-Oxetanone							15 ppm
10028-15-6	SA2240000	Ozone							5 ppm
376-53-4	AV2842000	Perfluoroadiponitrile							
376-89-6	ME8988000	Perfluoroglutaronitrile							
108-95-2	SL9170000	Phenol			•	•		200	
103-71-9	CY2520000	Phenyl isocyanate				•			
140-29-4	AL4130000	Phenylacetonitrile				•			30 mg/m^3
2687-12-9	CW3633000	1-Phenyl-3-chloro-1-propene							
638-21-1	SY6930000	Phenylphosphine							15 ppm
298-02-2	TC1092000	Phorate							0.6 mg/m^3

(continued overleaf)

TABLE 4 (*Continued*)

CAS No.	RTECS No.	Chemical Name	DHS	OPCW	US HPV	EU HPVC	AEGL-3 10 min (ppm)	ERPG-3 (ppm)	TEEL-3
7803-51-2	SY4830000	Phosphine	•				7.2		
10025-87-3	TD6832000	Phosphorus oxychloride	•	•	•	•	1.1		
1185-09-7		1,1,2,2-Tetrachloroethylsulfenyl chloride							
5216-25-1	XX3766000	α,α,α-*p*-Tetrachlorotoluene			•	•			
78-00-2	TU5040000	Tetraethyl lead			•	•			60 mg/m^3
597-64-8	WR6118000	Tetraethyltin							50 mg/m^3
2778-42-9	CV6454000	Tetramethyl-*m*-xylylene diisocyanate			•				
509-14-8	PD2464000	Tetranitromethane	•				2.2		
294-93-9	XJ3220000	1,4,7,10-Tetraoxacyclododecane							
108-98-5	DB3304000	Thiophenol			•				3.5 ppm
3982-91-0	XS4662000	Thiophosphoryl chloride				•			
7646-78-8	XU7588000	Tin tetrachloride			•	•			200 mg/m^3
7550-45-0	XV3248000	Titanium chloride	•			•	38		
584-84-9	CX0518000	2,4-Toluene diisocyanate	•		•		0.65		
1321-38-6		Toluene diisocyanate (mixture of isomers)							
921-03-9	UH4522000	Trichloroacetone							
76-02-8	AO0392000	Trichloroacetyl chloride							6 ppm
594-42-3	PC6468000	Trichloromethanesulfenyl chloride	•		•	•	1.6		
98-07-7	XX3976000	α,α,α-Trichlorotoluene			•	•			25 mg/m^3
998-30-1	WI6062000	Triethoxysilane							115 ppm
35037-73-1		Trifluormethoxyphenylisocyanate							
98-16-8	XX8456000	α,α,α-Trifluoro-*m*-toluidine			•	•			150 mg/m^3
2487-90-3	WI6188000	Trimethoxysilane			•			5	
141-59-3	SF2044000	2,4,4-Trimethyl-2-pentanethiol							
15112-89-7	WI5754000	Tris(dimethylamino)silane							
100-43-6	VA3262000	4-Vinylpyridine			•				

TABLE 5 Membership on Regulatory Lists and Guideline Values for Chemicals Toxic by the Dermal Route

CAS No.	RTECS No.	Chemical Name	DHS	US HPV	EU HPVC	AEGL-3 10min (mg/m^3)	TEEL-3
75-86-5	ON3024000	Acetone cyanohydrin	•	•	•	27	
309-00-2	JD4970000	Aldrin					25 mg/m^3
107-11-9	BA4858000	Allylamine	•			150	
28772-56-7	GZ7154000	Bromadiolone					1 mg/m^3
95465-99-9	TC3388000	Cadusafos					
8065-48-3	TC8792000	Demeton					10 mg/m^3
814-49-3	TB7994000	Diethyl chlorophosphate					8 mg/m^3
926-64-7	AL3304000	Dimethylaminoacetonitrile					
1122-58-3	UZ0910000	4-Dimethylaminopyridine					
77-77-0	KW9058000	Divinyl sulfone					
2104-64-5	TA6020000	Ethoxy-4-nitrophenoxy-phenylphosphine sulfide (EPN)					5 mg/m^3
13194-48-4	TC3416000	Ethoprop					50 mg/m^3
107-16-4	AL3556000	Glycolonitrile		•			4 ppm
105-31-7	MT5768000	1-Hexyn-3-ol					
78-97-7	ON2940000	Lactonitrile		•			150 mg/m^3
950-10-7	KF1386000	Mephosfolan					9 mg/m^3
126-98-7	UI2520000	Methacrylonitrile	•			32	
453-18-9	AJ2058000	Methyl fluoroacetate					5 mg/m^3
556-61-6	PC6272000	Methyl isocyanate		•			500 mg/m^3
3680-02-2	WW8330000	Methyl vinyl sulfone					
7786-34-7	HB6104000	Mevinphos					4 mg/m^3
5903-13-9	AE0938000	Nissol					
311-45-5	TB4298000	Paraoxon					
56-38-2	TC9772000	Parathion			•		10 mg/m^3
298-02-2	TC1092000	Phorate					0.6 mg/m^3
947-02-4	NP4102000	Phosfolan					9 mg/m^3
107-19-7	UO7966000	Propargyl alcohol		•	•		60 ppm
3689-24-5	XS8050000	Sulfotep					10 mg/m^3
13071-79-9	TC0350000	Terbufos					1 mg/m^3
2001-95-8	YZ0098000	Valinomycin					2.5 mg/m^3
3984-22-3	JX5551000	2-Vinyl-1,3-dioxolane					

factor. In addition, deployed sensors must be sufficiently sensitive to ensure that the potentially exposed population is protected while not producing false alarms that may occur if the sensor is overly sensitive (in this context, if a chemical is correctly identified by a sensor, but the concentration is below the concern level, then the detect should be considered a false positive because it will produce an unnecessary alarm). Other performance parameters, including response times, power requirements, service life, maintenance requirements, calibration needs, size, and detection, range all factor into the suitability of a sensor or sensor suite for use in homeland security applications to alert the population to a potential attack.

While it is not the purpose of this article to identify specific sensor technologies or current capabilities, two examples of the above discussion are provided. Table 2 identifies nine isocyanates that meet the Hodge and Sterner/Hine and Jacobsen criteria (i.e. 3-chlorophenyl isocyanate, 4-chlorophenyl isocyanate, 2,6-diisopropylphenyl isocyanate, 2-isocyanatoethyl methacrylate, methyl isocyanate, phenyl isocyanate, tetramethyl-*m*-xylylene diisocyanate, trifluormethoxyphenylisocyanate, and toluene diisocyanate). As a chemical class, therefore, isocyanates appear to be highly extremely toxic and shows high priority for sensor development (eight of the nine isocyanates are in the extremely toxic range). In addition, isocyanates and specifically methyl isocyanate have been involved in industrial accidents that highlight their danger and ability to cause harm to a large population (e.g. the 1984 release of methyl isocyanate in Bhopal, India). Current commercially available sensor technologies that are capable of distinguishing isocyanates and provide quantification, however, appear to be limited to paper tape technologies (the paper technology also appears to be specifically limited to diisocyanates). There is therefore a critical need to develop all types of sensors that are capable of identifying and quantifying isocyanates and deployable in homeland security applications.

In contrast to the isocyanates, the four hydrides presented in Table 2 (arsine, diborane, decaborane, and phosphine) appear to have over 20 adequate commercial sensors that are able to detect three of the four chemicals (no commercial sensors were found for decaborane). These hydrides show LC_{50} of 5.1–46 ppm for a 4-h exposure and all sensors have sensitivities that are well below the regulatory values for these chemicals, response times that are 1 min or less, and high selectivities.

In conclusion, there are more than 150 highly or extremely toxic commercially available compounds that were identified using a relatively restrictive definition of high and extreme toxicity. These compounds span a wide range of structures including isocyanates, halogens, acid chlorides, silanes, allylic/benzylic halides, nitrogen oxides, and phosphate esters. The diversity of chemical structures present a challenge for manufacturers to design suites of sensors that will be able to both identify and quantify these chemicals in homeland security applications. These include, for example, applications that will inform emergency responders and medical teams treating exposed populations about the identity of the chemical used so that the most effective medical countermeasures can be administered quickly. The identified chemicals additionally should provide an initial list that can be used to direct research efforts focused on developing new medical countermeasures, developing new "greener" synthetic methods and replacement chemicals that will eliminate the need for these toxic chemicals, and developing more robust security methods to protect the public from the threat of a terrorist event using these chemicals.

ACKNOWLEDGMENTS

The author wishes to thank Dr Richard Niemeier of the National Institute for Occupational Safety and Health, and Drs Patricia McGinnis and Peter McClure of Syracuse Research Corporation for enlightening discussions too numerous to count about toxicology and methods to identify toxic chemicals; Dr Benjamin Garrett of the Federal Bureau of Investigation for enlightening discussions about this article; Dr Lara Chappell for help in preparing some of the lists presented here, and Mr Brian Krafthefer of Honeywell International and Dr John Raymonda of CUBRC for discussions and information on sensors and sensor technology while developing a report for the National Technology Alliance (a US Government program).

REFERENCES

1. Shea, D. A., and Gottron, F. (2004). *Small-scale Terrorist Attacks Using Chemical and Biological Agents: An Assessment Framework and Preliminary Comparisons*, Congressional Research Service report for Congress. Order Code RL32391.
2. Karasik, T. (2002). *Toxic Warfare*, Rand Corporation. ISBN 0833032070. http://www.rand.org/publications/MR/MR1572/. September 21, 2005.
3. Cave, D., and Fadam, A. (2007). *The Reach of War; Iraq Insurgents Employ Chlorine in Bomb Attacks*, New York Times, New York, p. A1, 22 February 2007.
4. Partlow, J. (2007). *Baghdad Plan Has Elusive Targets: U.S. Patrols Still Unable to Tell Friend From Foe*, Washington Post, Washington, DC, p. A01, 26 February 2007.
5. Sharp, P. A. (2005). 1918 Flu and Responsible Science. *Science* **310**, 17.
6. Hauschild, V. D., and Bratt, G. M. (2005). Prioritizing industrial chemical hazards. *J. Toxicol. Environ. Health Part A* **68**, 857–876.
7. Hodge, H. C., and Sterner, J. H. (1949). Tabulation of toxicity classes. *Am. Ind. Hyg. Assoc. Q.* **10**, 93–96.
8. Hodge, H. C., and Sterner, J. H. (1956). Combined tabulation of toxicity classes. In W. S. Spector, Ed. *Handbook of Toxicology*, Vol. 1 – Acute Toxicities of Solids, Liquids and Gases to Laboratory Animals, W.B. Saunders Company, Philadelphia, PA.
9. Hine, C. H., and Jacobsen, N. W. (1954). Safe handling procedures for compounds developed by the petro-chemical industry. *Am. Ind. Hyg. Assoc. Q.* **15**, 141–144.
10. Department of Homeland Security (DHS) (2007). Chemical Facility Anti-Terrorism Standards. Interim Final Rule. *Fed. Regist.* **72**, 17687–17745.

2D-TO-3D FACE RECOGNITION SYSTEMS

Michael I. Miller, Marc Vaillant, William Hoffman, and Paul Schuepp
Animetrics, Conway, New Hampshire

1 INTELLIGENT VIDEO SYSTEMS

1.1 The Need for Intelligent Video Systems

Intelligent image and video interpretation systems of the future will be highly integrated with emergent video databases interacting with real-time access control and surveillance. The intelligent video surveillance software market, including video analysis, is experiencing meteoric growth. Airports, borders, ports, energy plants, historical buildings, monuments, manufacturing plants, retail establishments, and businesses all require access control and surveillance video solutions. Forrester predicts that 40% of businesses will need integrated security. The access control market is expected to reach nearly 14 billion dollars in 2009 [1]. Ultimately, these systems will integrate with and allow for the retrieval and cueing of the massive data stores such as the FBI's archives that contain both annotated as well as un-annotated video resources.

Figure 1 depicts an access control and video surveillance system handling the identities and monitoring dynamically the locations of individuals. According to ABI Research [2], the video surveillance market, already $13.5 billion as of 2006, will grow to $46 billion by 2012. The goal of spotting individuals of particular identities, and indexing and analyzing video archives is a fundamental challenge to the future of noncooperative FR. Solving the problem of dynamically identifying faces from live streaming or stored video will enable integrated, intelligent security solutions of the future.

The principal focus of this article is to describe the emerging 2D-to-3D technologies for extending FR systems historically applied to document verification and access control to the uncontrolled environments, which require pose and lighting invariant ID such as is required for tracking and recognizing faces in noncooperative surveillance and video applications.

1.2 The Barrier to Face

The surveillance environment has posed great challenges for FR technologies. Due to FR's nonintrusive nature, it has been pushed to the forefront as the biometric of choice. The International Civil Aviation Organization has embraced FR as its biometric standard. Yet, we must emphasize that, to date, the facial biometric has not adequately penetrated this marketplace. Presently, most systems do not incorporate FR biometrics. The major difficulty is that its advantage—its noninvasive, touch-free nature—is also its daunting challenge. Comparing it to fingerprint for the moment, imagine a "touch-free" fingerprint

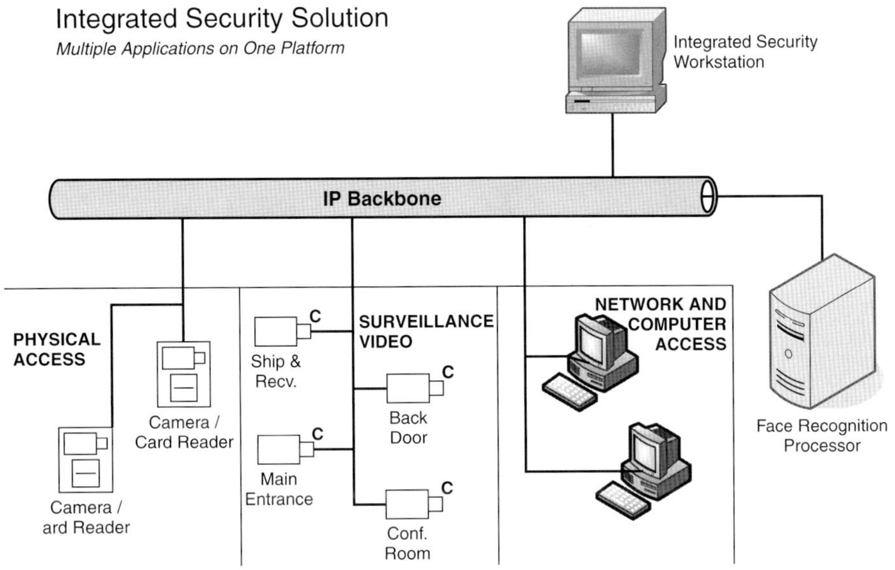

FIGURE 1 An intelligent video surveillance system.

system in which any random medium—perspiration, oil, sand, grit—you name it, could be between the finger and the sensor. Alternatively, imagine that the finger could be at any "distance to the sensor", and could be occluded by gloves or bandages. Could we expect to deploy fingerprint biometrics in such conditions and require "constant performance independent of environmental variables"—hardly.

Such issues directly confound the successful deployment of FR technologies including the huge range of camera qualities (CCTV, Webcams, high resolution imagery, etc.), the infinite variety and schema of environmental lightings, arbitrary subject's positioning, facial hair, ornaments such as eyeglasses and jewelry, and complex backgrounds. The left column of Figure 2 depicts control photographs associated with access and document verification. The right column shows uncontrol photographs with the confounding variations of lighting and complex backgrounds. Figure 3 shows the results from the Facial Recognition Grand Challenge (FRGC) 2005 report [3] depicting the gap in performance between controlled and uncontrolled data by FR systems. FR performance is studied worldwide by examining false reject rate (FRR) or verification rate (1-FRR) as a function of false accept rate (FAR). The table lists out the FRRs in percentages at an FAR of 0.001 meaning the systems only accepted people incorrectly 1 out of 1000 times. The difference in median performance of participants is 60% between the control and uncontrol.

The confounding challenges of the uncontrolled environment resulting in this gap must be solved in order to make effective use of existing video recordings and to embark upon "action oriented" intelligent analytical systems that will provide next generation security methods. This article examines how 2D-to-3D technologies provide the crucial technological link from the historical application of controlled front facial recognition for access control and document verification to the intelligent systems of video surveillance.

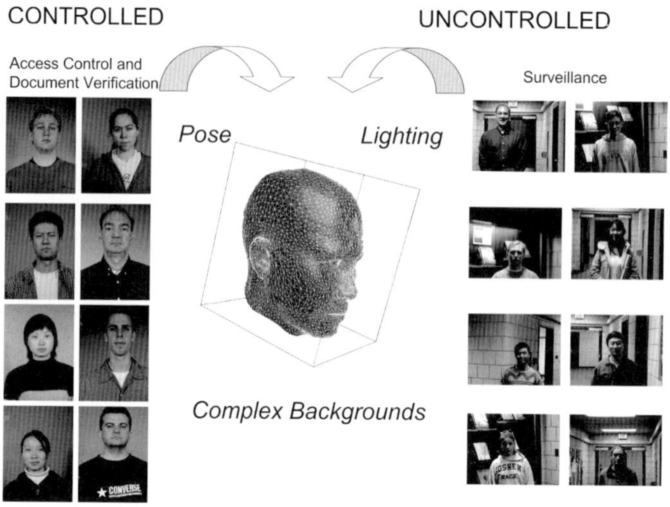

FIGURE 2 Control photographs associated with access control and document verification (left column), and uncontrol photographs (right column) with the confounding variations.

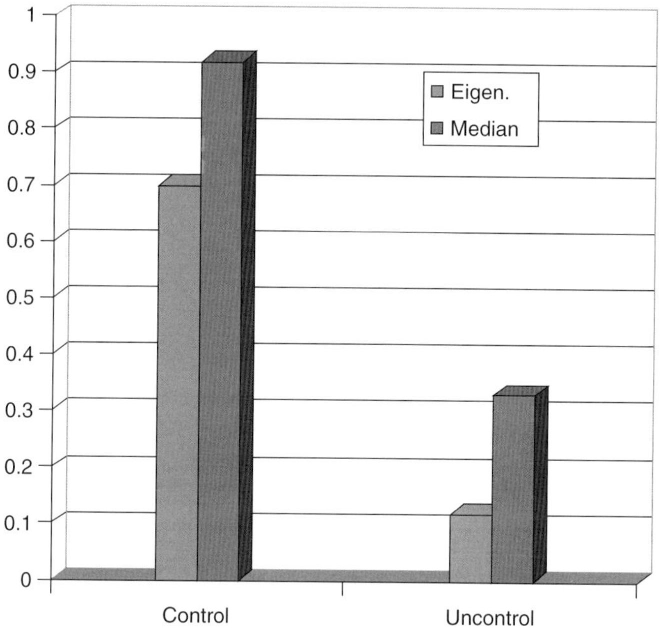

FIGURE 3 Gap in Performance of FR Systems, June 2005 FRGC for Control versus Uncontrol.

1.3 2D-to-3D Bridges the Performance Gap for Intelligent Video Systems

FR today works well in a "controlled setting" where the camera-person interaction is cooperative, illumination is monitored, and backgrounds are simple and uncluttered. Facial identification systems and face trackers have been in use for at least a decade. Typically, facial identification systems comprise detection and identification systems based on the manipulation of 2D likenesses of faces, which represent photometric and geometric variation robustly as manifest in the 2D likeness.

The "uncontrolled" surveillance environment introduces uncooperative subjects where facial pose is relatively arbitrary, lighting is infinitely variable, and backgrounds are arbitrarily complex. Next generation FR must accommodate these kinds of variations for successful transition from the controlled "checkpoint" access application to the "uncontrolled" surveillance video application. Purely 2D legacy FR technologies are limited in their deployment to the more controlled environments of access control and document verification. Figure 4 shows an example of 2D representational lattice model used in many of the legacy 2D FR systems.

Since 2D systems are limited to the manipulation of 2D geometric variations of in-plane geometric variation, they can be used for tracking and/or identification of faces while accommodating in-plane variation. However, they degrade as the target subjects are viewed out of plane. Since they rely on 2D likeness they cannot be robust to changes in both photometric and geometric variation, which depend on the 3D shape of the face and its interaction with variations in the external lighting illumination. 2D-to-3D technology provides a unified technological infrastructure for addressing all of these technical challenges, accommodating simultaneous high performance for imaging volumes at a physical access checkpoint (a more "controlled" scenario), as well as the "image at a distance" surveillance scenario.

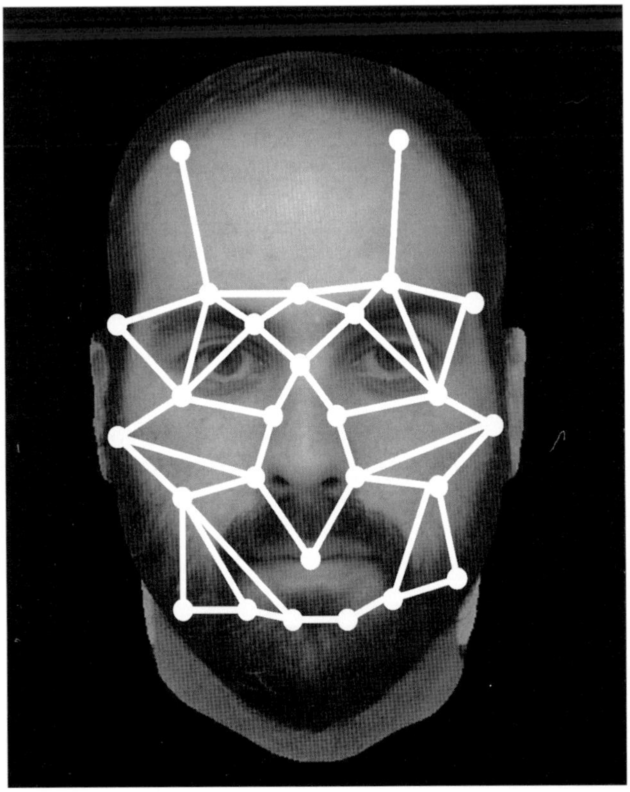

FIGURE 4 2D representation used in many legacy FR systems.

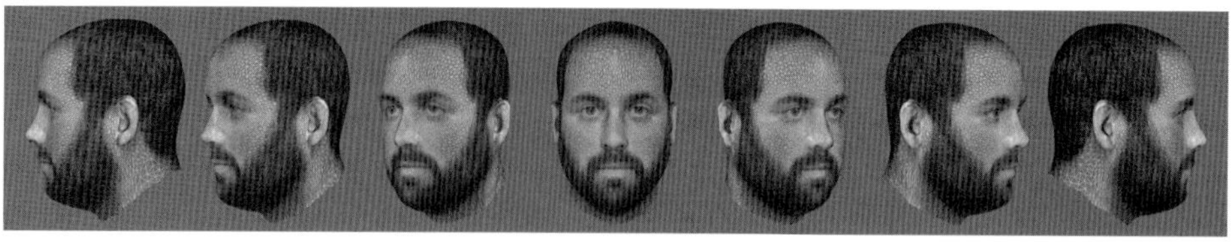

FIGURE 5 The multiple views associated with the 2D-to-3D geometric model.

Figure 5 depicts a 3D geometric model and texture representing the 2D photograph resulting from the 2D-to-3D technology. It is the core 3D data structure generated from a 2D image that can provide the opportunity for FR systems to pass between controlled frontal interactions between camera probing and arbitrary positions of uncalibrated surveillance cameras. The 3D geometric data structures unify (i) the uncalibrated nature of camera position to viewer allowing for identification to be invariant to pose (position and rotation), (ii) the infinite variety of variations introduced via the external variability of the lighting world can be accommodated via the representation of the light field associated with the observed luminance on the geometry, and (iii) the dynamic coherence of video is directly encoded into the rigid motions associated with the 3D object representation.

2 COMPUTATIONAL ANATOMY AND DIFFEOMORPHISMS FOR 2D-TO-3D MODEL GENERATION

Geometry based 3D facial ID systems are robust to geometric variation of the pose of individual faces and efficiently detects and identifies faces from projective imagery such as measured by conventional video cameras. The key technological advance required for their application using conventional projective imaging cameras lies in a system to automatically determine the 3D geometry of a person's face with a finite set of images or photographs or through the analysis of a persistent set of images from video data. The technological term being used to describe the generation of 3D geometries from single or multiple projective planar images is *2D-to-3D geometric model generation*. Knowledge of the 3D geometry of a subject allows for automated understanding of variables, such as photometric and chromatic characteristics of the environment or particular video, occlusion compensation, and pose. The use of a structured 3D model allows for more finite analysis of human faces for expression analysis and direction of gaze for any pose.

2.1 Computational Anatomy

Advances in 2D-to-3D technologies championed by companies such as Animetrics are based on recent inventions in computational anatomy [4–9] that uses mathematical algorithms to compute equations of motion, which form the basis for biometrics in the study of the shape and form of biological structures. These equations generate metrically precise geometries, which can be used for recognition and identification purposes.

In computational anatomy, 3D analysis of deformable structures provides a natural platform for the merging of coarse features associated with detection of objects, proceeding to the finest scales associated with full recognition of objects via deformable templates. The inference setting in computational anatomy has been almost exclusively in 3D; no special preference is given to particular poses of objects such as anatomical structures. The construction of geometric shapes is controlled by the dimension of the input imagery, composed of 0 (point), 1 (curves), 2 (surfaces), and 3 dimensional (volumes) imagery. The smoothness of the mappings is enforced through their control via the classic Eulerian ordinary differential equation for smooth flows of diffeomorphisms.

Computational anatomy studies anatomical manifolds and natural shapes in imagery using infinite dimensional diffeomorphisms, the natural extension of the finite dimensional matrix groups (rotation, translation, scale, and skew). Shapes and imagery are compared by calculating the diffeomorphisms on the image and volume coordinates $X \subset R^2, R^3$, which connect one shape to the other via the classical Eulerian flow equations:

$$\frac{d\phi_t}{dt} = v_t \circ (\phi_t) \tag{1}$$

The space of anatomical configurations is modeled as an orbit of one or more templates under the group action of diffeomorphisms:

$$I = \{I : I = I_{\text{temp}} \circ (\phi)\} \tag{2}$$

The unknown facial anatomical configurations are represented via observed imagery I^Dconsisting of points, curves, surfaces, and subvolumes. They are compared by defining a registration distance; $d : \left(I, I^D\right) \mapsto R^+$. Then the mapping, or correspondence, between anatomical configurations is generated by solving the minimum energy flow connecting the two coordinate systems and measuring the length or square root energy:

$$\inf_{\dot{\phi}=v(\phi)} \int_0^1 \|v_t\|_V^2 \, dt \text{ such that } d\left(I_{\text{temp}} \circ (\phi_1^{-1}), I^D\right) = 0 \tag{3}$$

where $\|v_t\|_V^2$ measures the energy of the vector in the smooth Hilbert space V with norm-square $\|v_t\|_V^2$. The computation performed becomes the matching of one object to the other by solving the minimization:

$$\inf_{\dot{\phi}=v(\phi)} \int_0^1 \|v_t\|_V^2 \, dt + \beta d^2 \left(I_{\text{temp}} \circ (\phi_1^{-1}), I^D\right) \tag{4}$$

where β controls the weight. The mathematics of such equations are examined in our work [4–9].

Shown in Figure 6 are the results of using the Computational Anatomy equations for deforming one surface manifold to another [20, 21]. The top row shows the flow of surfaces under diffemorphisms solving equation 4 mapping the template surfaces (panel 1) onto the target surface (panel 6). The bottom row (dashed line) shows the histogram of vertex distances between the original template and target surfaces (after rigid registration). The solid line shows the distance between vertices after diffeomorphic correspondence.

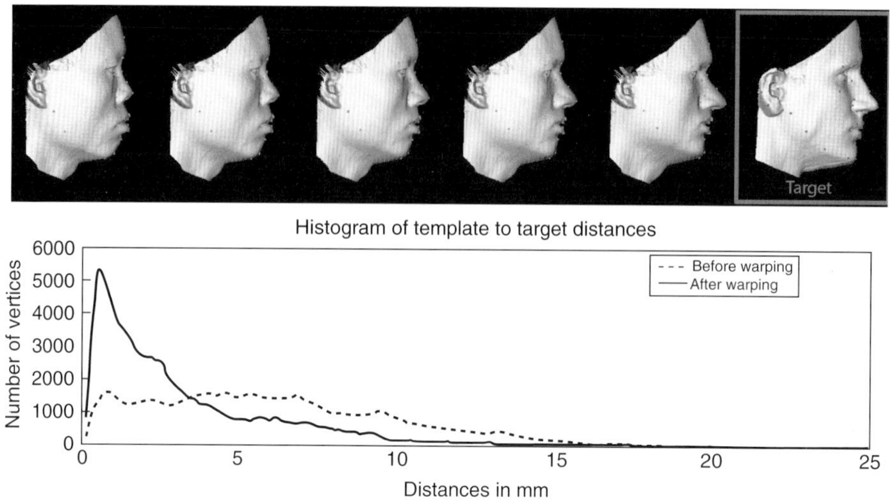

FIGURE 6 Top row shows a flow of surfaces under diffeomorphisms solving Eq. (4) mapping the template surface (panel 1) onto the target surface (panel 6). Bottom row dashed line shows the histogram of vertex distances between the original template and target.

2.2 One and Two-View Geometry Generation

The work in computational anatomy for tracking and deformation of objects is generally in 3D, rather than manipulating motions in the image plane, working with the full Euclidean motions in space and time [10–14]. The power of such 3D approaches is that they are not limited to estimated translation and rotational motion in the image plane, therefore they imply complete invariance to full 3D pose, scale, and translation. The 2D-to-3D model geometry generation technology enables the conversion of the two-dimensional photo into a three-dimensional geometry, which can then be rotated and seen from any view. To accomplish this, the three-dimensional equations of geometric motion associated with biological shape are integrated with the projective equations of flat 2D imagery. Through proper integration of analysis of the two-dimensional images into features and a geometrically correct process for smooth deformation, it is possible to generate realistic, accurate, fully structured, and defined three-dimensional geometric models. They are combined with the projective equations accommodating

$$P : (x, y, z) \rightarrow p(x, y, z) = \left(\frac{\alpha_1 x}{z}, \frac{\alpha_2 y}{z} \right) \tag{5}$$

associated with video imagery. The core problem, at least for deformable objects, is to infer the flow of the dense volume representations of the 3D geometric objects from 2D projective imagery. The inference problem is to estimate the low-dimensional matrix group for tracking associated with rigid body dynamics, as well as work on high-dimensional transformations for tracking deformable motions (such as facial expressions), and to incorporate information obtained from projective geometry.

Diffeomorphic mapping of anatomical configuration leverages progress in computational anatomy [4–9] supported by the National Institutes of Health for the study of anatomical structure. The methods for constructing geometric models from 2D manifold representations of the face via triangulated graphs estimates diffeomorphic mapping

FIGURE 7 The 2D-to-3D pipeline depicting the projective photograph (panel 1), face and eye detection, and two-dimensional feature analysis in the projective plane (panel 3), and the 3D geometric model generation in register and rotation with the given photograph (panel 4).

of templates onto pictures from single or multiple views of the individual invariant to camera pose. The algorithms incorporate both sparse and dense image information using mapping algorithms developed in computational anatomy. Figure 7 depicts the basic 2D-to-3D geometric model generation technology pipeline. The basic components required are two-dimensional image feature analysis (2DIFA), including eye detection and fiducial feature, and curve and subarea delineation. This is depicted in panel 3 of Figure 7. Once the 2DIFA is completed, the 3D model consisting of a triangulated mesh or quadrilateral representation is generated, which corresponds to the features generated with the photograph. The triangulated mesh model is depicted in panel 4 of Figure 7.

2.3 Statistical Validation of 2D-to-3D Model Generation

Figures 8, 9 and 10 show results from generating a single merged geometry from front and side view high resolution photographs. Shown in Figure 8 is an example of single textured geometry generated from a front and side view. The left column of Figure 8 shows the two input photographs; the middle column shows the textured model. The right column shows the front and side views of the single geometry. Figure 9 depicts a comparison of a subset of the anatomically defined fiducial marks superimposed over the manually defined features.

To examine the accuracy of the created mdoel geometries, 130 fiducial points have been attached to each model. These fiducial points are then compared to their positions in the image plane as viewed through the projective geometry. To illustrate, the points taken from the geometric model projected into the projective plane are shown in Figure 9. Superimposed over these model projected points are hand labelled manual points in the image plane for those features.

The geometric model generation produces a three-dimensional model from two-dimensional images by solving Eqs. 1–5. The accuracy of the models is essential for their accurate use for reorientation and normalization for improving identification rates with off-pose data. The geometric accuracy of the multiview geometry generation in the projective plane can be demonstrated via generation of hand featured databases with trained manual raters. Manual raters were trained to delimit in the order of 20 highly recognizable features in the projective imagery for front and 10 features for side view data. On each 3D model 130 fiducially identifiable points were defined on the surface of the volume. These points are the anatomical markers for the accuracy analysis. Each 3D model generated from the photographs was placed into the 3D space associated with the rigid motion generated for the solution of each variational problem.

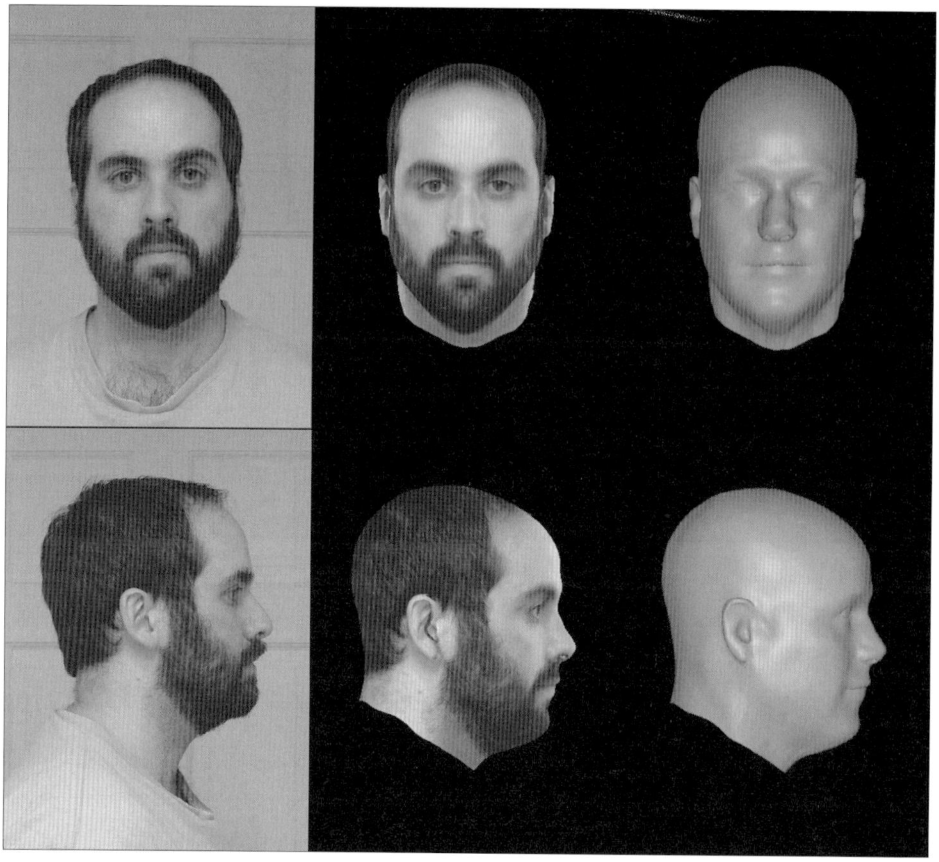

FIGURE 8 An example of single textured geometry generated from a front and side view: column 1 shows two photographs, column 2 the textured geometry, and column 3 the geometry in grayscale.

Using the projective equation, the 130 fiducial markers were projected into the image plane (if not occluded) and were compared to each of a subset of the 40 identifiable features that were manually marked in the respective photograph. The root mean square error (RMSE) was calculated for each of the manually labeled photographs on which the experiment was performed. Figure 10 shows a single textured photograph with an associated 3D geometry which is in register with the photograph Each geometry consists of a triangulated mesh of vertices and triangles with associated normals. Every geometry generated has a minimum of 130 labeled features that are in correspondence with features which can be defined in the projective image plane.

2.4 Root Mean Squared Error on Controlled and Uncontrolled Imagery

Validations have been performed on several standard databases. Figure 11 shows the results of the face recognition grand challenge (FRGC) controlled and uncontrolled data validation. Upwards of 500 total geometries were generated from the control and uncontrol photographs using Eqs. 1–5. Each were hand featured for comparison for statistical validation. For the control data, the average RMSE (square root of the sum of the squares)

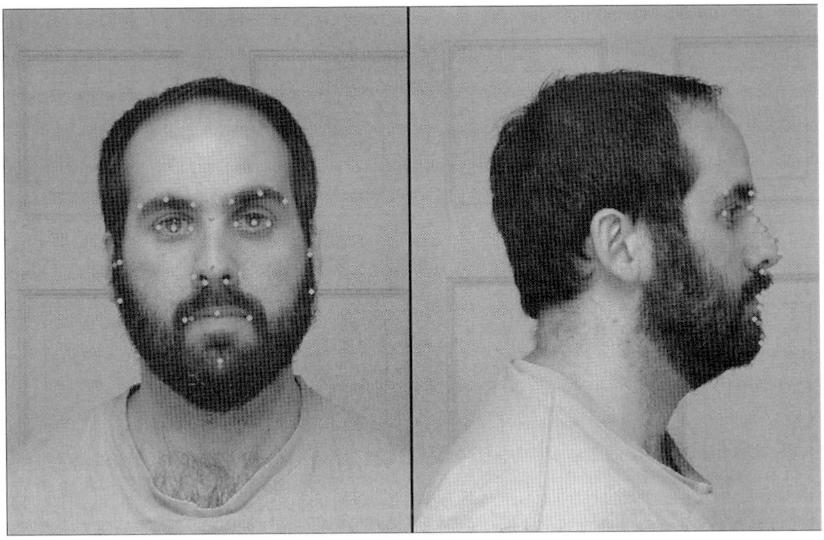

FIGURE 9 The comparison of a subset of the anatomically defined fiducial marks superimposed over the manually defined features; green points are taken from the geometry, projected into the projective plane; red points show the manually defined features.

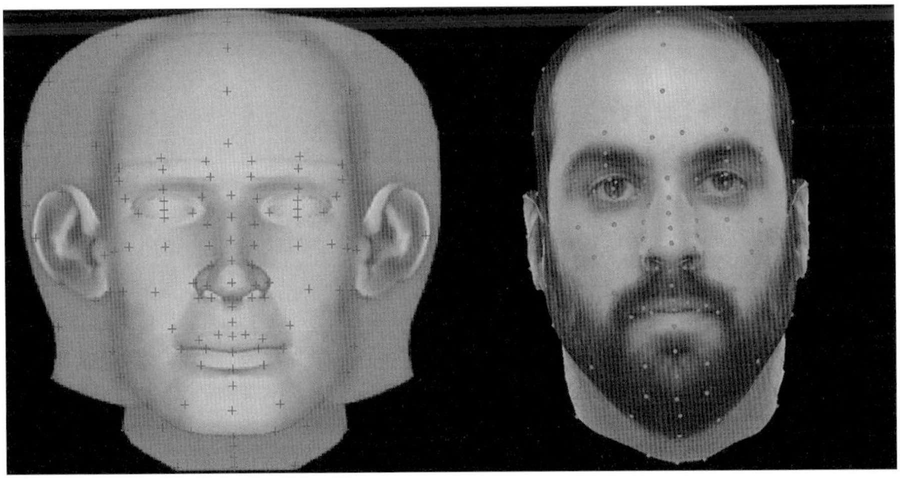

FIGURE 10 Model with fiducial points.

averaged over the landmarked features is about 3.0 pixels on a standard 64 pixels between the eyes image corresponding to approximately 1/20 of an eye distance (Fig. 12 left panel). For the uncontrolled FRGC data, comprised of high resolution images taken with intentional variation in scale and in environments with a high degree of clutter and lighting variation, the RMSE increases to 3.8 pixel normalized to the standard 64 pixels between the eyes (Fig. 12 right panel).

Figure 12 shows a similar RMSE analysis performed as a function of pose angle on the 15°, 25° and 40° Facial Recognition Technology Database (FERET) imagery. In this

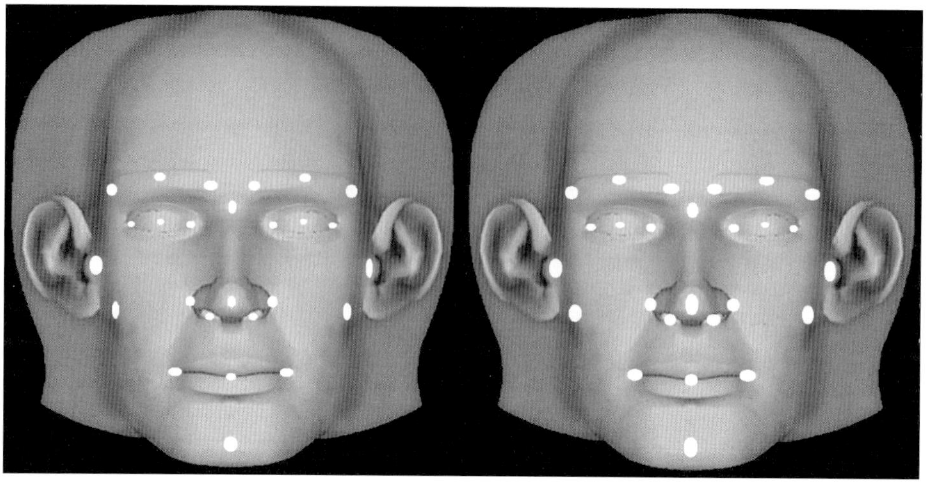

FIGURE 11 RMSE ellipses for FRGC controlled data (left) and uncontrolled data (right) for the 2D-to-3D model geometry generation.

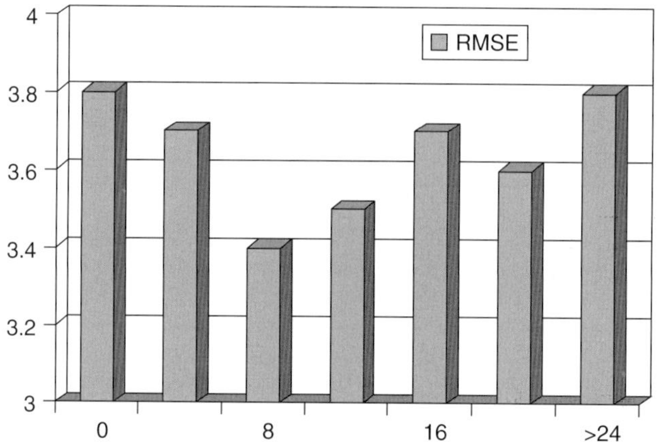

FIGURE 12 RMSE for square root of the sum of squares for 2D-to-3D model generation for FERET indexed as a function of y-orientation of the head.

case, the RMSE accuracy was examined to understand whether accuracy of the model generation is linked to the particular y-pose (plotted along x-axis) under study.

2.5 Rigid Motion Reconstruction Accuracy

Clearly, the errors in model generation accuracy are determined by the quantitative ability to estimate the rigid motions accurately. Not only must the shape of the 3D model be generated to be compatible with the photographs, but the rotation angles of the model fitting the photographs must also be captured accurately as well. The reliability and stability of pose estimation results from the chain of events: head detection rate, 2DIFA, and fine pose estimation. Table 1 shows the results obtained for the control and uncontrol FRGC data. The deviations of the acquired poses are shown in the last three columns.

TABLE 1 Deviation Results for Rotation Estimation of Controlled and Uncontrolled Imagery

Data set	Failure/ Degraded	Head Detection Success	Fine Pose Estimation Success (%)	X Std. Dev.	Y Std. Dev.	Z Std. Dev.
Controlled (608)	0	100%	100	4.38	1.71	0.66
Uncontrolled (492/508)	1	15	96.5/99.8	3.12	2.38	1.19

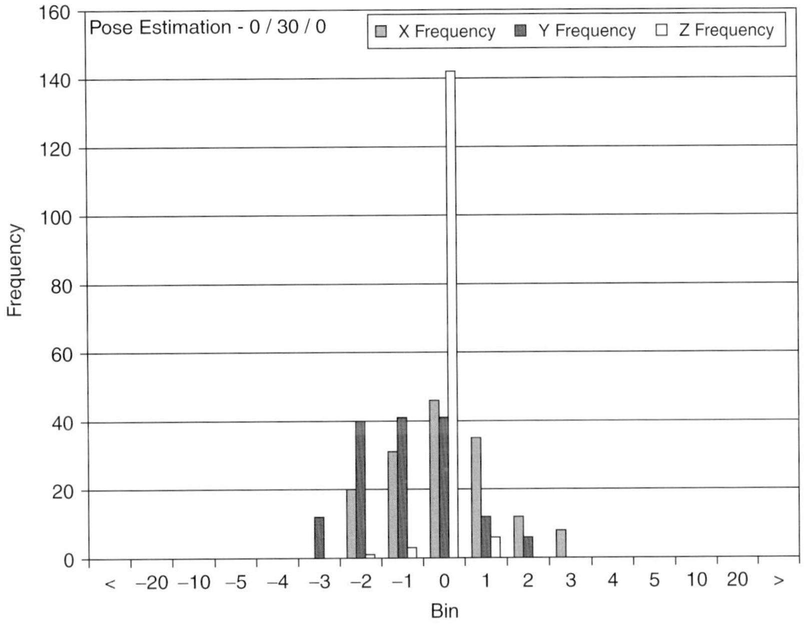

FIGURE 13 Histogram of rotation errors for each axis, X-pitch (red), Y-yaw (green), Z-roll (blue) for simulated imagery generated at 30° Y-yaw. The histogram counts (y-axis) show the errors in degrees (x-axis) obtained from the rigid motion acquisition compared to the known 0, 30, 0 rigid motions of the models.

The accuracy or bias of the capture of the rigid motion can be estimated by generating photographs of known rotations and positions, from which these parameters are acquired by building the geometric models to fit the photographs. This was done for rigid motions of 30° in X-pitch, Y-yaw, and Z-roll directions.

Figure 13 shows the results of studying accuracy of rigid motions for 3D models generated at known orientation of 0°, 30°, 0° X-pitch, Y-yaw, Z-roll, respectively. The rigid motion algorithm was used to acquire the pitch, yaw, and roll from the simulated projective imagery generated from the models at the known orientations. Errors are shown via the histogram.

3 THE 2D-TO-3D TECHNOLOGY FOR PHOTOMETRIC REPRESENTATION

The arbitrary nature of lighting presents a significant confounding variable for FR systems. Lighting representation provides the opportunity to create data structures and features, which provide conditions to make visual and computer-based recognition robust to arbitrary lighting conditions for identification purposes. The 3D model generated provides a direct substrate from which the luminance or lighting fields can be estimated as a "nuisance variable," essentially an extra data structure that itself represents the external environmental variables of the light sources and not the identity of the individual being imaged. 2D-to-3D technologies provide the opportunity to directly generate a 3D map of light intensities that are constructed in the infinite vector space of all lightings.

The 3D geometric models generated consist of two data structures: (i) the triangulated or quadrilateral mesh of vertices and normals representing the geometric positions of the 3D face and (ii) the second "texture" (determined by the albeido of the skin) data structure consisting of an red green blue (RGB) color vector $T(\cdot)$ at every vertex. Given any single or multiple set of photographs there is a lighting field—or luminance scalar field $L(\cdot)$—associated with each vertex of the model. The resulting observed photographs are the result of the interaction of the luminance field $L(\cdot)$ and the texture or albeido field $T(\cdot)$. To first order, the normal directions of the vertices of the models interact with the angular direction of the externals light source to determine the actual intensity of the measured image plane photometric values. The estimation of the luminance field $L(\cdot)$ directly removes the extra nuisance variation that is independent of the actual identity of the individuals—therefore is important for any ID system—and can be exploited for building photometrically invariant FR systems.

The 2D-to-3D solution represents the measured image $I(\cdot)$ in multiplicative form

$$I(x) = L(x)T(x), x \in FACE \tag{6}$$

where the texture or albeido is a vector field indexed over the face consisting of the RGB intensity vector. The multiplicative luminance field $L(\cdot)$ represents the interaction of the external lighting sphere with the surface of the face. The 3D geometry of the face provides the opportunity to manipulate the multiplicative luminance independently of the albeido or texture field, which is representative of the true identity of the individual.

Figure 14 shows examples illustrating the use of the 2D-to-3D technology to reconstruct both the geometric position and shape of the model as well as a luminance representation. The left column shows four example photographs. The middle column shows four luminance fields $L(\cdot)$ estimated on the reconstructed geometric model generated from the photographs in the left column. The right column shows the texture field resulting from the photographs on the left with the luminance resulting from variations in lighting removed.

4 2D-TO-3D GEOMETRIC MODEL NORMALIZATION

Once the 3D geometry is generated, it presents the opportunity to manipulate the projective imagery observed via conventional cameras. There are two kinds of normalization that are being actively pursued for FR technologies:

1. the first is geometric or pose normalization;
2. the second is termed *photometric normalization*.

We now examine each of these.

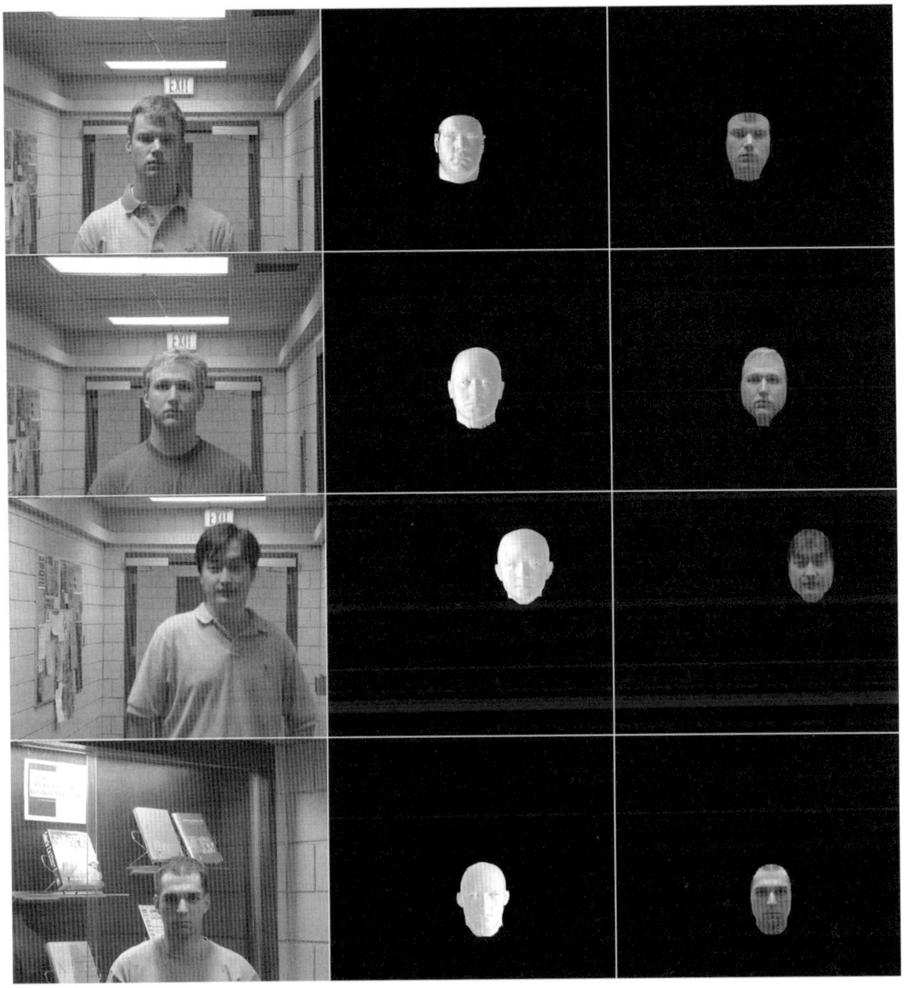

FIGURE 14 Column 1 shows four examples of photographs, column 2 shows four luminance fields estimated on the model associated with the photographs on the left, and column 3 shows the texture field resulting from the photographs on the left with the luminance normalized.

4.1 2D-to-3D Geometric Normalization

Given the 3D geometry fitted into the projective imagery, it is possible to manipulate the projective image to arbitrary neutral coordinates. Figure 15 shows the geometric pose normalization. The incoming photograph is shown in panel 1 with the reconstructed geometry model shown in register in panel 2. The model is then geometrically normalized to any scale and orientation as shown in panel 3. Panel 4 shows the re-textured model at this neutral geometric position.

4.2 2D-to-3D Photometric Normalization

Given a 3D geometry fitted into the projective imagery, it is possible to manipulate the projective image to arbitrary photometric coordinates. Figure 16 shows the use of

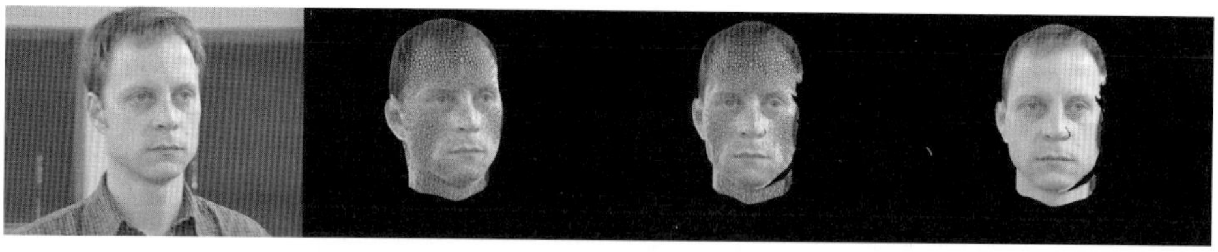

FIGURE 15 Geometric pose normalization. The incoming photograph is shown in panel 1, the reconstructed geometry in register is shown in panel 2, the geometrically normalized model is shown in panel 3, and the re-textured model at neutral pose and lighting.

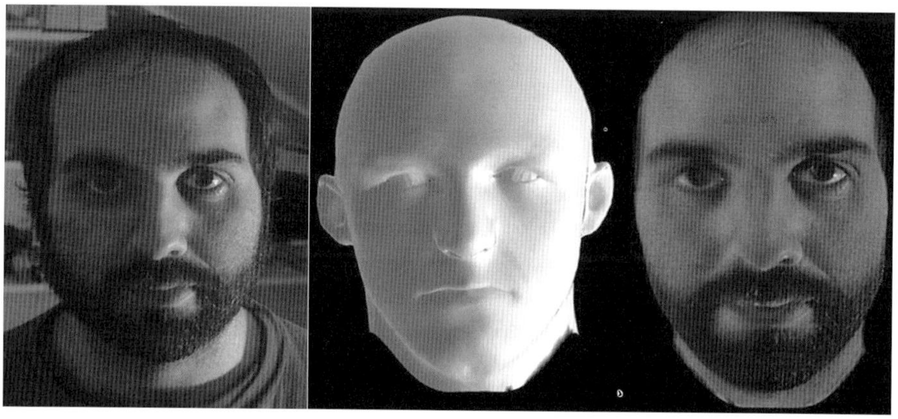

FIGURE 16 Panel 1 shows the photograph, panel 2 shows the reconstructed luminance on the reconstructed model, panel 3 shows the luminance normalized textured model.

photometric normalization. Panel 1 shows the photograph, panel 2 shows the reconstructed luminance field $L(\cdot)$ on the reconstructed model coordinates, and panel 3 shows the luminance normalized texture reconstruction $T(\cdot)$ on the model coordinates. Note that in panel 3 the strong shadowing is removed in the normalized texture field $T(\cdot)$.

4.3 Boosting Facial Recognition Systems via 2D-to-3D Geometric Model Generation

The 2D-to-3D geometry technology can be used to build pose and lighting invariant FR systems thereby boosting conventional 2D ID system performance in difficult environments. Clearly, conventional 2D ID systems degrade in applications where pose and lighting are highly variable; 2D-to-3D model generation technology has been applied in several large experiments exploring the boosting of conventional Face ID via pose variability normalization. Figure 17 shows an illustration of 2D-to-3D technology for generating enhanced galleries in several large DoD tests [15]. Each row shows the application of the 2D-to-3D geometric model generation for enhancing the gallery of pictures to contain representations that accommodate variation of pose and or luminance.

Gallery enhancement provides the opportunity to enhance existing legacy FR systems. Figure 18 shows the results of 2D-to-3D gallery enhancement of one of the conventional high performance 2D FR systems currently in use worldwide attained in a Unisys test

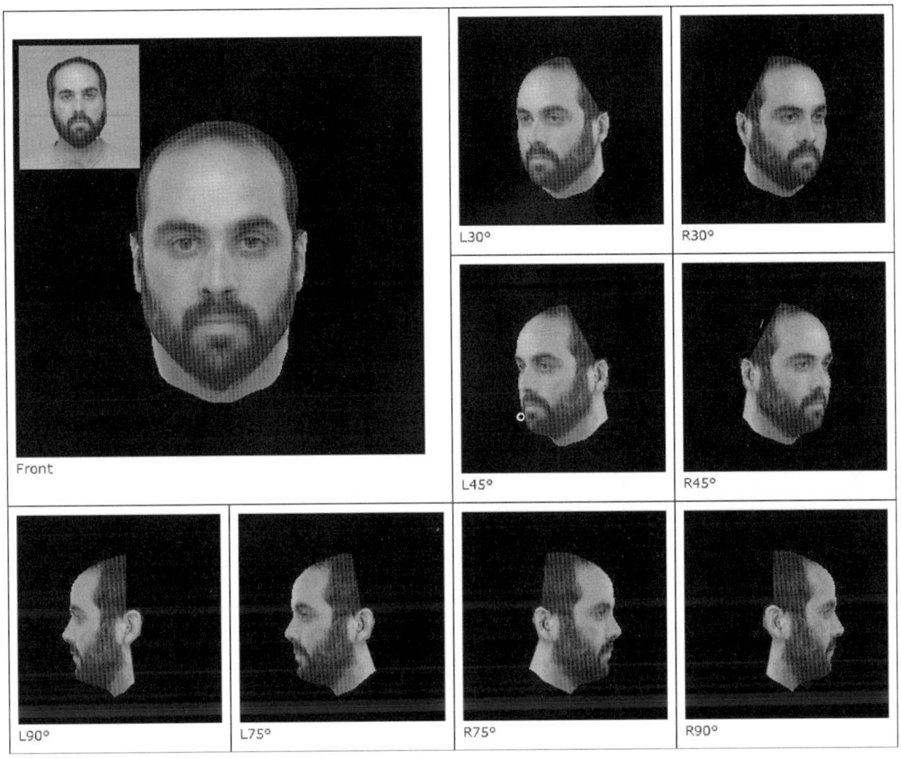

FIGURE 17 The use of 2D-to-3D technology for gallery enhancement. Each row shows the application of the geometric model.

for the DoD [15, 16]. The figure shows two performance curves of the FRR attained as a function of the FAR plotted along the x-axis.

The red curve is the 2D conventional ID system using the 2D-to-3D technology for gallery enhancement [15]. Note that the FRR was monotonically reduced for any fixed FRR. The FRR of 0.02 is depicted by the dashed line, the FRR is reduced by a factor of 10 from 0.01 to 0.001 by the gallery enhancement (from black to red).

5 POSE AND LIGHTING INVARIANT FACIAL RECOGNITION SYSTEMS BASED ON 2D-TO-3D GEOMETRY GENERATION

5.1 2D-to-3D Enabled Frontal Pose-Invariant Facial Recognition Systems

Conventional high performing 2D systems degrade dramatically in their extension from application of the controlled physical access checkpoint setting to the surveillance setting. In the surveillance setting, most faces are captured with some degree of pose, or nonfrontal view of the face. FR systems built upon 2D-to-3D technology have built in pose invariance and provide the robustness to handle the challenges of video surveillance.

Figure 19 depicts a 2D-to-3D enabled FR ID system [17, 18]. As depicted, the 2D-to-3D technology automatically represents all poses and lightings and can manipulate

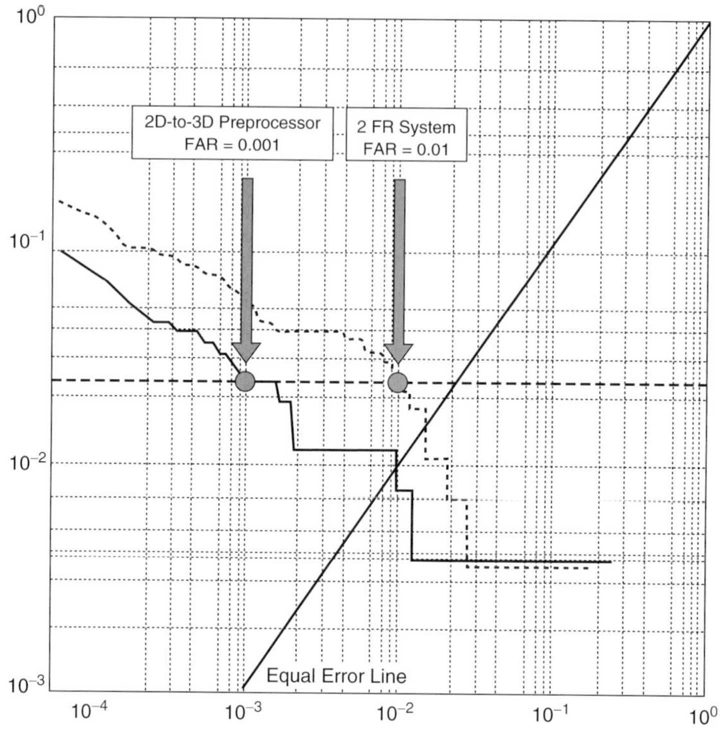

FIGURE 18 The application of 2D-to-3D technology for gallery enhancement. The curves shows the performance of the ID system as a function of FRR versus FAR. The dashed curve is a 2D FR system; the solid curve is the 2D FR system running with the gallery enhancement.

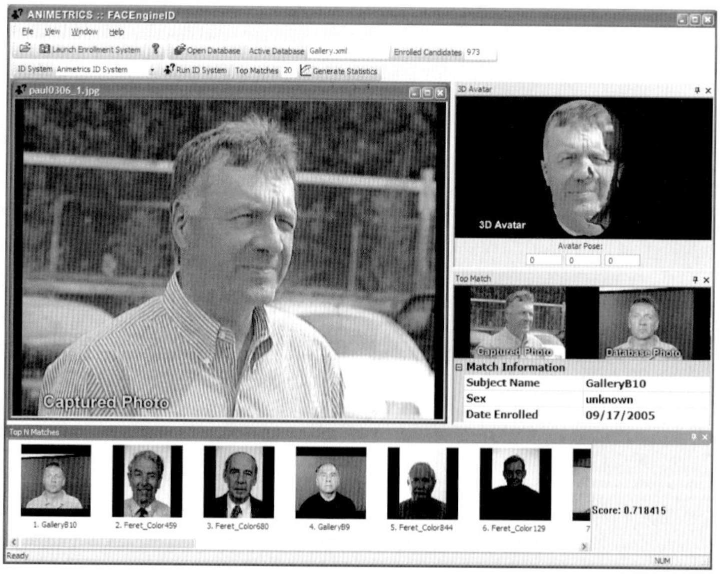

FIGURE 19 A pose and photometric invariant ID system [17, 18] built on 2D-to-3D technology is shown.

and normalize these faces to any pose position (i.e. neutral pose). This process is most beneficial for enhancing the face matching algorithm.

Figure 20 depicts the FR performance achievable with 2D-to-3D technologies when attempting to recognize face images at 40° of pose compared to that of a leading 2D conventional FR system [17]. The lower light solid shows a 94% verification rate at the FAR of 0.1%. This means that 94% of the time the system was able to recognize and verify that the person is who he says he is. At the same time, the system only accepted people incorrectly 1 out of 1000 times (i.e. the FAR). If the FAR is constrained to approximately 1 out of 100, or a 1% FAR, then it follows that the verification rate goes up.

Figure 20 shows a comparison of the 2D-to-3D enabled ID system (lower light solid) with a leading 2D conventional recognition system performance (bold solid). We see a dramatic improvement in performance. At a 95% verification rate of the 2D-to-3D technology, the conventional 2D system has degraded to 42%.

5.2 Lighting Invariant 2D-to-3D Facial Recognition Systems

The 3D models generated via the 2D-to-3D technologies support ID while accommodating lighting variation [18]. Figure 21 shows the FAR versus FRR performance for the FRGC uncontrol single gallery versus single probe experiment. The bold upper line shows performance for singleton galleries and the solid lower thin line for two in the gallery. In each case a single uncontrolled probe was compared to a single uncontrolled gallery

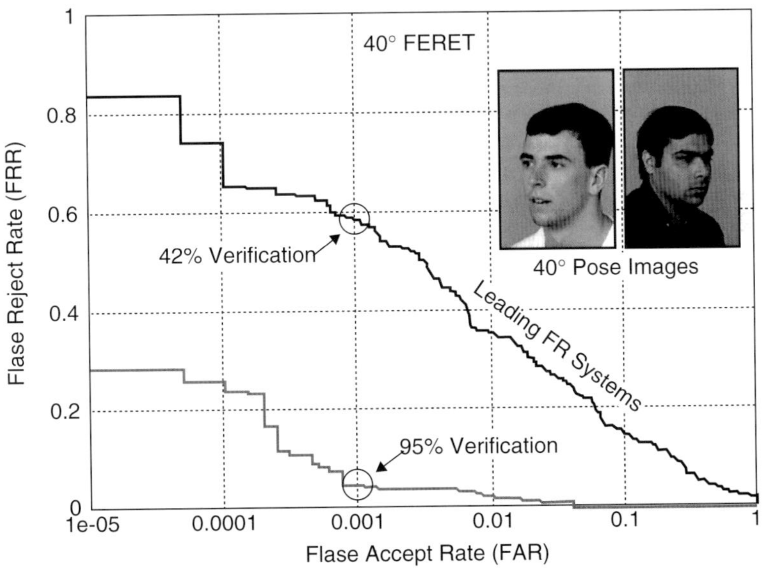

FIGURE 20 Pose invariance performance of 2D-to-3D ID systems [17, 18]. FRR as a function of FAR performance for the 40° probe FERET database is shown. The gallery is populated with front images of the same individuals.

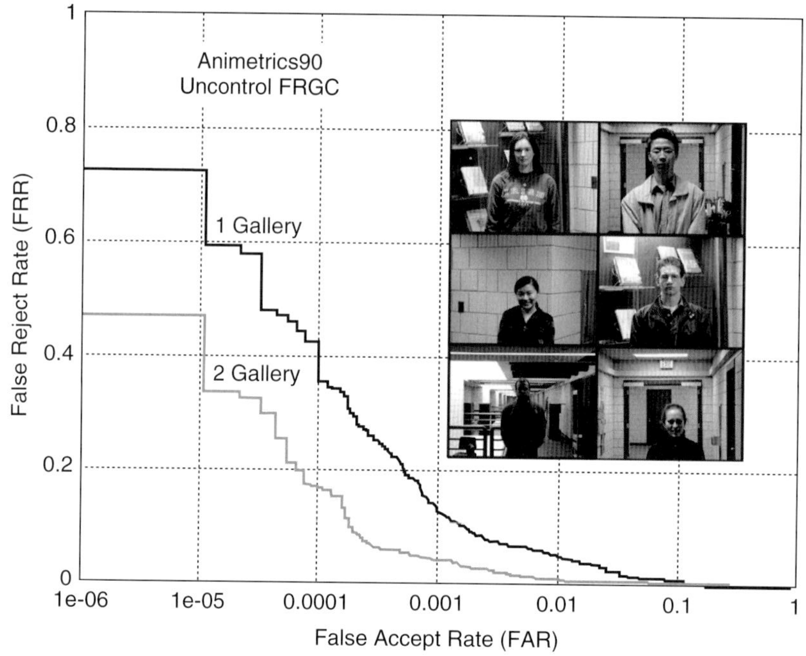

FIGURE 21 FRR versus FAR performance of the FRGC uncontrol single gallery versus single probe experiment. The red line shows performance for singleton gallery and the green line for two in the gallery.

token representing each class. The green line shows the performance of the 2D-to-3D enabled ID system given two gallery tokens for each class.

5.3 2D-to-3D Full Profile Identification Systems

The 3D geometric model provides for the integration of the profile biometric with the nonprofile biometric. Essentially, the 3D generated model is the organizing coordinate system for fusing all of the texture (albeido) and geometric information into a single common coordinate system. Figure 22 depicts the features in profile of the 3D model.

Once a 2D-to-3D model geometry and texture field is registered with the model coordinate system, the ID system can exploit this and generate pose-invariant ID. Figure 23 shows the results of the construction of a full profile ID system using the 2D-to-3D geometric model generation [19].

6 CONCLUSION

This article presents newly emerging 2D-to-3D geometric model generation technologies and their application to "noncooperative" FR surveillance systems. We demonstrate that the generated 3D models can be used to unify pose-invariant and lighting invariant ID systems. Results are demonstrated to use these 2D-to-3D model generation technologies for enhancing 2D systems via "pose normalization" as well as "gallery enhancement".

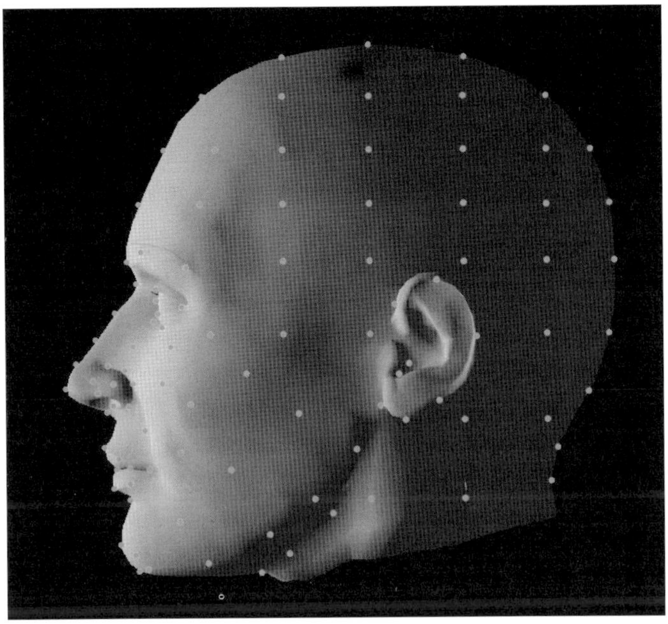

FIGURE 22 The model fiducial features for profile.

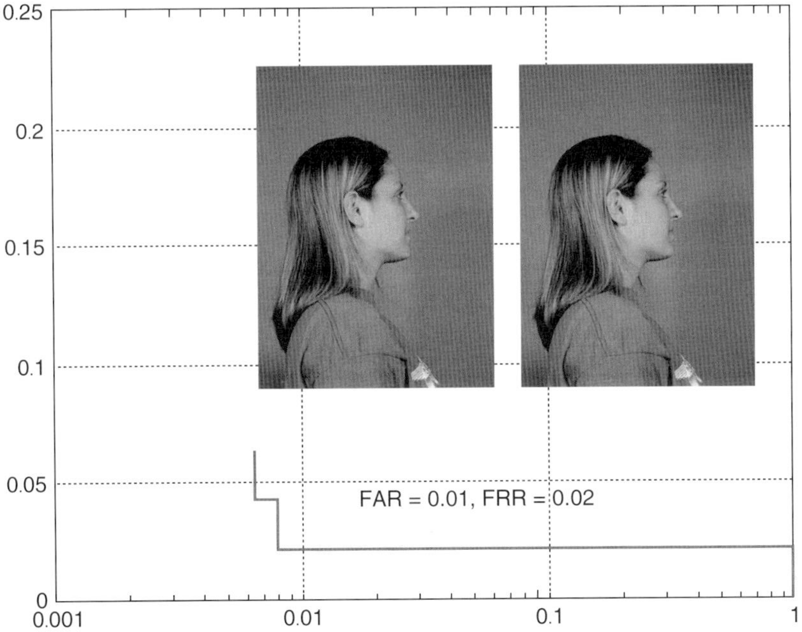

FIGURE 23 The FRR as a function of FAR performance of the profile single gallery versus single probe experiment.

As well, we demonstrate FR systems, which incorporate these 3D coordinate systems for complete 180° ID.

REFERENCES

1. Research and Consulting Outsourcing Services (RNCOS).. (2007). Access Control Technologies and Market Forecast World Over, p. 14.
2. ABI Research.. (2008). *Video Surveillance Systems: Explosive Market Growth and New Market Opportunities*, reference: http://www.securityinfowatch.com/article/printer.jsp?id=14777.
3. FRGC. (2005). Report http://www.frvt.org/FRGC/FRGC_Phillips_BC2005.pdf.
4. Grenander, U. and Miller, M. I. (1998). Computational anatomy: an emerging discipline. *Q. Appl. Math*. **56**, 617–694.
5. Dupuis, P., Grenander, U., and Miller, M. I. (1998). Variational problems on flows of diffeomorphisms for image matching. *Q. Appl. Math*. **56**, 587–600.
6. Miller, M. I., Banerjee, A., Christensen, G. E., Joshi, S. C., Khaneja, N., Grenander, U., and Matejic, L. (1997). Statistical methods in computational anatomy. *Stat. Methods Med. Res*. **6**, 267–299.
7. Miller, M. I. and Younes, L. (2001). Group actions, homeomorphisms, and matching: a general framework. *Int. J. Comput. Vision* **41**, 61–84.
8. Miller, M. I., Trouve, A., and Younes, L. (2002). On the metrics and Euler-Lagrange equations of computational anatomy. *Annu. Rev. Biomed. Eng*. **4**, 375–405.
9. Miller, M. I. (2004). Computational anatomy: shape, growth and atrophy comparison via diffeomorphisms. *NeuroImage* **23**, S19–S33.
10. Grenander, U. and Miller, M. I. (1994). Representations of knowledge in complex-systems. *J. Roy. Stat. Soc. B Met*. **56**, 549–603.
11. Grenander, U., Miller, M. I., and Srivastava, A. (1998). Hilbert-Schmidt lower bounds for estimators on matrix lie groups for ATR. *IEEE Trans. Pattern Anal. Mach. Intell*. **20**, 790–802.
12. Miller, M. I., Grenander, U., O'Sullivan, J. A., and Snyder, D. L. (1997). Automatic target recognition organized via jump-diffusion algorithms. *IEEE Trans. Image Process*. **6**, 157–174.
13. Miller, M. I., Srivastava, A., and Grenander, U. (1995). Conditional-mean estimation via jump-diffusion processes in multiple-target tracking recognition. *IEEE Trans. Signal Process*. **43**, 2678–2690.
14. Srivastava, A., Miller, M. I., and Grenander, U. (1997). Ergodic algorithms on special euclidean groups for ATR. *Progress in Systems and Control: Systems and Control in the Twenty-First Century*. Birkhauser, Boston, Vol. 22, pp. 327–350.
15. Mucke, E. (2004). *A Surveillance System Using Facial Image Enhancement with 3D Face Model Creation: Test Report to Unisys for Animetrics*, p. 12.
16. Miller, M. I., Vaillant, M., and Hoffamn, W. (2007). *SetPose: A 2D-to-3D Model Generation Technology for Library Enhancement and Pose Normalization*, http://www.animetrics.com/.
17. Miller, M. I., Vaillant, M., and Hoffamn, W. (2007). *Animetrics90: A 2D-to-3D Model Generation Pose invariant ID System*, http://www.animetrics.com/.
18. Miller, M. I., Vaillant, M., and Hoffamn, W. (2007). *Animetrics90: A 2D-to-3D Model Generation Lighting invariant ID System*, Whitepaper, http://www.animetrics.com/.
19. Miller, M. I., Vaillant, M., and Hoffamn, W. (2007). *Animetrics180: A 2D-to-3D Model Generation Full Profile ID System*, http://www.animetrics.com/.
20. Vaillant, M., and Glaunes, J. (2005). Surface matching via currents. *IPMI* **3565**, 381–392.
21. Vaillant, M., Qiu, A., Glaunes, J., and Miller, M. I. (2007). Diffeomorphic metric surface mapping in superior temporal gyrus. *Neuroimage*. **34**, 1149–1159.

EYE AND IRIS SENSORS

RIDA HAMZA AND RAND WHILLOCK
Honeywell International, Golden Valley, Minnesota

1 BIOMETRICS FOR HUMAN IDENTIFICATION

Biometrics is the study of automated methods for recognizing humans based on intrinsic physical or behavioral traits. In information technology, biometric authentications refer to technologies that measure and analyze physical characteristics in humans for authentication purposes. Examples of physical characteristics used for identification include fingerprints, eye retinas and irises, facial patterns, and hand measurements. The use of biometric indicia for identification purposes requires a particular biometric factor to be unique for each individual, readily measureable, and invariant over time. Although many indicia have been proposed, fingerprints are perhaps the most familiar example of a successful biometric identification scheme. As is well known, no two fingerprints are the same, and they do not change except through injury or surgery. Identification through fingerprints suffers from the significant drawback of requiring physical contact with the person. No method exists for obtaining a fingerprint from a distance, nor does any such method appear likely.

Recently, the iris of the human eye has been used as a biometric indicator for identification. We have witnessed wide-scale deployment of iris technology across many product categories.

The human iris has extremely data-rich physical structures that are unique to each human being. Biomedical literature suggests that iris features are as unique and distinct as fingerprints. These extraordinary structures consist of a multitude of patterns—arching ligaments, furrows, ridges, crypts, rings, corona, freckles, zigzag collaret, and other distinctive features—that make discrimination possible even between genetic twins. Even in the same person, no two irises are identical in texture or detail. The unique and highly complex texture of the iris of every human eye is essentially stable over a person's life.

As an internal component of the eye, the iris is protected from the external environment; yet, it is easily visible even from yards away as a colored disk behind the clear protective window of the eye's cornea and surrounded by the white tissue of the eye.

Although the iris stretches and contracts to adjust the size of the pupil in response to light, its detailed texture remains largely unaltered. While analyzing an iris image, distortions in the texture can be readily reversed mathematically to extract and encode an iris signature that remains the same over a wide range of pupillary dilations. The richness, uniqueness, and immutability of iris texture, as well as its external visibility, make the iris suitable for automated and highly reliable personal identification. Using a video camera, registration and identification of the iris can be performed automatically and unobtrusively, without any physical contact. All these desirable properties make iris recognition technology a reliable personal identification tool.

2 SCIENTIFIC OVERVIEW OF IRIS TECHNOLOGY

2.1 Technology Basics

The basic iris recognition approach that is the backbone of most fielded systems converts a raw iris image into a numerical code that can be easily manipulated. An automated iris recognition system is made up of several major components, which include at minimum: iris localization that determines the boundaries of the iris with the pupil and the limbus, iris map feature extraction, encoding, and an enrollment matching process. Figure 1 illustrates typical process stages of a recognition system.

In image acquisition, a digital image of the eye may be obtained at multiple resolutions, eye orientations and transitions, under variant illumination conditions, and in noise-laden environments. Segmentation defines the most suitable, usable area of the iris for feature extraction and analysis. The eyelids, deep shadow, and specular reflection are eliminated in the segmentation process. The texture patterns within the iris region are then extracted. The feature extraction process captures the unique texture of the iris pattern, and the encoder compresses the information into an optimal representation to expedite a matching process. The matching process computes the number of bits matched in the iris barcode against multiple templates of barcodes in a database.

The most crucial stage of iris recognition is isolating the actual iris region in the digitized image; herein, we refer to this process as segmentation or localization. In most of the existing systems, the iris region is approximated by geometric models, that is, two circles or ellipses, to simplify the image processing segmentation of the iris and pupil. Prior to encoding the extracted iris patterns, a normalization process adjusts the width of the pupillary boundary to the limbus zone to bring uniformity in cross-matching iris images taken at different resolutions and in different lighting and acquisition environments.

As the eye image goes through various deformations, normalization can be a crucial step in the overall analysis. Normalization scales the extracted iris region cropped from the image to allow for a fair comparison with the database templates, which may have been extracted from regions of different sizes. Dimensional inconsistencies among the captured iris images can occur for many reasons. The iris stretches or contracts as the pupil dilates with varying levels of illumination. Varying distance of image capture and image orientation may occur because the camera or the subject's head tilts. There may

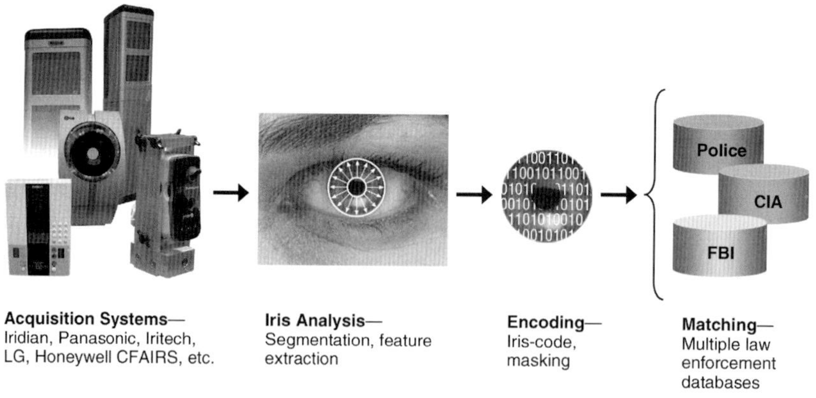

FIGURE 1 Typical iris recognition processes.

be local rotation of the eye within the eye socket. The subject or the subject's face might not be aligned directly with the acquisition device.

After the data related to the iris is normalized, the iris data is encoded into small byte iris codes. The encoding scheme converts a polar image map of an iris into a numeric code that can be manipulated for either storage or comparison to other stored iris codes. The encoded iris data can be sent to a storage module, if for example, it is a sample to be stored for comparison to other encoded iris data. The encoded iris data can also be sent to a module that compares it to stored data and looks for a match or a probable match.

Image enhancement may be applied to minimize illumination artifacts, that is, nonuniform brightness that illuminates some areas more than others within the iris annular region. Reducing the illumination artifacts can improve the quality of subsequent encoding and feature extraction steps. Perspective orientation effects may be addressed before feature extraction; however, this process could add more computational burden on the system. Some recent active contour segmentation algorithms do not appear to require these preprocessing steps to extract accurate features of the iris.

2.2 Research Review

Although prototype systems and techniques were proposed in the early 1980s, it was not until the mid-1990s that autonomous iris recognition systems became a reality. The deployment of an autonomous iris recognition system by Iridian (a.k.a. IriScan) was made possible by the work of Prof. John Daugman of Cambridge University. (The foundation of the iris recognition is well described in [1]). Since then, researchers in the field have made great progress. Most of these developments were built around the original Daugman methodology, with some key contributions such as the work of Wildes using a Hough transform for detecting the iris boundaries and eyelid edges [2–4] and pattern matching on the basis of an isotropic band-pass decomposition derived from applying Laplacian-pyramid spatial filters to the image data. Iris boundary detection takes place in a two-step process: (i) the sclera boundaries are detected using gradient-based filters tuned in vertical orientations. (ii) eyelid edges are detected using gradient filters that are oriented toward the horizontal edges. The horizontal orientation is determined by eyelash position within the eye image, based on the assumption that the head is in an upright position.

Hough-based methods require threshold values for edge detection, which may result in the removal or omission of critical information (e.g. edge points), resulting in failure to detect the iris or pupil regions. The brute-force approach of Hough transform is also computationally intensive, so it may not be suitable for real-time applications. Furthermore, since it works on local spatial features, the method may fail where the image is subject to local noise in the eye image. These features can make Wildes' method computationally expensive, since it relies on image registration, rescaling, and matching.

The Daugman integro-differential operator [1] is considered a variation of the Hough transform, since it makes use of derivatives of the image and performs a search to find geometric information that determines spatial parameters identifying the circles of the iris and pupil. The operator searches for a maximum partial derivative of the image over a circular model as a function of the circular radius and center. The eyelids can be detected in a similar fashion by performing the integration over elliptic arcs to simulate the eyelid curvatures. The advantage of the Daugman operator over Hough is that it does not require threshold values because it is based on raw derivatives. However, some images that suffer from heavy local spatial noise (e.g. spread specular reflections along the eye image, speckles due to digitization, and so forth) can raise challenges for both of these methods.

One approach to dealing with the eyelid occlusions that mask portions of the image makes use of linear fitting. However, the eyelid boundaries can be irregular and difficult to segment because of the presence of eyelashes. Another approach to this problem is to model the eyelids with parabolic curvatures and use the extracted configuration of model components to fine-tune the image intensity derivative information. This alternative can be computationally expensive, because it is based on edge detection and nonlinear curve fitting.

Boles' iris processing prototype builds a one-dimensional (1D) representation of the iris gray-level profiles [5]. An interesting aspect of this work is the ability of the wavelet transforms to eliminate the effect of glare due to reflection of the light source on the surface of the iris, a problem with which even some fielded devices cannot cope. Boles tested various resolutions and selected those that contained most of the energy from the iris signature and were thus least affected by noise.

Other research efforts have produced alternatives to the Daugman/Iridian (currently part of L-1 Identify solutions) methodology. Kim from Iritech introduced the multi-sector, multiresolution approach [6]. The work in [7] and [8] initiated the active contour concept and snake delineation for localization of nonideal irises. In the past few years, the authors have focused on developing new algorithms for a standoff iris recognition system [9–11].

The original foundation of the iris technology and algorithms derived from the information found in the literature do not address problems such as side looking eyes, self-occlusion, nonfrontal face images, etc. Recognition of this situation led to the development of alternative solutions [7, 10–12] to address the real-time operational requirements of a standoff iris recognition system. Unlike previous approaches, the conceptual design of [11, 13] employs an efficient and robust 1D fast segmentation approach built around our concept for the polar segmentation (POSE) system [9].

Major research activities have focused mainly on developing alternative segmentation and encoding techniques, while the conceptual design of an iris recognition system has not deviated much from the original foundation of the technology. In particular, the normalization step, the rubber sheet model proposed by Daugman [14] is the commonly used framework for mapping the deformed iris patterns because of dilation and occlusions into a fixed-size map. The technique maps each point in the Cartesian domain to its corresponding polar coordinates. The result is a fixed-size, unwrapped feature map. Some attempts have been made to base the normalization procedure using a Bayesian approach [15] by deriving a maximum *a posteriori* probability estimate of the relative deformation among iris patterns before the matching process.

Another unique iris pattern matching technique uses a 1D process that improves the encoding scheme for low-quality images. The technique proposes to provide a better representation of the iris features decomposed from the two-dimensional (2D) constructed iris polar representation map [16] when images are of poor quality, making it possible to process low-quality images that would be rejected by other methods. The matching process does not require circular shifting since the angular information is summed up and processed as a single score. Gaussian moments applied on a 1D representation of feature vectors have been advocated for the best representation of local features to indirectly quantify the variations in textures due to coronas, stripes, furrows, and so forth. 2D encoding schemes may appear to be much more reliable than a 1D process because of the inherited 2D relational spatial features in a 2D encoded signature. However, the 1D method allows encoding a lower level of iris image quality that is more suitable for a quick search mode.

To provide accurate recognition or identification of individual irises, it is important to encode the most discriminating information present in the polar presentation of the extracted iris. For some applications, only the most significant features of the iris patterns may need to be encoded so that comparisons between two subjects can be made easily and quickly.

The encoding scheme generates a template of a few bits that captures the essence of an iris pattern. The extracted numeric code is then used to compare the pattern to stored codes. Encoding the iris signature includes applying an algorithm such as wavelet, Gabor filters, or other techniques, described below, to extract textural information from images. These methods process detailed patterns of the iris to produce a bit-wise template of bits of information and exclude some of the corrupt areas using masking within the iris pattern. Encoding filters are chosen to achieve the best recognition rate and preserve the unique iris pattern information in a template.

Wang et al. [17–19] use local intensity variation and local ordinary measures to define a new iris encoding scheme. The authors [20], generate quantized phasor information represented by more than two bits, which are prioritized with the most significant bits over the least significant bits when conducting matching. The merit of this scheme is that it provides a quick way to match subjects and also generates the most probable match instead of the best match when facing poor quality iris images and patterns. In addition, the different bits can be weighted differently for matching using any of the information divergence measures.

Encoding can include the actual encoding of the extracted features processed at different resolution levels using different means of filtering and processing. The encoding mechanism can involve applying one or more selected filters to the segmented iris images. Filters used by the state-of-the art techniques include wavelet and bank filters. The wavelet approach may have an advantage over traditional Fourier transform in that the frequency data is localized. Gabor filters may also be able to present a conjoint representation of the iris pattern in a spatial and frequency domain. Log-Gabor filtering is more reliable than Gabor filtering. In a different study [21, 22], Haar filters were shown to be comparable to Gabor filters. Haar-based filters can be implemented more easily than Gabor filters because we can take advantage of the fact that the integral image representation can be computed using only addition operations. Just recently, Monro et al. [8] introduced a new patch encoding by applying zero-crossings of a 1D discrete cosine transform (DCT) coefficients of overlapped angular patches. The approach was fielded as part of the Smart Sensors Ltd. offerings.

The increased interest in iris technology has prompted new research in the field. Articles on iris recognition technology blossomed in a range of publications from scientific journals to Business Week and biometrics magazines. The iris challenger evaluation (ICE) program for 2006 provides a good perspective of where iris technology stands in comparison to other biometrics. ICE 2006 established the first independent performance benchmark of available algorithms. Although the evaluation was limited to only a few participants, the results of both ICE 2005 and ICE 2006 hold great promise for iris technology with a false reject rate between 0.01 and 0.03 at a false acceptance rate of 10^{-3} [23]. Furthermore, the reported performance of these algorithms did not account for failure to acquire in accordance with image quality assessments. Thus, the false nonmatching rate (FNMR) versus false matching rate (FMR) can be better even on larger datasets. The work reported in [24] probes the uniqueness and richness of iris patterns even when deployed to match billions of subjects—a requirement that must be met to deploy the technology for border control and the global war on terror.

3 CHALLENGES IN IRIS TECHNOLOGY

Iris recognition systems are designed to provide noninvasive and reliable authentication biometrics; however, current commercial solutions deployed by law enforcement and other agencies are impaired by stringent eye acquisition requirements, human intervention required for operation, difficulties with gazed irises, difficulties of encoding poorly captured iris images, lack of interoperability among different systems, and difficulties in processing moving irises.

3.1 Stringent Iris Acquisition Requirements

A major concern of existing iris recognition systems is that iris pattern features that uniquely identify humans can be easily missed because of the lack of accurate acquisition of the biometric data or to deviations in operational conditions. Iris recognition is a low-error, highly accurate method of retrieving biometric data, but iris scanning and iris image processing are costly and time consuming. Fingerprinting, facial patterns, and hand measurements have afforded lower cost, quicker solutions. During the past few years, iris recognition has matured sufficiently for it to compete economically with other biometric methods. Inconsistent iris image acquisition conditions, which most often occur in uncontrolled environments, sometimes cause systems to reject valid subjects or to validate imposters. In contrast, under controlled conditions, iris recognition has proven to be very effective. Iris recognition relies on more distinct features than other biometric techniques, and therefore provides a reliable solution by offering a much more discriminating biometric data set.

Constraints may be placed on the lighting levels, position of the scanned eye, and environmental temperature. These constraints can lead to a more accurate iris acquisition, but are not practical in many real world operations.

3.2 Standoff Iris Segmentation Challenges

Current iris recognition solutions do not reflect the full potential of the technology. The robustness of the standoff iris segmentation approach relies heavily on accurate iris segmentation techniques. Computing iris features requires a high quality segmentation process that focuses on the subject's iris and properly extracts its borders. Because iris segmentation is sensitive to the acquisition conditions, it is a very challenging problem. Most current devices try to maximize the segmentation accuracy by constraining the operation conditions and requiring subject cooperation. Such conditions make current iris systems unsuitable for many law enforcement, homeland security, and counterterrorism applications where subjects are not always cooperative.

When applied to nonconstrained operations, iris segmentation techniques of current systems fail miserably. Such operations may include subjects captured at variant ranges from the acquisition device or they may not align the eye directly with the imaging equipment. These conditions raise concerns about how to process accurate iris segmentation with limited operational constraints.

Most of the existing systems make use of first derivatives of image intensity to find edges to segment the borders of geometric models representing the boundaries of the iris. Other algorithms use costly procedures for geometric model searches throughout the digital image. Significant progress has mitigated this problem; however, these developments were mostly built around the original methodology, namely, circular/elliptical contour segmentation that has proven to be problematic in uncontrolled conditions. Some alternatives to the traditional approach still suffer similar issues with segmentation robustness under uncontrolled conditions.

A method that provides an iris recognition technique suited to moving "standoff" iris applications is much needed—that is, a system in unconstrained conditions that still provides an accurate, real-time result on the basis of the collected biometric data.

3.3 Lack of Database Cross Validation

Current commercial solutions for law enforcement and other agencies are impaired because data cannot be shared across systems. With the advent of economical successes of iris recognition devices, we have seen wide-scale deployment of the technology across many product categories. Most of these systems use a similar sequence of processes to that shown in Figure 1, but the diversity of implementation has resulted in multiple iris databases that lack interoperability. Cross validation with data collected by different law enforcement agencies is not supported, which decreases the usefulness of the iris as a robust biometric.

Some research effort has been made to build a common framework to enable interoperability and cross validation to implement a *versatile iris recognition tool* that is portable, interoperable, and deployable in any existing iris database maintained by law enforcement agencies. However, the market has not yet shown enough interest in support of such capabilities.

3.4 Image Quality Requirements and Preprocessing

Digital eye images are subject to a wide variety of distortions during acquisition, transmission, and reproduction. These distortions can degrade iris recognition performance. Performance of an iris recognition system depends heavily on a successful outcome at each stage of the iris recognition process. In turn, each stage may depend on the quality of the captured iris image. An objective image quality metric can play a variety of roles throughout iris processing. Many artifacts may affect one or more of these processes.

Under ideal conditions, a perfectly captured iris pattern clearly displays the texture of an iris that can be represented in a unique iris barcode. Many factors such as eye closure, obscuration, an off-angle eye, occlusions, imperfect acquisition with electronic noise, nonuniform illumination, different sensor wavelength sensitivity, pupil dilation, and specular light reflection can cause the captured iris map to be far from ideal. In particular, smearing, blurring, defocus, and poor resolution can result in poor quality images and can have a negative impact on iris segmentation and feature extraction and encoding. Encoding a blurry image may result in multiple false acceptances. It is crucial to avoid these poor quality images—especially during enrollment when false acceptance matches can occur because of smeared features and false rejection matches can occur because of obscuration, occlusions, or drastic dilation and gazing.

To counter this vulnerability, current systems have attempted to define quantitative measures that can automatically assess the quality of iris images before processing and develop an appropriate set of quantitative iris image quality metrics (IIQMs). The IIQMs are defined relative to image features based on acquisition performance. The quality of the image should correlate well with subjective iris processes. The IIQMs can be integrated into the preprocessing procedure to assess the quality of the iris image before the iris recognition process is initiated. On the basis of an evaluation with these metrics, the operator can accept the input image, reconfigure the processing to deal with degradations, or request a new capture of the iris.

4 FUTURE RESEARCH DIRECTIONS

Future research needs to bring iris recognition systems up to a more mature technology level capable of making iris biometrics a reliable, evidence-based tool while addressing the pitfalls of current iris devices.

4.1 Critical Needs Analysis

New research efforts are mainly directed toward solving the critical needs of the security market and to extending the technology to noncooperative standoff applications. Such applications are becoming crucial for law enforcement agencies in support of counterterrorism practices. The advancement of standoff iris recognition research presents opportunities to implement new systems well suited to high-security access control or "at-a-distance biometrics" applications with little or no constraints on subject positioning or orientation. An iris recognition operation may include subjects captured at various ranges from the acquisition device or include subjects who may not have an eye directly aligned with the imaging equipment. For such applications, it can be difficult to implement the level of control required by most of the existing systems for reliable iris recognition operations. Some emerging technical approaches to standoff iris recognition cope with asymmetry in acquired iris imaging, allowing systems to operate under uncontrolled operations as long as some of the iris annular is visible.

Recent contributions to the field are directed toward assessment of the quality of an eye image in real-time as a quality control procedure. These techniques may allow poor image acquisition to be corrected through recapture and facilitate the acquisition of a best possible image within the capture time window configured in the system. This acquisition results in a process for providing more good quality iris images to improve iris identification accuracy and the integrity of iris recognition systems. An objective of these advancements is to define rules to assess iris image quality and use these rules as discriminators for recovering information from poor quality iris images or reconfiguring the processing steps based on the image quality assessment.

Other research efforts focus primarily on developing tools that will enable interoperability among systems, making it possible to recognize enrolled irises from any commercially available acquisition device. The application of iris technologies that are more interoperable will make the technology more useful for law enforcement and broaden the use of the iris as a biometric.

4.2 Emerging Technical Approaches

In general, emerging techniques are focused on building a common framework to enable system interoperability, portability, and deployment in adverse environments with noncooperative subjects. Most recent research activities deviate from the original theme of the iris recognition approach based on regular linear fitting models, because of its limitations in handling noncooperative subjects. One prominent technical approach is the application of active contour techniques to model a more complex view of the iris at an off-angle. The following paragraphs describe three emerging concepts.

4.2.1 Standoff Iris Recognition System. Very few techniques address the true nature of iris irregularity in iris segmentation [7, 9, 12]. Apart from estimating the iris borders, the segmentation routine should also detect occlusions due to eyelashes, reflections, and eyelids. The authors have developed a standoff iris recognition system [13] and an algorithmic segmentation approach that address the real-time operational requirements of a standoff iris recognition system. The system can be regarded as "hands-off" or automatic iris processing. Consistent with the principles of noninvasive biometrics, this new segmentation technique is introduced along with a new encoding scheme resulting in an improved biometric algorithm. The feature extraction is based on a simplified polar segmentation (POSE) using a process with low computational load. The approach has proven useful for performing iris recognition under suboptimal image acquisition conditions. It is well suited for iris segmentation that detects all boundaries (inner, outer, eyelid and sclera, etc.) of the image iris simultaneously. This technical approach provides accurate segmentation and identifies good iris patterns under uncontrolled imaging conditions as illustrated in Figure 2.

Unlike existing art, which is often based on the brute force of a Hough transform, iterative active contour, or fitting model based on partial derivatives to tailor the iris edges to circular or regular matching templates, POSE employs an efficient and robust enhancement approach built around a POSE technique [5]. POSE differs from state-of-the art techniques in that it conducts a 1D segmentation process in the polar domain, replaces the exhaustive search for geometric models, and avoids the use of costly edge detection and curve fitting by executing a straightforward peak search on 1D signals. The technique can detect boundaries of the iris borders of any regular or irregular shape that might be due to eye gazing or suboptimal image acquisition conditions.

Recent improvements made to the POSE segmentation technique have resulted in a more robust and computationally efficient real-time iris recognition prototype [13]. The new POSE system takes the analysis of edges into the polar domain at an earlier stage

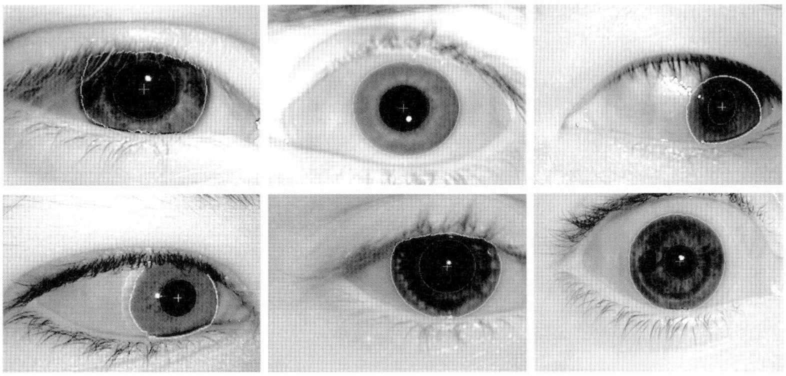

FIGURE 2 Honeywell POSE performance under uncontrolled imaging conditions.

and uses local patterns and an enhanced version of the original POSE proposition to detect iris features. The process is still based on mapping a located eye image into a polar domain with respect to a centered point in the pupil of the eye image. The centered point—not necessarily the exact center of the pupil—can be identified within the pupil region using any of a number of common eye finding algorithms. Once the inner and outer borders of an iris are estimated, the iris region can be extracted, normalized, and mapped into a compressed format where a barcode is generated for processing or database storage.

The POSE approach makes it possible to detect the irregular shape of iris curves using additional rules related to the nature of curve patterns and symmetry. This approach extracts the iris signature using a guided analysis to correctly normalize the stretching and compression of the patterns and bring uniformity into the interpretation of the patterns. The symmetry properties are used to cluster the edge points into at least two categories: (i) sclera and iris boundary points and (ii) iris and eyelid boundary points to be analyzed differently. This analysis helps to cluster obscured pixels and affected areas and weighs them with low scores or masks them out of the analysis. These weights are taken into consideration when patterns are matched against multiple codes in a database and are given weights based on the pattern visibility and exposure to the camera system during acquisition. This process assures that questionable data, such as near possible occlusions, is not weighted as highly as known good information.

Another standoff iris segmentation approach that addresses the real nature of non-ideal irises was introduced in [7]. Ross and Shah have formulated a snake segmentation approach that is based on geodesic active contours (GAC). The motive for using GAC models is to obviate the predication of the boundary limits, which in turn provides a better fit for tightened boundary around the iris borders. An anisotropic diffusion procedure is part of the solution for processing intraregion smoothing while inhibiting interregion smoothing of the iris rubber sheet image. The segmentation approach outlines the iris boundaries and the edges of the eyelid simultaneously to isolate the iris texture from its surroundings.

With the exception of the mandate for iteration in the technique, both propositions (i.e. POSE in [13] and [7]) should lead to similar solutions that can elicit the irregular limbic boundary of the iris (far from the circles and ellipses as previously modeled) to process smooth iris features. The reliability of the reaction of the latter technique to some challenging cases, for example, splitting and merging iris boundaries during evolution when processing an iris with more pronounced localized features (or iris images with heavy reflections) is yet to be proved. These artifacts can affect the stopping criteria before the segmentation reaches the actual iris boundaries—a common drawback of any snake approach when dealing with local minima.

4.2.2 Invariant Radial Encoding Scheme. The major downfall of earlier segmentation techniques is that the systems focus on segmentation of the inner and outer border of the iris to normalize the iris scaling and allow for cross matching of all records of iris barcodes. Many factors, including eyelid and eyelash occlusions that obscure the outer iris border and poor illumination that makes it difficult to distinguish from the sclera, may make it impossible to map the outer border accurately. Inaccurate mapping results in incorrect segmentation of the iris, which in turn negatively impacts the rest of the biometric recognition process. In addition, when applied to uncontrolled conditions, these segmentation techniques deteriorate for nonfrontal eyes. Such conditions may include subjects captured at various ranges and orientations from the acquisition device or subjects who may not have their eye directly aligned with the imaging equipment.

A new encoding scheme has been designed to avoid these pitfalls. The scheme uses a new approach to extract dyadic local iris features to account for the dilation nature of irises with low computational load. It does not rely on accurate segmentation of the outer bounds of the iris region, which is essential in prior techniques; rather, it relies on the identification of peaks and valleys in the iris contrast (i.e. the noticeable points of change in contrast in the iris). Regardless of a chosen filter, the encoding scheme does not use the exact location of the occurrence of peaks detected in the iris, but uses the magnitude variation of detected peaks relative to a referenced first peak. Since this algorithm does not rely on the exact location of pattern peaks and valleys, it does not require accurate segmentation of the outer boundary of the iris, which also eliminates the need for a normalization process.

Similar to previous approaches, to account for rotational inconsistencies and imaging misalignment when the information measure of two templates is calculated, one template is shifted left and right bit-wise (along the angular axis) and a number of distance measures are calculated from successive shifts. This bit-wise shifting in the angular direction corresponds to rotation of the original iris region by an angular resolution unit. From the calculated distance measures, only the lowest value is considered to be the best matching score of two templates.

4.2.3 Iris on the Move®. As stated above, one of the major benefits of iris technology is its noninvasive nature, which makes it practical to analyze subject identity without any physical contact. However, most fielded iris devices failed to meet this criterion, limiting their usefulness for security applications and border controls. Iris on the Move from Sarnoff Corporation [25] was the first to reinforce the noninvasiveness benefit and has worked toward an actual development to recognize an iris at a distance. Although the recognition algorithms and processes are based on previous techniques (e.g. Daugman and others), the system uses a conceptual design that takes advantage of advancements in optics and system engineering to capture an iris passing through a gate. The system is a ground-breaking iris recognition system for moving subjects and has opened the door for other players to contribute to the field. Other emerging techniques are leveraging the advancements in optics and iris algorithmic approaches to build cutting-edge iris recognition systems capable of capturing and processing an iris at farther distances—making the technology more practical for law enforcement applications.

5 CONCLUSION

Recent iris technology evolution has indeed demonstrated the reliability of iris biometrics as an effective means for human identification and recognition. Some prototype systems have even demonstrated standoff iris acquisition and recognition. As it stands, current iris technology offerings can enormously benefit law enforcement agencies. However, gains must be made to further improve performance and tap the full potential of iris recognition—especially for moving subjects. Subject motion presents technical challenges for iris segmentation and feature extraction. The task is made more difficult by factors including poorly captured or occluded iris images, obscuration, motion blur, and illumination artifacts. The field faces other challenges as well, including the unmet needs for full interoperability between different systems and acquisition of irises at greater distances.

REFERENCES

1. Daugman, J. How iris recognition works. *IEEE Trans. Circ. Syst. Video Technol*. **14**(1), 21–30.
2. Wildes, R. P. (1997). Iris recognition: an emerging biometric technology. *Proc. IEEE* **85**(9), 1348–1363.
3. Wildes, R. P., Keith, H., and Raymond, K. (1996). *Automated, Non-invasive Iris Recognition System and Method*, US Patent No. 5~572~596, US Government Printing Office, Washington, DC.
4. Wildes, R. P., Asmuth, J., Green, G., Hsu, S., Kolczynski, R., Matey, J., and McBride, S. (1994). A system for automated iris recognition. *Proceedings of the Second IEEE Workshop on Applications of Computer Vision*. Los Angeles, CA, pp. 121–128.
5. Boles, W. W. (1997). A security system based on human iris identification using wavelet transform. *First International Conference on Knowledge-based Intelligent Electronic Systems*. Adelaide, Australia, pp. 533–541.
6. Kim, D. H., and Ryoo, J. S. June 19, 2001 *Iris Identification System and Method of Identifying a Person Through Iris Recognition*, US Patent Application No. 62,247,813 B1.
7. Ross, A., and Shah, S. (2006). Segmenting non-ideal irises using geodesic active contours. In *Proceedings of 2006 Biometrics Consortium*, University of Virginia, Morgantown, REF 1-4244-0487. Available at: http://www.csee.wvu.edu/~ross/pubs/RossIrisGAC_BSYM2006.pdf.
8. Monro, D. M., and Rakshit, S. (2005). *JPEG 2000 Compression of Human Iris Images for Identity Verification*, ICIP report. Smart Sensors Ltd, Portishead, UK, research supported by the University of Bath.
9. Hamza, R. (2005). January 26, 2005 filed *A 1D Polar Based Segmentation Approach (POSE)*, US Non-Provisional Patent Application Serial NO. 11/043,366.
10. Hamza, R. (2005). Filed January 26, 2005. *Invariant Radial Iris Segmentation*, U.S. Provisional Application No. 60/647,270. US Patent Application Honeywell File No. H0011492.
11. Hamza, R. (2006). Filed March 3, 2006. *A Distance Iris Recognition System*, U.S. Provisional Application No. 60/778,770.
12. Proenca, H., and Alexandre, L. (2007). Toward noncooperative iris recognition: a classification approach using multiple signatures. *IEEE Trans. Pattern Anal. Mach. Intell.*, **29**(4), 607–612.
13. Hamza, R. (2007). Filed Feb 15, 2007. *Standoff Iris Recognition System*. U.S. Provisional Application No. 11/675,424.
14. Daugman, J. G.. 1994. US Patent No. 5~291~560, Biometric personal identification system based on iris analysis. US Government Printing Office, Washington, DC.
15. Thornton, J., Savvides, M., and Vijaya Kumar, B. V. K. (2007). A Bayesian approach to deformed pattern matching of iris images. *IEEE Trans. Pattern Anal. Mach. Intell.*, **29**(4), 596–606.
16. Du, Y., Ives, R., Etter, D., Welch, T., and Chang, C.-I. (2004). *A One-dimensional Approach for Iris Identification*, EE Dept, US Naval Academy, Annapolis, MD, http://www.usna.edu/EE/EE_Research/Biometrics/Papers/5404-57.pdf.
17. Ma, L., Tan, T., Zhang, D., and Wang, Y. (2004). Local intensity variation analysis for iris recognition. *Pattern Recognit.*, **37**(6), 1287–1298.
18. Ma, L., Wang, Y., and Tan, T. (2002). Iris recognition using circular symmetric filters. *Proceedings of the 16th International Conference on Pattern Recognition*. Quebec, Canada

Vol. 2:414-17. http://www.sinobiometrics.com/publications/lma/Iris Recognition Using Circular Symmetric Filters.pdf.

19. Sun, Z., Tan, T., and Wang, Y. (2006). Robust encoding of local ordinal measures: a general framework of iris recognition. *Proceedings of 2006 Biometrics Consortium Conference*. pp. 270–282.
20. Hamza, R. (2007). Filed on March, 2007. *Indexing and Database Search System*. US Patent Application 1100.1449101. Honeywell File No. H0014322.
21. Guo, G. (2006). *Face, expression, and iris recognition using learning base approaches*. Technical report #1575, August 2006, Computer Sciences Department, University of Wisconsin, Madison.
22. Lim, S., Lee, K., Byeon, O., and Kim, T. (2001). Efficient iris recognition through improvement of feature vector and classifier. *ETRI J.*, **23**(2), 61–70.
23. Phillips, P. J., Scruggs, W. T., O'Toole, A. J., Flynn, P. J., Bowyer, K. W., Schott, C. L., and Sharpe, M. *FRVT 2006 and ICE 2006 Large-Scale Results*, National Institute of Standards and Technology, NISTIR 7408. Arlington, VA.
24. Daugman, J. G. (2006). Probing the uniqueness and randomness of IrisCodes: Results from 200 billion iris pair comparisons. *Proc. IEEE* **94**(11), 1927–1935.
25. http://www.sarnoff.com/products_services/government_solutions/homeland_security/iom_brochure, 2007.

A TANDEM MOBILITY SPECTROMETER FOR CHEMICAL AGENT AND TOXIC INDUSTRIAL CHEMICAL MONITORING

H. WILLIAM NIU, DAVID E. BURCHFIELD, AND ANDREW W. SZUMLAS

Hamilton Sundstrand Corporation, Pomona, California

ANDREW ANDERSON

Sionex Corporation, Bedford, Massachusetts

1 INTRODUCTION

Increased efforts by government agencies to provide chemical warning systems for military and civilian locations have fueled the search for more sensitive and selective detectors. To better satisfy the demands required of sensors for these applications, we have developed an instrument that combines the technologies of differential and ion mobility. These two spectrometric techniques separate chemical warfare agent (CWA) and toxic industrial chemical (TIC) compounds in a complementary fashion, and allow the differential mobility spectrometer (DMS)–ion mobility spectrometer (IMS) to provide sensitive and selective detection for homeland security applications. This article describes the tandem DMS–IMS, its salient features, and the benefits that can be realized by application of the new sensor to homeland security applications.

2 SCIENTIFIC OVERVIEW OF TANDEM MOBILITY SPECTROMETRY

The detection capability of an analytical instrument is not only limited by detector sensitivity, but also by the environmental background presented in the sample. Hyphenated instrumentation uses combinations of two analytical techniques that enhance selectivity and interference rejection capability. An innovative example of this type of technology is the tandem DMS–IMS, which accomplishes the combination goals of hyphenated instrumentation.

2.1 Ion Mobility Spectrometry (IMS)

IMS is a technique that detects target compounds by measurement of their velocities (equivalent to measurement of the ions' gas-phase mobilities) as electrically charged target ions are drawn through a buffer gas [1, 2]. Measurement of ion mobilities as a function of time of flight in a drift tube under an applied electric field was developed in the early 1970s. The technique is simple in principle. A conventional IMS consists of an atmospheric pressure ionization (API) source, a drift tube, and a detector. As shown in Figure 1, ions are produced from air molecules by an ionizer such as an electrical discharge or radioactive source. Both positively and negatively charged ions are created

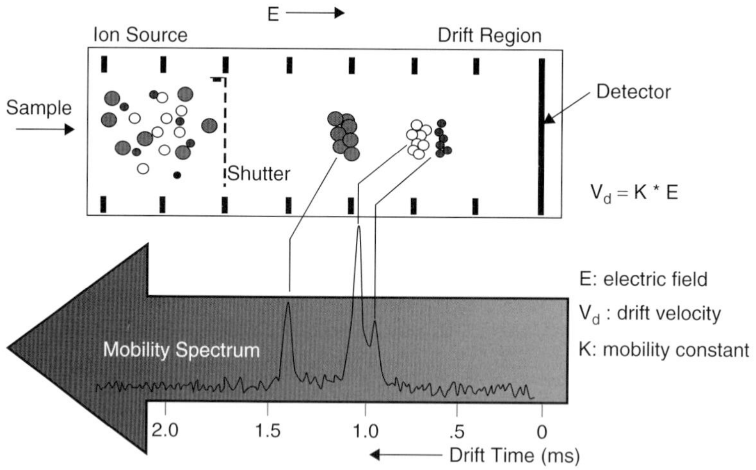

FIGURE 1 Schematic for a conventional ion mobility spectrometer. Ions arrive at the detector according to their mobilities (courtesy of Prof. Eiceman).

in the source and, if water is present at ppm or higher concentrations, these reactant ions typically percolate down to hydrated protons $\{(H_2O)_n \cdot H^+\}$ and O_2^-, respectively. These "reactant ions" can transfer charge to trace contaminants by either proton or electron transfer or can directly react with contaminants to generate ions. Any ions formed are then drawn through the analyzer under an electric field, timed as they transit a drift region, and the time to their arrival at a detector is converted to mobility. An ion's mobility is related to its mass and cross-sectional area in the gas phase. The IMS is analogous to a time-of-flight mass spectrometer. Application of IMS technology trades the exceptional selectivity of the mass spectrometer for instrument simplicity, especially elimination of a vacuum pump.

Because water plays an important part in ion-molecule chemistry, it is necessary to control the water vapor concentration within the spectrometer; optimum performance is best achieved by operating with relatively dry air containing low ppm levels of water vapor. For this reason, recirculated dry air is supplied to the spectrometer for most field applications.

On balance, IMS is attractive due to its high intrinsic sensitivity, conceptual simplicity, and potential for miniaturization. For over 20 years, ion mobility spectrometers have been successfully deployed for detecting CWAs for military applications and explosives at airports. This technique has also been applied to environmental monitoring, usually in conjunction with a gas chromatograph. In addition, IMS is highly sensitive and amenable to portable and inexpensive instrumentation.

2.2 Differential Mobility Spectrometry (DMS)

A variation of IMS is called *differential mobility spectrometry (DMS)* or *field asymmetric ion mobility spectrometry (FAIMS)* [3, 4]. In DMS, ions are carried down a drift tube by a gas flow while being subjected to an oscillating Radio Frequency (RF) electric field transverse to the flow direction. Depending on how the various ions' mobilities change under the high and low alternating fields, the RF waveform will drive some ions out of the flow, allowing only a narrow range of ion species to transit the drift tube and reach the detector. In this manner, the DMS operates as an RF ion filter analogous to a quadrupole mass spectrometer. The use of long dwell times, similar to selected ion monitoring in mass spectrometry, can generate high sensitivity (effectively 100% throughput) for ions of interest. The DMS can also be scanned continuously across the range of ions present or it can be step-scanned to extract target ions sequentially from a mixture.

A DMS analyzer consists of an ionization source, a tunable ion filter sandwich constructed of microfabricated electrodes, and two ion detectors (one for positive ions and one for negative ions). Figure 2 shows a DMS ion filter and illustrates in more detail the principle behind DMS. As ions move through the ion filter, an asymmetric oscillating field is applied perpendicular to the ion path. The asymmetric RF applies a very high electric field (of the order of 10–20 kV/cm) briefly, followed by a lower field of opposite polarity for a proportionally longer period. Any ions formed in the API source experience this asymmetric field and move with a zigzag motion down the filter. Most ions will adopt an off-axis trajectory, striking an electrode surface where they are neutralized and never reach the Faraday collectors. The resulting neutral molecules are carried out of the spectrometer by the gas flow. Only ions whose high-field mobilities exactly offset their low-field mobilities will make it through the separation region of the detector. The filter is tuned to an ion's specific differential mobility by superimposing a separate Direct

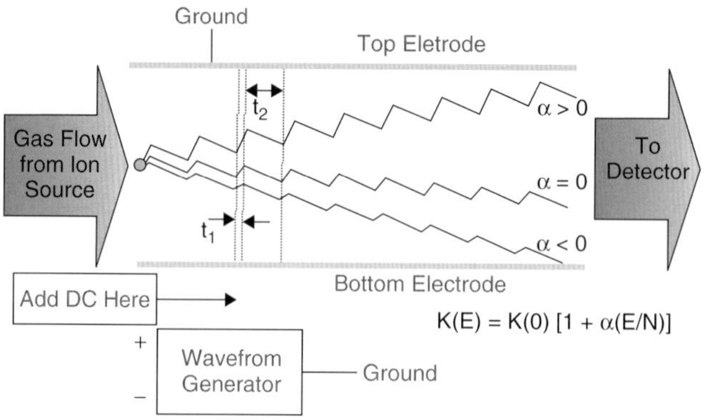

FIGURE 2 Field asymmetric DMS. Varying the compensation voltage determines the α or differential mobility value of the transmitted ions (courtesy of Prof. Eiceman).

Current (DC) bias on the RF field. This DC or "compensation" field is used to counter a specific target ion's drift toward the walls, restoring a narrow range of differential mobilities to the centerline of the drift tube and forcing all other ions with off-axis trajectories to be neutralized at a wall. Once through the ion filter, all remaining ions are drawn to the Faraday collectors where they present their charge to the amplifier, generating an analytical signal.

The analytical utility of the DMS is realized by sequentially selecting the ions that pass through the filter by scanning the compensation voltage (CV) over a suitable range, typically +10 to −45 V. Typical DMS spectra for dimethyl-methyl-phosphonate (DMMP), a simulant for the nerve agent Sarin, in dry air and lab air are shown in Figure 3.

DMS is a relatively new sensor technology. It has been applied similarly to conventional IMS for the detection of CWAs [5] and explosives [6]. A DMS is also commercially available from the Sionex Corp. (Bedford, MA) as a gas chromatograph detector. While related to IMS, DMS has superior sensitivity compared to IMS and similar classes of

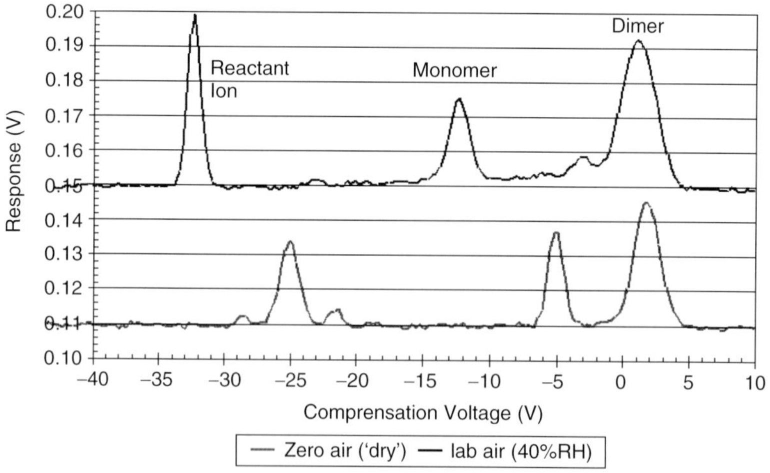

FIGURE 3 Typical compensation voltage (CV) spectrum for DMMP, a sarin simulant. The reactant ion peak, monomer, and dimer of DMMP are shown as a function of CV.

sensors. It can simultaneously detect positive and negative ions with response times on the order of seconds. Furthermore, the RF amplitude applied to the DMS filter can be varied to change the ion separation characteristics. The microfabricated DMS cell facilitates mass production with affordable cost.

2.3 Tandem DMS–IMS

Both IMS and DMS can be made portable for field applications. However, they are each relatively low-resolution devices that are susceptible to background interferences. For atmosphere monitoring at trace levels, a front-end separation device such as a gas chromatograph is usually employed to reduce the sample complexity associated with operation in an atmosphere with many constituents. However, gas chromatography is best performed with an inert carrier to preserve column lifetime and efficacy, which adds to the logistical burden of operation.

The tandem DMS–IMS concept links two mobility devices with different separation mechanisms to enhance selectivity while maintaining simplicity. In a conventional IMS, ions are formed in the source region and gated into a drift tube. Because the width of the ion packet is related to the time, the gate is turned on and the separation of those packets occur over a much longer time with the gate closed, the time-of-flight spectrometer only has about 1% throughput. A DMS has a much higher throughput, approaching 100% when tuned for a target compound. In the tandem arrangement, ions entering the instrument first pass through a DMS filter where target ions are separated from the bulk of the sample. The ions that pass through the DMS are presented to dual IMS drift tubes for detection. The CV of the DMS can be scanned across the entire range of ion species of interest while IMS spectra are collected. Alternatively, the DMS can be programmed to step-scan so that only specific targets are presented to the IMS detectors.

In the configuration presented in Figure 4, the Faraday ion collectors commonly employed in a DMS are replaced by a pair of IMS analyzers. Two small IMS drift cells serve as ion detectors while the DMS is used as a prefilter. In this concept, the analytical chain, which includes the source, DMS, detectors, pneumatics, and electronics, is shared and therefore the overall size is small. The analyzer shown has a total volume of less than 9 in.^3 and further miniaturization is possible.

The tandem DMS–IMS generates two-dimensional data, with separation realized along both differential mobility and ion mobility axes. An example spectrum generated by the DMS–IMS is shown in Figure 5. Here, the positive reactant ion peak (RIP) generated by the ion source is shown in the positive ion spectrum. A similar negative ion spectrum is generated simultaneously. The DMS separation dimension is given along the vertical axis in units of CV, and it is apparent from the spectrum that a CV of approximately −5 V is needed to allow the RIP to reach the IMS drift tube. The horizontal axis represents drift time in milliseconds. Through either dimension, a two-dimensional spectrum can be displayed by examination of a single drift time or CV as shown on the right-hand side of the figure. This characteristic allows each separation dimension to be examined individually.

Figure 6 illustrates the use of a DMS–IMS to detect a target species that generates ions of both positive and negative polarity. These two spectra show DMS–IMS spectra for the analyte ethyl parathion. For this and other analytes that exhibit this dual-polarity behavior, the truly simultaneous detection provided by the DMS–IMS can be leveraged to assist in detection. If one ion is more prominent than its counterpart in the other channel,

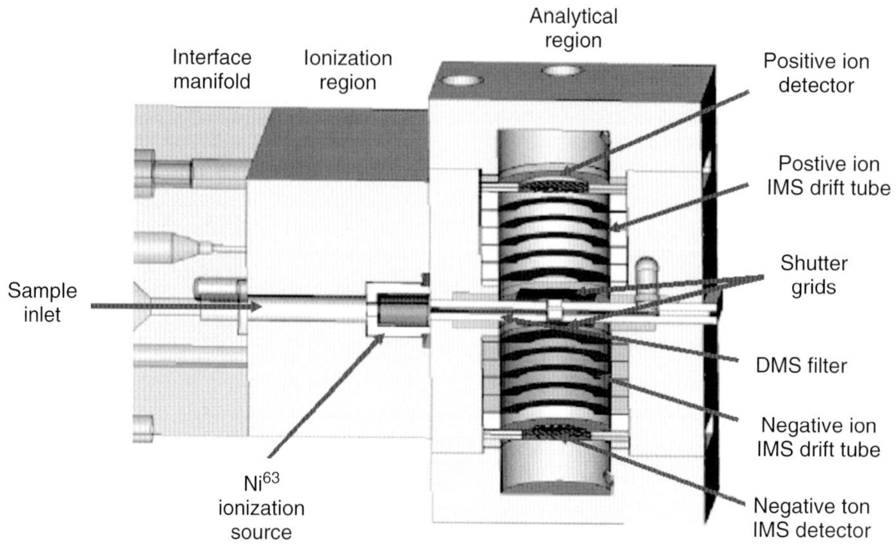

FIGURE 4 DMS–IMS schematic. The prototype sensor utilizes an orthogonal configuration to simultaneously detect both positive and negative ions.

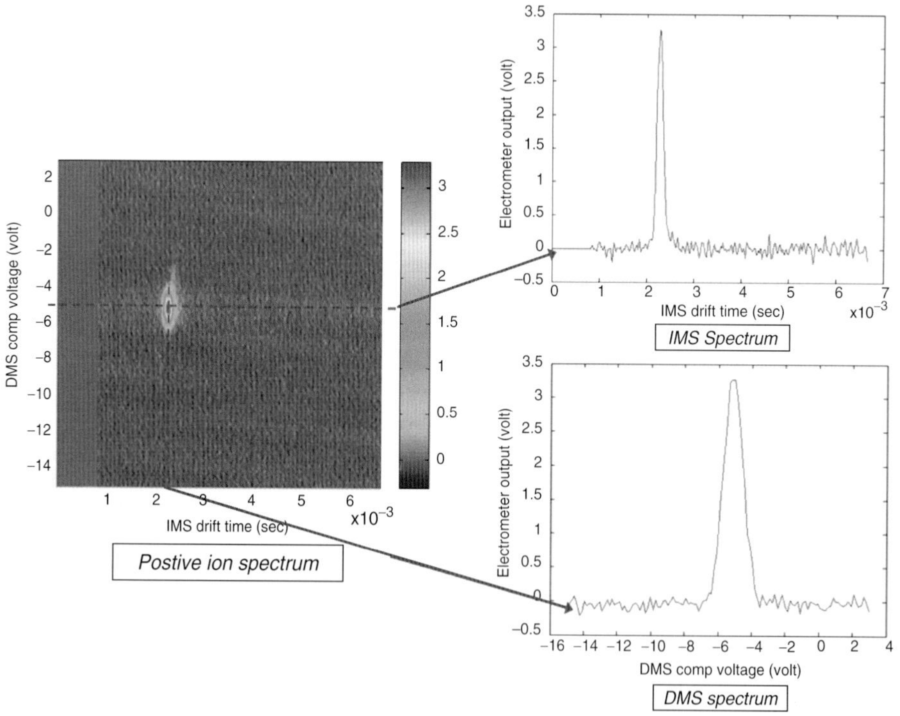

FIGURE 5 Positive ion DMS–IMS spectrum. This representative RIP spectrum displays the two-dimensional separation characteristics of the new sensor.

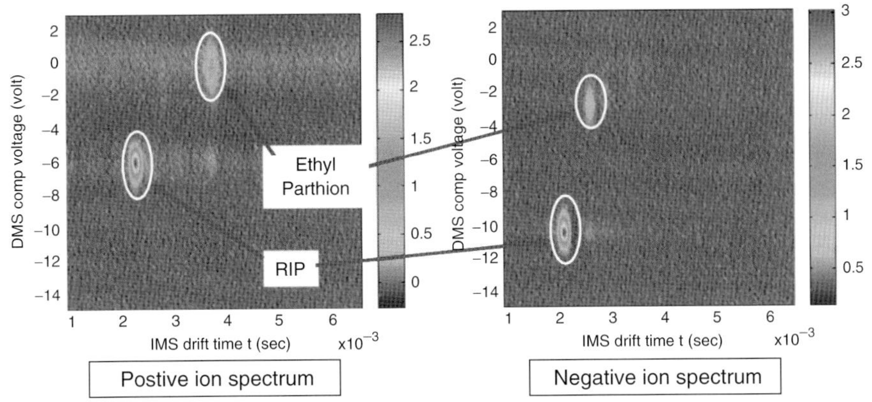

FIGURE 6 Demonstration of simultaneous positive and negative ion detection. Targets that form both positive and negative ions can be readily detected in the channel of the DMS–IMS that provides greater selectivity or sensitivity.

that ion can be utilized to provide quantitation and lower limit of detection. Additionally, the ion in the other channel can be used to confirm the presence of the ion used for quantitative analysis. In this fashion, the possibility of a false positive is greatly reduced, as two chemical species need to be present at specific locations in both the positive and negative ion channels. Targets that display positive and negative ion identifiers have proven to have quite unique spectral signatures in tests performed to date.

In addition to the benefits of dual-polarity ion detection, the DMS–IMS can also enhance the separation of TICs and CWAs due to the two different separation mechanisms. In general, the DMS prefilter is more effective in the separation of low molecular weight compounds than IMS, while drift-time analysis is more appropriate for the separation of compounds with higher molecular weights. Figure 7 shows a combined DMS–IMS spectrum formed from hydrogen chloride (HCl) and chlorine (panel A) spectra. The clear picture of the four analytical signals that can be attained through use of the DMS–IMS becomes ambiguous upon examination of the same system with either IMS (panel B) or DMS separation (panel C) alone.

3 RESEARCH AND FUNDING DATA

Interest in sensor technologies for homeland security applications has escalated in recent years. Search of two scientific databases shows the number of articles that contain "homeland security" begin to rise after the terrorist attacks of 2001 as shown in Figure 8. Interest in IMS rose concurrently, as ion mobility devices presented an available technology that had previously been applied to CWA detection by military users [7]. Extension of this technology to the detection of TIC compounds has provided a low-resistance path to fielding handheld detectors for first responders. The use of ion mobility for infrastructure protection has proven more challenging because of the demand of low false alarm rates. Coincident with the rise of IMS and its application to homeland security, the technique of DMS became more widely adopted as an analytical technique. A literature search using the keywords of DMS and FAIMS demonstrates the relative youth of the DMS analytical technique when compared to its more mature IMS counterpart. However, DMS

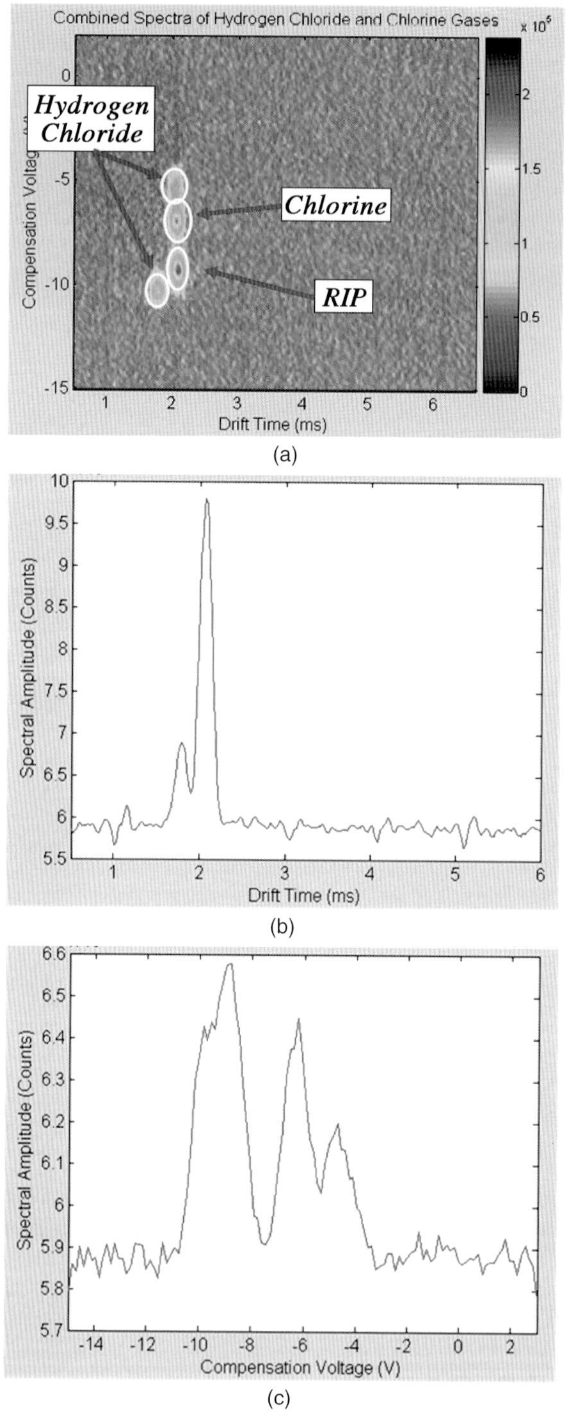

FIGURE 7 Example of resolution provided by a DMS–IMS sensor. (a) DMS–IMS spectrum obtained from combination of individual hydrogen chloride and chlorine spectra; (b) spectrum of hydrogen chloride and chlorine if using IMS separation only; (c) spectrum of hydrogen chloride and chlorine if only DMS separation were employed.

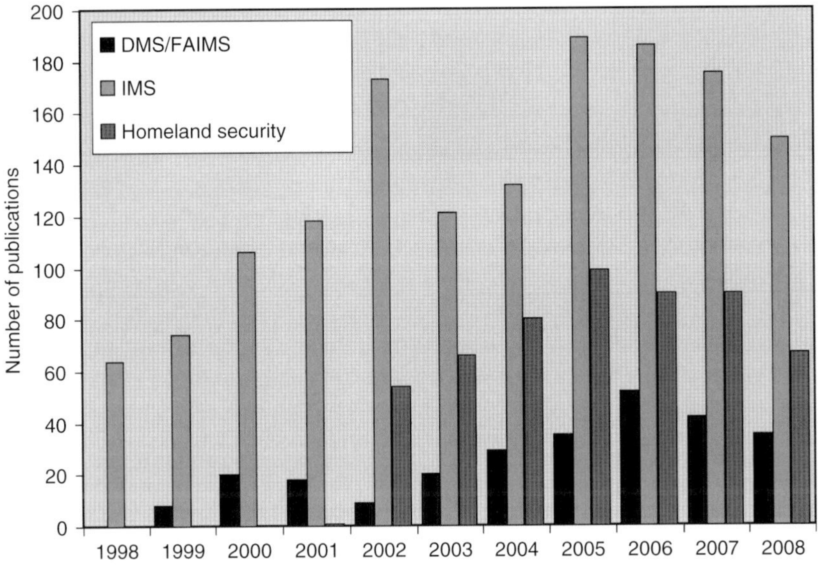

FIGURE 8 Number of publications in homeland security and mobility sensors in the past ten years. Number of publications per year as compiled from the Institute for Scientific Information (ISI) and Chemical Abstracts databases. Keywords searched were differential mobility spectrometry or FAIMS, ion mobility spectrometry, and homeland security.

has quickly gained acceptance as a viable sensor alternative. As an example, large-scale laboratory FAIMS-MS instruments are now commercially available.

Funding for sensors applied to homeland security or the protection of military personnel and installations has also increased in the current decade. The formation of the Department of Homeland Security (DHS) has provided a conduit for the direct funding of research applied to chemical countermeasures. In fiscal year 2009, the budget request for the DHS Science and Technology Directorate is \$869 million. At the same time, the military Joint Program Executive Office for Chemical and Biological Defense requested \$352 million for sensors.

4 CRITICAL NEEDS ANALYSIS

In homeland security applications, chemical detectors face a difficult operating environment with high demands on performance. Detectors fielded for infrastructure protection or emergency response must provide high sensitivity, excellent selectivity, longevity, and continuous, low cost operation. Furthermore, these detectors must operate in environments with complex chemical backgrounds. Examination of the qualities that are most important in homeland security applications suggest that the selectivity of the detector is of specific importance for detectors deployed to protect individuals at airports, subways, or other critical transportation infrastructure. The need for high selectivity is driven by the fact that if a detector reports a false positive, the subsequent societal disruptions and public panic that the alarm causes are extremely costly. Similarly, first responders require a chemical detector they can trust to provide the information they need to take decisive action in response to potential chemical threats or industrial accidents.

4.1 Infrastructure Protection

Chemical detectors play a key role in the protection of public and private infrastructure. Potential chemical attacks could be launched against public transportation hubs like subways, airports, and train stations. In addition, high-profile public and private buildings, as well as embassies abroad, have been identified as potential venues for chemical attacks. For these infrastructure protection applications, multiple chemical detectors would be placed at areas throughout the structure or facility based on either probability of release or modeling of chemical plume movement. The more detectors placed within the structure, the lower the likely response time after a chemical release. However, the more detectors that are desired, the lower their associated cost must be, with both initial and maintenance costs of the full system considered. To keep operational costs low, the detector and sampling system both need to be highly automated and reliable over years of operation.

Technologies similar to IMS and the DMS–IMS have been considered for infrastructure protection applications. A stand-alone DMS was evaluated by Hamilton Sundstrand for its efficacy in the detection of TICs and CWA simulants, and demonstrated excellent sensitivity and selectivity. However, complexity of the sample matrix in some facilities may cause the DMS to have false response issues similar to stand-alone IMS detectors, and it is likely that both DMS and IMS technologies will require a secondary separation technique to help resolve background interferents when placed in real-world operating conditions. Toward this end, a GC–IMS has also been evaluated, and generally displayed excellent analytical performance. However, this combination has the ultimate drawback of limited column reliability, need for a sorbent trap, and longer analysis time.

The DMS–IMS detector system has been designed specifically for extended protection of critical infrastructure. The sensor utilizes a membrane inlet system that selectively admits TIC and CWA compounds while simultaneously reducing the amount of background organic species that can make their way into the analyzer. The inlet system is designed for extended operation in various sample matrices, and since it isolates the detector from the external environment, clean analytical gas can be provided to the DMS–IMS analyzer through use of a simple filter system. The instrument's design supports long-lived, low-maintenance field operation, while it also provides the combination of two separation techniques necessary for these demanding applications.

4.2 Portable Applications

First responders often have the greatest need for a highly specific chemical detector. Emergency personnel on the scene of either industrial accidents or potential chemical attacks require a detector that can positively identify any gaseous hazards present in order to coordinate an appropriate response. In addition to selectivity, the responder's device needs to be highly portable, preferably in a handheld form-factor, battery powered, and no more than a few pounds in weight. Detectors that meet these criteria can assist emergency crews to accurately assess the incident and initiate a proper hazard response to save lives.

Different types of chemical detectors have been proposed or are currently available for use by first responders. The available technologies for this application include handheld IMS instruments, chemical-sensing arrays, and flame spectrophotometers [8]. The DMS–IMS technology is also highly amenable to portable applications as it combines IMS technology, currently the most widespread chemical detector for CWA detection, with an additional capability to reduce false positives. At the same time, instrument simplicity is maintained by the shared sensor platform, which allows the instrument to maintain the small form-factor and robust characteristics necessary for first responder devices.

5 FURTHER RESEARCH DIRECTION

At its current development stage, the DMS–IMS detector has proven effective for the detection of a broad range of both CWA and TIC compounds in a laboratory environment. The next stages of development require additional operational tests in real-world infrastructure protection environments. Areas of study to support these field tests include optimization of the instrument configuration and improvements to the detection algorithm. In certain cases, the use of a dopant within the system can suppress background species and enhance the separation possible with the DMS [9]. Dopant introduction, coupled with refinement of the detection algorithm, can assist in the proper identification of targets and reduce overlapping responses for targets and background chemical species. Although unattractive from a logistical standpoint, particular applications where high specificity for a select target list are desired may warrant addition of a gas chromatograph to the sample inlet to provide a third separation dimension.

Additional areas of interest include a widening of potential targets to additional TIC and CWA compounds. The use of the radioactive source employed to date has been driven largely by the need for simplicity. However, additional nonradioactive sources have recently become more prominent [10, 11] in the literature and may provide a new method for additional selectivity or target detection.

REFERENCES

1. Eiceman, G. A. and Karpas, Z. (2005). *Ion Mobility Spectrometry*, Taylor & Francis Group, Boca Raton, FL.
2. Creaser, C. S., Griffiths, J. R., Bramwell, C. J., Noreen, S., Hill, C. A., and Paul Thomas, C. L. (2004). Ion mobility spectrometry: a review. Part 1. Structural analysis by mobility measurement. *Analyst*. **129**, 984–994.
3. Miller, R. A., Nazarov, E. J., Levin, D. (2007). Differential mobility spectrometry (FAIMS): a powerful tool for rapid gas phase ion separation and detection. *J. Chromatogr. Libr*. **72**, 211–255.
4. Guevremont, R. (2004). High-field asymmetric waveform ion mobility spectrometry: A new tool for mass spectrometry. *J. Chromatogr. A*. **1058**(1–2), 3–19.
5. Krebs, M. D., Zapata, A. M., Nazarov, E. G., Miller, R. A., Costa, I. S., Sonenshein, A. L., and Davis, C. E. (2005). Detection of biological and chemical agents using differential mobility spectrometry (DMS) technology. *IEEE Sensors*. **5**(4), 696–703.
6. Eiceman, G. A., Krylov, E. V., and Krylova, N. S. (2004). Separation of ions from explosives in differential mobility spectrometry by vapor-modified drift gas. *Anal. Chem*. **76**(17), 4937–4944.
7. Eiceman, G. A. and Stone, J. A. (2004). Ion mobility spectrometers in national defense. *Anal. Chem*. **76**(21), 390A–397A.
8. Seto, Y., Kanamori-Kataoka, M., Tsuge, K., Ohsawa, I., Matsushita, K., Sekiguchi, H., Itoi, T., Iura, K., Sano, Y., and Yamashiro, S. (2005). Sensing technology for chemical-warfare agents and its evaluation using authentic agents. *Sens. Actuators B Chem*. **108**, 193–197.

9. Levin, D. S., Vouros, P., Miller, R. A., Nazarov, E. G., and Morris, J. C. (2006). Characterization of gas-phase molecular interactions on differential mobility ion behavior utilizing an electrospray ionization-differential mobility-mass spectrometer system. *Anal. Chem.* **78**(1), 96–106.
10. Andrade, F. J., Shelley, J. T., Wetzel, W. C., Webb, M. R., Gamez, G., Ray, S. J., and Hieftje, G. M. (2008). Atmospheric pressure chemical ionization source. 1. Ionization of compounds in the gas phase. *Anal. Chem.* **80**(8), 2646–2653.
11. Cody, R. B., Laramée, J. A., and Durst, H. D. (2005). Versatile new ion source for the analysis of materials in open air under ambient conditions. *Anal. Chem.* **77**(8), 2297–2302.

FURTHER READING

Sun, Y, Kwok, Y. O. (2005). *Detection Technologies for Chemical Warfare Agents and Toxic Vapors*. CRC Press, Boca Raton, FL.

DYNAMIC LOAD BALANCING FOR ROBUST DISTRIBUTED COMPUTING IN THE PRESENCE OF TOPOLOGICAL IMPAIRMENTS

Majeed M. Hayat

Department of Electrical and Computer Engineering and Center for High Technology Materials, University of New Mexico, Albuquerque, New Mexico

Jorge E. Pezoa and David Dietz

Department of Electrical and Computer Engineering, University of New Mexico, Albuquerque, New Mexico

Sagar Dhakal

Nortel Networks Inc., Richardson, Texas

1 INTRODUCTION

A new class of applications, based on sensor networks (SNs), has emerged in recent years. Examples of these applications are habitat monitoring, intrusion detection, defense and military battlefield awareness, structural health monitoring, and scientific exploration.

These applications share a particular feature, namely, the necessity of collaboration among the sensors. Consequently, the idea of performing distributed computing (DC) naturally appears in an SN environment. In fact, DC is being performed in wireless sensor networks (WSNs) in order to reduce the energy consumption of the set of sensors, make efficient use of network bandwidth, achieve desired quality of service, and reduce the response time of the entire system. These new applications have generated challenging scenarios, and the resource allocation solutions developed for traditional distributed computing systems (DCSs) are not appropriate under these new scenarios [1]. For instance, when DC is performed in a WSN, classical assumptions on node and communication-link reliability are no longer valid because (i) WSNs are typically deployed in harsh environments where sensors are prone to fail catastrophically and (ii) in order to save energy, sensors are allowed to turn themselves off at any time. In addition, assuming that the cost of transmitting data among the sensors is negligible or deterministic is not valid either. Also, large-scale donation-based distributed infrastructures, such as peer-to-peer networks and donation grids, exhibit a similar kind of behavior; the topology of the DCS changes over time as computing elements (CEs) enter into or depart from the DCS in a random fashion. Furthermore, any short-term or long-term damage that can be inflicted to the CEs in these infrastructures adds to this dynamic behavior. Clearly, the new scenarios for DC demand for solutions that can adapt to both fluctuations in the workload and changes in the number of CEs available in the DCS.

We review here modern resource reallocation techniques that are effective in modern dynamic DCSs [2–5]. In particular, we describe dynamic load balancing (DLB) policies that can be used to improve the performance of a DCS in the presence of random topological changes. These policies simultaneously attempt to improve the robustness of the DCS and to use the available CEs in the system efficiently. In this article, we have considered two different performance metrics: the average response time of a software application and the probability of executing an application successfully. These metrics are analytically modeled using the novel concept of stochastic regeneration. Our model takes into account the heterogeneity in the computing capabilities of the CEs, the random failure and recovery times of the CEs, and the random transfer times associated to communication network. Based upon this model, we devise DLB policies that optimize these two performance metrics. First, we discuss DLB policies suitable for scenarios where CEs can fail and recover at random instants of time, as in the case of DC in WSNs or donation-based DCSs. Then, we analyze policies for scenarios where CEs can fail permanently, which model long-term physical damages like those inflicted by weapons of mass destruction (WMD). Our theory is supported by Monte Carlo (MC) simulations and experimental results collected from a small-scale test-bed DCS.

2 THE LOAD BALANCING PROBLEM IN DISTRIBUTED COMPUTING

Large, time-consuming applications can be processed by a DCS in a parallel fashion. To this end, applications have to be divided into smaller units, called modules or tasks, which can be processed independently at any CE of the system. Tasks have to be intelligently allocated onto the nodes in order to efficiently use the computing resources available in the DCS. This task allocation is referred to in the literature as load balancing (LB). LB strategies can be divided into static and dynamic, centralized or decentralized, and sender initiated or receiver initiated [1]. Static LB is performed before the actual execution of the application in the system. At compiling time and based upon statistics of the DCS, the compiler divides the application accordingly into several tasks, which are allocated onto the CEs upon the execution of the application. In DLB, both applications and LB algorithms are being executed in the DCS. The DLB algorithm continuously monitors the queue length of the CEs, and upon the detection of an imbalance, it triggers the reallocation of tasks among the CEs. When the CEs are prone to fail, the LB algorithm must monitor also the working or failed state of the CEs. In this article, we focus on decentralized DLB policies because, in general, DLB policies outperform static LB policies in DCSs where both the system workload and the number of functioning CEs change dynamically with time [1].

In order to optimize a given performance metric, any DLB policy has to answer the following fundamental questions at each CE: (i) *When the jth CE has to trigger a LB action?* (ii) *How many tasks have to be processed at the jth CE?* and (iii) *How many tasks have to be reallocated from the jth CE to the other CEs?* The first question is answered by comparing the load at the jth CE with the average load in the system. To this end, information about the number of tasks queued at each CE must be collected by the jth CE. We denote by $\hat{Q}_{k,j}(t)$ the number of tasks queued at the kth CE as perceived by the jth CE at time t. Based upon this information, the jth CE computes its excess load at time t, denoted as $L_j^{\text{ex}}(t)$, using the formula:

$$L_j^{\text{ex}}(t) = Q_j(t) - \frac{\Lambda_j}{\sum_{l=1}^{n} \Lambda_l} \sum_{l=1}^{n} \hat{Q}_{l,j}(t) \quad (1)$$

A positive value for $L_j^{\text{ex}}(t)$ means that the jth CE is overloaded compared to the average load in the system, and as a consequence, an LB action must be triggered at time t. Note that the Λ_j parameters in Eq. (1) can be defined in several ways in order to establish different balancing criteria. If the Λ_j's are associated with the processing speed of the CEs, then the balancing criterion is determined by the relative computing power of the nodes. Alternatively, if the Λ_j's are associated to the failure rates of the nodes, then the balancing criteria is determined by the reliability of the CEs.

The second question is answered by employing Eq. (1) again. When the excess load at the jth CE is positive, the amount of tasks to be processed locally by the jth CE is given by the difference between its current and excess loads. The remaining tasks should be reallocated to other CEs. Otherwise, if $L_j^{\text{ex}}(t) \leq 0$ then the jth CE has to process all its tasks.

Finally, the third question is answered by computing which CEs are underloaded compared to the average load in the system, as is perceived by the jth CE. These underloaded processors are CE candidates to receive tasks. The number of tasks that each candidate CE receives is denoted by $p_{jk}L_j^{\text{ex}}(t)$, where p_{jk} is the partition of the excess load at the jth CE assigned to the kth candidate receiving CE. Partitions can be defined in several ways: evenly, according to the relative excess load of each candidate CE, and so on. In general, these partitions may not be effective and must be adjusted in order to compensate for the effects of the random communication times. Thus, the load to be transferred from the jth to the kth CE must be adjusted according to what is called the LB gain, which is denoted as K_{jk}. Finally, the number of tasks to be transferred from the jth to the kth CE is given by

$$L_{jk}(t) = \lfloor K_{jk} p_{jk} L_j^{\text{ex}}(t) \rfloor$$

where $\lfloor x \rfloor$ is the greatest integer less than or equal to x.

The LB gains are parameters that ultimately define the number of tasks to reallocate to each CE. We optimally select the value of such gains so that we can minimize the average response time of an application or maximize the probability of successfully serving an application. In order to optimize these metrics properly, first we need a precise model for the response time of an application. The response time is a complex random variable, which combines the randomness of the task processing times at the CEs, the failure and recovery times of the CEs, and the task transfer times in the communication network.

3 STOCHASTIC MODELING OF THE RESPONSE TIME

The idea of stochastic regeneration in DCSs is thoroughly reviewed; to do so, we freely draw from our earlier published works [2–4]. Our novel approach based on stochastic regeneration allows us to obtain recurrence equations that characterize both the average response time of an application and the probability of successfully serving an application [2–4]. The key idea is to introduce a special random variable, called the regeneration time, τ, which is defined as the minimum of the following six random variables: the time to the *first* task service by any CE, the time to the *first* occurrence of failure at any CE, the time to the *first* occurrence of recovery at any CE, the time to the *first* arrival of a queue-length information at any CE, the time to the *first* detection of a failure at any CE, or the time to the *first* arrival of a group of tasks at any CE. The key property of the regeneration time is that upon the occurrence of the regeneration event $\{\tau = s\}$, a *fresh* copy of the original stochastic process (from which the random response time is defined) will emerge at time s, with a *new* initial system configuration that transpires from the regeneration event.

In order to appreciate how the idea of regeneration works, let us consider the following simplified example. For a two-node DCS, let the following random variables W_j, X_j, Y_j, and Z_{jk} denote the service time of a task at the jth CE, the failure time of the jth CE, the recovery time of the jth CE, and the task transfer time from the jth to the kth CE, respectively, with $j, k \in \{1, 2\}$, $j \neq k$. Thus, the regeneration time is precisely defined as, $\tau = \min(W_1, W_2 X_1, X_2, Y_1, Y_2, Z_{12}, Z_{21})$. Let us denote by $T_{m_1,m_2}^{\mathbf{I},\mathbf{F}}(t_\text{b}; \mathbf{C})$ the random response time of an application that comprises $m_1 + m_2$ tasks, where t_b denotes the balancing instant and **I**, **F**, and **C** are vectors that monitor the number of tasks in the CEs, the working state of the CEs, and the number of tasks being transferred in the network, respectively. By exploiting the random variable τ, we can compute the average response time of the application as

$$\begin{aligned} \mathrm{E}\left[\mathrm{E}\left[T_{m_1,m_2}^{\mathbf{I},\mathbf{F}}(t_\text{b};\mathbf{C})\,|\tau\right]\right] = \mathrm{E}\Bigg[&\sum_{j=1}^{2} \mathrm{E}\left[T_{m_1,m_2}^{\mathbf{I},\mathbf{F}}(t_\text{b};\mathbf{C})\,\middle|\tau = W_j\right] P\left(\tau = W_j\right) \\ &+ \sum_{j=1}^{2} \mathrm{E}\left[T_{m_1,m_2}^{\mathbf{I},\mathbf{F}}(t_\text{b};\mathbf{C})\,\middle|\tau = X_j\right] \times P\left(\tau = X_j\right) \end{aligned}$$

$$+\sum_{j=1}^{2} \mathrm{E}\left[T^{\mathbf{I,F}}_{m_1,m_2}(t_\mathrm{b};\mathbf{C})\,|\tau = Y_j\right] P\left(\tau = Y_j\right)$$

$$\left.+\sum_{j=1}^{2} \sum_{k=1,j\neq k}^{2} \mathrm{E}\left[T^{\mathbf{I,F}}_{m_1,m_2}(t_\mathrm{b};\mathbf{C})\,|\tau = Z_{jk}\right] P\left(\tau = Z_{jk}\right)\right] \quad (2)$$

If we consider the regeneration event $\{\tau = W_1\}$ (the service of a task at the first CE), one can show that $\mathrm{E}\left[T^{\mathbf{I,F}}_{m_1,m_2}(t_\mathrm{b};\mathbf{C})\,|\tau = W_1\right] = \tau + \mathrm{E}\left[T^{\mathbf{I,F}}_{m_1-1,m_2}(t_\mathrm{b} - \tau;\ \mathbf{C})\right]$, which means conditional on the service of a task at the first CE the original problem starts afresh with a new initial task distribution (*m–1 tasks at the first CE and* m_2 tasks at the second CE) and a new balancing instant (at $t_{\mathrm{b}-\tau}$). Similar "recurrence" relationships can be proven for every possible regeneration event. After considering all possible regeneration events, we can obtain a set of coupled difference-differential equations that completely describe the dynamics of the average response time. Going back to our simplified example, a sample equation from the set is

$$\frac{\mathrm{d}}{\mathrm{d}t_\mathrm{b}}\mathrm{E}\left[T^{\mathbf{I,F}}_{m_1,m_2}(t_\mathrm{b};\mathbf{C})\right] = \sum_{j=1}^{2} \lambda_{d,j}\mathrm{E}\left[T^{\mathbf{I,F}}_{m_1-\delta_{j,1},m_2-\delta_{j,2}}(t_\mathrm{b};\mathbf{C})\right]$$

$$+\sum_{j=1}^{2} \lambda_{f,j}\mathrm{E}\left[T^{\mathbf{I,F}_j}_{m_1,m_2}(t_\mathrm{b};\mathbf{C})\right]$$

$$+\sum_{j=1}^{2} \sum_{k=1,k\neq j}^{2} \lambda_{jk}\mathrm{E}\left[T^{\mathbf{I,F}}_{m_1} + L_{21}\delta_{j,1}, m_2 + L_{21}\delta_{j,2}(t_\mathrm{b};\mathbf{C})\right] \quad (3)$$

where $\lambda_{d,j}$, $\lambda_{f,j}$ and λ_{jk} are the processing rate of the jth CE, the failure rate of the jth CE and the task transfer rate from the jth to the kth CE, respectively. The term L_{jk} is the number of tasks transferred from the jth to the kth CE and $\delta_{j,k}$ is the Kronecker delta. Given that L_{jk} depends on K_{jk}, we can formulate an optimization problem where t_b and the LB gains are judiciously selected so that the response time is minimized. Finally, structurally similar equations can also be obtained for the probability of successfully serving an application, as it is shown in [2–4].

4 SMALL-SCALE IMPLEMENTATION OF A DCS

We have implemented a small-scale test-bed DCS to experimentally test our DLB policies. The hardware architecture consists of the CEs and the communication network. Upon the occurrence of a failure, a CE is switched from the so-called working state to the backup state. If a node is in the backup state, then it is not allowed to continue processing tasks, but is allowed to work as a backup system that only receives and transmits tasks, so that no task in the system is missing. The occurrence of failures and recoveries at any CE is simulated by a software. The communication network employed in our architecture is the Internet, where the final links connecting the CEs are either wired or wireless. Our test bed also allows us to introduce, if needed, artificial latency by employing traffic shaper applications. The software architecture of our DCS is divided in three layers: application, LB, and communication. Layers are implemented in the software using POSIX threads. The application layer executes the application selected to illustrate the distributed processing of matrix multiplication. To achieve variability in the processing speed of the CEs, the randomness is introduced in the size of each row of the matrix by independently choosing its arithmetic precision with an exponential distribution. In addition, the application layer determines the failure and recovery instants at each CE by switching its state from working and to backup. The LB layer executes the LB policy defined for each type of experiment conducted. This layer monitors the queue length of the CEs and triggers the LB action. It also: (i) determines if a CE is overloaded with respect to the other nodes in the system; (ii) computes the amount of task to transmit to other CEs; and (iii) selects which CEs are candidates to receive the excess amount of tasks. Finally, the communication layer of each node handles the transfer of tasks as well as the transfer of queue-length information among the CEs. Each node uses the UDP transport protocol to transfer a queue-length packet to the other CEs, and the TCP transport protocol to transfer the tasks between the CEs. Also, when a CE is in the backup state, this layer receives and transmits tasks, if necessary.

5 DLB POLICIES FOR SYSTEMS WHERE CES FAIL AND RECOVER

5.1 The Preemptive DLB Policy

The so-called preemptive DLB policy allows a single load transfer between the CEs and no other balancing action is taken afterwards. The DLB action is preemptive in the sense that it will counteract for the combined effects of failures, recoveries, and communication times on the application response time. At time zero, CEs are assumed to be functioning and a CE, say the jth CE, transfers L_{jk} (t_b) tasks to the kth CE at time t_b, where $L_{jk}(t_b) = \left\lfloor K_{jk} p_{jk} L_j^{ex}(t_b) \right\rfloor$ and $L_j^{ex}(t_b)$ is computed using Eq. (1) with the Λ_j parameters defined to be equal to the processing rate of the CEs. The LB gains are optimally selected by solving recurrence Eq. (4) in [3].

Figure 1 (extracted from Figure 3 in [3]) depicts the average response time of the matrix multiplication application as a function of the LB gain. Notably, the theoretical, the MC-simulated, and experimental results show a fairly good agreement. In addition, for comparison purposes, the results for the no-failure case are also shown. From the theoretical curves we can observe that a proper task allocation can effectively reduce the average response time of an application. Note that the optimal number of tasks to reallocate is smaller in the scenario when CEs randomly fail and recover, as compared to the no-failure case. Intuitively, we can state that the optimal task reallocation in case of CE failure will always be less than the optimal task reallocation for the no-failure case.

5.2 The Responsive DLB Policy

In this policy, each CE initially executes a DLB action at time tb assuming that the CEs are totally reliable. In other words, the policy does not care about future CE failures and subsequent recoveries during the execution of the application. Therefore, the initial LB action is taken to achieve an "approximately" uniform division of the total workload of the DCS among all the nodes, assuming that all the CEs will remain functional. Once again, Eq. (1) is used to detect an imbalance and the Λ_j parameters are defined as the

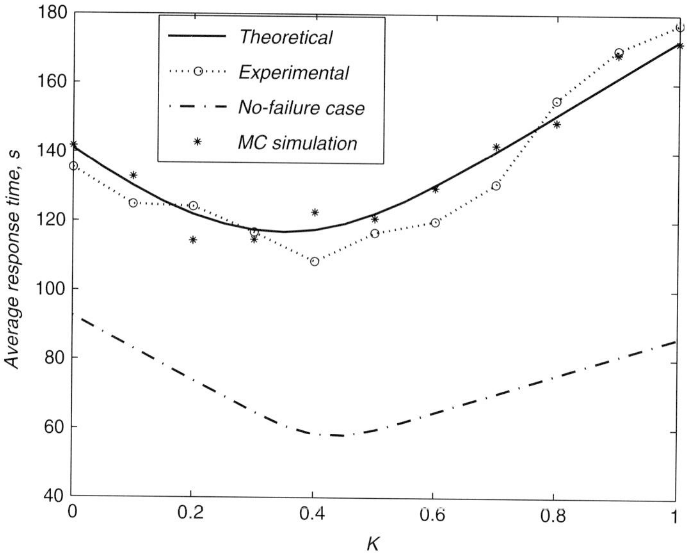

FIGURE 1 The average response time of the matrix multiplication application as a function of the LB gain K for the preemptive LB policy. Extracted from Figure 3 in [3].

processing rate of the CEs. The first difference with the preemptive DLB policy is that the LB gains are optimally selected solving a set of equations simpler than the one used by the preemptive DLB policy; these equations are stated in Eq. (9) in [2]. The second difference is that upon the occurrence of failure at any CE, the backup system of the failed CE executes an extra LB action in order to compensate for the time that will be wasted during the recovery process of the CE. Upon failure, the backup system of the failed CE reallocates tasks according to the following rule:

$$L_{jk}(t_f) = \left\lfloor \left(\frac{\lambda_{d,j}}{\lambda_{r,j}}\right) \left(\frac{\lambda_{d,k}}{\sum_{l=1}^{n} \lambda_{d,l}}\right) \left(\frac{\lambda_{r,j}}{\lambda_{r,j} + \lambda_{f,k}}\right) \right\rfloor \tag{4}$$

where the $\lambda_{r,j}$ is the recovery rate of the jth CE. Note that the first term at the right hand side of Eq. (4) is the average number of tasks that have not been served during the recovery time of the jth CE, the second term is the fair share of tasks of the receiving CE, and the third term is the steady-state probability of any CE to be functioning during the recovery time of the jth CE. Note that this, upon failure task reallocation, is fixed and determined by the parameters of the CEs.

Table 1 (extracted from Table II in [3]) lists the average response time achieved by the responsive DLB policy for some representative initial workload allocations. For comparison, we also list the average response time achieved by the preemptive DLB policy as well as the no-failure case. We can see that the responsive DLB policy outperforms the preemptive policy in all cases. In general, this behavior is observed for communication links where the transfer times per task are small (about 1 s per task). However, the preemptive DLB policy outperforms the responsive policy in scenarios where the communications times per tasks are larger than 1 s (more details shown in Table III in

TABLE 1 Experimental and Simulation Results for the Responsive DLB Policy Applied in Scenarios Where Nodes Can Fail and Recover

	Responsive DLB Policy		Preemptive DLB Policy	No Failure
Initial Workload	MC simulations	Experimental results	(Theoretical)	
(200,200)	277.9	263.4	275.0	141.9
(200,100)	202.4	188.8	210.1	106.9
(100,200)	203.1	212.9	210.1	106.9
(200,50)	170.8	171.4	177.1	89.3
(50,200)	170.8	177.6	177.1	89.3

For comparison, results for the no-failure case and the preemptive DLB policy are also listed. Extracted from Table II in [3].

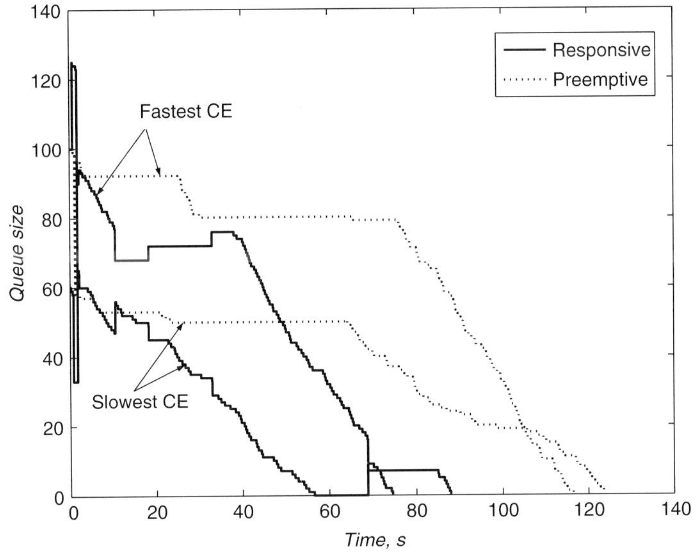

FIGURE 2 A DCS where CEs can randomly fail and recover: a sample realization of the queues dynamics. Extracted from Figure 4 in [3].

[3]). Finally, in order to compare the dynamics of the DCS under each policy, we show in Figure 2 (extracted from Figure 4 in [3]) the actual queues at each CE during one realization of the experiments performed for the preemptive and responsive DLB policies. We can observe that the long flat portions of the queues correspond to the recovery times of the CEs. The downward (upward) jumps in the queues correspond to the action of transferring (receiving) tasks after every failure instant.

6 DLB POLICIES FOR SYSTEMS WHERE COMPUTING ELEMENTS FAIL PERMANENTLY

In harsh environments where nodes fail catastrophically, or during a time much longer than the average application response time, an appropriate metric of reliability is the probability of successfully serving the application. Theorems 1 and 2 in [5] present the set of recurrence equations governing the dynamics of a DCS in the presence of this type of long-term failures. The DLB policies developed for this scenario employ Eq. (1) to determine the imbalances in the system. Unlike in the failure and recovery case, the Λ_j parameters are defined also as the average failure time of the CEs, or alternatively, they can be defined as $\Lambda_j = \lambda_{d,j}\left(1 - \lambda_{f,j} \Big/ \sum_{l=1}^{n} \lambda_{f,l}\right)$. For this last definition, the Λ_j parameters can be thought of as an *effective processing rate* of a CE, which penalizes the processing rate of a CE by its relative unreliability.

Figure 3 (extracted from Figure 2b in [5]) shows the probability of successfully serving an application, which we have denoted as $R^{\mathbf{I},\mathbf{F}}_{m_1,m_2}(t_b; \mathbf{C})$, as a function of the LB gains. We can observe again a remarkable agreement between theoretical, MC, and experimental results. We can see that the probability of successful completion seems to be convex as a function of the number of tasks to be reallocated. From our experience on the LB problem, we can say that the probability of successful completion (correspondingly, the average response time) has such concave (correspondingly, convex) shape when the communication network of the DCS exhibits notorious random transfer times.

Table 2 (extracted from Table 3 in [5]) lists the probability of success for different DLB policies. The policies differ in the balancing criterion used, that is, by the definition

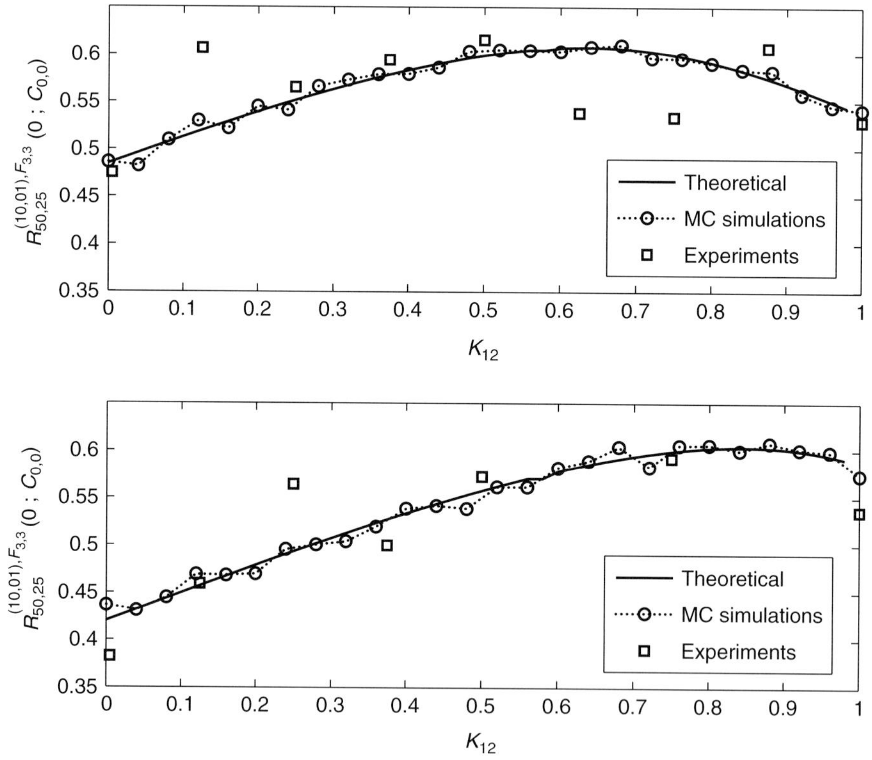

FIGURE 3 The probability of successfully serving an application as a function of the LB gains. The example considers a DCS where CEs can catastrophically fail at any random time. Extracted from Figure 2b in [5].

TABLE 2 The Probability of Successfully Serving an Application for Different Initial Workload Distribution

Initial Load (m_1,m_2,m_3,m_4,m_5)	Maximal-Service	Processing Speed	Complete	Optimum
(150,0,0,0,0)	0.2338	0.1845	0.2223	0.2632
(0,150,0,0,0)	0.2853	0.2908	0.2653	0.2908
(0,0,150,0,0)	0.2868	0.2678	0.2760	0.2867
(0,0,0,150,0)	0.2915	0.2965	0.2978	0.2978
(0,0,0,0,150)	0.2545	0.2940	0.3125	0.3125
(30,30,30,30,30)	0.2953	0.3105	0.3045	0.3170
(59,2,4,34,51)	0.2579	0.2583	0.2868	0.3183
(18,55,29,27,21)	0.2943	0.3098	0.3053	0.3148
(26,30,28,38,28)	0.3185	0.2978	0.3040	0.3185
(40,15,40,35,20)	0.2860	0.2873	0.2845	0.2963

Different balancing criteria have been employed to balance the DCS. Extracted from Table 3 in [5].

of the Λ_j parameters. The maximal-service policy defines the Λ_j's in terms of the average failure time of the CEs, the processing speed policy defines the parameters in terms of the processing rate of the CEs, and the complete policy defines the parameters in terms of the *effective processing rate* of the CEs. In addition, we have considered an example where the fastest CEs are at the same time the less reliable ones. From the data given in Table 2, it can be concluded that a similar performance level can be achieved independently of the balancing criterion employed, thereby showing the strength of our approach. Finally, as shown in Figure 4 (extracted from Figure 2c in [5]), caution must

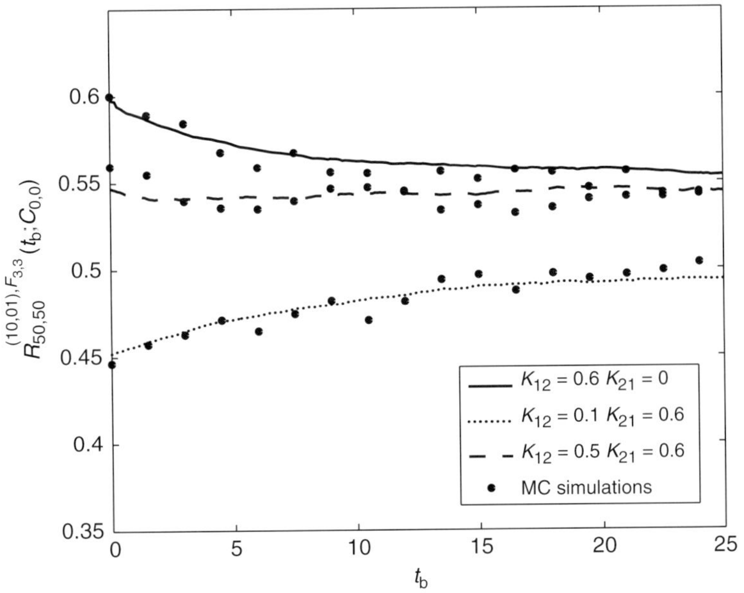

FIGURE 4 The probability of successfully serving an application as a function of the balancing instant. The example considers a DCS where CEs can catastrophically fail at any random time. Extracted from Figure 2c in [5].

be exercised in the selection of the balancing instant and the LB gains; otherwise, the probability of successfully serving an application can be notoriously degraded, as in the LB policy where $K_{12} = 0.1$ and $K_{21} = 0.6$.

7 RESEARCH DIRECTIONS

More fundamental research on modeling complex computing infrastructures is needed, in order to predict and understand the response of such infrastructures in the presence of network and/or CE failures. Models must include the possibility of failures at different layers. For instance, models must consider long-term physical attacks such as those inflicted by WMDs, communication-layer attacks like network flooding, and application-layer attacks like those produced by malicious software. New models will significantly enhance the capabilities of homeland security to (i) optimize computing networks' performance and robustness in the presence of failures; (ii) warn users and operators about a system dysfunction and provide quantitative description of its magnitude; (ii) design modern computing networks that have superior resilience and robustness in the presence of failures.

There is also the need for developing models to characterize, from the point of view of a computing network, the environment and the extent of damage produced by severe attacks on a DCS. The random nature of these attacks demands for a statistical framework, where spatial and temporal correlations of the attacks must be considered. In addition, early detection mechanisms of infrastructure network attacks are also required. These detection mechanisms must be informed also about the spatiotemporal correlations associated with the attacks, so that we can develop smart detection systems capable of warning the users affected potentially by a certain attack.

ACKNOWLEDGMENT

This work was supported by the Defense Threat Reduction Agency (Combating WMD Basic Research Program) and in part by the National Science Foundation (award ANI-0312611).

REFERENCES

1. Shirazi, B. A., Kavi, K. M., and Hurson, A. R. (1995). *Scheduling and Load Balancing in Parallel and Distributed Systems*. IEEE Computer Society Press, Los Alamitos, CA.
2. Dhakal, S., Hayat, M. M., Pezoa, J. E., Yang, C., and Bader, D. A. (2007). Dynamic load balancing in distributed systems in the presence of delays: a regeneration-theory approach. *IEEE Trans. Parallel and Distrib. Syst*. **18**, 485–497.
3. Dhakal, S., Hayat, M. M., Pezoa, J. E., Abdallah, C. T., Birdwell, J. D., and Chiasson, J. (2006). Load balancing in the presence of random node failure and recovery. *Proceedings of IEEE IPDPS*, Rhodes, Greece.
4. Dhakal, S., Pezoa, J. E., and Hayat, M. M. (2007). A regeneration-based approach for resource allocation in cooperative distributed systems. *Proceedings of ICASSP*, Honolulu, HI, III-1261–III-1264.

5. Dhakal, S., Pezoa, J. E., and Hayat, M. M. (2008). *Maximizing Service Reliability in Distributed Computing Systems with Random Failures: Theory and Implementation*. Submitted to IEEE Trans. Parallel and Dist. Systems Paper available at http://www.ece.unm.edu/lb.

FURTHER READING

Dhakal, S., Hayat, M. M., Elyas, M., Ghanem, J., and Abdallah, C. T. (2005). Load balancing in distributed computing over wireless LAN: effects of network delay. *Proceedings of IEEE WCNC*, New Orleans, LA.

Hayat, M. M., Dhakal, S., Abdallah, C. T., Birdwell, J. D., and Chiasson, J. (2004). Dynamic time delay models for load balancing. Part II: stochastic analysis of the effect of delay uncertainty. *Adv. in Time Delay Systems, LNCS* **38**, 355–368.

Dhakal, S., Paskaleva, B. S., Hayat, M. M., Schamiloglu, E., and Abdallah, C. T. (2003). Dynamical discrete-time load balancing in distributed systems in the presence of time delays. *Proceedings of IEEE CDC*, Maui, HI, 5128–5134.

The Resource Allocation Group at the Electrical and Computer Engineering, University of New Mexico, Albuquerque, NM. "The Resource Allocation GroupWebsite." Online since January (2003). Last update January 20th, 2009. http://www.ece.unm.edu/lb.

Sonnek, J., Chandra, A., and Weissman, J. (2007). Adaptive reputation-based scheduling on unreliable distributed infrastructures. *IEEE Trans. Parallel and Distrib. Syst*. **18**, 1551–1564.

Daley, D. J., and Vere-Jones, D. (1988). *An Introduction to the Theory of Point Processes*, Springer-Verlag Berlin, New York.

PASSIVE RADIO FREQUENCY IDENTIFICATION (RFID) CHEMICAL SENSORS FOR HOMELAND SECURITY APPLICATIONS

Radislav A. Potyrailo, Cheryl Surman, and William G. Morris
General Electric Global Research Center, Niskayuna, New York

1 INTRODUCTION

Development of new sensors is driven by the ever-expanding homeland security monitoring needs for the determination of chemical, biological, and nuclear threats

[1–4]. Over the last several decades, numerous principles of detection have been discovered, followed by the design and implementation of practical sensors and sensor arrays. New technologies for the detection of threats of importance to homeland security must be sensitive enough to detect agents below health risk levels, selective enough to provide minimal false-alarm rates, and rapid enough to enable an effective medical response [1]. This article provides a brief overview of chemical and biological threats, focuses in detail on modern concepts in chemical sensing, examines the origins of the most significant unmet needs in existing chemical sensors, and introduces a new philosophy in selective chemical sensing. This new approach for selective chemical sensing involves the combination of a sensing material that has different response mechanisms to different species of interest with a transducer that has a multivariable signal transduction ability to detect these independent changes. In the numerous laboratory and field experiments, the action of several response mechanisms in a single sensing film to different vapors was demonstrated, which resulted in the independent detection of these responses with a single sensor and correction for variable ambient conditions, including high levels of ambient relative humidity (RH).

2 BACKGROUND

Chemical sensors have found their niche among modern analytical instruments when real-time determination of the concentration of specific sample constituents is required. On the basis of a variety of definitions of sensors [5, 6], here we accept that a chemical sensor is an analytical device that utilizes a chemically responsive sensing layer to recognize a change in a chemical parameter of a measured environment and to convert this information into an analytically useful signal. In such a device (Fig. 1), a sensing material is applied onto a suitable physical transducer to convert a change in a property of a sensing material into a suitable form of energy. The obtained signal from the transducer is further processed to provide useful information about the concentration of species in the sample. The energy-transduction principles that have been employed for chemical sensing involve radiant, electrical, mechanical, and thermal types of energy [7]. As shown in Figure 1, in addition to a sensing material layer and a transducer, a modern sensor system often incorporates other important components such as sample introduction and data-processing components. A critical aspect of a modern sensor system is also its packaging design and implementation.

Compared to chemical sensing based on intrinsic analyte properties (e.g. spectroscopic, dielectric, and paramagnetic), indirect sensing using a responsive material expands the range of detected species, can improve sensor performance (e.g. analyte

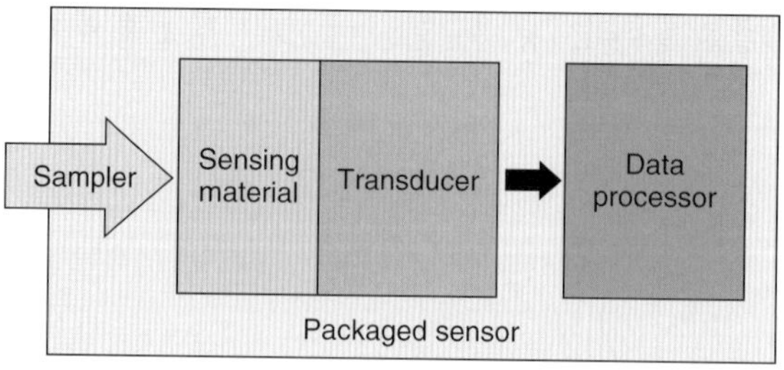

FIGURE 1 Main components of a modern chemical sensor system.

detection limits), and is more straightforwardly adaptable to miniaturization (e.g. through micro-electro-mechanical system (MEMS) or self-assembly). However, a possible challenge of the indirect sensing approach is a trade-off between the selectivity of response to an analyte of interest in a multicomponent complex sample and sensor reversibility.

Innovative ideas for sensors originate from new technological capabilities in sample manipulation [8], sensing materials [6, 9–11], transducer designs [12, 13], micro- and nanofabrication of transducers [14], wireless and proximity communication [15], and energy scavenging [16]. Although the design of a sensor for a particular application will be dictated by the nature and requirements of that application, it is useful to set down the features, listed in Table 1, that one would wish of an ideal sensor for chemical species. In real-world applications, the qualities of an ideal sensor are often weighted differently according to application. For example, low false-alarm rate and high probability of detection are among the most important requirements for homeland security applications [1, 17, 18]. The most important respects in which existing chemical sensors require additional improvements include long-term stability, selectivity, detection limits, and response speed [1].

A known problem in chemical sensing is cross sensitivity of individual chemical sensors to chemical compounds other than compounds of interest. For example, Figure 2

TABLE 1 Typical Requirements for an Ideal Sensor

Low false-alarm rate	Low initial cost
High probability of detection	Low operation cost
Broad dynamic range	Response reversibility
High sensitivity	Small size
High selectivity	Low power consumption
High long-term stability	Robustness
Maintenance simplicity	Self-calibration
High response speed	Ergonomic design

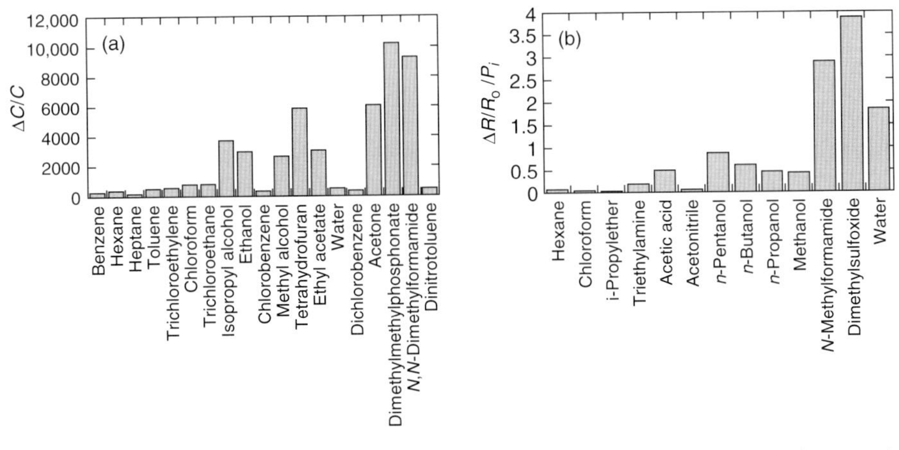

FIGURE 2 Typical cross sensitivity of response of different types of sensing materials to a variety of vapors: (a) capacitance response pattern of single-wall carbon nanotubes and (b) resistance response pattern of $LiMo_3Se_3$ nanowires. (a) Adapted from Reference [19] with permission and (b) adapted from Reference [20] with permission.

illustrates typical cross sensitivity of response of different types of sensing materials to a variety of vapors [19, 20]. Design of sensing materials that are selective to small molecules is challenging [6, 9, 10]. A common approach to address this problem is to build an array of partially selective sensors and to process the array response using multivariate analysis [21]. In such sensor arrays, individual transducers are coated with sensing materials and one response per sensing material (e.g. resistance, current, capacitance, work function, mass, temperature, optical thickness, and light intensity) is measured. An example of such a response of a sensor array is presented in Figure 3, where four sensors were combined into an array for analysis of toxic vapors of chlorinated organic solvents [22]. Although individual sensors in the array were only partially selective, the operation of the sensor array made possible discrimination not only perchloroethylene (PCE), trichloroethylene (TCE), and vinyl chloride (VC) but also three isomers of dichloroethylene (DCE), such as *cis*-1,2-DCE, *trans*-1,2-DCE, and 1,1-DCE. Such discrimination was accomplished with multivariate analysis, such as principal components analysis (PCA) [21]. Using identical transducers in the array simplifies array fabrication, whereas combining transducers based on different principles or employing transducers that measure more than one property of a sensing film [13] could improve array performance through hyphenation. Minimization of the number of sensors in an array is attractive because of simplification of data analysis, reduction of data-processing noise, simplification of sensing materials deposition, and simplification of device fabrication [6, 10, 12, 14, 21].

In designing a chemical sensor system with a single transducer or an array of transducers, attention should be paid to specific design requirements of each system component (sampler, sensing material, transducer, and data processor) and how these components are packaged. Table 2 highlights some of the key challenges and sensor-design aspects for each system component and for the packaged chemical sensor system. Table 3 presents

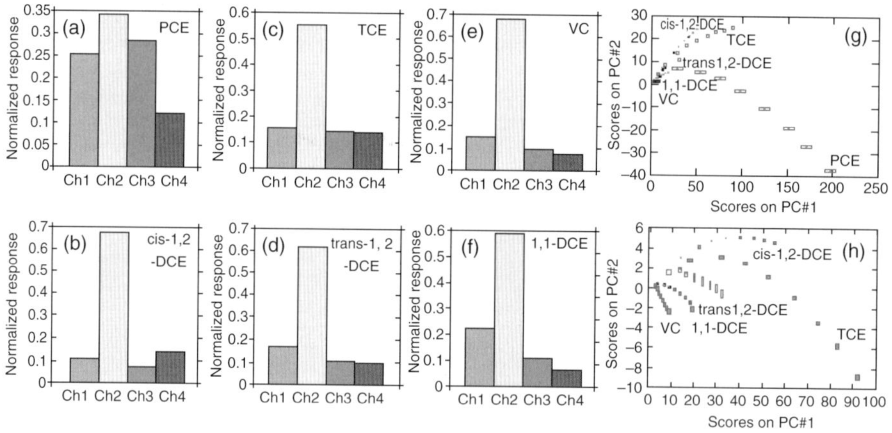

FIGURE 3 Example of operation of a four-sensor array for the detection of toxic vapors of chlorinated organic solvents: (a–f) Partially selective responses of four sensors in the array to 100 ppm of PCE, *cis*-1,2-DCE, TCE, *trans*-1,2-DCE, VC, and 1,1-DCE. (g, h) Score plots of the first two principal components for the data set obtained from the response of the four-sensor array to (g) six and (h) five toxic vapors of chlorinated organic solvents at different concentrations ranging from 0 to 100 ppm. Each data point in (g, h) is the mean of three measurements, rectangles represent one standard deviation. Adapted from Reference [22] with permission.

TABLE 2 Examples of Challenges and Sensor-Design Aspects Associated with Gas-Phase Chemical Sensing

Component	Challenges	Sensor-Design Aspects
Sampler	Particulate contamination Lack of representative sample for trace analysis	Preconcentration for detection limit improvement Integration into periodic self-cleaning sensor system
Sensing Material	Film poisoning Film aging Water condensation	Temperature-stabilized operation Temperature, gas-flow modulation to facilitate more reversible and selective response
Transducer	Signal-to-noise decrease with reduced transducer size Difficult readout from miniaturized transducer	Transducer design for higher order sensor response through hyphenated transduction techniques for selectivity, signal-to-noise, and stability improvement
Data processor	Reliable quantitation of multiple analytes in mixtures using partially selective sensing films and conventional sensor arrays Detection of low levels of analytes in the presence of high levels of interferences	Analysis of dynamic signatures for selectivity and stability improvement Analysis of multivariate signatures for selectivity, signal-to-noise, and stability improvement
Packaged system	Calibration drift Relatively large system size with sampler, battery, and data communicator	Self-calibration Power harvesting or wireless power

several examples of existing available individual sensors and sensor array systems for the detection of vapors of interest in homeland security applications.

3 THREATS, CHALLENGES, AND NEW TECHNICAL SOLUTIONS

To respond to chemical and biological threats, in addition to available analytical instrumentation [2] and sensor systems for homeland security applications (Table 3), new technologies with previously unavailable capabilities are needed. New technologies for the detection of threats of homeland security importance must satisfy at least three goals. They must be sensitive enough to detect agent concentrations at or below health risk levels, selective enough to provide acceptable false-alarm rates, and prompt enough to enable an effective medical response [1]. Out of these goals, *high selectivity* of chemical detection using existing sensor systems represents the most significant challenge. Significant *sensor sensitivity* improvements were demonstrated using new transducer designs and high-surface area sensing nanomaterials, whereas a significant increase in the *response speed* was demonstrated using high-surface area sensing nanomaterials [6, 10].

TABLE 3 Representative Examples of Commercially Available Individual Chemical Sensors and Sensor Array Systems

Individual Sensor or Sensor Array	Classes of Analytes	Company
Electrochemical sensors	Toxic industrial chemicals	City Technology Ltd, Portsmouth, Hampshire, UK www.citytech.com
Electrochemical and semiconductor sensors	Toxic industrial chemicals	Delphian Corp., Northvale, NJ, USA www.delphian.com
Metal oxide semiconductor sensors	Toxic industrial chemicals	Figaro Engineering Inc., Osaka, Japan www.figaro.co.jp
Array of interdigitated electrodes	Analytes of homeland security and defense relevance	Smiths Detection-Pasadena, Inc. (former Cyrano Sciences, Inc.), Pasadena, CA, USA www.smithsdetection.com
Array of surface-acoustic wave devices	Chemical warfare agents, toxic industrial chemicals	MSA, Pittsburgh, PA, USA www.msanorthamerica.com
Array of colorimetric sensing films	Toxic industrial chemicals	ChemSensing, Inc., Northbrook, IL, USA www.chemsensing.com

An attractive technical solution to the insufficient selectivity of existing sensors is to implement a new recently contemplated and experimentally demonstrated sensing concept for selective sensing that requires only a single sensor rather than a sensor array [23, 24]. This new concept involves the combination of a sensing material that has different response mechanisms to different species of interest with a transducer that has a multivariable signal transduction ability to detect these independent changes. In the numerous laboratory and field experiments, the action of several response mechanisms in a single sensing film to different vapors was demonstrated, which resulted in the independent detection of these responses with a single sensor and correction for variable ambient conditions.

3.1 Threats

In the foreseeable future, the United States and other nations will face an existential threat from the intersection of terrorism and weapons of mass destruction. Chemical agents (CAs), toxic industrial materials (TIMs), and biological agents (BAs) are among those compounds that are expected, by homeland security experts, to be utilized in future terrorist attacks [18].

3.1.1 Chemical Agents. Toxic chemical substances that are intended to kill, seriously injure, or seriously incapacitate people through their physiological effects are known as *chemical agents (CAs)* or *chemical warfare agents (CWAs)* [2, 3, 25]. During the twentieth century, over 50 different chemicals in liquid, gas, or solid form have been used and stockpiled as CAs. CAs can be organized into several categories (which can slightly vary in different literature sources) according to the manner in which they affect the human body. *Nerve agents* disrupt the mechanism by which nerves transfer messages to organs. *Blister (vesicant) agents* cause severe skin, eye, and mucosal pain and irritation. *Pulmonary (choking) agents* attack lung tissue, primarily causing pulmonary edema. *Blood agents* prevent the body from utilizing oxygen. Representative examples of CAs are listed in Table 4 [2, 3]. Other types of less lethal CAs include riot-control agents (e.g. pepper spray with capsaicin as an active ingredient and tear gas with *ortho*-chlorobenzylidene-malononitrile or chloroacetophenone as an active ingredient) and incapacitating agents (e.g. 3-quinuclidinyl benzilate, fentanyl-based Kolokol-1).

3.1.2 Toxic Industrial Materials. TIMs are industrial chemicals other than CAs that also have harmful effects on humans [2, 4]. TIMs are also often referred to as *toxic industrial chemicals (TICs)*. They have a LCt_{50} value (lethal concentration for 50% of the population multiplied by exposure time) less than 100 g min/m^3 in any mammalian species and are produced in quantities >30 ton/year at a given production facility. Although they are not as lethal as the highly toxic CAs, their ability to make a significant impact on the population is related to the amount of TIMs that can be released during a terrorist attack.

TIMs are ranked in three categories with respect to their hazard index ranking, indicating their relative importance. A *high hazard* ranking indicates a widely produced, stored, or transported TIM that has high toxicity and is easily vaporized. A *medium hazard* ranking indicates a TIM that may rank high in some categories but lower in others such as number of producers, physical state, or toxicity. A *low hazard* ranking indicates that this TIM is not likely to be a hazard unless specific operational factors indicate otherwise. Table 5 [2] summarizes TIMs by their hazard index. Many TIMs are toxic-by-inhalation (TIH) gases. The top four TIH gases that account for 55% of all highly hazardous chemical processes are listed in Table 6 [4].

3.1.3 Biological Agents. BAs, on the basis of their potential harmful nature, are classified into three categories [18, 25, 26]. *Category A* agents are those that can be easily disseminated or transmitted from person to person, cause high mortality rates, and have the potential for major public health problems as well as disruption of social life. *Category B* agents are next in the priority and include those that are moderately easy to disseminate and result in moderate morbidity rates and low mortality rates. *Category C* contains the third highest priority agents that include emerging pathogens that are easily available and can be produced in large quantity without a significant laboratory setting and have the potential to be engineered for mass dissemination. Selected BAs are summarized in Table 7 [18, 25, 26].

3.2 Challenges in Chemical Sensing of Homeland Security Threats

High selectivity of chemical detection of homeland security threats using existing sensor systems is the most significant challenge. The fundamental reason for the cross sensitivity of individual sensors to different detected species is the need to meet two conflicting

TABLE 4 Categories of Chemical Agents [2, 3, 25]

Nerve Agents	Blister (Vesicant) Agents	Pulmonary (Choking) Agents	Blood Agents
GA –Tabun	HD –Sulfur mustard	CG –Phosgene	CK –Cyanogen chloride
GB –Sarin	HN –Nitrogen mustard	DP –Diphosgene	AC –Hydrogen cyanide
GD –Soman	L –Lewisite	Cl –Chlorine	SA –Arsine
GF –Cyclosarin	MD –Methyldichloroarsine	PS –Chloropicrin	KCN –Potassium cyanide
VX –Methylphosphonothioic acid	PD –Phenyldichloroarsine	DM –Adamsite	NaCN –Sodium cyanide
Novichok	ED –Ethyldichloroarsine	BCME –*Bis*(chloromethyl) ether	

TABLE 5 TIMs Listed by Hazard Index [2]

High	Medium	Low
Ammonia	Acetone cyanohydrin	Allyl isothiocyanate
Arsine	Acrolein	Arsenic trichloride
Boron trichloride	Acrylonitrile	Bromine
Boron trifluoride	Allyl alcohol	Bromine chloride
Carbon disulfide	Allylamine	Bromine pentafluoride
Chlorine	Allyl chlorocarbonate	Bromine trifluoride
Diborane	Boron tribromide	*n*-Butyl chloroformate
Ethylene oxide	Carbon monoxide	sec-Butyl chloroformate
Fluorine	Carbonyl sulfide	*n*-Butyl isocyanate
Formaldehyde	Chloroacetone	*tert*-Butyl isocyanate
Hydrogen bromide	Chloroacetonitrile	Carbonyl fluoride
Hydrogen chloride	Chlorosulfonic acid	Chlorine pentafluoride
Hydrogen cyanide	Diketene	Chlorine trifluoride
Hydrogen fluoride	1,2-Dimethylhydrazine	Chloroacetaldehyde
Hydrogen sulfide	Ethylene dibromide	Chloroacetyl chloride
Nitric acid, fuming	Hydrogen selenide	Crotonaldehyde
Phosgene	Methanesulfonyl chloride	Cyanogen chloride
Phosphorus trichloride	Methyl bromide	Dimethyl sulfate
Sulfur dioxide	Methyl chloroformate	Diphenylmethane-4,4'-diisocyanate
Sulfuric acid	Methyl chlorosilane	Ethyl chloroformate
Tungsten hexafluoride	Methyl hydrazine	Ethyl chlorothioformate
	Methyl isocyanate	Ethyl phosphonothioic dichloride
	Methyl mercaptan	Ethyl phosphonic dichloride
	Nitrogen dioxide	Ethyleneimine
	n-Octyl mercaptan	Hexachlorocyclopentadiene
	Phosphine	Hydrogen iodide
	Phosphorus oxychloride	Iron pentacarbonyl
	Phosphorus pentafluoride	Isobutyl chloroformate
	Selenium hexafluoride	Isopropyl chloroformate
	Silicon tetrafluoride	Isopropyl isocyanate
	Stibine	Nitric oxide
	Sulfur trioxide	Parathion
	Sulfuryl chloride	Perchloromethyl mercaptan
	Sulfuryl fluoride	*n*-Propyl chloroformate
	Tellurium hexafluoride	Tetraethyl lead
	Titanium tetrachloride	Tetraethyl pyroposphate
	Trichloroacetyl chloride	Tetramethyl lead
	Trifluoroacetyl chloride	Toluene 2,4-diisocyanate
		Toluene 2,6-diisocyanate

requirements, such as sensor reversibility and sensor selectivity. High reversibility of sensor response should be achieved *via low energy* of interactions between the analyte and the sensing film, whereas high selectivity of sensor response should be achieved *via high energy* of interactions between the analyte gas and the sensing film. As a result, individual sensors and sensor arrays often cannot detect minute analyte concentrations in the presence of elevated levels of interferences, for example water vapor in air. Figure 4 [27, 28] illustrates typical degradation of the ability to detect low concentrations of

TABLE 6 Top Four TIH Gases that Account for 55% of All Highly Hazardous Chemical Processes [4]

TIH Gas	Percent from All Highly Hazardous Chemical Processes	Absolute Number Of Chemical Processes
Anhydrous ammonia	32.5	8343
Chlorine	18.3	4682
Sulfur dioxide	3	768
Hydrogen fluoride	1.2	315

TABLE 7 Biological Agents, Categorized Based on Their Potential Harmful Nature [18, 25, 26]

Category A	Category B	Category C
Bacillus anthracis (anthrax)	*Burkholderia pseudomallei*	Tickborne hemorrhagic fever viruses (*Crimean-Congo Hemorrhagic fever virus*, Tickborne encephalitis viruses, *Yellow fever virus*, Multidrug-resistant TB, *Influenza virus*, *Rabies virus*)
Clostridium botulinum	*Coxiella burnetti* (Q fever)	
Yersinia pestis	Brucella species (brucellosis)	
Variola major (smallpox) and other pox viruses	*Burkholderia mallei* (glanders)	
Francisella tularensis(tularemia)	Ricin toxin (from Ricinus communis)	
Viral hemorrhagic fevers (*Arenaviruses*, *Orthobunyavirus*, *Flaviruses*, *Filoviruses*)	Epsilon toxin of Clostridium perfringens	
	Staphylococcus enterotoxin B	
	Typhus fever (*Rickettsia prowazekii)*	
	Food and Waterborne Pathogens Bacteria (Diarrheagenic *Escherichia coli*, Shigella species, Salmonella, *Listeria monocytogenes*, *Campylobacter jejuni*, *Yersinia enterocolitica*), Viruses (*caliciviruses*, *Hepatitis A*), Protozoa (*Cryptosporidium parvum*, *Cyclospora cayatanensis*, *Giardia lamblia*, *Entamoeba histolytica*)	

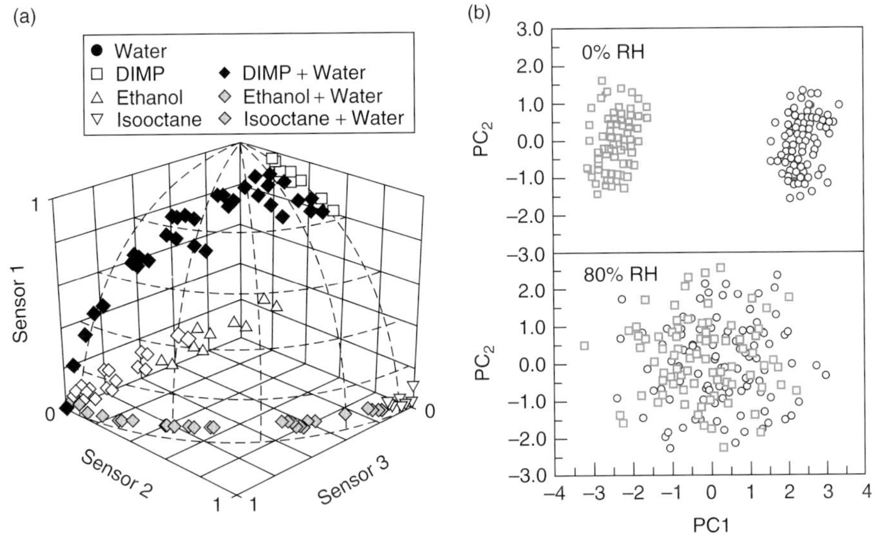

FIGURE 4 Typical effects of water vapor in air on the ability to selectively detect analyte vapors using conventional sensor arrays. (a) Normalized responses of three polymer-coated chemiresistor sensors to diisopropylmethylphosphonate (DIMP), ethanol, and isooctane, mixed with water vapor at concentrations from 0 to 100%. The responses to binary mixtures of solvent and water vapor form "trails" on the surface of the sphere from pure water vapor to the pure solvent vapor. (b) Principal components score plots of an array of 10 chemiresistors coated with diverse surface-functionalized single-wall carbon nanotubes sensing films upon exposure to two types of vapor mixtures (squares and circles) at 0 and 80% RH. (a) Adapted from Reference [27] with permission and (b) adapted from Reference [28] with permission.

analyte vapors in the presence of relatively high levels of water vapor. Thus, it is critical that new sensors for homeland security applications will be not disappointingly affected by variable levels of water vapor in air, with concentrations up to 50–95% of its saturated H_2O vapor pressure. There are also many other interferences of significance to homeland security applications in measured air. Some typical examples of additional interferences tested with sensors include vapors of diesel fuel, gasoline, floor stripper and polish formulations, disinfectant bleach, machine oil, and many others. However, concentrations of these interferences are much less, typically only up to 1–5% of their saturated vapor pressure [29]. Thus, the key successful phase in the development of new sensors is their ability to operate in the presence of high concentrations of water vapor in air.

3.3 Technical Solution—New Sensor Platform with Multivariable Signal Transduction

Over the years, wireless proximity-operated sensors have been reported based on diverse transducer designs such as resonant inductor–capacitor, magnetoelastic, thickness shear mode, and surface-acoustic wave transducers [24]. Radiofrequency identification (RFID) tags have been recognized as one of the disruptive technologies and are widely used ranging from the detection of unauthorized opening of containers to automatic identification of animals and to tracking of a wide variety of assets [30, 31]. Although usual RFID tags are ubiquitously employed as electronic labels and can cost only 5c in large quantities [32], known approaches of RFID sensing typically require a battery or a proprietary-redesigned integrated circuit (IC) memory chip with typically a one-bit analog input [30, 33] preventing the wide adoption of RFID devices for sensing. Although battery-powered active RFID sensors transmit data over large distances, unfortunately, the battery adds to the system maintenance and complexity and reduces system's life. Passive devices are attractive when there is a need for the smallest sensor size, when a sensor is deployed for a long-term application, when a high power RF transmission is prohibited, or when the sensor should be disposable or low cost.

Recently, ubiquitous passive RFID tags were adapted for unusually selective chemical sensing [23, 24]. By applying a carefully selected sensing material onto the resonant antenna of the RFID tag and measuring the complex impedance of the antenna, the impedance spectrum response was correlated to the concentration of a chemical compound of interest in the presence of high levels of background interferences. The digital data were also written into and read from the IC memory chip of the RFID tag. This IC memory chip stored sensor calibrations and user-defined information. Using this attractive sensing platform, a new concept for selective vapor sensing has been contemplated and experimentally demonstrated. This new sensing principle requires only a single sensor and involves the combination of a sensing material that has different response mechanisms to different vapors with a transducer that has a multivariable signal transduction ability to detect these independent changes. In numerous experiments, the action of several vapor-response mechanisms in a single sensing film to different vapors was demonstrated and the independent detection of these responses with such single sensor was performed.

Compared to other sensor technologies (summarized in exemplary references [2, 10, 12–14] and Table 3), developed RFID sensors have a significantly improved response selectivity to analytes, are able to detect several analytes with a single sensor, and are able to reject effects from interferences.

3.3.1 ***Principle of Chemical Sensing with RFID Sensors.*** The operation principle of chemical RFID sensors is illustrated in Figure 5 [24]. Reading and writing of digital information into the RFID sensor and measurement of complex impedance of the RFID sensor antenna are performed via mutual inductance coupling between the RFID sensor antenna and the pickup coil of a reader (Fig. 5a). A conventional digital RFID reader acquires digital data from an IC memory chip on the RFID sensor. This digital data have a unique factory programmed serial number (chip ID) as well as user-written data about the properties of the sensor (e.g. calibration curves for different conditions) and the object to which the sensor is attached (e.g. fabrication and expiration dates).

The origin of response of RFID sensor to chemical parameters is described in Figure 5 b and c. Upon reading of the RFID sensor with a pickup coil, the electromagnetic field generated in the RFID sensor antenna extends out from the plane of the RFID sensor (Fig. 5b) and is affected by the dielectric property of an ambient environment. When the resonant antenna of the RFID sensor is coated with a sensing film (Fig. 5c), the analyte-induced changes in the dielectric and dimensional properties of the sensing film affect the complex impedance of the antenna circuit through the changes in film resistance R_F and capacitance C_F between the antenna turns (inset of Fig. 5c). Such changes facilitate diversity is the response of individual RFID sensors and provide the opportunity to replace a whole array of conventional sensors with a single vapor-selective RFID sensor.

For selective analyte quantitation using individual RFID sensors, complex impedance spectra of the resonant antenna are measured as shown in Figure 5d. Several parameters

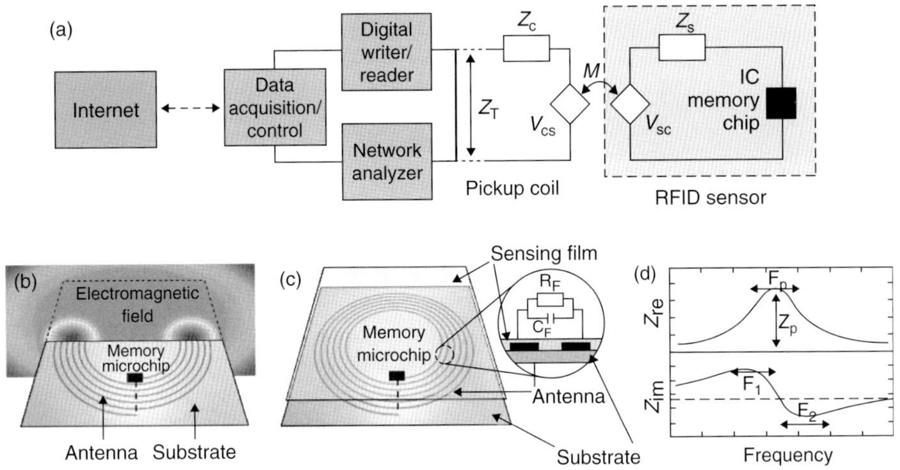

FIGURE 5 Operation principle of passive battery-free RFID sensors. (a) System schematic of writing and reading digital information into the sensor IC memory chip and measuring complex impedance of the sensor antenna. Z_C and Z_S are intrinsic impedance of the pickup coil and sensor, respectively; Z_T is total impedance; V_{CS} and V_{SC} are dependent voltage sources; and M is mutual inductance coupling. (b) Visualization of the electromagnetic field in the resonant-sensing antenna that is generated upon excitation with a pick up coil of the sensor reader. (c) Origin of response of RFID sensors to chemical parameters via a sensing film deposited onto the resonant antenna. Inset, analyte-induced changes in the film affect the complex impedance of the antenna circuit through the changes in film resistance R_F and capacitance C_F between the antenna turns. (d) Measured complex impedance spectrum (real part Z_{re} and imaginary part Z_{im} of complex impedance) and representative parameters for multivariate analysis: frequency of the maximum of the real part of the complex impedance (F_p), magnitude of the real part of the complex impedance (Z_p), resonant frequency of the imaginary part of the complex impedance (F_1), and antiresonant frequency of the imaginary part of the complex impedance (F_2). Adapted from Reference [24] with permission.

from the measured real and imaginary portions of the complex impedance are further calculated. Examples of calculated parameters include frequency of the maximum of the real part of the complex impedance (F_p), magnitude of the real part of the complex impedance (Z_p), resonant frequency of the imaginary part of the complex impedance (F_1), and antiresonant frequency of the imaginary part of the complex impedance (F_2). Additional parameters can also be calculated (e.g. zero-reactance frequency and quality factor). However, the use of F_p, F_1, F_2, and Z_p was found adequate for selective sensing. Upon a proper selection of a sensing film, the film-coated RFID sensor has responses different for each tested analyte or the analyte and interferences. By applying multivariate analysis of the full complex impedance spectra or the calculated parameters, quantitation of analytes and their mixtures with interferences is performed with individual RFID sensors. Examples of RFID tags adapted for sensing are presented in Figure 6.

3.3.2 Comparison of Analog and Digital RFID Sensor-Data Transfer. Wireless sensors are under development as passive (battery-free) and active (battery-powered) devices for diverse applications, where a connection between the sensor and the reader without an electrical contact is important. Battery-powered sensors have an obvious advantage of data transmission over large distances [30]. At the same time, a battery adds to the system

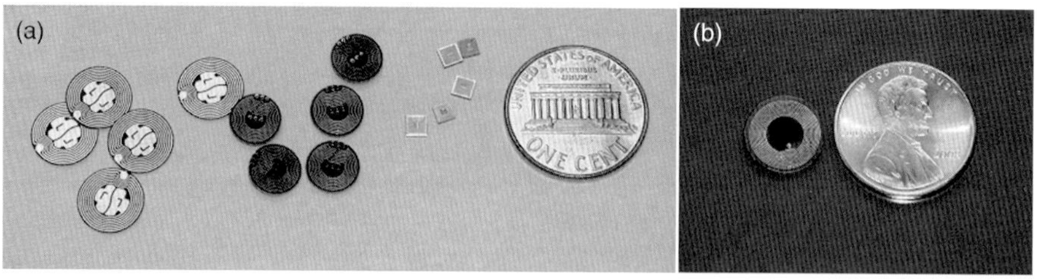

FIGURE 6 Examples of different sizes and form-factors of conventional RFIDs adapted for sensing: (a) RFID tags from different manufacturers adapted for sensing and utilized in initial experiments and (b) RFID sensor with an antenna structure specifically developed for sensing application.

maintenance and complexity, reduces system life, and limits the temperature range of sensor applications. Although a possible alternative is scavenging of ambient energy (solar, mechanical, thermal, etc.) [16], at present, power scavenging is not mature enough for its wide applicability [34]. Passive devices are more attractive in situations when there is a need for the smallest sensor size, when a sensor is deployed for a long-term application, when a high power RF transmission is prohibited, or when the sensor should be of low cost or disposable.

In passive sensors, two broad approaches for sensor-data transfer include analog and digital data transfer as summarized in Table 8 [24]. In digital data transfer, sensor circuits are limited to very simple designs because of the need for enough available electrical power to operate. As a result, these sensors often have only one-bit resolution [30, 33] and, therefore, they only operate as threshold switches without capability to provide continuous sensing information. For example, RFID temperature-threshold sensors [35] last only for one measurement because they change their state irreversibly upon reaching a temperature threshold. Performance of such RFID sensor depends on the design of the IC memory chip with an analog input. Even with an increase in the resolution of an analog-to-digital converter of an IC memory chip, these digital sensors will measure only a *single parameter per sensor* and, thus, will suffer from environmental interference effects similarly to other reported sensors. The RFID readers of digital sensors typically operate with their own proprietary digital protocols but cannot affect the quality of sensor performance. In sensors with digital signal transmission, the mutual inductance coupling between the pickup coil and the sensor needs to be controlled in order to avoid possible bit errors. Operation of a sensor coupled to the analog sensor input of the IC memory chip also requires significant power. Thus, the read range of such sensors is significantly less.

In analog data transfer, the sensing capability resides in design of both the RFID sensor antenna and RFID sensor reader. The synergistic combination of the resonant antenna design of the RFID sensor and the sensing film deposited onto the antenna provides the foundation for the selective and sensitive sensor response. However, it is the RFID sensor reader that is responsible for the high resolution and signal-to-noise ratio of the acquired signal. Typical resolution of the sensor reader is 16 bit (vs. 1-bit resolution of available digital sensors). RFID sensors with analog data transfer measure the complex impedance of the resonant sensor antenna, combine *several measured parameters* from the antenna with multivariate data analysis, and deliver unique capability for multianalyte sensing and rejection of interferences using only a single sensor.

TABLE 8 Key Features of Analog and Digital Sensors in Passive Inductively Coupled RFID Devices [24]

Sensor Performance Parameters	Capabilities of Analog Sensor-Data Transfer	Capabilities of Digital Sensor-Data Transfer
Opportunities in sensor–reader combination	Basis of RFID sensors is a standard low-cost RFID tag	Basis of RFID sensor is a custom IC chip with analog input on RFID tag
	Sensing performed by add-ons to standard RFID tag	Sensing performed by adding a separate sensor to IC chip
	Key monitoring capabilities are in RFID sensor reader that measures complex impedance of sensor	Key monitoring capabilities are in RFID tag with attached sensor
Measurement resolution	Typical 16-bit resolution provided by design of RFID sensor reader	Typical 1-bit resolution provided by design of IC chip
Multianalyte sensing and rejection of environmental interferences by a single sensor	Available by using multivariate analysis of measured complex impedance of resonant RFID antenna	Unavailable because only a single parameter per sensor is measured
Effects of variable mutual inductance coupling	Lead to sensor errors, however corrected using several methods	Lead to sensor bit errors, difficult to correct
Communication range	Several centimeters, limited to size of employed pickup coil	Several centimeters, limited by power for custom IC chip with analog input
Prospect as universal platform for physical, chemical, and biological sensing	Common sensor platform for measurements of physical, chemical, and biological parameters	Need for different separate sensors to be attached to RFID tag to detect different environmental parameters

3.3.3 RFID Dosimeter for Exposure to TIMs. Developed RFID sensors were tested for the detection of TIMs. Ammonia was selected as an analyte of choice based on its high hazard index (Table 5) and because ammonia tops the list of TIH gases (Table 6). As sensing materials, intrinsically conducting polymers (ICPs) were chosen for selective determination of vapors because of three key reasons [6, 9–11, 36]: (i) ICPs can exhibit several mechanisms of molecular recognition of gases that include changes in density of charge carriers, changes in mobility of charge carriers, polymer swelling, and conformational transitions of polymer chains. (ii) ICPs can be more sensitive than other sensing materials due to their inherent electrical transport property and energy migration. (iii) The inherent electrical transport property of ICPs is the material bulk transport property; thus, the readout of this material property can be more sensitive than potentiometric and amperometric methods that depend on local electronic structure.

As a sensing material in these experiments, polyaniline (PANI) was selected because it is a well-studied ICP for vapor sensing [11]. The response mechanism of PANI to NH_3 involves polymer deprotonation, whereas the response mechanism to H_2O involves formation of hydrogen bonds and swelling [37]. Results of sensor exposures to NH_3 and H_2O vapors are presented in Figure 7a–d. Deprotonation of the film upon NH_3

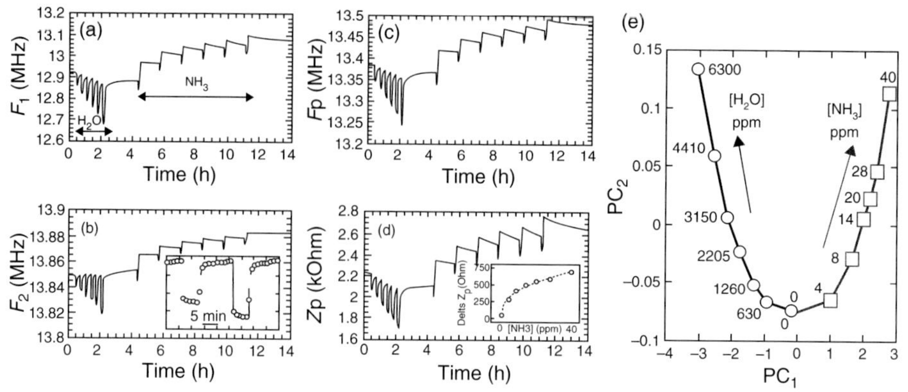

FIGURE 7 Selective analysis of NH_3 and H_2O vapors using a single sensor with the multivariable signal transduction. (a–d) Sensor responses F_1, F_2, F_p, and Z_p, respectively, upon ~10 min exposures of sensor to H_2O vapor (630, 1260, 2205, 3150, 4410, and 6300 ppm) and to NH_3 vapor (4, 8, 14, 20, 8, and 40 ppm). Note reversible response to H_2O vapor and nonreversible response to NH_3 vapor. Inset in (b), dynamic response to H_2O vapor. Inset in (d), univariate calibration curve for NH_3 determinations. (e) Scores plot of PC_1 versus PC_2 demonstrates discrimination between NH_3 and H_2O vapor responses.

exposures resulted in the increase in film impedance Z_p and shifts of the sensor resonance F_p, F_1, and F_2 to higher frequencies. The formation of hydrogen bonds and swelling of the polymer upon H_2O exposures resulted in the decrease in Z_p and shifts of F_p, F_1, and F_2 to lower frequencies. Measurements of multiple output parameters from a single sensor revealed different recovery kinetics of responses Z_p, F_1, F_2, and F_p during experiments with NH_3. Responses Z_p, F_1, and F_p showed a partial recovery from NH_3, while F_2 response was irreversible with only 1.2–3.5% signal recovery. This irreversible F_2 response is attractive to take dosimeter readings at a later time, for example at the end of an 8-h work shift. Response to H_2O vapor was reversible and at least 100-fold weaker over the response to NH_3. Univariate Z_p, F_1, F_2, and F_p calibration curves to NH_3 showed relatively high response sensitivity at low concentrations. This nonlinear behavior is typical to PANI films [38]. The detection limit (based on 3σ criterion) was calculated to be 15–80 ppb of NH_3 from Z_p, F_1, F_2, and F_p measurements. This achieved detection limit is much better over nanosensors with PANI nanowires [38] and single-walled carbon nanotubes [39].

For comparisons with earlier reported sensors, selectivity of developed RFID sensors was evaluated using PCA [21]. PCA is a robust tool for processing of multivariate signals that is used by 10 out of 13 surveyed electronic nose manufacturers [40]. As shown in the PCA scores plot (Fig. 7e), the action of several vapor-response mechanisms in a single sensing film was independently detected and quantified with a single multivariable signal transduction RFID sensor.

3.3.4 Reliable Quantitation of Toxic VOCs in the Presence of Variable Humidity. The ability of RFID sensors to operate in the presence of variable ambient RH and to reject effects of ambient humidity was further evaluated in detail. In one such vapor sensor, individual measured parameters are affected by RH as shown in a PCA scores plot versus experimental time (Fig. 8a). However, critical to the sensor performance,

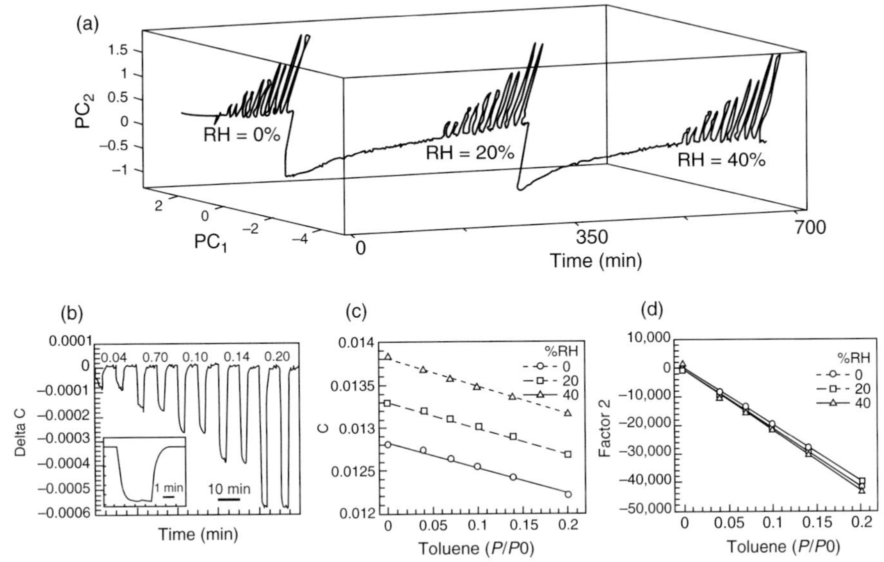

FIGURE 8 Initial demonstration of humidity-independent operation using a single sensor with the multivariable signal transduction. (a) Plot of PC_1 versus PC_2 versus time illustrates sensor response to five concentrations of toluene vapor (0.04, 0.07, 0.10, 0.14, and 0.20 P/P_o, two replicates each) at three humidity levels. (b) Response reproducibility and dynamics. Univariate (c) and multivariate (d) calibration curves for toluene detection at 0, 20, and 40% RH. P/P_o is partial vapor pressure during experiments.

sensing material applied onto the RFID antenna responds with the same magnitude to the model analyte vapor (toluene) in the presence of different humidity levels. The reproducibility and dynamics of the response to toluene at a single humidity level are presented in Figure 8b. Univariate calibration curves for toluene detection are presented in Figure 8c and show the preserved sensitivity and linearity of sensing film response at different humidity levels. A full correction of toluene response at different humidity levels was done using multivariate analysis of multiple responses from the single sensor. The resulting multivariate calibration curves at variable RH are identical (Fig. 8d) and provide a new capability to quantify vapors at different humidity. In these experiments, the detection of vapors was performed in the presence of up to 400-fold more concentrated water vapor. Measurements over extended period of time (45 h) were further performed to evaluate effects of even higher humidity levels, up to 76% RH where the sensor also did not change the response magnitude to the analyte at high humidity of the carrier gas.

Using the developed knowledge in the design of the sensing materials for RFID sensors, several types of new sensing materials were synthesized and determination of toxic vapors was performed down to 900 ppb detection limits and with eliminated humidity effects. Figure 9a illustrates an example of sensor response F_p to variable concentrations of TCE, water, and toluene vapors, demonstrating that water vapor response was not only negligible but also opposite in its response direction. Further, stability of sensor response to toluene vapor was tested at variable humidity levels of the carrier gas (0, 22, 44, 65, and 76% RH) as shown in Figure 9b. It was found that from conservative estimations based on the multivariable sensor response, the detection of toxic gases can be performed in the presence of 27,000-fold more concentrated water vapor.

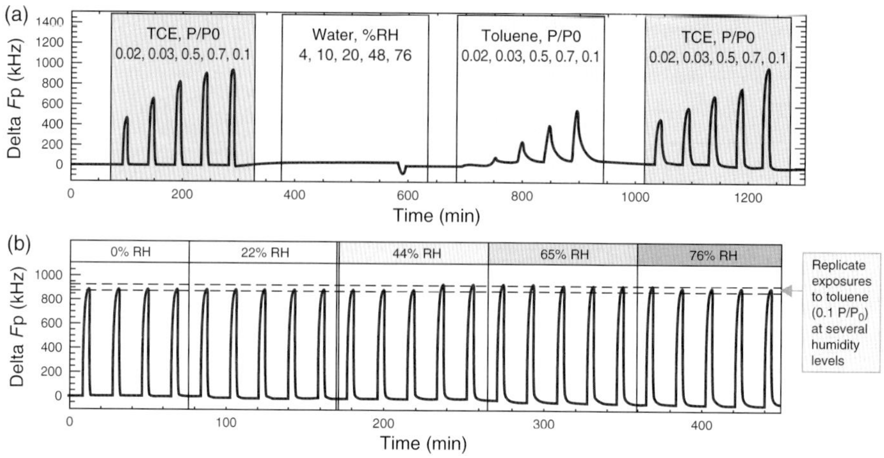

FIGURE 9 Advanced demonstration of humidity-independent operation using a single sensor with the multivariable signal transduction. (a) Typical response to different concentrations of TCE (trichloethylene) vapor, water vapor, and toluene vapor. (b) Stability of sensor response to 0.1 P/P_o of TCE vapor at variable humidity levels of the carrier gas.

3.3.5 Selective Detection of CWA Simulants. Effects of different vapors on the response of the developed sensors were further evaluated by selecting a difficult combination of vapors such as methanol (MeOH), ethanol (EtOH), water (H_2O), and acetonitrile (ACN) [23]. Acetonitrile was selected as a simulant for blood CWAs [41]. Figure 10a demonstrates the measured Z_p response for four analytes (H_2O, EtOH, MeOH, and ACN) for multiple concentrations and replicates ($n = 3$) [23]. Measurements of a single parameter of an RFID sensor, for example Z_p, cannot discriminate between different analytes. For example, if a signal Z_p is changed by

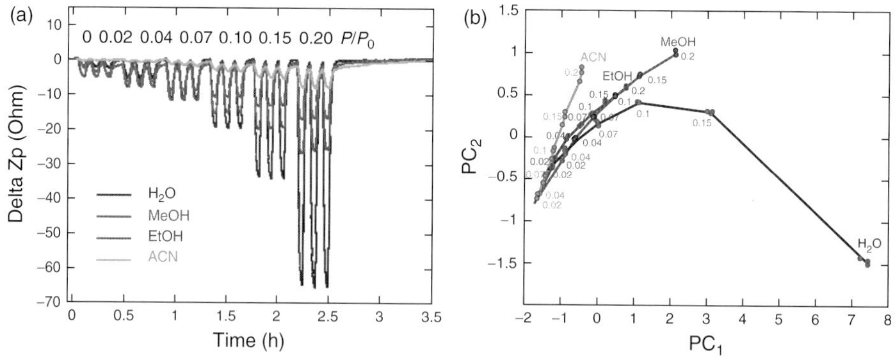

FIGURE 10 Selective detection of ACN as a CWA simulant using a single sensor with the multivariable signal transduction. (a) Measured Z_p response for four analytes (H_2O, EtOH, MeOH, and ACN), six concentrations (0, 0.02, 0.04, 0.07, 0.10, 0.15, and 0.20 P/P_o), and three replicates. (b) Results of the PCA multivariate analysis of measured parameters F_1, F_2, F_p, and Z_p of the RFID sensor to the changes in H_2O, EtOH, MeOH, and ACN at six concentrations each, and three replicates per concentration. Numbers are P/P_o values for individual vapors. Adapted from Reference [23] with permission.

~20 ohm, this change can be due to 0.1 P/P_0 of H_2O, 0.15 P/P_0 of MeOH, or 0.2 P/P_0 of EtOH. Thus, a single-parameter measurement of the RFID sensor cannot discriminate between different analytes and their concentrations. However, by applying the developed data-processing algorithm on the multivariable response of this single sensor, a good vapor discrimination has been achieved. Figure 10b illustrates that three out of four vapors were resolved using a selected sensing film and the complex impedance readout from the RFID sensor [23]. In particular, ACN vapor was well discriminated from H_2O, MeOH, and EtOH vapors. For detection of other analytes of interest to homeland security applications, diverse sensing materials can be applied [6, 9–11].

4 SUMMARY AND CONCLUSIONS

The detection of threats of importance to homeland security is needed with enhanced sensitivity for the detection of agents below health risk levels, with increased selectivity for operation in the presence of uncontrolled variable humidity and for providing minimal false-alarm rates and with expedited response rate for enabling an effective medical response. To meet these and many other new requirements, sensing technologies with previously unavailable capabilities are required.

These new capabilities will be difficult or even impossible to achieve using evolutionary improvements in existing sensing technologies. Thus, conceptually new technical solutions should be introduced in all key components of a sensor system as shown in Figure 1. Sensor systems with a sampler will benefit from new analyte preconcentrator concepts to rapidly collect and selectively release analyte species utilizing much less energy than is currently required. From the rigorous mathematical standpoint of data processing from a sensor, using a stable and high signal-to-noise multivariable response of a carefully designed sensor, it is possible to identify and quantify compounds that were not previously tested with the sensor [42]. To fully implement this capability, new sensing materials with more diverse response mechanisms to different species will be required, with well understood selection criteria [10]. These materials are under development using rational and combinatorial approaches [6, 10]. These sensing materials should be further coupled with new designs of multivariable transducers. Complete sensor systems will have more self-calibration and self-diagnostic abilities. At present, often the size of a transducer is much smaller than an associated battery needed for its operation. Thus, in sensors, too small to accommodate a conventional battery, power scavenging will become more important. This need will drive further the advanced packaging requirements to minimize needed power without degrading sensor performance. Of course, design, fabrication, and implementation phases of sensors will also be significantly impacted by progress in numerous disciplines and technologies as diverse as analytical chemistry, chemometrics, materials science, computer science, electrical engineering, artificial intelligence, and many others bound only by our imagination.

ACKNOWLEDGMENTS

This work has been supported by GE Corporate long-term research funds. We are grateful to J. Cella, K. Chichak, and T. Sivavec for materials synthesis.

REFERENCES

1. Fitch, J. P., Raber, E., and Imbro, D. R. (2003). Technology challenges in responding to biological or chemical attacks in the civilian sector. *Science* **302**, 1350–1354.
2. Fatah, A. A., Barrett, J. A., Arcilesi, J., Richard, D., Ewing, K. J., Lattin, C. H., and Helinski, M. S. (2000). *Guide for the Selection of Chemical Agent and Toxic Industrial Material Detection Equipment for Emergency First Responders, NIJ Guide 100-00*, Vol. 1, National Institute of Justice, Law Enforcement and Corrections Standards and Testing Program, Washington DC.
3. Shea, D. A. (2003). *High-Threat Chemical Agents: Characteristics, Effects, and Policy Implications*, CRS Report for Congress, Order Code RL31861, Congressional Research Service. The Library of Congress.
4. Hind, R. (2008). *Testimony before the Committee on Homeland Security on "Chemical Facility Anti-Terrorism Act of 2008"* http://www.greenpeace.org/raw/content/usa/press-center/reports4/chemsecuritytestimony.pdf.
5. Hulanicki, A., Glab, S., and Ingman, F. (1991). Chemical sensors: definitions and classification. Commission on General Aspects of Analytical Chemistry. *Pure Appl. Chem.* **63**(9), 1247–1250.
6. Potyrailo, R. A. (2006). Polymeric sensor materials: toward an alliance of combinatorial and rational design tools? *Angew. Chem. Int. Ed.* **45**, 702–723.
7. Middelhoek, S., and Noorlag, J. W. (1981/82). Three-dimensional representation of input and output transducers. *Sens. Actuators* **2**, 29–41.
8. Vilkner, T., Janasek, D., and Manz, A. (2004). Micro total analysis systems. *Recent Dev. Anal. Chem.* **76**, 3373–3386.
9. Hatchett, D. W., and Josowicz, M. (2008). Composites of intrinsically conducting polymers as sensing nanomaterials. *Chem. Rev.* **108**, 746–769.
10. Potyrailo, R. A., and Mirsky, V. M. (2008). Combinatorial and high-throughput development of sensing materials: the first ten years. *Chem. Rev.* **108**, 770–813.
11. Lange, U., Roznyatovskaya, N. V., and Mirsky, V. M. (2008). Conducting polymers in chemical sensors and arrays. *Anal. Chim. Acta* **614**, 1–26.
12. Röck, F., Barsan, N., and Weimar, U. (2008). Electronic nose: current status and future trends. *Chem. Rev.* **108**, 705–725.
13. Hierlemann, A., and Gutierrez-Osuna, R. (2008). Higher-order chemical sensing. *Chem. Rev.* **108**, 563–613.
14. Joo, S., and Brown, R. B. (2008). Chemical sensors with integrated electronics. *Chem. Rev.* **108**, 638–651.
15. Diamond, D., Coyle, S., Scarmagnani, S., and Hayes, J. (2008). Wireless sensor networks and chemo-/biosensing. *Chem. Rev.* **108**, 652–679.
16. Jiang, B., Smith, J. R., Philipose, M., Roy, S., Sundara-Rajan, K., and Mamishev, A. V. (2007). Energy scavenging for inductively coupled passive RFID systems. *IEEE Trans. Instrum. Meas.* **56**, 118–125.
17. Carrano, J. C., Jeys, T., Cousins, D., Eversole, J., Gillespie, J., Healy, D., Licata, N., Loerop, W., O' Keefe, M., Samuels, A., Schultz, J., Walter, M., Wong, N., Billotte, B., Munley, M., Reich, E., and Roos, J. (2004). Chemical and biological sensor standards study (CBS3). In *Optically Based Biological And Chemical Sensing For Defence*, Vol. 5617, J. C., Carrano, and A. Zukauskas Eds. SPIE—The International Society for Optical Engineering, Bellingham, WA, pp. xi–xiii.
18. Brower, J. L. (2006). The terrorist threat and its implications for sensor technologies. In *Advances in Sensing with Security Applications Amsterdam*, J. Byrnes, and G. Ostheimer Eds. Springer, The Netherlands, pp. 23–54.

19. Snow, E. S., Perkins, F. K., Houser, E. J., Badescu, S. C., and Reinecke, T. L. (2005). Chemical detection with a single-walled carbon nanotube capacitor. *Science* **307**, 1942–1945.

20. Qi, X., and Osterloh, F. E. (2005). Chemical sensing with LiMo3Se3 nanowire films. *J. Am. Chem. Soc* **127**, 7666–7667.

21. Jurs, P. C., Bakken, G. A., and McClelland, H. E. (2000). Computational methods for the analysis of chemical sensor array data from volatile analytes. *Chem. Rev.* **100**, 2649–2678.

22. Potyrailo, R. A., May, R. J., and Sivavec, T. M. (2004). Recognition and quantification of perchloroethylene, trichloroethylene, vinyl chloride, and three isomers of dichloroethylene using acoustic-wave sensor array. *Sensor Lett.* **2**, 31–36.

23. Potyrailo, R. A., and Morris, W. G. (2007). Multianalyte chemical identification and quantitation using a single radio frequency identification sensor. *Anal. Chem.* **79**, 45–51.

24. Potyrailo, R. A., Morris, W. G., Sivavec, T., Tomlinson, H. W., Klensmeden, S., and Lindh, K. (2008). RFID sensors based on ubiquitous passive 13.56-MHz RFID tags and complex impedance detection. *Wireless Commun. Mobile Comput.* DOI: 10.1002/wcm.711.

25. CDC (2008). *Emergency Preparedness and Response*. Centers for Disease Control and Prevention, Atlanta, GA, http://www.bt.cdc.gov.

26. Yadav, P., and Blaine, L. (2004). Microbiological threats to homeland security. *IEEE Eng. Med. Biol.* **23**, 136–141.

27. Patel, S. V., Jenkins, M. W., Hughes, R. C., Yelton, W. G., and Ricco, A. J. (2000). Differentiation of chemical components in a binary solvent vapor mixture using carbon/polymer composite-based chemiresistors. *Anal. Chem.* **72**, 1532–1542.

28. Peng, G., Trock, E., and Haick, H. (2008). Detecting simulated patterns of lung cancer biomarkers by random network of single-walled carbon nanotubes coated with nonpolymeric organic materials. *Nano Lett.* **8**, 3631–3635.

29. Meier, D. C., Evju, J. K., Boger, Z., Raman, B., Benkstein, K. D., Martinez, C. J., Montgomery, C. B., and Semancik, S. (2007). The potential for and challenges of detecting chemical hazards with temperature-programmed microsensors. *Sens. Actuators B* **121**, 282–294.

30. Finkenzeller, K. (2003). *RFID Handbook. Fundamentals and Applications in Contactless Smart Cards and Identification*, 2nd ed., Wiley, Hoboken, NJ.

31. Lehpamer, H. (2008). *RFID Design Principles*, Artech House, Norwood, MA.

32. Roberti, M. (2006). A 5-Cent breakthrough. *RFID J.* May, http://www.rfidjournal.com/article/articleview/2295.

33. Yang, C.-H., Chien, J.-H., Wang, B.-Y., Chen, P.-H., and Lee, D.-S. (2008). A flexible surface wetness sensor using a RFID technique. *Biomed. Microdevices* **10**, 47–54.

34. Philipose, M., Smith, J. R., Jiang, B., Mamishev, A., Roy, S., and Sundara-Rajan, K. (2005). Battery-free wireless identification and sensing. *IEEE Pervasive Comput.* **4**, 37–45.

35. Want, R. (2004). *Enabling Ubiquitous Sensing with RFID Computer*, pp. 84–86.

36. Sugiyasu, K., and Swager, T. M. (2007). Conducting-polymer-based chemical sensors: transduction mechanisms. *Bull. Chem. Soc. Jpn.* **80**, 2074–2083.

37. Nicolas-Debarnot, D., and Poncin-Epaillard, F. (2003). Polyaniline as a new sensitive layer for gas sensors. *Anal. Chim. Acta* **475**, 1–15.

38. Liu, H., Kameoka, J., Czaplewski, D. A., and Craighead, H. G. (2004). Polymeric nanowire chemical sensor. *Nano Lett.* **4**, 671–675.

39. Bekyarova, E., Davis, M., Burch, T., Itkis, M. E., Zhao, B., Sunshine, S., and Haddon, R. C. (2004). Chemically functionalized single-walled carbon nanotubes as ammonia sensors. *J. Phys. Chem. B* **108**, 19717–19720.

40. Snopok, B. A., and Kruglenko, I. V. (2002). Multisensor systems for chemical analysis: state-of-the-art in Electronic Nose technology and new trends in machine olfaction. *Thin Solid Films* **418**, 21–41.

41. Choi, N.-J., Kwak, J.-H., Lim, Y.-T., Bahn, T.-H., Yun, K.-Y., Kim, J.-C., Huh, J.-S., and Lee, D.-D. (2005). Classification of chemical warfare agents using thick film gas sensor array. *Sens. Actuators B* **108**, 298–304.

42. Grate, J. W., Wise, B. M., and Abraham, M. H. (1999). Method for unknown vapor characterization and classification using a multivariate sorption detector. Initial derivation and modeling based on polymer-coated acoustic wave sensor arrays and linear solvation energy relationships. *Anal. Chem.* **71**, 4544–4553.

FURTHER READING

Bartelt-Hunt, S. L., Knappe, D. R. U., and Barlaz, M. A. (2008). A review of chemical warfare agent simulants for the study of environmental behavior. *Crit. Rev. Environ. Sci. Technol.* **38**, 112–136.

Chauhan, S., Chauhan, S., D'Cruz, R., Faruqi, S., Singh, K. K., Varma, S., Singh, M., and Karthik, V. (2008). Chemical warfare agents. *Environ. Toxicol. Pharmacol.* **26**, 113–122.

Eubanks, L. M., Dickerson, T. J., and Janda, K. D. (2007). Technological advancements for the detection of and protection against biological and chemical warfare agents. *Chem. Soc. Rev.* **36**, 458–470.

Kendall R. J., Presley S. M., Austin G. P., and Smith P. N. Eds. (2008). *Advances in Biological and Chemical Terrorism Countermeasures*, CRC Press, Boca Raton, FL.

Sadik, O. A., Land, W. H. Jr, and Wang, J. (2003). Targeting chemical and biological warfare agents at the molecular level. *Electroanalysis* **15**, 1149–1159.

Szinicz, L. (2005). History of chemical and biological warfare agents. *Toxicology* **214**, 167–181.

PROTECTION, PREVENTION, RESPONSE AND RECOVERY

PROTECTION AND PREVENTION: AN OVERVIEW

John Cummings
Sandia National Laboratories, Albuquerque, New Mexico

1 BACKGROUND

Much of what we currently consider part of "homeland security" has its origins in the late 1990s—especially in the work leading to the publication of Presidential Decision Directive 63 (PDD-63) Protecting America's Critical Infrastructures [1]. This Presidential Directive built on the recommendations of the President's Commission on Critical Infrastructure Protection. In October 1997, the Commission issued its report calling for a national effort to assure the security of the United States' increasingly vulnerable and interconnected infrastructures, such as telecommunications, banking and finance, energy, transportation, and essential government services. PDD-63 assigned lead agencies for specific infrastructure sectors and functions. In addition, the Office of Science and Technology Policy (OSTP) was assigned the responsibility to coordinate research and development agendas and programs for the federal government through the National Science and Technology Council (NSTC).

OSTP established the Critical Infrastructure Protection R&D Interagency Working Group (CIP R&D IWG) shortly after PDD-63 was issued as a subgroup to the NSTC Committee on National Security (CNS) and Committee on Technology (CT). In effect, the CIP R&D IWG was jointly placed under the CNS and the CT, and it reported to both. The CIP R&D IWG developed a database of existing federal government CIP R&D programs in conjunction with the Office of Management and Budget (OMB) and through data calls to the federal executive branch departments and agencies. The IWG also developed a set of CIP R&D agendas (strategies and road maps) that formulated the conceptual framework for a national level, strategic Infrastructure Protection Research and Development (R&D) Plan [2] to mitigate both cyber and physical threats. The agendas recommended R&D investments by type and agency, based on gaps between desired R&D and actual R&D reported in the database. The OSTP continued to encourage agencies to conduct research and development to protect the nation's critical infrastructure after PDD-63, but most efforts were limited in scope and funding [3].

All of that changed after the events of September 11, 2001 and the anthrax incidents that followed shortly thereafter. The formation of the Office of Homeland Security (OHS) provided renewed focus on the nation's infrastructure and expanded the sectors and assets of concern beyond those addressed in PDD 63. A series of national strategies was published by OHS or its affiliates including one that focused on the critical infrastructure of the United States [4]. The National Research Council formed a Committee on Science and Technology for Countering Terrorism that wrote a report that focused on the scientific and technological means by which we can reduce the vulnerabilities of our society to terrorist attacks, and mitigate the consequences of those attacks when they occur [5]. The RAND National Defense Research Institute sponsored a series of workshops involving industry, academe, and government officials that gathered information on physical protection of the nation's most critical infrastructure sectors. Finally, in March, 2003 the Department of Homeland Security (DHS) came into being and began the task of coordinating the national effort to protect critical infrastructure and key assets.

In December, 2003 the White House published Homeland Security Presidential Directive 7 (HSPD-7; "Critical Infrastructure Identification, Prioritization, and Protection [6]") which outlined the requirements for protecting the nation's critical infrastructure. HSPD-7 defined these critical infrastructure as consisting of the following sectors and key resources: agriculture and food, water, public health and healthcare, emergency services, the defense industrial base, information technology, telecommunications, energy, transportation systems, banking and finance, chemical, postal and shipping, national monuments and icons, dams, government facilities, commercial facilities, and nuclear reactors, materials and waste.

HSPD-7 also required the secretary of DHS, in coordination with the director of the OSTP, to prepare on an annual basis a Federal Research and Development Plan in support of the directive. In 2004, DHS and OSTP published "The National Plan for Research and Development in Support of Critical Infrastructure Protection [7]". The plan is structured around nine "themes": detection and sensor systems; protection and prevention; entry and access portals; insider threats; analysis and decision support systems; response, recovery, and reconstitution; new and emerging threats and vulnerabilities; advanced infrastructure architectures and systems design; and human and social issues. The long-term vision of the National Plan involves three strategic goals: a national common operating picture for critical infrastructure; a next-generation computing and communications network with security "designed-in" and inherent in all elements; and resilient, self-diagnosing, and self-healing physical and cyberinfrastructure systems. An updated National R&D Plan has been written but is not yet published (Note: updates to the plan have been published and coordinated with other Homeland Security plans that respond to other Homeland Security Presidential Directives.).

The focus of this article is on protecting assets, networks, systems, and humans by preventing and/or mitigating *physical* attacks/damage. The possibility of combined cyber and physical attacks is of concern, and it can be considered here by examining cases involving degraded communications and IT systems that support physical protection/prevention systems.

A number of terms are used to describe what "protection and prevention" mean for homeland security and defense. These terms are often used loosely without an agreed upon set of definitions. In order to provide some rigor it may be useful to explicitly define these terms (definitions are from the Merriam-Webster online dictionary [8]):

Protect. To cover or shield from exposure, injury, damage, or destruction; to guard; to defend; to provide a guard or shield; to maintain the status or integrity of.

Prevent. To be in readiness for; to meet or satisfy in advance; to act ahead of; to go or arrive before; to deprive of power or hope of acting or succeeding; to keep from happening or existing; to hold or keep back; to hinder; to stop; to interpose an obstacle.

Mitigate. To cause to become less harsh or hostile; to mollify; to make less severe or painful; to alleviate; to extenuate; to relieve.

Respond. To say something in return; to react in response; to show favorable reaction; to be answerable; to reply.

Recover. To get back; to regain; to bring back to normal position or condition; to rescue; to make up for; to save from loss and restore to usefulness; to regain a normal position or condition.

Robust. Having or exhibiting strength or vigorous health; having or showing vigor, strength, or firmness; strongly formed or constructed; sturdy; capable of performing without failure under a wide range of conditions.

Resilient. Characterized or marked by resilience; capable of withstanding shock without permanent deformation or rupture; tending to recover from or adjust easily to misfortune or change; elastic.

Note that while the term *prevention* often means those efforts associated with intelligence and interdiction of terrorist plans, here it is focused on preparedness, deployment of defensive measures in advance of an attack, devaluation of targets, and so on.

The concepts of homeland security and homeland defense are often used interchangeably, but for our purposes we will align these terms with the agencies that are primarily accountable for them: DHS and the Department of Defense (DoD). Homeland security is focused on civilian and law enforcement areas of our national efforts to counter terrorist acts (including those complex issues associated with protecting our national borders). Homeland defense is focused on domestic use of military resources to defend the skies and seas (with some land-based actions focused on high-security events) and to provide assistance to other agencies in responding to and recovering from a major terrorist attack.

The threats and challenges from a homeland security and defense perspective range from relatively limited consequence suicide bombings to the possibility of weapons of mass destruction (WMD) being used to cause very large consequences. Threat actors range from radical Islamic fundamentalists to domestic extremist groups who support the use of violence—they are malevolent, intelligent, innovative, and dynamic. Challenges range from addressing root causes of unrest and support for violence to detecting WMD devices hidden in containers, vehicles, boats, and aircraft. There are literally millions of possible targets and thousands of miles of borders in the United States, so trying to protect and defend all of them is impossible. Terrorist acts focus on causing extreme fear in the civilian population so "hardening the will" of our citizens is a challenge. The primary goal of all our efforts is to maintain our core freedom and quality of life while providing an appropriate level of security. Understanding the dynamic risks from terrorism and deploying cost-effective solutions to address the highest risks is our challenge. The recent release of a national strategy for DoD involvement in civilian disasters [9] ensures a much more organized and stronger response to situations such as WMD in a major city that could overwhelm civilian-only resources.

In addition to significant science and technology advances, we also require validated assessment methodologies for defense facilities, critical infrastructure facilities, port and coast guard facilities, government facilities, and privately owned facilities. Currently, there are many methods, tools, techniques, and processes being used. In addition, we need training to help prevent attacks on facilities and to protect facilities—this is an important element of preparing for terrorist acts as well as for accidents and natural disasters that disrupt our infrastructure. A variety of methods, tools, techniques, and processes are used.

There are scientific and technological solutions that can address many of the most difficult challenges. High priority scientific and technological solutions include sensor and detection systems for explosives and WMD, protection systems to defend against the effects of those threats, enhanced risk analysis and management tools, and techniques to address the “insider” problem. Work on these solutions is well underway but many of the challenges are very difficult especially when placed in the context of functioning infrastructure and real assets. The grand challenges that lie ahead involve developing advanced concepts for our infrastructure and physical assets that foster self-awareness, self-healing, and graceful degradation should they be attacked.

2 THREATS, CHALLENGES, AND SOLUTIONS

Threats can be decomposed into perpetrators and potential kinds of attacks. We must consider a range of perpetrators (not just Al-Qaeda)—both foreign and domestic terrorists with a variety of motivations, intents, and capabilities—and we must consider a spectrum of attacks from suicide bombs to WMD. Both elements of the threat pose challenges and some of those can be addressed with science and technology solutions.

Protection from insiders is an age-old problem. Insiders are within our defenses and they are trusted. They know about our security measures and plans. They know where vulnerabilities exist or can be created by them, in our systems. Defending against malevolent insiders can involve a number of approaches based on technology. Part of the vetting process for hiring employees and support contractors can include database searches using smart algorithms. Access control can limit the effectiveness of most insiders by not allowing them complete freedom of movement and action within our defensive perimeter. Tagging and tracking using radio frequency identification (RFID) devices can alert security personnel to insiders who are not where they should be or who are doing what they are not supposed/allowed to do. Document protection and control is used to limit the amount of information about critical components and processes that is available to general employees and contractors. Advanced concepts may include sensors that use noninvasive means to measure malevolent “intent” and new designs of systems that are “smart” in terms of self-protection from harmful actions. Some of these are already available and are termed *skeptical* systems.

Many forms of attack involve intrusion by a human or vehicle through (or around) a defensive perimeter. Protection from intrusion may involve detection, possible elements to delay the intruder, and then response (e.g. a security force or an automated action). Fences, gates, a security guard force, and closed circuit television (CCTV) systems are fairly common elements of many systems. Physical intrusion detection may be by sensing motion, sound, or other parameters (temperature, human out-gassing, neuron activity, etc.). Metal and weapon detection systems are often employed at portals. Vehicles are often stopped or slowed using barriers and fences, bollards, serpentine roadway designs, architectural barriers, or vehicle traps. Access to a site or facility can be controlled with keys and locks, "smart" badges, passwords, biometric scans, and facial-recognition systems. Advanced designs and concepts for defense against intrusion are being developed and the use of advanced materials will improve the obstructing power of barriers. More intelligent and autonomous CCTV systems are under development and it is expected that they will provide significant improvements in cost-effective 24/7 detection of intrusion and attack.

Defense against explosive blasts is a growing concern. There are systems for detecting explosives as well as for disarming and deactivating explosive devices. Mitigation systems for buildings, windows, doors, and entry ways include blast design measures as well as debris and fragment shields. Defense against small arms fire, mortars, and rockets includes using barriers, shields, and armor, for body, buildings, and vehicles. Advanced designs and materials (e.g. carbon nanotubes) should reduce the cost of blast and fragment mitigation while improving performance. However, parallel advances in explosives and shrapnel technology of much higher performance make this area of R&D rather dynamic.

Chemical, biological, and radiological threats are generally manifested as airborne attacks with gases or aerosols. Heating, ventilation, and air conditioning (HVAC) systems can address these attacks on indoor facilities using detectors, filters, and alarms. There are HVAC systems with automated (smart) responses that can stop or redirect building airflows to mitigate effects. Drinking water and food are also possible vectors for these threats and current defenses involve detection and alarm. Future systems will employ new designs, concepts, and materials to detect, delay, mitigate, and possibly eliminate a broad range of these threats.

Less lethal and nonlethal force systems can be employed to delay or stop a physical attack on a facility. Common methods include the use of rubber and plastic bullets, beanbags, water cannons, tear gas, pepper spray, Mace spray, and stun guns. Newer concepts involve foams (sticky and slippery), slippery surfaces, acoustic devices, barometric and rapid gas exchange systems, laser dazzlers and microwaves. Many systems and techniques have been developed for use by law enforcement officers against criminals; evolutionary and revolutionary concepts are being developed and deployed constantly. An interesting observation in this area is the growing development and deployment of systems using robotic and automated response systems. These systems include those with lethal, less lethal, and nonlethal options. They offer the promise of cost-effective 24/7 response and are often used in combination with a guard force.

One of the methods of preventing an attack is to use cover, deception, and concealment systems so that the adversary is unaware of the location of the target. Techniques include camouflage and view-disruption systems. It is expected that nanotechnology may provide additional means of "hiding" systems and vulnerable components from view.

Water can be used as a weapon. This is clearly the case when considering major dams that are upstream of large population centers or important agricultural areas. Many of the methods, devices, techniques, and processes described previously to protect against intrusion and explosive blasts can be deployed to assist in defending such targets. There are specific conditions that must be taken into account for defense of dams such as waterborne and underwater attacks using explosives. Water barriers and nets are commonly used approaches.

Fire can also be used as a weapon. Small quantities of flammable liquids and accelerants are easily concealed and can be used in indoor facilities and vehicles of mass transportation. Major wildfires are common natural threats but they can also be intentionally initiated to endanger metropolitan and agricultural areas or to distract emergency responders. Since indoor fires occur somewhat regularly, detection and extinguishment systems are common in areas where large numbers of people gather. Large wildfires require a lot of effort from emergency responders, especially when conditions are dry and windy. Advancements in firefighting equipment (including enhanced communications capabilities) and new materials for extinguishment will enable faster limitation of the effects of fires.

Electromagnetic, microwave, laser, and directed-energy devices can be used as weapons. While not commonly deployed today, their use may increase in the future. Lasers have been used to "dazzle" pilots in ways that cause temporary impairment of eyesight. A number of devices are commercially available today, which use a range of electromagnetic frequencies and effects to disrupt electronic circuits and systems. These devices could be used to disable security or response systems and then combined with physical assaults on key facilities and assets. A range of defensive measures (optical filters for laser beams, radiation shielding, and hardening of electronic circuits, etc.) are available today for these threats and it is anticipated that countermeasures will evolve if these methods of attack are deployed by terrorists.

Response and recovery are important elements of resilient infrastructure (as is redundancy, of course). If the timescale of recovery is short enough, one of the key goals of terrorism (economic damage) can be significantly mitigated. This section of the Handbook does not address the needs of first responders for enhanced science and technology solutions. The focus here is rather heavily on infrastructure systems. Protective materials, paints, coatings, and films are being developed to assist in the rapid decontamination of surfaces that have been exposed to toxic chemicals, biological agents, and radiological materials. For rebuilding damaged facilities and assets we will need to be able to rapidly provide temporary structures. Long-term recovery means rebuilding with new concepts and materials that have security "designed-in" at the start. A broad range of research and development is underway in the building and manufacturing industries to incorporate advanced designs, concepts, and materials into the next generation of critical facilities and assets.

Critical infrastructure protection is common terminology for much of what is discussed in this section of the Handbook. While this overview article focuses on protection from and prevention of terrorist attacks, much of the current discussion in conferences and workshops has shifted to resilience of the nation's infrastructure. This difference is really a matter of definition and perspective. Resilience can imply rapid response and recovery from a catastrophe—the assumption being that the nation cannot protect everything from attack. A cost-effective strategy may be to be prepared to rebuild facilities and reconstitute capabilities very rapidly. Redundancy is an important element of resiliency. On the other hand, protection and prevention can be broadly viewed as containing a range of elements from robustness (hardening to withstand attacks) to mitigation of effects (limiting damage and impact) to resiliency (rapidly recovering from attacks). Whatever be the perspective, the science and technology base can and will provide solutions to many of the problems.

Information sharing and protection is a vital issue related to protection of the nation's critical infrastructure. The vast majority of the infrastructure is owned and operated by the private sector, cities, and municipalities. Most of the infrastructure is global in nature. Consequently, working together as partners to provide solutions that address the terrorism issue requires industry–government teams, US international teams, and openness in the science community to share research results while protecting sensitive information. Despite concerns about litigation and possible regulation, industry has often been reluctant to work closely with the government. Though this is not directly a science and technology issue, this area must be addressed in order to accelerate progress.

It is important to note that while this section of the Handbook is focused on research and development for solutions based primarily on the physical and engineering sciences, many of the solutions will come from the social sciences. In particular, terrorism is a violent form of action aimed at social, economic, cultural, ideological, and religious values. Addressing root causes of terrorism and removing some of the fear caused by it are extremely important for the ultimate solution for free and open societies.

3 FUTURE RESEARCH DIRECTIONS

Advanced designs, concepts, and materials for protection and prevention are areas of active research and development around the globe. New methods, devices, techniques, and processes are being developed every day. The key driver for many of these breakthroughs will be business continuity and reliability of infrastructure services. Another important driver for advancements will be protection from and prevention of criminal acts (including malevolent acts by insiders, theft, sabotage, and vandalism). Concerns about the impact of terrorist acts and natural catastrophes will drive much of the Federal investment in research and development for this area. Consequently, dual purpose R&D and multipurpose technology will help make future protection and prevention more cost-effective. It will also help increase the degree of deployment.

Information technology is one of the major contributors to current advancements and it will remain that way for the foreseeable future as well. Buildings and structures will contain significantly more embedded information technology that can sense the status of systems (and their constituent components) and provide automated responses. Designs will incorporate this capability into "smart" structures that will autonomously communicate with security and emergency officials, protect themselves, heal themselves, and degrade in ways that minimize loss of functionality and human casualties. Computer modeling and simulation tools will greatly aid in designing infrastructure systems that have security "built in" from the start and allow for cost-effective adaptability for the future.

Nanotechnology holds the promise of significant advancements in materials of all sorts. The construction of light-weight structures that are extremely strong is possible. Materials (e.g. based on carbon nanotubes) that can absorb blasts and contain fragments will be available at low cost. Special coatings, films, and surfaces that neutralize chemical and biological agents will be developed. Foams and sprays that can capture radiological particulates for removal and treatment may be manufactured. Materials with exotic properties will be developed that will provide new ways of protecting people, facilities, and assets. Other new materials will allow us to design enhanced systems that prevent malevolent acts from taking place at all or mitigate their consequences.

Biotechnology will provide us with new ways of defending against attacks using chemical or biological agents by allowing people to be vaccinated or inoculated in advance. There will be new designs and concepts for protection developed based on biological processes and systems. Many of the new materials envisioned for the future will be created by mimicking naturally formed materials or by using biological processes to manufacture them. Humans, agricultural crops, and livestock will recover more rapidly and fully from injuries and attacks by using newly developed techniques based on our improved biotechnology capabilities.

The nature of terrorism is such that terrorists will employ new methods based on improved science and technology just as we use new methods to provide enhanced prevention and protection. As in warfare, the effort is dynamic. It has both evolutionary and revolutionary elements. Our primary aim must be to develop and deploy a range of tools and techniques that address the full spectrum of response to terrorist acts—from understanding root causes to increasing interdiction to providing protection and recovery.

4 CONCLUSIONS

The history of violence and war goes back to the earliest humans and tribal groups. Yet terrorism brings a strong emotional element of irrational fear associated with the randomness of attacks and the focus on innocent civilians, including women and children. Nevertheless, following the cold war, the United States felt reasonably secure from international violence until the attacks of September 11, 2001. The nation's response since then has involved major reorganization of federal agencies to address the operational and R&D elements, increased partnership with other countries, and significant changes in the way private industry and infrastructure owners deal with security issues.

Prevention of terrorist attacks and protection of our nation's critical infrastructure from such attacks are important elements of homeland security and defense. As a free and open society we are vulnerable to acts of terrorism, partly due to our inherent cultural values. Consequently we must exploit the advantages we have in science and technology to develop and deploy solutions that allow us to defend our way of life and increase the difficulties for those who seek to cause terror.

Many of the possible solutions to the threats and challenges that we face from terrorists can be used for additional purposes. From an infrastructure perspective, business continuity and reliable delivery of goods and services act as economic drivers for improved security, especially as new systems are designed and constructed. From a human perspective, measures that address crime and violence can also serve to counter terrorist acts. Accidents (e.g. chemical and radioactive material releases) and naturally emerging diseases [e.g. severe acute respiratory syndrome (SARS) and Avian Influenza] are also reasons to support research and development apart from providing solutions for the terrorism problem.

Research and development can assist us in addressing issues from insider threats to suicide attacks with conventional explosives to dealing with WMD. New designs, concepts, materials, tools, devices, and technologies can provide cost-effective solutions that not only make us more secure but also more resilient if a catastrophe occurs for any reason (natural, accidental, or intentional and malevolent). Systems that are automated and built to be "smart" can defend, mitigate, and respond to attack and damage. Protection and prevention are important areas of homeland security and defense—they form part of the base of our science and technology response to acts of terrorism.

ACKNOWLEDGMENTS

The submitted manuscript has been authored by a contractor of the US Government under Contract No. DE-AC04-94AL85000. Accordingly, the US Government retains a nonexclusive, royalty-free license to publish or reproduce the published form of this contribution, or allow others to do so, for US Government purposes. Sandia is a multiprogram laboratory operated by Sandia Corporation, a Lockheed Martin Company, for the US Department of Energy's National Nuclear Security Administration under Contract DE-AC04-94AL85000.

REFERENCES

1. The White House. (1998). *Presidential Decision Directive 63: Protecting America's Critical Infrastructures*, May 28 1998, http://www.fas.org/irp/offdocs/pdd-63.htm.
2. MacDonald, B., and Rinaldi, S. M. (1998). *Critical Infrastructure Protection Research & Development Interagency Working Group Blue Book*. Office of Science & Technology Policy.
3. Robert T. M. (1997). *Critical Foundations: Protecting America's Infrastructures, Report of the Presidential Commission on Critical Infrastructure Protection, Washington, Dc*, October 1997, http://www.pccip.gov/.
4. The White House. (2003). *National Strategy for the Physical Protection of Critical Infrastructures and Key Assets*, February 2003, http://www.whitehouse.gov/pcipb/physical.html.
5. Committee on Science and Technology for Countering Terrorism. (2002). *Making the Nation Safer: The Role of Science and Technology in Countering Terrorism*. National Academies Press, Washington, DC.
6. The White House. Homeland Security Presidential Directive 7 (HSPD-7), (2003). *Critical Infrastructure Identification, Prioritization, and Protection*, December 17, 2003, http://www.whitehouse.gov/news/releases/2003/12/20031217-5.html.
7. DHS and OSTP. (2004). *The National Plan for Research and Development in Support of Critical Infrastructure Protection*.
8. Merriam-Webster online dictionary; http://www.m-w.com/.
9. Department of Defense. (2005). *Strategy for Homeland Defense and Civil Support*. Department of Defense, Washington, DC, June 2005.

FURTHER READING

Bullock, J., and Haddow, G. (2006). *Introduction to Homeland Security*, 2nd ed., Butterworth-Heinemann, Elsevier, Burlington, Mass, March 30 2006.

DHS. (2006). website: http://www.dhs.gov/dhspublic/.

DoD. (2006). website: http://www.defenselink.mil/.

Fay, J. (2007). *Encyclopedia of Security Management: Techniques and Technology*. Elsevier, Burlington, Mass, Butterworth-Heinemann.

Fennelly, L. J. Eds. (2003). *Effective Physical Security*, 3rd ed., Butterworth-Heinemann, Elsevier, Burlington, Mass.

Fischer, R. J., and Green, G. (2003). *Introduction to Security*, 7th ed., Butterworth-Heinemann, Elsevier, Burlington, Mass, November 2003.

Garcia, M. L. (2007). *The Design and Evaluation of Physical Protection Systems*. Butterworth-Heinemann, Elsevier, Burlington, Mass. February 23, 2001.

Homeland Security Institute. (2006). website: http://www.homelandsecurity.org/.

Howitt, A. M., and Pangi, R. L. Eds. (2003). *Countering Terrorism: Dimensions of Preparedness, BCSIA Studies in International Security*. The MIT Press, Boston, Mass, September 2003.

Kamien, D. (2005). *The McGraw-Hill Homeland Security Handbook*. McGraw-Hill, New York, NY, September 2005.

Lewis, T. G. (2006). *Critical Infrastructure Protection in Homeland Security: Defending a Networked Nation*. John Wiley & Sons, Hoboken, NJ.

National Commission on Terrorist Attacks. (2004). *The 9/11 Commission Report: Final Report of the National Commission on Terrorist Attacks Upon the United States*, July 22, 2004, http://www.9-11commission.gov/.

National Infrastructure Protection Plan. (2006). Department of Homeland Security, http://www.dhs.gov/xprevprot/programs/editorial_0827.shtm.

Sauter, M., and Carafano, J. (2005). *Homeland Security: A Complete Guide To Understanding, Preventing, and Surviving Terrorism, The McGraw-Hill Homeland Security Series*. McGraw-Hill, New York, NY.

Sennewald, C. (2003). *Effective Security Management*, 4th ed., Butterworth-Heinemann, Elsevier, Burlington, Mass.

PROTECTION AND PREVENTION: THREATS AND CHALLENGES FROM A HOMELAND DEFENSE PERSPECTIVE

JEFFREY D. MCMANUS

The Office of the Secretary of Defense, Washington, D.C.

1 INTRODUCTION

We recently passed an important anniversary in the United States on September 11, 2008. It is 8 years since the al-Qaeda attacks against the World Trade Center and the Pentagon. Unfortunately, this attack in which almost 3000 people were killed in a single day, has been joined by other significant terrorist incidents against other countries. These attacks have caused death to hundreds more innocent civilians in places like Mumbai, India (July 11, 2006); Bali, Indonesia (October 2, 2005); London, United Kingdom (July 7, 2005); and Madrid, Spain (March 11, 2004). A war has been declared upon our nation and waged against our people. The intent of transnational terrorists is to try to shape and degrade political will through extreme acts of violence in order to diminish resistance to their ideologies and agendas. Our future freedom and security depend on our efforts in meeting this challenge. It is imperative that we have a comprehensive preparedness strategy that focuses our combined national efforts on proactive prevention and protection activities *before* a terrorist incident occurs. This approach can greatly lessen the overall risk of both terrorist attacks as well as nonintentional hazards, and even assist the response and recovery efforts following those incidents that do occur.

2 BACKGROUND

It is the role of the Department of Defense and the US military to fight and win our nation's wars while protecting US sovereignty, territory, domestic population, and critical defense infrastructure against external threats and aggression. The structure and composition of military forces, built up during World War II and evolved over the 45-year Cold War, allowed for the detection, deterrence, or when necessary, defeat, of conventional threats to the United States. The principal threat was from the Soviet Union, capable of delivering nuclear weapons via missiles, bombers, or submarines, and having massive conventional forces poised in Eastern Europe for a strike west. The security of the US homeland depended on our nuclear deterrent and strong conventional forces. The Department of Defense and US military maintained a significant forward presence around the world that apart from providing a security shield, contained and deterred Soviet aggression, protected our allies and friends, and provided stability for the inevitable spread of personal and economic freedom.

The size, structure, and capabilities of US forces built up and maintained throughout the Cold War also provided key capabilities that could assist in specific situations here in the United States. These situations, called Defense Support to Civil Authorities (DSCA), allow for the use of US military forces within our homeland, under the direction of the President and the Secretary of Defense within the authorities provided by Title 10 U.S. Code (USC). However, the legal authorities and guidelines for DSCA are specific and limited. They preserve the spirit and intent of our founding fathers to limit the use of the military over our domestic population. Prior to any use of US military forces at home in a DSCA role, six criteria are assessed and evaluated:

- *Legality.* Is the activity compliant with existing law?
- *Lethality.* Is there no or very minimal potential for the use of lethal force by or against military forces?
- *Risk.* Is the activity safe for military forces?
- *Cost.* Is there an existing funding authority or reimbursement mechanism available to minimize impact to the Department of Defense budget?
- *Appropriateness.* Are military forces unique, capable, or necessary to provide the activity?
- *Readiness.* Will this activity not significantly interfere with other missions of the Department of Defense?

If answers to all of these questions are "Yes," then a DSCA mission can be approved and implemented.

These situations fall into four main categories. First is support following a civil disaster or emergency, such as a hurricane or an earthquake. These activities are governed under the Stafford Act (i.e. Title 42 USC, Sections 5121 and 5122). This allows utilization of military forces in response to a major disaster, whether natural or man-made. Second is support to counter drug operations, and is limited primarily to the detection and monitoring of drug traffickers attempting to cross US borders by air, sea, or land (i.e. authorities are outlined in Title 10 USC, Section 124). Third is support responding to either chemical or biological weapons emergencies, which is authorized under Title 10 USC, Section 382.

The final type of support would be that related to law enforcement functions, which is greatly restricted under the Posse Comitatus Act (i.e., 18 USC, Section 1385, also applicable is 10 USC, Section 375). This law prohibits US military forces from executing the civil laws of the United States, or stated another way, prohibits US military forces from performing direct law enforcement activities. These activities include interdiction of a vehicle, vessel, aircraft, or similar activity (with the exception of the specific drug detection and monitoring efforts described above); search or seizure; arrest or apprehension, such as stopping and frisking individuals; surveillance or pursuit of individuals; or undercover work, such as working as undercover agents, informants, investigators, or interrogators. The only exception from these restrictions would be if the President declares an emergency under the Insurrection Act (i.e. Title 10 USC, Sections 331–334). This would allow US military forces to respond to a civil disturbance and perform direct law enforcement functions.

3 CURRENT SITUATION

Following the terrorist attacks on September 11, 2001, the President working closely with Congress, created the Department of Homeland Security (DHS). *Homeland Security* is a concerted national effort to prevent terrorist attacks within the United States, reduce the vulnerability of the United States to acts of terrorism, and minimize the damage by and assist in the recovery from terrorist attacks that do occur. The DHS is the lead US federal agency for homeland security, with responsibilities for coordinating and integrating efforts related to homeland security, intelligence, law enforcement, and homeland defense. This environment is shown in Figure 1. It requires close coordination with other federal organizations which have critical roles in combating terrorism, such as the Directorate of National Intelligence and the intelligence community, the Department of Justice and the Federal Bureau of Investigation, and the Department of Defense. The DHS also has responsibilities beyond the prevention of terrorism, including leading the US government response to natural disasters and other emergencies.

There has been much progress made since September 2001. Many strategies, policies, and plans have been developed. New organizations have been established at the federal level, such as DHS, the Directorate of National Intelligence, the National Combating Terrorism Center, and US Northern Command. Each has consolidated authorities and functions that, while appropriate for the Cold War environment, were not meeting the requirements of the new security environment. It is important to recognize that we have more forms of protection in place than 8 years ago.

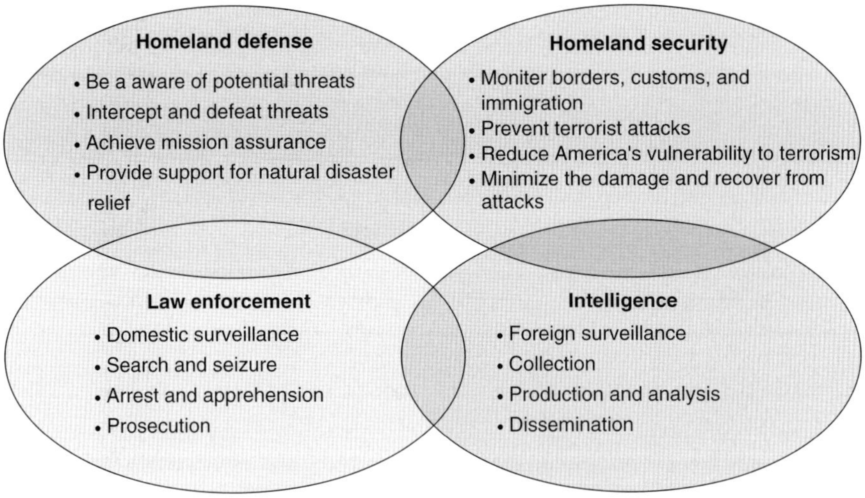

FIGURE 1 The homeland security environment.

However, while much has been done, there is still room for improvement. We have many national strategies and plans, but we still seem somewhat mired in a stove-pipe structure. While national plans have been developed through interagency processes, they still reflect somewhat limited perspectives unique to their communities of interest. For example, our national protection plans primarily reflect a security or law enforcement perspective. Remediation activities are being appropriately driven by a budgetary perspective, but risk management appears inconsistent and is therefore ineffective in maximizing the resource priority trade-offs. Response plans reflect an emergency response or consequence management perspective, but do not account for the longer term, sustained activities necessary for true recovery. Five years since September 11, 2001, our national effort seems to be "reactive" and "federal", that is, focused mostly on postincident efforts, and limited to the functions and capabilities that can be provided by federal departments and agencies.

President Bush signed and released the *National Strategy for Homeland Security* in July 2002. (A revised version was released in October 2007.) This strategy highlighted the new threat environment, discussed our vulnerabilities, and outlined six critical mission areas necessary to meet the new security challenge:

- intelligence and warning;
- border and transportation security;
- domestic counterterrorism;
- protecting critical infrastructure and key assets;
- defending against catastrophic threats;
- emergency preparedness and response.

This national strategy provided the structure and framework for the establishment of the DHS, and documented several necessary foundations for success: law, science and technology, information sharing and security, and international cooperation.

There are, however, deficiencies in the *National Strategy for Homeland Security*. The first is in the area of risk management because the strategy does not provide a comprehensive risk management structure. Risk is a function of criticality, vulnerability, and probability of incident occurrence. While vulnerabilities are addressed in the strategy, as are the evolving means of terrorist attack, the elements of criticality (i.e. consequence of loss) and probability of incident occurrence are not. These elements must be included to enable true risk management, that is, the process by which decision-makers accept, reduce, or offset risk. Another related deficiency is that the strategy does not adequately address nonhostile hazards. Events such as accidents or natural disasters can also result in catastrophic damage; for example, incidents such as Hurricane Katrina and the subsequent levee breaches in New Orleans. This paper later outlines the full scope of threats and hazards that should be considered when addressing homeland security.

The 2002 strategy also does not differentiate between mission areas (i.e. *what* we need to do), the domain environment (i.e. *where* we need to do them), and the preparedness continuum (i.e. *when* we need to do them). To be fair, the strategy does discuss protection and response activities. However, it does not fully articulate the steady-state prevention functions that can be implemented preincident to minimize general risks. It also does not articulate the postevent actions necessary over the longer term (i.e. once immediate response functions are completed) to get people and infrastructure back to their preincident situation. This paper will later outline the preparedness continuum, and focus on the proactive activities and challenges faced in the pre-event functions of prevention and protection.

4 THREATS AND HAZARDS

As discussed above, a key element in the risk management equation is that of threats and hazards. These are the incidents, whether intentional and man-made, or unintentional accidents and natural disasters, that cause damage or loss. Figure 2 provides an illustrative example. Vulnerabilities occur when a specific asset or function is both susceptible to damage by the threat or hazard and there is a probability for damage to occur. For example, the human body is vulnerable to injury from a bullet or a bomb, but not directly vulnerable to a cyber attack. Conversely, a computer system is vulnerable to cyber penetration and attack, but not vulnerable to a biological agent. These distinctions become very important when risk is being assessed for management through remediation or mitigation.

Threats are incidents intentionally caused by an adversary who has the intent, capability, and opportunity to cause loss or damage. This definition is related closely with that of Carl von Clausewitz and his definition for war: acts of violence used to compel an adversary to do one's will. Traditionally, war and the violence it contains has been the domain of nation-states, with an evolved code of conduct and laws for war. Policy makers and military leaders are used to dealing with "traditional" threats within the nation-state environment. The clearest example is the development, structure, and operation of "conventional" US military forces (e.g. the Army, Navy, Air Force, and Marine Corps) in the implementation of deterrence and containment of the Soviet Union. By and large, this structure also applies to the conventional threats of other countries, such as China, North Korea, or Iran. Two relatively recent and clear demonstrations of

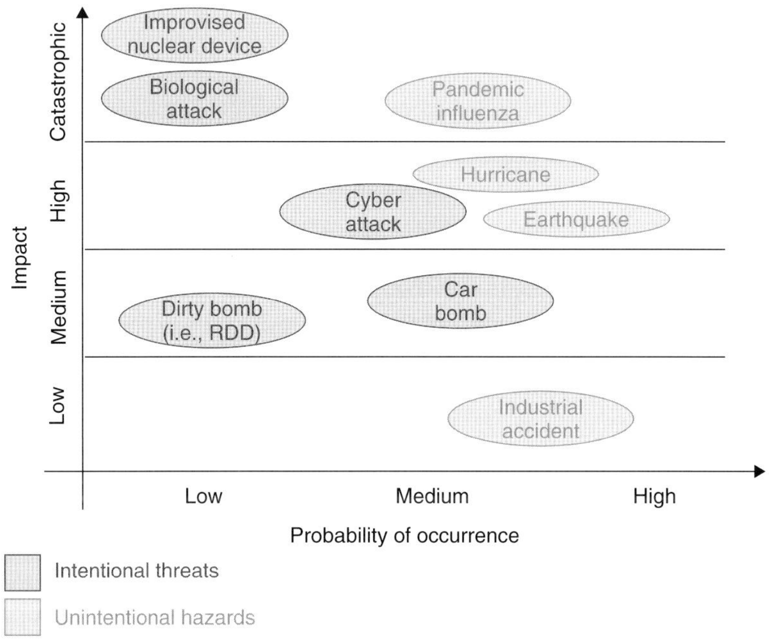

FIGURE 2 The relative impacts of threats and hazards.

the power of US conventional military force against opposing conventional forces were Operation Desert Storm in 1990–1991 and the 2003 invasion and liberation of Iraq.

The growing threats we face today, at both home and abroad, are asymmetric in both type and means. "Type" implies their nonstate character, for example, transnational groups such as al-Qaeda, Hezbollah, and Hamas. "Means" signifies that our enemies are not seeking to build or deploy conventional military forces or capabilities to match against our own. The reasons why are not difficult to understand because we have clearly demonstrated that no conventional military force in the world is any match against ours. If the purpose of war is to do acts of violence to compel an adversary to do your will, then the acts of violence are only effective if they can be performed, and their results can have great impact.

The means-to-date for organizations such as al-Qaeda and Hamas, although unconventional in deployment, has been utilization of relatively small, conventional explosives against civilian populations and infrastructures. The attack of September 11, 2001, was an unconventional use of a nonmilitary capability, to achieve a significant explosive event. The attacks since have also been asymmetric, directed against innocent civilian populations, but have utilized relatively small explosives:

- Mumbai (July 11, 2006);
- Karachi (March 2, 2006);
- Bali (October 2, 2005);
- London (July 7, 2005);
- Cairo (April 7, 2005);
- Jakarta (September 9, 2004);

- Beslan (September 4, 2004);
- Madrid (March 11, 2004);
- Istanbul (November 20, 2003);
- Casablanca (May 16, 2003);
- Jerusalem (November 21, 2002).

These terrorist attacks have been conducted against civilian populations, with secondary infrastructure impacts, and have killed tens and hundreds. However, the defining characteristic of the security environment we now face is the growing threat of a more substantial and diverse asymmetric attack which could cause catastrophic loss of life and result in mass panic, through the use of a *weapon of mass destruction* (*WMD*) in Pentagon parlance.

In the past, WMDs have been the exclusive domain of nation-states. It is important to note that the United States spent nearly 45 years and billions of dollars to deter their use by the Soviet Union. It is clear that today's terrorists groups will continue in their attempts to obtain, deploy, and ultimately use a WMD involving chemical, biological, radiological, nuclear, or high yield explosive capabilities. They will attempt to use these capabilities against targets in the United States. The bottom-line: terrorists intend to kill more of our people. They have ominously said so, and it is important that we listen:

- Dr. Ayman al-Zawahiri, al-Qaeda's second most influential leader said on July 27, 2006, "All the world is a battlefield open in front of us ... Our fight is a *jihad* [holy war] for the sake of God and will last until [our] religion prevails ... from Spain to Iraq ... We will attack everywhere ... ";
- Shaykh al-Fahd, a jihadist legal authority, wrote in May of 2003 in his *Treatise on the Legal Status of Using Weapons of Mass Destruction Against Infidels*.: "If people of authority engaged in *jihad* determine that the evil of the infidels can only be repelled by the means of weapons of mass destruction, they may be used."

Asymmetric WMD attacks can be addressed in three general categories. First are those whose consequences would be physically localized and relatively limited in scope. This category would include the use of high explosives, such as a car or truck bomb. A graphic example is the domestic terrorist attack in Oklahoma City, Oklahoma, on April 19, 1995, where a truck bomb was used to destroy the Alfred P. Murrah Federal Building. This attack caused the death of 168 people, its effects were immediate, and physical damage was localized. This category would also include an attack with a chemical weapon, or intentional destruction of a chemical storage site. This type of attack could cause the death of tens to hundreds of people, depending on location and scenario. The effects of a chemical attack would also be relatively immediate, depending on the specific chemical and its properties for dissipation. Physical damage, again, would be localized.

The second category of asymmetric WMD attacks are those whose consequences would be regional and/or persistent. This category would include a cyber attack against a critical infrastructure network. This type of attack probably would not cause many deaths, but could be very disruptive depending on the infrastructure targeted. A successful cyber attack against the financial network could cause loss of assets, and have extremely negative effects on business, markets, and the economy. Another cyber example would be an attack on the electric power grid, that caused a massive regional power outage similar in consequence to the power blackout of August 14, 2003, in the northeastern United States. This was the largest power blackout in US history, affecting an estimated 10 million people in the United States. and Canada; outage-related financial losses from this single event were estimated at $6 billion.

This category would also include a radiological weapon, also called a *dirty-bomb* or a radiological dispersion device ("RDD"). An RDD is a conventional explosive that is combined with radiological material. In physical damage this type of attack would be very similar to that of a high explosive. However, an RDD could also create additional panic and casualties, especially if used in a metropolitan area. Depending on the radiological material used, it could have contamination and radiation consequences. Therefore, recovery and cleanup could be extremely costly and time-consuming.

The last category of asymmetric WMD attack includes those whose consequences would be catastrophic and persistent. This category would include both biological and nuclear attacks. A biological attack, using something like the bacterial plague, could spread quickly if not contained and could kill thousands of people. The rate of spread would be amplified as a result of both, domestic and foreign travel. This type of attack would put a severe strain on the healthcare system, adversely impacting pharmacies, clinics, and hospitals. While this type of attack would not cause any physical damage to buildings or property, it could cause contamination that requires a costly cleanup.

This last category would also include a terrorist attack using an improvised nuclear device in a metropolitan city. It could kill tens to hundreds of thousands of people, cause incredible physical destruction over a 1–3 mile radius, further contaminate several thousand square miles, and displace more than half-a-million people. The economic impact would be hundreds-of-billions, with years for recovery. Impacts on the civilian population, healthcare system, economy, etc. are almost too grim to contemplate.

In addition to the threats outlined above risk management must also take into account hazards, which in many cases have a higher probability of occurrence than that of terrorist attack. Hazards are defined as nonhostile incidents, such as accidents, technological failure, or natural disasters, which cause loss or damage. Accidents can be due to negligence, bad maintenance, or truly bad luck, resulting in damage or death. A good example of an accident causing large-scale, regional impact was the electric power grid failure in the fall of 2003, mentioned earlier. This was the largest power blackout in US history, and it was caused when strained high voltage power lines shorted out after they came in contact with overgrown trees; it caused a cascading failure of the power network.

Another example of a hazard was the Bhopal disaster of December 1984, that killed at least 20,000 people and injured from 150,000 to 300,000 people. This incident followed the accidental release of 40 tons of a lethal chemical from a Union Carbide industrial plant in Bhopal, India. Later investigations showed that it was due to serious maintenance and safety problems at the plant. The situation degraded further when the transportation system of the city subsequently failed due to the overload and panic. People died when they were asphyxiated or trampled to death, or were seriously injured while trying to escape.

Natural disasters are also hazards that can cause significant death and physical destruction. The San Francisco earthquake of 1905, estimated at a magnitude greater than 7.5, killed at least several thousand people and left 300,000 people homeless. This earthquake and the resulting fire caused major damage to a principal US city and is remembered as one of the worst natural disasters in our history. Hurricane Katrina, and the subsequent levee breaches in New Orleans, is another example of a catastrophic natural disaster.

In August 2005, this Category 3 hurricane directly hit New Orleans, breaching levees and flooding 80% of the city. Over 1500 people were killed, hundreds of thousands displaced, and recovery from the storm is estimated to have cost $81 billion, making it the costliest natural disaster to date in US history. Tsunamis are another example of a natural disaster that can have devastating consequences. The Indian Ocean earthquake in December 2004, triggered massive tsunamis that killed an estimated quarter-of-a-million people and left millions homeless.

A final example of a hazard is an influenza pandemic, which the World Health Organization warns could occur in the next few years. The 1918 Spanish flu pandemic affected almost 25% of the population in the United States, with an estimated death toll of half-a-million. The ease and availability of both domestic and foreign travel would undoubtedly amplify the effect and speed of a pandemic. As with the biological attack scenario discussed previously, an influenza pandemic would add a significant strain on the healthcare system, medical personnel, hospital rooms, and medical supplies and distribution. It is unknown how much social disruption or panic could result, but it is clear that the effects of an influenza pandemic could be catastrophic.

5 ADDRESSING PRE-EVENT PREPAREDNESS

When addressing homeland security it is important to understand the relationship between missions, that is, what are the things we need to do, and the preparedness continuum, that is, when we need to do them. The preparedness continuum is a way of outlining the time elements of risk management. It can be articulated in three phases: the pre-event phase, the incident, and the postevent phase. This structure is shown in Figure 3.

The pre-event phase contains those activities and tasks that can and should be performed before an incident occurs. *Prevention* and *Protection* are both pre-event activities, and are discussed in detail below. The postevent phase contains those activities and tasks that can and should be performed after an incident occurs. *Response* and *Recovery* are both postevent activities. *Response* includes the actions taken immediately following an

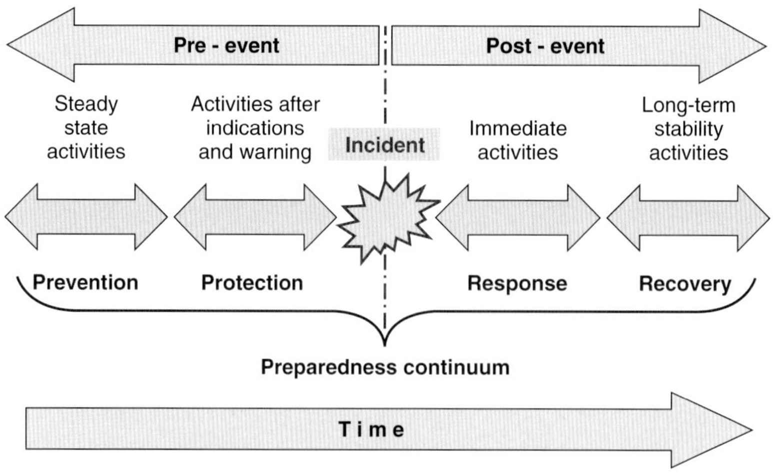

FIGURE 3 The preparedness continuum.

incident, to save lives and minimize damage to property. *Recovery* includes the actions necessary over the longer term, once immediate response functions are completed, to get people and infrastructures back to their preincident situation.

Prevention is the first element in the preparedness continuum, and is the steady-state baseline effort which should be conducted preincident to minimize general risks. There are many activities that contribute to prevention. First is a risk assessment for overall awareness that is, the first step in any assessment should be a prioritization of those functions or activities that are most important. This differentiates between the "must have" and the "nice to have." Once the most important or critical functions have been identified, vulnerabilities can be assessed to determine which threats or hazards could cause damage or loss. Finally, a risk assessment is complete when the probability of occurrence for a threat or hazard is considered.

For those areas considered at risk through assessment, there are many activities that can remediate risk. Remediation includes the actions taken to lessen or correct known vulnerabilities or weaknesses. Maintaining simple awareness of routine situations and, taking note and reporting suspicious activity is a form of remediation. Another way to lessen risk is to add resiliency to a function or network. This can be done through redundancy, by either increasing the number of elements that can perform a function or by increasing the routes between the functioning elements. For example, if the ability of a worker to get to a certain place of business is critical to operations, then a prevention plan for remediation could include having multiple means of travel, or having multiple routes that could be taken, or some combination of both.

Another way to increase resiliency is through stockpiling or storage of critical elements. These can include many examples, such as purchasing and maintaining a power generator in case primary power is lost, or maintaining both landline and cellular phones, or storing water and food. Prevention activities can be performed by organizations, companies, and groups, or by families and individual citizens. As Benjamin Franklin said, "*an ounce of prevention is better than a pound of cure*."

Prevention efforts can also have deterrent effects on enemies contemplating an attack. Since the purpose of terrorism is to cause mass casualties and panic to degrade political will, visible prevention activities such as those discussed above increase overall preparedness and can, therefore, dissuade possible attackers. At the least, attackers have been known to shift target locations if preattack surveillance shows increased levels of prevention and preparedness.

There are many challenges, however, that can hinder the establishment of effective prevention efforts. For example, added awareness through increased surveillance can have civil liberty ramifications. The same cameras used by authorities to detect criminal behavior can also be misused against legitimate civil activities. Another consequence from the use of surveillance systems can be profiling, or the broad targeting of individuals from a specific race or ethnic background. Appropriate overseeing is necessary to avoid these potential misuses of surveillance.

There are challenges related to resiliency as well. As discussed above, added resiliency can be very effective, whether through redundancy or stockpiling, but these activities can have significant financial costs associated with them. An appropriate risk management process can greatly assist in determining which systems require more resiliency or defining the correct balance between active and passive systems for deterrence. The bottom-line challenge concerning overall financial costs of prevention activities is determining the financial risk or business benefit to an intangible like security.

Protection is the second element in the preparedness continuum, and is the additional actions that can be taken preincident to further lessen risks, given appropriate and timely indications or warning. Since protection activities can be linked to indications or warnings of specific threats or hazards, knowing where to focus efforts can be somewhat easier than with the general prevention activities discussed above.

Like prevention, there are many types of activities that contribute to protection. First is changing tactics, techniques, or procedures, called "TTPs" in the military. That is, normal procedures can be modified or altered to lower the risks from a known threat or hazard. A recent example was the changes in the UK and US airport screening and passenger carry-on-items in August 2006, following the discovery of the bombing plot and subsequent arrests of terrorists in London. Knowing the terrorists meant to use initially separate liquid chemicals that could later be mixed on board an aircraft to create explosives allowed changes in both, passenger security screening as well as prohibiting specific items from passenger baggage. Another example of this type of activity would be boarding windows and sandbagging areas following warnings for an approaching hurricane.

Another type of protection activity following a warning would be isolation of, or evacuation from, the targeted asset. These activities could range from disconnecting the asset from the surrounding network to hardening the asset by active guarding or increased physical security measures. Complete isolation could be achieved through a total security lock-out or a full personnel evacuation.

Selecting another asset to perform the function of a targeted asset is another protection activity. This assumes a form of redundancy, such that the selected unthreatened asset can do the work of the asset that is identified at risk. An example would be diverting aircraft to a safe airport from an intended airport under threat of hostile attack or bad weather.

As with prevention, there are many challenges that can hinder implementation of effective protection efforts. In order to implement effective changes in techniques or procedures, the warnings must be specific and credible. In the example discussed above, airline security authorities knew the terrorists intended to smuggle multiple liquid chemicals on board aircrafts to later mix to create explosives. This allowed select materials to be banned from carry-on luggage (e.g. liquids) and specific screening procedures to be implemented. Without this "actionable" intelligence, the subsequent protection efforts would have been greatly hampered.

There are also challenges related to isolation or evacuation. Significant or total isolation is difficult to maintain for extended periods, due to the need to stockpile key components and material, such as equipment, fuel, food, and water, to sustain independent operations. Evacuations can also be challenging, placing increased burdens on transportation systems and networks. The example from the Bhopal disaster, discussed previously, showed what can happen when the transportation system of a city fails due to overload and panic during an evacuation.

There are challenges related to out-sourcing from targeted assets, or hardening or guarding a threatened asset. As discussed above, added protection activities can have significant financial costs associated with them. This cost burden can be compounded if there is uncertainty in the potential duration of the threat, requiring sustained protection activities over an extend period of time.

6 ACHIEVING PREVENTION AND PROTECTION SUCCESS

Our country and governmental system are based on the fundamentals of democracy, enabling us to live our lives, enjoy the benefits of liberty, and pursue happiness. We have a long tradition of freedom, a tradition that has stood the test of time, and endured many challenges. There is an inherent tension between freedom and security, a tension that was clearly recognized by our founding fathers, and best articulated by Alexander Hamilton in The Federalist Papers, Number 8:

> "... the continual effort and alarm attendant on a state of continual danger, will compel nations the most attached to liberty to resort for repose and security to institutions which have a tendency to destroy their civil and political rights. To be more safe, they at length become willing to run the risk of being less free."

It is important that we recognize this trade-off, and ensure that the efforts we put in place for prevention and protection to meet today's threat and hazard environment do not adversely degrade the very freedoms that are fundamental to our governmental system and our way of life.

Today's risks from terrorism and catastrophic hazards can be successfully managed if we implement a comprehensive approach. True risk management must consider three factors: consequence of loss, vulnerabilities, and probability of threat or hazard occurrence. These three elements must be combined and assessed to enable decision-makers to adequately accept, reduce, or offset risk.

An "all hazards" approach, one that addresses both threats and hazards as discussed above, must also be utilized to ensure we appropriately implement homeland security. The 15 National Planning Scenarios, developed by DHS in April 2005, for use in national, federal, state, and local homeland security preparedness planning, are structured using this all hazards approach. These scenarios span the spectrum of intentionally hostile threats, like chemical, biological, radiological, nuclear threats, or high yield explosives, to hazards like natural disasters and catastrophic accidents. They serve as an outstanding baseline for planning comprehensive prevention and protection activities.

A new updated strategic framework for homeland security must be implemented. This framework must address three areas. This paper discussed elements of one of them, that is, the prevention and protection activities of the broader preparedness continuum. However, the strategic homeland security framework must also include specific mission areas (i.e. what must be done for homeland security) and the domains environment (i.e. where the missions must be performed: on land, in the air, on the sea, and in cyber-space). This updated and comprehensive strategic framework for homeland security is shown in Figure 4.

This new Homeland Security Strategic Framework, based in large part on the preparedness continuum discussed in this paper, provides a context for synergy of effort between all levels of organizations in our country, from the largest government department down to the individual citizen and his respective community. It also places an emphasis on preincident activities, i.e., *prevention* and *protection*, which are proactive and can complement postevent planning. Our homeland security efforts can then be "powered down," meaning moving from the current "Top–Down" environment where large federal and state entities have the most responsibility, to a "Bottom–Up" environment where every

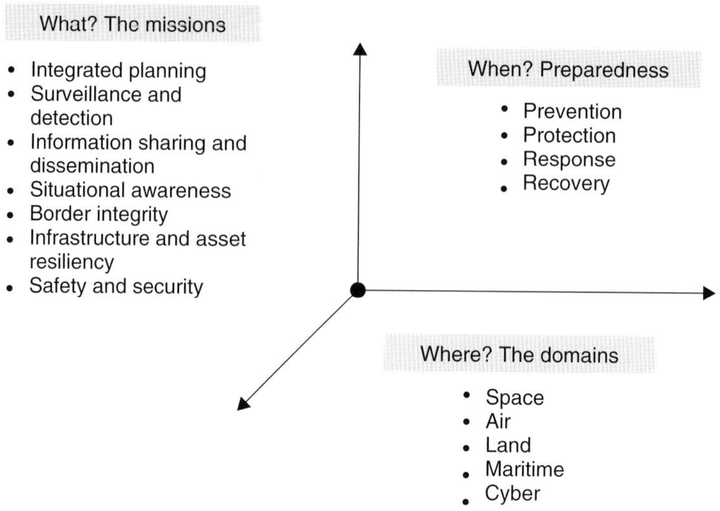

FIGURE 4 A comprehensive homeland security framework.

citizen, community, and city is empowered to contribute to a truly national effort to enhance our homeland's security.

Frederick the Great, who ruled the Kingdom of Prussia in Europe from 1740 to 1786 and was known as a great military strategist and commander, is quoted as having said: "... *he who tries to defend everywhere defends nowhere*." This quote points out the inherent difficulties in defending against a determined enemy and highlights the significant challenges faced in attempting to implement a successful strategy for prevention and protection. However, if we can enlist every citizen as part of the effort, and engage them to actively participate in prevention and protection activities broadly and where possible, I believe we can achieve a level of security for our homeland that is unprecedented and successfully counter both the natural elements as well as those few that would attempt to do us great harm. In this we may be able to prove Frederick the Great wrong—when *everyone* fully participates in the common defense, we can be truly strong *everywhere*.

FURTHER READING

Hamilton, A., Madison, J., and Jay, J. (1992). *The Federalist Papers*, specifically Numbers 8, 22–29, and 41. Buccaneer Books, Inc., Cutchogue, NY.

Homeland Security Presidential Directive No. 5, "Management of Domestic Incidents," The White House. (2003).

Homeland Security Presidential Directive No. 7, "Critical Infrastructure Identification, Prioritization, and Protection," The White House. (2003).

National Infrastructure Protection Plan, the Department of Homeland Security. (2006).

National Response Network, the Department of Homeland Security. (2008).

National Strategy for Homeland Security, The White House. (2002).

Quadrennial Defense Review, the Department of Defense. (2006).

Strategy for Homeland Defense and Civil Support, The Department of Defense. (2005).

CONSEQUENCE MITIGATION

PO-CHING DELAURENTIS, MARK LAWLEY, AND DULCY M. ABRAHAM
Purdue University, West Lafayette, Indiana

1 INTRODUCTION

The 1998 President's Commission on Critical Infrastructure Protection identified telecommunications, energy systems, water supply systems, transportation, banking and finance, and emergency and government services as essential core infrastructures for our modern society. A later national plan for critical infrastructure protection developed jointly by the Executive Office of the President, Office of Science and Technology Policy (OSTP) and the Department of Homeland Security (DHS), identified other key infrastructures such as agriculture and food, the defense industrial base, national monuments and icons, dams, commercial facilities, nuclear reactors, and materials and waste [1].

The increase in terrorist activities around the world, the possible use of weapons of mass destruction, and acts of nature such as earthquakes and floods pose threats to the security of various civil infrastructure systems. Further, increasing population concentrations have significantly stressed many infrastructure systems [2]. These factors combined with the cascading effects of system failures in power grids, telecommunication networks, transportation systems, and so on, intensify the need for effective disaster response planning and mitigation strategies.

Many definitions have been proposed for disaster/hazard/consequence mitigation. In this article, consequence mitigation refers to response strategies performed on existing infrastructures after an attack or disaster has occurred. It uses existing resources to take effective actions in order to minimize damage and propagation of damage and loss of life and property. Consequence mitigation strategies include contingency plans, rapid recovery plans, and operational tactics. Examples of these include damage control, toxic agent confinement, first responder deployment, and resource reallocation [3]. Figure 1 illustrates the distinction between mitigation strategies for pre- and postdisaster planning and response.

2 SCIENTIFIC OVERVIEW

There is only limited research in consequence mitigation. Most existing work focuses on public policy and government planning issues, with little emphasis on operational issues. Researchers distinguish between the goals of disaster prevention and consequence mitigation. Prevention focuses on preparation activities that reduce the likelihood of adverse events. These may include security improvements, upgrades in topology, addition of redundancies, and/or resiliency enhancements. In contrast, mitigation involves alleviating disaster effects, responding to their impacts, and recovering from the consequences.

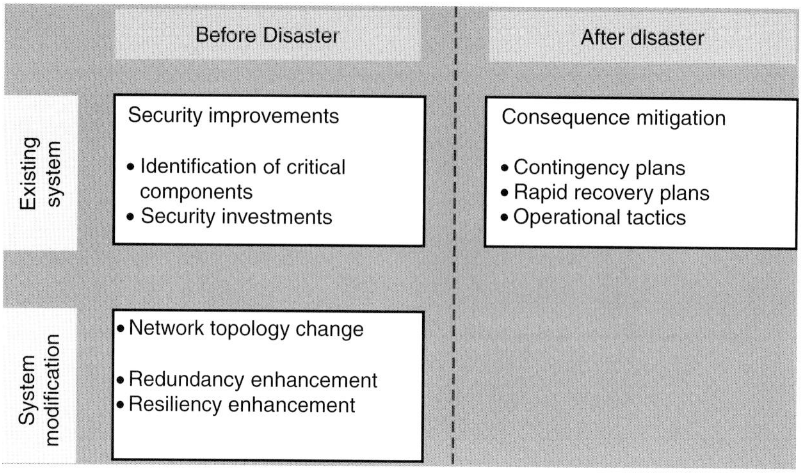

FIGURE 1 Disaster response and planning for infrastructure systems (adapted from Qiao et al. [3]).

Recovery may involve reconstruction of damaged facilities, restoration of physical and social networks, rehabilitation (including psychological recovery), and restitution (such as return to a previous physical or societal state). Both prevention and mitigation depend on predisaster planning and preparation.

The following sections provide an overview of recent mitigation research, mitigation tactics and technologies, and effectiveness measures for mitigation efforts.

2.1 Active Research Work and Recent Accomplishments

The National Critical Infrastructure Protection Research and Development (NCIP R&D) Plan was established in 2004 by the US DHS and the OSTP [1]. This plan identifies the critical R&D needs in securing the nation's infrastructures and key resources, and outlines several science, technology, and engineering themes that support all aspects of infrastructure protection.

The NCIP R&D Plan summarizes recent accomplishments in the development of strategies for consequence mitigation. It includes ongoing activities regarding protection of critical infrastructures undertaken by different agencies in the United States. Key highlights of consequence mitigation related efforts are listed in Table 1.

2.2 Mitigation Technologies and Tactics

There are different ways for mitigating disaster damages and potential hazards. One tactic aims at providing early detection and monitoring of the progression of a disaster so that damage control can be implemented at the earliest. Another method is the application of technologies that help minimize the damage caused by a disaster. Proper training for emergency workers who are called to respond to disasters is also deemed necessary for effective disaster mitigation.

2.2.1 Communication. A commonly suggested means for consequence mitigation is reliable wireless information technology (IT) infrastructure [6] that can be deployed in

TABLE 1 Ongoing and Recent Efforts of Consequence Mitigation by National Agencies in the United States

Agency	Highlights of Consequence Mitigation Effort
Department of Defense (DOD)	Development of technology for unexploded ordnance and dangerous materials detection inside assets and underground facilities
Department of Energy (DOE)	Development and deployment of a real-time global positioning system (GPS) with synchronized wide-area sensor network systems for electric grid monitoring and control
	Development of decontamination foam, which neutralizes chemical and biological agents in minutes
Department of Labor (DOL)	Developing protection, decontamination, and training guidance for hospital-based first receivers of victims of biological/chemical weapons of mass destruction
National Science Foundation (NSF)	Supporting research in nano- and biotechnology applications in protective materials and devices, new architectures for secure and resilient cyber and physical infrastructures, and sensors and sensor networks
US Department of Homeland Security, Federal Emergency Management Agency (FEMA)	*Best practice portfolio*—contains a collection of ideas, activities/projects, and funding sources that can help reduce or prevent the impacts of disasters, and is searchable by state, county, sector type, hazard, category/activity/project, and keywords [4]
US Environmental Protection Agency (EPA) and National Homeland Security Research Center (NHSRC)	Homeland security scientific research and technology development activities consist of the following: (i) threat and consequence assessment of human exposure to hazardous materials; (ii) decontamination and consequence management; (iii) water infrastructure protection; (iv) emergency response capability enhancement; (v) technology testing and evaluation [5]

various scenarios as an independent and secured emergency communication system [7]. Emergency management and control and responders may rely heavily on such infrastructure to communicate during disasters particularly when the public communications infrastructure is destroyed or severely damaged by the disaster. Developing frameworks for coordinated chains of communication is essential to provide quick response to victims as well as to ensure the safety of the responders.

2.2.2 Hazard Detection. Chemical/biological (C/B) or radiological attacks, in general, are difficult to detect and control especially in public spaces. The Program for Response

Options and Technology Enhancements for Chemical/Biological Terrorism in Subways (PROTECTS) is an initiative of the Department of Energy aimed at developing and applying technologies dealing with C/B terrorism. It covers both emergency planning and response phases of an incident and focuses on modeling and analyzing responses using engineering technologies [8]. A key technology developed by the PROTECTS program is the detection of the release of a C/B or radiological attack. Detection can be made by agent sensors, artificial intelligence, and video technologies. An artificial intelligence algorithm is used to recognize certain patterns of motion and sounds that are characteristics of a panicking crowd. The confirmation of a true incidence (not a false alarm) is then done by inspecting closed-circuit TV images. The use of this technology helps to keep casualties low since the situation can be observed remotely. Another frequently mentioned technology is sensor networks [9–11]. A sensor network is a collection of small, low-cost, and low-power devices with limited memory storage and wireless communication capabilities. One application is the personal digital assistant (PDA)-based multiple-patient triage device that can be used for on-scene patient triage, tracking, data recording, and monitoring of the physical environment. These data can help a medical facility prepare and appropriately allocate required resources (e.g. medical staff, beds, and operating rooms) before patients arrive, especially in a mass casualty scenario. Another sensor network application is the use of vital sign sensors that can be worn by disaster victims and first responders. These sensors can monitor a patient's physical condition and relay data to emergency medical personnel, enabling them to manage multiple patients simultaneously and be alerted if there are sudden changes in a patient's physiologic status [9]. Bioscience is another type of technology that can be used to detect hazards. For instance, the Sandia National Laboratories have been developing a state-of-the-art type of technology that uses special proteins to accurately and quickly detect specific bioterror threat agents [12].

2.2.3 Mitigating Technologies. Several technologies can be applied in mitigating hazard from a C/B attack: (i) inflatable barriers for blocking the spread of agent, (ii) water curtains, air curtains, and water or foam sprays for containing and detoxifying contaminated air, (iii) support tools for first responders and incidence commanders, such as hand-held devices that can receive information on-site, and (iv) training and exercises [8]. The Sandia National Laboratories have been active in developing effective tools to counter C/B attacks. One technology is containment foam that can be rapidly applied around an explosive device in order to reduce the blast effects and to capture the hazardous material that may be further spread [13]. Special types of coating materials can be used to contain radioactive materials by binding them, thereby preventing them from spreading. The lab has also developed bomb disablement tools that can be deployed by first responders to disable improvised explosive devices (IEDs) safely and remotely while preserving forensic evidence [12]. Successful technology applications are adopted by various US government agencies such as the DHS, the National Institute for Occupations Safety and Health (NIOSH), and the US Army.

2.2.4 Training. Emergency response/recovery technologies will not be effective without well-trained personnel [7]. Thus, organizations responsible for emergency or disaster mitigation tasks should provide proper training for first responders and incident management teams. Federal Emergency Management Agency (FEMA)'s Emergency Management Institute (EMI) provides a comprehensive list of training courses, both on-line and on-site, for emergency management officials [14]. In addition to educational and knowledge-based training, scenario-based physical exercises, such as simulated fires in a subway system, can also be an effective way for first responders and the incident commander to respond and implement optimal ventilation control strategies [8]. The Technical Support Working Group (TSWG), a US interagency forum, dedicates itself to developing technologies and equipment for the needs of the combating terrorism community. TSWG's training efforts include the development of "delivery architectures" (e.g. knowledge management systems and software architectures), "advanced distributed learning" (e.g. tools and guidelines for developing standard training materials), along with "training aids, devices and simulations" (e.g. virtual reality and computer-based simulations) [15].

2.2.5 Modeling and Simulation. Modeling and simulation tools can be used to understand what happens in disasters and thus what needs to be done in the mitigating and recovery stages. These tools can also be used as training exercises for the first responders, as well as the command and control personnel of Emergency Operations Centers (EOC) at all levels. In 2003 and 2004, the National Institute of Standards and Technology (NIST) held workshops on *Modeling and Simulation for Emergency Response* aimed at utilizing modeling and simulation technologies to better prepare for, mitigate, respond to, and recover from emergencies. Table 2 summarizes the range of emergency response modeling and simulation application tools discussed in the workshop. Details on the standard and related tools available can be found in Appendices A and B in NIST's report [16].

Bryson et al. [17] pointed out that there have been few studies in disaster recovery plan modeling in the management science (MS) and operations research (OR) community. They demonstrated an application of mathematical modeling techniques for decision-making support for disaster recovery planning, suggesting the need for dedicated efforts in integrating technical and MS/OR knowledge and techniques so that mitigation tactics and strategies can be more efficient and effective.

2.3 Effectiveness Measures

In addition to pursuing state-of-the-art technologies and techniques for disaster consequence mitigation planning and implementation, having precise and accurate methods for evaluating the effectiveness and efficiency of the mitigation actions is very important. Effectiveness measures can serve as a real-time feedback mechanism in the process of mitigating a disaster so that decision makers and responders can adjust plans and tactics accordingly. Examples of effectiveness measures include (i) public management—evacuation, restriction of entry, and quarantine; (ii) damage control—damage evaluation and pinpointing, agent confinement and elimination (e.g. toxic substance), and providing alternative sources of service; and (iii) recovery and rebuilding. However, the real challenge lies in defining and determining these effectiveness measures. Some methods that can help achieve the goals of effectiveness measures of consequence mitigation based on general emergency management include the following:

1. *Effectiveness measures developed from past experience*. After Hurricane Katrina devastated the New Orleans region in 2005, emergency response agencies began

TABLE 2 List of Modeling and Simulation Applications for Emergency Response Talks in NIST 2003 Workshop

Purpose	Application Tool	Developer
Planning applications	Virtual reality (VR) modeling and simulation—using geographic information systems (GISs) data, drawings, building plans, and city maps are processed using terrain generation software	Institute for Defense Analyses (IDA)
	JWFC simulation toolbox—providing multiechelon simulation across federal, state, and local government agencies	Joint Warfighting Center (JWFC)
	3D digital modeling, simulation, communication, and emergency response database (including facility models of high risk sites and libraries of the "best practice" processes)	Dassault Systems, Data Systems & Solutions (DS&S), Science Applications International Corporation (SAIC), and SafirRosetti
	Mass prophylaxis planning using ARENA simulation and queueing analysis in Excel	Department of Health and Human Services (DHHS), Agency for Healthcare Research and Quality (AHRQ)
Training	Top Officials (TOPOFF)—a national-level "real-time" weapons of mass destruction (WMDs) response exercise	The Department of Justice, the Department of State, and the Federal Emergency Management Agency (FEMA)
	2D simulations for command and control exercises of fire incidents	National Fire Programs, US Fire Administration, FEMA
	Automated exercise and assessment system (AEAS)—for emergency response and emergency management practitioners from the infrastructure owner/operator and local and state jurisdictional levels specific to WMD terrorist attacks	National Guard Bureau (NGB)
Real-time response	Tools and services for atmospheric plume predictions—in time for an emergency manager to decide if taking protective action is necessary to protect the health and safety of people in affected areas	National Atmospheric Release Advisory Center (NARAC)

to evaluate strategies for better evacuating and sheltering of residents and reducing loss of lives and properties. Whatever actions and plans were taken during that specific incident, for example, can serve as "the least effective" measure. Future mitigation plans and strategies for responses during hurricanes can then be improved by benchmarking against this measure.

2. *Use of risk and vulnerability assessment/mapping*. Risk and vulnerability assessment/mapping can be used for designing mitigation tasks and examining if current mitigation strategies and tactics are sufficient or effective. For example, matrices of multiple hazards can be used to analyze the relevance of different mitigation tools for responding to various hazards. Risk and vulnerability mapping can also be used as a tool to show the probabilities of disaster occurrences as well as the possible damage on physical properties, community infrastructure systems, and human lives [18].
3. *Simulation models*. Even though simulation may not provide the optimal solution, it can be useful as an effectiveness measure for disaster/hazard mitigation in integrating predisaster and postdisaster management and action plans and tactics to validate their effectiveness in a virtual environment. For example, the use of computer simulation for assessing the benefits of new technologies for disaster mitigation could reveal significant insight of such applications in a cost- and time-efficient manner [19]. Discrete event simulation and agent-based modeling, combined with human-in-the-loop and live simulation can provide significant insights on the effectiveness and efficiency of disaster responses.
4. *Use of detection device and remote sensing tools*. Detection devices such as sensors that can monitor or detect levels of C/B agents may be used after some hazardous material has been released. From these sensor readings, responders can determine the speed at which the agent is spreading or determine if the agent is confined where it was released. On the basis of this assessment they can plan their response. Then using real-time sensing, they can obtain feedback to assess whether their actions were effective in the situation.

3 ACTIVE RESEARCH AND FUNDING

Funding for research efforts in consequence mitigation has seen a surge since September 11, 2001. Table 3 lists several sources of funding for consequence mitigation.

4 CRITICAL NEEDS ANALYSIS

Consequence mitigation focuses on recovery after an attack or disaster has occurred. Technologies, such as those discussed above, support disaster mitigation efforts by providing more precise postdisaster information, such as pinpointing the location and amount of a toxic chemical released. Nevertheless, there is little discussion of the operational aspects of the disaster mitigation effort itself, that is, on the mitigation decision-making processes, triggering events, responder coordination, and so forth. Thus, research contributions in the following areas are essential for effective operational response after disaster.

4.1 Multiorganizational Coordination

In any disaster scenario, it is likely that multiple agencies across different jurisdictions in the affected area would participate in the mitigation effort. For example, if a C/B attack

TABLE 3 Research Funding Available in the United States and Other Countries

Agency	Funding Interest
US Sources of Funding	
National Science Foundation (NSF)—Division of Civil, Mechanical and Manufacturing Innovation, Infrastructure Management and Extreme Events (IMEE) [20]	For scientists, engineers, and educators focusing on large-scale hazards on civil infrastructure and society, and on issues of preparedness, response, mitigation, and recovery.
National Science Foundation (NSF)—hazards mitigation and structural Engineering (HMSE) [21]	For scientists, engineers, and educators working on fundamental research such as the design and performance of structural systems, and new technologies for improving the behavior, safety, and reliability of structural systems and their resistance to natural hazards.
Centers for Disease Control and Prevention (CDC)—engaging state and local emergency management agencies to improve states' ability to prepare for and respond to bioterrorism (funding opportunity number: CDC-RFA-TP06-601) [22]	Preparing the Nation's public health systems to minimize the consequences associated with natural or man-made, intentional or unintentional, disasters. Funding is only available to the National Emergency Management Association (NEMA).
National Institutes of Health—Small Business Innovation Research (SBIR) E-learning for HAZMAT and emergency response (SBIR [R43/R44]) [23]	For small businesses concerning the development of advanced technology training (ATT) products for the health and safety training and hazardous materials (HAZMAT) workers, emergency responders, and skilled support personnel.
US Department of Homeland Security—FEMA Hazard Mitigation Grant Programs (HMGP) [24]	For states and local governments for implementation of long-term hazard mitigation measures after a major disaster declaration.
US Department of Homeland Security—Predisaster Mitigation Grant Program [24]	For states, territories, Indian tribal governments and communities for hazard mitigation planning and the implementation of predisaster mitigation projects.
US Department of Homeland Security Office of Grants and Training—2006 Homeland Security Grant Program (HSGP) [25]	For state and urban areas to obtain critical resources to achieve the interim national preparedness goal and implement homeland security strategies.
US Department of Homeland Security Office of Grants and Training—Emergency Management Performance Grant (EMPG) [25]	The objective of this program is to help states develop effective emergency management systems that encourage the cooperation and partnership among government, business, volunteer, and community organizations and thus strengthening their preparedness for catastrophic events and emergency management capabilities.

TABLE 3 ***(Continued)***

Agency	Funding Interest
International Sources of Funding	
The European Mediterranean Disaster Information Network (EU-MEDIN)—applied multirisk mapping of natural hazards for impact assessment (ARMONIA) [26]	To provide the European Union (EU) with harmonized methodologies for producing integrated risk maps to achieve effective spatial planning in areas prone to natural disasters in Europe.
Disaster Hazard Mitigation Project Kyrgyz Republic [27]	To achieve (i) minimizing the exposure of humans, livestock, and riverine flora and fauna to radionuclides associated with abandoned uranium mine tailings and waste rock dumps in the Mailuu-suu area; (ii) improving the effectiveness of emergency management and response by national, local government agencies and local communities; (iii) reducing the loss of life and property in key landslide areas of the country.
Australian Government—the local grants scheme (LGS) and the national emergency volunteer support fund (NEVSF) [28]	For local governments to help communities develop and implement emergency management initiatives and enhance critical infrastructure protective measures, as well as to provide training of security awareness for local government staff.

were to occur at a subway station in a city, the local jurisdiction levels involved would include the subway management team (e.g., the city transportation administration), the city government, and the county in which the city is located. Public, private, and non-profit organizations would need to coordinate their actions with each other and across jurisdictions to create a dynamic emergency response system to ensure effective mitigation and response to a disaster. Since the most effective response processes and plans will transcend political and response agency boundaries, coordinated response plans must be developed through the collaborative efforts of all key responders.

4.2 Management Framework

A disaster response management framework is needed for supporting decision makers in a disaster scenario. Such an emergency management framework should include the use of real-time monitoring of the situation (such as the general environment, infrastructure systems, first responders, and victims) and real-time information and communication tools, both vertically in the organizational structure and horizontally among responders. Coordination of effective reporting from multiple sites (for instance, in an earthquake rescue effort) should also be incorporated in the framework so that mitigation decisions can be made to achieve effective and efficient actions taken to minimize negative impacts of a disaster [29]. Research on technology integration for real-time information exchange, decision-making support systems, and cross-infrastructure recovery efforts is very much needed. Furthermore, since response resources are likely to be limited or overwhelmed, it is important to prioritize response activities. Scarce resource allocation decisions need to be modeled, analyzed, and optimized beforehand to ensure that the most effective priorities and trade-offs occur during response.

5 RESEARCH DIRECTIONS

Four major themes can be identified for future research.

5.1 Computing Hardware, Software, and Research Tools

In the current environment, network of computers and devices embedded in infrastructure systems are susceptible to crippling cyber attacks. Thus, there is a significant need for next-generation Internet architectures that are designed to have security and inherent protection features at all levels of hardware and software [1]. The NCIP R&D plans propose the designing of cyber infrastructures that are resilient, self-diagnosing, and self-healing. In addition, future research will need to focus on developing integrated systems architecture/framework so that management and responders across jurisdictions and hierarchical response organizations can be well coordinated.

5.2 Interdependencies between Civil Infrastructures

Since modern civilization relies heavily on basic civil infrastructures that are interdependent of each other, improving coordination of security precautions taken by different utility systems is also an important task. In other words, it is necessary to look beyond the effects of an incident on a single system. Instead, future technologies and practice should be able to understand the perturbed behaviors of a complex, "system of systems" [30], which includes the unexpected cascading effects of infrastructure damage and failure. A good example is the concept of "resilient cities" that focuses on strengthening physical and social elements of an urban area and on bridging natural hazard and antiterrorism predisaster mitigation and postdisaster response strategies [31]. More effective use of technology transfer and broader industry/government support are needed to achieve the effectiveness of consequence mitigation for the industry/society as a whole [7].

5.3 Sensing Technologies for Assessment

Advanced sensing technologies are essential for assessing the causes and effects of disaster events. Future infrastructures must be designed with more embedded sensing, diagnostic, and predictive capabilities. Integrated sensor networks with powerful computational and communication capabilities are now becoming a reality, and infrastructural planners, engineers, managers, and operators must develop fuller understandings of how these can be incorporated into design and decision processes. For instance, a real-time building damage sensing network could be used to assess the condition of a building after an extreme event, such as an earthquake. This type of information would greatly help disaster responders responsible for evacuation, search, rescue, and building reenforcement efforts. Further, remote sensing technologies, such as satellite images, are especially helpful in surveying a large affected area [32].

5.4 Modeling and Analysis of Disaster Scenarios, Mitigation Plans, and Response Efforts

Simulation modeling can be very useful in assessing the effectiveness of mitigation plans for different disaster scenarios. For example, scenario analyses of various types and/or severity of terrorist attack can be performed using simulation. This type of "what if" analysis serves as a virtual exercise test bed for identifying critical mitigation resources and improving mitigation plans and decisions. Furthermore, optimization techniques are very useful for analyzing decision problems where priorities must be considered and trade-offs must be made. For example, reconfiguring damaged infrastructures and prioritizing repair tasks to optimally meet a population's demand during repair can be addressed by using advanced optimization techniques from Operations Research and Management Science. Technical challenges associated with developing these types of simulation and optimization models include model abstraction, validation, data scarcity, computational requirements, identifying appropriate objective functions, and interpreting the results, among others. A significant practical challenge is engaging laypeople in the development of the models and in ensuring that users of the model have adequate confidence in the results obtained through the models. Although this type of analysis has a long history in other large-scale planning and operation efforts such as manufacturing and transportation, it is in its infancy in disaster mitigation and response planning. Thus, there are significant opportunities for researchers to make seminal contributions in this area. One example is the application of a computer interface that integrates an agent-based discrete event simulation model and a geographic information system such that real-time data exchange and communication can coordinate and facilitate large-scale disaster response efforts [33]. Stochastic programming technique can be applied to address the issue of transporting first aid commodities and response personnel in an earthquake scenario [34]. Other examples of this type of work can be seen in [35–37].

REFERENCES

1. The Executive Office of the President Office of Science and Technology Policy (OSTP), and the Department of Homeland Security Science and Technology Directorate. (2004). *The National Plan for Research and Development in Support of Critical Infrastructure Protection*, US Department of Homeland Security, Washington, DC.
2. Hamilton, R. M. (2000). Science and technology for natural disaster reduction. *Nat. Hazard. Rev.* **1**(1), 56–60.
3. Qiao, J., Jeong, H. S., Lawley, M. A., and Abraham, D. M. (2006). Physical security aspects in water infrastructure. In *Advances in Homeland Security Volume 1, The Science of Homeland Security*, S. F. Amass, A. K., Bhunia, A. R., Chaturvedi, D. R., Dolk, S. Peeta, and M. Atallah, Eds. Purdue University Press, pp. 37–62.
4. FEMA. FEMA Mitigation Best Practices Search. (2008). http://www.fema.gov/mitigationbp/index.jsp.
5. U.S. Environmental Protection Agency. (2008). Homeland Security Research, http://www.epa.gov/nhsrc/index.htm.
6. Midkiff, S. F., and Bostian, C. W. (2002). Rapidly-deployable broadband wireless networks for disaster and emergency response. *The First IEEE Workshop on Disaster Recovery Networks (DIREN '02)*. June 24, 2002, New York.

7. Schainker, R., Douglas, J., and Kropp, T. (2006). *Electric Utility Responses to Grid Security Issues. IEEE Power and Energy Magazine*, March/April, 31–37.
8. Policastro, A. J., and Gordon, S. P. (1999). The use of technology in preparing subway systems for chemical/biological terrorism. *APTA 1999 Rapid Transit Conference*. Toronto, ON.
9. Lorincz, K., Malan, D. J., Fulford-Jones, T. R. F., Nawoj, A., Clavel, A., Shnayder, V., Mainland, G., Welsh, M., and Moulton, S. (2004). Sensor networks for emergency response: challenges and opportunities. Pervasive computing. *IEEE* **3**(4), 16–23.
10. ARC Research Network on Intelligent Sensors, Sensor Networks and Information Processing (ISSNIP) Applications. (2007). http://www.ee.unimelb.edu.au/ISSNIP/apps/index.html.
11. Zussman, G., and Segall, A. (2003). Energy efficient routing in ad hoc disaster recovery networks. *Proceedings IEEE INFOCOM 2003 Conference*. San Francisco, CA.
12. Sandia National Laboratories. (2006). *Defense Against Chemical and Biological Threats*. http://www.sandia.gov/mission/homeland/chembio/development/biotechnology/index.html.
13. Sandia National Laboratories. (2008). *Explosive Countermeasures*. http://www.sandia.gov/mission/homeland/programs/explosives/index.html/.
14. FEMA Emergency Management Institute. (2008). http://training.fema.gov/.
15. The Technical Support Working Group. (2008). http://www.tswg.gov/.
16. National Institute of Standards and Technology. (2008). *Modeling and Simulation for Emergency Response Workshops*. http://www.mel.nist.gov/div826/msid/sima/simconf/mns4er.htm.
17. Bryson, K.-M., Millar, H., Joseph, A., and Mobolurin, A. (2002). Using formal MS/OR modeling to support disaster recovery planning. *Eur. J. Oper. Res.* **141**, 679–688.
18. Godschalk, D. R., and Brower, D. (1985). Mitigation strategies and integrated emergency management. Public administration review. *Public Adm. Rev.* **45**, 64–71, Special Issue.
19. Robinson, C. D., and Brown, D. E. (2005). First responder information flow simulation: a tool for technology assessment. *Proceedings of the 37th Conference on Winter Simulation*, Orlando, FL.
20. NSF. *Infrastructure Management and Extreme Events (IMEE)*. (2008). http://www.nsf.gov/funding/pgm_summ.jsp?pims_id=13353&org=CMMI.
21. NSF. *Hazard Mitigation and Structural Engineering (HMSE)*. (2008). http://www.nsf.gov/funding/pgm_summ.jsp?pims_id=13358&org=CMMI&sel_org=CMMI&from=fund.
22. CDC Grant. (2007). http://www.grants.gov/search/search.do?oppId=10597&mode=VIEW.
23. Department of Health and Human Services. (2005). http://grants.nih.gov/grants/guide/rfa-files/RFA-ES-05-003.html.
24. FEMA. (2008). *FEMA Hazard Mitigation Grant Programs*. http://www.fema.gov/government/grant/hmgp/.
25. U.S. Department of Homeland Security, Office of Grants and Training (G&T). (2008). http://www.ojp.usdoj.gov/odp/about/overview.htm.
26. *ARMONIA –Applied multi Risk Mapping of Natural Hazards for Impact Assessment*. (2008). http://www.ist-world.org/ProjectDetails.aspx?ProjectId=46f1c981ed3b4821947cd3624b820fb4&SourceDatabaseId=7cff9226e582440894200b751bab883f.
27. Disaster Hazard Mitigation Project Kyrgyz Republic. (2004). http://web.worldbank.org/external/projects/main?pagePK=64283627&piPK=73230&theSitePK=40941&menuPK=228424&Projectid=P083235.
28. Australian Government. National Emergency Volunteer Support Fund (NEVSF). (2008). http://www.ema.gov.au/communitydevelopment.
29. Comfort, L. K., Dunn, M., Johnson, D., Skertich, R., and Zagorecki, A. (2004). Coordination in complex systems: increasing efficiency in disaster mitigation and response. *Int. J. Emerg. Manage.* **2**(1-2), 62–80.

30. Little, R. G. (2003). Toward more robust infrastructure: observations on improving the resilience and reliability of critical systems. *Proceedings of the 36th Annual Hawaii International Conference on System Sciences*. Big Island, HI.

31. Godschalk, D. R. (2003). Urban hazard mitigation: creating resilient cities. *Nat. Hazard. Rev.* **4**(3), 136–143.

32. Adams, B. J., Huyck, C. K., Gusella, L., Wabnitz, C., Ghosh, S. and Eguchi, R. T. (2006). Remote Sensing Technology for Post-Earthquake Damage Assessment- A Coming of Age. *Proceedings of the 8th U.S. National Conference on Earthquake Engineering*. April 18–22, San Francisco, CA.

33. Wu, S., Shuman, L., Bidanda, B., Kelly, M., Sochats, K. and Balaban, C. (2007). Disaster policy optimization: a simulation based approach. *Proceedings of the 2007 Industrial Engineering Research Conference*. Nashville, TN.

34. Barbarosoglu, G., and Arda, Y. (2004). A two-stage stochastic programming framework for transportation planning in disaster response. *J. Oper. Res. Soc.* **55**, 43–53.

35. Friedrich, F., Gehbauer, F. and Rickers, U. (2000). *Optimized Resource Allocation for Emergency Response after Earthquake Disasters, Safety Science 35.*

36. Altay, N., and Green, W. G., III. (2006). OR/MS research in disaster operations management. *Eur. J. Oper. Res.* **175**, 475–493.

37. Lee, E., Maheshwary, S., Mason, J., and Glisson, W. (2006). Large-scale dispensing for emergency response to bioterrorism and infectious-disease outbreak. *Interfaces* **36**(6), 591–607.

FURTHER READING

Disaster Planning and Public Policy

Berke, P. R., Kartez, J., and Wenger, D. (1993). Recovery after disaster: achieving sustainable development, mitigation and equity. *Disasters* **17**(2), 93–109.

Comfort, L., Wisner, B., Cutter, S., Pulwarty, R., Hewitt, K., Oliver-Smith, A., Wiener, J., Fordham, M., Peacock, W., and Krimgold, F. (1999). Reframing disaster policy: the global evolution of vulnerable communities. *Env. Hazard* **1**, 39–44.

Godschalk, D., Beatley, T., Berke, P., Brower, D. J., and Kaiser, E. J. (1999). *Natural Hazard Mitigation Recasting Disaster Policy and Planning*, Island Press, Washington, DC.

Freeman, P. K., Martin, L. A., Linnerooth-Bayer, J., Mechler, R., Pflung, G., and Warner, K. (2003). *Disaster Risk Management: National Systems for the Comprehensive Management of Disaster Risk and Financial Strategies for Natural Disaster Reconstruction*. Inter-American Development Bank, Sustainable Development Department, Environment Division, Integration and Regional Programs Department, Regional Policy Dialogue, Washington, DC.

Technical Research

Jacobson, S. H., Kobza, J. E., and Pohl, E., Eds. (2007). *IIE. Trans.* **39**(1), Special Issue on Homeland Security.

General Research for Natural Disasters

Cruz, A. M. (2008). *Engineering Contribution to the Field of Emergency Management*. http://training.fema.gov/EMIWeb/edu/docs/Engineering%20Contribution.pdf.

Heaney, J. P., Peterka, J., and Wright, L. T. (2000). Research needs for engineering aspects of natural disasters. *J. Infrastruct. Syst.* **6**(1), 4–14.

SECURITY ASSESSMENT METHODOLOGIES FOR U.S. PORTS AND WATERWAYS

D. Brian Peterman, Joseph DiRenzo III, and Christopher W. Doane
United States Coast Guard

1 INTRODUCTION

The challenge of securing critical infrastructure within the US Maritime Domain is daunting. With over 95,000 miles of maritime border, 361 maritime ports, thousands of critical maritime facilities, and millions of recreational and commercial maritime users, the US Maritime Domain is too vast to be completely protected. Compounding the problem is the Maritime Transportation System that provides over $700 billion a year to the nation's economy; a system that is highly sensitive to interruption. The only feasible solution to this maritime security dilemma is to prioritize maritime security efforts using a rigorous process to assess risk at the national, regional, and port levels. Since the horrific attacks of September 11, 2001, the US Coast Guard has developed, refined, and continuously updated maritime risk assessment tools and methods to inform maritime security operations against the ever-changing worldwide asymmetric terrorist threat. Settling upon the equation, risk = threat × vulnerability × consequence, the capability to assess maritime risk is now several iterations ahead of where the country was after the attacks. This assessment process will remain a "living process" requiring constant modification to stay apace of changes in maritime use and terrorist adaptations to security measures.

2 BACKGROUND

The attractiveness of the US Maritime Domain to terrorists is captured in the US National Strategy for Maritime Security, "the infrastructure and systems that span the maritime domain ... have increasingly become both targets of and potential conveyances for dangerous and illicit activities" [1, p. 2]. The challenge of protecting the US maritime is daunting for multiple reasons. The country's maritime borders, rivers, and waterways are extensive; the maritime border alone is over 12,400 miles [2, p. 6; 3]. These shorelines are host to thousands of facilities either critical to the national economy and/or processing highly volatile or toxic materials dangerous to nearby populations. The waterways also contain other critical infrastructure, such as locks and dams, whose damage or destruction would have dire effect on the Maritime Transportation System in addition to population. "The challenge is immense as it involves nearly 13 million registered US recreational vessels, 82,000 fishing vessels and 100,000 other small vessels" [4, p. i]. The potential security challenge posed by recreational boats is made even more complicated with each state maintaining vastly different databases that do not electronically connect or share information easily or by automation.

The only solution for reducing risk to such a vast and complex system as the US Maritime Domain without undermining the Maritime Transportation System is to employ a risk-based strategy, identifying and focusing security efforts on high risk targets while preparing to minimize the consequences of attacks that do occur. This concept is clearly stated in the country's National Strategy for Homeland Security, "We accept that risk—a function of threats, vulnerabilities, and consequences—is a permanent condition. We must apply a risk-based framework across all homeland security efforts in order to identify and assess potential hazards (including downstream effects), determine what levels of relative risk are acceptable, and prioritize and allocate resources among all homeland security partners, both public and private, to prevent, protect against, and respond to and recover from all manner of incidents" [5, p. 41].

2.1 Risk Assessment Methods

An understanding of "risk" and its associated components of "threat," "vulnerability," and "consequence" is essential to any discussion of risk assessment methods. Defining these four critical terms is not easy, nor has there been agreement within academic, law enforcement, or military circles as to their meaning. Adding to this difficulty is the issue of "real risk" versus "perceived risk" (public perception is an additional reality that must be respected). Assessing risk is a process that involves as much art as it does science. The art element is driven by the high degree of uncertainty created by questions such as the following: What is acceptable risk? What is the enemy's true capability? What is the enemy's intent? In simple terms, the terrorists hold every one of the controlling factors in the successful accomplishment of an attack. They can pick the time, place, location, and method of the attack. Terrorists also have the ability to call off or delay an attack if security forces are in position to disrupt their attack knowing that these forces cannot remain at high alert indefinitely.

The analysis of risk informs several critical decisions: determining how limited resources and capabilities will be deployed; identifying what security actions offer the greatest risk reduction for the least investment; and what new technologies or capabilities offer the best investment for limited funds. The same discussion occurs as mitigation strategies are developed, analyzed, and selected. In selecting a specific course of action, risk models help quantify potential results. What also must be evaluated are unintended consequences of the action.

The RAND Corporation's team of Dr. Henry Willis, Dr. Andrew Morral, Terrence Kelly, and Jamison Jo Medby presented one of the definitive papers regarding the "risk" equation at the Society for Risk Analysis' Annual Meeting in 2004. Dr. Henry Willis, who is an Associate Policy Research at RAND's Pittsburgh's office, expanded on this equation on February 7, 2007 in testimony before the Committee on Appropriations Subcommittee on Homeland Security, United States House of Representatives. Willis noted at the very beginning of his testimony that, "There is no single correct method for measuring terrorism risk ... Terrorism risk is a function of three factors: a credible *threat* of attack on a *vulnerable* target that would result in unwanted *consequences*. Risk only exists if terrorists want to launch an attack, if they have the capability to do so successfully in a way that avoids security and compromises the target, and if the attack results in casualties, economic loss, or another form of unwanted consequences" [6] (*see* Risk Analysis Framework for Counterterrorism).

The US military uses very specific processes to evaluate potential targets that can be invaluable to an overall discussion of risk. "Target analysis is the detailed examination of potential targets to determine their military importance, priority of attack, scale of effort, and the lethal or nonlethal weapons required to attain a specified effect. It is a systematic approach to establishing the adversary vulnerabilities and weaknesses to be exploited. This is accomplished through the methodical examination of all information pertaining to a given target" [7, p. A-1]. This target analysis examines vulnerability to attack and what the overall effect would be if an attack occurred.

To conduct this analysis, US Special Forces use a method called CARVER. The name is an acronym for the variables the method uses for evaluation. The use of this method is applicable to infrastructure evaluation as part of an overall approach to risk evaluation. A numerical value/rating scheme is applied to each of the following CARVER variables.

C = Criticality. "Criticality or target value is the primary consideration in targeting. Criticality is related to how much a target's destruction, denial, disruption, and damage will impair the adversary's political, economic, or military operations, or how much a target component will disrupt the function of a target complex" [7, p. A-3].

A = Accessibility. "In order to damage, destroy, disrupt, deny or collect data on a target, Special Operations Forces (SOF) must be able to reach it with the necessary equipment, either physically or via indirect means" [7, p. A-4].

R = Recuperability. "In the case of Direct Action (DA) missions, it is important to estimate how long it will take the adversary to repair, replace, or bypass the damage inflicted on a target. Primary considerations are spare parts availability and the ability to reroute production" [7, p. A-4].

V = Vulnerability. "A target is vulnerable if SOF has the means and expertise to attack it. At the strategic level, a much broader range of resources and technology is available to conduct the target attack" [7, p. A-4].

E = Effect. "The target should be attacked only if the desired military effects can be achieved. These effects may be of a military, political, economic, informational, or psychological nature" [7, p. A-4].

R = Recognizability. "The target must be identifiable under various weather, light, and seasonal conditions without being confused with other targets or components" [7, p. A-4].

For each CARVER element a criterion provides a value that is included in the overall assessment of either the ease or difficulty in attacking a particular target and achieving the desired effect. For example, under "Critically" a score of "5" might equal "significant damage to overall mission capability." However, an assigned score of "1" could mean "loss would not effect overall mission performance." [Note: These types of notional factors are applied across all six criteria listed for CARVER]

But CARVER is only a process, or some would say a "tool," in an overall evaluation regime. An even more critical factor in any maritime risk assessment process is the *capability* of a terrorist group to use and exploit the maritime environment for an attack. This is an important component discussion above the standard risk = threat × vulnerability × assessment. It is often overlooked because it is extremely difficulty to judge. But when you think about it, is it not the most basic question—what are terrorist groups, both national and international truly capable of? (*see* Vulnerability Assessment).

2.2 Assessing Maritime Risk

The capability and intent for some terrorist groups to exploit the maritime environment has been well documented. Certainly the Tamil Tigers have conducted their bold maritime attacks, including use of the underwater environment, and tactical acumen rivaling many of the world's military and security forces. What is debated in both government and academic circles is the ability of a terrorist group to obtain a chemical, biological, radiological, or nuclear (CBRN) weapon of mass destruction (WMD). What is truly available on the black market? Do terrorists groups have the technical expertise to assemble, deploy, and trigger a WMD inside a port? These are the true questions of risk that should be part of any overall discussion.

So what has been and is being done to assess maritime risk due to terrorism?

Primary responsibility for conducting these assessments falls on the US Coast Guard as the Department of Homeland Security's Executive Agent for Maritime Security [1, p. 23] and the Federal Maritime Security Coordinator in the port [8, p. 7]. Before entering into a detailed discussion of the Coast Guard's use of risk assessment methods for analyzing port security requirements, an understanding of the service's historic role in securing US ports and employing risk management is necessary.

At the start of World War I the Coast Guard was assigned the responsibility for the security of US ports under the Magnuson Act of 1917. For both World Wars and the Korean War, the service expanded its numbers significantly to meet the tremendous demands of securing the nation's 12,400 miles of maritime border and hundreds of ports. During World War II, the Coast Guard grew to more than 240,000 active duty personnel with more than 53,000 assigned to port security duties [9] (the current strength of the Coast Guard is just over 40,000 active duty personnel) [10].

After the Korean War, the port security threat to the United States steadily diminished and the port security program essentially became a contingency mission assigned to the Coast Guard Reserve. The perception was that port security operations would only be needed in support of the Department of Defense. Planning centered around two functions: securing a limited number of domestic ports conducting military out load operations (ports of embarkation) and providing port security at foreign ports to support military offload operations (ports of debarkation). By September 11, 2001 the Coast Guard had been reduced to a force of just over 35,000 active duty personnel who had little to no involvement in port security.

During the 18 months preceding the September 11th attacks the Coast Guard had gone on record stating that it lacked the resources necessary to meet all of its missions; a resource demand that essentially did not include port security [11, p. 11]. Following the attacks, port security suddenly became a primary mission of the active duty force. This resurgence of port security as a primary mission significantly exacerbated the already sizeable resource shortfall within the Coast Guard.

On September 11, 2001 the port security task facing Coast Guard planners and operators was formidable. As previously discussed, the vastness, complexity, and economic importance of the US Maritime Domain and the Maritime Transportation System are truly immense. Protecting everything was clearly not an option. In addition, the nearly 12,000,000 recreational boaters operating on US waters created ideal camouflage for terrorists seeking to disguise preparations for small boat attacks.

To add to the maritime security difficulty, industry had shifted to a "just-in-time" inventory system, maintaining minimum inventory and counting on being resupplied just in time with the critical supplies and material needed to continue operations. As much of these materials are transported through the Maritime Transportation System, any inordinate delay in the flow of commerce due to security measures would have a telling effect on the economy. It was clear to Coast Guard planners that, barring a significant force increase to the levels of World War II, maritime security operations would have to be focused, efficient, and effective in order to maximize the use of existing security forces while minimizing the impact on legitimate users; their solution, risk management. The greatest risks in the maritime relative to security needed to be identified and effective risk mitigation measures implemented.

Fortunately, the Coast Guard had been using risk management principles for almost a decade prior to 2001. To inform its Marine Safety and Environmental Protection programs, the service had employed Risk-Based Decision Making [12, p. 4-1] and was using Operational Risk Management [13, p. 1] to assist field commanders in assessing the risks associated with a specific action or activity. These two initiatives created a risk management culture within the Coast Guard that significantly facilitated the service's adoption of risk-based operations to meet the new demands of maritime security following the 2001 attacks.

The Operational Risk Management model follows a seven-step task-oriented process: identify mission tasks, identify hazards, assess risks, identify options, evaluate risk versus gain, execute decision, and monitor situation. This model calculated risk using the equation

$$\text{Risk} = \text{Severity} \times \text{Probability} \times \text{Exposure}$$

Severity is the potential consequences in terms of degree of damage, injury, or mission impact. Probability is the likelihood of the potential consequence. Exposure is the amount of time, number of occurrences, number of people, and/or equipment involved in the task [13, p. 4]. Coast Guard commanders were taught to assess risk associated with specific actions or missions using this Operational Risk Management model (*see* Terrorism Risk: Characteristics and Features).

The service's Risk-Based Decision Making model contains five major components: decision structure, risk assessment, risk management, impact assessment, and risk communication. Decision structure includes the concept of obtaining input from external stakeholders in the risk analysis process to determine available options and influencing factors. Risk assessment uses the equation:

$$\text{Risk} = \text{Probability} \times \text{Consequence}$$

The assessment also looks at factors such as: reasons for the assessment; type of results needed; available resources; complexity of the assessment; type of activity or system analyzed; and type of incidents targeted. Risk management recognizes that the benefit of action to manage or reduce risk must outweigh the cost, be acceptable to other stakeholders, and not cause other significant risk. Impact assessment requires tracking of the risk mitigation actions taken to ensure that the desired benefits are realized. Risk communications incorporates two-way communication with stakeholders to identify key issues, provide information, and obtain consensus [12, p. 4–5].

3 POST-9/11 MARITIME RISK ASSESSMENT

With this background in risk management, Coast Guard leadership and planners readily accepted risk-based decision making as the means to solve the maritime security challenge. Shortly after the 2001 attacks the Coast Guard developed its first security risk assessment tool called the *Port Security Risk Assessment Tool*s (PSRAT) [14, p. 16]. This tool used the risk equation:

$$\text{Risk} = \text{Threat} \times \text{Consequence} \times \text{Vulnerability}$$

The tool was provided to Coast Guard Captains of the Port to assess maritime risk in their areas of responsibility. Individual facilities, infrastructure, and waterways seen as potential terrorist targets were assessed for risk using the tool. The tool generated a dimensionless Risk Index Number (RIN) for the target based upon the assessed values for threat, consequence, and vulnerability; the higher the RIN, the greater the risk.

The RIN scores for the various facilities and other infrastructure were compiled and ranked at the national level. This allowed the service to develop its first list of nationally critical maritime infrastructure based upon risk. Coast Guard security forces and those of other agencies then focused their security operations in support of these critical infrastructures.

While PSRAT was a good start, it had shortcomings. The level of effort put into the analysis varied depending upon staff availability and workload at each port. As a result, consistency in the analysis between ports was not ideal. In addition, the vulnerability and consequence analysis was conducted primarily by law enforcement personnel, not scientist and engineers, creating inaccuracies. A significant weakness was poor linkage between attack methods and resulting consequences. A second run of the PSRAT model was conducted with modified guidelines, improvements in the results were noted, but significant inconsistencies were still apparent.

Coast Guard planners gathered to improve their port security risk assessment and developed the Maritime Security Risk Analysis Model (MSRAM) in 2006 [14, p. 16]. Although based upon the same risk equation, this tool offered improved threat information including potential damage information, target classification, more rigorous training, and data treatment. Still, the 2006 MSRAM analysis produced results with some clear errors. The problems were addressed and a second MSRAM analysis was conducted in 2007 with much improved result [14, p. 16].

In MSRAM, threat is treated as the probability a certain type of terrorist attacks, such as explosive-laden small boats, standoff attacks, sabotage, and so on might occur [14, p. 19]. This threat analysis includes concepts, such as terrorist intent, terrorist capability, and timeframe in which terrorists will acquire the capability. In addition to terrorist competency and possession of necessary material for a given attack method, the assessment of capability also considers the ability of the terrorists to produce that capability in a given geographic region. For example, terrorists have proven their ability to conduct suicide attacks with explosive-laden small boats in the Middle East; but do they have the infrastructure in place to conduct such attacks in the United States? (*see* Risk-Based Prioritization).

The consequence assessment looks at both primary consequences and secondary economic impacts that might result from a selected attack method. Primary consequences include factors such as: death, injury, primary economic impact, national security impacts, symbolic effect, and environmental impact that would probably result from a successful attack of a given type. Secondary economic impacts consider factors such as: recoverability, redundancy, and secondary economic impacts [14, p.19].

Primary economic impacts include direct factors such as the value of the infrastructure that would be lost in the attack. Secondary economic impacts consider factors such as the monetary impact through lost revenue if the attacked target remains inoperable. Recoverability is an assessment of how quickly an attacked function can be restored at the site of the attack. Redundancy looks at how many of the same type facility or infrastructure exist [14]. For example, the consequences of losing the only bridge in a region may be far more significant that a region with several bridges spanning the same waterway.

A significant difficulty in calculating consequence stems from equating the consequence values associated with different factors. Equating economic value or national security valuc to lives lost is not an exact science, but requires objective analysis and stakeholder consensus. Overtime, the Coast Guard worked with a variety of other agencies and entities to develop these equivalencies.

Vulnerability is the likelihood that a selected attack method will succeed in producing the consequences feared. Vulnerability considers factors, such as achievability, security systems in place, and target hardness. Achievability considers attack aspects such as whether the depth of water surrounding a piece of infrastructure would support a small boat attack. Security systems include the security capabilities at the selected facility as well as the security capabilities of local law enforcement and the Coast Guard. Target hardness evaluates the ability of physical barriers, such as wall material and thickness, to resist a given attack type.

3.1 Maritime Risk Assessment Challenges

In calculating risk, it is critical that the consequence and vulnerability values used are those associated with the selected attack method. Consider an explosive-laden small boat attack on a waterfront chemical facility. The degree of physical destruction, subsequent chemical release, and resulting death and injury must be those resulting realistically from the amount of explosive a small boat can carry and the proximity to the facility a small boat can achieve; not necessarily a worse case release.

The accuracy of the risk assessment is wholly dependent upon the accuracy of the inputs to the tool. Accurately establishing the values for each of the risk factors requires expert and deliberate analysis. For the explosive-laden small boat attack on a waterfront chemical facility example: the blast effect of various explosive loads must be calculated, the ability of the facility's wall to withstand the blast also must be calculated, and the effect of the blast energy that reaches the chemical needs to be understood; Will the chemical ignite? If so, with what thermal or blast effect? Will the chemical be released in a toxic plume? If so, in what concentration, for what duration and distance? Answers to these questions require in-depth analysis by scientists and engineers (*see* High Consequence Threats: Chemical and High Consequence Threats: High-Grade Explosives).

3.2 Application of Maritime Risk Assessment

As a result of the risk analysis supported by MSRAM, the national maritime critical infrastructure list was significantly refined. This in turn allowed port-level security forces to prioritize their security operations with increased confidence. Although inconsistencies and need for improvements were still apparent, the Coast Guard has developed enough confidence in the tool to require its use in developing Area Maritime Security Plans as required in the Maritime Transportation Security Act of 2002 (MTSA) [15, p. 36].

3.3 Opportunities for Improving Maritime Risk Assessment

Perhaps the most significant opportunity for improvement in the Coast Guard's port security risk analysis does not reside with the risk analysis tool, but with the input values. As noted earlier, the need for accurate inputs is critical; however, the inputs for vulnerability and associated consequences continue to be provided by port-level operators with little expert analysis by scientist and engineers. This more in-depth expertise is needed to accurately calculate risk. Given the size and complexity of the US Maritime Domain, and the limited size of maritime security forces and the national budget, an accurate risk assessments is a must to ensure the most effective and efficient use of these resources.

Identification of effective risk mitigation activities is another use for the MSRAM. Once the areas of greatest security risk are identified, the obvious next step is to determine how to best reduce vulnerability. Although it is possible to reduce potential consequences (move the facility away from populated areas, dilute the chemical concentration, institute a robust response organization, etc.), the current focus is to reduce the vulnerability. Ideally, port planners should be able to use the risk assessment tool to "plug in" various security options and measure the risk reduction gained to determine the best returns on investment.

Unfortunately, in its present form, the MSRAM tool is too cumbersome and lacks sufficient sensitivity for effective use as a risk mitigation assessment tool. As currently configured, the ranges for consequence values are essentially orders of magnitude [14], whereas, in many cases, the ability of available security forces to mitigate potential consequence is far smaller. Therefore, the risk changes created by different force employment operations cannot be discerned with the tool.

4 CONCLUSION

Risk assessment is not a one time effort, rather it is an ongoing process that identifies initial risk and adjusts risk as new threat information is received, risk is reduced through security activities, infrastructure is added or removed, and new technologies emerge. The ever-changing nature of risk demands that risk assessment be dynamic and responsive.

For example, in the absence of intelligence of specific attack planning for a port, we must use default analysis of previous targeting choices and the stated strategic objectives of our enemy to make our best guess at the "threat" value for our risk equation. The risk mitigation activities developed with the resultant risk assessment forms the core of our steady state maritime security strategy.

But what happens when attack planning intelligence is received for a port? The natural tendency is to put as many prevention resources around the intended target as possible to deter or defend against the specified threat. This would be a good response if there were adequate strategic law enforcement reserves to do targeted point defense while regular forces continue steady state protection, but unfortunately this is rarely the case. As a result, more risk is assumed throughout the system as we rush to protect a specific target (*see* Emerging Terrorist Threats).

If time is available, it would be useful to enter specified threat information into the risk assessment tool to reassess how the entire port security system might be impacted. This analysis may show that it is prudent to focus all protection resources against the threat, but it may also show that it might be worth absorbing a lower consequence attack than dropping our guard around higher consequence infrastructure. The thought of our enemy using operational deception to draw us away from high value targets must never be ignored. Whatever risk mitigation strategy is selected, the risk tool should again be employed to measure the expected effects on risk throughout the port security system.

A fundamental premise of port security is that we will never be able to eliminate risk. The adaptive threats posed by our enemies and limited resources available to counter those threats requires that we work toward reducing, not eliminating the threat. The risk assessment tool is critical in helping to assess where we will achieve the greatest risk "buy down" through applied security activities. While we can use the risk assessment tool to inform the budget process for increasing security capabilities and capacities were the investment is warranted, we cannot and should not try to protect everything with our limited resources. If we do, we expose the highest risk infrastructure to greater risk while spreading precious resources thin in the process of protecting lower risk areas. Deciding where to focus security efforts is one of the most difficult problems faced by a Captain of the Port. MSRAM is one of the few tools available to the operational commander to help make those decisions. It is therefore critical that MSRAM be a robust, agile tool that not only assesses risk, but also assesses the effects of security measures. The current tool is capable in the former task, but needs more work on the latter (*see* Risk-Based Prioritization).

It is important that local operational commanders have maneuvering room to develop locally focused security plans because all ports have fundamental differences. As noted previously, the MTSA designated the Coast Guard Captains of the Port as the Federal Maritime Security Coordinator for the ports in their areas of responsibility. They are supported by an Area Maritime Security Committee for each port that includes representatives from interested federal, state, and local governments as well as the private sector. Each Area Maritime Security Committee has helped to develop a comprehensive Area Maritime Security Plan tailored to the security requirements of the individual port. MSRAM assessment plays a role in developing this Plan. The Plan is exercised at least once a year and modified as needed. The Plan is required to be reviewed and reapproved by the Coast Guard once every five years.

Through the Area Maritime Security Committee and resultant partnerships, the Coast Guard Captain of the Port seeks as many resources as possible to bolster port security.

Maritime resources from State and local law enforcement agencies play a large role in supplementing Coast Guard resources. Nationally, about 25% of the port security effort comes from non-Coast Guard law enforcement agencies [16]. Some state and city governments have passed local laws making violation of federal security zones a state/local offense as well, thus allowing nonfederal law enforcement officials to enforce federally designated security zones. This not only improves patrolling but also facilitates prosecution because violators can be held accountable in federal, state, or local courts.

4.1 Consequence Management

Although this article is about port facility protection and prevention, a brief discussion of consequence management is prudent. Much of the Coast Guard's counterterrorism efforts to date have focused on protection and prevention. More effort is now going into consequence management and the reason is clear from the risk equation. Our enemies wish to attack us in order to achieve a certain serious consequence. If we can reduce the consequences of an attack, we reduce the effects our enemy can achieve. By building a robust consequence mitigation capability, we get into our enemy's planning cycle and can potentially prevent an attack by showing the enemy that his desired effects will not be achieved. If we show the enemy that even a tactically successful attack on a port will only have limited consequences, it might deter our enemy from attacking. Thus, post-attack consequence mitigation planning and capacity building can have a powerful deterrent effect and must be a key part of our overall counterterrorism strategy. MSRAM and future risk assessment tools must include the capability to assess the power of consequence mitigation to help Captains of the Port plan consequence management (*see* High Consequence Threats: Chemical).

4.2 Summation

The challenges faced by those responsible for US port security are daunting and the ability to accurately assess risk at the national, regional, and port levels are critical to deterring and preventing attacks as well as mitigating the consequences of an attack. Current assessment tools are well ahead of where we started, but much more must be done to improve their scope, make them more agile, enhance their display, and make them available to the people who need to know.

REFERENCES

1. U.S. President (2005). *National Strategy for Maritime Security*. White House, Washington, DC, pp. 1–27.
2. Allen, T. W. (2008). *Statement of Admiral Thad W. Allen Commandant on the Fiscal Year 2009 President's Budget Before the Committee on Appropriations Subcommittee on Homeland Security U.S. House of Representatives*, pp. 1–18.
3. U.S. Coast Guard (2007). *The U.S. Coast Guard Strategy for Maritime Safety, Security and Stewardship*. Headquarters U.S. Coast Guard, Washington, DC, p. 34.
4. Department of Homeland Security (2008). *Small Vessel Security Strategy*. SECDHS, Washington, DC, pp. 1–31.
5. U.S. President (2007). *National Security Strategy for Homeland Security*. White House, Washington, DC, pp. 1–53.

6. Willis H. RAND Corporation (2007). *Testimony before the Committee on Appropriations Subcommittee on Homeland Security, United States House of Representatives*, pp. 1–7.
7. U.S. Office of the Chairman of the Joint Chiefs of Staff (2003). *Joint Tactics, Techniques and Procedures for Special Operations Targeting and Mission Planning*. CJCS, Joint Publication (JP), Washington, DC, pp. I-1–IV-18.
8. U.S. Congress (2002). *An Act to Amend the Merchant Marine Act, 1936, to Establish a Program to Ensure Greater Security for United States Seaports, and for Other Purposes: Maritime Transportation Security Act of 2002*. 107th Congress, 2002, Public Law 107–295, pp. 1–71.
9. Doane, C., DiRenzo J. III (2004). The history of port security: A coast guard mission since 1917. *Maritime Reporter and Engineering News* 8, 28–47.
10. U.S. Coast Guard Fact Sheet (2008). Available from http://www.uscg.mil/top/xabout/
11. Loy, J. M. (2000). *Statement of Admiral James M. Loy before the House of Representatives, Subcommittee on Coast Guard and Maritime Transportation, Committee on Transportation and Infrastructure*, U.S. Congress, Washington, DC, p. 8–32. Available from http://commdocs.house.gov/committees/trans/hpw106-95.000/hpw106-95_1.HTM.
12. U.S. Coast Guard. Office of the Commandant (2004). *Contingency Preparedness Planning Manual, Volume I: Planning Doctrine and Policy*, COMDTINST M3010.11C, Headquarters U.S. Coast Guard, Washington, DC, pp. 1-1–10-2.
13. U.S. Coast Guard. Office of the Commandant (1999). *Operational Risk Management*, COMDTINST 3500.3, Headquarters U.S. Coast Guard, Washington, DC, pp. 1–13. Available from http://www.uscg.mil/directives/ci/3000-3999/CI_3500_3.pdf.
14. Downs, B. (2007). Balancing resources to risk. *Presentation to SCOTS/NCHRP 20–59*, August 2007, Irvine, CA, pp. 1–57. Available from www.transportation.org/sites/security/docs/Downs—USCG%20Balancing%20Resources%20to%20Risk.pdf.
15. General Accounting Office (2007). *Port Risk Management: Additional Federal Guidance Would Aid Ports in Disaster Planning and Recovery*, GAO Report GAO-07-412, Washington, DC, pp. 1–53. Available from www.gao.gov/new.items/d07412.pdf
16. U.S. Coast Guard Atlantic Area (2008). *Operation Neptune Shield Scorecard for June 2008 (Security Sensitive Information).*

FURTHER READING

Daniels, W. and DiRenzo, J. III (2005). *Maritime Anti-Terrorism at the Crossroads of National Security and Homeland Defense, National Defense*, http://nationaldefense.ndia.org/issues/2005/feb/Maritime_Anti-Terrorism.htm.

DiRenzo, J. IIII and Doane, C. (2007). *Small Vessel Security Summit Initiates Constructive Dialogue, Maritime and Border Security News*, http://www.imakenews.com/ejkrause/e_article000864979.cfm?x=b11,0,w.

Linacre, N. A., Koo, B., Rosegrant, M. W., Msangi, S., Falck-Zepeda, J., Gaskell, J., Komen, C., John, M. J., and Birner, R. (2005). *Security Analysis for Agroterrorism: Applying the Threat, Vulnerability, Consequence Framework to Developing Countries*.

Parfomak, P. W. and Frittelli, J. (2007). *Maritime Security: Potential Terrorist Attacks and Protection Priorities.* Congressional Research Service, Washington, DC.

Willis, H. H., Morrall, A. R., Kelly, T. K., and Medbey, J. J. (2004). Risk-based allocation of counterterrorism resources. *RAND. Presented at the Society for Risk Analysis Annual Meeting Palm Springs.* Palm Springs, CA.

Willis, H. H., Morrall, A. R., Kelly, T. K., and Medbey, J. J. (2005). *Estimating Terrorism Risk.* RAND, Santa Monica, CA.

DEFENDING AGAINST MALEVOLENT INSIDERS USING ACCESS CONTROL

DALE W. MURRAY

DoD Security System Analysis Department, Sandia National Laboratories, Albuquerque, New Mexico

BETTY E. BIRINGER

Security Risk Assessment Department, Sandia National Laboratories, Albuquerque, New Mexico

1 INTRODUCTION

The greatest challenge to any security system is protecting against the malevolent insider because the he or she may be authorized to own 'the keys to the kingdom'. An insider is defined as anyone with knowledge of operations, sensitive information, and/or security systems and who has unescorted access to the facilities or critical assets. A malevolent insider is an insider who has decided to become an adversary. Protecting against the malevolent insider threat requires an integrated security system that minimizes the potential for hiring an adversary and deters the on-staff employee from becoming an adversary. The security system must integrate such protection functions as personnel security, physical security, cyber security, and operations security to *make it easy for the insider to do the right thing and very difficult for the insider to do the wrong thing*. If the insider decides to do the wrong thing, the security system should be able to protect against the adversarial acts or, if the malevolent insider is able to overcome the security system, the system should photograph, record, or otherwise document the event in order to provide evidence for prosecution. The physical security system prevents the malevolent insider from causing an undesired event by detecting the action early enough and delaying the insider adversary long enough so that an appropriate response can interrupt the scenario before the undesired event is caused. Access control is an important element of defending against the malevolent insider because it supports the detection, delay, and response functions of the physical security system. The current state of access control technology, as applied to the insider adversary, focuses on identity verification, contraband detection, and tamper indication.

2 PROTECTION AGAINST THE MALEVOLENT INSIDER

The determination of how much security is enough cannot be made without some judgment about the level, access, and sophistication of the threat that the system must protect against. A description of the insider threat spectrum must be completed in order to design or assess the effectiveness of a security system. An insider is defined as anyone with knowledge of operations, sensitive information, and/or security systems who has unescorted access to the facility or security interests. The passive insider adversary commits no overt acts; he or she only provides information. The active insider adversary may be nonviolent, unwilling to use force against personnel, or he or she may be violent, willing to provide active, violent participation including force against personnel. The active, violent insider is a very difficult adversary to protect against. More than one insider adversary is possible but emphasis is placed on addressing the single insider adversary, the most probable insider threat. Another underlying assumption is that the nonviolent insider will give up if detected.

2.1 Insider Threat Analysis

The motivations for the insider adversary can be the same as those for the outsider adversary. Motivation is an important indicator for both the level of malevolence and the likelihood of insider attack. Motivations might be ideological (the insider holds a fanatical conviction), financial (the insider wants or needs money), vengeful (the insider is a disgruntled employee, contractor, or customer), egotistical ("Look what I can do"), psychotic (the insider is mentally unstable but capable), voluntary (the insider is a volunteer), or coercive (the insider or the insider's family is threatened).

Insider adversaries have advantageous characteristics that distinguish them from other adversaries:

- Operational/system knowledge that can be used to advantage
- Authorized access to the facility, information system, sensitive information, and security systems, without raising the suspicion of others
  - Can conduct tests and rehearsals
  - Can test the system with normal "mistakes"
- Opportunity to choose the best time to commit an act or extend acts over a long period of time
- Capability to use tools available at work location
- Recruitment/collusion with others, either insiders or outsiders [1]

All employment positions at a facility should be included in the insider threat analysis. Any employee may pose a potential insider threat, even trusted managers and security personnel. Insider positions at a typical building might include managers, staff, information system administrators, security personnel, administrative staff, contractors, custodians, maintenance personnel, vendors, and past employees. Visitors hold some, but not all, of the advantages of an insider.

A higher level of protection may be required for high risk positions. High risk positions are those that afford employees access to the most sensitive information or critical assets as a part of the normal job assignment. The insider threat analysis describes general personnel job categories in terms of *knowledge, access*, and *authority* related to each of the undesired events or related critical assets. Efforts should be made to assure that all appropriate personnel assignments are included. The purposes are to identify the job categories that could provide the greatest advantage for the insider adversary to cause the undesired events, and to understand the potential capabilities of an insider adversary. The types of insider knowledge that provide a significant advantage to the insider adversary include knowledge of security/control features, work schedules and assignments, locations and characteristics of critical assets, specific details of facility operations, and known weaknesses and gaps in protection. Insider access that can be used to cause an undesired event includes the usual authorized work access, special temporary access to other areas, and the access to other employees as a source of expanded information. Insider authority that the insider adversary can exploit is described as management authority over others, personal influence over others, the authority to do assigned tasks, and the ability to get temporary authority to do any task.

2.2 Role of Access Control in Protecting against the Insider Threat

Protection features from personnel security, physical security, cyber security, and operations security must be integrated to mitigate the insider threat. Usually, these protection features function independently and are not integrated towards a common objective, but not a single one of these functions, acting alone, can answer the insider threat. The physical security features must function together to detect, delay, and appropriately respond to the insider threat to prevent the undesired event.

Access control plays a prominent role in the physical protection system because it supports all three functions of detection, delay, and response. Access control can detect insider noncompliance with procedures, insider attempts to access areas for which he or she is not authorized, and insider attempts to enter an area with contraband (weapons, explosives, or any restricted item) on his or her person, or in packages. Access control can delay the insider adversary by denying access to unauthorized areas or locking doors to areas he or she is not authorized to enter, thus forcing the insider either to switch to an overt method of attack or be deterred from becoming an adversary.

An important objective for the physical protection system is to minimize the opportunity for malevolent insider acts. A common method of achieving this objective is to compartmentalize the facility. Compartmentalization can be achieved by limiting access to sensitive information or actual assets to only those needing it for job duties, further restricting access to the assets that could result in a high consequence undesired event, enforcing a multiperson presence in critical areas, and monitoring activities to detect potential malevolence and identify who is responsible for the act. Entry/exit control can be used to enforce compartmentalization and access. Entry control enforces authorization checks with a picture-badge inspection, electronic credential, personal identification number (PIN), or a biometric check, such as fingerprints, eye-retinal pattern, and the like. Entry control schemes might include contraband detection to prevent the introduction of weapons, explosives, or other tools that could be used in a malevolent attack. Exit control is used to detect the unauthorized removal of high value assets.

3 ACCESS CONTROL TECHNOLOGY

Access control is the function performed by the security system to allow authorized individuals into a secured area while preventing entry by unauthorized individuals. Other functions that are performed by an access control system are enforcing a two- (or more) person rule, antipassback (*passback* refers to entry credentials being passed back by the first user to a second person for use), contraband detection, and tamper indication. Some of these functions are related directly to addressing the outsider while some address the insider adversary. Before discussing the way in which the access control points address insider issues, some general background on access control is required.

3.1 Balancing Intrusion Detection and Access Control Portals

Perimeter and building security can be extremely effective at detecting persons attempting to penetrate a security area through the detection layers, but authorized persons and materials must still have access to the area. Thus it is very important to have portals in the security area that meet the same level of security as the intrusion detection system. In order to accomplish that, the access control and contraband detection system must be capable of accurately verifying the identity and determining the authorization of persons attempting to enter via the portal. It also has to be able to determine that authorized persons are not carrying materials and objects that are prohibited in the area.

The portal should not be so secure that it greatly exceeds the security of the intrusion detection system. It is a waste of money and resources to create a highly secure portal that can be easily bypassed by entering covertly through the intrusion detection layers. In addition to wasting money and resources, very high security portals are more time-consuming to pass through and have a higher likelihood of falsely rejecting an authorized individual than lower security portals. Depending on the application and the need for very high security, additional time to enter and a higher probability of falsely rejecting authorized persons may be acceptable trade-offs.

3.2 Verifying Identity

In the process of access control it is important to verify the identity of an individual seeking access before granting entry into the security area. There are three criteria used for identity verification. These criteria are not limited to electronic identity verification; they are the means we all use to recognize each other and have been used throughout human history so it is not surprising that they have been carried over into electronic security systems. These criteria are easily remembered as something you know, something you have, or something you are.

The most common means of verifying identity in an electronic access control system based on "something you know" is the PIN. At access control portals for physical access to a security area, it is usually difficult to provide a full alphanumeric keyboard; typical access control systems use keypads for numeric entry of PINs. But, PINs used alone to verify identity does not provide a high level of security since there is often nothing to prevent an adversary from guessing PINs. If an adversary is allowed an unlimited number of guesses, it is simply a matter of time until an enrolled PIN is guessed. One means of reducing the chance of success at guessing a PIN is to make the number of possible combinations greatly exceed the number of PINs actually enrolled and used. For instance, a four-digit PIN allows only 10,000 unique combinations. If there are 1000 PINs enrolled, an adversary has a 1-in-10 chance of guessing correctly with each attempt. By simply increasing the number of digits in the PINs to five, there will be 100,000 possible combinations, thereby reducing the chance of correctly guessing on a single try to 1 in 100.

There are a number of examples of "something you have" used for identity verification. Keys and credentials are the most frequently used objects for gaining access into secured areas and systems. In the case of electronic security systems, the coded credential is required. Coded means that there is some way to store information on the credential—usually a card—that can be easily read by an electronic interface. Coded credentials include bar codes, magnetic strips, proximity badges, and smart cards. As with keys, anyone in possession of the credential can use it to gain entrance to a security area if the credential is the only criterion used by the system. For this reason a coded credential combined with a PIN is considerably more secure. Guessing the correct four-digit PIN for the credential is a 1 in 10,000 probability event; thus the adversary's task when both are used is to gain an individual's coded information and the associated PIN in order to gain entry.

Electronic systems currently in use can make measurements of some physical or behavioral feature of humans—in other words, "something you are". The measurements obtained by the electronic interface are called a *biometric*; the electronics that make and analyze the measurements are called *biometric systems*. Physical features used for identification include (but are certainly not limited to) height, weight, finger print, blood vessel pattern on the retina, iris structure, facial structure, or shape of the hand. Behavioral features include speech patterns (speaker recognition systems can be physical; some work on tones that are determined by dimensions of the vocal tract), signature dynamics, typing dynamics, and walking gait. The security of the biometric technique used to identify the individual depends on how unique the feature is to the individual. Obviously a person's height or weight is not unique to the individual; many people have the same height or weightwhereas the structure of the human eye and fingerprints are unique to the individual.

In very high security applications the highest level of security is provided by combining all three criteria into a single application. Figure 1 shows one such system. This combination of all three identification criteria creates a very high confidence that the identity of the individual can be verified.

3.3 Detecting Contraband

Nearly anything that can be used by an adversary to perform a malicious act can be considered to be contraband. Sometimes materials like drugs and alcohol that are simply detrimental to production or safety are considered contraband. Guns and other weapons

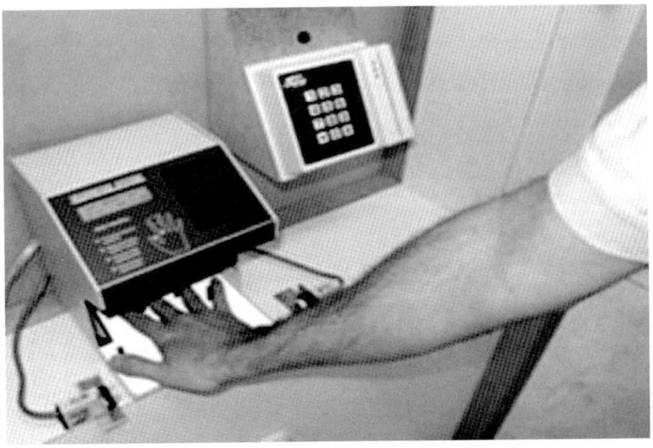

FIGURE 1 A magnetic stripe card reader used in combination with a PIN pad and a hand geometry biometric system.

FIGURE 2 An automobile door handle being swipe-sampled for explosives detection.

are an obvious class of contraband items. For the protection of sensitive information, cell phones, cameras, and recording devices are often considered contraband. Tools that can be used by an adversary can also be considered contraband. Contraband detection is a clear example of using technology to counter the insider because only persons who are authorized into a security area are screened for contraband. Not all contraband is readily detectable by technology. Examples of contraband that can be detected by modern contraband detection equipment include explosives, metallic objects like tools and weapons, and radiological materials. Systems are in development for the detection of chemical and biological agents. Figure 2 shows a contraband detection tool in use.

3.4 Tamper-Indicating Devices

Another means to counter the insider with an access control technology is the use of tamper-indicating devices (TIDs). These devices, commonly known as *seals*, have been used to authenticate documents and other items since the earliest records in history.

The purpose is to indicate unauthorized access by insiders, into secure containers or rooms. The role of a TID is to provide an indication of an unauthorized covert opening of a container, package, door, or other object to which a TID has been affixed. Seals are broadly divided into two categories: passive and active. A passive seal requires an inspection to determine whether an unauthorized entry has occurred between inspection intervals. Passive seals may or may not be operated electronically. Passive seals are further divided into groups based on the way they are applied. These groups are pressure-sensitive (or tape) seals, loop seals, and bolt seals. Pressure-sensitive seals are sheets of tamper-indicating materials with adhesive backing, as shown in Figure 3. These are applied like masking tape to a closed container or room. Any attempt to open the space will cause the seal to tear, stretch, or become detached. Loop seals, as shown in Figure 4, are used in conjunction with a hasp so that the container lid or room door cannot be opened without removing the loop seal. The loop is designed to become damaged and cannot be reapplied after removal. Bolt seals are similar in operation to loop seals in that they are also used in conjunction with a hasp. The bolt is designed to become damaged and is very difficult to reapply after removal. Figure 5 shows examples of bolt seals.

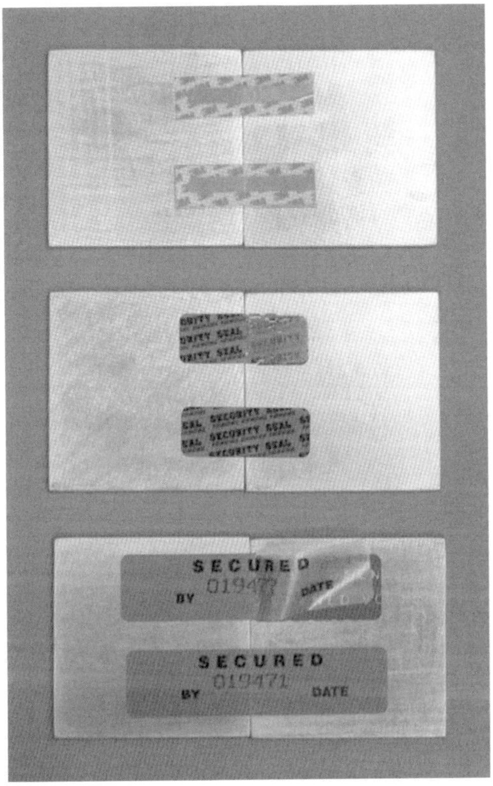

FIGURE 3 Examples of tape or pressure-sensitive seals.

An active seal is electronically operated and provides a near-real-time indication, or alarm, of an unauthorized access or entry. Most active seals are in development at this time.

4 ACCESS CONTROL FEATURES APPLIED AGAINST THE INSIDER THREAT

Because the process of access control is designed to identify individuals for the purpose of separating those authorized to enter (insiders) from those not authorized (outsiders), the process is primarily intended to address the outsider threat. However, there are some features of access control systems that can be used to counter the insider threat.

4.1 Entry/Exit Logs

Entry control systems can provide detailed logs of all entries into all security areas. If the system uses exit readers, a complete log of the comings and goings of all personnel is maintained. While this feature cannot prevent an insider from performing a hostile act, it may act as a deterrent. Knowing that the system can identify personnel at the scene of a hostile act may deter insiders who consider the chance of being apprehended too great a risk. Determined insiders who do not fear being caught will not be deterred.

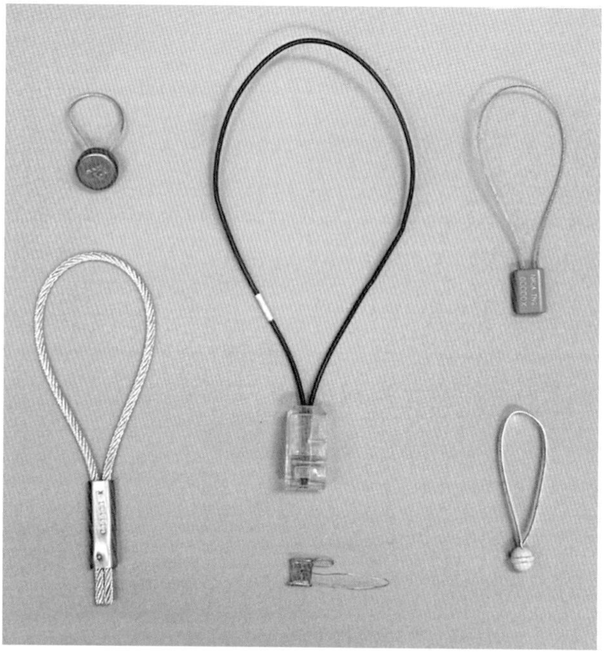

FIGURE 4 Examples of loop seals.

4.2 Antipassback Features

The antipassback feature is used to make the job of an insider acting in collusion with one or more outsiders more difficult. In a collusion scenario an insider may use his or her credential and PIN to allow outsiders to enter a security area. The insider shares the PIN with an outsider and either creates a counterfeit of the coded credential or passes back the real credential to the outsider. The outsider can thus gain entry and join the insider in the secure area to perform a hostile act.

Antipassback is a feature that is often standard on commercial access control systems. This feature prevents the same credential codes and PINs from being used in tandem. This is intended to prevent an insider from assisting an outsider by "passing back" his or her credential to the outsider for use in defeating the access control system and entering the secured area. The insider must first exit the system for the codes and PINs to be valid for another entrance. This can be accomplished through either a timed antipassback feature, which prevents the coded credential and PIN from being used for some programmed period of time after it is used for entry, or an absolute antipassback feature, in which the credential and PIN must be used in an exit reader before being used for another entrance into the area. These antipassback features can be useful for complicating the task of an insider allowing an outsider accomplice into the area, but do not address the insider acting alone.

Another factor in determining the effectiveness of antipassback features is the portal that the access control system is controlling. Simple one-door portals will not be very effective because the insider can just hold the door open while one or more outsiders enter. More effective is a turnstile that is sufficiently small to make the introduction of more that one person at a time difficult and obvious. Electronic sensors developed to detect the entrance of more than one person through a portal at a time show varying degrees

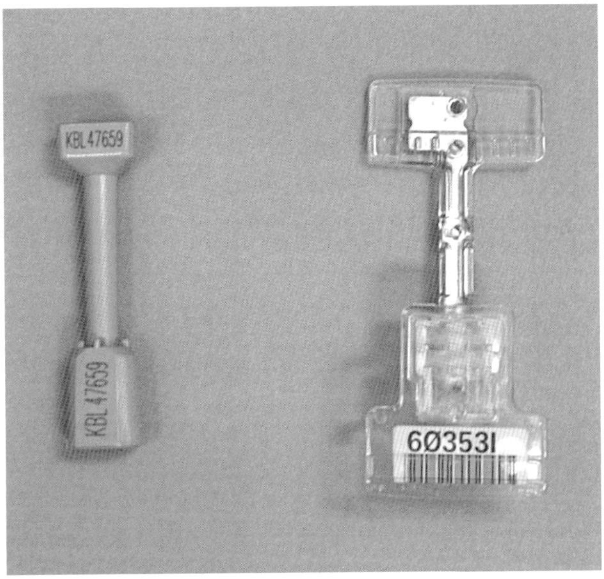

FIGURE 5 Examples of bolt seals.

of success. The best portal design for use with antipassback features is the two-door portal or mantrap. This system requires the insider to enter the portal and close the door behind him or her, before using the access control system to gain entrance through the second door into the secure area. Electronic sensors can be used to determine that a single individual is inside the portal while it is in use or the portal can be equipped with scales to weigh the contents to ensure that only a single individual is inside.

4.3 Two-Person Rule Enforcement

The two-person rule is a security technique used to address the insider acting alone. A two-person rule specifies that no individual is allowed into a security area alone. This technique is typically reserved for the most secure areas. A two-person rule can be implemented either as a procedure or, in an electronic system, the entry control point can be programmed to force the implementation. In the electronic system approach two authorized persons must submit credentials and enter the associated PINs within a short period of time to open the portal. The system must also be equipped with out-bound or exit readers. The system keeps track of how many persons are inside the area and can prevent the exit of a person if that exit violates the two-person rule. Alternatively, the system can be programmed to allow the exit and transmit an alarm to the monitoring station to alert the guard force that a two-person rule violation has occurred. This powerful technique is highly effective at addressing the insider acting alone.

4.4 Compartmentalization Feature

Another feature of access control systems that is helpful in countering the malevolent insider is the use of compartmentalization. This feature is simple to implement using electronic access control systems. These systems are fully capable of developing compartments (security areas) within a facility that prevent unauthorized access. Access control systems in wide use today can control access to interior security areas at the individual cardholder level. The system can be programmed to permit (or deny) access to individuals for specified days of the week and/or a certain time of day at a particular location. Combined with a two-person rule this feature is effective in preventing any single malevolent insider from gaining sufficient information to do significant damage.

5 RESEARCH DIRECTIONS

While these features and techniques provide varying degrees of effectiveness, they cannot prevent all malevolent acts by insiders. Neither antipassback nor the two-person rule can guarantee that the insider does not have free movement once inside the area. Antipassback effectiveness can be greatly enhanced by the use of two-door portals capable of detecting the presence of others when the insider is operating the access control system. Advances in sensor technology are required to enable high confidence decisions that a single individual is present inside the portal. One way to increase the effectiveness of the two-person rule would be to develop a system that can actively track all individuals inside a secure area. Alarms could be generated if either there are no longer two persons inside the area (a strict two-person rule violation) or if the personnel are no longer in direct line-of-sight with each other (a violation of the spirit of the two-person rule). Continued development of active seals that can generate an alarm when tampering occurs is recommended to detect the insider who defeats the access control system.

Currently there is little direct funding being invested in the effort to counter the insider with access control. However the considerable investment in improving access control technologies carries a side benefit of helping counter the insider threat. Substantial funding is currently used to develop reliable facial recognition techniques. Facial recognition holds the potential for not requiring cooperation from the person whose face is being captured and analyzed. This ability is dependent on further development and the use of multiple and covert cameras. Without covert coverage the insider adversary can take actions to avoid having his face seen by the cameras. Alternately an alarm could be generated when a sufficiently advanced image analysis detects someone who is actively avoiding having his face imaged. There are systems currently in development for detecting suspicious behavior and these algorithms could be modified for this purpose. Advances in video camera coverage inside a facility include a potential for insider tracking to identify and counter malevolent insiders. Past interior tracking designs relied on radio frequency (RF) tags and had only limited success, due to failed reads and the required cooperation from the potential insiders.

Improved contraband detection is another area where funding is strong. While metal detectors and X-ray package search systems are mature technologies, explosives detection equipment can be improved in such performance areas as sensitivity, reliability, and materials detected. Technologies for the detection of chemicals and biological agents are still in their infancy and will require time and funding to mature.

As with any security system there is no silver bullet that will solve all insider access control problems. Overall system effectiveness will be enhanced by integrating features. Just as using more than one of the three identity verification methods increases effectiveness, the use of access control techniques to counter the insider threat will be most effective when several features are combined. However effective it is, access control is the primary detection system for the malevolent insider threat.

ACKNOWLEDGMENTS

The submitted manuscript has been authored by a contractor of the US Government under Contract No. DE-AC04-94AL85000. Accordingly, the US Government retains a nonexclusive, royalty-free license to publish or reproduce the published form of this contribution, or allow others to do so, for US Government purposes. Sandia is a multi-program laboratory operated by Sandia Corporation, a Lockheed Martin Company, for the US Department of Energy's National Nuclear Security Administration under Contract DE-AC04-94AL85000.

REFERENCE

1. James, P. (2006). Chapter 22, Insider Analysis. *The Nineteenth International Training Course for the Physical Protection of Nuclear Facilities and Materials*. Sandia National Laboratories and the International Atomic Energy Agency, Albuquerque, NM.

FURTHER READING

Turner, J. T., and Gelles, M. G. (2003). Chapter 12, insider threat: risk management consideration. *Threat Assessment, A Risk Management Approach*. Haworth Press, Binghamton, NY.

LESS-LETHAL PAYLOADS FOR ROBOTIC AND AUTOMATED RESPONSE SYSTEMS

Hobart Ray Everett, Greg Kogut, Larry Drymon, Brandon Sights, and Kelly Grant

Space and Naval Warfare Systems Center, San Diego, California

1 BACKGROUND

The numerous types of less-lethal weapons developed over the past several decades represent a natural application payload for the growing number of unmanned ground vehicles (UGVs) now being employed in military, security, and law-enforcement scenarios. For the most part, however, these UGVs are all teleoperated, and thus suffer from real-time control problems because of communication latencies, poor situational awareness, and unacceptable burden imposed upon the operator. The incorporation of teleoperated lethal/less-lethal weapons on such systems further exacerbates the situation, resulting in unacceptable performance in all but the most simplistic cases.

In 1996, US Department of Defense Directive 3000.3 defined nonlethal weapons as follows [1]: "Weapons that are explicitly designed and primarily employed so as to incapacitate personnel or materiel, while minimizing fatalities, permanent injury to personnel, and undesired damage to property and the environment." When it later became apparent that many of these so-called nonlethal systems could under certain circumstances prove fatal, the more appropriate terminology "less lethal" was generally adopted instead.

A large number of less-lethal munitions have evolved over the years [2], most of which can generally be classified as either ballistic projectiles or directed energy. The most well known of the ballistic category is the kinetic energy (i.e. blunt force) variety, typical examples being rubber balls, plastic bullets, paint balls, bean bags, sponges, rings, and dowels. Ballistic incapacitants include nets that can be fired at human targets to ensnare and entangle (Fig. 1), and what is known as "sticky foam", extremely tacky and tenacious materials, are also used to entrap or impair. Examples of ballistic area-effect weapons are malodorants such as tear gas (CS) and pepper (OC), as well as flash-bang grenades that emit both a blinding flash (several thousand candela) and deafening sound (170 dB).

The most familiar of the directed-energy category is the taser, a ballistically delivered trailing-wire pair of barbs, which administers an incapacitating electric shock. A more

FIGURE 1 A would-be intruder and his weapon are temporarily entangled by FMI's Snare Net munition during 1999 tests in Waltham, MA.

recent enhancement of this concept uses a laser-induced plasma channel (LIPC) to eliminate the need for wires, ionizing the air to form a conductive path for the electrical charge. Acoustical directed-energy systems such as the long range acoustic device (LRAD), on the other hand, focus debilitating high-intensity sound waves toward the target. Optical versions, such as strobe lights and laser dazzlers, employ intense beams of visible light to cause temporary blindness and disorientation. The radar-based "pain ray" active denial system similarly projects millimeter-wave radio-frequency (RF) energy, generating an extremely uncomfortable burning sensation on the surface of the skin.

Pursuit of weapon payloads for military UGVs initially involved the adaptation of conventional (i.e. lethal) ballistic munitions in a force-projection role, allowing the remote operator to maintain a safer standoff distance from the enemy. In 1983, for example, the Hawaii Laboratory of the Naval Ocean Systems Center (NOSC) incorporated an M-16 machine gun on a teleoperated dune buggy for the US Marine Corps [3]. The PROWLER (Programmable Robot Observer with Logical Enemy Response), developed in 1983–1985 by Robot Defense Systems (RDS) of Thornton, CO, featured two turret-mounted M-16s on a six-wheeled standard manufacturing all-terrain-vehicle chassis. In 1986, NOSC installed a 0.50-caliber M-2 machine gun on a high mobility multi-wheeled vehicle (HMMWV)-based UGV for the Marines under the Ground/Air TeleRobotic Systems (GATERS) program [4].

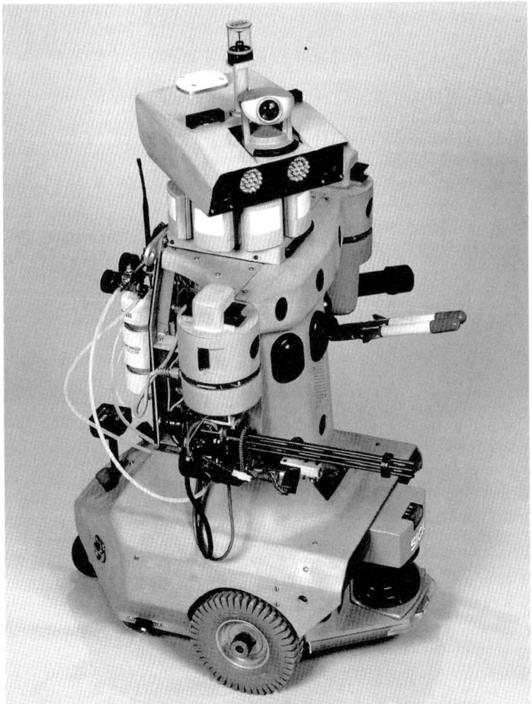

FIGURE 2 ROBART III is a laboratory surrogate used by the Space and Naval Warfare Systems Center San Diego (formerly NOSC) to develop automated nonlethal response capabilities for security and force-protection missions.

FIGURE 3 The FMI Snare Net ballistic incapacitant deploys after launch from a less-lethal payload mounted on the company's Lemming man-portable robot in 1999.

When a negative public reaction began to surface in the late 1980s, followed by Congressional direction prohibiting the incorporation of lethal weapons on robots, attention shifted to nonlethal systems. In 1992, for example, the NOSC laboratory robot ROBART III (Fig. 2) featured a pneumatically powered Gatling-style weapon that could fire either tranquilizer darts or rubber bullets [5]. Other less-lethal payloads included 130-dB acoustic disrupters and high-intensity xenon (later light-emittins-diode (LED)) strobes. The principal focus of this research effort was to pursue a more automated force-projection capability, as is further discussed later.

One of the first robotic implementations of a ballistic incapacitant was a net launching payload developed by Foster-Miller, Incorporated (FMI), funded in part under the DARPA Tactical Mobile Robot (TMR) program. Figure 3 illustrates a fully deployed net immediately after being fired from a laser-sighted payload on the front of an FMI Lemming man-portable robot. Interfaced via a 12-V power connection and an RS-232 serial link, the modular launch payload could be adjusted in elevation, with training in azimuth a function of robot heading. The FMI Snare Net round was originally designed for a standard 37-mm grenade launcher, the short-range version of which was effective from 1.5 to 8 m. For longer distances out to 150 m, an optical proximity-fuzed fire-and-forget variant could be used to delay net deployment until within 3 m of the target.

With the proliferation of man-portable UGVs supporting coalition forces in Iraq and Afghanistan, considerable interest has been recently expressed in the addition of nonlethal means for protecting the robot from tampering and sabotage. A more futuristic application scenario currently under consideration involves an autonomous UGV protecting its human partner during proximal human–robot teaming. Under this Warfighter's Associate concept [6], the robot infers to a large degree what its behavior should be by evaluating the perceived threat and observing the actions of its human partner.

2 TECHNICAL CHALLENGES

The technical challenges associated with the use of less-lethal weapon payloads on mobile robotic systems can be subdivided into two general categories: (i) controlling the robot and (ii) controlling its weapon.

2.1 Robotic Control

From a mobility perspective, the type of control strategy employed on a UGV runs the full spectrum defined by teleoperated at the low end through fully autonomous at the upper extreme. A teleoperated machine of the lowest order has no onboard intelligence and blindly executes the drive and steering commands from a remote operator. A fully autonomous system, on the other hand, keeps track of its position and orientation and typically uses some type of world modeling scheme to represent and avoid perceived objects in its surroundings. The most common form of control employed in robotic systems used by military and law enforcement today is teleoperation, for the simple reason that it is the least complex and therefore the least expensive.

An unfortunate trade-off, however, is that teleoperated systems require continuous high-bandwidth communications links, which not only are difficult to maintain in practice, but also introduce unwanted latencies into the control loop. This situation is further aggravated by other end-to-end systemic delays that can arise from video digitizing, compression, encoding and decoding, as well as operator and vehicle response times [3]. Collectively these various sources can contribute to overall latencies ranging anywhere from several hundred milliseconds to tens of seconds, the effects of which are well documented [7]. Elliott and Eagleson reported " . . . latencies as small as a few hundred milliseconds will prevent the operator from controlling a device in a natural way" [8]. Held et al. noted that feedback delays cause operators to consistently overcompensate in their joystick movements [9], while Day concluded latency typically generates oscillation in the operator's control responses [10].

Another drawback to teleoperated systems is significantly reduced situational awareness on the part of the remote operator, relative to a human performing the same task directly. In a comprehensive study, Drury et al. concluded that robot operators exclusively devote an average of 30% of their time to acquiring situation awareness, sometimes ignoring performance-critical feedback in the process [11]. In the late 1980s, many researchers were attempting to address this problem through a technique known as *remote presence* [3], employing helmet-mounted stereo displays and binaural hearing (Fig. 4). A robotic head–neck assembly in the driver's seat, for example, was slaved to the operator's helmet, allowing him or her to view whatever the robot was seeing from the vehicle perspective [12].

While indeed providing operators with a 3-D sense of immersion in the remote environment, this telepresence approach introduced a new set of problems that often outweighed its advantages. For starters, the use of stereo cameras drove the communication bandwidth requirements even higher than that for monocular vision. Even worse, the previously mentioned systemic latencies induced adverse physiological effects in a large percentage of users, since camera motion was no longer controlled by a joystick but now slaved to the motions of the operator's head. This problem arises due to the vestibulo-ocular reflex that actively stabilizes our eyes in response to accelerations sensed by the inner ear. Any unexpected disparities in feedback from our visual and vestibular systems can quickly lead to headache, disorientation, sweating, fatigue, and even nausea. If the remote cameras do not immediately mimic the operator's head movements, the effect is much more noticeable than the equivalent lag of a joystick-commanded camera motion.

A third negative aspect of teleoperation is the driving burden imposed upon the operator. Even the very best video cameras are no match for the human eye in terms of acuity

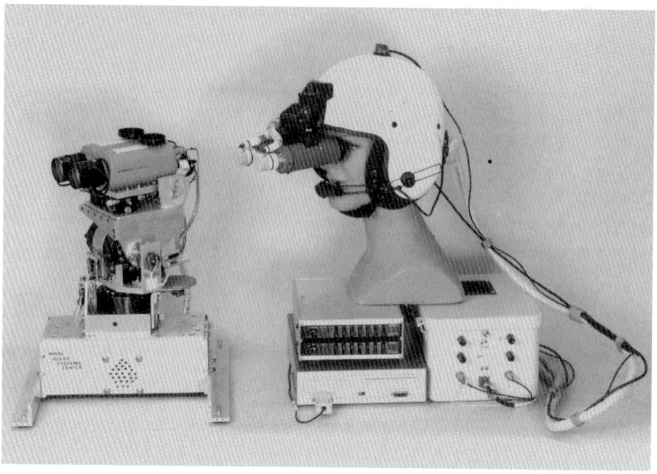

FIGURE 4 The ground air telerobotic system (GATERS) helmet-mounted display employed an electromagnetic Polhemus sensor (top of helmet) to track the operator's head movements.

and dynamic range, and improved resolution is often achieved by zooming in the camera. The subsequently reduced field of view drastically restricts the operator's peripheral vision, however, down from somewhere around 180° (both eyes) to perhaps 40° or less. This "tunnel-vision" perspective forces operators to drive in an unnatural manner, with increased reliance on foveal vision relative to normal "manned-vehicle" driving, which is supported more by our subconscious perception.

Early NOSC attempts to reduce operator's burden through the application of onboard intelligence led to the reflexive-teleoperated control scheme developed on ROBART II [13]. The robot's sonar and optical collision-avoidance sensors, originally intended only for autonomous navigation, were also employed during manual operation to minimize the possibility of operator error. In this telereflexive mode, the control inputs for drive and steering still originate from a human operator as before. Both the speed and direction of the platform, however, are automatically altered as needed by the onboard software to keep the robot traveling at a safe speed without running into surrounding objects. A good analogy here is that of riding a horse versus riding a motorcycle: pointing the latter at a tree will result in a collision, whereas most horses will instinctively alter course to avoid contact.

In summary, the four fundamental shortcomings of teleoperated control include the following: (i) requires a continuous high-bandwidth communications link; (ii) suffers from feedback and control latencies; (iii) provides limited situational awareness; and (iv) imposes an excessive burden upon the human operator.

2.2 Weapon Control

While manually driving a remote vehicle is a very tedious and potentially fatiguing operation just by itself, adding a teleoperated weapon payload makes a bad situation even worse. Unless firing at a relatively motionless target from a static platform, the degree of accuracy and responsiveness required for effectively aiming a weapon is significantly greater than that needed for driving. Introducing any target or vehicular motion further exacerbates the situation to the point where precise fire control is for the most part impossible. In addition, weapon-payload tasks are often complicated by potential time constraints and the need to follow definitive rules of engagement. If the remote operator has to simultaneously manipulate three different joysticks (i.e. one for drive and steering, another for camera pan and tilt, and yet a third for the weapon), the chances of successfully performing coordinated actions in a timely and effective fashion are minimal.

For this reason, work was begun in 1993 on ROBART III to extend the concept of reflexive-teleoperation into the realm of sensor-assisted camera and weapon control [5]. As a near-term strategy, the surveillance-camera pan-and-tilt axes could optionally be slaved to those of the weapon, so the camera automatically looked wherever the operator was aiming. The mobility base could similarly be slaved, causing the robot to turn and face the direction toward which the weapon was pointed. If a forward drive speed was commanded at this point, the operator merely had to keep the weapon trained on the intruder, and the robot would automatically give chase. Although this was clearly a step in the right direction in terms of reduced operator burden, it did nothing to address improved situational awareness or communication latencies. The most difficult task of all, controlling the weapon payload, was still teleoperated.

A more effective approach involves applying local sensor-assisted automation to those control tasks that can be off-loaded from the human operator. A breakdown of potential weapon-payload tasks is as follows:

Target detection. The two most distinguishing human characteristics that can be exploited by automated detection systems are that humans tend to move around and also give off heat. For this reason, the most common sensors employed in an initial detection role are Doppler radar, conventional video, and thermal imagers. Appearance cues such as skin temperature, color, texture, and aspect ratio are often taken into account as well to filter out nuisance alarms.

Target acquisition. The process of selecting a specific target from the set of all detected possibilities, and then aiming the weapon accordingly. A variety of factors come into play here, such as the perceived threat level, proximity to high-value assets, distance from the UGV, and likelihood of escape [7].

Target tracking. Maintaining a valid fire-control solution as the acquired target and/or UGV continue to move, which is a bit more complicated than simply pointing the weapon at the last observed target location.

Target discrimination. Rejecting inappropriate targets based on a variety of distinguishing features (i.e. noncombatant and unarmed), currently one of the biggest challenges for automated systems.

Target prosecution. The arming and firing of the weapon itself.

In the human-supervised sensor-assisted control scheme of ROBART III, the first three tasks of detection, acquisition, and tracking are automated, freeing the operator to focus on target discrimination and prosecution [6]. Since all detection and tracking functions take place locally, the debilitating effects of communications latency are eliminated, and any secondary systemic delays can be explicitly modeled for optimal performance. Two field-ready examples of more recent automated targeting systems are presented in the next section.

3 CURRENT LESS-LETHAL IMPLEMENTATIONS

The Force Protection Joint Experiment (FPJE) is a focused series of four technology-integration assessments to identify, scope, and mitigate risk for the Joint Force Protection Advanced Security System (JFPASS) Joint Capability Technology Demonstration (JCTD). Orchestrated by technical representation from the Army, Navy, and Air Force, the Joint Experiment serves as an effective venue for the evaluation of emerging force-protection technologies. A specific focus area is to integrate and assess automated response capabilities using lethal/nonlethal means to include semiautomated unattended weapons and semiautonomous security robots.

3.1 Unattended Weapon System

The network-integrated remotely operated weapon system (NROWS) is a stand-alone weapon platform designed to provide a remote lethal/less-lethal response to intruders (Fig. 5). The system employs autonomous surveillance, detection, and automated target tracking to enable timely response to hostile activity, but target prosecution remains

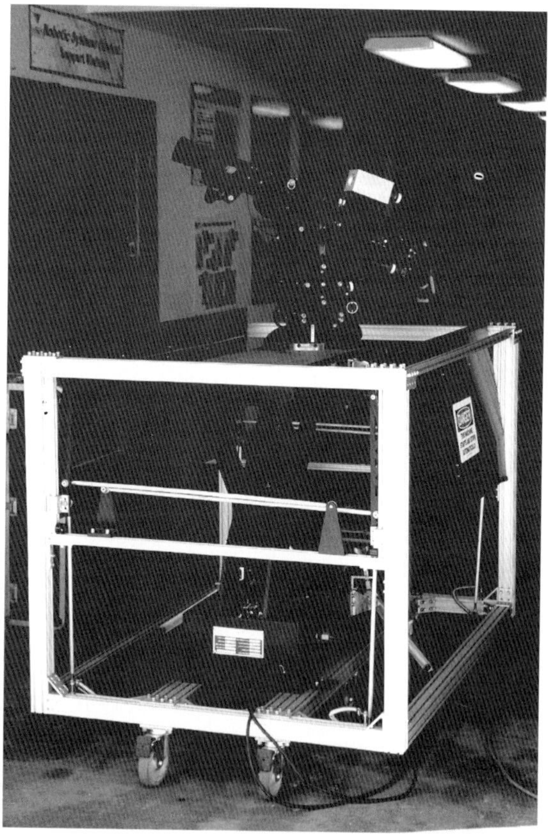

FIGURE 5 The NROWS unattended weapon, shown here in the deployed position, can be housed in a protective "pop-up" enclosure until needed.

FIGURE 6 Shown here on an iRobot Warrior, the multibarrel Metal Storm weapon can electronically fire (i.e. no moving parts to jam) a variety of lethal/less-lethal rounds.

under operator control. A transmission control protocol/internet protocol (*TCP/IP*) network architecture allows for flexible integration and operation of multiple platforms from a single control station. Communications between the remote unit and its command-and-control (C2) station can be established over a direct land line or via wireless link, providing a real-time unattended weapons pod that effectively extends delay/denial response capabilities.

On the basis of a precision weapon-aiming platform built by TeleRobotics Corporation (TRC) of Sausalito, CA, the system can apply a variety of small and medium automatic weapons in support of security or combat operations. These weapons require no modifications of any kind, facilitating use of standard government-issue munitions such as M4/16 and M240/249 machine guns. Less conventional armaments such as the multibarrel stacked-round Metal Storm weapon (Fig. 6) may also be adapted. NROWS platforms can be housed in a variety of configurations, including roof or wall mounts, inside protective ballistic shrouds, or in a "pop-up" unit suitable for installation at or directly below ground level. With the latter method, the weapon system can remain hidden until needed, making it less vulnerable to enemy surveillance and preemptive attack.

3.2 Robotic Response System

The mobile detection assessment response system (MDARS) provides a robotic capability to conduct semiautonomous random patrols and surveillance activities, including barrier-assessment and theft-detection functions. The current platform features diesel-powered four-wheel hydrostatic drive, global positioning system (GPS) navigation, and ladar-based obstacle avoidance. Soon to enter production by general dynamics robotics systems, the MDARS robot is equipped with an intrusion detection system (IDS) that can acquire and track multiple human targets out to 300 m. The IDS sensor suite includes Doppler radar and both regular vision and forward looking infrared (FLIR) cameras, which also provide a digital video feed to the C2 console.

The addition of a less-lethal weapon module, long envisioned as a preplanned product improvement under the MDARS Modernization Program, was completed in 2007 for

FIGURE 7 Two air-powered FN303 less-lethal weapons are mounted on TRC precision aiming platforms attached to either side of the MDARS superstructure.

evaluation at the Joint Experiment. This add-on capability allows the remote system to audibly challenge intruders and attempt to delay or repel those showing hostile intent. The current weapon is a semiautomatic air-powered FN Herstal FN303 that is highly effective to 50 m, with a maximum range of 100 m. There are several different types of fin-stabilized 18-mm rounds available for a variety of missions, including impact, impact plus marking (i.e. indelible or washable paint and inert powder), or impact plus irritant, such as pepper spray.

Two FN303 weapons are mounted on TRC precision aiming platforms (Fig. 7), which in turn are attached to either side of the MDARS superstructure that supports the IDS module. This configuration allows both payloads to be trained on an intruder directly to the front or rear, or for individual tracking of separate targets. Each TRC system uses a single-board controller to receive incoming commands, position the pan-and-tilt axes, control bore-sighted cameras, and monitor arming and firing of the attached weapon. The controller accepts XML remote-procedure calls (RPCs) from the C2 software pertaining to weapon movement, arming, and firing.

When the robot enters sentry mode, the onboard software moves the weapons from their stowed to ready position and begins looking for intruder data packets from the IDS module. These packets contain a list of detected targets, each with an identifier, range and bearing from the robot, status flags, and a confidence percentage indicating the probability that the target is human. Each potential intruder is analyzed by the onboard software to determine the optimal target to prosecute. The remote operator can influence this process if desired by individually designating the console-displayed intruder tracks as either primary or friendly. The system attempts to point both weapons at a primary-flagged intruder if possible, or the nearest intruder otherwise, and will not track a target designated as friendly.

Since the intruder data packet contains a "snapshot" representation of where potential intruders were last seen, the predicted locations are estimated by calculating the speed and heading of each target, based on up to four previous range and bearing values. In

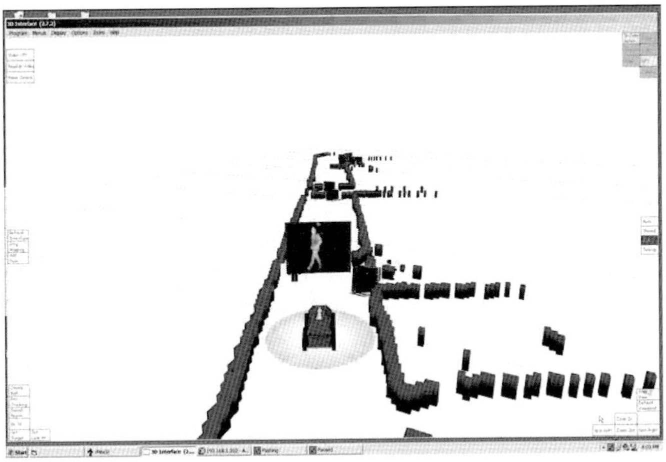

FIGURE 8 Originally developed by Curtis Nielsen at Brigham Young University: this 3-D display provides enhanced situational awareness to the operator, such as laser-detected surroundings, robot location, robot pose, camera orientation, and video imagery.

this fashion, the projected fire-control solutions are computed in advance, and the system is locked on and ready to shoot as soon as the operator gives the signal to engage. The weapon-payload response can be optionally configured to initially fire warning shots at the ground in front of the intruder, followed by a second volley directed at chest level if he or she continues to advance.

4 FUTURE PLANS

While efforts to enhance the automation of detection, acquisition, tracking functions for UGV weapon payloads continue, achieving performance compared to the sophisticated fire-control systems already employed on military aircraft and tanks will be challenging. The most prevalent UGVs on today's battlefield are man-portable, where size, weight, and power constraints impose significant hurdles to adapting the complex armament that much larger vehicles are able to accommodate. These same man-portable systems are also more prone to tampering and sabotage, being easy to tip over, disable, or even capture, and so near-term incorporation of less-lethal defensive measures is of considerable interest.

Longer term efforts are investigating augmented-virtuality displays for improved situational awareness, allowing operators to more quickly assess the tactical situation (Fig. 8). Under the Warfighter's Associate concept, the control paradigm has progressively shifted from low-level joystick to high-level mouse input, and even speech recognition is finally showing real-world potential. Laboratory demonstrations of proximal human–robot interaction have shown that desired robot actions can be reliably inferred from the perceived state and behavior of its human counterpart. This innovative approach may one day allow an intelligent robot to acquire and prosecute a confirmed threat, based on where its human partner is observed to be aiming and firing.

REFERENCES

1. Policy for Non-Lethal Weapons (1996). *Official Website for DoD Issuances*, available online at: http://www.dtic.mil/whs/directives/.
2. Department of Defense Nonlethal Weapons and Equipment Review: A Research Guide for Civil Law Enforcement and Corrections (2004). *Special Report*. National Institute of Justice, http://www.ncjrs.gov/pdffiles1/nij/205293.pdf.
3. Hightower, J. D., and Smith, D. C. (1983). Teleoperator technology development. *Proceedings of the 12th Meeting of the United States-Japan Cooperative Program in Natural Resources*. San Francisco, CA.
4. Aviles, W. A., Hughes, T. W., Everett, H. R., Umeda, A. Y., Martin, S. W., Koyamatsu, A. H., Solorzano, M. R., Laird, R. T., and McArthur, S. P. (1990). Issues in mobile robotics: the unmanned ground vehicle program teleoperated vehicle. *Proceedings SPIE Mobile Robots V*, Boston, MA, pp. 587–597.
5. Everett, H. R., and Gage, D. W. (1996). *A Third Generation Security Robot*, Vol. 2903, SPIE Mobile Robot and Automated Vehicle Control Systems, Boston, MA, pp. 20 21, 118–126.
6. Everett, H. R., Pacis, E. B., Kogut, G., Farrington, N., and Khurana, S. (2004). Towards a Warfighter's Associate: eliminating the operator control unit. *SPIE Proceedings 5609: Mobile Robots XVII*, Philadelphia, PA.
7. Kogut, G., Drymon, L., Everett, H. R., Pacis, E. B., Nguyen, H., Stratton, B., Goree, J., and Feldman, B. (2005). Target detection, acquisition, and prosecution from an unmanned ground vehicle. In *SPIE Proceedings 5804*, G. R. Gerhart, C. M. Shoemaker, and D. W. Gage Eds. Unmanned Ground Vehicle Technology VII, Orlando, FL, pp. 560–568.
8. Elliott, E. D., and Eagleson, R. (1997). Web-based teleoperated systems using EAI. *Proceedings of the IEEE International Conference on Systems, Man, and Cybernetics*, Orlando, FL. Vol. 1, pp. 749–754.
9. Held, R., Efstathiou, A., and Greene, M. (1966). Adaption to displaced and delayed visual feedback from the hand. *J. Exp. Psychol.* **72**, 887–891.
10. Day, P. N. (1999). An investigation into the effects of delays in visual feedback on real-time system users. In M. A., Sasse, and C. Johnson Eds. *Human-Computer Interaction —INTERACT '99*. IOS Press, Oxford, pp. 674–675.
11. Drury, J. L., Scholtz, J., and Yanco, H. A. (2003). Awareness in human-robot interactions. *Proceedings of the IEEE Conference on Systems, Man, and Cybernetics*. Washington, DC.
12. Martin, S. W. and Hutchinson, R. C. 1989. Low-cost design alternatives for head-mounted displays. *Proceedings, SPIE 1083*. Three Dimensional Visualization and Display Technologies, Los Angeles, CA.
13. Laird, R. T., and Everett, H. R. 1990. Reflexive teleoperated control. Proceedings, association for unmanned vehicle systems, *17th Annual Technical Symposium and Exhibition (AUVS '90)*, Dayton, OH, pp. 280–292.

FURTHER READING

Unmanned Ground Vehicles http://www.spawar.navy.mil/robots/, 2007.

Non-lethal Bibliography http://www.au.af.mil/au/aul/bibs/soft/nonlethal.htm, 2007.

Ankin, A. C. "Governing Lethal Behavior", Technical Report GIT-GVU-07-11, available online at http://www.gatech.edu/ai/robot-lab/online-publications/formalizationv35.pdf, 2008.

DEFENDING AGAINST DIRECTED ENERGY WEAPONS: RF WEAPONS AND LASERS

MICHAEL J. FRANKEL
EMP Commission, Washington, D.C.
EDWARD T. TOTON
Toton, Incorporated, Reston, Virginia

1 INTRODUCTION

Like many inventors of a new technology, the first caveman to heave a rock in anger at his fellow troglodyte must have quickly realized he was on to something. The concentration and direction of expended energy on a target where it will have the most impact, while minimizing the expenditure of wasted energy elsewhere where it will not, is the essence of directed energy weapons (DEWs). So, heaved rocks and other forms of kinetic weapons such as bullets are directed energy threats. High explosives, which may in more usual circumstances squander much of their kinetic energy in a spherically symmetric expansion into mostly empty space, may also be configured as DEWs through shape charge configurations or other specialized charge designs. But, the more common perception of DEWs today deals with devices that radiate electromagnetic energy in the form of high or low energy "light"—lasers and radio frequency (RF) weapons—and both charged and neutral particle beams.

A great deal of research dollars have been expended—and continue to be expended—by the US government in pursuit of the development of DEWs. The first suggested use of such devices in a military context dates back to the 1950s when charged particle beams were studied as a proposed defense against atomic weapons [1].[1] In the 1970s, there was particular interest in charged particle beam weapons in which directed beams of electrons might be used to protect ships from incoming missiles by engaging and thermally exploding energetic materials in incoming antiship missiles [2].[2] These efforts involved huge accelerators to produce the particles, elaborate steering mechanisms, and a host of technical challenges to propagate a beam, which had to overcome the natural tendency of similarly charged particles to run away from each other's vicinity. In the 1980s, the Strategic Defense Initiative Organization (SDIO) provided the sponsorship and funding to demonstrate a range of DEW: X-ray lasers (powered by nuclear explosions) and other nuclear-pumped directed energy schemes, continuous wave and pulsed chemical lasers, tunable free-electron lasers, neutral particle

[1]Compact accelerators capable of firing high energy electrons at incoming nuclear warheads were proposed by Nobel Laureate Robert Wilson in a now almost forgotten episode that presaged the inauguration of the "Star Wars" program in the 1980s by almost 30 years.
[2]cf. Navy's Chair Heritage program.

beams, high powered microwave weapons, and of course rocks [3].[3] Additionally, the technical feasibility of even more exotic directed energy schemes, such as γ-ray lasers, was examined, at least analytically. Resources have also been devoted to development efforts that may protect US systems against the threat of hostile use of DEWs by adversaries. These efforts range from the development of directed energy attack sensors to the development of hardened systems that can withstand an assault by DEW threat systems. Presently, the Department of Defense continues to explore the potential of DEWs for a variety of both tactical and strategic defense missions. However, much of this information has been classified and, in many ways, is not directly relevant to the threat to the civilian infrastructure described below.

2 DIRECTED ENERGY WEAPONS AND HOMELAND SECURITY

The focus of this article is homeland security and what can be done to protect ourselves from directed energy threats. Much of the directed energy gadgetry developed in pursuit of robust weapons systems, the gigantic accelerators and huge pulse power and energy storage systems, and technologically sophisticated and complex tracking and targeting systems would not seem to represent the most likely, easily transportable and concealable, threats available to either terrorists or rogue regimes intent on creating local, homeland, disruption. What remains then to consider are two directed energy threats to homeland security: lasers and radio frequency (RF) weapons.

2.1 Lasers

Lasers (*light amplification by stimulated emission of radiation*) are devices that produce highly concentrated beams of light, either continuously or in a pulsed mode operation. The mechanism of lasing involves the creation of a so-called electronic population inversion in a lasing medium in which excited atoms all transition to the same lower energy state, each atom giving up the same bit of energy in the form of a light wave. The coherent concentration of these individual but identical bits of light energy gives rise to the intensely energetic monochromatic beam that emerges from a laser. Lasing has been achieved in solid, liquid, and gaseous media and at a wide spectrum of frequencies ranging from the infrared through the visible, ultraviolet, and even × rays.

2.2 Radio Frequency (RF) Weapons

RF weapons deliver energy to a target in the form of electromagnetic radiation. The essential elements of such an electromagnetic weapon are a power source, a physical device to convert the power into electromagnetic radiation, and an antenna to broadcast the radiation in the desired direction. Traditionally, RF weapons are classified into two classes: narrow band, often described as high power microwave (HPM) weapons and wideband, usually described as ultra-wideband (UWB) weapons. HPM weapons are characterized by a narrow frequency band and generally operate in the gigahertz range. Depending upon the design details and type of power source, they may operate either

[3]The modern incarnation of designer rocks involve a number of complex interceptor systems including such as a theater high altitude area defense (THAAD) system, which is designed as a "hit to kill" intercept.

continuously or in pulsed mode. They will also generally require an expensive and highly specialized type of vacuum tube to convert the energy from a power source into guided microwave frequency waves. On the other hand, UWB RF sources will discharge their electromagnetic energy in a brief pulse containing a wide spectrum of energies ranging from tens of megahertz to tens of gigahertz. Instead of a technologically complex tube, they may use an intense spark gap as a radiation source. Depending on the design, these may also be repetitively pulsed. UWB weapons tend to be high power, but low energy, threats, which facilitates their design in small packages, unlike many HPM threats. Other nomenclature usage—such as transient electromagnetic discharge (TED) devices in place of UWB—may be found in the literature.

3 CONSEQUENCES OF EMPLOYMENT: WHAT DIRECTED ENERGY WEAPONS CAN DO

3.1 Lasers

In a terrorist employment, the chief concern is that lasers might be used as antipersonnel weapons to temporarily blind, or even permanently damage, the eyesight of individuals at critical operational moments. In particular, concern over the potential vulnerability of airline pilots to illumination by a mobile ground-based laser during takeoff and landing of commercial airliners has been a cause of considerable disquiet. At one end of the range of potential effects of aiming such focused laser energy on a pilot is the possible infliction of permanent eye damage, since intense light energy is focused by the eye's lens on the retina. The rods and cones situated there, the physiological task of which is to distinguish and transduce the different colors of ordinary light, can be destroyed at a sufficiently intense illumination. At lower intensities a pilot may experience temporary blindness from the intense glare or be "dazzled" by an intense sparkle as laser light scatters off the cockpit glass (Figure 1). The magnitude of this threat is enhanced by the ease of acquisition of such laser weapon threats. It is not only the relatively compact and high power systems that are readily available through many industrial catalogues and at educational institutions can represent an active threat, but the ubiquitous laser pointers favored by large numbers of faculty at almost any institution of higher learning in the United States and available at almost any store that sells educational supplies can also

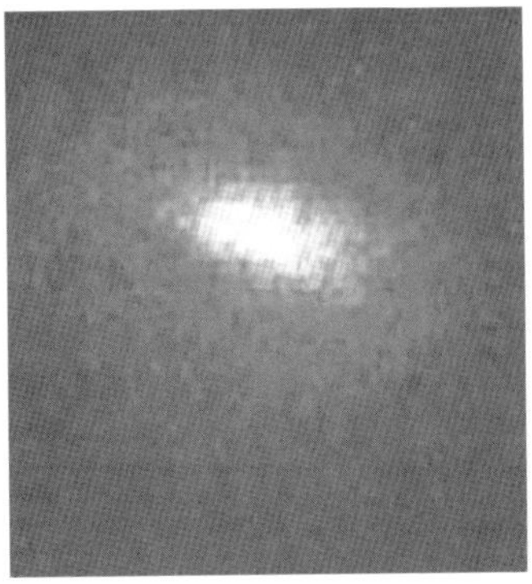

FIGURE 1 Laser-induced cockpit glare. A pilot's eye view.

represent an active threat. Most troubling is the identification of more than 400 incidents since 1990 by the Federal Aviation Administration wherein aircraft cockpits have been illuminated by lasers leading to pilot distraction during critical phases of flight [4, 5]. While no aircraft losses resulted from these incidents, the National Transportation Safety Board has documented cases of pilot eye damage and incapacitation during critical phases of flight [6]. In the words of a recent report by the Congressional Research Service:

> A recent rash of incidents involving lasers aimed at aircraft cockpits has raised concerns over the potential threat to aviation safety and security. While none of these incidents has been linked to terrorism, security officials have expressed concern that terrorists may seek to acquire and use higher powered lasers to, among other things, incapacitate pilots. There is also growing concern among aviation safety experts that the ubiquity and low cost of handheld laser devices could increase the number of incidents where pilots are distracted or temporarily incapacitated during critical phases of flight [7].

Unlike energy emanating from more usual sources, which tends to spread itself out and dissipate its intensity with distance from the source, the highly collimated and energetic laser beam tends to propagate long distances with relatively little spread. Thus, its almost undiminished effects may be felt at very long distances from its origin in the absence of absorption by the atmosphere. On a clear day with low water vapor in the air, laser beams may propagate for very long distances. NASA has documented one incident of interference with pilot vision by a green laser at a distance of 90 miles from the source [8].

3.2 RF Weapons

It is often asserted that RF weapons affect only electronic devices, and do not directly affect people. This is not entirely true since microwave systems have also been developed for nonlethal antipersonnel applications.[4] But our concern over terrorist employment leads us to focus on the threat to electronic systems. RF systems, both of the HPM and UWB variety, work by coupling unwanted electrical energy contained in the strong electromagnetic field projected by the RF weapon to runs of wire and circuit elements in vulnerable electronic equipment. The effects of such RF weapon illumination on equipment will generally vary with the intensity of the illumination. At relatively low intensities, computer chips may experience "upset" and "latch-up" conditions, in which the internal electronic states in chips would have been affected by the imposed currents and voltages and a manual restart, or reboot, may be necessary. At higher intensities, the electronics may be degraded to the point they cannot be restarted and would have to be repaired or replaced. At still higher intensities, manifestations of physical damage, such as arcing or melting of circuit elements, may be observed (Figure 2).

RF technologists often speak of "front door" and "back door" vulnerability of equipment. Front door refers to the coupling of the RF energy from a beam directly into the vulnerable circuit element—such as memory chips and capacitors—through an intended signal path, usually via an antenna that collects the radiated energy and conducts it in the form of an impressed electric current to the vulnerable element. Back door refers to

[4]As anyone who owns a microwave oven is aware, RF energies can also heat things up. The Pentagon's nonlethal weapons development effort has demonstrated a "people stopper" that can be used for crowd control, which works by inflicting an intensely painful burning sensation when specially designed microwave devices are focused on human skin. Other organizations have also explored antipersonnel use of RF.

FIGURE 2 Flashover observed in integrated circuit board during current injection testing (courtesy W. Radasky, Metatech Corp.)

the coupling of radiant energy through unintended paths, such as joints or cracks, and from there to circuit elements by coupling to metallic elements that act as expedient antennae. Most HPM and UWB threats attack systems through the "back door" as their frequency content is in the gigahertz regime and, at those frequencies and corresponding wavelengths, the physical size of typical commercial equipment is on the order of the threat wavelength.

The advance of electronics technology has also conspired to increase the potential vulnerability of our electronic systems. "Old" electronic systems that utilized vacuum tubes and relatively large-scale metallic elements were much more robust and resistant to interference by stray electromagnetic energy than modern digital electronic systems. The latter are characterized by submicron scale circuit features and "smart" capabilities enabled by a plethora of low-voltage computer chips whose continued functioning may not withstand stray circuit voltages. The shift from analog to digital processing inside modern electronics and the shift to processing frequencies above 1 GHz also greatly increases the probability of upset.

Of particular concern is modern supervisory control and data systems (SCADAs), which are increasingly deployed within every nook and cranny of our modern electronic-based critical infrastructures. Their function is to automatically monitor and remotely control the operation of geographically far flung systems such as the electric power grid, the pumps on our oil and gas pipelines, and our national telecommunications infrastructure. These ubiquitous control units confer great economic benefit and enormous operational agility, but are essentially unhardened computers—electronically quite similar to the personal computers found on every desk—which have been demonstrated to be highly sensitive to RF radiation. In the 1980s, catastrophic failure of a SCADA system controlling flow in a natural gas pipeline located 1 mile from the naval depot of Den Helder in the Netherlands caused a large gas explosion in a 36 in. diameter pipeline. This failure was caused by electromagnetic interference traced to an L-band naval radar coupling into the wires of the SCADA system. Radio frequency

FIGURE 3 Consequences of SCADA system failure. Smoke from gas pipeline explosion, Bellingham, Washington.

energy caused the SCADA system to open and close at the radar scan frequency, a relay that was, in turn, controlling the position of a large gas flow-control valve. The resulting changes in valve position created pressure waves that traveled down the pipeline and eventually caused the pipeline to fail. A similar gas pipeline explosion in Bellingham Washington was also traced to the failure of SCADA control (Figure 3). In another incident ultimately traced to the operation of a Navy AN/SPS-49 radar system 25 miles off the coast of San Diego, the radar-induced failure of a SCADA system controlling the opening and closing of water and gas flow valves led to the subsequent letter of complaint by the San Diego County Water Authority to the Federal Communications Commission, warning of a potential "catastrophic failure" of the aqueduct system. The potential consequences of a failure of this 825 million gallon per day flow rate system ranged from "spilling vents at thousands of gallons per minute to aqueduct rupture with ensuing disruption of service, severe flooding, and related damage to private and public property."

The critical infrastructures on which the functioning of modern society is dependent are susceptible to disruption, degradation, or destruction by RF weapons fired at a distance to the line-of-sight exposed electronic components. The electric power grid depends on unhardened SCADA monitoring systems to detect and respond to problems. The SCADAs along with the critically important autonomous relays that protect the very long lead replacement time circuit elements such as the very high voltage transformers (which are no longer manufactured domestically and typically require more than a year's delay to fill an order) are increasingly of digital electronic design and susceptible to RF weapon damage. In the case of the autonomous relays, over the last 20 years, solid-state digital systems have increasingly replaced the older electromechanical designs, which has significantly increased their vulnerability to the threat of an RF weapon environment. A worrisome scenario envisions a series of simultaneous attacks on geographically separated elements of the electric grid by such electromagnetic weapons, which may produce a cascading failure of large sections of the grid. The data and operational control centers that enable the accounting and daily flow of trillions of dollars through our financial and banking systems are enabled by computers and other electronic elements that are intrinsically susceptible to RF weapon environments. Similarly, transportation control centers, as well as switches that control railroad track function, contain digital electronic systems susceptible to RF energies. Thus, a terrorist in possession of such a means would find himself in a very target rich environment.

4 THE ATTRACTION OF DEW FOR TERRORISTS

From the perspective of terrorist employment, directed energy weaponry offers a number of attractive features. First, they are easily available. Compact high power laser and RF emitter systems can be purchased almost anywhere in the country for legitimate industrial or educational uses. They may be obtained at moderate cost and without the risks of discovery associated with the purchase, manufacture, or need for concealment from the attention of security and law enforcement authorities, that attends acquisition, transport, and employment of conventional high explosive systems. Secondly, high power lasers and high power RF systems are easily transportable. They are compact enough to be easily carried and concealed by a small truck or van. Yet smaller systems, still powerful enough to inflict damage are man-portable. Although probably not a terrorist weapon of choice, we have already noted how even a laser pointer can produce effects that may endanger the safety of airline travel. An RF weapon in a suitcase is available for purchase today (Figure 4). Thirdly, compact DEWs are easy to conceal and use covertly from a stand-off distance. A terrorist employment could drive an RF weapon system mounted in the back on enclosed van, park at a significant distance from a line-of-sight target, and take down an electronic target in complete silence, escaping immediate notice and facilitating an easy retreat and opportunity for additional attacks on other facilities. It might even take considerable time before it is realized that a deliberate attack was perpetrated. Fourthly, while an explosive may be fired and used only once, a directed

FIGURE 4 RF weapon in a suitcase. Available for purchase from Diehl Stiftung and Co. KG, Germany.

energy system may have a "deep magazine." Unless powered by a special one-shot output device such as a compressed flux explosive generator, a directed energy device can be fired again and again, as long as the power system contains energy. Many RF weapon environments are high power and low energy environments, which reduces the need for a large energy source.

5 THE RISK OF TERRORIST EMPLOYMENT

Risk is formally accounted as the multiplicative concatenation of the (conditional) probabilities of threat, times vulnerability, and times consequence. We have discussed some of the potential consequences of terrorist employment of such weapons in the previous section. The vulnerability of modern digital electronics to electromagnetic disruption is also well established, not only by the accidental demonstrations such as the SCADA failures previously referenced (along with countless other instances) but also by scientific testing programs, which have established the susceptibility levels and thresholds for electromagnetic shock for commercial electronic devices [9]. Testimony at a Congressional hearing indicated that electronic equipment may be degraded or damaged when impressed field levels are on the order of kilovolts to tens of kilovolts per meter [9]. Similarly, the vulnerability of human eyesight to high power lasers has been amply demonstrated over many years. The threat is also clearly real. If we consider it simply from a capabilities-based perspective, it is clear that the technology knowledge is not just widely disseminated but acquisition of the physical devices themselves is about as easy as filling an order from an internet catalogue. A recent government-sponsored effort explored the ease with which a home-brew RF weapon might be assembled from parts available for purchase from a local radio shack [10].[5] Congressional reports have also speculated about the possibility of terrorist interest in exploiting the capabilities of such devices [11].[6] That is, the existence of threat, vulnerability, and consequence is well established, and the risk is undeniable. Quantitative assessment of such factors relative to other risks and risk prioritization remains a task for the Department of Homeland Security.

6 MANAGING THE RISK: RECOMMENDATIONS

Characterizing risk in terms of threat, vulnerability, and consequence is the essence of risk analysis. What to do about it is the essence of risk management [12, 13]. There are a number of actions that may be taken to reduce our vulnerability to DEW employment, and reduce the consequences as well. The following recommendations are offered:

- Perform a site vulnerability assessment by knowledgeable RF professionals.
- Reduce line-of-sight vulnerabilities to directed energy interference. Data centers or operational centers should not be sited within glass walled buildings with clear views of nearby roads. If you can see the road, the road can see you and may

[5]The hearings were devoted to a different sort of electromagnetic pulse phenomenon, that due to a nuclear source which a different frequency spectrum than that associated with RF weapons, but the radiation field intensities for expected damage are not dissimilar.

[6]Subsequent government-sponsored tests at Aberdeen proved inconclusive.

point something back. Similarly, unimpeded views into open upper floor and open windows should not be offered to potential observers in other buildings, even if the other building is at a kilometer distance.

- Consider the routing of cables and other conducting metal runs that enter a facility. These may become conduits into internal equipment for unwanted electrical energies imposed by RF weapons on exposed runs of cable. Where possible, these should be buried or otherwise hidden from view.
- Compile and maintain lists of critical equipment whose loss would not only shut down operations or endanger safety but might also be difficult to quickly replace. Consider maintaining an inventory of spares for critical components.
- For "high value" targets, deploy attack warning sensors to improve situational awareness that an attack has actually occurred or is underway.
- Consider the cost–benefit and utility of selective hardening of some critical components. For high value operations such as computer data centers, consider placing the computers within an interior, RF shielded, room.
- Continue current government investment in protective materials development efforts for both laser and RF threats. Such materials might include easily applicable "spray on" hardening for RF protection or optically reactive materials for laser protective applications. For private industry, begin such investment.
- For critical facilities, enforce access control and maintain "keep out" distances. Do not allow uninspected bag, briefcases, or shipments into critical facilities.

Risk may not be eliminated, but prudent preparations and some advance planning may serve to reduce it to tolerable levels.

REFERENCES

1. Schweber, S. S. (2007). Defending against nuclear weapons, a 1950s proposal. *Phys. Today* 36–41.
2. Snow, D. M. (1980). Lasers, charged-particle beams, and the strategic future. *Polit. Sci. Q.* **95**(2), 277–294.
3. Missile Defense Agency http://www.mda.mil/mdaLink/pdf/thaad.pdf.
4. Nakagawara, V. B., Dillard, A. E., McLin, L. N., and Connor, C. W. (2004). *The Effects of Laser Illumination on Operational and Visual Performance of Pilots During Final Approach*, Department of Transportation, DOT/FAA/AM-04/9, June.
5. *U.S. Secretary of Transportation Norman Y. Mineta Announces New Laser Warning and Reporting System for Pilots, Measures to Safeguard Pilots and Passengers, Support Timely Enforcement*, Department of Transportation Press Release DOT 08-05.
6. National Transportation Safety Board *Safety Recommendation Letter–Safety Recommendations A-97-13 through -15*. Washington, DC.
7. Elias, B. (2005). *Lasers Aimed at Aircraft Cockpits: Background and Possible Options to Address the Threat to Aviation Safety and Security*, CRS Report for Congress, January 26, 2005, http://fas.org/sgp/crs/RS22033.pdf.
8. National Aeronautics and Space Administration, Aviation Safety Reporting System (NASA/ASRS). Report Numbers 285090 and 290037. Moffett Field, CA.
9. Radasky W. A. (2004). Special issue on High Power Electromagnetics (HPEM) and Intentional Electromagnetic Interference (IEMI). *IEEE Trans. Electromagn. Compat*. **46**(3).

10. Jakubiak, S. (1999). *Statement before the House Military Research and Development Subcommittee, hearing on EMP Threats to the U.S. Military and Civilian Infrastructure*, October 7, 1999.
11. Shriner, D. (1998). *The Design and Fabrication of a Damage Inflicting RF Weapon by 'Back Yard' Methods*, Testimony before the Joint Economic Committee, United States Congress, February 25, 1998.
12. Wilson, C. (2004). *High Altitude Electromagnetic Pulse (HEMP) and High Power Microwave (HPM) Devices: Threat Assessments*, CRS Report for Congress, August 20, 2004, http://digital.library.unt.edu/govdocs/crs/permalink/meta-crs-6028:1.
13. Scouras, J., Cummings, M. C., McGarvey, D. C., Newport, R. A., Vinch, P. M., Weitekamp, M. R., Colletti, B. W., Parnell, G. S., Dillon-Merrill, R. L., Liebe, R. M., Smith, G. R., Ayyub, B. M., and Kaminskiy, M. P. (2005). *Homeland Security Risk Assessment. Volume I. An Illustrative Framework*, RP04-024-01a. Homeland Security Institute, Arlington, VA, November 11, 2005.

FURTHER READING

Beason, D. (2005). *The E-bomb: How America's New Directed Energy Weapons Will Change the Way Future Wars Will Be Fought*, Da Capo Press.

McCarthy, W. J. (2000). *Directed Energy and Fleet Defense: Implications for Naval Warfare*, Occasional Paper No. 10, Center for Strategy and Technology, Air University, Maxwell AFB, https://research.maxwell.af.mil/papers/ay2000/csat/csat10.pdf.

THE SENSOR WEB: ADVANCED TECHNOLOGY FOR SITUATIONAL AWARENESS

KEVIN A. DELIN

SensorWare Systems, Inc., Pasadena, California

EDWARD SMALL

Sacramento Metropolitan Fire District, Sacramento, California and FEMA Urban Search and Rescue Team, CA Task Force 7, Sacramento, California

1 INTRODUCTION

The need for situational awareness in the dynamic environment of emergency and rescue operations is well understood. Data must be continually collected, analyzed, assimilated, and disseminated to both local operational personnel and remote commanders. The basic principles of "Facts, Probabilities, Own Situation, Decision, and Plan of Operation" for fire and rescue strategies are just as relevant today as they were when originally described in 1953 [1]. Simply stated, situational awareness informs decision making and decreases reaction time to changing conditions, even allowing for anticipation of events in certain instances.

Failure to effectively collect, synthesize, and distribute facts to personnel involved at all levels of a field operation will result in service delays, or worse, death. Emergency services personnel cannot begin operations without having the ability to monitor for hazards and account for personnel. Because emergency and rescue operations are labor intensive, however, continuous and effective monitoring for hazardous conditions often becomes less of a priority, or disappears entirely. It is therefore of critical importance to find a technological means to generate situational awareness for those personnel working in the hazardous area, both as a means to speed the course of the operation and to protect the personnel from danger.

Here, one such new piece of equipment, the Sensor Web, is examined. This technology can aid and substitute for human efforts in understanding the changing, and often chaotic, conditions during emergency service operations. First, the Sensor Web technology will be briefly described. Then, a series of representative field applications, including actual operations, will be given, which illustrate the unique capabilities of the Sensor Web as applied to emergency services. Finally, future directions of the technology are considered.

2 SENSOR WEB TECHNOLOGY

The Sensor Web is an embedded, intelligent infrastructure for sensors. Physically, it consists of spatially distributed sensor/actuator platforms (called "pods") that wirelessly communicate with one another (Fig. 1). Originally developed at National Aeronautics and Space Administration (NASA) for planetary exploration of unknown environments, the Sensor Web is also well suited for providing situational awareness in the chaotic and unpredictable environments associated with emergency and rescue operations. Despite its sophistication, such a system would cost no more than traditional, less capable wireless solutions and could actually reduce total operational costs by providing continual, automated in-field analysis thereby freeing up rescue personnel for other, more demanding, tasks.

2.1 Sensor Web Protocols

The wireless communication between pods should be thought of as an information bus, in the same way that buses connect the individual components (hard drive, optical drive, memory, logic units, etc.) of a computer. Consequently, while the individual pods are certainly networked with each other, the Sensor Web is not, in and of itself, a network but rather a spatially distributed macroinstrument [2]. The distinction is crucial: a network consists of components that *route information* along communication paths to specific destination points, while a macroinstrument consists of components that *share information* with each and every other piece at all times without any intervening routing. The Sensor Web's applicability to situational awareness, in fact, derives from having

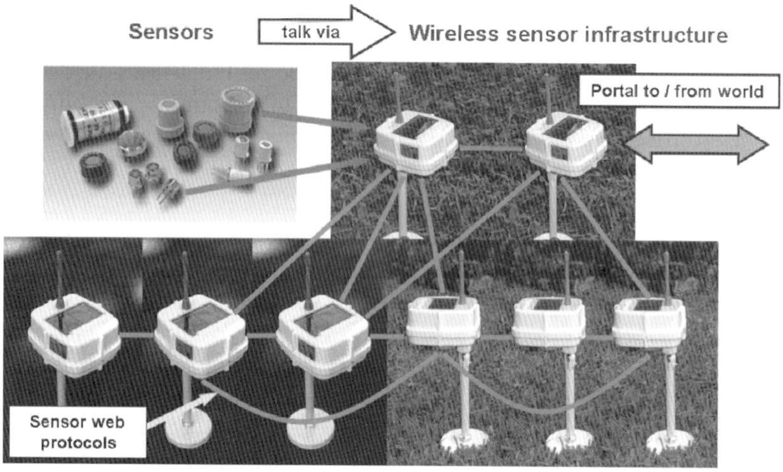

FIGURE 1 Schematic representation of a Sensor Web. Sensors are connected to the pods (white boxes). The pods communicate wirelessly to form an amorphous network where all pods are equivalent and any pod can be a portal to the outside world.

its sensor measurements taken, distributed, and interpreted *collectively* over this unique, massively redundant, communication architecture.

It can be stated in another way, that every pod pushes its data out onto the Sensor Web in an omni-directional manner. There is never a purposeful routing of data toward a specific pod or a special portal or gateway. In this way, every pod is made aware of conditions throughout the entire Sensor Web during each measurement cycle. In fact, while the computation hardware in a pod can be quite sophisticated, it is the sharing of information among the pods that gives the Sensor Web its macrointelligence. This is similar to how perception is created in the brain from a complex, interacting set of neurons that share electrochemical signals [3] rather than from individual intelligence at each neuron.

The Sensor Web communication architecture of data sharing is distinct from both hub-and-spoke and mesh network types. In hub-and-spoke networks, individual spoke nodes can be synchronized to the hub but the information must always be routed through the hub to get to other points. Mesh networks are typically based on asynchronous Internet-like (TCP/IP) protocols and require information routing as well. In marked contrast to these two network types, the Sensor Web communication architecture, by design, is synchronous (all measurements across the system are taken at the same time) and requires no routing.

The Sensor Web communication protocols are simple and robust. Each measurement cycle begins with the pods taking in sensor data. After a measurement is taken, each individual pod in the system broadcasts its information (data it has taken or received from others) in an omni-directional manner to all pods in communication range. Each pod then processes and analyzes the information it has received and the cycle repeats. In this way, the information is hopped pod-to-pod and spread throughout the entire Sensor Web. The entire system becomes a coordinated whole by possessing this internal, continuous data stream, drawing knowledge from it, and reacting to that knowledge.

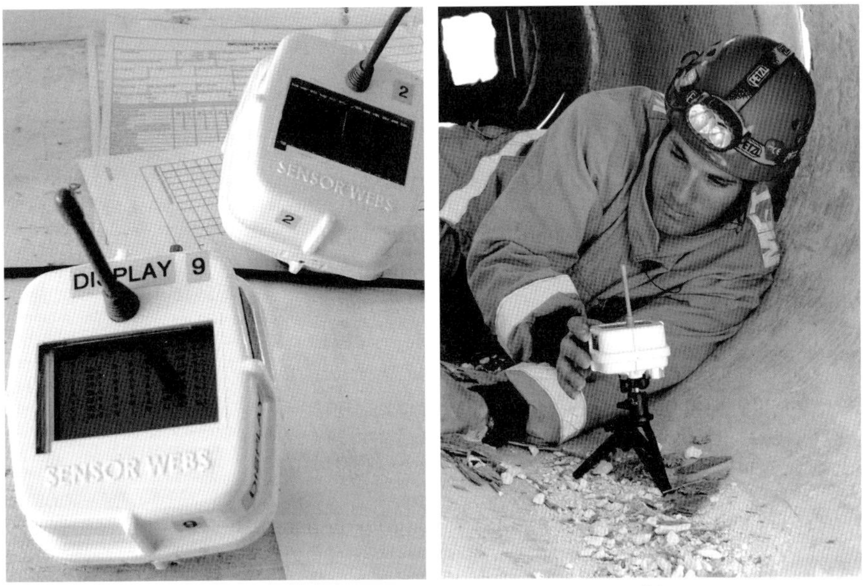

(a) (b)

FIGURE 2 Sensor Web pods. (a) A standard pod is in the background and shows both the 900-MHz antenna and a solar panel to harvest additional energy for the pod's rechargeable batteries. In the foreground is a special display pod where a flat panel display has replaced the solar panel and reveals the conditions at other pods. This allows mobile personnel to monitor the Sensor Web without having to access a computer. Labels have been attached to the pods to clearly mark their software-assigned identification. (b) A responder deploys a Sensor Web pod. The pods can be mounted using a variety of hardware including stands, magnetic bases, and spikes.

2.2 Sensor Web Pods

A key feature of the Sensor Web is that its component parts, the sensor platforms or pods, are all alike (Fig. 2). In general, they only differ by the sensors attached to them. A single pod, known as the mother, holds the single, system-wide clock that will synchronize all pods. The mother, however, holds no special hardware; indeed any pod may be designated as the mother simply by labeling it as such. Unlike hub-and-spoke and mesh networks, the Sensor Web is a truly amorphous network with no central point and no specially designated portal or gateways.

Each Sensor Web pod consists of five basic modules:

1. *The radio.* Although any radio frequency can be chosen, the 900 MHz license-free Industrial, Science and Medical (ISM) band has been used in manufactured Sensor Web systems to date. This frequency requires no licensing of end-users and does not compete with the more common frequencies found at emergency sites (minimizing jamming). In addition, radios operating at this frequency do not require line-of-sight communication (even going through concrete walls) and have an upper range of about 200 m (compliant with government power regulations). Because each pod will essentially function as a repeater and retransmit data it receives from other pods, the effective radio range is extended far beyond the limits imposed by the specifications and regulations associated with a single radio.

2. *The microprocessor.* This component contains the system's protocols, communicates with the attached sensors, and carries out data analysis as needed.
3. *The power system.* The combination of solar panels, rechargeable batteries and micropower electronic design have kept Sensor Web pods operating in the field for years without requiring maintenance.
4. *The pod packaging.* The package is lightweight, durable, inexpensive, and sealed against such elements as rain, standing water, water sprays, dust storms, and caustic chemicals. In addition, it enables an easy and rapid mounting.
5. *The sensor suite.* This module is completely determined by the specific application. It is the ability to accommodate a wide range of sensor types that makes the Sensor Web so versatile. For the types of operations discussed here, typical sensors include those for monitoring gases and environmental conditions such as air temperature and humidity.

2.3 Sensor Web Properties

As just described, the Sensor Web is a distributed macroinstrument based on its unique protocols that allow for data sharing via nonrouted, synchronous inter-pod communication. These characteristics create valuable Sensor Web properties when the technology is applied to emergency and rescue operations (Fig. 3).

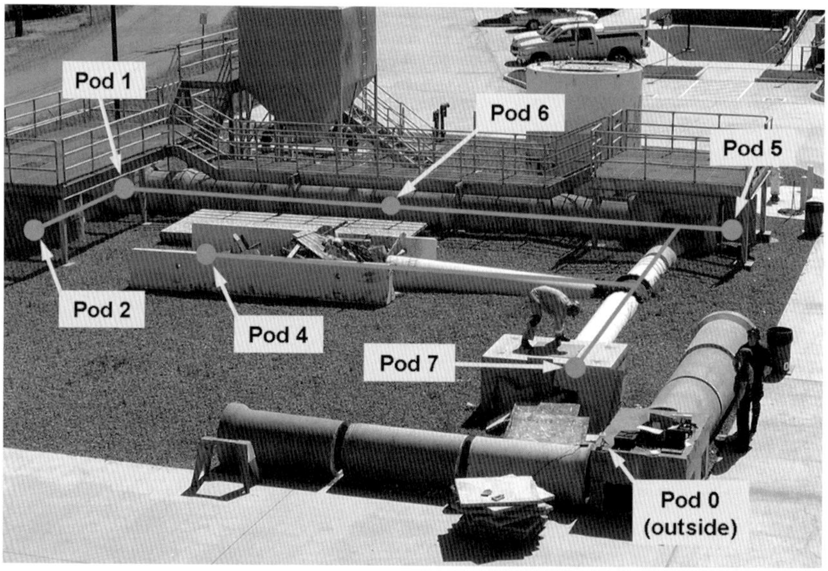

FIGURE 3 A confined space training facility for Urban Search and Rescue. The Sensor Web system was deployed inside the structure as indicated schematically by the dots. The size of the structure is indicated by the size of the firefighter on top of the structure near pod 7. Lines connecting dots indicate physical paths, not pod connectivity. When deployed, the pods were able to wirelessly communicate through the concrete barriers; for example, pod 4 was in direct communication with pods 1, 2, and 6. When a chain saw was intentionally left running between pods 5 and 6, personnel were able to observe the migration of carbon monoxide through the structure with one front toward pods 6, 1, and 2 and a secondary front toward pod 7. All pods in the system went into alarm together when the gas first reached threshold levels at pod 6.

1. The redundant, nonrouted data sharing allows for any pod to be a portal to the outside world. Since every piece of the Sensor Web contains all the same information, anyone with access to a single pod will be provided with the same situational awareness picture. This means that the first responder at the rescue scene looking at the flat panel display of a pod, the incident commander directing operations who has plugged his laptop computer into another pod, and remote government personnel examining the information of a third pod being sent out over the Internet, will all simultaneously have a common, unified picture of operations.
2. A single, system-wide clock provides for an immediate synchronous snapshot that can be intuitively understood by field personnel. Because the Sensor Web is synchronous, all the measurements are taken at the same time and, in essence, become a pixel in the overall picture taken by the Sensor Web. This picture is an immediate integration and valuable for in-field personnel to understand plume motion, for example, without the need for time-consuming posthoc analysis. Moreover, when combined with the massively redundant, nonrouted communication paths, this snapshot is known to all pods in the Sensor Web which allows for *anticipatory* warnings and alarms. In this way, field personnel are alerted to dangerous conditions anywhere in the work zone immediately by the Sensor Web rather than by a remote incident commander.
3. Synchronous system behavior reduces latency between data taking and data reporting. Each pod has the same situational awareness picture at the end of every measurement cycle. Because each pod can act as a portal into the system, this implies that all end-users are kept up-to-date on every measurement cycle.
4. The massively redundant, nonrouted communication paths provide for a highly robust structure with no central point of system failure. Since all pods are potential portals, the loss of any particular pod does not affect the entire operation. Since any pod can be designated as the single, system-wide clock, the worse case scenario of a damaged mother pod is rapidly recoverable simply by assigning (via a software label) the designation of "mother" to another pod. This worse case failure can even be corrected automatically without human intervention if, after a predetermined amount of time without receiving signal from the mother, the pod with the current lowest serial number in the Sensor Web promotes itself to mother.
5. The redundant, nonrouted, data sharing allows for recoverable single point sensor failure. Because every pod contains a microprocessor and knows all sensor measurements across the system at each measurement cycle, each pod can locally analyze global conditions across the entire Senor Web. It is therefore possible for the system to immediately evaluate a seemingly anomalous measurement against the background of neighboring measurements to determine, on a statistical basis, false positives or to distinguish between a merely worsening trend versus a true critical situation requiring evacuation. It is also possible for the system to "suggest", within the same measurement cycle, a missing sensor measurement by combining spatial interpolation of neighboring measurements with recent local measurement trends.
6. The massively redundant, nonrouted communication paths allow the system to be easily and rapidly deployed. No special skill is required to set up a Sensor Web as all the pods are essentially the same (in contrast to other networking schemes which requires special hardware gateways or router tables). As a result, once the mother

pod is switched on to provide a clock for the system, the pods may be dispersed as dictated by the needs of the situation. Data taking is immediate and the pods may even be reshuffled on the fly and, as long as they stay within communication range, maintain the overall Sensor Web macroinstrument. This is particularly valuable if pods are assigned to specific rescue squads that are moving independently through a building. Because emergency service operations do not allow the time to leisurely recall complex equipment function, the simplicity and speed of deployment may be among the Sensor Web's most compelling and important features.

3 SENSOR WEBS APPLIED TO FIELD OPERATIONS

To date, there have been over 30 Sensor Web field deployments with systems spanning distances up to 6 miles and running continuously for over 3 years. The systems have been tested extensively in numerous, challenging environments from the remote ice slopes of Antarctica to the searing heat of the central New Mexico desert to the corrosive salt air of the Florida coast [4–6]. Real time, streaming output of some of these systems may be viewed over the Internet using a variety of user interface displays [6]. Here, the Sensor Web capabilities as applied specifically to emergency and rescue operations will be examined. As will be shown, the Sensor Web's properties and physical robustness allow it to efficiently bring key environmental parameters together in a continuous operational picture and disseminate this picture to in-field and remote personnel.

A critical issue for any type of deployment is determining a pod's location. For extended, outdoor operations, this is most easily accomplished by using an external Global Positioning System (GPS) unit during deployment, noting pod placement coordinates, and putting these coordinates into the pod's memory as it is deployed. Pods can then share their individual coordinates with each other just as they do with sensor data. If pod power usage, size, and cost are not an issue, it is also possible to place a GPS unit inside each pod. Sometimes, however, GPS coordinates may not be a practical option because (i) pods may be shielded from strong GPS signals (as when placed inside a building), (ii) building geometry provides a more transparent understanding of pod placement (e.g. "pod at the west end of the first floor corridor"), or (iii) typical GPS resolution may not be accurate enough (as in the case of placing a pod against a wall and determining which room it is in). In such cases, simple hand mapping by in-field personnel has been found to be effective and easily performed, even under rapid deployment circumstances. In addition, these hand-mapped "coordinates" typically provide rescue personnel with a clearer, intuitive picture of how the Sensor Web is deployed in the area. Pods can still autonomously perform spatial data analysis in these cases because a relative pod placement map can be formed by each pod through a shared knowledge of every individual pod's nearest neighbors.

3.1 Atmospheric Monitoring

The predominant cause of death in a confined space incident is from a hazardous atmosphere. Typically, in a confined space or structural collapse operation, one highly competent person is dedicated to the position of "environmental officer". This person continuously monitors the atmosphere by way of a gas sensor with remote sampling capability, or by periodically requesting a reading from the entrant who is carrying a gas sensor, or both. This person also directs the forced ventilation efforts to enhance victim survivability and to maintain as tenable, and explosion-free, an atmosphere as possible.

There are several difficulties associated with the present technique. First, because all of the gases monitored have different vapor densities, they tend to stratify at different levels in the space, or may become trapped in dead spaces and fail to diffuse into the atmosphere. For this reason, atmospheric sampling must occur at 4-ft intervals, both vertically and horizontally. This significantly slows the progress of field personnel motion as they move into or even exit a confined space operation. Second, while the use of a sampling pump and tubing is a common method of obtaining remote gas samples within the confined space, there is the issue of the time it takes for the atmospheric sample to be drawn through the sample tubing. Since this may take up to 3 sec/ft of tubing and 50-ft sections of tubing are not uncommon, gas monitoring can require 1.5–2.5 min of delay per sample. Third, with the portable devices typically used by rescue personnel, evacuation of the space is often at the discretion of those working. Experience has shown that most in-field personnel consider an alarm to be an annoyance, and remain focused on completing their task; the atmosphere is only a "little hazardous". Finally, documentation of the atmospheric monitoring for most types of confined space operation is required by state/federal law, yet maintaining this documentation during the operation takes away from the actual emergency or rescue services.

Sensor Webs have been built and successfully used for monitoring confined space atmospheres. Here, the pods are equipped with the four gas sensors necessary for this application (e.g. oxygen, carbon monoxide, hydrogen sulfide, and explosive limits). The pods also contain sensors for air temperature and humidity. The measurement cycle for this Sensor Web is programmed for 30 sec. These systems have been used for several years now in confined space operations and collapsed structure training exercises. It has been found that the rescue personnel readily adapt to the new technology and have no difficulty infusing the technology into standard procedures (Fig. 4).

Benefits of using the Sensor Web for atmospheric monitoring are as follows:

1. Providing a permanent sensing infrastructure that frees the rescue personnel from having to take measurements. As a squad penetrates the confined space or collapsed structure, it deploys pods along the ingress path, effectively growing the Sensor Web. Once in place, the Sensor Web allows personnel to move freely into and out of the operations area. This is especially important during lengthy operations where squads will be replaced periodically. Moreover, since all pods would alarm when any pod detects a hazard, other responders working in other portions of the building will be aware if there is a gas leak that could affect them as well. As a result, responders will quickly know when to exit and how to modify egress paths by detecting remote atmospheric changes due to gas leaks.
2. Reducing the latency of obtaining measurements compared to drawing gas through a tube. Measurements are now available to the environmental officer at the sampling frequency (here, 30 sec.) throughout the field of operation.
3. Providing the environmental officer and incident commanders with a full picture of atmospheric conditions without diverting the rescuer's attention from other crucial tasks (Fig. 5). During an actual collapsed structure training operation, the Sensor Web revealed trends of oxygen displacement from the expired carbon dioxide of rescuers in confined areas, as well as increases in temperature and humidity from rescuers and equipment. This allowed the incident commander to move ventilation

(a) (b)

FIGURE 4 Sensor Web used for atmospheric monitoring. (a) A 10-pod Sensor Web system, including laptop and several varieties of pod mounting hardware fit compactly in a case, ready for rapid field deployment. (b) A pod being lowered into a confined space to determine atmospheric conditions. The pod will be left in this space for the duration of the operation allowing personnel to freely move into and out of the area.

fans from other parts of the operation into the affected confined areas to allow rescuers to proceed without stopping. While the Sensor Web system provides warning of imminent hazardous conditions, just like the single station gas detectors, the greater value may be in its ability to display trends in environmental conditions and disseminate that information to commanders with authority to act.

4. Providing an accurate and immediate record of conditions recorded every 30 sec by the Sensor Web and output to a laptop connected to the mother pod.

3.2 Structural Integrity Monitoring

The dangers present in and around structures compromised by natural or man-made disasters are rarely static. Building conditions can continue to deteriorate by the actions of earthquake aftershocks, wind, rain, and snow loading, and the intentional or unintentional actions of rescue workers.

The 4-gas Sensor Web pods described above also have an accelerometer built into them. The accelerometer functions as a tiltmeter to determine a change of state in the pod's orientation. The pod, attached to shoring or a building wall, can monitor changes while they are occurring and warn of impending failure. Such pods therefore perform double-duty, monitoring both atmosphere and structural integrity with the attendant reporting benefits for both parameters (Fig. 6).

Shoring stress tests revealed that the Sensor Web provided warning 60 s (two measurement cycles) before shoring failure. Such an advanced warning, distributed throughout

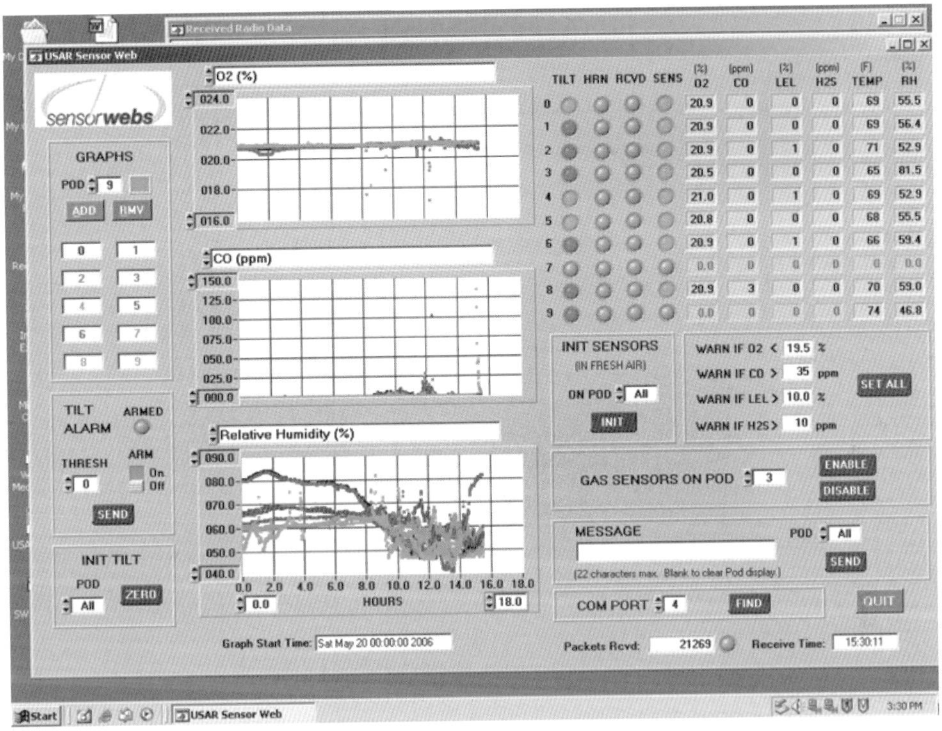

FIGURE 5 A screen-capture of the user interface for the Sensor Web. Time-based trending data are immediately available as are current readings from the system. The interface also serves as a command portal into the system. Note the spike in CO concentration at hour 15. This was a result of an acetylene torch being ignited inside the collapsed structure.

the entire space of operations, would greatly reduce the risk of personnel being caught in a further collapse.

3.3 Decontamination Monitoring

During biological decontamination operations, a chemical agent is introduced at the proper temperature and humidity to destroy the intended pathogen. Such operations are typically very complex logistically and labor intensive. The target structure to be decontaminated is first sealed with tarps that must hold the caustic decontaminant gas over many hours. Typical operation will have the caustic gas pumped into and out of the structure continually with small fans placed throughout the structure to ensure even distribution of the gas. The entire building's atmosphere needs to be monitored during the decontamination procedure to ensure that proper conditions exist (decontaminant concentration, temperature, and humidity) to kill the pathogen.

Presently, wet chemistry techniques are used to monitor the operation. Long plastic tubing is distributed throughout the structure from a central hub. Atmospheric samples are pumped out of the hub at regular intervals and examined chemically to maintain appropriate conditions in the building.

There are several disadvantages with this technique. First, the distribution of the plastic tubes can take nearly a full day, even in a modest-sized home. Second, sampling

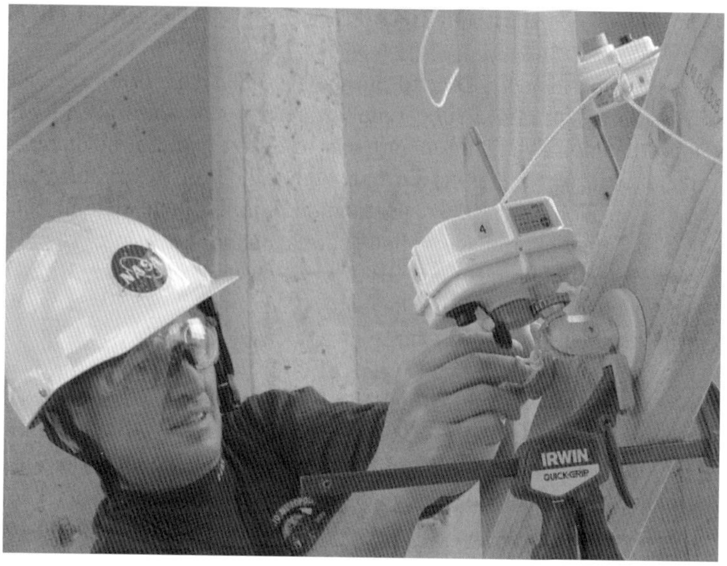

FIGURE 6 Pods are attached to shoring.

in this manner means that the chemical analysis will only yield the average conditions of the building due to gas mixing in the central hub. As a result, there still might be pockets in the structure where an appropriate concentration of decontaminant is not obtained, allowing for the pathogen to live. Lastly, water can collect in the plastic tubing due to the high humidity conditions typically needed during operations. This water can effectively absorb the gas (ClO_2 in the case of anthrax decontamination), which will create spurious results during chemical analysis. The tubes can also fill with ice during winter-time decontaminations and therefore make sample retrieval impossible and force operators to abort the entire operation.

A Sensor Web with ClO_2 sensors attached to each pod has been successfully used to alleviate these problems (Fig. 7). Inclusion of the technology into the operations was easy and personnel were able to learn the system in literally under 15 min. At least a dozen actual decontaminations have been performed using this system. In addition to the tremendous labor reduction in eliminating the tube system for gas sample retrieval, the Sensor Web allows operators to follow the ClO_2 gas plume throughout the structures in real time. On more than one occasion, this enabled operators to immediately find leaks in the building's tarps as well as building pockets where ClO_2 gas concentrations were lower than expected.

4 FUTURE DIRECTIONS

The Sensor Web is a general information infrastructure for sensors. It collects, analyzes, and reacts to whatever conditions are important to the end user, and can be coupled with existing methods of obtaining the necessary data. Any low-bandwidth sensor can be attached to the Sensor Web and enhance the situational awareness properties of the system. Initial experiments using a Sensor Web to track in-field personnel during operations

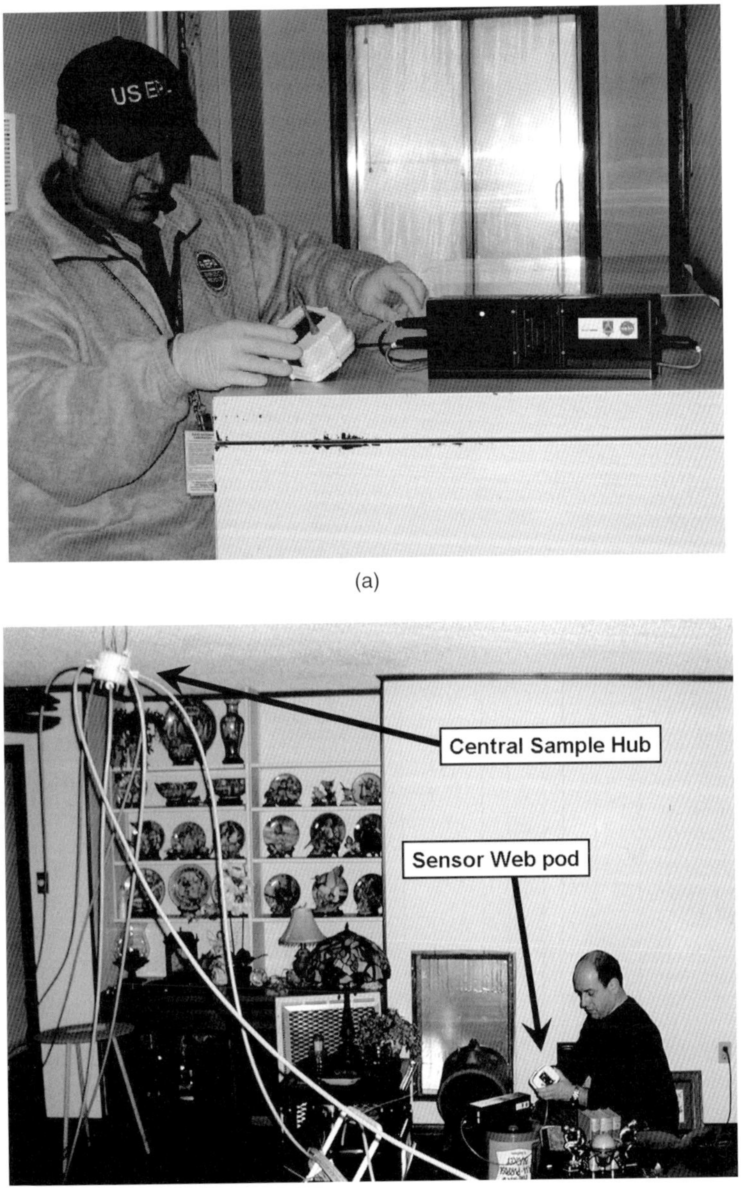

(a)

(b)

FIGURE 7 Sensor Web used for decontamination operations. (a) Environmental Protection Agency (EPA) decontamination team member deploying a Sensor Web pod. The black box attached to the pod is the ClO_2 sensor. (b) A Sensor Web is deployed in a decontamination area that also uses the traditional sampling method to monitor ClO_2 concentration. Note the traditional method requires plastic tubing that runs through the entire house and connects to a central hub where atmosphere samples will be pumped out for chemical analysis.

show promise. Such a capability would benefit locating fire fighters lost in buildings as well as provide an accurate count of personnel going into and out of a rescue operation area.

Other applications for the Sensor Web immediately present themselves, especially involving infrastructure protection. The properties of the Sensor Web make it ideal for sentinel security systems where a disturbance at one pod will be known by all. Moreover, the massively redundant connectivity of the macroinstrument makes the Sensor Web amenable to over-the-horizon monitoring of rail and highways.

Finally, the reactive capabilities of the Sensor Web to dynamic environments are only now being explored. Future systems may, for example, control ventilation during rescue operations based on changing atmospheric conditions and consequently free up additional labor that can be better applied to the actual rescue tasks at hand. Clearly, the situational awareness capability inherent in the Sensor Web can only increase with advances in the underlying technology and with no additional cost compared to that of more traditional, less capable wireless solutions.

REFERENCES

1. Layman, Lloyd. (1953). *Fire Fighting Tactics*, National Fire Protection Association, Quincy, MA.
2. Delin, K. A. (2002). The Sensor Web: A macro-instrument for coordinated sensing. *Sensors* **2**, 270–285.
3. Koch, C., and Laurent, G. (1999). Complexity and the Nervous System. *Science* **284**, 96–98.
4. Delin, K. A. (2005). Sensor webs in the wild, *Wireless Sensor Networks: A Systems Perspective*, N. Bulusu, and S. Jha, Eds. Artech House, Norwood, MA, pp. 259–272.
5. Delin, K. A., Jackson, S. P., Johnson, D. W., Burleigh, S. C., Woodrow, R. R., McAuley, J. M., Dohm, J. M., Ip, F., Ferré, T. P. A., Rucker, D. F., and Baker, V. R. (2005). Environmental studies with the sensor web: principles and practice. *Sensors* **5**, 103–117.
6. See links at www.sensorwaresystems.com.

Critical Information Infrastructure Protection

Country Surveys
International Organization and Forums

CRITICAL INFORMATION INFRASTRUCTURE PROTECTION, OVERVIEW

Myriam Dunn Cavelty

Center for Security Studies (CSS), ETH Zurich, Switzerland

1 INTRODUCTION

For a number of years, policymakers at the highest levels have been expressing their concern that insecure information systems threaten economic growth and national security. As a result of these concerns, a complex and overlapping web of national, regional, and multilateral initiatives has emerged. Despite the sometimes substantial differences between these governmental protection policies, they offer a wealth of empirical material from which a variety of lessons can be distilled for the benefit of the international community.

2 BACKGROUND

The importance of protecting infrastructures has greatly increased in the global security political debate of late, due in particular to the traumatic terrorist attacks in New York and Washington (2001), Madrid (2004), and London (2005). In all of these cases, the perpetrators exploited elements of the civilian infrastructure for the purpose of indiscriminate murder. In the case of the 11 September 2001 attacks in the US, they used the transport infrastructure by turning airplanes into weapons. In Europe, trains, underground railways, and train stations as well as computers were targeted. This approach not only demonstrated the brutal nature of the "new terrorism", but also reinforced the view that traditional concepts of domestic security were no longer commensurate to contemporary requirements and needed to be adapted.

Reprinted with permission from Elgin M. Brunner and Manuel Suter. **International CIIP Handbook 2008/2009, Series Editors**: Andreas Wenger, Victor Mauer and Myriam Dunn Cavelty, Center for Security Studies, ETH Zurich.

Long before these attacks, the protection of strategically important installations in the domestic economic and social sphere had already been an important part of national defense concepts [1]. The term "Critical Infrastructure Protection" (CIP), however, refers to a broader concept with a distinctly new flavor. First of all, it is no longer restricted to concrete defense against immediate dangers or criminal prosecution after a crime has been committed, but increasingly refers to preventive security measures as well. Furthermore, contemporary modern societies have become significantly more vulnerable, and the spectrum of possible causes of disruptions and crises has become broader and more diffuse. This is why CIP has become a crystallization point for current security policy debates [2].

3 FROM THREATS TO RISKS

The genesis and establishment of the concept of CIP is the result of two interlinked and at times mutually reinforcing factors: The expansion of the threat spectrum after the Cold War, especially in terms of malicious actors and their capabilities on the one hand, and a new kind of vulnerability due to modern society's dependency on inherently insecure information systems on the other.

During the Cold War, threats were mainly perceived as arising from the aggressive intentions of states to achieve domination over other states. Among other things, the end of the Cold War also heralded the end of unambiguous threat perceptions: Following the disintegration of the Soviet Union, a variety of "new" threats were moved onto the security policy agendas of most countries [3]. The main distinguishing quality of these "new" challenges is the element of uncertainty that surrounds them: uncertainty concerning the identity and goals of potential adversaries, the time frame within which threats are likely to arise, the contingencies that might be imposed on the state by others, the capabilities against which one must prepare, and uncertainty about the type of challenge one had to prepare for [4]. Clearly, the notion of "threat" as something imminent, direct, and certain no longer accurately describes these challenges. Rather, they can be characterized as "risks", which are by definition indirect, unintended, uncertain, and situated in the future, since they only materialize when they occur in reality [5].

As a result of these diffuse risks and due to difficulties in locating and identifying enemies, part of the focus of security policies has shifted away from actors, capabilities, and motivations towards general vulnerabilities of entire societies. The catchphrase in this debate is "asymmetry", and the US military has been a driving force behind the shaping of this threat perception in the early 1990s [6]. The US as the only remaining superpower was seen as being predestined to become the target of asymmetric warfare. Specifically, those adversaries who were likely to fail against the American war machine might instead plan to bring the US to its knees by striking against vital points at home that are fundamental not to the military alone, but to the essential functioning of industrialized societies as a whole [7]. These points are generally defined as critical infrastructures (CI). They are deemed critical because their incapacitation or destruction would have a debilitating impact on the national security and the economic and social welfare of a nation.[1]

[1]The definition of what to include in a definition of critical infrastructure varies slightly from country to country. This Handbook shows in detail how each country defines the critical infrastructure and what sectors are included.

Fear of asymmetrical measures against such "soft targets" was aggravated by the second factor: the so-called information revolution. Most of the CI relies on a spectrum of software-based control systems for smooth, reliable, and continuous operation. In many cases, information and communication technologies (ICT) have become all-embracing, connecting other infrastructure systems and making them interrelated and interdependent. These technologies are in general regarded as inherently insecure: Security has never been a system design driver, and pressure to reduce time-to-market is intense, so that a further explosion of computer and network vulnerabilities is to be expected, leading to the emergence of infrastructures with in-built instability, critical points of failure, and extensive interdependencies [8]. At the same time, the spread of ICT was (and is) seen to make it much easier to attack asymmetrically, as big, specialized weapons systems or an army are no longer required. Borders, already porous in many ways in the real world, are nonexistent in cyberspace.

4 EVOLUTION OF THE CRITICAL (INFORMATION) INFRASTRUCTURE PROTECTION (CIIP) ISSUE

Commensurate with this threat perception, the US was the first nation to address the new vulnerability of the vital infrastructures in a broad and concerted effort. New risks in designated sectors[2] like information and communications, banking and finance, energy, physical distribution, and vital human services were identified by the Presidential Commission on Critical Infrastructure Protection (PCCIP) [9].[2] The PCCIP concluded in 1997 that the US was so dependent on these infrastructures that the government had to view them through the lens of a "national security focus", since serious consequences for the entire nation were to be expected if these elements were unavailable for any significant amount of time.

According to this approach, critical infrastructures should be understood to include material and IT assets, networks, services, and installations that, if disrupted or destroyed, would have a serious impact on the health, security, or economic well-being of citizens and the efficient functioning of a country's government. Such infrastructures could be damaged by structural threats as well as by intentional, actor-based attacks. The first risk category would, for example, include natural catastrophes, human-induced catastrophes (e.g., dam failure, nuclear reactor accident), personnel shortages through strikes or epidemics, organizational shortcomings due to technical or personal failures, human error, technical outages, and dependencies and supply shortages. In the second category, the spectrum of possible attackers is extensive, ranging from bored teenagers, disaffected or dissatisfied employees, organized crime, fanatics and terrorist cells, to hostile states.

There is an equally broad range of attack options, including hacker attacks as well as the physical destruction of civilian or military installations. The main focus of early CIP efforts was, however, directed towards the as-yet largely unknown risks emanating from cyberspace: The global information infrastructure appeared to facilitate anonymous attacks from anywhere in the world, while at the same time serving as a source for hacker tools for everyone. Based on this threat perception, a CIP policy crystallized under US President Bill Clinton that was largely directed towards information security. In many ways, other countries followed this lead with a similar focus on the information aspect.

[2]A sector is defined as "A group of industries or infrastructures which perform a similar function within a society", see: [9].

However, since the terrorist attacks of 11 September 2001, there has been a noticeable return of the classical threat concept to the CIP debate. Especially from the US point of view, efforts have been made since then to tackle a series of structural threats within the framework of an increasingly actor-oriented counter-terrorism strategy. In the US, CIP became a key component of Homeland Security and is currently discussed predominantly with a view to developing strategies against terrorism. In this context, the physical aspects of CIP have gained more attention, while the importance of information aspects has diminished slightly in comparison. In the meantime, this CIP focus on counterterrorism has also become a hallmark of recent debates in the EU, which has recently begun to develop a CIP policy that consists mainly of coordinating the measures adopted by member states. The same is true for other parts of the world.

5 DISTINCTION BETWEEN CIP AND CIIP

More than ten years after the beginning of the CIP debate, there still is little clarity with regard to a clear and stringent distinction between the two key terms "CIP" and "CIIP". In official publications, the term CIP is frequently used even if the document is only referring to the information aspects of the issue.

The reason for this is that the two cannot and should not be discussed as completely separate concepts. In our view, CIP is more than CIIP, but CIIP is an essential part of CIP. Focusing exclusively on cyber-threats while ignoring important traditional physical threats is just as dangerous as the neglect of the virtual dimension—what is needed is a sensible handling of both interrelated concepts. Nonetheless, there is at least one characteristic for distinguishing between the two: While CIP comprises all critical sectors of a nation's infrastructure, CIIP is only a subset of a comprehensive protection effort, as it focuses on measures to secure the critical *information* infrastructure. The definition of exactly what should be subsumed under CI, and what should come under the heading of CII, is another question: Generally, the CII is that part of the global or national information infrastructure that is essentially necessary for the continuity of a country's critical infrastructure services. The CII, to a large degree, consists of, but is not fully congruent with, the information and telecommunications sector, and includes components such as telecommunications, computers/software, the internet, satellites, fiber-optics, etc. The term is also used for the totality of interconnected computers and networks and their critical information flows.

Due to their role in interlinking various other infrastructures and also providing new ways in which they can be targeted, information infrastructures do play a very specific role in the debate, as we have already mentioned. They are regarded as the backbone of critical infrastructures, given that the uninterrupted exchange of data is essential to the operation of infrastructures in general and the services that they provide. Centralized SCADA (Supervisory, Control, and Data Acquisition) systems are widely employed to monitor and control infrastructures remotely. But SCADA-based systems are not secure: once-cloistered systems and networks are increasingly using off-the-shelf products and IP-based networking equipment, and require interconnection via the internet, which opens the door to attackers from the outside in addition to those on the inside.

As the information revolution is an ongoing and dynamic process that is fundamentally changing the fabric of security and society, continuing efforts to understand these changes are necessary. This requires research into information-age security issues, the identification of new vulnerabilities, and new ways for countering threats efficiently and effectively. One such effort is the International CIIP Handbook, which aims to make a contribution towards this ambitious goal. The entire publication is available on the internet (http://www.crn.ethz.ch).

The CIIP Handbook focuses on national governmental efforts to protect critical (information) infrastructure as well as those of international organizations. The overall purpose of the International CIIP Handbook is to provide an overview of CII protection practices in a broad range of countries.

6 CRITICAL SECTORS

Specific countries identify critical sectors in their territory and provide definitions of CII and CIIP. Some countries, such as Australia, Canada, Germany, the Netherlands, New Zealand, the UK, or the US, provide clear definitions of what constitutes CIP, while other countries—for example Brazil, Korea, or Russia—, offer no definition. Everywhere, CIIP is understood more or less explicitly as a subset of CIP, including protection, detection, response, and recovery activities at both the physical and the virtual levels. However, a clear distinction between CIP and CIIP is lacking in most countries, and one finds both terms being used interchangeably. As was pointed out in the introduction, this reflects the continuing difficulties that arise from having to distinguish between physical and virtual aspects of critical infrastructures.

In designating critical sectors, all countries have followed the example of the Presidential Commission on Critical Infrastructure Protection (PCCIP), which was the first official publication to correlate critical infrastructures with specific business sectors or industries [10]. The choice of the "sector" as a unit of analysis is a pragmatic approach that roughly follows the boundaries of existing business/industry sectors. This approach reflects the fact that the majority of infrastructures is owned and operated by private actors. In addition, the decision on which infrastructures and sectors to include in the list of critical assets requires input from private-sector experts, besides experts and officials at various levels of government. More often than not, expert groups address the issue, either in larger or smaller groups [11]. A component or a whole infrastructure is usually defined as "critical" due to its strategic position within the whole system of infrastructures, and especially due to the interdependency between the component or the infrastructure and other infrastructures. However, as we show below, there is also a more symbolic understanding of criticality that influences the designation of critical assets.

It is broadly acknowledged, however, that the focus on sectors is too artificial to represent adequately the realities of complex infrastructure systems. For a more meaningful analysis, it is therefore deemed necessary to evolve beyond the conventional "sector"-based focus and to look at the services, the physical and electronic (information) flows, their role and function for society, and especially the core values that are delivered by the infrastructures. Therefore, experts groups often focus on four steps in the identification of what is critical: 1) critical sectors, 2) sub-sectors for each sector on the basis of organizational criteria, 3) core functions of the sub-sectors, and 4) resources necessary for the functioning of the sub-sectors [12].

Table 1 shows which country defines which sectors as critical. One must be careful, however, to avoid misleading comparisons: While for instance Australia, Canada, the Netherlands, the UK, and the US are very precise in identifying critical sectors and

TABLE 1 Critical Sectors in Different Countries

Country Sector	AUS	A	BR	CAN	EST	F	FIN	GER	HUN	IND	IT	JAP	KOR	MAL	NL	NO	NZ	POL	RU	SE	SING	SPA	SWIT	UK	USA	TOTAL
Banking and Finance	√		√	√	√	√	√	√	√	√	√	√	√	√	√	√	√	√	√	√	√	√	√	√	√	**24**
Central Govern-ment/Government Services		√		√	√	√		√	√		√	√	√	√	√	√	√	√	√	√		√	√	√	√	**20**
Chemical and Nuclear Industry				√						√				√	√			√				√	√		√	**8**
Emergency/Rescue Services	√		√	√	√	√			√	√	√		√	√		√	√	√	√				√	√	√	**17**
Energy/Electricity	√	√		√	√	√	√	√	√	√	√	√	√		√	√	√	√			√	√	√	√	√	**21**
Food/Agriculture	√			√	√	√	√	√	√		√	√			√	√					√	√	√	√	√	**16**
Health Services	√		√	√	√	√	√		√		√			√	√	√					√	√	√	√	√	**16**
Information Services/Media	√	√				√	√		√		√		√		√	√			√	√	√		√		√	**14**
Military Defense/ Army/Defense Facilities						√			√	√			√	√		√			√					√	√	**9**
National Icons and Monuments	√																								√	**2**
Sewerage/Waste Management	√										√			√	√	√		√					√	√	√	**9**
Telecommunications	√	√	√	√	√	√	√	√	√	√	√	√	√		√	√	√	√	√	√	√	√		√	√	**22**
Transportation (land, sea, air)/Logistics/ Distribution	√	√	√	√	√	√	√	√	√	√	√	√	√	√	√	√	√	√		√	√	√	√	√	√	**24**
Water Infrastructure					√										√										√	**3**
Water (Supply)	√		√		√	√	√	√	√		√	√		√	√	√				√	√	√	√	√	√	**18**

sub-sectors as well as products and services that these sectors provide, other countries, such as Austria, Brazil, Poland, Russia, have no official list of critical sectors.

Variations between countries can be explained by differences in conceptualizations of what is critical, but also by country-specific peculiarities and traditions. Sociopolitical factors as well as geographical and historical preconditions determine whether or not a sector is deemed to be critical.

The most frequently mentioned critical sectors in all countries are listed below. These are the core sectors of modern societies, and possibly the areas where a large-scale interruption would be most devastating:

- Banking and Finance,
- Central Government/Government Services,
- (Tele-) Communication/Information and Communication Technologies (ICT),
- Emergency/Rescue Services,
- Energy/Electricity,
- Health Services,
- Food,
- Transportation/Logistics/Distribution, and
- Water (Supply).

A comparison across time shows that the concept of criticality has undergone change, and that the criteria for determining which infrastructures qualify as critical have expanded over time; the PCCIP, for example, defined eight sectors as critical for the US, while today, critical infrastructures in the US already include 18 sectors.

We can thus distinguish between two differing, but interrelated perceptions of criticality [13]:

- *Criticality as systemic concept.* This approach assumes that an infrastructure or an infrastructure component is critical due to its structural position in the whole system of infrastructures, especially when it constitutes an important link between other infrastructures or sectors, and thus reinforces interdependencies.
- *Criticality as a symbolic concept.* This approach assumes that an infrastructure or an infrastructure component is inherently critical because of its role or function in society; the issue of interdependencies is secondary—the inherent symbolic meaning of certain infrastructures is enough to make them interesting targets.[3]

The symbolic understanding of criticality allows the integration of non-interdependent infrastructures as well as objects that are not man-made into the concept of critical infrastructures, including significant personalities or natural and historical sites with a strong symbolic character. Additionally, the symbolic approach allows essential assets to be defined more easily than the systemic one, because in a sociopolitical context, the defining element is not interdependency as such, but the role, relevance, and symbolic value of specific infrastructures [15].

The emphasis on the interconnectedness of various sectors, in connection with this symbolic understanding, creates a specific set of problems for decision-makers: Basically,

[3]For an example (critical assessment without interdependencies), see: [14].

everything is networked, and even a discrete event of little apparent significance could potentially set off unpredictable cascading effects throughout a large number of sectors. When the concept of criticality, and accordingly the scope of what is to be secured, is expanded from interconnected physical networks like the electrical grid and road networks to include everything with emotional significance, ranging from schools to national monuments, almost everything becomes potentially critical. In this situation, decision-makers must be careful not to follow the natural impulse to increase security ad absurdum and aim to protect everything that could possibly be at risk, because total protection will never be possible, and the effects on society will likely be negative. Prioritization must be based on careful risk assessment that comprises calculations of the likelihood of a given threat source displaying a particular potential vulnerability, and the resulting impact of that adverse event [16].

At the same time, one must be aware of the fact that current methodologies for analyzing CII—and first and foremost among them risk analysis—are insufficient in a number of ways: One of the major shortcomings is that the majority of them do not pass the "interdependency test". In other words, they fail to address, let alone understand, the issue of interdependencies and possible cascading effects. Besides, the available methods are either too sector-specific or too focused on single infrastructures and do not take into account the strategic, security-related, and economic importance of CII.[4] Moreover, there are both theoretical and practical difficulties involved in estimating the probabilities and consequences of high-impact, low-probability events—and this is the scenario we are dealing with in the context of CI(I)P. It also appears that there is no way of apolitically cataloguing objects, vulnerabilities, and threats on a strategic policy level, such as the economy at large, in a meaningful way.

Clearly, therefore, long-term research into CIP and CIIP matters is needed. A holistic and strategic threat and risk assessment at the physical, virtual, and psychological levels is the basis for a comprehensive protection and survival strategy, and will thus require a comprehensive and truly interdisciplinary research and development agenda that encompasses fields ranging from engineering and complexity sciences to policy research, political science, and sociology.

7 PAST AND PRESENT INITIATIVES AND POLICY

Many of the national CIIP efforts, including, specific committees, commissions, task forces, and working groups as well as key official reports and fundamental studies, and important national programs, were triggered or at least accelerated by the *Presidential Commission on Critical Infrastructure Protection* (PCCIP) set up by US President Bill Clinton in 1996, and to some extent by the preparations for anticipated problems on the threshold of the year 2000 (Y2K problem). This led to the establishment of (interdepartmental) committees, task forces, and working groups. Their mandate often included scenario work, the evaluation of a variety of measures, or assessments of early-warning systems. These efforts resulted in policy statements—such as recommendations for the establishment of independent organizations dealing with information society issues—and reports laying down basic CIIP policies.

[4]This issue is addressed in additional detail in by Myriam Dunn, who argues that shortcomings include the lack of data to support objective probability estimates, persistent value questions, and conflicting interests within complex decision-making processes [17].

In the aftermath of 11 September 2001 ("9/11"), several countries launched further initiatives to strengthen and allocate additional resources to their CIP/CIIP efforts. Prior to 9/11, for many people, critical infrastructure protection was synonymous with cyber-security. The attacks of 9/11, however, highlighted the fact that terrorists could cause enormous damage by attacking critical infrastructures directly and physically, and thus demonstrated the need to re-examine physical protection, especially in the US [18]. The perception that the cyber-dimension had been unduly prioritized before 9/11 subsequently led to a shift in focus from the virtual to the physical domain, and from CIIP to CIP. Subsequently, CIP became a key component of *Homeland Security* and is currently discussed predominantly with a view to developing strategies against Muslim terrorism. The physical aspects of CIP have been moved to the forefront, while the importance of information aspects has diminished. This CIP focus on counterterrorism has also become a hallmark of debates in the EU, which has recently begun to develop a CIP policy that consists mainly of coordinating the measures adopted by member states.

CIIP policies are at various stages of implementation—some are already being enforced, while others are just a set of suggestions—and come in various shapes and forms, ranging from a regulatory policy focus concerned with the smooth and routine operation of infrastructures and questions such as privacy or standards, to the inclusion of cyber-security into more general counter-terrorism efforts. Most countries consider CIIP to be a national-security issue of some sort. In parallel, however, they often pursue a business continuity strategy under the "information society" label. The law enforcement/crime prevention perspective is also found in all countries. Furthermore, data protection issues are a major topic for civil rights groups. While all of the perspectives can be found in all countries, the emphasis given to one or more of the perspectives varies to a considerable degree.[5]

All countries examined have recognized the importance of public-private partnerships (PPP). Governments actively promote information-sharing with the private sector, since large parts of critical infrastructures are owned and operated by the business sector. Different types of such partnerships are emerging, including government-led partnerships, business-led partnerships, and joint public-private initiatives. In Switzerland, Korea, the UK, and the US, strong links have already been established between the private business community and various government organizations. One of the future challenges in many countries will be to achieve a balance between security requirements and business efficiency imperatives. Satisfying shareholders by maximizing company profits has often led to minimal security measures. This is because like many political leaders, business leaders tend to view cyber-attacks on infrastructures as a tolerable risk.

Despite the general consensus on the positive aspects of PPPs, their implementation remains difficult. It has been shown that it is relatively easy for the government and private actors in a PPP to agree on the existence of a problem and on the need for a remedy. It is, however, much harder to agree on actual measures to be taken, on the actors responsible for implementing them, on the party that will assume legal responsibility for such measures, and on the party that will bear the costs for implementing them [20].

[5]This issue is also addressed by Isabelle Abele-Wigert, who shows how practical and academic dialog is hampered by vastly differing terminology and viewpoints of what constitutes the problem [19].

8 ORGANIZATIONAL OVERVIEW

Only in a few countries central governmental organizations have been created to deal specifically with CIIP. For example, the US, France, Switzerland, Singapore, and Korea have all made provisions in this regard. Mostly, responsibility lies with multiple authorities and organizations in different governmental departments. Very often, responsibility for CIIP protection is given to well-established organizations or agencies that appear suitable for the task. Depending on their key assignment, these agencies bring their own perspective to bear on the problem and shape policy accordingly.

In countries such as France and New Zealand, CIIP efforts are mainly led by the defense establishment, whereas in other countries, such as the UK or Switzerland, approaches to CIIP are jointly led by the business community and public agencies. Furthermore, in Australia as well as in the US and New Zealand, CIIP is integrated into the overall counterterrorism efforts, where the intelligence community plays an important role. In India, Korea, Japan, Singapore, and Estonia, the fostering of the information society and economic growth through safe information infrastructures is at the forefront.

The establishment of these organizational units and their location within the government structures are influenced by various factors such as civil defense tradition, the allocation of resources, historical experience, and the general threat perception of key actors in the policy domain. Depending on their influence or on the resources at hand, various key players shape the issue in accordance with their view of the problem. Different groups, whether they be private, public, or a mixture of both, do not usually agree on the exact nature of the problem or on what assets need to be protected with which measures. There are at least four (overlapping) typologies for how CIIP issues are viewed: an IT-security perspective, an economic perspective, a law enforcement perspective, and a national-security perspective. While all typologies can be found in all countries, the emphasis given to one or more of them varies to a considerable degree. Ultimately, the dominance of one or several typologies has implications for the shape of the protection policies and, subsequently, for determining appropriate protection efforts, goals, strategies, and instruments for solving problems.

In the end, the distribution of resources and the technical and social means for countering the risk are important for the outcome. We can observe that the different actors involved—ranging from government agencies and the technology community to insurance companies—have divergent interests and compete with one another by means of scenarios describing how they believe the threat will manifest itself in the future [21]. Furthermore, the selection of policies seems to depend largely upon two factors: One is the varying degree to which resources are available to the different groups. The other factor is the impact of cultural and legal norms, because they restrict the number of potential strategies available for selection [22]. In general, we can identify two influential discourses: On the one hand, law enforcement agencies emphasize their view of the risk as "computer crime", while on the other hand, the private sector running the infrastructures perceives the risk mainly as a local, technical problem or in terms of economic costs [23]. Because the technology generating the risk makes it very difficult to fight potential attackers in advance, protective measures focus on preventive strategies and on trying to minimize the impact of an attack when it occurs. Here, the infrastructure providers are in a strong position, because they alone are in the position to install technical safeguards for IT security at the level of individual infrastructures.

Norms are also important in selecting the strategies. Most importantly, the general aversion of the new economy to government regulation as well as legal restrictions limits the choice of strategies [24]. Besides these cultural differences with regard to strategy, the nature of cyber-attacks naturally positions law enforcement at the forefront: It is often impossible to determine at the outset whether an intrusion is an act of vandalism, computer crime, terrorism, foreign intelligence activity, or some form of strategic attack. The only way to determine the source, nature, and scope of the incident is to investigate. The authority to investigate such matters and to obtain the necessary court orders or subpoenas clearly resides with law enforcement. As a consequence of the nature of cyber-threats, the cyber-crime/law enforcement paradigm is emerging as the strongest viewpoint in most countries.

9 EARLY WARNING AND PUBLIC OUTREACH

The earlier a potential risk is identified, the greater the chance to act in a timely, resource-efficient, and strategically adequate manner. Therefore, timely warning of attacks is an indispensable component of ensuring that a breakdown of important infrastructure, or even only of certain components of ICT, will be limited to an incident that is short, rare, controllable, geographically isolated, or with as little consequences as possible for the national economy and security.

Early-warning systems are designed for the following purposes: understanding and mapping the hazard; monitoring and forecasting impending events; processing and disseminating understandable warnings to political authorities and the population; and undertaking appropriate and timely actions in response to the warnings. In CIIP, early warning is focused mainly on IT security incidents. The general trend in CIIP early warning points towards establishing central contact points for the security of information systems and networks. Among the existing early-warning organizations are various forms of *Computer Emergency Response Teams* (CERTs), e.g., special CERTs for government departments, CERTs for small and medium-sized businesses, CERTs for specific sectors, and others. CERT functions include handling of computer security incidents and vulnerabilities or reducing the probability of successful attacks by publishing security alerts.[6] Internationally, CERTs primarily exchange information at the Forum of Incident Response and Security Teams (FIRST).

In some countries, permanent analysis and intelligence centers have been developed in order to make tactical or strategic information available to the decision-makers within the public and private sectors more efficiently. Tasks of early-warning system structures include analysis and monitoring of the situation as well as the assessment of technological developments. Examples can be found in Canada (Integrated Threat Assessment Center) [26]. and in Switzerland (Reporting and Analysis Center for Information Assurance, MELANI) [27].

Often, these entities manage outreach, cyber-security awareness, and partnership efforts to disseminate information to key constituencies and build collaborative actions with key stakeholders. Generally, many private enterprises, public entities, and home users lack the resources to manage cyber-security risks adequately. Many entrepreneurs and home users are unaware of the extent to which their individual cyber-security preparedness affects overall security, and internet users must be made aware of the importance of sound cyber-security practices and require more user-friendly tools to

[6]The issue is further addressed by Thomas Holderegger, who examines early-warning players in the CIIP sector and specifies their tasks and responsibilities, with a specific focus on the role of the nation state [25].

implement them. Public outreach efforts therefore entail cataloguing existing best practices, developing strategies to market those practices to specific audiences, creating incentive plans to ensure acceptance of those practices, contributing to the development of a national advertising campaign, and developing a strategy to communicate the importance of cyber-security and their role in enhancing it to public and private CEOs across the country.

10 LEGAL ISSUES

Although many countries have been concerned with the protection and security of information (infrastructures) and related legislation for some years, they have begun to review and adapt their cyber-security legislation after 9/11. Because national laws are developed autonomously, some countries have preferred to amend their penal or criminal code, whereas others have passed specific laws on cyber-crime.

The following is an overview of important common issues currently discussed in the context of legislation procedures:

- Data protection and security in electronic communications (including data transmission, safe data storage, etc.);
- IT security and information security requirements;
- Fraudulent use of computer and computer systems, damage to or forgery of data, and similar offences;
- Protection of personal data and privacy;
- Identification and digital signatures;
- Responsibilities in e-commerce and e-business;
- International harmonization of cyber-crime law;
- Minimum standards of information security for (e-)governments, service providers, and operators, including the implementation of security standards such as BS7799, the code of practice for information security management ISO/IEC 17799, the Common Criteria for Information Technology Security Evaluation ISO/IEC 15408, and others;
- Public key infrastructure and its regulation.

Across all boundaries, there are two main factors that influence and sometimes even hinder efficient law enforcement—one with a national, the other with an international dimension:

- *Lack of know-how or of functioning legal institutions.* Even if a country has strict laws and prohibits many practices, the enforcement of such laws is often difficult. Frequently, the necessary means to prosecute misdemeanors effectively are lacking due to resource problems, inexistent or emerging cyber-crime units, or a lack of supportive legislation, such as the storing of rendition data [28].
- *Lack or disparity of legal codes.* While most crimes, such as theft, burglary, and the like are punishable offenses in almost every country of the world, some rather grave disparities still remain in the area of cyber-crime [29].

11 INTERNATIONAL ISSUES

From the discussion of legal issues, it becomes obvious that like other security issues, the vulnerability of modern societies—caused by dependency on a spectrum of highly interdependent information systems—has global origins and implications. To begin with, the information infrastructure transcends territorial boundaries, so that information assets that are vital to the national security and the essential functioning of the economy of one state may reside outside of its sphere of influence on the territory of other nation-states. Additionally, "cyberspace"—a huge, tangled, diverse, and universal blanket of electronic interchange—is present wherever there are telephone wires, cables, computers, or electromagnetic waves, a fact that severely curtails the ability of individual states to regulate or control it alone. Any adequate protection policy that extends to strategically important information infrastructures will thus ultimately require transnational solutions.

There are four possible categories of initiatives that may be launched by multilateral actors: deterrence, prevention, detection, and reaction.

- *Deterrence.* or the focus on the use of multilateral cyber-crime legislation: Multilateral initiatives to deter the malicious use of cyberspace include initiatives a) to harmonize cyber-crime legislation and to promote tougher criminal penalties (e.g. the Council of Europe Convention on Cybercrime) [30], and b) to improve e-commerce legislation (e.g., the efforts of the United Nations Commission on International Trade Law (UNCITRAL) for electronic commerce) [31].
- *Prevention.* or the design and use of more secure systems and better security management, and the promotion of more security mechanisms: Multilateral initiatives to prevent the malicious use of cyberspace center around a) promoting the design and use of more secure information systems (e.g., the Common Criteria Project) [32]; b) improving information security management in both public and private sectors (e.g., the ISO and OECD standards and guidelines initiatives) [33]; c) legal and technological initiatives, such as the promotion of security mechanisms (e.g., electronic signature legislation in Europe).
- *Detection.* or cooperative policing mechanisms and early warning of attacks: Multilateral initiatives to detect the malicious use of cyberspace include a) the creation of enhanced cooperative policing mechanisms (e.g., the G-8 national points of contact for cyber-crime); and b) early warning through information exchange with the aim of providing early warning of cyber-attacks by exchanging information between the public and private sectors (e.g., US Information Sharing & Analysis Centers, the European Early Warning & Information System, and the European Network and Information Security Agency (ENISA)).
- *Reaction.* or the design of stronger information infrastructures, crisis management programs, and policing and justice efforts: Multilateral initiatives to react to the malicious use of cyberspace include a) efforts to design robust and survivable information infrastructures; b) the development of crisis management systems; and c) improvement in the coordination of policing and criminal justice efforts.

The most important legislative instrument in this area is the Council of Europe Cybercrime Convention (CoC). This convention is the first international treaty on crimes committed via the internet and other computer networks. Its main objective is to pursue a common law enforcement policy aimed at the protection of society against cyber-crime, especially by adopting appropriate legislation and fostering international cooperation [34]. An additional protocol to the CoC outlaws racist and xenophobic acts committed through computer systems.

While other politically powerful entities such as the G8 also try to foster collaboration and a more efficient exchange of information when it comes to cyber-crime and terrorism, the CoC goes one step further. It lays out a framework for future collaboration between the prosecution services of the signature states. It achieves this mainly by harmonizing the penal codes of the CoC signatory states. As a result, crimes such as hacking, data theft, and distribution of pedophile and xenophobic material, etc., will be regarded as illegal actions per se, thus resolving the problem of legal disparities between nations that was mentioned above. This also allows the authorities to speed up the process of international prosecution. Since certain activities are defined as illegal by all CoC member states, the sometimes long and painful task of cross-checking supposed criminal charges committed in a foreign country becomes obsolete if the offence is already included in the national penal code. Consequently, reaction times will be shortened and the parties to the CoC will establish a round-the-clock network within their countries to handle aid requests that demand swift intervention [35]. While the implementation of the CoC will most likely be a slow and sometimes thorny process, the idea of finding a common denominator and harmonizing the response to at least some of the most crucial problems is certainly a step in the right direction.

REFERENCES

1. Luiijf, E. A. M., Burger H. H., and Klaver M. H. A. (2008). Critical infrastructure protection in The Netherlands: a quick-scan. In *EICAR Conference Best Paper Proceedings 2003*, U. E. Gattiker, P. Pedersen, and K. Petersen, Eds. http://cipp.gmu.edu/archive/2_Netherlands CIdefpaper_2003.pdf [last accessed in June 2008].
2. Dunn, C., and Kristensen, S. K. (2008). *Securing the Homeland: Critical Infrastructure, Risk, and (In)Security*, Routledge, London.
3. Buzan, B., Wæver, O., and de Wilde, J. (1998). *Security: A New Framework for Analysis*, Lynne Rienner, Boulder.
4. Goldman, E. O. (2001). New threats, new identities and new ways of war: the sources of change in national security doctrine. *J. Strateg. Stud.* **24**, 12–42.
5. (a) Bailes, A. J. K. (2007). Introduction: a world of risk. In *SIPRI Yearbook 2007: Armaments, Disarmament and International Security*, pp. 1–20; (b) Beck, U. (1999). *World Risk Society*, Polity Press, Cambridge.
6. Rattray, G. (2001). *Strategic Warfare in Cyberspace*, MIT Press, Cambridge.
7. Berkowitz, B. D. (1997). Warfare in the information age. In *In Athena's Camp: Preparing for Conflict in the Information Age*, A. John, and R. David, Eds. RAND, Santa Monica, pp. 175–190.
8. Rathmell, A. (2001). Controlling computer network operations. *Infor. Secur. Int. J.* **7**, 121–144.
9. President's Commission on Critical Infrastructure Protection (PCCIP) (2008). *Critical Foundations: Protecting America's Infrastructures*, Washington, October 1997: Appendix B, Glossary, B-3. http://www.ihs.gov/misc/links_gateway/download.cfm?doc_id=327&app_dir_id=4&doc_file=PCCIP_Report.pdf [last accessed in June 2008]. Publication quoted in the following as PCCIP.
10. President's Commission on Critical Infrastructure Protection (PCCIP) (2008). *Critical Foundations: Protecting America's Infrastructures*, Washington, October 1997. http://www.ihs.gov/misc/links_gateway/download.cfm?doc_id=327&app_dir_id=4&doc_file=PCCIP_Report.pdf [last accessed in June 2008]. Publication quoted in the following as PCCIP.

11. Dunn, M. (2004). Part II: analysis of methods and models for CII assessment. In *The International Critical Information Infrastructure Protection (CIIP) Handbook 2004*, M. Dunn, and W. Isabelle, Eds. Center for Security Studies, Zurich, pp. 219–297.
12. Dunn, M. (2004). Part II: analysis of methods and models for CII assessment. In *The International Critical Information Infrastructure Protection (CIIP) Handbook 2004*, M. Dunn, and W. Isabelle, Eds. Center for Security Studies, Zurich, p. 227f.
13. Metzger, J. (2004). The concept of critical infrastructure protection (CIP). In *Business and Security: Public-Private Sector Relationships in a New Security Environment*, A. J. K. Bailes, and F. Isabelle, Eds. Oxford University Press, Oxford, pp. 197–209.
14. United States General Accounting Office (GAO) (2008). Testimony before the Subcommittee on National Security, Veterans Affairs, and International Relations; House Committee on Government Reform. *Homeland Security: Key Elements of a Risk Management*, Statement of Raymond J. Decker, Director Defense Capabilities and Management. 12 October 2001, p. 6. http://www.gao.gov/new.items/d02150t.pdf [last accessed in June 2008].
15. Metzger, J., op. cit.
16. Stoneburner, G., Goguen, A., and Feringa, A. (2002). Risk management guide for information technology systems. In *Recommendations of the National Institute of Standards and Technology*. NIST Special Publication 800-30. US Government Printing Office, Washington, p. 8. http://csrc.nist.gov/publications/nistpubs/800-30/sp800-30.pdf [last accessed in June 2008].
17. Dunn, M. (2006). Understanding critical information infrastructures: an elusive quest. In *International CIIP Handbook 2006. Analyzing Issues, Challenges, and Prospects*, M. Dunn, and V. Mauer, Eds. Center for Security Studies, Zurich, Vol. II, pp. 27–53.
18. Moteff, J. (2005). Risk Management and Critical Infrastructure Protection: Assessing, Integrating, and Managing Threats, Vulnerabilities and Consequences. CRS Report for Congress. February 4, p. 3.
19. Wigert, I. (2006). Challenges governments face in the field of critical information infrastructure protection (CIIP): stakeholders and perspectives. In *International CIIP Handbook 2006, Analyzing Issues, Challenges, and Prospects*, M. Dunn, and V. Mauer, Eds. Center for Security Studies, Zurich, Vol. II, pp. 55–68.
20. Andersson, J. J., and Malm, A. (2006). Public-private partnerships and the challenge of critical infrastructure protection. In *International CIIP Handbook 2006, Analyzing Issues, Challenges, and Prospects*, M. Dunn, and V. Mauer, Eds. Center for Security Studes, Zurich, Vol. II, pp. 139–167.
21. (a) Bendrath, R. (2003). The American cyber-angst and the real world—any link? In *Bombs and Bandwidth: The Emerging Relationship Between IT and Security*, R. Latham, Ed. The New Press, New York, pp. 49–73; (b) Bendrath, R. (2001). The cyberwar debate: perception and politics in US critical infrastructure protection. In *The Internet and the Changing Face of International Relations and Security*, Information & Security: An International Journal. A. Wenger, Ed. Vol. 7, pp. 80–103.
22. Dunn, M. (2004). Cyber-threats and countermeasures: towards an analytical framework for explaining threat politics in the information age. *Conference paper, SGIR Fifth Pan-European IR Conference*. The Hague, 10 September 2004.
23. Bendrath, R. The cyberwar debate, op. cit., p. 97.
24. Bendrath, R. The cyberwar debate, op. cit., p. 98.
25. Holderegger, T. (2006). The aspect of early warning in critical information infrastructure protection (CIIP). In *International CIIP Handbook 2006. Vol. II. Analyzing Issues, Challenges, and Prospects*, M. Dunn, and V. Mauer, Eds. Center for Security Studies, Zurich, pp. 111–135.
26. http://www.itac-ciem.gc.ca/index-eng.asp, 2008.
27. http://www.melani.admin.ch/, 2008.

28. Goodman, S. E., Hassebroek, P. B., King, D., and Azment, A. (2002). International coordination to increase the security of critical network infrastructures. Document CNI/04. *Paper presented at the ITU Workshop on Creating Trust in Critical Network Infrastructures*. Seoul, 20–22 May 2002.

29. Gelbstein, E., and Kamal, A. (2002). *Information Insecurity. A Survival Guide to the Uncharted Territories of Cyber-Threats and Cyber-Security*, United Nations ICT Task Force and United Nations Institute for Training and Research, New York, November 2002. http://www.un.int/unitar/patit/dev/old%20site/curriculum/Information_Insecurity_Second_Edition_PDF.pdf.

30. Council of Europe (2008). *Convention on Cybercrime*, http://conventions.coe.int/Treaty/EN/Treaties/Html/185.htm [last accessed in June 2008].

31. http://www.uncitral.org/english/workinggroups/wg_ec/index.htm, 2008.

32. http://www.commoncriteriaportal.org [last accessed in June 2008].

33. (a) The International Organization for Standardization ISO (2000). *Developed a code of practice for information security management (ISO/IEC 17799:2000)*. http://www.iso.org/iso/en/prods-services/popstds/informationsecurity.html; (b) The Organisation for Economic Co-operation and Development (OECD) (2008). *Promotes a "Culture of Security" for Information Systems and Networks*, http://www.oecd.org/document/42/0,2340,en_2649_33703_15582250_1_1_1_1,00.html [last accessed in June 2008].

34. *Convention on Cybercrime*, op. cit.

35. Taylor, G. (2008). *The Council of Europe Cybercrime Convention. A Civil Liberties Perspective*, http://www.crime-research.org/library/CoE_Cybercrime.html [last accessed in June 2008].

AUSTRALIA

Manuel Suter and Elgin Brunner
Center for Security Studies (CSS), ETH Zurich, Switzerland

1 CRITICAL SECTORS

Australia takes an all-hazards approach to the protection of critical infrastructures, whether information-based or not. The definition of critical infrastructure (CI) accepted

Reprinted with permission from Elgin M. Brunner and Manuel Suter. **International CIIP Handbook 2008/2009, Series Editors**: Andreas Wenger, Victor Mauer and Myriam Dunn Cavelty, Center for Security Studies, ETH Zurich.

by Australia is "those physical facilities, supply chains, information technologies and communication networks that, if destroyed, degraded or rendered unavailable for an extended period, would significantly impact on the social or economic well-being of the nation or affect Australia's ability to conduct national defense and ensure national security" [1]. The national information infrastructure (NII) is a subset of the critical infrastructure. As in many countries, the majority of the elements of the critical infrastructure are owned or operated as commercial enterprises.

In Australia, the CIP program is led by the Attorney-General's Department (AGD), primarily through the Trusted Information Sharing Network for Critical Infrastructure Protection (TISN). The TISN brings together the nine sectors considered to be critical to Australia. These are [2]:

- Communications (Telecommunications (Phone, Fax, Internet, Cable, Satellites) and Electronic Mass Communications),
- Energy (Gas, Petroleum Fuels, Refineries, Pipelines, Electricity Generation and Transmission),
- Banking and Finance (Banking, Finance, and Trading Exchanges),
- Food Supply (Bulk Production, Storage, and Distribution),
- Emergency Services,
- Health (Hospitals, Public Health, and Research and Development Laboratories),
- Mass Gatherings (Icons (e.g., Sydney Opera House) and places of mass gatherings)
- Transport (Air Traffic Control, Road, Sea, Rail, and Inter-modal (Cargo Distribution Centers)),
- Utilities (Water, Waste Water, and Waste Management).

2 PAST AND PRESENT INITIATIVES AND POLICIES

National Counter-Terrorism Plan (2003, revised 2005)
E-Security National Policy Statement (2007)

2.1 Guiding Principles of Australia's CIP Policy

Critical Infrastructure Protection (CIP) requires the active participation of the owners and operators of infrastructure, regulators, professional bodies, and industry associations, in cooperation with all levels of government, and the public. To ensure this cooperation and coordination, all of these participants should commit to the following set of common, fundamental principles of CIP [3]. These principles are to be read as a whole, as each sets the context for the one following.

- CIP is centered on the need to minimize risks to public health, safety, and confidence, to ensure Australia's economic security and maintain the country's international competitiveness, and to ensure the continuity of government and its services;
- The objectives of CIP are to identify critical infrastructure, analyze vulnerability and interdependence, and protect Australia from, and prepare for, all hazards;

- Because all critical infrastructures cannot be protected from all threats, appropriate risk management techniques should be used to determine their relative severity and duration, the level of protective security, and to set priorities for the allocation of resources and the application of the best mitigation strategies for business continuity;
- The responsibility for managing risk within physical facilities, supply chains, information technologies, and communication networks primarily rests with the owners and operators;
- CIP needs to be undertaken with an all-hazards approach, with full consideration of interdependencies between businesses, sectors, jurisdictions, and government agencies;
- CIP requires a consistent cooperative partnership between the owners and operators of critical infrastructure and governments;
- The sharing of information relating to threats and vulnerabilities will assist governments, and owners and operators of critical infrastructure, in managing risk better;
- Care should be taken, when referring to national security threats to critical infrastructure, including terrorism, to avoid causing undue concern in the Australian domestic community and to potential tourists and investors overseas;
- Stronger research and analysis capabilities can ensure that risk mitigation strategies are tailored to meet Australia's unique critical infrastructure circumstances.

2.2 CIP and Counter-Terrorism Policy

The National Counter-Terrorism Committee (NCTC) has primary responsibility for the oversight of the protection of critical infrastructures from terrorism. In general, however, CIP is a shared responsibility of the corporate sector and the Australian federal, state, and territory governments. In the field of CIIP, the Attorney-General's Department coordinates arrangements [4].

The Australian government takes actions in the following fields:

- Identifying Australia's critical infrastructure and determining broad areas of risk;
- Assisting businesses in mitigating their risk through business-government partnerships, e.g., the Trusted Information Sharing Network (TISN) and Infrastructure Assurance Advisory Groups (IAAGs), and through state and territory governments;
- Promoting domestic and international best practices in CIP.

2.3 e-Security

Resulting from a review of the e-Security environment, the former government released an e-Security national policy statement in 2007. The former government followed this up with funding of AUS$74 million over four years for e-Security initiatives [5]. The catalyst for this action was the increasing interconnectedness of the electronic environment and the need to address e-Security threats to different segments on the Australian economy holistically [6]. In consequence, the agenda appoints a new interdepartmental committee with responsibility across the entire range of government, the E-Security Policy and Coordination (ESPaC), to coordinate e-Security policy throughout the different areas.

In order to assign the roles and responsibilities of relevant Australian government agencies clearly, three priorities are defined by the agenda:

- Reducing the e-Security risk to Australian government information and communication systems;
- Reducing the e-Security risk to Australia's national critical infrastructure;
- Enhancing the protection of home users and SMEs from electronic attacks and frauds.

These new priorities and the focus on the interconnectivity of the different areas have had a considerable impact on the administrative arrangements in the field of e-Security (see chapter Organizational Overview). A second departure from the 2001 agenda is the emphasis on initiatives to address sophisticated and targeted attacks that are difficult to detect and fight by conventional measures. Thirdly, it has been decided to review the new agenda every two years instead of every four years, given the rapid evolvement of new e-Security threats [7].

One of the major initiatives was a significant expansion of the Australian Government Computer Emergency Readiness Team [8] (GovCERT.au) within the AGD.

3 ORGANIZATIONAL OVERVIEW

In Australia, the CIP program is led by the Attorney-General's Department (AGD), in close collaboration with the owners and operators of critical infrastructures. CIP efforts are primarily coordinated through the Trusted Information Sharing Network for critical infrastructure protection (TISN) [3], which provides the framework for public-private collaboration in the field of CIP and CIIP (see the chapter on Organizational Overview).

The AGD collaborates also closely with other public agencies. In 2007, as a result of the budgetary announcement by the former government and the revised E-Security National Agenda, the role and responsibilities of several public agencies increased. The most important administrative change concerned the establishment of a whole of government E-Security Policy and Coordination Committee (ESPaC) in line with the push towards a holistic approach to addressing the security of the electronic environment. While the newly created ESPaC is a standing interdepartmental committee with responsibility for e-Security policy, all agencies involved in CIIP collaborate closely. For instance, the Defence Signals Directorate (DSD), the Australian Security Intelligence Organisation (ASIO), and the Australian Federal Police (AFP) are engaged in formal Joint Operating Arrangements supporting threat and vulnerability assessment and the analysis of, and the response to, critical incidents affecting the integrity of Australia's information infrastructure.

3.1 Public Agencies

3.1.1 Attorney-General's Department (AGD). Attorney-General's Department (AGD), provides expert support to the Government in the maintenance and improvement of Australia's system of law and justice and its national security and emergency management systems. The mission of the Attorney-General's Department is achieving a just and secure society [9].

Within the department, the Security and Critical Infrastructure Division (SCID) is responsible for the administration and development of legislation and the provision of legal and policy advice with respect to counter-terrorism, national security, telecommunications interception and critical infrastructure protection. The Division coordinates Australian Government activities in critical infrastructure protection, building on the work to protect Australia's National Information Infrastructure that began in 1999, and provides policy and legal policy advice on these issues. The division performs a leadership role in the development of a business-government partnership for critical infrastructure protection with Australian industry [10].

3.1.2 E-Security Policy and Coordination (ESPaC) Committee. The E-Security Policy and Coordination (ESPaC) Committee was established in 2007 and replaced two former committees—the Electronic Security Coordination Group (ESCG), run by the Department of Broadband, Communications and the Digital Economy, and the Information Infrastructure Protection Group, run by the AGD. The incorporation of these agencies into the new ESPaC committee "ensures effective e-security coordination across the three areas of critical infrastructure, home and SMEs, and government" [3].

The tasks of the ESPaC Committee correspond to those of its predecessors (the Electronic Security Coordination Group and the Information Infrastructure Protection Group): awareness raising, promoting e-Security skills, advancing research and development, and coordinating the government policies related to e-Security.

The ESPaC Committee is chaired by the AGD and is comprised of representatives from the following government agencies: the Australian Communications and Media Authority; the Australian Government Information Management Office; the Australian Federal Police; the Australian Security Intelligence Organization; the Department of Broadband, Communications and the Digital Economy; the Defence Signal Directorate; the Department of Defence; the Department of the Prime Minister and Cabinet; and the Office of National Assessments.

The Information Infrastructure Protection Group (IIPG) was an interdepartmental committee of the Australian government responsible for providing policy coordination and/or technical response in relation to threats to the National Information Infrastructure (NII). It was replaced by the ESPaC [11].

3.1.3 Department of Broadband, Communications and the Digital Economy (DBCDE). The Department of Broadband, Communications and the Digital Economy (DBCDE), formerly the Department of Communications, Information Technology and the Arts, (DCITA) participates in the Australian government's CIP activities through the Trusted Information Sharing Network (TISN). It chairs and provides secretariat support to the IT Security Expert Advisory Group (ITSEAG). The ITSEAG provides advice to the TISN on current and emerging security issues affecting owners and operators of critical infrastructure, including:

- Voice over Internet Protocol (VoIP) enterprise systems,
- Supervisory Control and Data Acquisition (SCADA) systems,
- Wireless services.

DBCDE also provides the secretariat for the Communications Sector Infrastructure Assurance Advisory Group (CSIAAG) of the TISN, which has developed an all-hazards risk management framework for the national critical communications infrastructure [12].

3.1.4 Australian Government Computer Emergency Readiness Team (GovCERT.au). GovCERT.au was established in 2005 within the Attorney-General's Department to enhance Australia's preparedness with regard to attacks on information security. GovCERT.au is responsible for:

- Liaising with the Computer Emergency Response Teams of foreign governments;
- Coordinating inquiries from foreign governments about cyber-security issues that affect Australia's critical infrastructure and business sector;
- Coordinating the Australian government's policy on how to prepare for, respond to, and recover from computer emergencies affecting the national information infrastructure;
- Managing the Australian government's Computer Network Vulnerability Assessment Program [13], which provides cash grants to critical infrastructure owners and operators to undertake security assessments of their IT systems and networks, including physical and personnel security aspects relating to those networks.

GovCERT.au is the Australian government's point of contact for foreign governments on Computer Emergency Response issues affecting the national information infrastructure.

GovCERT.au receives information about IT security issues from foreign governments that needs to be passed on to Australian critical infrastructure owners and operators. GovCERT.au does not handle day-to-day computer incidents [14].

3.1.5 Australian Government Information Management Office (AGIMO). The Australian Government Information Management Office (AGIMO), part of the Department of Finance and Administration, provides strategic advice, activities, and representation relating to the application of ICT to government administration, information, and services.

AGIMO's functions and responsibilities include:

- Promoting improved government services through technical interoperability and the integration of business processes across Australian government services and with state/territory and local authorities;
- Developing and enhancing government e-procurement processes;
- Promoting comprehensive telecommunications arrangements for the entire government;
- Identifying and promoting the development of the ICT infrastructure necessary to implement emerging strategies for the entire government;
- Developing an e-Government Authentication Framework to assist people in verifying electronic communications.

In cooperation with other government bodies, AGIMO manages international contacts and represents Australia in world forums on ICT-related issues. AGIMO also manages the .gov.au domain in consultation with state and territory governments [15].

3.1.6 Defence Signals Directorate (DSD). The Defence Signals Directorate (DSD) is Australia's national authority on information security and signals intelligence. DSD plays an integral role in the protection of Australia's official communications and information systems. It does so by providing expert assistance to Australian agencies in relation to cryptography, network security, and the development of guidelines and policies on information security.

The activities of the DSD's Information Security Group (INFOSEC) include information and incident collection, analysis and warning services, setting awareness and certification standards, and defensive measures, including protective security measures, response arrangements, and contingency planning. In addition to its support for Australian government departments and authorities, INFOSEC also plays an important role working with industry towards the development of new cryptographic products [16].

3.1.7 Australian Security Intelligence Organisation (ASIO). The Australian Security Intelligence Organisation (ASIO) is Australia's national security service. Its functions are set out in the Australian Security Intelligence Organisation Act 1979 (the ASIO Act). ASIO's main role is to gather information and produce intelligence that will enable it to warn the government about activities or situations that might endanger Australia's national security. The ASIO Act defines security as the protection of Australia and its people from espionage, sabotage, politically motivated violence, the promotion of communal violence, attacks on Australia's defense system, and acts of foreign interference. Some of these terms are further defined in the ASIO Act [17].

3.1.8 The Australian Federal Police (AFP). The introduction of the Cybercrime Act (2001) prompted the Australian Federal Police (AFP) to join forces with state and territory police to create a national organization to address the threat of cyber-crime. The distinction between cyber-crime and cyber-terrorism is blurred because many of the tools and techniques are common to both activities. Consequently, the creation of the Australian High Tech Crime Centre (AHTCC) was a major and important CIIP measure. The AHTCC provides a national coordinated approach to dealing with instances of high-tech crime affecting the Australian jurisdiction, including the investigation of electronic attacks against the National Information Infrastructure [18].

3.2 Public-Private Partnerships

3.2.1 The Trusted Information Sharing Network for Critical Infrastructure Protection (TISN). Because the vast majority of the critical infrastructure is owned or operated on a commercial basis, public-private collaboration is a key component of CIIP. The Attorney-General's Department writes: "As with most businesses, those who own or run critical infrastructure know the best way to protect it, how to manage an incident and how to get things up and running again. While the Government believes that regulations are not the best way to protect all types of critical infrastructure in some areas regulations are needed for special reasons. For example, in the transport industry regulations are needed so Australia can meet international obligations" [19]. The Trusted Information Sharing Network for Critical Infrastructure Protection (TISN) is the most important initiative to encourage the cooperation between the private and the public actors.

Building on the recommendations of the first Consultative Industry Forum (CIF) [20], the former government announced the formation of the Business-Government Task Force on Critical Infrastructure. The task force recommended replacing the CIF with a "learning network" to share information about critical infrastructure protection. In 2002, the government announced the creation of a Trusted Information-Sharing Network for Critical Infrastructure Protection (TISN) [21].

The TISN is organized according to Australia's critical infrastructure sectors. Each of the sector groups, the so-called Infrastructure Assurance Advisory Groups (IAAGs), is chaired by a representative of the critical infrastructure from that sector [3]. Membership is restricted to owners and operators of CI and government. Logistical support for the group is provided by government agencies that deal with the sector on a day to day basis, e.g., the Health Department with the Health group. The Attorney-General's Department provides support to Emergency Services, Banking, and Mass Gatherings. Each sector group is represented by their chair at the Critical Infrastructure Advisory Council (CIAC). The CIAC reports to the attorney-general. It is a way for critical infrastructure owners and operators to communicate with the Australian government at a high level. It also feeds into Australia's counter-terrorism arrangements.

Two permanent Expert Advisory Groups have been set up to advise the Critical Infrastructure Advisory Council—one for IT Security and the other for Critical Infrastructure Protection Futures [22].

4 EARLY WARNING AND PUBLIC OUTREACH

There are two key organizations that provide comprehensive early-warning services for cyber-attacks in Australia. The Defence Signals Directorate (DSD) has the remit to assist federal and state/territory IT networks, and the Australian Computer Emergency Response Team (AusCERT) provides some similar services to private sector operators of CI. In addition, the Australian government has launched the OnSecure website, run by the DSD.

4.1 Information Security Incident Detection Reporting and Analysis Scheme (ISIDRAS)

The DSD manages the Information Security Incident Detection Reporting and Analysis Scheme (ISIDRAS). The function of the ISIDRAS is the collection of information on security incidents that affect the security or operability of government computer and communication systems.

The ISIDRAS facilitates high-level analysis of information security incidents with the aim of improving knowledge of both threats and vulnerabilities to Australian government information systems and about how to protect these systems more effectively. ISIDRAS provides regular reporting of incidents. Government agencies that have detected a security breach can report the incident by completing an Australian Government IT Security Incident Reporting Form or via the OnSecure Website (which is a joint initiative between the Defence Signal Directorate and the Australian government Information Management Office to assist government agencies in dealing with information security breaches) [23]. Information derived from these reports is used as a basis for threat assessments and security advice.

4.2 Australian Computer Emergency Response Team (AusCERT)

The Australian Computer Emergency Response Team (AusCERT) is an independent non-profit organization located at the University of Queensland. It provides an important information security service to the private sector and to some government agencies on a fee-for-service basis. AusCERT's aims are to reduce the probability of successful attacks, to reduce the direct costs of security to organizations, and to lower the risk of consequential damage [24]. In May 2003, the Australian government announced the launch of AusCERT's National Information Technology Alert Service (NITAS) [25], which is sponsored by the federal government. NITAS provides a free service to subscribers, most of whom are owners and operators of the NII [26].

5 LAW AND LEGISLATION

5.1 Electronic Transactions Act 1999

The Electronic Transactions Act of 1999 creates a light-handed regulatory regime for using electronic communications in transactions. It facilitates electronic commerce in Australia by removing existing legal impediments under Commonwealth law that may prevent a person from using electronic communications. The act gives business and the community the option of using electronic communications when dealing with government agencies [27].

5.2 Cybercrime Act 2001

The Cybercrime Act of 2001 amended the Criminal Code Act 1995. It also amended the Crimes Act 1914 and the Customs Act 1901 to enhance the applicability of the existing search-and-seizure provisions relating to electronically stored data. It gives federal law enforcement agencies the authority to investigate and prosecute groups who use the internet to plan and launch cyber-attacks (such as hacking, computer virus propagation, or denial-of-service attacks) that could seriously interfere with the functioning of the government, the financial sector, and industry. The offenses and investigation powers were drafted in a manner to make them consistent with the draft of the Council of Europe's Cybercrime Convention.

The act covers:

- Unauthorized modification of data to cause impairment;
- Unauthorized impairment of electronic communication;
- Unauthorized access to, or modification of, restricted data;
- Unauthorized impairment of data stored on a computer disk, etc.;
- Possessing, producing, supplying, or obtaining data with intent to commit a computer offense;
- Causing an unauthorized computer function with intent to commit a serious offense.

The offenses were drafted in a way that recognizes the inter-jurisdictional character and extend to situations where:

- The conduct occurs wholly or partly in Australia;
- The result of the conduct occurs wholly or partly in Australia; or
- The offender was an Australian citizen or Australian company.

5.3 Security Legislation Amendment (Terrorism) Act 2002

The Security Legislation Amendment (Terrorism) Act 2002 [28] amended the Criminal Code Act 1995 to:

- Create a new offense of engaging in a terrorist act and a range of related offenses;
- Modernize Australia's treason offense; and
- Create offenses relating to membership in or other specified links with a terrorist organization.

An organization can be listed in regulations if the attorney-general is satisfied that the organization is a terrorist organization and that the organization has been identified in a decision of the United Nations Security Council relating to terrorism. A court may also find that an organization is a terrorist organization [29].

The act also specifically outlawed cyber-terrorism: "The action or threat of action which seriously interferes with, seriously disrupts, or destroys, an electronic system including, but not limited to information, telecommunications and financial systems [...]. The action is done or the threat is made with the intention of: advancing a political, religious or ideological cause; and coercing, or influencing by intimidation, the government of the Commonwealth or a State, Territory or foreign country (or part of)" [30].

5.4 Spam Act 2003

Australia's anti-spam legislation was introduced in 2003 in response to concerns about the impact of spam on the effectiveness of electronic communication and the costs imposed on end users. The Spam Act 2003 [31] prohibits the sending of spam, which is defined as a commercial electronic message sent without the consent of the addressee via e-mail, short message service (SMS), multimedia message service (MMS), or instant messaging. The requirements under the Spam Act apply to all commercial electronic messages, including both bulk and individual messages. The Australian Communications and Media Authority (ACMA) has enforcement responsibility for the Spam Act.

In June 2006, the former Department of Communications, Information Technology and the Arts (now Department of Broadband, Communications and the Digital Economy) released a review of the Spam Act [32] which found that the measures in the Act had been successful in curbing spam, but that it remained a significant problem.

ACKNOWLEDGMENT

We acknowledge the contribution of the expert, Alex Webling of the Attorney-General's Department within the Australian government, who validated the content of this chapter.

REFERENCES

1. Attorney-General's Department National Security. (2008). Website http://www.ag.gov.au/www/agd/agd.nsf/Page/Nationalsecurity_CriticalInfrastructureProtection.
2. http://www.tisn.gov.au/agd/WWW/TISNhome.nsf/Page/Business-Govt_partnership, 2008.
3. http://www.tisn.gov.au/, 2008.

4. Commonwealth of Australia (2005). *National Counter-Terrorism Plan*, 2nd ed., September. http://www.nationalsecurity.gov.au/agd/WWW/rwpattach.nsf/VAP/(5738DF09EBC4B7EAE52BF217B46ED3DA)~NCTP_Sept_2005.pdf/$file/NCTP_Sept_2005.pdf.
5. http://www.homelandsecurity.org.au/files/2007_e-security_agenda.pdf, 2008.
6. Australian Government *E-Security National Agenda 2007*, http://www.dbcde.gov.au/__data/assets/pdf_file/71201/ESNA_Public_Policy_Statement.pdf, 2008.
7. Yates A. (2007). *National Security Briefing Notes*, July. http://www.homelandsecurity.org.au/files/2007_e-security_agenda.pdf.
8. http://www.tisn.gov.au/agd/WWW/rwpattach.nsf/VAP/(930C12A9101F61D43493D44C70E84EAA) ~GovCERT.au+October+2007.PDF/$file/GovCERT.au+October+2007.PDF, 2008.
9. The Attorney-General's Department *About the Department*, http://www.ag.gov.au/www/agd/agd.nsf/Page/About_the_Department, 2008.
10. http://www.ag.gov.au/www/agd/agd.nsf/Page/Organisational_StructureNational_Security_and_Criminal _JusticeSecurity_and_Critical_Infrastructure, 2008.
11. *E-Security National Agenda*, (2007). op. cit., p. 1.
12. http://www.dbcde.gov.au/communications_for_business/security/critical_infrastructure_security, 2008.
13. http://www.tisn.gov.au/agd/WWW/tisnhome.nsf/Page/CIP_Projects#section2, 2008.
14. http://www.ag.gov.au/govcert, 2008.
15. http://www.agimo.gov.au, 2008.
16. http://www.dsd.gov.au, 2008.
17. http://www.asio.gov.au, 2008.
18. http://www.ahtcc.gov.au, 2008.
19. http://www.ag.gov.au/www/agd/agd.nsf/Page/Nationalsecurity_CriticalInfrastructure Protection, 2008.
20. Attorney-General's Department, Protecting Australia's National Information Infrastructure. (1998). *Report of the Interdepartmental Committee on Protection of the National Information Infrastructure*. Canberra, December 1998.
21. http://www.cript.gov.au, 2008.
22. http://www.tisn.gov.au/agd/WWW/rwpattach.nsf/VAP/(930C12A9101F61D43493D44C70E84EAA)~TISN+diagram+v.2+Dec+07.pdf/$file/TISN+diagram+v.2+Dec+07.pdf, 2008.
23. http://www.onsecure.gov.au, 2008.
24. NII Report, (1998). op. cit. p. 2 http://www.auscert.org.au.
25. http://www.nationalsecurity.gov.au/www/attorneygeneralHome.nsf/0/64534A395BA69AF4CA256D24007BDCA2, 2008.
26. http://www.national.auscert.org.au, 2008.
27. http://www.ag.gov.au/agd/WWW/securitylawHome.nsf/Page/e-commerce_Electronic_Transactions_Act_-_Advice_for_Commonwealth_Departments, 2008.
28. Security Legislation Amendment (Terrorism) Act 2002, No. 65, (2002). *An Act to Enhance the Commonwealth's Ability to Combat Terrorism and Treason, and for Related Purposes*, http://scaleplus.law.gov.au/html/comact/11/6499/pdf/0652002.pdf.
29. http://www.nationalsecurity.gov.au/agd/www/NationalSecurityHome.nsf/Page/RWPA41035442ED47EF7 CA256D6A001215A5, 2008.
30. Security Legislation Amendment (Terrorism) Act (2002). op. cit.
31. http://www.dbcde.gov.au/__data/assets/pdf_file/0015/34431/Main_Features_of_the_spam_act.pdf, 2008.
32. http://www.dbcde.gov.au/__data/assets/pdf_file/0008/40220/Report_on_the_Spam_Act_2003_Review-June_2006.pdf, 2008.

AUSTRIA

MANUEL SUTER AND ELGIN BRUNNER
Center for Security Studies (CSS), ETH Zurich, Switzerland

1 CRITICAL SECTORS

Contemporary sources of dangers and risks to the state, the society, and the individual may be found in the fields of politics, the economy, the military, society, the environment, culture and religion, and information technology (IT). Information and communication technology has acquired a dimension of its own in security policy because it links all other security aspects, thus becoming a power factor in its own right and leaving room for many options. Austria as a modern society and as a small state is particularly vulnerable in the area of information. This includes both the military and the civilian sectors, and increasingly business and industry as well [1].

Accordingly, Critical Information Infrastructure Protection is of crucial importance for Austria. Responding to a parliamentary inquiry [2], the Austrian federal chancellor defined critical infrastructures as "natural resources; services; information technology facilities; networks; and other assets which, if disrupted or destroyed would have serious impact on the health, safety, or economic well-being of the citizens or the effective functioning of the Government" [3]. This definition conforms to the definition elaborated by the EU (see chapter on the EU in this book).

The same inquiry also raised the question of whether there was a list of critical infrastructures in Austria [3]. In its answer, the Ministry of Internal Affairs clarified that there is a list of civilian objects worthy of protection, but they are not explicitly denoted as critical infrastructure. However, it can be assumed that Critical Infrastructure Protection in Austria manly refers mainly to these objects. The list of civilian objects worthy of protection includes about 180 items, which are categorized in the following classes:

- Institutions of the legislative, executive, and judiciary powers,
- Infrastructure facilities of energy supply companies,
- Information and communication Technologies,
- Infrastructure facilities that ensure the provision of vital goods,
- Transport and traffic infrastructures.

Reprinted with permission from Elgin M. Brunner and Manuel Suter. **International CIIP Handbook 2008/2009, Series Editors**: Andreas Wenger, Victor Mauer and Myriam Dunn Cavelty, Center for Security Studies, ETH Zurich.

2 PAST AND PRESENT INITIATIVES AND POLICIES

Following the Security and Defense Doctrine of 2001, which can be considered to be the guideline for Austria's security and defense policy, security in all its dimensions is the basic prerequisite for the existence and functioning of a democracy as well as for the economic welfare of the community and its citizens. Therefore, security must be conceived and implemented within a comprehensive security policy.

There have been several organizational and procedural efforts since the 1990s to manage CIP/CIIP in Austria. The issue of CIIP has been addressed by the government, especially by the Ministry of Internal Affairs; the Ministry of Defense; the Ministry of Traffic, Innovation, and Technology; and the Federal Chancellery, which has taken the leadership and is the central point in different projects.

On the European level, Austria takes part in all relevant EU activities regarding the protection of critical infrastructures, such as the European Program for the Protection of Critical Infrastructure (EPCIP) and the Critical Infrastructure Warning Information Network (EUCIWIN). Austria, like most other EU member states, shares the opinion that the protection of critical infrastructures has to follow the principle of subsidiarity, which means that the protection of the critical infrastructure is primarily the task of the member states. Activities of the EU are seen as complementary measures.

2.1 Security and Defense Doctrine 2001

According to the principle of comprehensive security, the Security and Defense Doctrine [4] recommends the development of the existing Comprehensive National Defense Program into a system of Comprehensive Security Provision by focusing on the new risks and threats and by amending legal provisions [5]. One can therefore deduce that this will also include all measures referring to CIIP.[1] This doctrine clearly stresses that for small states, full and unimpaired access to the information they require is a basis for their freedom of action in security matters [5].

The implementation of Austria's security policy within the framework of the Comprehensive Security Provision relies on systematic co-operation among various policy areas on the basis of appropriate sub-strategies.

2.2 IT Strategy and the "Platform Digital Austria"

The IT strategy of the government was formulated in July 2001, based on a decision of the Council of Ministers of 6 June 2001 referring to the New Structuring of the IT Strategy of the Government. The strategy consisted of the following three service types: Administration and Public Relations, Techniques and Standards, and Project Management and International Affairs. A special body, the ICT Board, was established to guarantee

[1]The concept of "Comprehensive National Defense" as developed from 1961 onwards was embedded in the Constitution in 1975. Under Article 9a of the Austrian Constitution, the role of Comprehensive National Defense is to "maintain [Austria's] independence from external influence as well as the inviolability and unity of its territory, especially to maintain and defend permanent neutrality". Together with the constitutional amendment, the Austrian parliament unanimously adopted a resolution in 1975 "on the fundamental formulation of Comprehensive National Defense in Austria" (defense doctrine). These were the foundations of the national defense plan, which was adopted by the Austrian government in 1983 and identified the "protection of the country's population and fundamental values from all threats" as a basic goal of Austrian security policy.

strategic co-ordination of ICT within the framework of the public government. This board was composed of the chief information officers of all the Federal Ministries and was located at the Federal Chancellery. It was responsible for coordinating the IT activities with each ministry, the local authorities, and the municipalities.

In 2005, the strategy was restructured. The basic elements of the 2001 strategy were retained, and the existing organizations were consolidated. However, in order to ensure sustainability, the ICT Board and the E-Cooperation board (the body responsible for coordination of e-government) were summarized in one unit, the ICT-Strategy Unit of the federal government [6]. This group forms the central unit of the new "Platform Digital Austria", where all IT and e-government efforts are coordinated. The ICT-Strategy Unit is responsible for public relations; the ICT budget, controlling, and sourcing; law, organization, and international activities; program and project management; and technical infrastructure.

2.3 Citizen Card and e-government

The Austrian Citizen Card Concept does not define a single citizen card for electronic identification, but only specifies the minimum technical requirements in a neutral way. Because of the open, technologically neutral approach, a variety of entities can issue citizen cards. These include both public bodies (including federal ministries and universities) and private bodies (certification authorities, banks) and can even involve other technologies such as mobile phone signatures [7]. In order to make use of the possibilities offered by electronic identification, citizens need to register their card as citizen card, download software, and buy a reader for the chip. After this registration process, they can use their card for e-identification and for electronic signatures. Different governmental services can be accessed by using the citizen card, and the e-government activities will be extended continuously [8].

2.4 Zentrales Ausweichsystem (ZAS)

After a fire at the Austrian Central Bank at the end of the 1970s, the government decided to establish an alternative replacement system for the data stock of the government. This system is located in the so-called Einsatzzentrale Basisraum (EZB) in St. Johann/Salzburg. Due to its coordinative function in the procurement of IT technologies, the Federal Chancellery has been responsible for the development of the EZB.[2]

The Zentrale Ausweichsystem (ZAS) has been a central part of the governmental crisis prevention system since the 1980s and has been fully operational on a day-to-day basis ever since. Some fundamental and very important systems (like the law information system/RIS) are run by this system. In addition, the ZAS serves as an archive for important backup data, such as the data from the public record office and from the Schengen Information System.[3]

2.5 Austrian Information Security Handbook

By order of the Federal Chancellery, the Center for Secure Information Technology Austria (A-sit, see below) publishes the Austrian Information Security Handbook (known

[2]The ZAS is located on an installation of the Austrian military; therefore, not much is publicly known about the institution itself.

[3]Cf. [9].

until 2005 as the IT-Security Handbook). This handbook gives an overview of IT security in general and informs readers in a broad and comprehensive way about fundamental aspects and measures in the field of IT. The handbook was updated in 2003, 2004, and 2007 based on the idea that security is a continuous process. It consists of two parts: "IT Security Management", which offers concrete instructions in this field; and "IT Security Measures", which describes standard security measures for IT systems requiring a medium security level [10].

2.6 Official Austrian Data Security Website

The Official Austrian Data Security Website [11], which is coordinated by the Federal Chancellery, serves as an information desk for citizens in important matters such as data security, the Schengen Information System, Europol, etc. It also informs the public about the work of the Commission on Data Protection, whose reports are available on the website. It also serves as a complaint board for citizens who want to report violations of their data privacy.

3 ORGANIZATIONAL OVERVIEW

At the public level, no single central authority is responsible for CII/CIIP, which is considered to be a cross-agency task. However, the Federal Chancellery fulfills a coordinating task. CIIP is mainly addressed by the Ministry of Internal Affairs, the Ministry of Defense, and the Ministry of Traffic, Innovation, and Technology. In addition, the Center for Secure Information Technology Austria (A-SIT) and the Stopline.at—Initiative, both organized as public-private partnerships perform important tasks in the field of CIIP.

3.1 Public Agencies

3.1.1 Ministry of Internal Affairs (BMI). Several divisions of the Ministry of Internal Affairs (BMI) deal with CIIP, especially with aspects of data security and cyber-crime. For example, the head office for the public safety at the Federal Crime Police Office operates a reporting center for child pornography [12].

Another important agency belonging to the BMI is the Federal Agency for State Protection and Counter-Terrorism (BVT), which is responsible for the coordination of personal security and the security of installations. In addition, it evaluates and develops the ability to provide protection on a permanent basis with regard to possible new threat scenarios.

The BMI also serves as the point of contact for European Processes concerning Critical Infrastructure Protection.

3.1.2 Ministry of Defense. In the framework of the Ministry of Defense, Department II (also known as the "control department") is responsible for all aspects of information warfare. It fulfills its duties in close cooperation with the Leadership Support Command[4] and the two military intelligence services.[5] One of these, the Abwehramt, which

[4]The Austrian armed forces and the Ministry of Defense are currently undergoing reform, so that a change in responsibilities is possible.

[5]"Heeresnachrichtenamt" and "Heeresabwehramt".

is responsible for the protection of the armed forces, also has a special department called Electronic Defense.[6]

The Austrian Federal Constitution and the Defense Law determine the cooperation between the army and civil authorities in crisis situations if the latter are not able to guarantee the maintenance of public order and inner security themselves. Part of this is the protection of civilian installations against interference by unauthorized third parties, including the protection of critical information infrastructures.

The final report of the Politico-Military Commission [14], which was released in autumn 2004, recommends that the Austrian armed forces be given an important role in the protection of the vital, civilian ICT, as well as the capacity to provide redundant systems in case of catastrophes or threats [15].

These protective measures have been tested in several exercises held in close co-operation with the civilian institutions. The largest maneuver of this kind in Austria took place in the federal states of Carinthia and Styria from 13–16 April 2004. The Schutz 2004 maneuver was planned and executed as a security assistance mission under the leadership of the civil authorities.

3.1.3 Ministry for Traffic, Innovation, and Technology (BMVIT). The Ministry for Traffic, Innovation, and Technology (BMVIT) is responsible for the safety of the public critical infrastructure. It operates a coordinating center for private owners and operators of critical infrastructure, and a center for security research. One of its recent activities has been to order an ICT master plan that would analyze the strengths and weaknesses and the state of the art of Austria's critical infrastructure. Another part of this mandate consisted in presenting options for measures, targets, missions, and visions [16]. The BMVIT also coordinates the Austrian Security Research Program, in which critical infrastructure protection will play an essential part [17].

3.1.4 Commission on Data Protection (DSK). The Commission on Data Protection (DSK) serves as independent control authority that deals with data processing in the public and private sectors.[7] The DSK is located at the Federal Chancellery. All citizens have the right to appeal to this commission if their rights in the field of data security are violated. The commission verifies these claims and takes measures to remedy confirmed violations. The Council on Data Protection has exclusive consultative agendas and periodically publishes the Report on Data Security.

3.2 Public-Private Partnerships

3.2.1 Center for Secure Information Technology Austria (A-SIT). The Center for Secure Information Technology Austria (A-SIT) was founded in May 1999 as an association supported by the Austrian National Bank, the Ministry of Finance, and the University of Technology in Graz. Its tasks include general monitoring issues of IT security[8] and the evaluation of encryption procedures [19], as well as supporting the introduction of the Citizen Card, supporting public institutions, and developing a security policy for all

[6]The chief of this department, Colonel Walter J. Unger, published several articles concerning IT security and cyberterrorism. See e.g.: [13].

[7]For more information, see [18].

[8]A-SIT offers tools and demonstration examples on its homepage: http://demo.a-sit.at.

important electronic payment systems for the Austrian National Bank. It is also a member of the Computer Incident Response Coordination Austria (CIRCA).

3.2.2 ***Stopline.at.*** Stopline.at is an online center that can be addressed by all internet users—also anonymously if they wish—who come across child pornography or right-wing extremist content on the internet. The relevant laws describing the respective crimes are §207a StGB (Austrian penal code) regarding child pornography, and the Austrian National Socialist prohibition law and the law against displaying National Socialist regalia as well as symbols of right-wing radicalism, respectively.

Stopline.at was founded as a private initiative by the Austrian internet service providers and has become reporting office that is authorized and accepted by the public authorities. Stopline.at cooperates closely with the Federal Ministry of the Interior (Federal Office of Criminal Investigation and Federal Office for the Protection of the Constitution and Counter-Terrorism).

4 EARLY WARNING AND PUBLIC OUTREACH

4.1 Computer Incident Response Coordination Austria (CIRCA)

The Computer Incident Response Coordination Austria (CIRCA) is Austria's main organization in the field of IT early-warning systems. It is a public-private partnership whose main actors are the Federal Chancellery, the Federation of the Austrian Internet Service Providers (ISPA), and A-SIT. Other members are representatives of the social partners (economic interest groups), the federal states, and of other critical infrastructure providers. It is a web of trust between Internet Service Providers (ISPs), IP network operators from the public and private sectors, and enterprises in the field of IT security. The electronic communication network of the private sector is run by ISPA, whereas the Federal Chancellery has the lead in the public sector.

The aim of this Austrian security net is to provide an early-warning system against worms, viruses, distributed denial-of-service attacks, and other threats that endanger IP networks and their users. Therefore, CIRCA issues alerts and risk assessments and provides information about precautionary measures. Its strategy is both proactive and reactive, and involves a continuous exchange of information and news between the Federal Chancellery and CIRCA [20].

4.2 Computer Emergency Response Team (CERT.at)

In March 2008, the Austrian domain registry nic.at launched the Austrian Computer Emergency Response Team (CERT.at) [21]. The purpose of the CERT is to coordinate security efforts and incident response for IT security problems on a national level in Austria. The level of support given by CERT.at varies depending on the type and severity of the incident or issue, the type of constituent, the size of the user community affected, and CERT.at's resources at the time. Special attention is given to issues affecting critical infrastructure. In addition, the CERT also releases educational material for SMEs and the general public.

The CERT.at cooperates with local and international CERTs as well as with other information security teams. It therefore shares information about incidents and security breaches with its partners. Nevertheless, it strictly protects the privacy of its customers.

5 LAW AND LEGISLATION

There is a broad variety of legal acts and laws dealing with CII/CIIP in a very broad sense. Most of them refer to the processing, collection, transfer, and protection of (personal) data through or by public agencies (e.g. the police and security agencies).

The general responsibilities of governmental authorities are laid out in the Bundesministeriengesetz (Federal Ministry Law), which defines the agendas of each ministry.

The following can be regarded as the central and most relevant legislative acts:

5.1 Information Security Law and Information Security Order

With the Information Security Law[9] and the Information Security Order,[10] Austria guarantees the secure use of classified information within the jurisdiction of the federal government according to international law. They regulate the access, transmission, identification, electronic processing, registration, and preservation of classified information. In accordance with international law, information regarding security arrangements within the EU or with other states qualifies as classified information. The Information Security Law specifies four types of classified information:

- *Limited.* if the unauthorized transmission of information would be contrary to the interests mentioned in Article 20, paragraph 3 of the Federal Constitution;
- *Confidential.* If the information has to be kept secret according to additional federal laws and if maintaining secrecy is in the public interest;
- *Restricted.* If the information is confidential and its publication would harm the interests mentioned in Article 20, paragraph 3 of the Federal Constitution;
- *Top Secret.* If the information is secret and its publication could seriously damage the interests mentioned in Article 20, paragraph 3 of the Federal Constitution.

Consequently, every type of classification corresponds to a certain security infrastructure (building, organizational structures, and personnel).

The Data Security Law therefore only grants access to Confidential, Restricted, and Top Secret information to individuals who have completed an advanced security examination according to paragraphs 55 to 55b of the Security Police Act. In the civilian sphere, this security examination is conducted by the Federal Office for Constitutional Protection and Counter-Terrorism.

5.2 Data Security Law 2000

The Data Security Law (DSG)[11] contains extensive regulations on the processing of personal data. With this law, Austria adopted the EU guideline for data security of the year 1995. The DSG 2000 stresses the importance of data-security measures and measures to enhance confidentiality for personal data. As a rule, the user of personal data is responsible for ensuring that the information is used in a correct manner, that no unauthorized persons have access to data, that the data is not destroyed, and that its secure storage is guaranteed. The DSG lists the following as civil rights:

[9]BGBl I Nr. 23/2002.
[10]BGBl II Nr. 548/2003.
[11]BGBl 165/99; see the explanations given by the Ministry of Internal Affairs. http://www.bmi.gv.at.

- The fundamental right to a secure processing of personal data,
- the right of information,
- the right to have incorrect or wrong data corrected,
- and the right to have data deleted.

Another important part of the DSG's activities is the duty to report. This means that with certain exceptions (e.g., for reasons of national security), all applications for personal data must be reported. Additionally, the Data Security Website contains all necessary information, forms, and addresses for rapid reporting.

5.3 Security Police Law

The Security Police Law (SPG)[12] defines the duties and authority of the civilian security services. Several articles and/or sections refer to the collection, transfer, storage, and deletion of personal data,[13] as well as measures to prevent the unauthorized use of data. It also provides special rights for individuals whose privacy has been violated by the security services.[14]

Together with this law, the office of a "legal protection agent"[15] was established as a controlling institution. The main duty of the legal protection agent is to protect the rights of citizens by ensuring that investigations of threats as well as observation and surveillance stay within legal rules.

5.4 Military Competence Law

In analogy to the Security Police Law, the Military Competence Law (MBG)[16] regulates the tasks and duties of the Austrian armed forces, including the two military intelligence services.[17] The MBG regulates the collection, transfer, and deletion of personal data. Paragraph 55 regulates the rights of citizens in cases where data security measures have been disregarded. The MBG also provides for the establishment of the institution of a "legal protection agent" who monitors the legality of measures undertaken by the intelligence services.[18]

5.5 Telecommunication Law

The Telecommunication Law[19] (TKG) includes extensive and detailed regulations referring to data security in general, and specific regulations regarding communication exchange. Furthermore, these regulations stipulate confidentiality of telecommunication.[20] The law also states that the suppliers of communication lines are responsible for securing all data. Paragraph 89 obliges the suppliers of communication lines to place all technical means necessary for the surveillance of telecommunication at the disposal of the security agencies.

[12]BGBl 566/91 idF BGBl 85/2000.
[13]Cf. especially section 4 of the law, "Verwenden personenbezogener Daten im Rahmen der Sicherheitspolizei".
[14]Cf. especially section 6 of the law, "Besonderer Rechtsschutz".
[15]"Rechtsschutzbeauftragter" (ombudsman in charge of protecting the rights of citizens).
[16]BGBl 86/ 2000.
[17]Second Section of the Law on Intelligence Services.
[18]Paragraph 57 of the law.
[19]Telekommunikationsgesetz, BGBl 100/ 1997 idF BGBl 134/ 2002.
[20]Fernmeldegeheimnis, Datenschutz. Chapter 12, paragraphs 87–101.

5.6 Penal Code (StGB)

Several articles of the Austrian Penal Code (StGB) refer to CII/CIIP. Some new regulations were introduced to the Penal Code in 2002 [22]:

Paragraph 118a. Unlawfully accessing a computer system: This crime is punishable with a prison sentence up to six months or a fine. The law applies not only to illegal access, but also to unauthorized registration on a computer system or to those who offer these possibilities to another person, make them public, or use them to gain benefit. The law also applies to cases where users who are authorized to use part of the system have accessed other parts that are off-limits to them. But an essential element is that a violation of security measures has to have occurred. Thus, if no security measures are in place, unauthorized access is not a crime. It is worth mentioning that the perpetrator will only be prosecuted with authorization from the injured party.

Paragraph 119. Infraction of the confidentiality of telecommunications: This crime is defined in a similar way to violations of the privacy of correspondence. The punishments and the requirement for the prosecution are the same as in paragraph 118a.

Paragraph 119a. Improper interception of data: This constitutes a crime that is punished and prosecuted. It is essential that the intercepted data not be intended for the intercepting person. It does not matter whether the perpetrators intend to use the data for themselves, to make it public, or to offer it to another party. The law makes no distinction between the methods applied.

Paragraph 126b. Disruption of the operability of computer systems: The elements of the crime of "Disruption of the operability of computer systems" are directly connected with paragraph 126a. The law outlaws the disruption of systems by introducing or sending data. The authorization of the injured party is not needed for prosecution, because this law applies to the diffusion of viruses, worms, etc.

Paragraph 126c. Abuse of computer programs or access data: This article is a very complex one. It prohibits the abuse of computer programs or access data, such as passwords. It is generally intended to cover Trojans and spy programs, as well as accessing and distributing passwords and access codes for various purposes. However, the maximum punishment is not higher than in the other articles.

5.7 Penal Procedure (StPO)

The Penal Procedure (StPO) regulates the special investigation methods for combating organized crime. These methods are provisions for optical and acoustic surveillance by civilian security institutions. The law also regulates the installation of a legal protection agent who monitors the legality of the special investigation methods. According to the StPO, the Minister of Justice is obliged to report annually on the use of special investigation methods to the Council for Data Protection (DSR),[21] the Commission on Data Protection (DSK), and the Austrian parliament.[22]

[21]The Council for Data Protection (Datenschutzrat, DSK) is a consultative body, which advises the government in questions concerning data protection. http://e-campus.uibk.ac.at/planet-et-fix/M6/3_Datenschutzrecht/3_Institutionen/K633_20datenschutzrat.htm.

[22]Cf. Bundesministerium für Justiz. "Gesamtbericht über den Einsatz besonderer Ermittlungsmethoden im Jahr 2001" (Vienna 2002).

In 2004, Austria introduced the European arrest warrant into its Penal Procedure System. It is an EU regulation that simplifies the extradition of persons for trial or for the enforcement of sentences. It comprises a catalog of 32 crimes where no close examination is required for extradition. A major problem is that these 32 offenses are not defined properly. One of these crimes is "cyber-crime", which has given rise to a lot of controversy, because each of the 25 member states may define it in a different way. In the Austrian penal code, for example, there is no such offence as "cyber-crime".

5.8 Electronic Signature Law (SigG)

Since 1999, the Electronic Signature Law (SigG)[23] has regulated the admission of electronic signatures in the Austrian legal system. The controlling board is the Austrian Telecom Control Commission, which gives the suppliers the necessary certificates. It also informs its constituency about security measures related to electronic signatures [23]. Since 24 September 2002, it has been fully operational with the Public-Key-Infrastructure (PKI).

ACKNOWLEDGMENT

We acknowledge the contribution of the expert, Otto Hellwig of Technische Universität Graz, who validated the content of this chapter.

REFERENCES

1. Resolution by the Austrian parliament (2001). *Security and Defense Doctrine: Analysis*, —Draft expert report of 23 January. http://www.austria.gv.at/2004/4/18/doktrin_e.pdf.
2. http://www.parlament.gv.at/pls/portal/docs/page/PG/DE/XXII/J/J_04641/imfname_067709.pdf, 2008.
3. http://www.parlament.gv.at/pls/portal/docs/page/PG/DE/XXII/AB/AB_04595/imfname_069768.pdf, 2008.
4. http://www.bundeskanzleramt.at/2004/4/18/doktrin_e.pdf, 2008.
5. Federal Chancellery Austria. (2001). *Security and Defense Doctrine*, http://www.bundeskanzleramt.at/2004/4/18/doktrin_e.pdf.
6. Digital Austria *Administration on the Net: An ABC Guide for E-Government in Austria*, p. 28. http://www.digitales.oesterreich.gv.at/DocView.axd?CobId=19394, 2008.
7. http://www.buergerkarte.at, 2008.
8. http://www.help.gv.at/sigliste/sig_bund.jsp?cmsid=281, 2008.
9. Der Standard (2007). *Österreichs Hochsicherheits-Datenspeicher wird 25 Jahre alt*, (15 September 2007).
10. The complete handbook is available at http://www.a-sit.at/pdfs/OE-SIHA_I_II_V2-3_2007-05-23.pdf, 2008.
11. http://www.dsk.gv.at/indexe.htm.
12. http://www.bmi.gv.at/kriminalpolizei, 2008.

[23]Signaturgesetz BGBl 1999/ 190.

13. (a) Unger, W. J., and Vetschera, H. (2005). Cyber War und Cyber Terrorismus als neue Formen des Krieges. *Österreichische Militärische Zeitschrift* **43**2, 203–211; (b) Unger, W. J. (2004). Angriff aus dem Cyberspace I-III. In *Truppendienst*, No. 2, pp. 143–147, 271–275, 382–386; No. 3; No. 4.
14. Bundesheerreformkommission http://www.bmlv.gv.at/facts/bh_2010/archiv/pdf/endbericht_bhrk.pdf.
15. Bundesheerreformkommission (2004). *Endbericht 2004*, p. 49f, 2008.
16. http://www.bmvit.gv.at, 2008.
17. http://www.kiras.at/wDeutsch/index.php and http://www.bmvit.gv.at/innovation/sicherheitsforschung/index.html, 2008.
18. http://www.dsk.gv.at/indexe.htm, 2008.
19. http://www.a-sit.at/asit/asit.htm, 2008.
20. http://www.circa.at, 2008.
21. http://www.cert.at/english/missionstatement/content.html, 2008.
22. http://www.cybercrimelaw.net/countries/austria.html, 2008.
23. http://www.signatur.rtr.at.

BRAZIL

MANUEL SUTER AND ELGIN BRUNNER
Center for Security Studies (CSS), ETH Zurich, Switzerland

1 CRITICAL SECTORS

Broadly defined, the Brazilian critical infrastructures include the areas of oil, electric energy, and telecommunications [1]. More specifically, the SecGov 2006 conference [2] held in Brasilia in November 2006 and sponsored by the Institutional Security Cabinet (Gabinete de Segurança Institucional—GSI) had the goal of discussing topics and questions on Critical Infrastructure Security in Brazil, Information and Communication Security and Terrorism. Eight discussion panels took place on the following topics:

- Public Safety,

Reprinted with permission from Elgin M. Brunner and Manuel Suter. **International CIIP Handbook 2008/2009, Series Editors**: Andreas Wenger, Victor Mauer and Myriam Dunn Cavelty, Center for Security Studies, ETH Zurich.

- Energy,
- Finance,
- Transport Systems,
- Water Supply,
- Public Health,
- Telecommunications,
- Terrorism.

Although the Brazilian government has not formally defined what the critical infrastructures are, at least the first seven topics are unofficially considered to represent critical sectors[1].

As regards critical *information* infrastructure, the focus lies on telecommunications and the internet. Based on the understanding that critical infrastructure protection on a nationwide level has consequences that can impact a nation socially, politically, and economically, a new approach was developed and proposed specifically by the federal telecommunications regulatory body, Anatel (see the chapter on Organizational Overview), to be applied to the telecommunications infrastructure, in order to understand the related risks and to develop a suitable program based on four main points: contextualization, a protection strategy, a set of methodologies, and software tools to support them. These methodologies include the development of tools for identifying critical (information and communication) infrastructures and the potential threat landscape, for scenario creation, and for diagnosing [3]. Moreover, information security is no longer understood as an exclusive problem of the sectors related to IT, or even of a particular organization, industry, or government; it is understood instead as consisting of regional and global strategies that facilitate an organized response to the threats and vulnerabilities associated with technology use [4].

2 PAST AND PRESENT INITIATIVES AND POLICIES

As mentioned above, Brazilian policies for information infrastructure protection focus on two particular aspects: the internet and telecommunications. Both of these, it is argued, play an important role in social (and digital) inclusion and are essential for national cohesion. The policies adopted in order to create trust in critical network infrastructures [5] show that the two sectors cannot be separated, since the interests of telecom and internet providers in operating secure networks are clearly inter-related, and the latter depend almost entirely on the former for backbone infrastructure and access networks. The Brazilian government has initiated several initiatives in association with internet diffusion, network protection, and communications security.

2.1 Brazilian Internet Steering Committee (CGI)

The initiatives of internet governance are mainly conducted under the auspices of the Brazilian Internet Steering Committee (Comitê Gestor da Internet no Brasil—CGI). This committee is a multi-stakeholder organization composed of members of government

[1] Information provided by an expert.

agencies, backbone operators, representatives of the internet service provider industry, users, and the academic community, and was jointly created in 1995 by the Ministry of Communications and the Ministry of Science and Technology.[2] The committee's main tasks are:

- To propose policies and procedures related to the regulation of internet activities;
- To recommend standards for technical and operational procedures for the internet in Brazil;
- To establish strategic directives related to the use and development of the internet in Brazil;
- To promote studies and technical standards for the network and security of services in the country;
- To coordinate the allocation of internet addresses and the registration of domain names;
- To collect, organize, and disseminate information on internet services, including indicators and statistics.

The committee maintains three working groups—on network engineering, on computer security, and on the training of human resources—in order to provide technical, administrative, and operational input for the committee's decisions and recommendations. Moreover, several projects in areas of fundamental importance for the operation and development of the internet in Brazil are coordinated. In order to execute its activities, the non-profit civil organization Brazilian Network Information Center NIC.br was created.

The Brazilian Internet Steering Committee has lately issued the second edition of its Survey on the Use of Information and Communication Technologies in Brazil—ICT Enterprises and ICT Households, reflecting the concern and the commitment of the committee in monitoring and sharing information about the evolution of the internet, which is considered an essential tool for social and economic development, as well as for the democratic participation of citizens and countries in the information society [6].

Of particular importance are also some of the recently initiated combined actions to improve internet security, such as instruction for users. Several initiatives are undertaken by the Computer Emergency Response Team Brazil (CERT.br), which is maintained by the Internet Steering Committee. The Internet Security Best Practices [7] document has been published since 2000 in order to help increase users' security awareness. While this document was written specifically for internet end users and has been constantly updated to reflect the evolving nature of attack and protection technologies, another document has been developed by CERT.br that is aimed specifically at companies: the Best Practices for Internet Network Administrators [8]. This document is addressed to security professionals and network professionals who do not have a dedicated security team at their disposal [9].

[2]The Brazilian Internet Steering Committee was created by interministerial ordinance no. 147 of 31 May 1995 and altered by presidential decree no. 4829 of 3 September 2003. Since July of 2004, the representatives of the civil society are chosen democratically to participate directly in the deliberations and to debate priorities for the internet, along with the government.

2.2 Brazilian electronic government program (e-gov)

The Brazilian government sees itself as having an important role to play both as a promoter and as a user of information and communication technologies. Therefore, the government has made the adoption of advanced information communication technologies for its administrative processes and delivery of services to its citizens a high priority. In 2000, it launched an electronic government initiative under the auspices of the Information Society Program of the Ministry of Science and Technology with three overall aims relating to the goal of digital inclusion: to universalize services, to make the government accessible to everyone, and to advance the infrastructure. A presidential decree established the Executive Committee of Electronic Government on 18 October 2000. Three years later, the presidency of Brazil published another important decree creating eight technical committees of e-government, with tasks including the implementation of free software, the advancement of digital inclusion, the integration of systems, legal systems and software licenses, the administration of websites and online services, network infrastructure, government to govern (G2G), and knowledge and strategic information management [10]. The Brazilian e-government model aims at integrating the different government organs in order to guarantee multiple channels of access for the citizens, institutions, local executives, and civil servants through manifold devices such as the traditional office counter and telephone, but also internet and digital TV [11]. In concrete terms, the driving principles of Brazil's electronic government are defined as follows [12].

- The priority of electronic government is to promote citizenship;
- Digital inclusion is inseparable from electronic government;
- Free software is a strategic appeal to implement electronic government;
- Knowledge management is a strategic instrument for the articulation and administration of the public policies of electronic government;
- Electronic government needs to rationalize the use of resources;
- Electronic government needs to relate to the integrated outline of policies, systems, templates, and norms;
- The activities of electronic government must be integrated with other levels of government.

3 ORGANIZATIONAL OVERVIEW

Major public efforts in Brazil concerning CIIP include the Information Security Steering Committee, the national policies for ICT under the auspices of both the Ministry of Science and Technology and the Ministry of Communication, and the Brazilian Network Information Center. Brazil has a complex and very sophisticated infrastructure of institutions involved in developing information security policy. Information security issues lie within the jurisdiction of the Institutional Security Cabinet (Gabinete de Segurança Institucional—GSI), which is an essential organ of the Presidency of the Brazilian Republic and assigned with the competence to coordinate the activities respective to information security [13]. The GSI's activities are defined by decree no. 5083 of 17 May 2004 [14]. It does not handle security issues directly, but works through other related organizations. Under its auspices, the Information Security Committee was formed.

As public-private partnerships, Anatel (the federal telecommunications regulatory body), Serpro (the federal data processing service), and CERT.br (the Computer Emergency Response Team Brazil) strive to further and deepen the cooperation between the public and the private sectors.

3.1 Public Agencies

3.1.1 Information Security Steering Committee (CGSI). The Brazilian Information Security Steering Committee (Comitê Gestor da Segurança Informação - CGSI) was created by decree no. 3505 on 13 June 2000 [15]. It is composed by representatives from every ministry [16]. The participants discuss information security issues and define the future policy directions of the Brazilian federal administration in working groups. This committee oversees the federal government's commitment under decree no. 3505, which stipulates that there must be an information security policy for every department of the Brazilian federal government [17]. Information security is defined by the committee as including the protection of the information systems from denial of service to authorized users, and against intrusion or unauthorized modification of data and information. It is seen as broadly including the security of human resources, of documents and material, of areas and installations of communication and computing, as well as being designed to prevent, detect, deter, and document eventual threats and their development [18].

3.1.2 National Policies for ICT. The Ministry of Science and Technology maintains a program dedicated to information and communications technologies (ICT). This program formulates a national policy and addresses issues such as software, microelectronics, network services, legal questions, and digital inclusion [19]. The focus lies on the technological and developmental aspects of the information and communication technologies.

Likewise, the Ministry of Communications maintains programs addressing digital inclusion, radio-diffusion, postal services, and telecommunications. These programs all aim to democratize access to these different means of communication and information and to reduce social and regional inequalities therein [20].

3.1.3 Brazilian Network Information Center (NIC.br). As mentioned above, the Brazilian Internet Steering Committee (CGI) was created by interministerial ordinance no. 147 of 31 May 1995 and altered by presidential decree no. 4829 of 3 September 2003. It is a public agency by nature, but its members include representatives of the private corporate and third sectors, as well as of academia. It is responsible for promoting the technical quality, innovation, and dissemination of internet governance and services, and has created the Brazilian Network Information Center (NIC.br) [21] in order to execute its activities. These activities include services—registro.br, CERT.br, and PTT.br—as well as projects such as antispam.br, statistics and indicators, and the internet security card.

This is to say that, as a set of services, the center coordinates Brazilian domain registration and IP assignments, it sponsors the CERT.br, and aims at providing the necessary infrastructure for the direct interconnection between the diverse networks that operate in the metropolitan regions (Ponto de Troca de Tráfego—PTT). Moreover, the committee's Center of Studies on Information and Communication Technologies (Centro de Estudo sobre as Tecnologias da Informação e da Comunicação—CETIC.br) is responsible for the collection, analysis, and dissemination of data about the use and penetration of the internet in Brazil [22].

The projects maintained by the Internet Steering Committee's executive branch, the Network Information Center, include, as mentioned, an anti-spam website designed to serve as an impartial and technically based source of reference concerning spam. This site represents the effort to inform both the users and network administrators about spam, its implications, and the forms of protection and combat. Furthermore, two other projects work on the statistics and indicators about Brazilian internet development and growth and on a document containing recommendations about how to navigate the internet more securely and about how individuals can protect themselves against so-called 'cyber-threats'.

3.2 Public-Private Partnerships

3.2.1 ***Anatel.*** Anatel (Agência Nacional de Telecomunicações), the federal telecommunications regulatory body modeled on the Federal Communications Commission of the US, was established with the mission of enabling a new model for Brazilian telecommunications, starting with the privatization of the Telebrás system. After privatization has been achieved between 1995 and 2003, the main role of Anatel became that of regulation, concession, and supervision of the telecommunications services in the country [23]. Among the important issues under discussion are the mechanisms for achieving cooperation between the Brazilian government and the private sector under the auspices of Anatel. Initial steps have been taken to address cyber-security issues facing the Brazilian telecom sector infrastructure through cooperation between private companies and this regulatory body.[3] Moreover, and as mentioned earlier, the methodology proposed and used by Anatel in order to identify critical infrastructure—called MI^2C—was used for the purpose of defining the critical parts of the Brazilian telecommunications infrastructure [24].

3.2.2 ***SERPRO.*** The SERPRO (Serviço Federal de Processamento de Dados) is a private company owned by the Brazilian government with the mandate of providing networking services for information technologies to government agencies in Brazil. Serpro supports thousands of federal government IT systems and runs a large IP-based government intranet system. There are extensive physical and logical security arrangements in place. Serpro has a security committee of about 35 people who develop government system security policies. The coordinator of the committee is a member of the above-mentioned Federal Government's Security Committee (CGSI). Moreover, Serpro cooperates on security issues with the Brazilian Internet Steering Committee and its Computer Emergency Response Team.[4] Serpro maintains different programs grouped under the three labels of governmental, entrepreneurial, and citizenship matters, which are closely linked to the Brazilian e-government program.

3.2.3 ***CERT.br partnerships.*** The Computer Emergency Response Team Brazil, maintained by the Internet Steering Committee, has a close partnership with the Software Engineering Institute (SEI) of Carnegie Mellon in matters of education. Within this partnership, the governmental cell is provided with educational courses in computer security creation and management, technical formation of information security, and the fundamentals and details of incident handling. Moreover, due to this cooperation, Brazil is a

[3]Cf. Robert Bruce et al., op. cit.
[4]Cf. Robert Shaw., op. cit.

member to the Carnegie Mellon Software Engineering Institute (SEI) CERT coordination center, which is useful in networking and coordinating information security issues internationally.

Moreover, CERT.br is a member of the global Forum of Incident Response and Security Teams (FIRST) [25], which, by bringing together a variety of Computer Security Incident Response Teams (CSIRTs) from government, commercial, and educational organizations worldwide, aims to foster cooperation and coordination in incident prevention, to stimulate rapid reaction to incidents, and to promote information-sharing among members.

CERT.br is also a research partner of the Anti-Phishing Working Group (APWG), which is the global pan-industrial and law enforcement association focused on eliminating the fraud and identity theft that result from phishing, pharming, and e-mail spoofing of all types [26].

4 EARLY WARNING AND PUBLIC OUTREACH

4.1 CTIR Gov

The Computer Security and Incident Response Team (Centro de Tratamento de Incidentes de Segurança em Redes de Computadores da Administração Pública Federal, CTIR Gov) is subordinate to the Institutional Security Office of the Presidency of the Republic (GSI) and deals with incidents on networks belonging to the federal public administration of Brazil. An executive order of 30 June 2003 created a working group responsible for determining the various aspects related to the installation and operation of a Computer Emergency Center. The mission of CTIR Gov is to coordinate responses to computer security incidents, to assure the necessary information exchange, and thereby to offer its constituency services that are both reactive (by responding as soon as notification arrives) and proactive (designed to prevent incidents and to reduce their impact). The reactive services aim to reveal the patterns and tendencies by continuous observation of events in order to serve as input to security recommendations, which are later issued to the constituency. The proactive services, which include information assets analysis and constitutive structures from the various information technology environments in the Federal Public Administration, provide a broad view of available resources, their usefulness, and associated risks [27].

4.2 CERT.br

CERT.br [28], formerly known as NBSO/Brazilian CERT, is the Brazilian National Computer Emergency Response Team, maintained by the NIC.br—the executive branch of the Brazilian Internet Steering Committee. CERT.br is a service organization that is responsible for receiving, reviewing, and responding to computer security incident reports and activity related to networks connected to the Brazilian internet. Besides doing incident handling activities, CERT.br also works to increase awareness in the community and to help new Computer Security and Incidence Response Teams (CSIRTs) to establish their activities. The range of services of CERT.br includes [29].

- To provide a focal point for reporting computer security incidents that provides coordinated support in response (and indication to others) to such reports;

- To establish collaborative relationships with other entities such as law enforcement, service providers, and telecom companies;
- To support tracing intruder activity;
- To provide training in incident response, specially for CSIRT staff and for institutions starting the creation of a CSIRT.

Additionally, CERT.br maintains a list of all Brazilian CSIRTs [30]. CERT.br also participates in the coordination of the Brazilian Honeypots Alliance and uses the data collected thereby to identify malicious activity originating in the Brazilian internet space, and to notify the administrators of the networks involved in malicious activities identified.

4.3 Brazilian Honeypots Alliance

The objective of the Brazilian Honeypots Alliance/Distributed Honeypots Project [31] is to increase the capacity for incident detection, event correlation, and trend analysis in the Brazilian internet space. To achieve these goals, the project is working to:

- Set up a network comprising distributed low-interaction honeypots, covering most of the Brazilian IP address space;
- Build a data analysis system that allows to study the attacks trends and correlations;
- Work with CSIRTs to disseminate the information.

The project is jointly coordinated by the CERT.br and the CenPRA (Centro de Pesquisas Renato Archer), a research institution of the Ministry of Science and Technology [32]. The honeypots network has 25 partner institutions including representatives from academia, the government, industry, and the military, which provide hardware and network blocks and maintain their own honeypots. Statistics about malicious activities observed in the honeypots are generated daily [33]. The collected data is used for intrusion detection purposes.

4.4 RNP/CAIS

In order to coordinate separate initiatives and secure the integration of regional networks into a national network, the Ministry of Science and Technology created the National Education and Research Network (Rede Nacional de Ensino e Pesquisa—RNP) in 1989 and assigned to it the task of building a national internet network infrastructure for academic purposes. Ten years later, in 1999, the Ministry of Science and Technology and the Ministry of Education jointly started the inter-ministerial Program for the Implantation and Maintenance of the RNP, with the aim of elevating the academic network to a new position. This RNP2 backbone was officially inaugurated in 2000. Since 2002, the RNP has had an agreement with the government to reach certain goals aimed at fostering the activities of technological research in network development and the operation of advanced network means and services that benefit national education and research [34]. The RNP was the basic platform for the early development of internet technology in Brazil, and because of its historic role, it continues to play an important role in security issues.[5]

[5]Cf. Robert Shaw, op. cit.

The RNP's Security Incidents Attendance Center (Centro de Atendimento a Incidentes de Segurança—CAIS), which was created in 1997, is more specifically concerned with network security within the RNP. The mission of CAIS is to resolve and prevent security incidents on the networks of RNP2, to divulge information and security alerts, and also to participate in international organizations for networking purposes. Hence, CAIS acts in the detection, solution, and prevention of security incidents on the Brazilian academic network, in addition to developing, promoting, and spreading security practices for the networks. Its concrete activities range from the providing of incident response services and the promotion of the creation of new security groups nationwide to the testing and recommendation of security tools and policies [35].

5 LAWS AND LEGISLATION

Decree no. 3505 of 13 June 2000 establishes the information security policy to be used throughout the government and across all related partners in a number of different areas, including:

- Classification and treatment of information;
- Research in technologies to support national defense;
- Accreditation and certification of products and services;
- Assurance of interoperability of systems;
- Establishing rules and standards relating to cryptography;
- Systems for the confidentiality, availability, and integrity of information.[6]

This decree was updated on 21 June 2004. The update makes the Secretaria de Comunicação de Governo e Gestão Estratégica da Presidência da República a full member of the Comitê Gestor de Segurança da Informação (CGSI) [36].

5.1 Brazilian Penal Code

Two amendments to the Brazilian Penal Code dating from 2000 created two new offenses relative to information security. Articles 313-A and 313-B of law no. 9983 of 14 July 2000, respectively, criminalize

- The "entry, or facilitation on the part of an authorized employee of the entry, of false data, improper alteration or exclusion of correct data with respect to the computer systems or the data bank of the public administration for purposes of achieving an improper advantage for himself or for some other person, or of causing damages", as well as;
- The "modification or alteration of the information system or computer program by an employee, without authorization by or the request of a competent authority".[7]

[6]Cf. Robert Bruce et al., op. cit.
[7]This law is an amendment of decree-law no. 2848 of 7 December 1940 in the Penal Code.

5.2 Cybercrime Laws

Brazil's criminal law states that gaining unauthorized access to a computer system or violation of the secrecy of a computer system belonging to either a financial institution or securities dealer is a crime under article 18 of law no. 7492 of 16 June 1986, which defines crimes against the national financial system [37].

Senate bill PLS 00152 [38] of 1991 defines the crimes involving wrongful use of computer (and also contains other provisions). This legislation defines as a crime the violation of data by means or clandestine of hidden access to a computer program or system, as well as the violation of the secrecy of data by gaining access to information contained in the system or physical medium of a third party.

Moreover, Brazil has several laws prohibiting the interception of telephone, data, or telematic communications. These laws ensuring privacy and criminalizing data interception are outlined both in the Brazilian Federal Constitution and in public law.[8]

5.3 Brazilian Cybercrime Bill

The Brazilian Congress is currently discussing a more specialized Cybercrime Bill. Under the responsibility of Senator Eduardo Azeredo, this bill is said to be inspired by the Convention on Cybercrime of the Council of Europe [39] and attempts to bring together three draft bills dating from 1996 and 1999. In both the criminal and the military criminal codes, 11 offenses are to be typified [40].

- Dissemination of malicious codes aimed at stealing passwords (phishing);
- Credit card fraud;
- Cell phone cloning;
- Offenses against honor (libel, slander, and defamation, with the stipulation of increased penalties);
- Dissemination of malicious codes aimed at causing harm (viruses, trojans, worms, etc.);
- Unauthorized access to computer network;
- Unauthorized access to information;
- Unauthorized possession, transportation, or provision of such information;
- Unauthorized disclosure of a database;
- Compound larceny with the use of computer systems;
- Disruption of public utility services;
- Attacks against a computer network—DoS, DDos, DNS, etc.

While this bill is still under discussion, in the meantime, cyber-crimes in Brazil are being judged in analogy to the Brazilian Penal code.

ACKNOWLEDGMENT

We acknowledge the contribution of the experts, Mariana Balboni of the Brazilian Internet Steering Committee, Regina Maria De Felice Souza of Agência Nacional de Telecomunicações, and João Henrique de A. Franco and Sérgio Luis Ribeiro of CPqD Telecom & IT Solutions, who validated the content of this chapter.

[8]For more details, cf. Goodman and Brenner, op. cit.

REFERENCES

1. Claudio, P. *Internet in Developing Countries: the Case of Brazil*, http://www.research.ibm.com/people/p/pinhanez/publications/netbrasil.htm, 2008.
2. http://www.secgov.com.br.
3. Regina Maria De Felice Souza. (2007). *Critical Telecommunication Infrastructure Project*, InfoCitel Electronic Bulletin No. 33, http://www.citel.oas.org/newsletter/2007/marzo/infraestructura_i.asp.
4. Forum of Incident and Response Teams (FIRST). 2nd COLAIS. (2006). *2nd Latin American Conference for Security Incident Response*. Rio de Janeiro, 6–12 October 2006, http://www.rnp.br/en/events/colaris/.
5. Robert S. (2002). Creating trust in critical network infrastructures: the case of Brazil. *ITU Workshop on Creating Trust in Critical Network Infrastructures*. Seoul, Republic of Korea, 20–22 May 2002.
6. Brazilian Internet Steering Committee, Brazilian Network Information Center. (2007). Preface to the *Survey on the Use of Information and Communication Technologies in Brazil—ICT Households and ICT Enterprises 2006*, 2nd ed., São Paulo, http://www.cetic.br/tic/2006/indicadores-2006.pdf.
7. http://cartilha.cert.br (in Portugese), 2008.
8. http://www.cert.br/docs/seg-adm-redes, 2008.
9. Christine, H., and Steding-Jessen, K. (2006). Information security in Brazil. In *Survey on the Use of Information and Communication Technologies in Brazil 2005—ICT Households and ICT Enterprises*, 1st ed., Brazilian Internet Steering Committee/Mariana Balboni, Ed. http://www.cetic.br/tic/2005/indicadores-2005.pdf.
10. http://www.governoeletronico.gov.br, 2008.
11. http://portal.etsi.org/docbox/Workshop/@METIS_Kick-off/Presentations/E-gov%20interoperability%20in%20Brazil%20ePing.ppt, 2008.
12. http://www.governoeletronico.gov.br, 2008.
13. http://www.presidencia.gov.br/estrutura_presidencia/gsi/sobre, 2008.
14. http://www.presidencia.gov.br/estrutura_presidencia/casa_civil/atos/destaque/estreg_gsi/view?searchterm=5083, 2008.
15. http://www.planalto.gov.br/gsi/cgsi, 2008.
16. The current list of representatives is available here: http://www.planalto.gov.br/gsi/cgsi/, 2008.
17. Bruce, R., Dynes, S., Brechbuhl, H., Brown, B., Goetz, E., Verhoest, P., Luiijf, E., and Helmus, S. 30 June (2005). *International Policy Framework for Protecting Critical Information Infrastructure: A Discussion Paper Outlining Key Policy Issues*. TNO Report 33680, Tuck School of Business, Dartmouth, Delft.
18. http://www.planalto.gov.br/gsi/cgsi, 2008.
19. http://www.mct.gov.br/index.php/content/view/2143.html, 2008.
20. http://www.mc.gov.br, 2008.
21. http://www.nic.br, 2008.
22. http://www.cetic.br, 2008.
23. http://www.globaliswatch.org/files/pdf/GISW_Brazil.pdf, 2008.
24. MI^{2}C is described in Bezerra, E. K., Nakamura, E. T., Ribeiro, S. L. (2005). Critical telecommunications infrastructure protection in Brazil. *First IEEE International Workshop on Critical Infrastructure Protection*. http://ieeexplore.ieee.org/xpl/freeabs_all.jsp?arnumber=1572288.
25. http://www.first.org, 2008.
26. http://www.antiphishing.org, 2008.
27. http://www.ctir.gov.br, 2008.
28. http://www.cert.br/index-en.html, 2008.

29. http://www.cert.br/mission.html, 2008.
30. This list is accessible at http://www.cert.br/contact-br.html, 2008.
31. http://www.honeypots-alliance.org.br, 2008.
32. http://www.cenpra.gov.br, 2008.
33. http://www.honeypots-alliance.org.br/stats, 2008.
34. http://www.rnp.br/en/rnp/history.html, 2008.
35. For a more detailed list, see http://www.rnp.br/cais/sobre.html, 2008.
36. http://www.planalto.gov.br/casacivil/site/exec/arquivos.cfm?cod=560&tip=doc, 2008.
37. This section is based on: Goodman, M. D. and Brenner, S. W. (2002). The emerging consensus on criminal conduct in cyberspace. *UCLA Journal of Law and Technology*, **6**(1), http://www.lawtechjournal.com/articles/2002/03_020625_goodmanbrenner.php.
38. http://legis.senado.gov.br/pls/prodasen/prodasen.layout_mate_detalhe.show_materia?p_cod_mat=1463, 2008.
39. See also: http://www.coe.int/t/e/legal_affairs/legal_co-operation/combating_economic_crime/3_technical_cooperation/cyber/567-LEG-country%20profile%20Brazil%20_30%20May%2007_pdf, 2008.
40. Senado Federal, Gabinete do Senador Eduardo Azeredo. (2007). Octopus interface conference—cooperation against cybercrime. *Cybercrime legislation in Brazil, Presentation by Senator Eduardo Azeredo*. Strasbourg, 11th June 2007.

CANADA

Manuel Suter and Elgin Brunner
Center for Security Studies (CSS), ETH Zurich, Switzerland

1 CRITICAL SECTORS

In Canada, critical infrastructure (CI) consists of the physical and information technology facilities, networks, and assets essential to the health, safety, security, or economic well-being of Canadians, and the effective functioning of government [1]. Canada's federal government (i.e., the government of Canada, and each of the provincial and territorial

Reprinted with permission from Elgin M. Brunner and Manuel Suter. **International CIIP Handbook 2008/2009, Series Editors**: Andreas Wenger, Victor Mauer and Myriam Dunn Cavelty, Center for Security Studies, ETH Zurich.

governments) structures its respective critical infrastructure programs as it deems appropriate. The government of Canada classifies critical infrastructure within the ten sectors listed below:

- Energy and Utilities,
- Communications and Information Technology,
- Finance,
- Health Care,
- Food,
- Water,
- Transportation,
- Safety,
- Government,
- Manufacturing [1].

The government of Canada recognizes that the nation's critical infrastructure could potentially be affected by both physical and cyber threats, whether natural or human-induced. Recognizing the complex nature of the threat environment, the government has adopted an all-hazards approach to protect critical infrastructure.

2 PAST AND PRESENT INITIATIVES AND POLICIES

Canada began implementing dedicated CIP and CIIP policies in 2001 in response to the new risk environment and the increasing interconnectedness of both physical and cyber-based infrastructures. In 2003, the government of Canada brought together the office responsible for critical infrastructure and emergency preparedness and the various agencies responsible for national security into one department, Public Safety and Emergency Preparedness Canada (PSEPC). This department, which is now called Public Safety Canada, was created to keep Canadians safe from a range of risks, including natural disasters, crime, and terrorism. Its responsibilities include ensuring a coordinated response to threats and developing initiatives and programs aimed at strengthening Canada's critical infrastructure [2].

Given the interdependencies and connectedness between critical infrastructures, the interruption of any one service could have a cascading effect and disrupt other essential services or systems. For example, during the North American Power Outage of 2003, large segments of rural and urban communities were in the dark: traffic and street lights were out; banking and government services were interrupted, and fuel distribution was disrupted. The disruption in one sector—electricity—affected a score of others, interrupting the delivery of important services to Canadians.

In light of this increasing interdependency, Public Safety Canada has taken a leadership role in promoting a national partnership among private and public-sector critical infrastructure stakeholders. This leadership has led to the development of the National Strategy and Action Plan for Critical Infrastructure (National Strategy and Action Plan).

2.1 The National Strategy and Action Plan for Critical Infrastructure

To address the need for coordinated action, federal, provincial, and territorial governments have drafted a National Strategy and Action Plan that will enhance the resiliency of Canada's critical infrastructure. Its goal is to build a safer, more secure, and more resilient Canada. To achieve this goal, the National Strategy sets out a model for public-private sector partnership, an information sharing framework, and a risk-based approach to protecting critical infrastructure. The Action Plan identifies near-term deliverables that will be used to establish national priorities, goals, and requirements so that funding and resources are applied in the most effective manner [3].

Achieving meaningful progress under the National Strategy and Action Plan calls for critical infrastructure partners to have:

- Risk-based plans and programs in place addressing and anticipating risks and threats;
- Access to robust information-sharing networks that include relevant intelligence and threat analysis; and
- Plans in place to identify and address dependencies and interdependencies to allow for more timely and effective response and recovery [3].

The public-private partnership described in the National Strategy and Action Plan provides the bedrock for effective critical infrastructure protection. Governments and private-sector partners each bring core competencies that add value to the partnership and enhance Canada's protective posture.

The government can support industry efforts and assist in broad-scale protection through activities such as:

- Providing owners and operators with timely, accurate, and useful information on risks and threats;
- Ensuring that industry is engaged as early as possible in the development of risk management activities and emergency management plans; and
- Working with industry to develop and prioritize key activities for each sector.

The federal government will establish sector networks for each of the ten critical infrastructure sectors, which will provide standing fora for public-private sector partners to engage in information exchange and address critical infrastructure priorities (e.g., identify and address risks, develop plans, and conduct exercises). The federal government will also establish a National Cross-Sector Forum, which will be composed of representatives from each of the ten sector networks. The Forum will identify and address cross-sector interdependencies, and provide advice and recommendations to the minister of public safety [3].

Risk management under the National Strategy and Action Plan builds on the Emergency Management Act [4], which requires federal ministers to identify risks, address these risks through plans and conduct exercises. This risk management approach includes:

- Risk profiles that identify and assess risks;
- Plans to protect the most vulnerable areas of critical infrastructure;
- Exercises to validate plans and protective measures; and

- Risk management tools and guidance.

The National Strategy and Action Plan for Critical Infrastructure represents the first milestone in the road ahead. This document identifies a clear set of goals and objectives and outlines the guiding principles that will underpin our efforts to secure infrastructure vital to our public health and safety, national security, governance, economy, and public confidence. Most importantly, it establishes a foundation for building and fostering a cooperative environment where governments and industry can work together to protect our critical infrastructure and secure the foundations of our country and way of life.

2.2 Information-Sharing

Information-sharing is one of the most significant issues in CIP and CIIP. Canada has been working to identify better ways to achieve this goal. Information-sharing can be viewed as a means to manage actions that can help deter, prevent, mitigate, and respond to the impact of a threat, as well as a tool to manage risk.

Government of Canada information-sharing practices related to CIP and CIIP are based on the principles articulated in the Access to Information Act (ATIA) [5], which include the public's right to access information held by the government of Canada along with specific exceptions to that right. For example, when confidential information is provided to the government of Canada by a foreign government, that information is protected by a specific and mandatory exemption in the Access to Information Act (ATIA) and cannot be disclosed.

Building on Canada's current system of safeguards, the Emergency Management Act[1] includes important amendments to the ATIA that protect specific critical infrastructure and emergency management information shared by private-sector owners and operators of Canada's critical infrastructure. This type of information will enable the government of Canada to develop comprehensive emergency management plans and mitigation and preparedness measures, improve warning capabilities, and develop better defenses and responses.

To support the information-sharing requirements in the National Strategy and Action Plan for Critical Infrastructure, Canada has developed two guides called "Information Sharing and Protection under the Emergency Management Act" [6] and "Identifying and Marking Critical Infrastructure Information Shared in Confidence with the Government of Canada" [7], both of which elaborate on the information protection measures in the Emergency Management Act. These guides form a framework that provides a clear structure for the process of establishing information-sharing relationships, and encourage consistent approaches among participants, while ensuring that such processes are workable for and relevant to all key stakeholders. The primary goals of Canada's information-sharing framework are to assess threats and vulnerabilities, improve warning and reporting capabilities, and analyze attacks to develop better defenses and responses.

3 ORGANIZATIONAL OVERVIEW

In Canada, the lead department dealing with CIP and CIIP is Public Safety Canada. As mentioned above, the department was created in 2003 out of the integration of the former

[1] See the chapter on Law and Legislation.

Department of the Solicitor General, the National Crime Prevention Centre, and the former Office of Critical Infrastructure Protection and Emergency Preparedness (OCIPEP).

The premise for Public Safety Canada's CIP and CIIP efforts is accurate and timely threat information. The Integrated Threat Assessment Centre (ITAC) helps to arrange the information collected by various intelligence sources. In addition, the Permanent High-Level Forum on Emergencies was established to ensure cooperation between the federal and local governments.

In Canada, the private sector owns and operates more than 80 per cent of the nation's critical infrastructure. This underscores the need for effective relationships between the government of Canada and the private sector, and between all levels of government and the organizations involved in preventing and responding to the various potential threats.

3.1 Public Agencies

3.1.1 Public Safety Canada. Public Safety Canada provides policy advice and support to the minister of public safety on issues related to public safety, including national security and emergency management, policing and law enforcement, interoperability and information-sharing, border management, corrections and conditional release, Aboriginal policing, and crime prevention. The Public Safety Canada portfolio also includes the Royal Canadian Mounted Police (RCMP), the Canadian Security Intelligence Service (CSIS), the Correctional Service of Canada, the National Parole Board, the Canada Firearms Centre, the Canada Border Services Agency, and three review bodies [8].

Public Safety Canada continues the mandate given to the Office of Critical Infrastructure Protection and Emergency Preparedness (OCIPEP) to combine critical infrastructure protection and emergency management responsibilities in one organization. This approach reflects the new risk environment, where the physical and virtual dimensions of infrastructures are increasingly interconnected. Combining critical infrastructure protection and emergency management resources and policy tools with acquired knowledge and experience in emergency management should ensure a stronger, more integrated and effective national security posture. Critical infrastructure protection and emergency management are not seen as separate endeavors, but as part of the assurance and protection continuum.

Public Safety Canada is the focal point for coordinating, analyzing, and sharing information related to physical and virtual threats to the Canadian critical infrastructure. Once it has received notification, the Government Operations Center, located in Public Safety Canada, assesses the threat to Canada and further distributes the bulletin and assessment to critical infrastructure owners and operators as well as emergency management contacts in Canada.[2]

3.1.2 Integrated Threat Assessment Centre (ITAC). The Integrated Threat Assessment Centre (ITAC) [9] was created to facilitate the integration of intelligence from various sources into comprehensive threat assessments. These are based on intelligence and trend analysis evaluating both the probability and potential consequences of threats. Such assessments are aimed at assisting the government of Canada to coordinate activities in response to specific threats more effectively in order to prevent or mitigate risks to public safety.

Several federal government departments feed into ITAC, including: Public Safety Canada, the CSIS, the Department of National Defence, the Canada Border Services

[2]Information provided by an expert.

Agency, Foreign Affairs Canada, Transport Canada, the RCMP, the Communications Security Establishment, the Privy Council Office, and the Ontario Provincial Police [10]. The focus of the threat assessments is on events and trends related to domestic and international terrorism. Although the assessments are related to national security issues, they are produced at various levels of classification, allowing for a broader distribution. ITAC assessments are currently distributed to the federal government and foreign partners through ITAC; law enforcement agencies receive the assessments through the RCMP.

3.1.3 Federal Provincial High-Level Forum on Emergencies. Major emergencies require extremely close cooperation between the federal government, provinces and territories, municipalities, and first responders. The government of Canada has therefore invited provinces and territories to establish a permanent high-level forum on emergencies in order to allow for regular strategic discussion of emergency management issues among key national players[2].

3.2 Public-Private Partnerships

The Canadian private sector, which owns and operates more than 80 per cent of the nation's infrastructure, plays a key role in securing cyberspace. National sector associations such as the Canadian Electricity Association (CEA), the Canadian Bankers Association (CBA), the Canadian Telecommunications Emergency Preparedness Association (CTEPA), and others have been active in promoting enhanced CIP/CIIP efforts. Currently, Canada's CI sectors are working to enhance information-sharing among their members, with government, and between sectors.

It is increasingly recognized that information on threats, vulnerabilities, corrective measures, and best practices should be shared widely across sectors and with governments. Canadian industry and governments at all levels are working together to improve information-sharing and analysis efforts. Industry sectors have identified a variety of challenges, including such issues as timeliness and relevancy of threat information. As industry efforts to increase cooperation and information-sharing mature, so will the national ability to respond to and manage cyber-incidents and attacks.[3]

4 EARLY WARNING

4.1 Canadian Cyber Incident Response Centre (CCIRC)

Public Safety Canada's Canadian Cyber Incident Response Centre (CCIRC) [12] provides national and international leadership in cyber-readiness and response. CCIRC is Canada's national focal point for coordinating cyber-security incident response and monitoring the cyber-threat environment 24 hours a day, seven days a week.

CCIRC leverages the IT security capabilities of the federal government to provide the following services to critical infrastructure sectors:

- Incident response, coordination, and support;
- Monitoring and analysis of the cyber-threat environment;
- IT security-related technical advice;

[3]Cf. [11].

- National awareness and education (training, standards, best practices).

When warranted, Public Safety Canada issues cyber-alerts and advisories, as well as other cyber-related information products to respond to potential, imminent, or actual threats, vulnerabilities, or incidents affecting Canada's critical infrastructure. This information is made available to all levels of government, as well as to non-government organizations.

CCIRC will build upon existing international relationships and is designed for improved interoperability with its allied partners.

4.2 Government Operations Centre (GOC)

Public Safety Canada is home to the Government Operations Centre (GOC) [13]. The GOC operates 24 hours a day, seven days a week. Its purpose is to provide strategic-level coordination and direction on behalf of the government of Canada in response to an emerging or occurring event affecting the national interest. It also receives and issues information dealing with any emerging or occurring threat to the safety and security of Canadians and Canada's critical infrastructure.

Information received by the GOC is quickly verified, analyzed, and distributed to the appropriate response organizations. This is made possible through Public Safety Canada's close linkages with other government departments and agencies; provincial, territorial, and municipal governments; and the private sector.

Calling upon resources and experts in various fields, the GOC helps to ensure that the right resources are in the right place at the right time. It coordinates the response to calls for help from other government departments and agencies; provincial, territorial, and municipal governments; and the private sector.

5 LAW AND LEGISLATION

5.1 Canadian Criminal Code Sections

- 342.1 (1): Every one who, fraudulently and without colour of right, (a) obtains, directly or indirectly, any computer service, (b) by means of an electromagnetic, acoustic, mechanical or other device, intercepts or causes to be intercepted, directly or indirectly, any function of a computer system, (c) uses or causes to be used, directly or indirectly, a computer system with intent to commit an offense under paragraph (a) or (b) or an offense under section 430 in relation to data or a computer system, or (d) uses, possesses, traffics in or permits another person to have access to a computer password that would enable a person to commit an offense under paragraph (a), (b) or (c) is guilty of an indictable offense and liable to imprisonment for a term not exceeding ten years, or is guilty of an offense punishable on summary conviction [14].
- 342.2 (1): Every person who, without lawful justification or excuse, makes, possesses, sells, offers for sale or distributes any instrument or device or any component

thereof, the design of which renders it primarily useful for committing an offense under section 342.1, under circumstances that give rise to a reasonable inference that the instrument, device or component has been used or is or was intended to be used to commit an offense contrary to that section, (a) is guilty of an indictable offense and liable to imprisonment for a term not exceeding two years; or (b) is guilty of an offense punishable on summary conviction [14].

- 430. (1.1): Every one commits mischief who willfully (a) destroys or alters data; (b) renders data meaningless, useless or ineffective; (c) obstructs, interrupts or interferes with the lawful use of data; or (d) obstructs, intercepts or interferes with any person in the lawful use of data or denies access to data to any person who is entitled to access thereto [15].

5.2 Emergency Management Act 2007

The Emergency Management Act (EMA) came into force in August 2007 and replaced its predecessor, the Emergency Preparedness Act 1985, with new and more comprehensive measures that strengthen the federal role in emergency management and critical infrastructure protection.

The purpose of the new Emergency Management Act (EMA) is to strengthen the readiness posture of the government of Canada to prepare for, mitigate the impact of, and respond to all hazards in Canada by emphasizing the need for a common and integrated approach to emergency management activities in the government of Canada. It recognizes that emergency management in an evolving risk environment requires a collective and concerted approach between all jurisdictions, including the private sector and non-governmental organizations. The act reflects a comprehensive, all-hazards approach to emergency management.

The EMA sets out the duties and responsibilities of the minister in providing national leadership by coordinating emergency management for the government of Canada. In particular, this involves:

- Coordinating the federal response to emergencies in Canada and the US;
- Establishing standardized elements for emergency plans within the government of Canada;
- Monitoring, evaluating, and testing the robustness of EM plans of government institutions;
- Enhancing cooperation with other jurisdictions and entities by promoting common standards and information-sharing.

The EMA also outlines the responsibilities of other federal ministers in carrying out their emergency management responsibilities.

The act addresses a common concern within the private sector: the confidentiality of the information shared with government, and specifically its protection from disclosure in response to a request under the Access to Information Act. Such releases could harm the competitive position and business reputation of service providers and prevents the building of trusted partnerships between industry and government. The importance of information-sharing was recognized in the EMA with the inclusion a consequential amendment to the Access to Information Act that exempts from disclosure critical infrastructure and emergency management information that is shared in confidence with the government [4].

5.3 The Department of Public Safety and Emergency Preparedness Act 2005

The Department of Public Safety and Emergency Preparedness Act [16] is Public Safety Canada's enabling legislation that sets out the general powers, duties, and functions for the department. The act establishes the public safety minister's powers and authorities to secure public safety and emergency preparedness, and to provide leadership at the national level.

ACKNOWLEDGMENT

We acknowledge the contribution of the experts, Claudia Zuccolo, Marta Khan, Michel De Jong, and Suki Wong of Public Safety Canada, who validated the content of this chapter.

REFERENCES

1. http://publicsafety.gc.ca/prg/em/nciap/about-en.asp, 2008.
2. http://www.publicsafety.gc.ca/abt/index-eng.aspx, 2008.
3. Public Safety Canada (2008). *Working Towards a National Strategy and Action Plan for Critical Infrastructure. Draft for Consultation*, http://www.publicsafety.gc.ca/prg/em/cip/_fl/nat-strat-critical-infrastructure-eng.pdf.
4. http://www.publicsafety.gc.ca/media/nr/2007/bk20070807-eng.aspx, 2008.
5. http://laws.justice.gc.ca/en/showdoc/cs/A-1///en?page=1, 2008.
6. http://www.publicsafety.gc.ca/prg/em/cip/_fl/information-sharing-and-protection-under-the-ema-eng.pdf, 2008.
7. http://www.publicsafety.gc.ca/prg/em/cip/_fl/labelling-sensitive-cip-information-eng.pdf, 2008.
8. http://www.publicsafety.gc.ca/abt/wwa/index-eng.aspx, 2008.
9. http://www.itac-ciem.gc.ca/index-eng.asp, 2008.
10. http://www.itac-ciem.gc.ca/prtnrs/index-eng.asp, 2008.
11. Public Safety Canada *Working Towards a National Strategy*, op. cit., 2008.
12. http://www.publicsafety.gc.ca/prg/em/ccirc/index-eng.aspx, 2008.
13. http://www.publicsafety.gc.ca/prg/em/goc/index-eng.aspx, 2008.
14. http://laws.justice.gc.ca/en/showdoc/cs/C-46/bo-ga:l_IX-gb:s_335//en#anchorbo-ga:l_IX-gb:s_335, 2008.
15. http://laws.justice.gc.ca/en/showdoc/cs/C-46/bo-ga:l_X-gb:s_422//en#anchorbo-ga:l_X-gb:s_422, 2008.
16. http://laws.justice.gc.ca/en/P-31.55/index.html, 2008.

ESTONIA

MANUEL SUTER AND ELGIN BRUNNER
Center for Security Studies (CSS), ETH Zurich, Switzerland

1 CRITICAL SECTORS

Estonia is one of the most rapidly developing information societies in Central and Eastern Europe. Estonia attracted a lot of attention in 2005 when it carried out its first round of internet-based voting in the local government elections of 2005 (in 2007, Estonia even became the first country in the world to feature e-voting in parliamentary elections). These elections were the results of constant and ambitious efforts to foster the information society.

The uninterrupted functioning of information and communication infrastructures (ICTs) provides the basis for such a highly developed information society. Information security and the protection of critical information infrastructures are therefore essential parts of Estonia's security policy.[1]

There are following critical sectors as defined by the Emergency Preparedness Act (consolidated text July 2002): "Vitally important sectors and the ministries administering these are the following:

- Maintenance of public order, fire extinguishing and rescue work, organization of protection of data banks—the Ministry of Internal Affairs;
- Functioning of the energy and gas system, organization of supply with staple goods; organization of telecommunications and postal services, and transport—the Ministry of Economic Affairs and Communications;
- Organization of supply with foodstuffs—the Ministry of Agriculture;
- Functioning of the financial system—the Ministry of Finance;
- Organization of health care, social insurance and social welfare, provision of psycho-social help, assistance to refugees and the evacuated, labor force calculation—the Ministry of Social Affairs;
- Organization of protection of cultural property—the Ministry of Culture;

[1]Information security and CIIP became even more important after the online attacks on Estonian government sites of April/May 2007, which attracted worldwide attention.

Reprinted with permission from Elgin M. Brunner and Manuel Suter. **International CIIP Handbook 2008/2009, Series Editors**: Andreas Wenger, Victor Mauer and Myriam Dunn Cavelty, Center for Security Studies, ETH Zurich.

- Organization of environmental protection and monitoring—the Ministry of the Environment.

The Ministry of Internal Affairs is the leading ministry in the field of crisis management" [1].

2 PAST AND PRESENT INITIATIVES AND POLICY

"I would not consider it an exaggeration to say that "e" has put Estonia back on the world map" [2]. This statement by Meelis Atonen, the then Minister of Economic Affairs and Communication in the preface of the policy paper Estonian IT Policy: Towards a More Service-Centered and Citizen Friendly State: Estonian Information Policy 2004–2006, outlines the importance of IT for Estonia.

Accordingly, the Estonian government has promoted various initiatives to strengthen the IT-sector. The first policy paper, Principles of Estonian Information Society, was set out in 1998. It was followed by the above-mentioned paper, which defined the principles of the Estonian information policy for 2004–2006; and since January 2007, the Estonian IT policy has been defined by the Estonian Information Society Strategy 2013.

Due to these strategies and their efficient implementation, Estonia succeeded in making considerable progress on the way towards an information society (for example, Estonia successfully launched new ID cards in 2002 that can also be used for issuing digital signatures and for using web-based services of the state) [3].

2.1 National Security Concept of the Republic of Estonia 2004

With the nation's rapid transformation into an information society, information security and the protection of communication infrastructure became important issues of national security. The National Security Concept 2004 therefore refers explicitly to the risks stemming from threats to information security. It is stated that "the constantly increasing rate at which electronic information systems are adopted in Estonia, and their connection with and dependence upon worldwide information systems, increases the threat of computer crime as well as the vulnerability of information systems, including spheres of primary importance to national security" [4].

2.2 National Information Security Policy

One of the aims of the policy paper for the Estonian information policy 2004–2006 was to define basic principles of a common IT security policy [5]. These basic principles were elaborated by a joint working group representing both the public and the private sectors and formulated in the National Information Security Policy.

The purpose of the Estonian Information Security Policy is to contribute to the development of a secure and security-aware information society. More specifically, the policy includes the following goals: elimination of non-acceptable risks to electronic communication networks and communication systems; defense of basic human rights; raising awareness about IT security and providing the respective training; participation in international initiatives related to e-security; and increasing the competitiveness of the Estonian economy [6].

In order to achieve these goals, the Estonian information security policy comprises five domains:

- *Cooperation and coordination at the national and international levels.* this domain includes initiatives such as the development and maintenance of a computer incident response capacity as well as participation in the European Network and Information Security Agency (ENISA);
- *Crisis management and cybercrime.* this domain includes preparations of crisis management plans and all initiatives designed to fight national and international cybercrime;
- *Education and training.* activities related to awareness-raising in government agencies as well as in the private sector and among the general public;
- *Legislation and regulation related to IT security.* specification, elaboration, and implementation of procedures, documentation, and means for ensuring information security;
- *Activities for the protection of people and assets.* protection of human rights and particularly of personal data.

As the Ministry of Economic Affairs and Communication states in the yearbook 2005 on Information Technology in Public Administration of Estonia [7], the Information Security Policy is designed to address IT security issues in the public sector as well as in the private sector.

In the same yearbook, information security is also clearly defined as part of critical infrastructure protection efforts: "The information security policy contributes to critical information infrastructure protection and takes into account information security aspects in other fields of critical information protection. The various fields of information security policy provide support and basic data for the protection of critical infrastructure and vice versa" [8].

2.3 Estonian ID Card

The Estonian ID card is not only a plastic card for the identification of its owner, but also contains a chip with a personal data file and two certificates enabling secure electronic authentication and digital signature [9]. It can be used for internet-based services provided by the Estonian government as well as for several services offered by the private sector.

Ninety per cent of the residents of Estonia already carry the new ID card. However, only a minority uses the ID card as an identification and authentication tool for digital services. The Estonian government, in cooperation with private-sector partners, tries to promote the usage of the ID card (see the chapter on Organizational Overview for the public-private initiative Computer Protection 2009).

2.4 Estonian Information Society Strategy 2013

Since January 2007, the new Estonian Information Society Strategy 2013 entered into force, setting out the general framework, objectives, and respective action fields for the development of the information society in Estonia. The strategy emphasizes the importance of cooperation between the public and private sectors and the need for coordination among all ministries involved.

Three objectives are mapped out by the strategy:

- *Development of a citizen-centered and inclusive information society.* the percentage of internet users in Estonia is to be further increased;
- *Development of a knowledge-based economy.* ICT uptake by enterprises is to be promoted and the competitiveness of the ICT sector to be increased;
- Development of citizen-centered, transparent, and efficient public administration by improving the efficiency of the public sector and providing user-friendly e-services in the public sector.

One of the principles for the development of the information society as defined in the document also refers to the importance of information security. It is stated that "the development of the information society must not undermine people's sense of security" [10]. Non-acceptable risk must be avoided, and personal data and identities must be secured.

2.5 The Estonian IT Interoperability Framework

The Estonian IT interoperability framework [11] is a set of standards and guidelines aimed at ensuring the provision of services for public administration institutions, enterprises, and citizens both in the national and in the European context. The latest version of the document (available in Estonian) [12] comprises the IT security interoperability framework, which specifies the activities related to CIIP and the use of the system of security measures by organizations.

2.6 The Estonian Cybersecurity Strategy

The Estonian Cybersecurity Strategy lays out the priorities and activities aimed at improving the security of country's cyberspace. The Cybersecurity Strategy concentrates on the following areas: the responsibilities of state and private organizations, vulnerability assessments of critical national information infrastructure, the response system, domestic and international legal instruments, international cooperation, and training and awareness-raising issues.[2]

3 ORGANIZATIONAL OVERVIEW

In Estonia, there is no single central authority responsible for CIIP. Several ministries and their respective subunits are directly involved. However, the main tasks of CIIP are assigned to the Ministry of Economic Affairs and Communication (MEAC) [13]. The MEAC plays a leading role with regard to information security, since two central agencies for the national IT policy are subordinated to the MEAC: The Department of State Information System (RISO), which is the central body for overall ICT coordination; and the Estonian Informatics Centre (RIA), which constitutes the implementing body under the MEAC. Other important public agencies that are dealing with CIIP are located within the Ministry of the Internal Affairs and within the Ministry of Defense. These two

[2]Information provided by an expert.

ministries are responsible for internal security and crisis management. With the project Computer Protection 2009, there is also an important public-private partnership, which aims to foster the security of the Estonian information society.

3.1 Public Agencies

3.1.1 The Department of State Information System (RISO). Within the Ministry of Economic Affairs and Communication (MEAC), the Department of State Information System (RISO) [14] is responsible for overall ICT coordination. With regard to IT security, it ensures the involvement of the private sector and cooperation among the different IT managers of the governmental agencies. In order to improve the security of the governmental communication network, RISO also coordinates the actions of the county and local governments and launches and supports broad public awareness campaigns.

The department also prepares appropriate legislation drafts and defines standard procedures for e-government. These regulative measures are usually developed in coordination with other ministries and with the private sector.

At the international level, RISO participates in the European Network and Information Security Agency (ENISA), and is involved in other cross-border initiatives, such as the development of the International Telecommunication Union (ITU) Global Cybersecurity Agenda.

3.1.2 The Estonian Informatics Centre (RIA). The Estonian Informatics Centre (RIA) was established to develop and manage data communication services for governmental organizations. Thus, the RIA is responsible for the technical security of the state's communication and information infrastructure. That includes measures to ensure the security of the governmental portals (which consists of three platforms on the internet);[3] preventive measures to maintain the security of the governmental data communication network; and monitoring and improving the overall security of IT in Estonia.

In 2005, the Estonian Computer Emergency Response Team (CERT) was established at RIA, in compliance with the obligation to form a national center for IT security, as laid out in the policy paper "Principles of the Estonian Information Policy 2004–2006". With the establishment of the Estonian CERT, the RIA has consolidated its role as the responsible body for the technical facets of CIIP. (For more information on the Estonian CERT, see the chapter on Early Warning and Public Outreach).

3.1.3 The Estonian National Communications Board. The Estonian National Communications Board manages and regulates the postal sector as well as the market for electronic communications in Estonia. It is responsible for the management of limited communication resources (e.g., radio frequencies) as well as for the regulation of the electronic communications market in Estonia [15]. In this function, the National Communication Board oversees the companies operating in the field of electronic communications and ensures the compliance of these companies with security requirements.

[3]http://www.riik.ee, which is the e-government platform; http://www.eesti.ee, which is the information platform; and https://www.esti.ee, which is the citizens' portal.

3.1.4 The Security Agencies. The task of the security agencies—the Security Police Board (belonging to the Ministry of Internal Affairs) and the Information Board (located within the Ministry of Defense)—is to ensure national security and maintain constitutional order through non-military preventive measures [16]. The functions of the Security Police Board are to prevent espionage, protect state secrets, and combat terrorism and corruption. The Information Board, in turn, collects intelligence concerning foreign countries and is responsible for the security of electronically transmitted information.

3.2 Public-Private Partnerships

3.2.1 Computer Protection 2009. Computer Protection 2009 is a joint project of the Look@World Foundation and the Ministry of Economics and Communications. The Look@World Foundation was established in 2001 by ten leading companies in Estonia with the goal to foster the development of the IT society in Estonia.

The Computer Protection 2009 project (also called Look@World 2) aims to foster the security of the Estonian information society, so that in 2009, Estonia will be the country "with the most secure information security in the world" [17]. To achieve this ambitious goal, the signing partners of the initiative started broad promotion programs to raise public awareness of IT security. In particular, they try to encourage citizens to use their ID card for electronic personal authentication.

The main activities of the foundation, however, include sharing of information among companies, public agencies, and citizens on how to adequately recognize threats to information security and to protect oneself against them [18]. Improving and promoting internet security-related dialog and cooperation between the public and private sector is a distinctive concern of the Computer Protection 2009 initiative.

4 EARLY WARNING AND PUBLIC OUTREACH

4.1 CERT Estonia

Established in 2005 at the Department for Handling Information Security Incidents of the Estonian Informatics Centre (RIA), the Computer Emergency Response Team of Estonia is responsible for the management of security incidents in the.ee computer networks. Its main task is "to assist internet users in Estonia in the implementation of preventive measures in order to reduce possible damage from security incidents and to help them in responding to security threats" [19]. This means that CERT Estonia offers support for incident handling and acts as an early-warning center for IT security.

The process of incident handling comprises the collection of information on incidents, analysis of attacks, and coordination of the response activities. However, since not all incidents are of the same importance, it is also important to assign priorities to each incident according to the severity level and scope. In assessing this prioritization, CERT Estonia takes the following aspects into account: the number of affected users; the type of incident; the target of an attack as well as the attack's point of origin; and the required resources for handling the incident [19]. Of course, attacks on critical infrastructures that may jeopardize people's lives would be considered to be incidents of highest priority.

In the domain of early warning, CERT Estonia cooperates with various national and international partners. The broad network enables the CERT to recognize new threats and vulnerabilities in a timely manner. Warnings are mainly issued in cases of attacks with higher level of severity, extremely widespread threats, and highly severe vulnerabilities [20].

In addition, once CERT Estonia has managed to successfully launch the above-mentioned services, it also intends to contribute to the promotion of awareness-raising in the field of IT security [19].

4.2 Infosecurity Portal

When the Computer Protection 2009 initiative was launched, a gateway to IT security-related information and discussions was established that is available in Estonian and in Russian [21]. The portals contain numerous links, articles, and news, and enable the users to obtain and share information about threats related to the internet. The goal of the portal is to help citizens to familiarize themselves with the world of information security.

5 LAW AND LEGISLATION

Since 1997 Estonia enacted a series of laws with regard to CIIP in general. Estonia was also among the first countries to sign the Council of Europe's Convention on Cybercrime in 2001, and fully enacted it in 2004.

The following legal instruments are relevant to information security and CIIP:

- *Emergency Preparedness Act.* This act provides the legal basis for the organization of emergency preparedness of and for crisis management by the national government, government agencies, and local governments [1];
- *State Secrets Act.* defines state secrets, access to state secrets, and the basic procedure for the processing of state secrets [22];
- *Personal Data Protection Act.* This act determines the principles for processing personal data (Chapter 1). Paragraph 6 defines the principle of security, which is binding for all processors of personal data: "security measures to prevent the involuntary or unauthorized alteration, disclosure or destruction of personal data shall be applied in order to protect the data" [23];
- *Public Information Act.* The purpose of this act is to ensure that the public and every person has the opportunity to access information intended for public use, based on the principles of a democratic and social rule of law and an open society, and to create opportunities for the public to monitor the performance of public duties. The act defines the state information system, describes the organization of databases belonging to a state information system, and lays out the legal basis for providing and using data services. The act provides for an approach integrating different areas of government through legislation that defines the administrative system of the state information system and other support systems for the state information system, as well as the status of databases established within the information system of the public sector [24];
- *Electronic Communications Act.* This act defines the requirements for the publicly available electronic communications networks and communications services. With regard to CIIP, the security requirements are of particular interest: "A communications undertaking must guarantee the security of a communications network and prevent third persons from accessing the data [...] without legal grounds" [25];

- *Information Society Services Act.* This act provides the requirements for information society service providers, the organization of supervision, and liability for violation of this act [26].

5.1 Penal Code

Several articles of the Estonian penal code refer to information security:

- *Article 206 (Computer sabotage).* Unlawful replacement, deletion, damaging or blocking of data or programs in a computer, as well as unlawful entry of data or programs in a computer is punishable by fines or imprisonment.
- *Article 207.* Damaging or obstruction a connection to a computer network or computer system is punishable.
- *Article 208.* Spreading computer viruses is punishable by fines or imprisonment.
- *Article 217.* Unlawful use of a computer, computer system, or computer network by way of removing a code, password, or other protective measures is punishable by fines or imprisonment.
- *Article 284.* Unlawfully handing over the protection codes of a computer, computer system, or computer network, if committed for the purpose of personal gain and in a manner which causes significant damage or results in other serious consequences, is punishable by fines or imprisonment.

ACKNOWLEDGMENT

We acknowledge the contribution of the experts, Thomas Viira of the Estonian Informatics Center and Jaak Tepandi of the Institute of Informatics, Tallinn University of Technology, who validated the content of this chapter.

REFERENCES

1. http://www.rescue.ee/index.php?page=143&PHPSESSID=220faec9593e393510ea6f39fef5a197, 2008.
2. Meelis, A., Ministry of Economic Affairs and Communication Preface. In " *Estonian IT Policy: Towards a More Service-Centered and Citizen-Friendly State. Principles of the Estonian Information Policy 2004–2006*, http://www.riso.ee/en/files/Policy.pdf, 2008.
3. http://www.id.ee/?id=11019, 2008.
4. http://web-static.vm.ee/static/failid/067/National_Security_Concept_2004.pdf, 2008.
5. http://www.riso.ee/en/files/Policy.pdf, 2008.
6. http://www.riso.ee/en/information-policy/security, 2008.
7. Ministry of Economic Affairs and Communications of Estonia (2005). *Information Technology in Public Administration of Estonia 2005*, p. 33. http://www.riso.ee/en/pub/yearbook_2005.pdf, 2008.
8. Ministry of Economic Affairs and Communications of Estonia (2005). *Information Technology in Public Administration of Estonia 2005*, p. 48. http://www.riso.ee/en/pub/yearbook_2005.pdf, 2008.

9. Ministry of Economic Affairs and Communications of Estonia (2006). *Information Technology in Public Administration of Estonia 2006*, p. 23. http://www.riso.ee/en/pub/2006it/, 2008.
10. Estonian government *Estonian Information Society Strategy 2013*, p. 4. http://www.riso.ee/en/files/IYA_ENGLISH_v1.pdf, 2008.
11. http://www.riso.ee/en/information-policy/interoperability, 2008.
12. http://www.riso.ee/wiki/Pealeht, 2008.
13. http://www.mkm.ee/index.php, 2008.
14. http://www.riso.ee/en/, 2008.
15. http://www.sa.ee/atp/?id=3476, 2008.
16. Estonian government (2004). *National Security Concept of the Republic of Estonia 2004*, p. 16.
17. http://www.riso.ee/en/node/80, 2008.
18. Ministry of Economic Affairs and Communications of Estonia (2006). *Information Technology in Public Administration of Estonia 2006*, p. 42, 2008.
19. http://www.ria.ee/28201, 2008.
20. Ministry of Economic Affairs and Communications of Estonia (2005). *Information Technology in Public Administration of Estonia 2005*, p. 36, 2008.
21. The Estonian version is available at http://www.arvutikaitse.ee; the Russian one can be found at http://www.infosecurity.ee, 2008.
22. http://www.legaltext.ee/text/en/X30057K7.htm, 2008.
23. http://www.legaltext.ee/text/en/X70030.htm, 2008.
24. http://www.esis.ee/ist2004/106.html, 2008.
25. http://www.legaltext.ee/text/en/X90001K2.htm, Chapter 10, Paragraph 101, 2008.
26. http://www.legaltext.ee/text/en/X80043.htm, 2008.